BIOLOGICAL MEMBRANES

This graduate text, suitable for students of physiology and biophysics, and medical students specializing in neurophysiology and related fields, provides a comprehensive discussion of biological mass transfer and bioelectrical phenomena. Emphasis has been given to the applicability of physics, physical chemistry and mathematics to the quantitative analysis of biological processes, with all the necessary mathematical grounding provided in Chapter 1.

The quantitative analysis is broken into four key stages:

- formulation of a biological/biophysical model,
- derivation of the associated mathematical description of the model,
- solution of the mathematical expression, and
- interpretation of the mathematical solution to a biological explanation.

This book guides the student through these stages, which are central to the understanding of cell membrane functions.

BIOLOGICAL MEMBRANES

Theory of transport, potentials and electric impulses

OVE STEN-KNUDSEN

Professor Emeritus of Biophysics
University of Copenhagen

CAMBRIDGE UNIVERSITY PRESS
Cambridge, New York, Melbourne, Madrid, Cape Town, Singapore, São Paulo

Cambridge University Press
The Edinburgh Building, Cambridge CB2 8RU, UK

Published in the United States of America by Cambridge University Press, New York

www.cambridge.org
Information on this title: www.cambridge.org/9780521810180

First published 2002
This digitally printed version 2007

A catalogue record for this publication is available from the British Library

Library of Congress Cataloguing in Publication data
Sten-Knudsen, Ove.
Biological membranes : mass transfer, membrane potentials, and electrical impulses /
by Ove Sten-Knudsen.
p.   cm.
Includes bibliographical references and index.
ISBN 0 521 81018 3 (hb)
1. Neural conduction – Mathematical models.   2. Action potentials
(Electrophysiology) – Mathematical models.   3. Mass transfer.   4. Biophysics.   I. Title.
QP363 .S775   2002
571.6′4 – dc21   2001052488

ISBN 978-0-521-81018-0 hardback
ISBN 978-0-521-03635-1 paperback

To
Nan-Marie
Helge, Nina
and
Henrik

Contents

Contents

Foreword

I was delighted when my friend Ove Sten-Knudsen asked if I would write a Foreword to this English translation of his book on biological membranes. I confess that I then had no idea of the immense scope and magisterial quality that I found when I saw the proofs. I wish that such a book had been available in the days when I was concerned with membranes and ion movements: I often had to struggle through the derivation of equations that are here worked through step by step. This will be invaluable not only to students but to the many biologists who work on membranes and use mathematics but are not themselves mathematicians in the full sense of the word. It will also be a major convenience to have the whole background collected in a single volume, instead of being scattered in numerous articles and books.

Another feature that gives me great pleasure is the biographical notes on the authors of classical papers in the field. Whenever one of those great men is mentioned for the first time, there is a footnote telling us his dates, where he worked and his main achievements. This is a welcome contrast to the usual practice of merely giving a name with no indication that it refers to an actual human being.

The book's title hardly does justice to its content: as well as dealing with the properties shared by all cell membranes, it includes very full accounts of the fundamentals of nerve conduction and of synaptic transmission. These were established half a century ago and nowadays they are too easily taken for granted while emphasis is put on the more modern studies of ion channels that followed from them.

This book will be a godsend to all who aim for a quantitative understanding of membrane phenomena.

Sir Andrew Huxley, OM, FRS
Cambridge
April 2002

Preface

I often say that when you can measure what you are speaking about, and express it in numbers, you know something about it; but if you cannot measure it, when you cannot express it in numbers, your knowledge is of a meager and unsatisfactory kind: It may be the beginning of knowledge, but you have scarcely in your thought advanced to the stage of science.

Lord Kelvin, 1883

The phenomena of mass transfer and electric activity across biological cell nembranes involve a variety of complex processes. To be able to understand and describe these basic mechanisms it is essential to have a thorough insight into the nature of mass transfer by migration, diffusion, and electrodiffusion, and of concepts and fundamental principles of physics and physical chemistry. It is difficult to understand and describe processes in a biological membrane without leaning on the knowledge of mechanisms that operate in simpler systems. In describing transport processes through a cell membrane, we use the concepts of simple passive transport theory. Command of these concepts is the basis on which one performs a quantitative analysis of processes that are observed experimentally.

The purpose of the present book is to give an overall account of diffusion, electrodiffusion, equilibrium potentials, diffusion potentials and membrane potentials, and, in this connection, to take relevant examples from membrane physiology/biophysics to illustrate the use of these tools in a quantitative analysis of the experimental results. It is hoped that the text may be of use to undergraduate studies in general physiology/biophysics and to some postgraduate researchers.

The mathematical background needed to read the book corresponds to the level of a college student graduating in science. These prerequisites for reading the book are presented in **Chapter 1**. They serve to refresh the memory of the reader, making frequent use of cross-reference in the following chapters. In the following chapters, the mathematical build-up is developed in detail without requiring an additional mathematical course. In this respect the book is self-contained. The presentation of the mathematics is deliberately written in a user-friendly manner, intended for readers who are not professional mathematicians or physicists. Therefore, the mathematical derivations are written out in detail, which may appear unnecessarily detailed to a reader who is skilled in handling mathematics. It is my experience that in this way most readers with no basic training in mathematics can be brought to understand – and later to apply – rather complicated mathematics. On the other hand I do not want to conceal the fact that at places the reader *is* confronted with problems that unavoidably are rather complicated to handle. But the reader will never meet the rather condescending phrase: 'it is easily seen that . . .'.

The theory of mass transfer by diffusion, migration, and by the superposition of these processes, is developed in **Chapter 2**. The chapter starts by introducing principles of migration flux and diffusion flux that define the associated parameters mobility and diffusion coefficient. This leads to the law of mass conservation (the diffusion equation). The exposition is based upon macroscopic considerations. Later these processes are reconsidered from the microscopic point of view of *Brownian motion* and *Boltzmann statistics*. I regard the introduction of these complementary aspects to be particularly useful to understand mass transfer across cell membranes. To solve the time-dependent diffusion equation no use is made of integral transforms, a mathematical technique that is beyond the scope of this book. Ludwig Boltzmann used an inspired transformation that by simple allowed him means to find a particular solution of the diffusion equation in terms of the Error function. The presentation in this book succeeded in using adaptations of this solution to solve both the new diffusion problems in **Chapter 2** and to handle problems in the remaining chapters.

Chapter 3 deals with the theory of transport of ions in aqueous solutions, as well as the origin of electric potential differences across cell membranes, i.e. membrane potentials. The basis for this is the theory of *electrodiffusion*: diffusion superimposed upon migration of ions with an electric field as driving force. The chapter starts by introducing the necessary concepts: electric potential and field, electric conductance of a single ion and of a mixture of ions. The Nernst–Planck equations, the equations of electrodiffusion most frequently occurring in electrophysiology, follow as a natural sequel of the general transport equation in Chapter 2. The condition of *electroneutrality* is assumed to hold almost throughout. The *equilibrium potential* across a membrane is introduced by applying the Nernst–Planck equation to the transport across an ion selective permeable membrane. The Donnan system is treated partly by making use of a thermodynamic argument and partly by using the Poisson–Boltzmann equation to determine the potential and concentration profiles. The treatment of the *diffusion potential* begins by considering the concentration differences of a single salt. This is followed by the descriptions of the general diffusion regimes of Planck and by Henderson which both relate to a membrane that is separated by mixtures of electrolytes of different composition. Membrane potentials in living biological cells will be described as complex diffusion potentials. Of common interest are three types of membrane whose properties are described in detail: (1) the anion/cation selective permeable membrane; (2) the constant field membrane (Goldman equation), and (3) the mosaic membrane, each type having its own merits. The chapter ends by comparing the membrane potentials obtained in experiments on a living single frog-muscle cell with those predicted from theory.

Most readers of this book are assumed to be biologists seeking an in-depth understanding of biological transport and bioelectrical phenomena. **Chapter 4** comprises a bridge between the theoretical treatment and some fundamental biological experiments performed in nerve (and muscle) fibre. Thus the chapter is essential for demonstrating the applicability of physics and physical chemistry to biology. Furthermore, it contains analyses not been treated in the previous chapters (e.g. cable analysis). The chapter starts by summarizing the basic properties of nerve excitability and impulse transmission. The emphasis is the meticulously planned and epoch-making work on the axon of the squid (*Loligo*) by A.L. Hodgkin & A.F. Huxley, with participation of B. Katz and R.D. Keynes. The experimental results were interpreted and followed by a quantitative analysis accounting for the origin of the action potential in the giant nerve of the squid. The lessons to be learned from this work extend far beyond its relation to nerve activity.

Chapter 5 presents the important investigations of Katz and co-workers on the processes involved in the neuromuscular transmission, where statistical arguments were used to account for their observations.

It should be emphasized that the aims of **Chapter 4** and **Chapter 5** are not to provide an up to date review of the field. The examples chosen only serve the didactic intentions of this book.

The basis of the present book is a similar text published in Danish in 1995* aimed at students of general physiology, biophysics and postgraduate researchers in neuromedicine. The present book is the result of many revisions and it has been supplemented with new sections.

I wish to thank my colleagues professor Rodney Cotterill, Dr phil & scient, Professor Erik Hviid Larsen, Dr scient, and Professor Ulrik V. Lassen, Dr med.,

---

* O. Sten-Knudsen (1995): *Stoftransport, membranpotentialer of elektriske impulser over biologiske membraner*. Akademisk Forlag, København.

for their encouragement to write the manuscript and for their help and advice in connection with its submission for publication; likewise Senior Lecturer Jørgen Warberg, Dr med., Head of Department of Medical Physiology for hospitality and assistance, and to Senior Lecturer Per Hedegaard, Lic. scient., and Mr Peter Busk Laursen, UNI-C, for instruction in handling some special typesetting problems. Ms. Nina Sten-Knudsen has given great help in making the line drawings. My particular thanks goes to Senior Researcher Else Marie Bartels, Ph.D., D.Sc., for giving me the inestimable help in reading the proofs of the book and in the completion of the index and the list of references.

I am most grateful to the Press Syndicate of The Cambridge University Press and their staff for publishing the book, in particular Dr Shana Coates, Editor (Biological Sciences), for in her friendly manner of directing the manuscript (and its author) from beginning to completion, and Mrs Beverly Lawrence for her expert copy-editing of a difficult manuscript and for excellent communication via e-mail, and also to Ms Carol Miller, production controller.

My thanks are also due to the authors referred to in the legends for kindly allowing me to copy figures from their papers and for confirming their permissions in writing, as well as to editors of the journals: *J. Physiol.*, *J. gen. Physiol.*, *Biophys. J.*, *Proc. R. Soc. Lond.*, *Progr. Biophys.*, *Nature*, *J. Neurophysiol.*, *Amer. J. Physiol.*, *Amer. J. Med.*, *Arch. Sci. Physiol.* and to Liverpool University Press, McGraw-Hill Education, Charles C. Thomas Publisher and University of California Press.

The Carlsberg Foundation has kindly supported the work by grants to providing the author with the adequate EDP equipment and to cover expenses related to the publication of the book.

Finally, it gave me great pleasure that Sir Andrew Huxley, OM, FRS, kindly agreed to write the Foreword to the book. I also greatly appreciated the concomitant constructive comments to the text.

Ove Sten-Knudsen
Gentofte, May 2002

Chapter 1

Mathematical prelude

For more than two thousand years some familiarity with mathematics has been regarded as an indispensable part of the intellectual equipment of every cultured person.

(Richard Courant, 1941)

1.1 Introduction

In biological research there is a steadily increasing trend to describe functions and mechanisms *quantitatively* by applying ideas and concepts from physics and physical chemistry. This tendency is found in large areas of biology, extending from ecology over the function of the integrated organism to processes taking place at the cellular and molecular level. This development will doubtless continue.

However, a quantitative treatment of any phenomenon in physics or physical chemistry requires an adequate command of the mathematical tools that are needed to formulate and solve the particular problem that is subject to such close scrutiny. For that reason, mastery of certain elements of mathematical analysis is an indispensable element in the arsenal of tools that are loaded into the knapsack of the serious student of general physiology or cell biology.

The sections that follow in this chapter are not presented as a self-contained mathematical text. The intention is to present a summary – short in some places, more detailed in others – of the mathematical concepts and techniques that are used in this book. It is presumed that the reader is already familiar with these concepts. Thus, a cursory reading of this chapter may have the effect of acting as a reminder of items that are known but perhaps not immediately recalled from memory.

1

1.2 Basic concepts of differential calculus

1.2.1 Limits

A collection of numbers

$$a_1; a_2; a_3; a_4; \ldots a_n;$$

that follow each other according to a given law is called a *sequence* of numbers. If the number of elements n increases without bound the sequence is an *infinite* sequence. The elements of the sequence are said to *converge* to a *limit L* if the elements beyond that of a_μ behave in such a way that the difference

$$|L - a_n| \quad \text{for } n > \mu$$

is smaller than any arbitrarily small positive number ε. If the elements a_n do not pile up in this manner, the sequence is made up of elements that *diverge*. When the elements of a sequence are added they constitute a *series*

$$S_n = a_1 + a_2 + a_3 + a_4 + \cdots a_n,$$

which may be *finite* or *infinite* according to whether the number of elements n is bounded or not. An infinite series may converge to a definite value S_n when n increases beyond the boundary. This value $S_\infty = L$ is called the *limit* of the series. This is generally written as

$$S_n \to L, \quad \text{for } n \to \infty, \quad \text{or} \quad \lim_{n \to \infty} S_n = L.$$

1.2.2 Functions

Let x and y represent two arbitrary quantities that are coupled together in such a way that to each value of x there exists a definite value of y. We say then that the quantity y is a *function* of the quantity x. Usually this is represented as

$$y = f(x), \tag{1.2.1}$$

where x is called the *independent variable* and y is called the *dependent variable**. Of course one could equally well have considered the *inverse function*

$$x = g(y), \tag{1.2.2}$$

where y is now the independent variable and x is the dependent variable. The condition that the inverse function $x = g(y)$ is so well-behaved that there exists in the interval $a \leq x \leq b$ one and only one value of x for a given value of y, is

---

* To facilitate the readability of this text, mathematical and physical variable quantities are printed in *italics*. Similarly, mathematical operators are printed in Roman type.

that the function $y = f(x)$ is increasing or decreasing *monotonically* in the same domain. Thus, the function $y = x^2$ is monotonically decreasing in the region $-a \leq x \leq 0$, and to every value of y there corresponds only one value $x = -\sqrt{y}$. In the region $0 \leq x \leq a$ the function $y = x^2$ increases monotonically, and to every value of y there corresponds likewise only one value $x = \sqrt{y}$. With increasing values for x in the region $-a \leq x \leq a$ the function $y = x^2$ both decreases monotonically as well as increasing, and for a given value of y we have the corresponding values $x = -\sqrt{y}$ and $x = \sqrt{y}$. A function that suddenly *jumps* from one value to another is said to be a *discontinuous function*. Thus, the function

$$y = f(x) = \begin{cases} 2 & \text{for } x \geq 1 \\ 1 & \text{for } x < 1 \end{cases}$$

is a discontinuous function for $x = 1$, since

$$f(1 + \varepsilon) - f(1 - \varepsilon) = 1$$

no matter how small we make the positive quantity ε. A *continuous function* is, roughly speaking, a function that does not do such things. Thus, the function

$$y = f(x) = \begin{cases} x^2 & \text{for } x \geq 1 \\ x & \text{for } x \leq 1 \end{cases}$$

in continuous at the point $x = 1$ since

$$f(1 + \varepsilon) - f(1 - \varepsilon) = (1 + \varepsilon)^2 - (1 - \varepsilon) = 3\varepsilon + \varepsilon^2 \to 0 \quad \text{for } \varepsilon \to 0,$$

although the formula displays changes for $x = 1$.

1.2.3 The derivative

Consider the function $y = f(x)$ that is continuous in the range $a < x < b$. If the quantity, denoted the *difference quotient*, for the function $y = f(x)$ at the point x

$$\frac{f(x + h) - f(x)}{h}, \tag{1.2.3}$$

converges towards a *definite limit* as h approaches zero in an arbitrary manner 0, the value of this limit

$$\lim_{h \to 0} \left[\frac{f(x + h) - f(x)}{h} \right] \overset{\text{def}}{\equiv} f'(x), \tag{1.2.4}$$

is called the *first derivative* of the function $y = f(x)^*$. Another name for $f'(x)$
is the *differential quotient* of $f(x)$. We can illustrate this limiting process geometrically as follows: Eq. (1.2.3) represents the value of the slope of a straight
line that is anchored at the curve point P_0 with coordinates $(x, f(x))$ and makes
another section with the curve at the point P_1 at $(x + h, f(x + h))$. This line
is called a *secant* to the curve. When we let h decrease in an arbitrary manner,
the point P_1 approaches the point P_0 from either side according to the sign of
h, and when $h \to 0$ the slope of the secant attains a limiting value that is equal
to the slope of the line that, at the point P_0, has only one point in common with
the curve $y = f(x)$, namely the *tangent* of the curve at P_0, or

$$\lim_{P_1 \to P_0} (\text{Slope of secant anchored at } P_0) = (\text{Slope of tangent at } P_0)$$

always provided there is a tangent with a well-defined direction at the point
P_0 on the curve. This occurs if the limit of the ratio $(f(x + h) - f(x))/h$
in Eq. (1.2.4) converges to the definite value $f'(x)$ when $h \to 0$. In many
physical applications involving the derivative it may useful to keep in mind this
geometrical representation of $f'(x)$.

The expression $y' = f'(x)$ goes back to the work of J.-L. Lagrange[†]. Another
way of writing the derivative $f'(x)$ is

$$f'(x) \overset{\text{def}}{\equiv} \frac{dy}{dx}, \tag{1.2.5}$$

which was introduced by G.W. Leibniz (1646–1716)[‡], has many practical ad-
vantages, and is almost always used in applied mathematics.

The quantity (dy/dx) is not a fraction in the usual sense but a compact *symbol*
meaning that the function $y = f(x)$ has been subjected to the operation that is
defined by Eq. (1.2.4). To emphasize the character of dy/dx as a mathematical
operation many people prefer to use the typographical convention

$$\frac{dy}{dx} \overset{\text{def}}{\equiv} \frac{dy}{dx}, \tag{1.2.6}$$

to distract one's thoughts from a fraction. This notation will be used in this
book.

* The symbol $\overset{\text{def}}{\equiv}$ is used in this text to emphasize that it is a definition.
† J.-L. Lagrange (1736–1813) was a Professor at École Polytechnique. He was one of the greatest
  mathematicians of the eighteenth century, who made fundamental contributions to the development of differential and integral calculus, calculus of variation, theory of numbers and to
  mechanics (Mécanique analytique) and astronomy.
‡ This is a remainder of the derivative being obtained from the difference quotient which he wrote
  as
$$\frac{f(x + \Delta x) - f(x)}{\Delta x} = \frac{\Delta y}{\Delta x}, \quad \text{for } \Delta x \to 0.$$

As an illustration we consider the function $y = f(x) = x^2$. We have

$$\frac{(x+h)^2 - x^2}{h} = \frac{(x^2 + 2hx + h^2) - x^2}{h} = \frac{2hx + h^2}{h} = 2x + h.$$

Hence

$$\lim_{h \to 0} \frac{(x+h)^2 - x^2}{h} = 2x.$$

Thus, the limit exists, giving

$$f'(x) = \frac{dy}{dx} = 2x.$$

Continuing this argument to $y = f(x) = x^n$, where n is any real number, one gets

$$\frac{d}{dx}(x^n) = n\,x^{n-1}.$$

Naturally the operations of Eq. (1.2.3) and Eq. (1.2.4) can be applied to the function $f'(x)$. If the limit exists it is called the *second derivative* of the function $f(x)$. The notation for this limit is

$$f''(x) \overset{\text{def}}{\equiv} \frac{d}{dx}\left(\frac{dy}{dx}\right) \overset{\text{def}}{\equiv} \frac{d^2 y}{dx^2}. \tag{1.2.7}$$

Some mathematicians have never become reconciled to Leibniz's notation and have instead replaced the operator $d(\ )/dx$ by the symbol D to denote the operation*

$$D f(x) \overset{\text{def}}{\equiv} \lim_{h \to 0}\left[\frac{f(x+h) - f(x)}{h}\right] \overset{\text{def}}{\equiv} f'(x).$$

The D notation will not be used in this text.

The requirement for the limit of Eq. (1.2.4) to exist is that the function $f(x)$ is *continuous*. However, this condition is not sufficient, because a continuous function may exhibit a sudden break at a point x_0. In this case $f'(x_0 - \varepsilon)$ and $f'(x_0 + \varepsilon)$ both exist no matter how small we make ε, but they may differ drastically from each other in value, leaving $f'(x)$ to have a discontinuity at the point x_0.

---

* This was introduced in 1808 by Brisson and gained a footing owing to the extensive use of the operator D made by A.L. Cauchy (1789–1857).

1.2.3.1 A few derived functions

Using the operations that are defined by Eq. (1.2.4) on the elementary mathematical functions one obtains explicit expressions for the derivatives of the
functions in question. Below are a few important elementary examples*

(a) If $f(x) = A$, where A is a constant, $f'(x) = 0$.
(b) If $f(x) = Au(x)$, $f'(x) = Au'(x)$.
(c) If $f(x) = u(x) + v(x)$, $f'(x) = u'(x) + v'(x)$.
(d) If $f(x) = u(x) v(x)$, $f'(x) = u'(x) v(x) + u(x) v'(x)$.
(e) If $f(x) = \dfrac{u(x)}{v(x)}$, $f'(x) = \dfrac{u'(x) v(x) - u(x) v'(x)}{v(x)^2}$.
(f) If $f(x) = x$, $f'(x) = 1$.
(g) If $f(x) = x^n$, $f'(x) = nx^{n-1}$.
(h) If $f(x) = \sin x$, $f'(x) = \cos x$.
(i) If $f(x) = \cos x$, $f'(x) = -\sin x$.
(j) If $f(x) = \tan x$, $f'(x) = 1/\cos^2 x$.

1.2.4 Approximate value of the increment Δy

In physics many relations are described in terms of the *rate* of change of a
quantity. This change may depend upon time, position in space, or both. With
hardly a single exception it is sufficient initially to express this change with an
approximate accuracy that may be improved later as occasion requires. In this
context, differential calculus is a very useful tool. One proceeds as follows. The
curve in Fig. 1.1 shows an arbitrary differentiable function $y = f(x)$. The line
AB denotes the tangent to the curve on the point (x, y) having a slope that is
equal to the value of the derivative $f'(x)$ taken at the point (x, y). Let $x + h$
be a neighboring point to x that corresponds to assigning a finite increment
$h = \Delta x$ to the value x of the independent variable. We denote the value of the
function at the neighboring point $x + h$ as $f(x + h) = y + \Delta y$, where Δy is
the increment in $y = f(x)$ due to the change h in the argument. According to
Eq. (1.2.3) and Eq. (1.2.4), that defines the derivative $f'(x)$, the increment can
be written as

$$\Delta y = f(x + h) - f(x) = f'(x)h + \varepsilon \Delta x, \qquad (1.2.8)$$

or

$$y + \Delta y = f(x + \Delta x) = f(x) + f'(x)\Delta x + \varepsilon \Delta x, \qquad (1.2.9)$$

---

* For more about hyperbolic functions, see Appendix I.

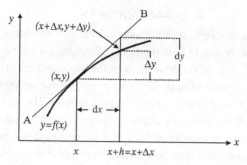

Fig. 1.1. Approximation of the increment Δy of a function $y = f(x)$ by a linear function. The figure also illustrates the geometrical meaning of the differentials dy and dx.

where $\varepsilon = \varepsilon(\Delta x)$ depends on the magnitude of Δx and approaches zero when $h = \Delta x \to 0$.

We now regard the variable x as fixed and let the increment $h = \Delta x$ vary in an arbitrary manner. Equation (1.2.9) now states that the increment Δy to the value y of $f(x)$ at a given value of x is made up of two terms:

(i) a term $f'(x)h = f'(x)\Delta x$ that is proportional to the increment $h = \Delta x$ with $f'(x)$ as the proportionality coefficient that is a constant at a fixed value of x, and

(ii) a correction term $\varepsilon h = \varepsilon \Delta x$, which can be made as small as we wish relative to h by making the increment $h = \Delta x$ sufficiently small. Thus, the smaller we make the interval in question $h = \Delta x$ around x the more precisely will the function $f(x + h)$, being a function of h, be represented by its linear part

$$f(x + h) \approx f(x) + f'(x)h, \qquad (1.2.10)$$

where both $f(x)$ and $f'(x)$ are two fixed numbers for a given value of x. From a geometrical viewpoint this approximate description of the value $f(x + h)$ of the function $y = f(x)$ at the point (x, y) means that the curve of $f(x)$ is *replaced* by the tangent and that the expression for the increment of the function

$$\Delta y = \Delta f = f(x + h) - f(x),$$

corresponding to the increment Δx of the independent variable, can be written approximately as

$$\Delta y = \Delta f \approx f'(x)\Delta x, \qquad (1.2.11)$$

provided Δx is sufficiently small to make the term $\varepsilon \Delta x$ negligible relative to the term $f'(x)\Delta x$.

1.2.5 Differential

The approximate description of the increment Δy by the linear part $f'(x)h = f'(x)\Delta x$ can also be used to put the term *differential* on a firmer logical basis. The original meaning of differentials as infinitely small quantities – different from zero – very soon turned out to have no precise meaning. One of the founders of differential calculus G.W. Leibniz (1646–1716) tried, without success, around 1680 to define the differential quotient as the ratio between two infinitely small increments dy and dx that were considered just before both quantities assumed the value zero. More than 100 years passed before the Bohemian priest B. Bolzano (in 1817) sharpened the definitions of such concepts as limits, continuity, etc., and then described the derivative by the limiting process in Eq. (1.2.4). However, Leibniz's notation has turned out to be the most suitable for handling calculations in physics and chemistry. For that reason, it is of value to attempt to give an unambiguous description of the identity

$$f'(x) \stackrel{\text{def}}{\equiv} \frac{dy}{dx},$$

in such a way that the expression dy/dx need not be regarded only as a symbol for the limiting process

$$\frac{dy}{dx} = \lim_{h \to 0} \frac{f(x+h) - f(x)}{h},$$

but can also be considered as a quotient between two actual, well-defined, quantities.

Starting from the definition of the derivative $f'(x)$ as a limiting process, as in Eq. (1.2.4), we then assign a fixed value to the independent variable x and consider the increment $h = \Delta x$ as the variable (see Fig. 1.1). The quantity $h = \Delta x$ is then called the *differential* of x, and is designated as dx. We then *define* the quantity

$$dy \stackrel{\text{def}}{\equiv} f'(x)\,dx, \tag{1.2.12}$$

as the *differential* dy of the function $y = f(x)$ corresponding to the differential dx of the independent variable. Thus, by means of this definition the derivative $f'(x)$ is regarded as the ratio between two quantities dy and dx, which can have any value provided their ratio is constant and equal to $f'(x)$. Comparing Eq. (1.2.9) with Eq. (1.2.10) shows that the differential dy is equal to the linear portion of the increment Δy that corresponds to the increment dx of the independent variable x (compare Fig. 1.1).

The introduction of the differentials dy and dx due to S.-F. Lacroix (1765–1843) and A.L. Cauchy (1789–1857) does not represent a new idea. But their merit is to make more precise the wording of "infinitesimal quantity": these quantities are now of finite magnitude, and not quantities "just differing from zero". Hence, when considering a particular problem, they may be chosen to be small enough so that one can, with confidence, replace the increment Δy of the function with its differential dy and write

$$\Delta y \approx dy = f'(x)\,dx = \left(\frac{dy}{dx}\right)dx, \qquad (1.2.13)$$

and

$$f(x+dx) \approx f(x) + f'(x)\,dx = f(x) + \left(\frac{dy}{dx}\right)dx. \qquad (1.2.14)$$

The validity of the above approximation depends on the special character of the physical situation in question. In general, the error introduced will be insignificant for the solution of the physical problem as long the infinitesimal quantities introduced are smaller than the actual error of measurement that are related to the physical situation.

1.2.5.1 The chain rule

One often finds that the dependent variable y is a function of the independent variable u that again is a function of the independent variable x, e.g.

$$y = u^3 \quad \text{and} \quad u = \sin x.$$

This situation is described by saying that y is a *function of a function* or that y is a *compound function* of x. In general we write this as

$$y = f(x) = F(u) = F\{u(x)\}.$$

If both derivatives

$$\frac{dF}{du} \quad \text{and} \quad \frac{du}{dx}$$

exist it can be shown that

$$f'(x) = F'(u)\,u'(x),$$

or, in terms of Leibniz's notation,

$$\frac{dy}{dx} = \frac{dF}{dx} = \frac{dF}{du}\frac{du}{dx}, \qquad (1.2.15)$$

which illustrates both the flexibility and suggestive strength of this notation. It appears as if the symbols dy and dx are quantities that can be considered and manipulated as if they were real numerical quantities. In fact, they can. According to Eq. (1.1.10) we have

$$dF = \frac{dF}{du}du, \quad \text{and} \quad du = \frac{du}{dx}dx,$$

so that

$$dF = \frac{dF}{du}\frac{du}{dx}dx,$$

which on division on both sides by dx becomes Eq. (1.2.15). In the above example we have $dy/du = 3u^2$ and $du/dx = \cos x$. Hence

$$\frac{dy}{dx} = \frac{dy}{du}\frac{du}{dx} = 3\sin^2 x \cos x.$$

For the function $y = \sin^3 \alpha x$ we obtain $dy/dx = 3\alpha \sin^2 \alpha x \cos \alpha x$, since

$$\frac{d(\sin \alpha x)}{dx} = \frac{d(\sin \alpha x)}{d(\alpha x)}\frac{d(\alpha x)}{dx} = \alpha \cos \alpha x.$$

If $y = \sin \sqrt{x} = \sin u$, where $u = \sqrt{x} = x^{1/2}$ we have

$$\frac{dy}{dx} = \frac{d\sin u}{du}\frac{du}{dx} = \cos \sqrt{x}\frac{d}{dx}(\sqrt{x}) = \cos \sqrt{x}\left(\frac{1}{2}\right)x^{-\frac{1}{2}} = \frac{1}{2}\frac{\cos \sqrt{x}}{\sqrt{x}}.$$

1.2.5.2 The derivative of the inverse function

It has previously been stated that if a continuous function $y = f(x)$ is either increasing or decreasing monotonically in an interval (say $a \leq x \leq b$) then the *inverse* function $x = g(y)$ also exists as a single-valued function that is continuous and monotonic in the same interval. If the function $y = f(x)$ is differentiable in the interval, the function increases monotonically if $f'(x) > 0$ in the interval and, correspondingly, can decrease monotonically if $f'(x) < 0$. Knowledge of the differentiability of a function in a given interval provides a tool for deciding whether the function also possesses an unambiguous inverse function as expressed in the following statement.

If the function $y = f(x)$ is differentiable in the interval $a < x < b$ and $f'(x) > 0$ everywhere or $f'(x) < 0$ everywhere, then the inverse function $x = g(y)$ also has a derivative $x' = g'(y)$ in the whole interval. The derivative of the original function $y = f(x)$ and that of the inverse function $x = g(y)$ are for the values of x and y belonging together connected by the following relation:

$$f'(x) \cdot g'(y) = 1, \tag{1.2.16}$$

or written in the form

$$\frac{dy}{dx} = \frac{1}{(dx/dy)}.$$ (1.2.17)

This is demonstrated by applying the definition for the derivative (Eq. (1.2.4)) on $y = f(x)$ and its inverse function $x = g(y)$. For the function $y = f(x)$ we have

$$f'(x) = \lim_{\Delta x \to 0} \frac{f(x + \Delta x) - f(x)}{\Delta x}.$$

The numerator is written as $f(x + \Delta x) - f(x) = \Delta y$. As $x = g(y)$ we can express Δx by means of the increment Δy, since $\Delta x = g(y + \Delta y) - g(y)$. Hence, the above difference quotient can also be written as

$$\frac{\Delta y}{g(y + \Delta y) - g(y)} = \frac{1}{[g(y + \Delta y) - g(y)]/\Delta y}.$$

Since the two functions $f(x)$ and $g(y)$ are continuous we have $\Delta y \to 0$ when $\Delta x \to 0$, and vice versa. This implies that

$$f'(x) = \lim_{\Delta x \to 0} \frac{f(x + \Delta x) - f(x)}{\Delta x} = \lim_{\Delta y \to 0} \frac{1}{[g(y + \Delta y) - g(y)]/\Delta y} = \frac{1}{g'(y)},$$

provided that $f'(x) \neq 0$ and $g'(y) \neq 0$ in the interval $a \leq x \leq b$.

1.3 Basic concepts of integral calculus

Integral calculus emerged from the need to determine areas of surfaces differing from those of rectangles and to find equations for curves where the behavior of their tangents were known. The basic method was known to the Greek mathematicians*, for example in their attempts to find the area of a circle, which was confined between the n-sided regular inscribed and circumscribed polygons, whose areas are known from Euclidian geometry. As n increases, the difference between the two areas becomes smaller. We can make this difference as small as we please by choosing n sufficiently large, and so the value of the area can be estimated to any degree of accuracy that is required. This method of exhaustion is essentially that of integral calculus.

---

* It was known in particular by Archimedes (287–212 BC), who, in addition to his great contributions to mathematics, is also regarded as the founder of the laws of equilibrium in rigid and fluid bodies. He was also an imaginative inventor.

1.3.1 Definite and indefinite integral

Let $y = f(x)$ be a function represented by a finite, positive value in the interval $a \leq x \leq b$. The *definite integral* of the function $y = f(x)$ from $x = a$ to $x = b$ is defined by the following operation. The interval $a \leq x \leq b$ is divided in n subintervals

$$\Delta x_1, \Delta x_2, \ldots, \Delta x_i, \ldots, \Delta x_n.$$

Let $f(x_i)$ be the value of the function somewhere in the subinterval Δx_i. One then introduces the sum

$$S_n = f(x_1)\Delta x_1 + f(x_2)\Delta x_2 + \cdots + f(x_i)\Delta x_i + \cdots + f(x_n)\Delta x_n, \quad (1.3.1)$$

or

$$S_n = \sum_{i=1}^{i=n} f(x_i)\Delta x_i = \sum_{i=1}^{i=n} \Delta A_i, \quad (1.3.2)$$

where $\sum$ stands for "sum of elements of the form ...", in this case

$$f(x_i)\Delta x_i = \Delta A_i,$$

where ΔA_i is the area of the rectangle with sides $f(x_i)$ and Δx_i.

If this sum S_n assumes a definite value, *the limit* of S_n, when all intervals Δx_i approach zero as the number of intervals $n \to \infty$, the function $f(x)$ is said to be *integrable* in the interval between $x = a$ and $x = b$. The value of this limit for S_n is denoted the *definite integral* of $y = f(x)$ from $x = a$ to $x = b$. The symbolism that reflects this operation is

$$\lim_{\substack{n \to \infty \\ \Delta x_i \to 0}} \sum_{i=1}^{i=n} f(x_i)\Delta x_i \stackrel{\text{def}}{\equiv} \int_a^b f(x)\,dx. \quad (1.3.3)$$

The symbol $\int$ – an elongated S – was introduced by Leibniz* to make an association to the "sum of infinitely large number of infinitely small subelements", and the symbol has retained its value of convenience ever since and is called the *integral sign*. We denote $x = a$ as the *lower limit* of the definite integral and $x = b$ as the *upper limit*. The arithmetic definition above also holds if $a > b$, as the only change that arises is that the differences $\Delta x_i = f(x_{i+1}) - f(x_i)$ now become negative when the interval is traversed from a to b. This suggests the relation

$$\int_a^b f(x)\,dx = -\int_b^a f(x)\,dx, \quad (1.3.4)$$

* In a manuscript dated 29th October 1675.

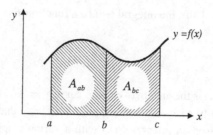

Fig. 1.2. Illustration of the definite integral as an area.

and the definition

$$\int_a^a f(x)\,dx = 0. \tag{1.3.5}$$

From a geometric point of view, Eq. (1.3.1) gives the value of the area between the curve $y = f(x)$ and the x-axis that is delimited by the lines $x = a$ and $x = b$. An example of such an area A_{ab} is shown in Fig. 1.2 together with the adjacent area A_{bc} that is delimited by the curve $y = f(x)$ and by the lines $x = b$ and $x = c$. Denoting the total area between the lines $x = a$ and $x = c$ as A_{ac}, we have: $A_{ab} + A_{bc} = A_{ac}$, or

$$\int_a^b f(x)\,dx + \int_b^c f(x)\,dx = \int_a^c f(x)\,dx. \tag{1.3.6}$$

On account of Eq. (1.3.4) and Eq. (1.3.5) this relation will hold for any mutual positions the three points a, b and c may assume.

In Fig. 1.2 it is assumed that the function $f(x)$ is positive in the whole range considered. However, the integral that is defined by Eq. (1.3.1) as the limit of the sum of elements $f(x_i)\Delta x_i$ is independent of such an assumption. If $f(x) < 0$ in part of the range from a to b it only results in making the summation elements in question negative, thereby assigning a negative value to the area where the curve of $f(x)$ is located *below* the x-axis. Thus, the total area that is enveloped by an arbitrary curve $y = f(x)$, will in general comprise positive as well as negative areas.

Let $y = f(t)$ represent a function of the independent variable that, for reasons of convenience, we shall denote by t. Next we consider the integral of this function taken from a fixed point $t = a$ to another point $t = x$, which we allow to vary on the t-axis. The value of this integral is then determined by the value that is assigned to x. Thus, the integral will be a function $F(x)$ of its upper limit $t = x$, namely

$$F(x) = \int_a^x f(t)\,dt. \tag{1.3.7}$$

The function $F(x)$ is the area between the curve $y = f(t)$ and the t-axis that is delimited by the fixed line $t = a$ and the line $t = x$ that may vary as we please. For that reason an integral $F(x)$ with a variable upper limit is called an *indefinite integral*. The condition for the existence of an indefinite integral $F(x)$ is that the function $y = f(t)$ is *continuous*.

1.3.2 The fundamental law

The fundamental law of integral and differential calculus* states: *the derivative of the indefinite integral $F(x)$ of the function $y = f(t)$ with respect to x is equal to the value of $f(t)$ for $t = x$, namely*

$$F'(x) = \frac{dF}{dx} = f(x), \tag{1.3.8}$$

that is *the process of integration that leads from the function $f(x)$ to $F(x)$ can be reversed by taking the derivative of the function $F(x)$ with respect to x.*

This important theorem can be demonstrated by applying the limiting procedure Eq. (1.2.4) to the difference quotient (Eq. (1.2.3)) of the indefinite integral, i.e.

$$F'(x) = \lim_{h \to 0} \left[\frac{F(x+h) - F(x)}{h} \right].$$

From Eq. (1.3.4) and Eq. (1.3.6) it follows that the denominator can be written as

$$
\begin{aligned}
F(x+h) - F(x) &= \int_a^{x+h} f(t)\,dt - \int_a^x f(t)\,dt \\
&= \int_x^a f(t)\,dt + \int_a^{x+h} f(t)\,dt \\
&= \int_x^{x+h} f(t)\,dt.
\end{aligned}
$$

The right-hand side of Fig. 1.3 can be visualized as the area between the curve $y = f(t)$ and the t-axis that is delimited by the lines $t = x$ and $t = x + h$. Furthermore it is seen that this area is contained between the two rectangles of

---

* This theorem was discovered around 1670 by Isaac Newton (1642–1727) and by G.W. Leibniz (1646–1716), independently of each other.

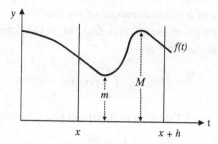

Fig. 1.3. To the derivation of the fundamental law of integral and differential calculus.

areas $h\,m$ and $h\,M$, where m and M are the smallest and largest value respectively of $y = f(t)$ in the interval $x \le t \le x + h$. Thus, we have

$$m \le \frac{F(x+h) - F(x)}{h} \le M.$$

As the function $y = f(t)$ is continuous both m and M will approach the value $f(x)$ when $h \to 0$. At the same time the difference quotient $(f(x+h) - f(x))/h$ will approach $F'(x)$. Thus, the above limit becomes

$$F'(x) = \lim_{h \to 0} \left[\frac{F(x+h) - F(x)}{h} \right] = f(x).$$

This version of the derivation of Eq. (1.18) is due to Cauchy* (1823).

Thus, to obtain an *indefinite integral* or a *primitive function* of the function $y = f(x)$ one has to find a function $F(x)$, whose derivative is equal to $f(x)$, i.e. find a function with the property

$$F'(x) = f(x). \tag{1.3.9}$$

1.3.3 Evaluation of a definite integral

Having at our disposal *one* primitive function $F(x)$ – an indefinite integral – that satisfies Eq. (1.3.8), we can construct any number of primitive functions, such as the function

$$G(x) = F(x) + C, \tag{1.3.10}$$

where C is a constant that will also satisfy Eq. (1.3.8), because the derivative of the function $y = C$ is equal to zero. This property leads to an important rule

---

* Augustin Louis Cauchy (1789–1857) was one of the greatest mathematicians. He was the founder of the modern theory of functions of complex variables, and was responsible for further development of the theory of differential equations, difference equations and infinite series.

1. Mathematical prelude

for finding the value of a definite integral of the function $f(x)$ taken between
the limits a and b, if a primitive function $G(x)$ of $f(x)$ is known.

Consider the primitive function

$$F(x) = \int_a^x f(t)\,dt,$$

of the function $y = f(x)$. Equation (1.3.10) can then be written as

$$G(x) = \int_a^x f(t)\,dt + C. \qquad (1.3.11)$$

This expression is also valid for $x = a$, namely

$$G(a) = \int_a^a f(t)\,dt + C.$$

But according to Eq. (1.3.5) we have

$$\int_a^a f(t)\,dt = 0$$

and hence

$$G(a) = \int_a^a f(t)\,dt + C = 0 + C.$$

Inserting $C = G(a)$ in Eq. (1.3.11) and putting $x = b$ gives

$$G(b) = \int_a^b f(t)\,dt + G(a),$$

or

$$\int_a^b f(t)\,dt = G(b) - G(a), \qquad (1.3.12)$$

no matter which of the many possible forms for $G(x)$ one may choose to use. We
then have the following important result: *to calculate the value of the definite
integral*

$$\int_a^b f(x)\,dx,$$

*we have only to find a function $G(x)$ with the property $G'(x) = f(x)$ and then
form the difference $G(b) - G(a)$.*

To simplify the notation it has been found to be convenient to remove the
limits from the integral sign in Eq. (1.3.11) and modify the graphics for the

indefinite integral to

$$G(x) = \int f(x)\,dx + C,\qquad (1.3.13)$$

where $\int \cdots dx$ means: find a function $F(x)$ with the property $F'(x) = f(x)$, and have the additive constant C in mind. Sometimes it may be useful to remember the above formula in this way

$$G(x) = \int \left(\frac{dF}{dx}\right) dx + C = F(x) + C,\qquad (1.3.14)$$

in particular in those cases where it is almost directly obvious that the function $f(x)$ can be written as the derivative of a function $F(x)$. The indefinite integral on the form $\int dx$ sometimes leads to difficulties in understanding until one realizes that the integrand in this case is $f(x) = 1$, which again is the derivative of the function $F(x) = x$. Hence we have: $\int dx + C = x + C$.

1.3.4 The mean value theorem

There are several ways for estimating the value of a definite integral. We shall consider the simplest. Let $y = f(x)$ represent a continuous non-negative function – either positive or zero – in the interval $a \le x \le b$, i.e. $f(x) \ge 0$. For the definite integral it holds that

$$\int_a^b f(x)\,dx = \lim_{\substack{n\to\infty \\ \Delta x_i \to 0}} \sum_{i=1}^{i=n} f(x_i)\Delta x_i \quad \ge 0,$$

as the sum contains only positive elements. Let M denote a number such that $M \ge f(x)$ for every value of x in the interval $a \le x \le b$. Furthermore, let m denote another number such that $m \le f(x)$ for every x in the interval $a \le x \le b$. Hence we have

$$\int_a^b m\,dx \le \int_a^b f(x)\,dx \le \int_a^b M\,dx.$$

This double inequality is illustrated geometrically in Fig. 1.4.

But we have

$$\int_a^b m\,dx = m\int_a^b dx = m(b-a), \quad \text{and also} \quad M\int_a^b dx = M(b-a),$$

and hence

$$m(b-a) \le \int_a^b f(x)\,dx \le M(b-a).$$

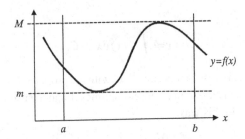

Fig. 1.4. Illustration of the mean value theorem for the definite integral.

Therefore, the value of the definite integral can be represented as the product of $(b - a)$ and some number μ that is located between m and M:

$$\int_a^b f(x)\,dx = \mu(b - a), \qquad m \le \mu \le M, \tag{1.3.15}$$

where we can regard μ as the *mean value* of $f(x)$ in the interval $a \le x \le b$. The function $y = f(x)$ is continuous in the interval considered, and will therefore assume all values between the largest and smallest value of $f(x)$ in the interval. Therefore, we can put $\mu = f(\xi)$ where ξ is located somewhere in the interval. The last expression can therefore also be written as

$$\int_a^b f(x)\,dx = (b - a)f(\xi), \qquad a \le \xi \le b. \tag{1.3.16}$$

This formula is called the *mean value theorem of the integral calculus*.

1.4 The natural logarithm

1.4.1 Definition of the natural logarithm

After this recapitulation of the fundamentals of the integral calculus we consider the function

$$y = f(x) = x^n.$$

If n is different from -1 there exists an indefinite integral

$$G(x) = \int x^n\,dx = \frac{1}{n+1}x^{n+1} + C, \tag{1.4.1}$$

since $G'(x) = x^n$. If $n = -1$, the function assumes the form

$$f(x) = \frac{1}{x} = x^{-1}.$$

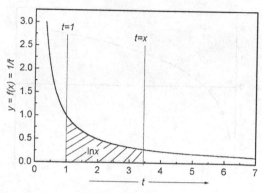

Fig. 1.5. Illustration of the geometric meaning of the natural logarithm $y = \ln x$ as an area.

The right-hand side of Eq. (1.4.1) then becomes indeterminate since $1/(n+1) = 1/(-1+1) = 1/0$. Thus, in this case the integral of $f(x) = 1/x$ cannot be expressed by Eq. (1.4.1). It turns out to be impossible to find an indefinite integral of the function $y = 1/x$ that is expressed in terms of elementary functions, i.e. polynomials, fractional rational functions (the ratio between two polynomials) or algebraic functions (e.g. the square root of a polynomial). Because of the frequent occurrence of the integral $\int dx/x$, mathematicians found it convenient to define a *new* function by means of this integral. This function is called the *natural logarithm* and is denoted as $\ln x$. This function is *defined* by the integral

$$\ln x = \int_1^x \frac{1}{t}\, dt, \tag{1.4.2}$$

i.e. as the area between the rectangular hyperbola $y = 1/t$ and the x-axis that is delimited between the line $t = 1$ and the line $t = x$ (Fig. 1.5). The variable x can be any positive number, but $x = 0$ is excluded because the integral diverges as the integrand $y = 1/t$ becomes infinite when $x \to 0$.

1.4.2 Elementary properties of the logarithm

The function $y = \ln x$ is useful for several reasons. The first follows from the fundamental theorem Eq. (1.3.8). We have

$$f'(x) = \frac{d \ln x}{dx} = \frac{1}{x}. \tag{1.4.3}$$

Thus, the derivative of $y = \ln x$ is always positive, but it decreases for increasing values of x. In accordance with this we see that the area under the rectangular

1. Mathematical prelude

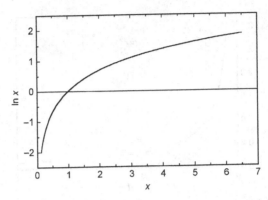

Fig. 1.6. The course of the function $y = \ln x$. The number $x = e$ satisfies the relation $\ln e = 1$.

hyperbola $y = 1/x$ taken between the two lines at x and $x + \Delta x$ decreases monotonically with increasing values of x. The course of the function $y = \ln x$ is illustrated in Fig. 1.6. Below we shall recapitulate the three basic properties of the logarithmic function.

1.4.2.1 Logarithm of a product

The main property of the logarithmic function is given by the formula

$$\ln a + \ln b = \ln(ab). \qquad (1.4.4)$$

To demonstrate this theorem we consider the function $F(x) = \ln x$ together with another function

$$G(x) = \ln(ax) = \ln w = \int_1^w \frac{1}{t} dt, \qquad (1.4.5)$$

where $w = ax$. Taking the derivative of $G(x)$ with respect to x yields (see Eq. (1.2.15))

$$G'(x) = \frac{d \ln w}{dw} \frac{dw}{dx} = \frac{1}{w} \frac{d(ax)}{dx} = \frac{1}{ax} a = \frac{1}{x}.$$

We also have

$$F'(x) = \frac{1}{x}.$$

The two functions $F(x)$ and $G(x)$ have exactly the same derivative and, consequently, can only differ from each other by a constant number. Thus,

$$G(x) = F(x) + C,$$

or

$$\ln(ax) = \ln x + C, \tag{1.4.6}$$

where C is a constant whose value does not depend upon x. To determine the value of C we put $x = 1$ in Eq. (1.4.6)

$$\ln(a1) = \ln 1 + C.$$

From the definition in Eq. (1.4.2) we have

$$\ln 1 = \int_1^1 \frac{1}{t} \, dt = 0. \tag{1.4.7}$$

Hence

$$\ln a = 0 + C,$$

and inserting this into Eq. (1.4.6) gives

$$\ln(ax) = \ln a + \ln x.$$

Putting then $x = b$ yields

$$\ln(ab) = \ln a + \ln b, \tag{1.4.8}$$

which is the addition theorem for the natural logarithm: *the logarithm of a product is equal to the sum of the logarithm of each factor.*

1.4.2.2 Logarithm of a quotient

To apply the above result we write the quotient a/b as $a \cdot (1/b)$. We have then

$$\ln \frac{a}{b} = \ln a + \ln \frac{1}{b}. \tag{1.4.9}$$

To see the meaning to be assigned to $\ln(1/b)$ we put $a = b$ in Eq. (1.4.8) and obtain

$$\ln 1 = 0 = \ln b + \ln \frac{1}{b},$$

from which it follows that

$$\ln(1/b) = -\ln b.$$

Hence

$$\ln \frac{a}{b} = \ln a - \ln b. \tag{1.4.10}$$

Thus, *the logarithm of a quotient is equal to the logarithm of the numerator minus the logarithm of the denominator.*

1.4.2.3 Logarithm of an exponential

Putting $a = b = x$ in Eq. (1.4.8) gives

$$\ln(xx) = \ln x^2 = \ln x + \ln x = 2 \ln x.$$

Then making the sequence of products $x^2 x = x^3$ etc., yields

$$\ln x^3 = 3 \ln x,$$

and

$$\ln x^n = n \ln x. \tag{1.4.11}$$

This relation that holds for any integer can be proven to hold for any number of rationals, irrationals as well as complex numbers. In other words: *the logarithm to a number x that is raised to any power r, is equal to the product of exponent and the logarithm of the number x.*

1.5 The exponential function

1.5.1 Definition of the exponential function

The function

$$F(x) = \ln x$$

assumes the value zero for $x = 1$ and grows monotonically towards ∞ with a decreasing slope ($F'(x) = 1/x$). Therefore, a number x must exist having the property $\ln x = 1$. This number was denoted e by L. Euler*. Thus, the number e is defined by the relation: $\ln e = 1$. The number e turns out to be an irrational[†] number whose numerical value is $e = 2.718\ 2818\ 28 \ldots$ (see Section 1.6). For values $0 < x < 1$ we have $-\infty < \ln x < 0$. This follows from $F(x) = \ln x = -\ln(1/x)$ and $1/x \to \infty$ when $x \to 0$. Thus, the function grows monotonically from $-\infty$ to ∞. This implies that to each value of x in the range $0 < x < \infty$ there exists one and only one value of y. Hence, the function

$$y = \ln x \tag{1.5.1}$$

---

* Leonhard Euler (1707–1783) was a Swiss mathematician. He first held various positions in St. Petersburg, and in 1741 was appointed Professor in mathematics at the academy of Frederick II of Prussia. In 1766 he returned as director of the Academy of Sciences in St. Petersburg. Euler was one of the most prolific of writers in mathematics, who made important contributions to almost every branch of mathematics. He was awarded the Prize of the Parisian Academy of Sciences on ten occasions.

† That is neither an integer nor a fraction.

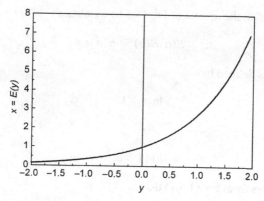

Fig. 1.7. The course of the inverse function $x = E(y)$ – the exponential function – of the logarithmic function $y = \ln x$ shown in the range $-2 \le y \le 2$ (compare Eq. (1.5.2)).

is associated with a *single valued inverse function*, that for now we shall denote by

$$x = E(y). \qquad (1.5.2)$$

In Eq. (1.5.1) x is the independent variable and y is the dependent variable. Equation (1.5.2) describes the same functional relationship, but now it is y that is considered as the independent variable and x as the dependent variable. The graphical representation of Eq. (1.5.2) is shown in Fig. 1.7. It could have been obtained by rotating Fig. 1.6 90° anti-clockwise and then reflecting the graph with respect to the y-axis.

We shall now consider the most important property of the E-function that is given by the relation

$$E(a)E(b) = E(a + b), \quad \text{for all values of } a \text{ and } b. \qquad (1.5.3)$$

To prove this relation we write the functions

$$E(a) = x \quad \text{and} \quad E(b) = z, \qquad (1.5.4)$$

and their inverse functions

$$a = \ln x \quad \text{and} \quad b = \ln z. \qquad (1.5.5)$$

From Eq. (1.4.8) and Eq. (1.5.5) we have

$$\ln(xz) = \ln x + \ln z = a + b. \qquad (1.5.6)$$

From Eq. (1.5.4) we have $xz = E(a)E(b)$, which is the inverse function of

1. Mathematical prelude

$\ln(xz)$. Taking then the inverse of Eq. (1.5.6) we obtain

$$xz = E(a)E(b) = E(a + b). \tag{1.5.7}$$

The number e is defined by

$$\ln e = 1.$$

Thus, we have

$$E(1) = e. \tag{1.5.8}$$

Hence, it follows from Eq. (1.5.7) that

$$e^2 = ee = E(1)E(1) = E(1 + 1) = E(2),$$
$$e^3 = e^2 e = E(2)E(1) = E(2 + 1) = E(3),$$
$$\cdot =$$
$$\cdot =$$
$$e^n = E(n),$$

where n is a positive integer. If n is a negative integer we proceed as follows. Since $\ln 1 = 0$ we have

$$E(0) = 1.$$

We can therefore write

$$E(n)E(-n) = E(n + (-n)) = E(0) = 1.$$

Hence

$$e^{-n} = \frac{1}{e^n}. \tag{1.5.9}$$

These relations can be proved to hold also for values of n that are rational, irrational and complex.

The function

$$y = e^x \tag{1.5.10}$$

is called the *exponential function* (sometimes known as the natural exponential function) and has the inverse function

$$x = \ln y. \tag{1.5.11}$$

The functions e^{-x} and $1 - e^{-x}$ for $0 \leq x \leq \infty$ turn up in many practical applications. The course of these functions is illustrated in Fig. 1.8.

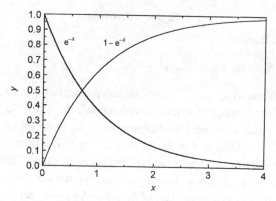

Fig. 1.8. Graph of the functions $y = e^x$ and $1 - e^{-x}$ for $x \geq 0$.

1.5.2 Derivative and integral

To find the derivative of the exponential function e^{-x} we consider the inverse function

$$y = e^x.$$

Since $x = \ln y$ is a continuous, monotonically increasing function in the range $0 < y < \infty$ the derivative dy/dx of the inverse function $y = e^x$ is*

$$\frac{dy}{dx} = \frac{1}{dx/dy}. \tag{1.5.12}$$

Inserting Eq. (1.4.3) we obtain the remarkable result

$$\frac{dy}{dx} = \frac{1}{1/y} = y = e^x, \tag{1.5.13}$$

or, in other words: *the derivative of the exponential function e^x is the function itself.*

To find the derivative of the function

$$y = e^{\alpha x} = e^u$$

we put $u = \alpha x$. It follows then from the chain rule in Eq. (1.2.15)

$$\frac{dy}{dx} = \frac{d}{dx}(e^{\alpha x}) = \frac{dy}{du}\frac{du}{dx} = e^u \alpha = \alpha e^{\alpha x} \tag{1.5.14}$$

and

$$F(x) = \int e^{\alpha x} dx = \frac{1}{\alpha} e^{\alpha x}, \tag{1.5.15}$$

since $F'(x) = (1/\alpha)\alpha e^{\alpha x} = e^{\alpha x}$.

* See Section 1.2.3.2, Eq. (1.2.17).

1.6 Taylor's theorem

A series of the form

$$S_n = a_0 + a_1 x + a_2 x^2 + a_3 x^3 + \cdots + a_n x^n + \cdots \qquad (1.6.1)$$

where the coefficients $a_0, a_1, a_2, \ldots a_n$ are independent of x, is called a *power series* in x. If the number of terms n is unbounded, the series S_∞ is called an *infinite series*. Such series are called *convergent* if S_∞ assumes a definite value when $n \to \infty$ for values of x in the range $a \le x \le b$. If, instead, the value of S_∞ becomes indeterminate, the series is *divergent* in the range. Such series play an important role in many branches of pure and applied mathematics, because it is possible to express a large class of functions by power series, and by this process we may often facilitate the mathematics of handling complex functions.

1.6.1 Taylor and Maclaurin series

Of particular interest is the case where the coefficients in the power series are composed of the derivatives of the function $F(x)$ in question.

1.6.1.1 Expansion of a polynomial

First we consider the simple situation where the function $F(x)$ is a finite *polynomial*

$$F(x) = a_0 + a_1 x + a_2 x^2 + a_3 x^3 + \cdots + a_n x^n, \qquad (1.6.2)$$

of the n-th degree. We differentiate on both sides of Eq. (1.6.2) once, twice, etc., with respect to x, to obtain

$$\frac{\mathrm{d}F}{\mathrm{d}x} = 1 \times a_1 + (\text{sum of powers of } x \text{ up to degree } n-1)$$

$$\frac{\mathrm{d}^2 F}{\mathrm{d}x^2} = 1 \times 2 \times a_2 + (\text{sum of powers of } x \text{ up to degree } n-2)$$

$$\frac{\mathrm{d}^3 F}{\mathrm{d}x^3} = 1 \times 2 \times 3 \times a_3 + (\text{sum of powers of } x \text{ up to degree } n-3)$$

$$\frac{\mathrm{d}^n F}{\mathrm{d}x^n} = 1 \times 2 \times 3 \cdots n \times a_n = n!\, a_n.$$

Putting $x = 0$ in Eq. (1.6.2) and in the above set of equations yields

$$a_0 = F(0), \quad a_1 = F'(0), \quad a_2 = \frac{1}{2!} F''(0), \quad a_3 = \frac{1}{3!} F'''(0) \cdots a_n = \frac{1}{n!} F^{(n)}(0).$$

Thus, for the polynomial $F(x)$ of the n-th degree the coefficients $a_0 \ldots a_n$ can be expressed in terms of the derivatives of $F(x)$ taken at $x = 0$. Inserting these values for $a_0 \ldots a_n$ in Eq. (1.6.2) results in

$$F(x) = F(0) + xF'(0) + \frac{x^2}{2!}F''(0) + \frac{x^3}{3!}F'''(0) \cdots \frac{x^n}{n!}F^{(n)}(0), \quad (1.6.3)$$

which states that the value of the polynomial $F(x)$ at the point x can be expressed from the values of the function and its derivatives at the point $x = 0$.

1.6.1.2 Expansion of an arbitrary function

The above expansion of $F(x)$ in terms values at $x = 0$ can be generalized to apply to the value of $F(x + h)$ in terms of the values of $F(x)$, where h is an arbitrary increment to the value x. In $F(x)$ we replace x by $\xi = x + h$ and consider the function $F(\xi) = F(x + h) = G(h)$ as a function of h, and regard for the moment the variable x as fixed and h as the independent variable. It then follows that $d\xi = dh$, and therefore also that

$$G'(h) = F'(\xi), \; G''(h) = F''(\xi), \; \ldots G^{(n)}(h) = F^{(n)}(\xi).$$

Putting $h = 0$ in the above set gives

$$G'(0) = F'(x), \; G''(0) = F''(x), \; \ldots G^{(n)}(0) = F^{(n)}(x).$$

As the function $G(h)(= F(x + h))$ is a polynomial in h and also in the n-th degree, $G(h)$ is related to $G(0)$ according to Eq. (1.6.3), namely

$$G(h) = G(0) + hG'(0) + \frac{h^2}{2!}G''(0) + \frac{h^3}{3!}G'''(0) \cdots \frac{h^n}{n!}G^{(n)}(0).$$

Replacing $G(h)$ by $F(x + h)$ and $G(0)$ by $F(x)$ we obtain the *Taylor series*

$$F(x + h) = F(x) + hF'(x) + \frac{h^2}{2!}F''(x) + \frac{h^3}{3!}F'''(x) \cdots \frac{h^n}{n!}F^{(n)}(x) \quad (1.6.4)$$

for a polynomial of the n-th degree that expresses the value in the neighboring position $x + h$ in terms of the values at position x. This expansion, which terminates with the $n + 1$-th term, is exact. The suggested structure of Eq. (1.6.4) led mathematicians to search for a similar structure in the case where $F(x)$ is an arbitrary function, not necessary a polynomial. In this case the above expansion represents an *approximation* to the function by a polynomial $P(x)$ that is given by the right-hand side of Eq. (1.6.4). Furthermore, the number of terms in the series does not remain finite as in Eq. (1.6.4) where all the derivatives beyond $F^{(n)}(x)$ vanish. The validity of the expansion and the deviation R_n – the remainder – of the polynomial value $P_n(x)$ from the actual value of $F(x)$ is contained in the following statement, which we quote without proof.

If the function $F(x)$ has continuous derivatives up to the $(n + 1)$-degree in the region between x and $x + h$, then

$$F(x + h) = F(x) + hF'(x) + \frac{h^2}{2!}F''(x) + \frac{h^3}{3!}F'''(x) \cdots$$

$$+ \frac{h^n}{n!}F^{(n)}(x) + R_n, \tag{1.6.5}$$

where the remainder R_n is given by

$$R_n = \frac{1}{n!}\int_0^h (x + \xi)^n F^{n+1}(x + \xi)\,d\xi. \tag{1.6.6}$$

More tractable forms of R_n due to Lagrange and to Cauchy will not be considered here, apart from noting that the behavior of the remainder R_n for $n \to \infty$ is decisive for the utility of Eq. (1.6.4). If $R_n \to 0$ for $n \to \infty$ then the series

$$F(x + h) = F(x) + hF'(x) + \frac{h^2}{2!}F''(x) + \frac{h^3}{3!}F'''(x) \cdots \frac{h^n}{n!}F^{(n)}(x) \tag{1.6.7}$$

represents a good approximation of the function $F(x)$ that is made better the more terms one chooses to include and which converges to $F(x)^*$ as $n \to \infty$. This series is known as a Taylor series[†]. With the same provisos we likewise have

$$F(x) = F(0) + xF'(0) + \frac{x^2}{2!}F''(0) + \frac{x^3}{3!}F'''(0) \cdots \frac{x^n}{n!}F^{(n)}(0), \tag{1.6.8}$$

which is usually called a Maclaurin series[‡].

1.6.1.3 The binomial series

Consider the function

$$F(x) = (1 + x)^\alpha, \tag{1.6.9}$$

where $x > -1$ and α is an arbitrary number that can be positive or negative, rational or irrational. To expand $(1 + x)^\alpha$ in a Maclaurin power series in x we

---

[*] Several test procedures are available to decide whether a given infinite series diverges or converges to a limiting value within a given range $a < x < b$.

[†] Stated in 1712 by Brook Taylor, a student of Newton. His derivation of the formula appeared in 1715 as a part of his development of the calculus of finite differences.

[‡] Given in 1742 by Colin Maclaurin, who rightfully stated that it was simply a special case of Taylor's formula. Some mathematicians are reluctant to credit Maclaurin by attaching his name to Eq. (1.6.8), but in practice it is convenient to have an easy way to distinguish between Eq. (1.6.7) and Eq. (1.6.8).

first calculate the derivatives

$$
\left.\begin{aligned}
F'(x) &= \alpha\,(1+x)^{(\alpha-1)}, \\
F''(x) &= \alpha \times (\alpha - 1)(1+x)^{(\alpha-2)}, \\
F'''(x) &= \alpha \times (\alpha - 1) \times (\alpha - 2)(1+x)^{(\alpha-3)}, \\
&\ddots, \\
F^{(\nu)}(x) &= \alpha \times (\alpha - 1) \times (\alpha - 2) \times (\alpha - 3)\cdots \\
&\quad (\alpha - \nu + 1)(1+x)^{(\alpha-\nu)}.
\end{aligned}\right\}
\qquad (1.6.10)
$$

For $x = 0$ we have

$$
\left.\begin{aligned}
F'(0) &= \alpha, \\
F''(0) &= \alpha \times (\alpha - 1), \\
F'''(0) &= \alpha \times (\alpha - 1) \times (\alpha - 2), \\
&= \cdot, \\
F^{(\nu)}(0) &= \alpha \times (\alpha - 1) \times (\alpha - 2) \times (\alpha - 3)\cdots(\alpha - \nu + 1).
\end{aligned}\right\}
\qquad (1.6.11)
$$

Inserting these values together with $F(0) = 1$ into Eq. (1.6.8) gives the Maclaurin series for the *binomial* $(1 + x)^\alpha$

$$
(1 + x)^\alpha = 1 + \alpha x + \frac{\alpha(\alpha - 1)}{2!} x^2 + \frac{\alpha(\alpha - 1)(\alpha - 2)}{3!} x^3 + \cdots
$$
$$
+ \frac{\alpha(\alpha - 1)(\alpha - 2)\cdots(\alpha - n + 1)}{n!} x^n + R_n. \qquad (1.6.12)
$$

When α is *not* a *positive integer* the number of terms becomes infinite. A closer examination shows that if $|x| < 1$ then $R_n \to 0$ for $x \to \infty$, and the expression $(1 + x)^\alpha$ can be expanded in the infinite binomial series

$$
(1 + x)^\alpha = 1 + \frac{\alpha}{1!} x + \frac{\alpha(\alpha - 1)}{2!} x^2 + \cdots = \sum_{\nu=0}^{\infty} \binom{\alpha}{\nu} x^\nu \qquad (1.6.13)
$$

where the general binomial coefficient is usually written as

$$
\frac{\alpha(\alpha - 1)\cdots(\alpha - \nu + 1)}{\nu!} = \frac{\alpha!}{(\alpha - \nu)! \cdot \nu!} = \binom{\alpha}{\nu}, \quad \text{for } \nu > 0. \qquad (1.6.14)
$$

We may also consider

$$
(p + q)^\alpha = p^\alpha (1 + x)^\alpha, \quad \text{where } x = \frac{q}{p}, \qquad (1.6.15)
$$

which combined with Eq. (1.6.11) gives

$$(p+q)^\alpha = p^\alpha + \alpha p^{(\alpha-1)}q + \frac{\alpha(\alpha-1)}{2!}p^{(\alpha-2)}q^2$$

$$+ \frac{\alpha(\alpha-1)(\alpha-2)}{2!}p^{(\alpha-3)}q^3 + \cdots$$

$$+ \frac{\alpha(\alpha-1)(\alpha-2)\cdots(\alpha-\nu+1)}{n!}p^{(\alpha-\nu)}q^\nu. \qquad (1.6.16)$$

In this general form* of the binomial expansion the infinite series will converge to $(p+q)^\alpha$ if $|q| < |p|$. If $\alpha = n$ is a positive integer the expansion is a polynomial that stops after the $n+1$-th term.

1.6.2 Series of the logarithmic and exponential functions

The series expansions of these functions were obtained early by the pioneers of the calculus.

1.6.2.1 The logarithm

We have from Eq. (1.4.3)

$$\ln(1+x) = \int_1^{1+x} \frac{du}{u}.$$

Changing the variable of integration to $t = u - 1$ we have $du/dt = 1$, and the limits are changed to $t = 0$ for $u = 1$, and $t = x$ for $u = 1 + x$. Hence the above integral can be written

$$\ln(1+x) = \int_0^x \frac{1}{1+t} \, dt.$$

For $|t| < 1$ the integrand $1/(1+t) = (1+t)^{-1}$ can be expanded as a binomial series

$$(1+t)^{-1} = 1 + \frac{-1}{1!}t + \frac{-1(-1-1)}{2!}t^2 + \frac{-1(-1-1)(-1-2)}{3!}t^3 + \cdots$$

$$= 1 - t + t^2 - t^3 + \cdots.$$

Substituting the right-hand side in the above integral and integrating terms by turn gives

$$\ln(1+x) = x - \tfrac{1}{2}x^2 + \tfrac{1}{3}x^3 + \tfrac{1}{4}x^4 + \cdots, \qquad \text{for } -1 < x < 1 \qquad (1.6.17)$$

---

* This was discovered in 1664 by Newton, and was the first of his many outstanding contributions to mathematics and physics.

or replacing x by $-x$

$$\ln(1-x) = -x - \tfrac{1}{2}x^2 - \tfrac{1}{3}x^3 - \tfrac{1}{4}x^4 + \cdots, \quad \text{for } -1 < x < 1. \quad (1.6.18)$$

These equations are of limited value as they are only valid for $-1 < x < 1$. The usefulness of this expansion is improved by taking the differences between the two series for the same value of n. This gives (compare Eq. (1.4.10))

$$\ln\frac{1+x}{1-x} = \ln z = 2\left(x + \frac{x^3}{3} + \frac{x^5}{5} + \frac{x^7}{7} + \cdots\right), \quad (1.6.19)$$

which now allows the calculation of any positive number z, since

$$z = \frac{1+x}{1-x}, \quad (1.6.20)$$

irrespective of its magnitude, always has a solution for x that lies between $x = -1$ and $x = +1$. For example, if $z = 3.0$ this yields $x = 0.5$. Furthermore, the expansion converges faster than the two other series.

1.6.2.2 The exponential function

The function $F(x) = e^x$ represents one of the simplest expansions. We have from Eq. (1.5.13)

$$F'(x) = e^x, \quad F''(x) = e^x \quad F'''(x) = e^x \ldots F^{(n)}(x) = e^x,$$

and

$$F'(0) = 1, \quad F''(0) = 1, \quad F'''(0) = 1 \ldots F^{(n)}(0) = 1.$$

Thus, the Maclaurin series for e^x is

$$e^x = 1 + x + \frac{x^2}{2!} + \frac{x^3}{3!} + \cdots \frac{x^n}{n!}, \quad (1.6.21)$$

which converges for all values of x. Putting $x = 1$ gives

$$e = 1 + 1 + \tfrac{1}{2!} + \tfrac{1}{3!} + \tfrac{1}{4!} + \cdots \approx 2.71828 \quad (1.6.22)$$

as the numerical value of the (irrational) number e.

Similarly we obtain for the trigonometric series

$$\sin x = x - \frac{x^3}{3!} + \frac{x^5}{5!} + \cdots + (-1)^n \frac{x^{2n+1}}{(2n+1)!}, \quad (1.6.23)$$

and

$$\cos x = 1 - \frac{x^2}{2!} + \frac{x^4}{4!} + \cdots + (-1)^n \frac{x^n}{n!}. \qquad (1.6.24)$$

1.6.3 Approximate expressions of functions

In many mathematical calculations involving a function $F(x)$ the complexity of $F(x)$ may make it difficult or impossible to find a solution that is exact and able to be expressed in terms of elementary functions. Instead of wasting time trying a probably hopeless attempt, it is often advantageous first to seek a simpler type of solution by replacing $F(x)$ with an approximate formula that facilitates the handling of the mathematical problem but still reproduces a solution having an acceptable degree of accuracy. To this end the most used tools are the binomial series and the Taylor and Maclaurin series. For example, if it suffices to consider only small values of x in $F(x)$ one takes the Maclaurin series and writes

$$F(x) \approx F(0) + x\, F'(0), \qquad (1.6.25)$$

to represent $F(x)$ as *a first-order approximation*. We can say that "$F(x)$ has been linearized". Thus, for small values of x one has from Eq. (1.6.14) and Eq. (1.6.18)

$$\ln(1+x) \approx x, \quad \text{and} \quad e^x \approx 1 + x. \qquad (1.6.26)$$

If the first-order approximation appears to be too crude we could add the next term of the Maclaurin series and write

$$F(x) \approx F(0) + x\, F'(0) + \frac{x^2}{2!} F''(0) \qquad (1.6.27)$$

to represent $F(x)$ as *a second-order approximation*.

The expansion of an arbitrary function by polynomials is of particular value in the case of integration of an intractable function, since integration of x^n for $x \neq 1$ does not represent a problem. For example to integrate

$$\int \sin(x^2)\, dx,$$

we replace x in Eq. (1.6.20) by x^2 to give

$$\sin(x^2) = x^2 - \frac{x^6}{3!} + \frac{x^{10}}{5!} - \cdots.$$

Integration of both sides with respect to x then yields

$$\int \sin(x^2) = \frac{1}{2}x^3 - \frac{x^7}{6 \cdot 3!} + \frac{x^{11}}{10 \cdot 5!} - \cdots.$$

1.6.4 Evaluation of an undetermined expression 0/0

Consider a function

$$F(x) = \frac{f(x)}{g(x)}.$$

It sometimes happens that there is a value $x = a$ such that

$$f(a) = g(a) = 0.$$

In this case the function $F(x)$ takes the so-called *indeterminate* form

$$F(a) = \frac{0}{0}$$

for $x = 0$. As it stands, the right-hand side does not make sense unless it is regarded as the *limiting value* of a quotient in which the numerator and denominator both tend to zero as $x \to a$, i.e.

$$F(a) = \lim_{x \to a} \left(\frac{f(x)}{g(x)} \right). \tag{1.6.28}$$

In many cases the limit assumes a definite value. This can be found by expanding $f(a + h)$ and $g(a + h)$ in a Taylor series. We write

$$\frac{f(a + h)}{g(a + h)} = \frac{f(a) + h \, f'(a) + h^2 \, f''(a)/2! + h^3 \, f'''(a)/3! + \cdots}{g(a) + h \, g'(a) + h^2 \, g''(a)/2! + h^3 \, g'''(a/3!) + \cdots}.$$

Since $f(a) = g(a) = 0$ we have

$$\frac{f(a + h)}{g(a + h)} = \frac{f'(a) + h \, f''(a)/2! + h^2 \, f'''(a)/3! + \cdots}{g'(a) + h \, g''(a)/2! + h^2 \, g'''(a/3!) + \cdots}.$$

Now

$$\lim_{x \to a} \left(\frac{f(x)}{g(x)} \right) = \lim_{h \to 0} \left(\frac{f(a + h)}{g(a + h)} \right) = \lim_{h \to 0} \left(\frac{f'(a) + h \, f''(a)/2! + \cdots}{g'(a) + h \, g''(a)/2! + \cdots} \right). \tag{1.6.29}$$

Thus

$$\lim_{x \to a} \left(\frac{f(x)}{g(x)} \right) = \frac{f'(a)}{g'(a)}. \tag{1.6.30}$$

1. Mathematical prelude

This result is known as L'Hospital's* rule. Consider, for example,

$$F(x) = \frac{x}{1 - e^x} = \frac{f(x)}{g(x)}.$$

Here $f(0) = g(0) = 0$. We have $f'(x) = 1$ and $g'(x)) = -e^x$, giving $f'(0) = 1$ and $g'(0) = -1$. Hence

$$\lim_{x \to 0} \left(\frac{x}{1 - e^x} \right) = \frac{1}{-1} = -1.$$

1.7 Basic techniques of integration

Whereas we have a definite set of rules that applies to calculation of the derivative $F'(x)$ of function $F(x)$, no such standard procedures exist for us to perform the *inverse* process, i.e. to find a function $F(x)$ whose derivative $F'(x)$ is equal to a given function $f(x)$. In the simplest cases one can manage by relying on one's fingertip knowledge of the derivatives of the elementary functions (e.g. knowing that the derivative of x^n is equal to $n\,x^{n-1}$, it is also evident that the integral of x^n is $x^{n+1}/(n+1)$). However, in the majority of cases we have to manipulate the integrand in some way or other to transform the given integral in one or several steps to forms that are recognized as belonging to the class of elementary, well-known integrals. Two such general useful methods are available.

1.7.1 The method of substitution

In this method we introduce the integration of a new variable. Let

$$F(x) = \int_a^x f(t)\,dt$$

represent an indefinite integral of the function $f(x)$, i.e. $F'(x) = f(x)$. We then express the function $F(x)$ by a new variable u, that is defined by means of the equation

$$x = \phi(u),$$

where $\phi(u)$ is any continuous, single-valued function of u. The problem then is expressing the integral of a function $f(x)$ with respect to x as the integral of

---

* The Marquis de L'Hospital, a pupil of Leibniz, published the first textbook on differential calculus.

a function of u with respect to the new variable u. We have

$$F\{\phi(u)\} = \int_a^{\phi(u)} f(t)\,dt.$$

To differentiate this expression with respect to u we make use of the chain rule in Eq. (1.2.15) and Eq. (1.3.9)*. This gives

$$\frac{dF}{du} = \frac{d}{d\phi}\left(\int_a^{\phi(u)} f(t)\,dt\right)\frac{d\phi}{du} = f\{\phi(u)\}\frac{d\phi}{du},$$

from which it follows that

$$F\{\phi(u)\} = \int f\{\phi(u)\}\,\phi'(u)\,du. \qquad (1.7.1)$$

Since

$$F\{\phi(u)\} = F(x) = \int f(x)\,dx$$

the formula of transformation takes the form

$$\int f(x)\,dx = \int f\{\phi(u)\}\,\phi'(u)\,du = \int h(u)\frac{dx}{du}\,du, \qquad (1.7.2)$$

where $h(u) = f(\phi(u))$.

Looking at the terms connected with the first equality sign, the rule is: if the integrand has the special form of a composite function $f\{\phi(u)\}$ times the derivative of the $\phi'(u)$ of the internal function one can replace $\phi(u) = x$ and instead evaluate the integral $\int f(x)\,dx$ with respect to x (i.e. formally replacing (substituting) $\phi'(u)\,du$ by dx).

EXAMPLE The integral $\int \sin(u^2)\,u\,du$ can also be written as $\frac{1}{2}\int \sin(u^2)\,2u\,du$ and is thereby put into the above standard form $f\{\phi(u)\}\,\phi'(u)$. Replacing $x = u^2$ we have

$$\int \sin(u^2)\,u\,du = \frac{1}{2}\int \sin(u^2)\,2u\,du = \frac{1}{2}\int \sin x\,dx = -\frac{1}{2}\cos x = -\frac{1}{2}\cos u^2.$$

One could also arrive at this result by direct inspection by realizing that

$$-\frac{d}{du}(\cos u^2) = -\frac{d}{du^2}(\cos u^2)\frac{du^2}{du} = \sin(u^2)\,2u,$$

and so

$$\int \sin(u^2)\,2u\,du = -\int\left(\frac{d}{du}(\cos u^2)\right)du = -\cos(u^2).$$

---

* The fundamental theorem of integral calculus.

In many cases the integrand $f(x)$ will enter as a composite function that can be written as $f(x) = h(u)$, where $u = \psi(x)$, where it appears to be a fairly easy matter to integrate $h(u)$ with respect to u. In this case it may be advantageous to *shift* to this new variable of integration u that is related to the x variable by $x = \phi(u)$. Connecting together the terms on the left-hand side and right-hand side of Eq. (1.7.2) the rule reads: if the variable of integration x of a composite function $f(x) = h(u)$, where $u = \psi(x)$, is changed to the variable u the integrand $h(u)$ must be multiplied with the derivative dx/du. Thus

$$\int h\{\psi(x)\}\,dx = \int h(u)\,dx = \int h(u)\left(\frac{dx}{du}\right)du \qquad (1.7.3)$$

or, if we are dealing with a definite integral between the limits a and b the corresponding limits after change of variable are $\psi(a)$ and $\psi(b)$, thus

$$\int_a^b h\{\psi(x)\}\,dx = \int_{\psi(a)}^{\psi(b)} h(u)\left(\frac{dx}{du}\right)du. \qquad (1.7.4)$$

EXAMPLE To evaluate $\int (ax+b)^n\,dx$ one introduces the substitution $u = \psi(x) = ax+b$, i.e. $x = \phi(u) = (u-b)/a$ and $dx/du = 1/a$. Using Eq. (1.7.3) we have

$$\int (ax+b)^n\,dx = \int u^n\left(\frac{dx}{du}\right)du = \frac{1}{a}\int u^n\,du$$

$$= \frac{1}{a}\frac{1}{n+1}u^{n+1} = \frac{1}{a(n+1)}(ax+b)^{n+1} \qquad (n \neq -1).$$

For $n = -1$ we have

$$\int \frac{1}{ax+b}\,dx = \frac{1}{a}\int \frac{1}{u}\,du = \frac{1}{a}\ln u = \frac{1}{a}\ln(ax+b).$$

1.7.2 Partial integration

We consider the basic formula* for the derivative of the product of two functions $f(x)$ and $g(x)$

$$\frac{d}{dx}(f(x)g(x)) = f(x)g'(x) + f'(x)g(x).$$

* See Section 1.2.3.1

Both sides are then integrated with respect to x^*

$$\int \frac{d}{dx}(f(x)g(x))dx = \int f(x)g'(x)dx + \int f'(x)g(x)dx$$

$$f(x)g(x) = \int f(x)g'(x)dx + \int f'(x)g(x)dx,$$

or

$$\int f(x)g'(x)dx = f(x)g(x) - \int g(x)f'(x)dx, \qquad (1.7.5)$$

which is the formula for the integration technique that is called *partial integration*. When the integrand is split into the product of two functions $f(x)$ and $g'(x)$, it may turn out that it is easier to integrate the product of $g(x)f'(x)$ than the product of $f(x)g'(x)$. Whereas the replacement of $f(x)$ by $f'(x)$ on the right-hand side does not in principle present any difficulty, the usefulness of the procedure depends on the possibility of replacing $g'(x)$ by $g(x) = \int g'(x)dx$. For that reason, $g'(x)$ is in general chosen to be easily integrable.

In the case of a definite integral between the limits $x = a$ to $x = b$ the formula becomes

$$\int_{x=a}^{x=b} f(x)g'(x)dx = [f(x)g(x)]_{x=a}^{x=b} - \int_{x=a}^{x=b} g(x)f'(x)dx. \quad (1.7.6)$$

A slightly different notation might make Eq. (1.7.5) easier to memorize. We put

$$f(x) = u, \quad \text{and} \quad f'(x) = \frac{du}{dx},$$

$$g(x) = v, \quad \text{and} \quad g'(x) = \frac{dv}{dx},$$

whereby Eq. (1.7.5) takes the form

$$\frac{d(uv)}{dx} = u\frac{dv}{dx} + v\frac{du}{dx}.$$

Integration of both sides with respect to x gives

$$\int u\frac{dv}{dx}dx = uv - \int v\frac{du}{dx}dx.$$

These integrals can also be evaluated by changing the integration variable x to u on the left-hand side and to v on the right-hand side $(du/dx)dx = du$ and

---

* Disregarding any additive constants of integration.

$(dv/dx)dx = dv$, thus

$$\int u\, dv = uv - \int v\, du,$$
(1.7.7)

or in the case of dealing with a definite integral between the limits $x = a$ and $x = b$

$$\int_{x=a}^{x=b} u\, dv = [uv]_{x=a}^{x=b} - \int_{x=a}^{x=b} v\, du.$$
(1.7.8)

EXAMPLE In this example

$$\int \ln x\, dx = \int \ln x \cdot 1 \cdot dx$$

we have particularly emphasized the integrand on the right-hand side to show that we put $f(x) = \ln x$ and $g'(x) = 1$. Thus, $f'(x) = 1/x$ and $g(x) = x$. Hence, according to Eq. (1.7.7) we obtain

$$\int \ln x\, dx = x\, \ln x - \int \frac{x}{x}\, dx = x\, \ln x - \int 1\, dx = x\, \ln x - x.$$

In the next example

$$\int x^2 e^x\, dx,$$

we put $u = x^2$, i.e. $du = 2x\, dx$, and $dv = e^x\, dx$, i.e. $v = e^x$. Thus, the integral is

$$\int x^2 e^x\, dx = uv - \int v\, du = x^2 e^x - \int 2x\, e^x\, dx.$$

Repeating the procedure on the last integral by putting $u = 2x$, i.e. $du = 2dx$, and again $dv = e^x dx$, i.e. $x = e^x$ gives

$$\int 2x\, e^x\, dx = 2x\, e^x - \int 2e^x\, dx = 2x\, e^x - 2e^x,$$

and hence

$$\int x^2 e^x\, dx = e^x(x^2 - 2x + 2).$$

1.8 Functions of several variables

At an early stage in the development of the mathematical analysis the need emerged to consider functions that depended on several variables in contrast to the functions hitherto considered that all were of the form $y = f(x)$, where the value of the independent variable y depends only on the value of a *single* variable x. This demand arose naturally by applying mathematics to describe events occurring in nature, where many of the phenomena studied turned out to be governed by two, three or more independent variables. A simple example presents the ideal gas, whose volume V, pressure p and temperature T are connected by the gas law

$$pV = n\mathcal{R}T,$$

where $\mathcal{R}$ is the gas constant $(8.314\ \mathrm{J \cdot mol^{-1} \cdot K^{-1}})$ and n the number of moles contained in the volume in question. It appears that the gas volume V depends on the two variables p and T, since*

$$V = n\mathcal{R}\frac{T}{p}.$$

We split the equation in the following two ways

$$V = \begin{cases} \dfrac{n\mathcal{R}}{p} \cdot T \\[2ex] n\mathcal{R}T \cdot \dfrac{1}{p} \end{cases},$$

or

$$V \propto \begin{cases} T, & \text{if } p = \text{constant} \\[2ex] \dfrac{1}{p}, & \text{if } T = \text{constant} \end{cases}.$$

Thus, the gas volume V is proportional to the temperature T, if the pressure p is kept *constant,* and is inversely proportional to the pressure p at a *constant temperature* T. In both cases the volume V is *a function of one variable,* i.e. T or p. Thus, the effect on the volume V of small changes ΔT at constant pressure p and of small changes Δp at constant temperature T can be handled in Section 1.2.5. At constant p we have

$$(\Delta V)_p = \frac{n\mathcal{R}}{p}\frac{\mathrm{d}}{\mathrm{d}T}(T)\,\Delta T = \frac{n\mathcal{R}}{p}\,\Delta T,$$

and at constant T we have

$$(\Delta V)_T = n\mathcal{R}T\frac{\mathrm{d}}{\mathrm{d}p}\left(\frac{1}{p}\right)\Delta p = -\frac{n\mathcal{R}T}{p^2}\,\Delta p,$$

where the subscripts $(\Delta V)_p$ denote that volume change occurs while the pressure p is held constant, and similarly with $(\Delta V)_T$.

---

* Naturally this selection of V as the independent variable is quite arbitrary. One could equally have chosen T or p as independent variables.

1.8.1 Geometrical representation

A function of one variable can be visualized as a curve in a two-dimensional plane. Correspondingly a function of two variables can be represented geometrically as a *surface*. To illustrated this we consider the relation

$$x^2 + y^2 + z^2 = r^2, \qquad (1.8.1)$$

where (x, y, z) are coordinates* in a rectangular three-dimensional coordinate system and r is a constant. Every point P in space with coordinates that satisfy Eq. (1.8.1) have the same distance r to the origin $(0,0,0)$ of the coordinate system. Geometrically this means that the totality of such points P constitutes the *surface* of the sphere with radius r and center in $(0,0,0)$. Consider now

$$z = \sqrt{r^2 - (x^2 + y^2)}$$

where x and y are prescribed by the constraint to satisfy

$$r^2 - (x^2 + y^2) \geq 0,$$

i.e. the points (x, y) are only localized inside the circle in the x–y plane with radius r and center at $x = 0$ and $y = 0$. A straight line drawn through the point (x, y) in the x–y plane and parallel to the z-axis will cut the surface of the sphere of radius r in a point whose distance to the x–y plane is equal to z and whose magnitude $0 \leq z \leq r$ depends upon the position of the point (x,y) in the x–y plane. Thus, the variable z is a function

$$z = f(x, y)$$

of the two independent variables x and y, and is represented as a *surface* in the space above the x–y plane (in the present case the surface of the hemisphere with radius r and center at $(0,0,0)$). Similarly, any function of two variables x and y

$$W = z = f(x, y)$$

can be represented as a surface in a rectangular coordinate system that contains all the z-values that arise from the set of points (x, y) in the x–y plane.

1.8.2 Partial derivatives

The concept of the derivative or differential quotient of a function of one variable $y = f(x)$ shall now be extended to comprise functions of several variables. For the sake of simplicity we shall consider a function of two variables only,

---

* These are also called Cartesian coordinates.

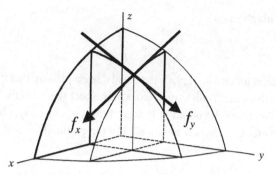

Fig. 1.9. Illustration of the partial derived functions $\partial f(x, y)/\partial x \overset{\text{def}}{=} f_x$ and $\partial f(x, y)/\partial y \overset{\text{def}}{=} f_y$. For further explanation see text.

$W = f(x, y)$ as sketched in Fig. 1.9. This function can be constrained to behave as a function of one variable, i.e. of either x or y. If we assign a definite, constant value y_0 to the variable y we are now dealing instead with a function $W = f(x, y_0)$ of *one* variable, where y_0 now serves as a parameter. We can represent this function $f(x, y_0)$ geometrically by letting the surface be cut by the plane $y = y_0$, i.e. a plane perpendicular to the y-axis at the position y_0. Since all the points in this plane contain the value $y = y_0$ the curve of intersection between the plane $y = y_0$ and the surface $W = f(x, y)$ represents a plane curve $W = f(x, y_0)$ situated in the plane and represented by the equation $u = f(x, y_0)$, i.e. a function that varies only with x. Consider then a fixed point x_0 and we can differentiate this function* as shown in Section 1.2 by the limit

$$\lim_{h \to 0} \left[\frac{f(x_0 + h, y_0) - f(x_0, y_0)}{h} \right], \tag{1.8.2}$$

by means of which we arrive at the *partial derivative of $f(x, y)$ with respect to x in the point* (x_0, y_0). Geometrically this partial derivative represents the *slope* of the curve $W = f(x, y_0)$ at the point $x = x_0$ or, in other words the *slope of the point $x = x_0$ on the surface $W = f(x, y)$* in the direction of the x-axis. For a function of one variable $f'(x)$ is used to denote the derived function. Correspondingly, the above limit is expessed as

$$\lim_{h \to 0} \left[\frac{f(x_0 + h, y_0) - f(x_0, y_0)}{h} \right] \overset{\text{def}}{=} f_x(x_0, y_0), \tag{1.8.3}$$

to indicate the partial derivative with respect to x of the function $f(x, y)$ of two variables. To emphasize the limiting process one very often uses the analogy to

---

* Provided the derivative exists.

the Leibniz's notation

$$\frac{\partial f}{\partial x} \quad \text{or} \quad \frac{\partial}{\partial x} f, \tag{1.8.4}$$

by using a special letter ∂ to replace the usual d to emphasize that it is a function of several variables that is differentiated with respect to one of the variables. In physics and chemistry the notation is used almost exclusively, although some authors still prefer Cauchy's D notation*. For the partial derivative the notation is

$$\frac{\partial f}{\partial x} = D_x f.$$

Similarly the partial derivative of $f(x, y)$ with respect to y in the point (x_0, y_0) is defined by the limit

$$\lim_{k \to 0} \left[\frac{f(x_0, y_0 + k) - f(x_0, y_0)}{k} \right] \begin{array}{l} \stackrel{\text{def}}{\equiv} f_y(x_0, y_0) \\[2mm] \stackrel{\text{def}}{\equiv} \dfrac{\partial f}{\partial y} \\[2mm] \stackrel{\text{def}}{\equiv} D_y f(x_0, y_0) \end{array} \Bigg\}. \tag{1.8.5}$$

The above partial derivatives are defined at the fixed point (x_0, y_0). But these operations could just as well have been carried out at any point (x, y), where the function $f(x, y)$ is defined and differentiable. Hence, in general one writes

$$\begin{aligned} \frac{\partial f(x, y)}{\partial x} &= \frac{\partial f}{\partial x} = f_x(x, y) = D_x f(x, y) \\ \frac{\partial f(x, y)}{\partial y} &= \frac{\partial f}{\partial y} = f_y(x, y) = D_y f(x, y) \end{aligned} \Bigg\} \tag{1.8.6}$$

from which it follows that the derivatives are also functions of x and y.

Derivatives of second and higher orders enter into calculations. Thus, the notation for second-order derivatives is

$$\begin{aligned} \frac{\partial^2 f(x, y)}{\partial x^2} &= \frac{\partial^2 f}{\partial x^2} = f_{xx}(x, y) = D_{xx} f(x, y) \\ \frac{\partial^2 f(x, y)}{\partial y^2} &= \frac{\partial^2 f}{\partial y^2} = f_{yy}(x, y) = D_{yy} f(x, y) \end{aligned} \Bigg\}. \tag{1.8.7}$$

Since the first-order derivative is generally also a function of several variables, one can calculate, for example, the rate of change with y of the partial derivative

* $Df \stackrel{\text{def}}{\equiv} f'(x) \stackrel{\text{def}}{\equiv} dy/dx$. See Section 1.2.

$f_x(x, y)$ with respect to the variable x. Thus

$$f_{xy} = \frac{\partial f}{\partial y}\left(\frac{\partial f}{\partial x}\right) = \frac{\partial^2 f}{\partial y \partial x}.$$

Furthermore, it can be proven that if both $f_x(x, y)$ and $f_y(x, y)$ are *continuous* functions then the order of differentiation of $f(x, y)$ with respect to x and y is immaterial. Therefore, we have

$$f_{xy}(x, y) = \frac{\partial f}{\partial y}\left(\frac{\partial f}{\partial x}\right) = \frac{\partial f}{\partial x}\left(\frac{\partial f}{\partial y}\right).$$

EXAMPLE Consider the function of two variables

$$f(x, y) = x^3 y.$$

To calculate the derivative of this function with respect to x we treat y as a constant. This gives

$$\frac{\partial f(x, y)}{\partial x} = y\frac{\mathrm{d}}{\mathrm{d}x}(x^3) = 3x^2 y,$$

and by calculating the partial derivative of y we consider x as a constant. Thus,

$$\frac{\partial f(x, y)}{\partial y} = x^3 \frac{\mathrm{d}}{\mathrm{d}y}(y) = x^3.$$

We have likewise

$$\frac{\partial^2 f}{\partial x^2} = 3y\frac{\mathrm{d}}{\mathrm{d}x}(x^2) = 6xy, \quad \text{and} \quad \frac{\partial^2 f}{\partial y^2} = x^3 \frac{\mathrm{d}}{\mathrm{d}y}(1) = 0.$$

1.8.3 Total differential

For a function of one variable $y = f(x)$ the differential of y

$$\mathrm{d}y = \left(\frac{\mathrm{d}f(x)}{\mathrm{d}x}\right)\mathrm{d}x$$

is defined as the linear part of the increment Δy that corresponds to the increment $\mathrm{d}x$ of the independent variable x (the differential* of x). The smaller we make $\mathrm{d}x$ the lesser becomes the deviation between the actual increment $\Delta y = f(x + \mathrm{d}x) - f(x)$ and the differential $\mathrm{d}y$. We shall now apply this argument to functions of several variables.

---

* See Section 1.2.5.

We consider a point on the surface of the function $W = f(x, y)$ at the position (x, y). The values of the derivatives at this point are

$$\frac{\partial f(x, y)}{\partial x} \quad \text{and} \quad \frac{\partial f(x, y)}{\partial y}.$$

These represent the slopes on the surface at the point (x, y) in the directions of the x- and y-axis, respectively. We now imagine the position of x undergoing a change $x + dx$, while keeping the position y constant. As a result W undergoes a change $(\Delta W)_y$* whose linear part can be written

$$(dW)_y = \frac{\partial f(x, y)}{\partial x} \, dx.$$

Similarly, a change dy in the direction of the y-axis at constant x gives

$$(dW)_x = \frac{\partial f(x, y)}{\partial y} \, dy.$$

The linear combination dW of these two increments

$$dW = \frac{\partial f(x, y)}{\partial x} \, dx + \frac{\partial f(x, y)}{\partial y} \, dy, \qquad (1.8.8)$$

is called the *the total differential* of $W(x, y)$. It represents a natural extension of the definition of the differential dy of a function of one variable; that is, the linear part of the increment Δy that moves along the tangent at (x, y) as result of the change in position from x to $x + dx$. Similarly, the increment in the function $W(x, y)$

$$\Delta W = f(x + dx, y + dy) - f(x, y),$$

deviates from that of dW but for sufficiently small values of dx and dy we can disregard this difference and replace ΔW by dW.

With functions like $W = f(x, y, z)$ with three or more variables it is no longer possible to visualize the function geometrically. But there are of course no objections to introducing the partial derivatives with respect to the variables x, y, z,

$$\frac{\partial f}{\partial x}, \quad \frac{\partial f}{\partial y} \quad \text{and} \quad \frac{\partial f}{\partial z}, \qquad (1.8.9)$$

as the limit corresponding to Eq. (1.8.3), and to use the usual rules but bearing in mind that when calculating the derivative with respect to one variable – e.g. x – the remaining variables should be treated as constants.

---

* $(\Delta W)_y$ means a change in W resulting from a change in x while y is kept constant.

The form of the *total differential* of a function with three or more variables is

$$dW = \left(\frac{\partial f}{\partial x}\right) dx + \left(\frac{\partial f}{\partial y}\right) dy + \left(\frac{\partial f}{\partial z}\right) dz + \cdots, \qquad (1.8.10)$$

that contains all the terms for the linear contribution to the increment ΔW (x, y, z) resulting from the changes of the variables x, y and z by the differentials dx, dy and dz.

This expression works as the crank arm in numerous mathematical treatments of problems in physics, chemistry and biology.

1.8.4 The chain rule once more

The rule for calculation of the derivative of a composite function such as $y = F\{\psi(x)\}$ is shown in Section 1.2.5.1. With functions of several variables it very often happens that their arguments enter also as composite functions. To calculate the derivatives in such cases the rules are analogous to those that hold for one variable, although it is in the nature of things that the formula may take a more complicated form. A sketch of the derivation follows below.

Let the function

$$U = U(\xi, \eta),$$

e.g. $U = e^{\xi/\eta}$, represent a function of the two variables ξ and η, such that both are functions of the two variables x and y, which we represent as

$$\xi = \xi(x, y) \quad \text{and} \quad \eta = \eta(x, y),$$

e.g. $\xi = x + \alpha y$ and $\eta = x - \alpha y$. Thus, the function

$$U = U\{\xi(x, y), \eta(x, y)\}$$

is a function of the two variables x and y and must therefore also possess partial derivatives $\partial U/\partial x$, $\partial U/\partial y$ ... with respect to these variables. Now let the independent variables ξ and η be subjected to the changes $\Delta\xi$ and $\Delta\eta$. The resulting change in $U(\xi, \eta)$ can be written as

$$\Delta U = \left(\frac{\partial U}{\partial \xi}\right) \Delta\xi + \left(\frac{\partial U}{\partial \eta}\right) \Delta\eta + \delta_1\Delta\xi + \delta_2\Delta\eta,$$

i.e. as the sum of the linear contributions to the increment plus the quantities $\delta_1\Delta\eta$ and $\delta_2\Delta\eta$ where $\delta_1, \delta_2 \ll 1$ vanish concurrently as $\Delta\xi$ and $\Delta\eta$ approach zero. In this expression the increments $\Delta\xi$ and $\Delta\eta$ result from the change of the two variables x and y by the increments Δx and Δy. The changes of ξ and

1. Mathematical prelude

η written in terms of the linear contributions are then

$$\Delta\xi = \left(\frac{\partial\xi}{\partial x}\right)\Delta x + \left(\frac{\partial\xi}{\partial y}\right)\Delta y + \epsilon_1\Delta x + \gamma_1\Delta y,$$

and

$$\Delta\eta = \left(\frac{\partial\eta}{\partial x}\right)\Delta x + \left(\frac{\partial\eta}{\partial y}\right)\Delta y + \epsilon_2\Delta x + \gamma_2\Delta y,$$

where the numbers $\epsilon_1, \epsilon_2, \gamma_1, \gamma_2$, vanish as Δx and Δy approach zero. The increments $\Delta\xi$ and $\Delta\eta$ are then inserted in the above expression for the increment ΔU. This gives

$$\Delta U = \frac{\partial U}{\partial\xi}\left[\left(\frac{\partial\xi}{\partial x}\right)\Delta x + \left(\frac{\partial\xi}{\partial y}\right)\Delta y + \epsilon_1\Delta x + \gamma_1\Delta y\right]$$
$$+ \frac{\partial U}{\partial\eta}\left[\left(\frac{\partial\eta}{\partial x}\right)\Delta x + \left(\frac{\partial\eta}{\partial y}\right)\Delta y + \epsilon_2\Delta x + \gamma_2\Delta y\right],$$

which also can be written as

$$\Delta U = \left[\left(\frac{\partial U}{\partial\xi}\right)\left(\frac{\partial\xi}{\partial x}\right) + \left(\frac{\partial U}{\partial\eta}\right)\left(\frac{\partial\eta}{\partial x}\right)\right]\Delta x$$
$$+ \left[\left(\frac{\partial U}{\partial\xi}\right)\left(\frac{\partial\xi}{\partial y}\right) + \left(\frac{\partial U}{\partial\eta}\right)\left(\frac{\partial\eta}{\partial y}\right)\right]\Delta y + \epsilon\Delta x + \gamma\Delta y,$$

where it can be shown that both ϵ and γ vanish concurrently with Δx and Δy. Now let only x be subjected to a change Δx. This gives

$$\Delta U_y = \left[\left(\frac{\partial U}{\partial\xi}\right)\left(\frac{\partial\xi}{\partial x}\right) + \left(\frac{\partial U}{\partial\eta}\right)\left(\frac{\partial\eta}{\partial x}\right)\right]\Delta x + \epsilon\Delta x,$$

where ΔU_y denotes the increment of U for a change Δx of x with an unchanged value of y. Division by Δx on both sides gives

$$\frac{\Delta U_y}{\Delta x} = \left(\frac{\partial U}{\partial\xi}\right)\left(\frac{\partial\xi}{\partial x}\right) + \left(\frac{\partial U}{\partial\eta}\right)\left(\frac{\partial\eta}{\partial x}\right) + \epsilon.$$

When $\Delta x \to 0$, the left-hand side approaches $\partial U/\partial x$ according to Eq. (1.8.3) and on the right-hand side $\epsilon \to 0$. Hence

$$\frac{\partial U}{\partial x} = \left(\frac{\partial U}{\partial\xi}\right)\left(\frac{\partial\xi}{\partial x}\right) + \left(\frac{\partial U}{\partial\eta}\right)\left(\frac{\partial\eta}{\partial x}\right), \tag{1.8.11}$$

and correspondingly for the derivative with respect to y

$$\frac{\partial U}{\partial y} = \left(\frac{\partial U}{\partial\xi}\right)\left(\frac{\partial\xi}{\partial y}\right) + \left(\frac{\partial U}{\partial\eta}\right)\left(\frac{\partial\eta}{\partial y}\right). \tag{1.8.12}$$

This is the rule for the calculation of the first derivatives $\partial U/\partial x$ and $\partial U/\partial y$ when $U = U\{\xi(x, y), \eta(x, y)\}$ is a composite function of the variables x and y.

To calculate the second partial derivative $\partial^2 U/\partial x^2$ we differentiate Eq. (1.8.11) partially with respect to x

$$\frac{\partial^2 U}{\partial x^2} = \frac{\partial}{\partial x}\left(\frac{\partial U}{\partial x}\right) = \frac{\partial}{\partial x}\left(\frac{\partial U}{\partial \xi}\frac{\partial \xi}{\partial x}\right) + \frac{\partial}{\partial x}\left(\frac{\partial U}{\partial \eta}\frac{\partial \eta}{\partial x}\right)$$

$$= \frac{\partial}{\partial x}\left(\frac{\partial U}{\partial \xi}\right)\frac{\partial \xi}{\partial x} + \frac{\partial U}{\partial \xi}\frac{\partial^2 \xi}{\partial x^2} + \frac{\partial}{\partial x}\left(\frac{\partial U}{\partial \eta}\right)\frac{\partial \eta}{\partial x} + \frac{\partial U}{\partial \eta}\frac{\partial^2 \eta}{\partial x^2},$$

in accordance with the rule for differentiating a product in Section 1.2.3.1. To proceed we make use of the result of Eq. (1.8.11) but replace U with $\partial U/\partial \xi$ and $\partial U/\partial \eta$ in the first and third terms. This gives

$$\frac{\partial^2 U}{\partial x^2} = \left[\frac{\partial}{\partial \xi}\left(\frac{\partial U}{\partial \xi}\right)\frac{\partial \xi}{\partial x} + \frac{\partial}{\partial \eta}\left(\frac{\partial U}{\partial \xi}\right)\frac{\partial \eta}{\partial x}\right] \cdot \frac{\partial \xi}{\partial x}$$

$$+ \left[\frac{\partial}{\partial \xi}\left(\frac{\partial U}{\partial \eta}\right)\frac{\partial \xi}{\partial x} + \frac{\partial}{\partial \eta}\left(\frac{\partial U}{\partial \eta}\right)\frac{\partial \eta}{\partial x}\right] \cdot \frac{\partial \eta}{\partial x}$$

$$+ \frac{\partial U}{\partial \xi}\frac{\partial^2 \xi}{\partial x^2} + \frac{\partial U}{\partial \eta}\frac{\partial^2 \eta}{\partial x^2}$$

$$= \frac{\partial^2 U}{\partial \xi^2}\left(\frac{\partial \xi}{\partial x}\right)^2 + \frac{\partial^2 U}{\partial \xi \partial \eta}\frac{\partial \xi}{\partial x}\frac{\partial \eta}{\partial x}$$

$$+ \frac{\partial^2 U}{\partial \xi \partial \eta}\frac{\partial \xi}{\partial x}\frac{\partial \eta}{\partial x} + \frac{\partial^2 U}{\partial \eta^2}\left(\frac{\partial \eta}{\partial x}\right)^2 + \frac{\partial U}{\partial \xi}\frac{\partial^2 \xi}{\partial x^2} + \frac{\partial U}{\partial \eta}\frac{\partial^2 \eta}{\partial x^2}$$

$$= \frac{\partial^2 U}{\partial \xi^2}\left(\frac{\partial \xi}{\partial x}\right)^2 + \frac{\partial^2 U}{\partial \eta^2}\left(\frac{\partial \eta}{\partial x}\right)^2 + 2\frac{\partial^2 U}{\partial \xi \partial \eta}\frac{\partial \xi}{\partial x}\frac{\partial \eta}{\partial x}$$

$$+ \frac{\partial U}{\partial \xi}\frac{\partial^2 \xi}{\partial x^2} + \frac{\partial U}{\partial \eta}\frac{\partial^2 \eta}{\partial x^2}. \tag{1.8.13}$$

In a similar way the expression for the second partial derivative of U with respect to y becomes

$$\frac{\partial^2 U}{\partial y^2} = \frac{\partial^2 U}{\partial \xi^2}\left(\frac{\partial \xi}{\partial y}\right)^2 + \frac{\partial^2 U}{\partial \eta^2}\left(\frac{\partial \eta}{\partial y}\right)^2 + 2\frac{\partial^2 U}{\partial \xi \partial \eta}\frac{\partial \xi}{\partial y}\frac{\partial \eta}{\partial y}$$

$$+ \frac{\partial U}{\partial \xi}\frac{\partial^2 \xi}{\partial y^2} + \frac{\partial U}{\partial \eta}\frac{\partial^2 \eta}{\partial y^2}. \tag{1.8.14}$$

A simpler case of a composite function of two variables is the function

$$U = U(\zeta) = U\{\zeta(x, y)\}, \tag{1.8.15}$$

1. Mathematical prelude

where $U(\zeta)$ is a function of a single variable ζ, which, however, is a function of the two variables x and y.

An increment of the independent variable $\Delta\zeta$ results in an increment dU, giving

$$\Delta U = \left(\frac{dU}{d\zeta}\right) \Delta\zeta + \delta\Delta\zeta. \tag{1.8.16}$$

As before, we express the increment $\Delta\zeta$ by the changes Δx and Δy of the two variables x and y

$$\Delta\zeta = \left(\frac{\partial\zeta}{\partial x}\right) \Delta x + \left(\frac{\partial\zeta}{\partial x}\right) \Delta x + \epsilon\Delta x + \gamma\Delta x,$$

which, inserted in Eq. (1.8.16), gives

$$\Delta U = \Delta U = \left(\frac{dU}{d\zeta}\right) \left(\frac{\partial\zeta}{\partial x}\right) \Delta x + \left(\frac{dU}{d\zeta}\right) \left(\frac{\partial\zeta}{\partial y}\right) \Delta y + \epsilon\Delta x + \gamma\Delta y.$$

A change in x by the amount Δx but with the other variable y kept constant gives

$$\Delta U_y = \left(\frac{dU}{d\zeta}\right) \left(\frac{\partial\zeta}{\partial x}\right) \Delta x + \epsilon\Delta x,$$

and division by Δx on both sides yields

$$\frac{\Delta U_y}{\Delta x} = \left(\frac{dU}{d\zeta}\right) \left(\frac{\partial\zeta}{\partial x}\right) x + \epsilon,$$

which for $\Delta x \to 0$ gives

$$\frac{\partial U}{\partial x} = \left(\frac{dU}{d\zeta}\right) \left(\frac{\partial\zeta}{\partial x}\right), \tag{1.8.17}$$

and similarly for the derivative with respect to y

$$\frac{\partial U}{\partial y} = \left(\frac{dU}{d\zeta}\right) \left(\frac{\partial\zeta}{\partial y}\right), \tag{1.8.18}$$

which bears a close resemblance to the basic formula Eq. (1.2.15).

EXAMPLE An important field of application of these equations occurs in the cases where the connection between a function $U(x, y)$ and some of its partial derivatives is known in rectangular coordinates, but where for some reason or the other it is preferable to write this connection in terms of another system of coordinates, e.g. polar coordinates (r, θ), whose relation to the rectangular coordinates (x, y) are given by

$$x = r \cos \theta \quad \text{and} \quad y = r \sin \theta.$$

Inserting the polar coordinates in the primary expression gives

$$U(x, y) = U(r \cos \theta, r \sin \theta),$$

where $U(x, y)$ now appears as a composite function of the independent r and θ. Using the chain rule we have

$$\frac{\partial U}{\partial r} = \frac{\partial U}{\partial x}\frac{\partial x}{\partial r} + \frac{\partial U}{\partial y}\frac{\partial y}{\partial r} = \frac{\partial U}{\partial x}\cos \theta + \frac{\partial U}{\partial y}\sin \theta$$

$$\frac{\partial U}{\partial \theta} = \frac{\partial U}{\partial x}\frac{\partial x}{\partial \theta} + \frac{\partial U}{\partial y}\frac{\partial y}{\partial \theta} = \frac{\partial U}{\partial x}(-r \sin \theta) + \frac{\partial U}{\partial y}(r \cos \theta).$$

Solving these two equations for $\partial U/\partial x$ and $\partial U/\partial y$ we obtain

$$\frac{\partial U}{\partial x} = \frac{\partial U}{\partial r}\cos \theta - \frac{\partial U}{\partial \theta}\frac{\sin \theta}{r}$$

$$\frac{\partial U}{\partial y} = \frac{\partial U}{\partial r}\sin \theta + \frac{\partial U}{\partial \theta}\frac{\cos \theta}{r}.$$

1.9 Some ordinary differential equations

Let $y(x)$ represent an unknown function of the independent variable x; and let the respective derived functions of y with respect to x be y', y'', $\ldots$, $y^{(n)}$. Any functional connection between the function $y(x)$ and the variable x in which at least one of the derivatives also enters is called an *ordinary differential equation*. A differential equation is characterized by its *order*, that is the order of the highest derivative involved. The functional relation

$$F(x, y, y', \ldots, y^{(n)}) = 0$$

represents an ordinary differential equation of the n-th order. The equation

$$\frac{d^2 y}{dx^2} + A\frac{dy}{dx} + By = 0,$$

where A and B are constants, is an example of a second-order *linear* differential equation with constant coefficients. The *linear* notation arises from the property that all three quantities y, y' and y'' are of the first order, which has the important implication that no matter how they are mutually added or subtracted the result is still a linear combination of y, y' and y''. For that reason the equation

$$P(x)\frac{d^2y}{dx^2} + Q(x)\frac{dy}{dx} + R(x)y = 0,$$

where $P(x)$, $Q(x)$ and $R(x)$ are functions of the independent variable x, is still a second-order linear differential equation. In contrast, the equations

$$\frac{dy}{dx} + \frac{x}{y} = 0$$

and

$$\frac{d^2y}{dx^2} + y\frac{dy}{dx} + y^2 = 0 \quad \text{and} \quad \frac{d^2y}{dx^2} + \left(\frac{dy}{dx}\right)^{1/2} = 0$$

are first- and second-order nonlinear differential equations, since the terms y^{-1}, $y(dy/dx)$, y^2 and $(dy/dx)^{1/2}$ enter into the equations in a nonlinear manner.

To *solve a differential equation* amounts to finding the particular relation $y = f(x)$ that exists between the unknown function y and the independent variable x. The methods to be used differ according to the character of the differential equation in question, but one or several integrations are almost always involved, except in the simplest cases. Thus, in the case of the differential equation

$$\frac{d^2y}{dx^2} - x\frac{dy}{dx} + y = 0,$$

the solution $y = x$ emerges quickly; but in general the solution procedure may be more laborious.

Differential equations were introduced very early after the discovery of differential and integral calculus*, and developed almost explosively both in pure mathematics and to describe numerous processes in physics, physical chemistry, biology and in nearly any branch of technology.

1.9.1 Four first-order differential equations

The general form of the ordinary first-order differential equation is

$$\frac{dy}{dx} + P(x)y = Q(x), \tag{1.9.1}$$

of which we shall consider some simple – but important – cases.

---

* In 1676, Leibniz introduced the concept *æquatio differentialis* to denote the connection between the differentials dx and dy and the two variables x and y. Newton's great discoveries in his *Philosophiae Naturalis* resulted almost exclusively from formulating and solving his the physical problems in terms of differential equations (see, for example, Chandrasekhar, S. (1995) *Newton's Principia for the Common Reader*, Clarendon Press, Oxford).

1.9.1.1 The equation $y' + \alpha y = 0$

Putting $P(x) = \alpha$, where α is a constant, and $Q(x) = 0$ we are left with

$$\frac{dy}{dx} + \alpha y = 0. \tag{1.9.2}$$

This is the so-called *homogeneous, first-order linear differential equation*. The word *homogeneous* arises because the term on the right of the equality sign is *zero*. We divide the equation on both sides by y and multiply by dx. This gives*

$$\frac{dy}{y} + \alpha \, dx = 0.$$

This enables us to replace each of the two terms on the left-hand side by the equivalent indefinite integrals without any effect apart from replacing the left-hand side by a constant A_1 that differs from zero, thus:

$$\int \frac{dy}{y} + \int \alpha \, dx = A_1,$$

or (compare Eq. (1.4.3))

$$\ln y = -\alpha x + A_1. \tag{1.9.3}$$

The magnitude of the constant A_1 depends upon the character of the physical problem. One requirement could be that two values (y_0, x_0) should enter as part of the problem. Inserting these values in Eq. (1.9.3), we obtain

$$\ln y_0 = -\alpha x_0 + A_1,$$

or $A_1 = \ln y_0 + \alpha x_0$. Inserting this value back in Eq. (1.9.3) gives

$$\ln y - \ln y_0 = \alpha(x_0 - x),$$

or, according to Eq. (1.4.10),

$$\ln \frac{y}{y_0} = \alpha(x_0 - x),$$

or (from Eq. (1.5.10) and Eq. (1.5.11))

$$y = y_0 \, e^{\alpha(x_0 - x)}. \tag{1.9.4}$$

If the values of (y_0, x_0) have the special form $(y_0, 0)$ we find

$$y = y_0 \, e^{-\alpha x}, \tag{1.9.5}$$

i.e. y decreases exponentially with x from the *initial values* $y = y_0$ for $x = 0$.

* The variables are said to be *separated*.

1.9.1.2 The equation $y' + \alpha y = K$

Putting $P(x) = \alpha$, and $Q(x) = K$ where both α and K are constants we have

$$\frac{dy}{dx} + \alpha y = K. \tag{1.9.6}$$

This slightly more complicated version is the *inhomogeneous, first-order linear differential equation*. The word *inhomogeneous* indicates that the term to the right of the equality sign is different from zero. A solution can be found by putting*

$$y = u(x)e^{-\alpha x}, \tag{1.9.7}$$

i.e. a solution of the homogeneous equation (Eq. (1.9.2)) that is multiplied by an unknown function $u(x)$, whose form is subsequently investigated. This is done by inserting Eq. (1.9.7) and Eq. (1.9.6), which gives[†]

$$\frac{du}{dx}e^{-\alpha x} - u(x)\alpha\, e^{-\alpha x} + \alpha u(x)e^{-\alpha x} = K,$$

or

$$\frac{du}{dx} = K\, e^{\alpha x}.$$

Integration then gives

$$u(x) = \int K\, e^{\alpha x}dx = \frac{1}{\alpha}K\, e^{\alpha x} + A_1,$$

where A_1 is a constant of integration. This expression for $u(x)$ is inserted into Eq. (1.8.7) whereby we obtain the following expression as the solution of Eq. (1.8.6)

$$y = \frac{K}{\alpha} + A_1\, e^{-\alpha x}. \tag{1.9.8}$$

Let the physical process under consideration be of such a character that $y = 0$ when $x = 0$. These values inserted in Eq. (1.9.8) give $A_1 = -K/k$, as $e^0 = 1$, and replacement of A_1 by this value in Eq. (1.9.8) yields

$$y = \frac{K}{a}[1 - e^{-ax}]. \tag{1.9.9}$$

---

* This artifice (variation of the parameter), was developed in its general form in 1774 by J.-L. Lagrange (1736–1813) – one of the greatest mathematicians of all times. The simplest version as applied to Eq. (1.9.6) had already been used in 1739 by L. Euler (1707–1783).
† See Section 1.2.3.1 (d) and Section 1.5.2, Eq. (1.5.14).

For $\alpha > 0$ the value inside the square brackets [ ] moves towards 1 when $x \to \infty$, and y is said to display an asymptotic, exponential growth from $y = 0$ towards $y = K/\alpha$ with x approaching $+\infty$.

1.9.1.3 The equation $y' + \alpha y = Q(x)$

We shall now seek a solution of

$$\frac{dy}{dx} + \alpha y = Q(x), \qquad (1.9.10)$$

where the quantity $Q(x)$ on the right-hand side of the equality sign is a function of x. We proceed as before by assuming a solution of the form

$$y = u(x) e^{-\alpha x}, \qquad (1.9.11)$$

and insert the expression for y into Eq. (1.9.10). This gives

$$\frac{du}{dx} = Q(x) e^{\alpha x}.$$

It then follows that

$$u(x) = \int e^{\alpha x} Q(x) \, dx + A_1,$$

where A is a constant of integration. Inserting this value for $u(x)$ in Eq. (1.9.11) gives

$$y e^{\alpha x} = \int e^{\alpha x} Q(x) \, dx + A_1, \qquad (1.9.12)$$

as the solution of the inhomogeneous equation containing the "driving function" $Q(x)$. To put more meat into this expression one must know (i) the explicit form of $Q(x)$ and (ii) a set (x_0, y_0) of initial values in order to be able to adjust the value of the constant A_1 to the character of the physical problem in question.

1.9.1.4 The equation $y' + P(x)y = Q(x)$

As in Section 1.9.1.2 we first seek a solution of the homogeneous equation

$$\frac{dy}{dx} + P(x)y = 0. \qquad (1.9.13)$$

The variables are separated by division on both sides by y and multiplication with dx. This gives

$$\frac{dy}{y} = -P(x) \, dx,$$

which is equivalent to

$$\int \frac{dy}{y} = -\int P(x)\,dx + A_1,$$

or

$$\ln y = -\int P(x)\,dx + A_1,$$

where A_1 is a constant of integration. We write the expression of the homogeneous solution as

$$y = A_2\,e^{-\int P(x)\,dx}, \tag{1.9.14}$$

where A_2 is another form of the constant of integration. We again apply the method of variation of the parameter by writing

$$y = u(x)\,e^{-\int P(x)\,dx}, \tag{1.9.15}$$

and search for the form of $u(x)$ that makes Eq. (1.9.15) satisfy Eq. (1.9.13). Differentiation of Eq. (1.9.15) gives

$$\begin{aligned}
\frac{dy}{dx} &= \frac{du}{dx}\,e^{-\int P(x)\,dx} + u(x)\frac{d}{dx}\left(e^{-\int P(x)\,dx}\right) \\
&= \frac{du}{dx}\,e^{-\int P(x)\,dx} + u(x)(-1)e^{-\int P(x)\,dx} \cdot \frac{d}{dx}\left(\int P(x)\,dx\right) \\
&= \frac{du}{dx}\,e^{-\int P(x)\,dx} - u(x)P(x)e^{-\int P(x)\,dx},
\end{aligned}$$

which, inserted into Eq. (1.9.13), yields

$$\frac{du}{dx}\,e^{-\int P(x)\,dx} - u(x)P(x)e^{-\int P(x)\,dx} + u(x)P(x)e^{-\int P(x)\,dx} = Q(x).$$

From this it follows that

$$\frac{du}{dx} = e^{\int P(x)\,dx}\,Q(x),$$

or

$$u(x) = \int e^{\int P(x)\,dx}\,Q(x)\,dx + A_3,$$

which, inserted into Eq. (1.9.15), leads to

$$y\,e^{\int P(x)\,dx} = \int e^{\int P(x)\,dx}\,Q(x)\,dx + A_3, \tag{1.9.16}$$

which is the general solution of the linear first-order differential equation in its most complete form. Equation (1.9.16) is also written in this form

$$y = e^{-\int P(x)\,dx} \left(\int e^{\int P(x)\,dx} Q(x)\,dx + A_3 \right), \qquad (1.9.17)$$

or, to be more explicit about the order of doing the integration,

$$y = e^{-\int P(x)\,dx} \left(\int^{\xi=x} \exp\left\{ \int^{\eta=\xi} P(\eta)\,d\eta \right\} Q(\xi)\,d\xi + A_3 \right).$$

1.9.2 Two second-order differential equations

Differential equations that contain a second derivative are called *second-order equations*. They turn up in numerous different problems in physics – both in their linear and nonlinear form. The simplest cases – but useful in the present text – are as follows.

1.9.2.1 The equation $y'' + \kappa^2 y = 0$

The homogeneous second-order equation

$$\frac{d^2 y}{dx^2} + \kappa^2 y = 0, \qquad (1.9.18)$$

where κ is a constant, is among other things the equation of motion of the harmonic oscillator, i.e. the elastic, undamped vibration of a particle that moves under the influence of a restoring force that is proportional to the displacement from the equilibrium position. The solution must be of such a character that the function $y = f(x)$ differentiated twice with respect to x is equal to a (negative) multiple of itself, i.e. to $-\kappa^2 y$. For the function $y = \sin \kappa x$ it holds* that

$$\frac{dy}{dx} = \kappa \cos \kappa x \quad \text{and} \quad \frac{d^2 y}{dx^2} = \kappa[-\kappa \sin \kappa x] = -\kappa^2 \sin \kappa x,$$

which implies that $\sin \kappa x$ is a solution of Eq. (1.9.18). But in this case $A \sin \kappa x$, where A is a arbitrary constant, can also be a solution because of the linearity of Eq. (1.9.18). Similarly $\cos \kappa x$ is also a solution, since

$$\frac{dy}{dx} = -\kappa \sin \kappa x \quad \text{and} \quad \frac{d^2 y}{dx^2} = -\kappa[\kappa \cos \kappa x] = -\kappa^2 \cos \kappa x.$$

Thus, the complete solution of Eq. (1.9.18) is of the form

$$y = A \sin \kappa x + B \cos \kappa x. \qquad (1.9.19)$$

---

* See Section 1.2.3.1.

1.9.2.2 The equation $y'' - \kappa^2 y = 0$

Another important example is the solution of

$$\frac{d^2 y}{dx^2} - \kappa^2 y = 0. \qquad (1.9.20)$$

This equation appears, for example, in the description of the distribution of the electric currents in a cylindrical nerve or muscle fiber or the distribution of the electric charges near an electrified interface.

In this case the second derivative of $y = f(x)$ must be equal to a multiple of itself. Consider the function $y = e^{\alpha x}$. We have from Eq. (1.5.14)

$$\frac{dy}{dx} = \alpha e^{\alpha x}, \quad \text{and} \quad \frac{d^2 y}{dx^2} = \alpha[\alpha e^{\alpha x}] = \alpha^2 e^{\alpha x}.$$

Inserting the values for y and $d^2 y/dx^2$ in Eq. (1.9.20) gives

$$\alpha^2 e^{\alpha x} - \kappa^2 e^{\alpha x} = 0.$$

It follows that

$$\alpha^2 = \kappa^2, \quad \text{where } \alpha = \pm \kappa.$$

Thus, $e^{\kappa x}$ and $e^{-\kappa x}$ are separate solutions of Eq. (1.9.20). Hence, the linear combination

$$y = A e^{\kappa x} + B e^{-\kappa x}, \qquad (1.9.21)$$

where A and B are arbitrary constants, represents the complete solution of Eq. (1.9.20).

The two constants have to be adjusted in such a way that the function reproduces correctly the physical problem that is under consideration. The function may have the property of constraint that $y \to 0$ for $x \to \infty$. This requires that the constant A in Eq. (1.9.21) must be zero to prevent this undesired behavior of y for $x \to \infty$, and the solution must then be of the form $B e^{-\kappa x}$. If, in addition, the requirement is that y assumes a prescribed value at $x = 0$, then we must have $B = Y_0$, and the solution that satisfies this requirement – also called a *boundary condition* – takes the form

$$y = Y_0 e^{-\kappa x}. \qquad (1.9.22)$$

This solution will be used in Chapter 3 and in Chapter 4.

1.10 A note on partial differential equations

Let $f(x, y)$ be a function of the two independent variables x and y*, whose partial derivatives of f with the variables x and y we write as

$$f_x = \frac{\partial f}{\partial x}, \quad f_y = \frac{\partial f}{\partial y}, \quad f_{xy} = \frac{\partial^2 f}{\partial x \partial y}, \quad f_{yx} = \frac{\partial^2 f}{\partial y \partial x}$$

$$f_{xx} = \frac{\partial^2 f}{\partial x^2}, \quad f_{yy} = \frac{\partial^2 f}{\partial y^2}, \dots . \tag{1.10.1}$$

Any functional relation between the function $f(x, y)$ and the independent variables x, y together with the various partial derivatives f_x, f_y etc.

$$F(x, y, f, f_x, f_y, f_{xx}, f_{yy}, \dots) = 0 \tag{1.10.2}$$

represents a *partial differential equation*. As with ordinary differential equations, the *order* of the partial differential equation is given by the highest order of the derivative involved. Similarly, we have the designation *linear* versus *nonlinear* equations, and their *degree*. Thus, the general linear first-order partial differential equation is of the form

$$A(x, y)\frac{\partial f}{\partial x} + B(x, y)\frac{\partial f}{\partial y} + C(x, y)f = D(x, y),$$

where $A(x, y)$, $B(x, y)$, $C(x, y)$ and $D(x, y)$ are known functions of x and y, and

$$A(x, y)\frac{\partial f}{\partial x} + \beta\frac{\partial^2 f}{\partial y^2} = 0 \quad \text{and} \quad \frac{\partial^2 f}{\partial x^2} + \frac{\partial^2 f}{\partial y^2} = 0$$

are examples of linear second-order partial equations, which in addition are called *homogeneous* as the term on the right-hand side is zero. In contrast

$$\left(\frac{\partial^2 f}{\partial x^2}\right)^2 + \left(\frac{\partial^2 f}{\partial y^2}\right)^2 = 1 \quad \text{and} \quad f\frac{\partial^2 f}{\partial x^2} + \left(\frac{\partial f}{\partial y}\right)^2 = f^2$$

are two examples of *nonlinear* equations.

All linear differential equations obey the *principle of superimposition*. This means that if the $F(x, y)$, $G(x, y)$ and $H(x, y)$ taken separately are solutions of a given partial differential equation, then any linear combination

$$\alpha F(x, y) + \beta G(x, y) + \gamma H(x, y)$$

of these three equations is also a solution. This property is of extreme importance when one attempts to build up a sum of singular solutions to reproduce a given physical problem.

---

* Quite generally f could be a function $f(x_1, x_2, x_3, \dots x_n)$ of n independent variables.

It was seen in Section 1.9 that the solution obtained by the integration of an ordinary differential equation gave rise to *arbitrary constants* whose numbers were determined by the order of the equation (i.e. one constant for a first-order equation; two constants for a second-order equation, etc.). The numerical values of these constants were then evaluated by fitting the solution to conform to the properties of the physical problem in question, e.g. by the requirement that the solution contains the particular value y_0 corresponding to the value $x = x_0$ of the independent variable.

In the partial differential equations that appear from time to time in this book, one of the independent variable is the *time t*, e.g. the equation

$$\frac{\partial C}{\partial t} = D\frac{\partial^2 C}{\partial x^2}, \tag{1.10.3}$$

where D is a constant, represents the one-dimensional spread of matter of concentration C by a process of diffusion. However, before tackling the solution of such an equation several essential conditions for the process must be formulated in mathematical terms from the outset, i.e. the *size* of the space involved and the *initial distribution* of matter at the start of the process, e.g. as an equation of the form

$$C(x, t) = H(x); \quad \text{for } t = 0. \tag{1.10.4}$$

This condition is called the *initial condition* of the problem. In addition, the matter concentration will be constrained to assume certain values in the space such as

$$C(x, t) = C_1; \quad \text{for } x = x_1, \quad \text{and} \quad t > 0, \tag{1.10.5}$$

$$\frac{\partial C(x, t)}{\partial x} = 0, \quad \text{for } x = x_2, \quad \text{and} \quad t > 0. \tag{1.10.6}$$

Equation (1.10.5) and Eq. (1.10.6) are called the *boundary conditions* of the problem.

To solve a partial differential equation that fulfills both the initial condition and the set of boundary conditions – two in the above case of a second-order equation – may pose great problems. With a few exceptions it is not possible to find a general solution, as it is only part of the information needed to obtain a particular solution that is contained in the differential equation itself. The additional information required arises from the complementary demands stipulated by the initial and boundary conditions.

As a illustration we consider the one-dimensional wave equation

$$\frac{\partial^2 U}{\partial t^2} = v^2 \frac{\partial^2 U}{\partial x^2}, \qquad (1.10.7)$$

where we seek a solution without making use of these auxiliary conditions. We introduce two new coordinates that are defined by

$$\left. \begin{array}{l} \xi = x + vt \\ \eta = x - vt \end{array} \right\}$$

and consider $U = U(\xi, \eta) = U(x + vt, x - vt)$ as a function of these coordinates. We have*

$$\frac{\partial^2 U}{\partial x^2} = \frac{\partial^2 U}{\partial^2 \xi} + 2\frac{\partial^2 U}{\partial \xi \partial \eta} + \frac{\partial^2 U}{\partial^2 \eta},$$

$$\frac{\partial^2 U}{\partial t^2} = v^2 \left\{ \frac{\partial^2 U}{\partial^2 \xi} - 2\frac{\partial^2 U}{\partial \xi \partial \eta} + \frac{\partial^2 U}{\partial^2 \eta} \right\},$$

whereby the wave equation transforms to

$$-4v^2 \frac{\partial^2 U}{\partial \xi \partial \eta} = 0.$$

Since $v \neq 0$, we have

$$\frac{\partial^2 U}{\partial \xi \partial \eta} = \frac{\partial}{\partial \xi} \left(\frac{\partial U}{\partial \eta} \right) = 0. \qquad (1.10.8)$$

Integration with respect to ξ yields

$$\frac{\partial U}{\partial \eta} = g(\eta),$$

as $\partial g(\eta)/\partial \xi = 0$. The function $g(\eta)$ is an entirely arbitrary function that is independent of the variable ξ. Integration of the above expression with respect to η gives

$$U(\xi, \eta) = \int g(\eta)\, d\eta + F(\xi) = F(\xi) + G(\eta),$$

where $F(\xi)$ is an arbitrary function that does not depend on η. The general

---

* See Section 1.8.4, Eq. (1.8.11) and Eq. (1.8.13). But the details of this derivation is not necessary for the understanding of the following.

solution of the wave equation is therefore*

$$U(x, t) = F(x + vt) + G(x - vt),$$

which represents two arbitrary configurations, one of them G moving with the velocity v along the positive direction of the x-axis, whereas the other configuration F moves with the same velocity in the opposite direction. The general solution of an ordinary differential equation of order n results in general from n successive integrations each returning a separate arbitrary constant. As illustrated above, the general solution of a partial differential equation unconstrained by boundary conditions returns instead *arbitrary functions* that only reflect the general behavior that is dictated by the equation – in the above case a uniform motion along the x-axis. Therefore, when dealing with solutions of practical problems in physics and chemistry it is not very profitable to look for the general solutions of the equations. Instead one looks for solutions that most easily make the adjustment of the boundary conditions. Thus, in general it is the character of the boundary value problem that determines the tactics chosen to solve the equation concerned. The solution of a given problem may turn out to be complicated and difficult, and in many cases may require a knowledge of a larger selection of mathematical disciplines and techniques than we expect to be at the disposal of the readers of an introductory text such as this.

In the following chapters we need to find special solutions to a few partial differential equations. However, this should not give rise to any misgivings from our prospective readers. We believe that familiarity and practice in handling the mathematics that is summarized in this chapter will be enough to take a fairly large step ahead†.

The equations in question are of the type:

$$\frac{\partial C}{\partial t} = D \frac{\partial^2 C}{\partial x^2}, \tag{1.10.9}$$

which is the simple *one-dimensional diffusion equation*, along with

$$\frac{\partial C}{\partial t} = D \frac{\partial^2 C}{\partial x^2} + v \frac{\partial C}{\partial x}, \tag{1.10.10}$$

* This solution was found in 1747 by Jean leRond d'Alembert (1717–1783), French mathematician and philosopher, who made important contributions to the development of differential and integral calculus, and to the physics of mechanics, hydrodynamics, meteorology and astronomy. He collaborated with Diderot as editor of *Encyclopédie ou dictionaire raissonné des sciences, des arts et des métiers* (1751–1765).

† In this context I feel a spiritual kinship with the British sergeant who answered a young physicist who, owing to his training, was transferred to a section working on radio receivers, and one day asked a question about one of the circuits and got the truly immortal reply: "All you need to know is Ohm's law, but you need to know Ohm's law bloody well". (Longair, M.S. (1984) *Theoretical Concepts in Physics*, Cambridge University Press, *cit.* p. 350.)

which is the equation for *diffusion with migration/convection superimposed*, and

$$\tau \frac{\partial V}{\partial t} = \lambda^2 \frac{\partial^2 V}{\partial x^2} - V, \qquad (1.10.11)$$

which is the *equation for a leaky cable*.

This may appear a substantial undertaking. But – dear reader – do not feel disheartened. From a pure mathematical viewpoint we have only to tackle one type of equation, namely Eq. (1.10.9), since the other two equations can be put into the same form by means of rather simple transformations. Furthermore, the special solutions that are required from time to time in this text can *all* be derived from a *single* solution of Eq. (1.10.9), that is obtained by means of the transformation*

$$\xi = \frac{x}{\sqrt{t}},$$

using only simple differential and integral calculus. Thus, the mathematical developments in this book can be understood without the need to draw on any of the standard methods (e.g. use of Fourier series, Fourier integrals, integral transform technique, etc.) of solving boundary value problems in mathematical physics. The cost of this tactic is less elegance[†] in the presentation of the mathematics and more pages of algebra – a price that we hope will be counterbalanced by the lighter demands of skills in using mathematical routines that do not go beyond those that are referred to in this chapter. But . . .

Even kings cannot take a short cut to mathematics.

(C.F. Gauss, 1777–1855)

---

[*] See Chapter 2, Section 2.5.5.1.

[†] I find it justified to sacrifice elegance in the interest of a simple and more easy reading. On this point I adhere to the dictum of the great L. Boltzmann, according to whom matters of elegance ought to be left to the tailor and to the cobbler. (Einstein, A. (1944) *Relativity, the Special and General Theory*, 13th edition, Methuen & Co. Ltd, London (*cit* p. vi).)

Chapter 2

Migration and diffusion

The relation of size to the time required for diffusion is one of the most fundamental of all considerations in the 'design' of animals and plants. On a humbler level, it is also of primary importance in planning physiological experiments, or interpreting physiological results.

(A.V. Hill, 1965)

2.1 Introduction

Membrane structures are present everywhere in biological systems. They are part of many intracellular structures, e.g. mitochondria, and each cell is surrounded by a surface membrane that separates the structure of the interior of the cell from its environment but allows for an exchange of small molecules and energy. The range of rates by which the different substances are transported across the membrane is large: some substances penetrate through the membrane rather easily, while others are almost, or completely, impermeable. Furthermore, the membrane permeability for a given substance may differ greatly from one cell type to the other.

During the past 40 years a large number of observations have been collected to show that the function of the cell membrane is not only to serve as a passive barrier for transport of matter between the cell and its environment. Many substances do not cross the membrane *passively*, i.e. moving in a direction that is determined by the set of physical conditions such as concentration, electric potential, hydrostatic pressure, etc., that are characteristic of the cell interior and the surrounding solution.

Some mass transport through the cell membrane occurs as *active transport*. This is a type of transport that takes place in specialized protein molecules – or transport molecules – in the cell membrane. The transport molecules are

62

designed specifically to transport a single, perhaps two, or even several species simultaneously, e.g. Na^+, or Na^+ and K^+. The transport protein drives, via an *energy requiring* process, the species in question *unidirectionally* through the membrane, for example from the inside to the outside. The energy available for the transport is derived primarily from the metabolic processes inside the cell, whereby chemical free energy is temporarily bound in phosphorous compounds like adenosine triphosphate (ATP), which serves as fuel for the transport molecule. Ordinarily the actively transported molecule is transferred from an environment where its free energy is low to an environment of higher free energy*, i.e. in the direction *opposite* to that taken spontaneously by the fraction of the species that does not enter into the transport system. Active transport plays an important part in the regulation of the interior constituents of the cell and is of decisive importance for the survival of the cell. Thus, cell membranes are complex, dynamic structures in possession of a chemical reactivity designed to bring about a selective transport of substances in and out of the cell according to the requirements of the cell and the organism as a whole. Therefore, the cell membrane is not a passive sack that simply delimits the interior of the cell, but rather an organ with a *regulatory capacity*.

In order to decide whether a particular transport process is an *active* process, it is imperative to have a clear understanding of the basic physical principles and mechanisms that govern *passive transport processes*: a movement reflecting the spontaneous tendency of the species in question to run downhill energetically, by moving from a state of high free energy to that of a lower one, provided that opposing forces are not sufficient to prevent the downhill movement. In most cases passive transport processes are those whose immediate progression is solely determined by the set of conditions that make up the environment in the two phases surrounding the membrane, such as differences in concentration, electric potential and pressure.

As stated above, the passive transport process will progress in every system provided the distribution of the substance (e.g. between the extra- and intracellular phases) differs from the thermodynamic equilibrium distribution. Classical thermodynamics answers precisely whether a system is in equilibrium and – if this is not the case – also the direction of the movements within the system necessary to attain equilibrium. But thermodynamics does not give any information about the *time* that is required for the system to move from an initial state to its equilibrium state. From a practical point of view, the time it takes to

---

* A colloquial verbalism for such a movement is "in the energetically uphill direction", i.e. for an uncharged molecule a movement from a state of low to high concentration. It was the presence of such an uphill transport that coined the word "active transport".

produce a given change of a quantity is of the utmost importance. It may be of small comfort to know that a given process will progress in a given direction if the change is attained first after, say, 1000 years. For that reason it is necessary to have a *kinetic* description at one's disposal.

In the following pages an account will be given of some of the physical concepts and mechanisms entering into the description of the passive transport of solutes through a *homogeneous medium* or through layers of homogeneous media (simple passive transport). Even though the passive transport through biological cell membranes appears in many cases to take place by rather complicated mechanisms, this does not invalidate the theory of passive transport as a useful tool in the description of passive transport across cell membranes. It is a difficult – and a not very promising – undertaking to describe the behavior in a complicated system like a cell membrane without familiarity with the mechanisms that operate in a simple system and with the physical laws that govern these mechanisms. Furthermore, in describing the transport of a substance across a cell membrane we use all those concepts that have proven useful in the domain of simple passive transport theory. For that reason we shall give an account of the essential elements of that part of the simple passive transport theory that can serve as the necessary background for describing passive mass transfer through a cell membrane, and can function as a point of departure for understanding the character of those modifications it has been found necessary to introduce.

In the treatment in this book we will restrict ourselves to systems that are *isothermal* and – with a few exceptions – at a *uniform pressure*. The point of departure is a small number of equations, which are justified empirically through studies of the behavior of macroscopic systems. These general equations are then used to account for the transport of dissolved substances in concrete, well-defined physical situations resembling those encountered in biology. In this description of the movement of a given substance – expressed in amounts such as kilogram, mole or any other convenient unit – none of the details at the molecular level which underlies the transport process is considered. As a natural supplement to this description a sketch is given of a simple model which illustrates how a *single* dissolved particle is expected to move in a solvent as a result of thermal agitation (*random walk*). Next we consider a *swarm of particles*, all of which execute their incessant thermal movements to and fro in a purely random manner. If the number of particles in the swarm moving in one direction is larger than the number moving in the opposite direction in a given time the result is a *net transport* of particles in the given direction. This situation will occur if the *density* of the particles in the swarm *varies* in a given direction. A net transport also will occur if all the particles are driven by an *external force*, which adds an extra unidirectional velocity component to the random thermal movements.

2.2 Flux

To characterize the intensity of the net transport of a given particle species in the system a convenient measure is the *transport-flow density* or *flux J* of the transported component. The flux is defined as the amount of substance that per unit time (s) passes a unit area (m²) placed at right angles to the direction of the flow. Flux has therefore the following dimensions*

$$J \triangleq \text{(amount of substance)} \, \text{m}^{-2} \, \text{s}^{-1}, \qquad (2.2.1)$$

where the amount of the substance is given in units (e.g. kg, dm³, mol, number of molecules, etc.) that are convenient in the given physical situation. The flux is analogous to the electric current density, which is given in amperes per square meter (A m^{-2}) or coulombs per square meter per second (C m^{-2} s^{-1}).

2.3 Types of passive transport

Mass transport in an isothermal system may take place by means of the following mechanisms.

(i) Because of the irregular thermal movements (the molecular chaos), which affect both the molecules of the solvent and those of the solute, all molecules in the system will incessantly change their positions and move around in a purely random manner in the volume accessible to them. If the particle concentration is uniform everywhere throughout the system, there will be no net transport of the dissolved particles in any particular direction. If – on the other hand – there is a non-uniform distribution of the dissolved particles, there will be a *net transport* of particles in the *direction* of the *concentration drop*. The net transport will take place because the number of particles that in a certain time move *out* from a region with a high concentration will be larger than the number of particles which the region *receives* in the same time from an adjacent region having a *lower concentration*. A mass transport whose origin is an uneven distribution of the dissolved molecules whose movement is due to this thermal chaos is called a **diffusion process**.

---

* In this text the symbol $\triangleq$ means: "has the following units".

(ii) If each dissolved particle is under the influence of an *external force*, every particle will experience an extra velocity component along the direction of the force that superimposes upon their thermal movement. The result is that each particle – and hence the swarm of particles – will move as a whole with a certain velocity in the direction of the force. In this case a net particle transport will occur even if the the concentration is uniform throughout the system. This type of transport, which is due to the presence of an external field of force acting on each of the dissolved particles, is called **migration**. The most frequently occurring forces are those of *gravity* and other *g-forces*, and also *electric forces* in the case of electrically charged particles. Transport by migration will take place even if the system contains only a single dissolved particle. It is only meaningful to talk of transport by diffusion, however, when the number of dissolved particles is sufficiently high to ensure that the particles make a continuum in space and time, i.e. provided that the influence of thermal fluctuations on the particle concentration can be ignored.

(iii) Finally, mass transport can take place because the system as a whole is not at rest, but is *flowing* in a given direction due to a pressure fall throughout the system. This type of transport is called **hydrodynamic flow** or **convection**. Mass transport often occurs as a combination of several of the above mentioned mechanisms. A combination is diffusion superimposed on migration. If the transported particles are ions and the driving force is an electric field, the transport process is called *electrodiffusion*.

In the following we describe explicit expressions for the passive transport mechanisms mentioned above, as well as the factors that determine the magnitude of the flux.

2.4 Migration

First, we define some useful concepts.

2.4.1 Friction coefficient and mobility

Consider a particle of mass m, which is suspended in a solution (Fig. 2.1) and acted upon by an external force X (e.g. gravity) in the positive direction of the x-axis* driving the particle in that direction. Let the particle velocity at time

---

* In the calculations that follow we consider only processes that can be regarded as occurring in one dimension, i.e. the physical quantity in question only changes relative to one direction – the positive direction of the x-axis – and depends only upon one space coordinate x. Vectors such as force, velocity, flux, etc., have – according to sign – the same (+) or opposite (−) direction as the positive direction of the x-axis.

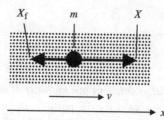

Fig. 2.1. The relation between an external driving force X acting on a particle with mass m and the frictional force X_f, origination from the motion of the particle with a velocity v in the medium.

t be v (m/s). During its motion, the particle has to push and move the molecules of the solute and, as a result, be acted upon by frictional force X_f directed in an opposite direction to the external driving force X. To a first approximation, the frictional force is proportional to the particle velocity v with respect to the surrounding medium, which is assumed to be at rest:

$$X_f = -\beta v,\qquad(2.4.1)$$

where the minus sign is due to the opposite directions of X_f and v.

The coefficient β in Eq. (2.4.1) is the *friction coefficient*; it depends on the size and shape of the particle and on the properties of the suspending medium. The resulting force in the direction of the x-axis is, according to Newton's second law, equal to the particle mass m times its acceleration dv/dt at time t. Hence, the particle's equation of motion can be written as

$$m\frac{dv}{dt} = X + X_f,$$

or, by invoking Eq. (2.4.1),

$$m\frac{dv}{dt} = X - \beta v,\qquad(2.4.2)$$

After a certain time the particle will have attained constant velocity (the stationary velocity) where the frictional force X_f and the driving force X have the same magnitude. As $v = $ constant entails $dv/dt = 0$, we have

$$X = \beta v,\qquad(2.4.3)$$

where v is the stationary velocity caused by the external driving force X. Equation (2.4.3) is rewritten as

$$v = \frac{X}{\beta} = BX,\qquad(2.4.4)$$

where the new constant

$$B = \frac{1}{\beta} = \frac{v}{X},$$ (2.4.5)

is the particle's *mechanical mobility* or just *mobility*. It follows from Eq. (2.4.5) that B equals the stationary velocity the particle attains when acted upon by unit driving force. Hence the dimensions of B in the SI-system are

$$B \overset{\Delta}{=} \text{m s}^{-1} \text{N}^{-1} \quad \text{(meters per second) per newton.}$$

2.4.2 Migration flux

Consider a suspension of identical particles uniformly distributed in a medium, and all moving with the same stationary velocity v under the influence of a driving force X. We shall now derive the relation connecting the flux J, the particle concentration C and the stationary velocity v. We consider a surface element of area A (m^2), which is placed at right angles to the direction of the mass transport J (see Fig. 2.2). How many particles will pass this area per unit time? If all the particles move downwards with a stationary velocity v, then in time Δt each particle will have moved a distance $v \Delta t$. All particles that are contained inside the prism with volume $Av \Delta t$ will therefore in the time Δt pass the surface element A. If the number of particles per cubic meter is N, the prism will contain $N A v \Delta t$ particles. This number equals the number of particles $\Delta M^{(\Delta t)}$ which pass the surface element A in time Δt. Therefore

$$\Delta M^{(\Delta t)} = N A v \, \Delta t.$$

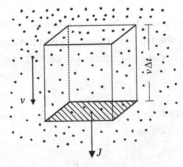

Fig. 2.2. Illustration of the concept of flux or transport-flow density.

The flux J is the number of particles that pass a unit area per unit time. Dividing both sides of the above expression by Δt and A gives the migration flux as

$$J = Nv. \tag{2.4.6}$$

The dimensions of N and v are (particles) m^{-3} and m s^{-1}, respectively, and the product Nv has the dimension: (particles) m^{-3} $\text{m s}^{-1} = $ (particles) m^{-2} s^{-1}, in agreement with the definition in Eq. (2.2.1).

The flux also can be expressed in terms of the driving force X acting on each particle. Inserting Eq. (2.4.4) in Eq. (2.4.6) gives

$$J = BNX, \tag{2.4.7}$$

namely

$$\text{flux} = \text{mobility} \times \text{concentration} \times \text{driving force.}$$

Equation (2.4.7) is valid also if other concentration measures are used: dividing on both sides with Avogadro's constant* $\mathcal{N}_A$ gives

$$N/\mathcal{N}_A \overset{\Delta}{=} \text{(number of molecules) } \text{m}^{-3}/\text{(number of molecules) } \text{mol}^{-1}$$
$$\equiv \text{mol m}^{-3},$$

and

$$J/\mathcal{N}_A \overset{\Delta}{=} \text{(number of molecules) } \text{m}^{-2} \text{s}^{-1}/\text{(number of molecules) } \text{mol}^{-1}$$
$$\equiv \text{mol m}^{-2} \text{s}^{-1}.$$

When each particle is subjected to a driving force X, the amount of substance $M^{(t)}$ that in time t is transported through a cross-sectional area A perpendicular to the direction of motion is, according to Eq. (2.2.1) and Eq. (2.4.7),

$$M^{(t)} = JAt = ABCXt. \tag{2.4.8}$$

The above argument can be extended to the case where the concentration varies throughout the system, i.e. $C = C(x)$. Making Δt sufficiently small, the concentration $N(x)$ or $C(x)$ will approximately be constant in the region between the positions x and $x + v\,\Delta t$. Hence, we obtain

$$J(x) = B\,C(x)\,X, \tag{2.4.9}$$

where the migration flux $J(x)$ is a function of the position x in the space. The consequences of this situation will be considered in Section 2.5.3.

* Avogadro's constant $= \mathcal{N}_A = 6.023 \times 10^{23}$ molecules per mole. See Appendix S.

2.5 Diffusion

A prerequisite for diffusion is that the substance in question is distributed non-uniformly. In this section we describe the factors that determine the magnitude of the diffusion flux and derive the equations that govern the diffusion processes. These equations then are used to calculate how that diffusion proceeds under a given set of constraints where we develop some useful mathematical tools. Finally, we give a description of the mechanisms that underlie the diffusion processes at the molecular level.

2.5.1 Phenomenological description

To illustrate diffusion one can take a strongly colored, soluble crystal, e.g. $KMnO_4$, and drop it into a beaker of water. It will be observed that not only do the salt molecules dissolve but, as times goes on, the colored molecules will spread throughout the entire water phase until finally they appear to be uniformly distributed in the accessible water volume. When this state is reached one will no longer observe any macroscopic change in the distribution of the dissolved molecules. To further illustrate the characteristic features of a diffusion process, we use a diffusion cell (Fig. 2.3). In (A) a pure water phase and a colored solution are kept separate in each of two glass chambers. In (B) the two chambers are carefully moved until contact is established between the two phases. It then will be observed that the colored substance gradually moves into the water phase in such a way that the concentration in the chamber receiving the colored molecules is always highest at the zone of transition to the other chamber, even as the concentration in this zone gradually decreases as the colored molecules diffuse into the upper chamber. If we wait long enough we will observe that the colored

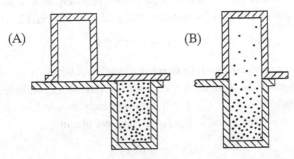

Fig. 2.3. Diffusion cell. In (A) the lower chamber contains a colored solution, while the upper chamber contains pure water. In (B) the two chambers are moved carefully to establish contact between the interiors to allow the dissolved molecules to diffuse into the upper chamber.

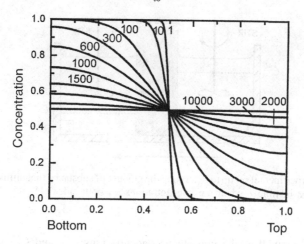

Fig. 2.4. The spatial distribution (concentration profiles) of a diffusing colored substance in a diffusion cell in Fig. 2.3(B) in the direction bottom→top. The curves show the concentration profiles at various times, whose mutual magnitudes relative to the steepest curve are: 1:10:100:300:600:1000:1500:2000:3000:10 000.

molecules are distributed *uniformly* in both chambers, i.e. the concentration is the same all over in the two chambers. Figure 2.4 shows the set of concentration profiles measured at various times $t_0 < t_1 < t_2 < t_3, < \cdots t_\infty$, e.g. by using an optical scanning technique.

This experiment illustrates the essential features of the diffusion process.

(a) A necessary condition for diffusion to take place is a non-uniform spatial distribution of the diffusing molecules.

(b) Mass transport by diffusion takes place in the *direction* of *decreasing* concentration of the diffusing substance.

(c) The *steeper* the concentration profile, the higher the *rate* of transport.

To study the connection between the rate of transport and the steepness of the concentration profile we can use an experimental set up shown in Fig. 2.5. The two phases (1) and (2), contain the diffusing, non-ionic molecules, concentrations $C^{(1)}$ and $C^{(2)}$, which are kept uniform in each phase by stirring. M is a membrane consisting of a homogeneous material of thickness h, which allows for passage of the dissolved particles, but with properties such that the stirring does not cause mixing within the membrane. If $C^{(1)}$ is higher than $C^{(2)}$, the transport will proceed in the direction phase$^{(1)} \to$ phase$^{(2)}$. When the transport process is stationary – i.e. when the flux through the membrane is independent of time – we measure the amount $M^{(\Delta t)}$ that in time Δt passes from phase$^{(1)}$

2. Migration and diffusion

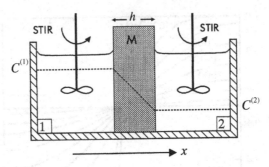

Fig. 2.5. Experimental set up to determine the amount of substance that diffuses through a membrane in time t when the concentrations on both sides (1) and (2) are held constant.

to phase$^{(2)}$ as well as the two mean concentrations $C^{(1)}$ and $C^{(2)}$ in the two phases corresponding to the time interval Δt. This measurement is repeated for other values of $C^{(1)}$ and $C^{(2)}$, and for different areas A and thicknesses h of the membrane. We also use substances with different molecular weights. The results thus obtained can be summarized by the following expression:

$$M^{(\Delta t)} = \text{constant} \times A \, \frac{C^{(1)} - C^{(2)}}{h} \, \Delta t, \qquad (2.5.1)$$

where $C^{(1)} - C^{(2)}$ is the concentration drop in the direction of transport and the constant is characteristic of the diffusing substance and of the membrane.

This expression resembles strongly that describing heat conduction through a slab (Fourier's law: concentrations $C^{(1)}$ and $C^{(2)}$ replace the temperatures $T^{(1)}$ and $T^{(2)}$ and $M^{(\Delta t)}$ the transported quantity of heat Q), or the expression for the movement of electric charge in a conductor (Ohm's law: $C^{(1)}$ and $C^{(2)}$ replace the electric potentials $\psi^{(1)}$ and $\psi^{(2)}$ and $M^{(\Delta t)}$ the transported electric charge q, where the quantity $q/\Delta t = $ electric current i).

It may be helpful to rewrite Eq. (2.5.1) as

$$M^{(\Delta t)} = DA\frac{C^{(1)} - C^{(2)}}{\Delta x} \, \Delta t = -DA\frac{C^{(2)} - C^{(1)}}{\Delta x} \, \Delta t = -DA\frac{\Delta C}{\Delta x} \, \Delta t, \qquad (2.5.2)$$

where D is a constant factor and $\Delta C = C^{(2)} - C^{(1)}$ is the concentration increment over the length $h = \Delta x$ in the direction of transport – in the direction from *high* concentration $C^{(1)}$ to *low* concentration $C^{(2)}$. The increment ΔC is a *negative* quantity.

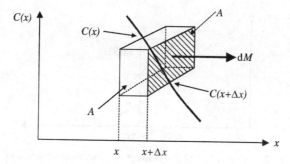

Fig. 2.6. Derivation of Fick's law.

2.5.2 Diffusion flux (Fick's law)

The above result will now be used to describe transport by diffusion in the case where the concentration profile varies as a function of distance as sketched in Fig. 2.4 and Fig. 2.6. In this situation, the concentration at any point x will change with time t. Thus, the concentration is a function $C(x, t)$ of the two variables x and t.

At a given time t, consider the layer between two parallel planes of equal areas A and A' at positions x and $x + \Delta x$, where the distance Δx is small enough to ensure that the concentration drop from $C(x)$ at position x to $C(x + \Delta x)$ at position $x + \Delta x$ can be approximated as being linear. Furthermore, the time Δt, is so small that the change of the concentration profile is negligible: $C(x, t) \approx C(x, t + \Delta t)$; and $C(x + \Delta x, t) \approx C(x + \Delta x, t + \Delta t)$. The amount of substance $M^{(\Delta t)}$ that in time Δt has diffused through the areas A and A' is, according to Eq. (2.5.2),

$$M^{(\Delta t)} = DA \frac{C(x, t) - C(x + \Delta x, t)}{\Delta x} \cdot \Delta t = -DA \frac{C(x + \Delta x, t) - C(x, t)}{\Delta x} \cdot \Delta t.$$

Dividing on both sides by $A \cdot \Delta t$ gives

$$\frac{M^{(\Delta t)}}{A \, \Delta t} = J = -D \frac{C(x + \Delta x, t) - C(x, t)}{\Delta x},$$

where J is the flux across the layers at x and $x + \Delta x$. If $\Delta x \to 0$, the quotient on the right-hand side approaches the partial derivative of $C(x, t)$ with respect to x^*. Hence, the flux $J(x, t)$ at position x and at time t can be expressed as

$$J(x, t) = -D \frac{\partial C(x, t)}{\partial x}, \quad \text{Fick's law.} \quad (2.5.3)$$

---

* See Chapter 1, Section 1.8.2, Eq. (1.8.2).

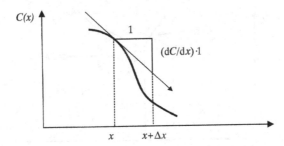

Fig. 2.7. Illustration of the concept of concentration gradient.

Equation (2.5.3) is called Fick's law to honor the German physiologist Adolf Fick*, who set up the equation in 1855[†]. The quantity $\partial C(x, t)/\partial x$ represents the slope of the concentration profile at the point x and at the time t; in this context $\partial C/\partial x$ often is referred to as the *concentration gradient* in the direction of the x-axis[‡]. The *concentration gradient* is equal to concentration change ΔC per unit increase in distance if the concentration profile had continued with the slope it had at the point x, i.e. $\Delta C = (\partial C/\partial x) \times 1$ or, in other words, equal to the differential dC_t for constant t, when the differential of the independent variable x has the value $dx = 1.0$ (cf. Fig. 2.7). The *sign* of the concentration gradient gives the direction of increasing concentration: if $\partial C/\partial x$ is positive then the concentration will increase with increasing values of x, whereas a negative sign means that the concentration will decrease with increasing values of x. Thus, a positive value of $J(x)$ requires a negative value of $\partial C/\partial x$. Equation (2.5.3) states therefore that the *diffusion flux in the positive direction of the x-axis is proportional to the concentration drop per unit length* ($-dC/dx$) *of the concentration profile at the position x.*

2.5.2.1 The diffusion coefficient

The coefficient D in Eq. (2.5.2) is called the *diffusion coefficient* of a diffusing substance. The units of D follow from Eq. (2.3.2) as

$$(\text{amount}) \cdot m^{-2} \cdot s^{-1} = D \cdot (\text{amount}) \cdot m^{-3} \, m^{-1}$$

* Adolf Fick (1829–1901), was Professor of physiology in Würzburg. He was one of the first advocates for applying mathematical and physical concepts in physiological research. "Fick's principle" is treated in any physiological textbook. He also showed among other things that carbohydrates are the most important sources of energy during muscular work.

† Equation (2.5.3) is often introduced as a postulate. The tactic used here reflects our general trend to reduce the number of postulates as much as possible and to use instead formulations built upon experimental data.

‡ The words "concentration gradient" sometimes cause verbal confusion when they are also regarded as synonymous to a spatial concentration *difference* $\Delta C = C^{(1)} - C^{(2)}$ between the points (1) and (2) in space.

or

$$D \stackrel{\Delta}{=} m^2 \, s^{-1} \quad \text{(in SI units)}.$$

However, the values of D are frequently still given in the c.g.s. units of $cm^2 \, s^{-1}$, i.e. $D_{SI} = 10^{-4} \times D_{c.g.s.}$.

The diffusion coefficient D is a characteristic of the molecular species diffusing in a given surrounding. Its magnitude depends not only upon the molecular properties, such as size and shape of the diffusing molecule, but also on the molecular properties of the *medium* in which the diffusion takes place, such as viscosity. In some cases, e.g. large concentrations of the diffusing molecules, D will vary with concentration (see also Section 2.5.3). Most low molecular weight substances in aqueous solutions have diffusion coefficients varying between 10^{-9} and $10^{-10} \, m^2 \, s^{-1}$.

2.5.2.2 A simple application of Fick's law

This example serves only as an exercise to keep control of the dimensions. Consider a liquid column contained in a tube with a 10 cm diameter. A substance is dissolved in the liquid in such a way that its concentration profile decreases linearly along the axis of the tube. The magnitude of the diffusion coefficient of the substance is $5 \times 10^{-5} \, cm^2 \, s^{-1}$. The concentration of the substance is $1 \, mol \, dm^{-3}$ and $0.5 \, mol \, dm^{-3}$ at the positions x and $x + 10 \, cm$, respectively. How much of the substance diffuses through a cross section between these two points in the course of $2\frac{1}{2}$ min, provided there has been no change in the concentration profile?

We have

$$\Delta C = C(x_2) - C(x_1) = -0.5 \, \text{mole dm}^{-3} = -500 \, \text{mole m}^{-3},$$

and $\Delta x = x_2 - x_1 = 10 \, cm = 0.1 \, m$. Hence

$$\frac{dC}{dx} = \frac{\Delta C}{\Delta x} = -\frac{500 \, \text{mol m}^{-3}}{0.1 \, m} = -5000 \, \text{mol m}^{-3} \, m^{-1}.$$

Furthermore

$$D = 5 \times 10^{-5} \, cm^2 \, s^{-1} = 5 \times 10^{-9} \, m^2 \, s^{-1}.$$

Inserting into Eq. (2.5.2) gives

$$J = -(5 \times 10^{-9} \, m^2 \, s^{-1}) \, (-5 \times 10^3 \, \text{mol m}^{-4}) = 2.5 \times 10^{-5} \, \text{mol m}^{-2} \, s^{-1}.$$

From Eq. (2.4.8); $M = J \, At$, where A is the cross section of the tube $= \frac{\pi}{4} d^2$, where d is the tube diameter (10 cm $= 0.1$ m). Furthermore: $t = 2\frac{1}{2} \, min = 150 \, s$.

2. *Migration and diffusion*

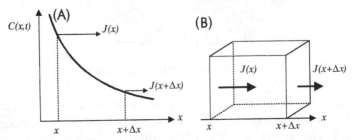

Fig. 2.8. (A) Example of a concentration profile $C(x, t)$ at time t. The concentration drop in the direction of the x-axis is higher at position x than at $x + \Delta x$, resulting in a greater diffusion flux at x than at $x + \Delta x$, and an accumulation of matter takes place in the region between x and $x + \Delta x$. (B) An element of volume in the shape of a prism between the planes at x and $x + \Delta x$ with the surface areas A with fluxes $J(x)$ and $J(x + \Delta x)$.

So, finally

$$M = 2.5 \times 10^{-5} \, \text{mol m}^{-2} \, \text{s}^{-1} \times (3.1416/4) \, 0.01 \, \text{m}^2 \, 150 \, \text{s}$$
$$= 2.95 \times 10^{-5} \, \text{mol}.$$

2.5.3 The diffusion equation

A dissolved substance that is distributed non-uniformly in space will diffuse from regions of high concentration to regions of low concentration. The concentration differences, therefore, will gradually equalize as illustrated in Fig. 2.4. This process is bound to obey a certain regularity, which we shall derive. We consider (Fig. 2.8(A)) a one-dimensional* diffusion process with concentration $C(x, t)$ decreasing in the direction of increasing values of x. The shape of the concentration profile $C(x, t)$ is arbitrary, but is such that the steepness at the position x is greater than that at the neighboring point $x + \Delta x$, i.e. the concentration drop per unit length is greater at x than at $x + \Delta x$. Accordingly, the flux $J(x)$ at x will be greater than the flux $J(x + \Delta x)$ at $x + \Delta x$, and, during time Δt, more matter will enter by diffusion at x than will leave by diffusion at $x + \Delta x$. The resulting accumulation of material will increase the average concentration $\langle C(x, t) \rangle$ inside the region between x and $x + \Delta x$. The greater the difference $J(x, t) - J(x + \Delta x, t)$, the greater the concentration increase ΔC in time Δt. According to Fick's law the difference $J(x, t) - J(x + \Delta x, t)$ is proportional to $(\partial C/\partial x)_x - (\partial C/\partial x)_{x+\Delta x}$, and the greater this difference the

---

* This apparent self-contradictory statement implies a diffusion process which proceeds everywhere in one given direction. If we place a three-dimensional Cartesian coordinate system in such a way that the positive direction of the x-axis coincides with the direction of the diffusion flux, it follows that along the directions of the y- and z-axis we have: $\partial C/\partial y = \partial C/\partial z = 0$. Thus, the concentration can everywhere be described with the *single variable* x as $C(x)$.

greater will be the change in slope of the concentration profile between x and $x + \Delta x$. We therefore would expect that there will be a relation between the rate of the concentration change at point x and the curvature of the concentration profile at that point. This relation, which we shall now derive, is called *the diffusion equation*. We consider two situations.

2.5.3.1 Diffusion with mass conservation

The simplest case to consider – and fortunately the one encountered most frequently – is a diffusion process which occurs under conditions of **mass conservation**, i.e. the diffusing species is nowhere *produced* nor *destroyed**. First, we express the condition of mass conservation in a mathematical form. Consider, as illustrated in Fig. 2.8(B), two planes placed at right angles to the x-axis at the points x and $x + \Delta x$. We will calculate the concentration change that occurs inside the volume element between x and $x + \Delta x$ during an infinitesimal time interval between the time t and $t + \Delta t$. During the interval Δt an amount $M_x^{(\Delta t)}$ of the diffusing substance will pass through a surface area A at the plane at x:

$$M_x^{(\Delta t)} = J(x)A\,\Delta t.$$

In the same time interval an amount $M_{x+\Delta x}^{(\Delta t)}$ will be removed from the volume element $\Delta V = A\,\Delta x$ through the surface area A at $x + \Delta x$:

$$M_{x+\Delta x}^{(\Delta t)} = J(x + \Delta x)A\,\Delta t.$$

The amount $\Delta M^{(\Delta t)}$ accumulated inside the volume element $\Delta V = A\,\Delta x$ in time Δt is

$$\Delta M^{(\Delta t)} = M_x^{(\Delta t)} - M_{x+\Delta x}^{(\Delta t)} = J(x)A\,\Delta t - J(x + \Delta x)A\,\Delta t,$$

where it is assumed that the concentration profile $C(x, t)$ changes so slowly with time that $J(x, t) \approx J(x, t + \Delta t)$ and $J(x + \Delta x, t) \approx J(x + \Delta x, t + \Delta t)$. The amount $\Delta M^{(\Delta t)}$ can also be expressed as the increment in concentration $\Delta C^{(\Delta t)}$ times the volume element $\Delta V = A\Delta x$

$$\Delta M^{(\Delta t)} = \Delta C^{(\Delta t)}A\Delta x = [C(x', t + \Delta t) - C(x', t)]A\,\Delta x,$$

where $C(x', t + \Delta t)$ and $C(x', t)$ are mean concentrations in the volume element, i.e. the concentrations at a position x' somewhere between x and $x + \Delta x$. These two equations for $\Delta M^{(\Delta t)}$ give

$$[C(x', t + \Delta t) - C(x', t)]A\,\Delta x = -[J(x + \Delta x) - J(x)]A\,\Delta t,$$

---

* In physics this assumption is used so often and in many different contexts that it is referred to as **the principle of mass conservation**.

or, by dividing both sides by $A \, \Delta t \, \Delta x$

$$\frac{C(x', t + \Delta t) - C(x', t)}{\Delta t} = -\frac{J(x + \Delta x) - J(x)}{\Delta x}.$$

Now let $\Delta t \to 0$ and $\Delta x \to 0$. Because $x' \to x$, the left-hand side will approach the partial derivative of $C(x, t)$ with respect to t, whereas the right-hand side approaches the partial derivative of $J(x, t)$ with respect to x. Hence, mass conservation at the position x can be expressed as

$$\frac{\partial C(x, t)}{\partial t} = -\frac{\partial J(x, t)}{\partial x}. \tag{2.5.4}$$

Inserting the expression for Fick's law $J = -D \partial C / \partial x$, where for typographical reasons $C(x, t)$ is replaced by C, gives the diffusion equation

$$\frac{\partial C}{\partial t} = -\frac{\partial}{\partial x}\left(-D\frac{\partial C}{\partial x}\right) = D\frac{\partial^2 C}{\partial x^2} + \frac{\partial D}{\partial x}\frac{\partial C}{\partial x}$$

$$= D\frac{\partial^2 C}{\partial x^2} + \left(\frac{\partial D}{\partial C}\frac{\partial C}{\partial x}\right)\frac{\partial C}{\partial x} = D\frac{\partial^2 C}{\partial x^2} + \frac{\partial D}{\partial C}\left(\frac{\partial C}{\partial x}\right)^2.$$

When D is independent of x we have $\partial D / \partial x = 0$ and also $\partial D / \partial C = 0$, and the diffusion equation in one dimension assumes the following form*

$$\frac{\partial C}{\partial t} = D\frac{\partial^2 C}{\partial x^2}. \tag{2.5.5}$$

The *diffusion equation* is a second-order partial differential equation. It describes how the concentration spreads out in space and time in conformity with the requirement of *mass conservation:* the rate of concentration change at x is equal to the second derivative[†] of C with respect to x at the same point times the diffusion coefficient D.

If the concentration depends upon all the the Cartesian coordinates x, y, z as $C(x, y, z, t)$, the diffusion flux will have the components

$$J_x(x, y, z, t), \quad J_y(x, y, z, t) \quad \text{and} \quad J_z(x, y, z, t),$$

corresponding to the directions of the x-, y- and z-axes. A similar argument of mass conservation for a prism with sides Δx, Δy and Δz leads to the following

---

* Often called Fick's second law.
[†] It can be shown that the magnitude of $d^2 y / dx^2$ is a measure of how strongly the curve $y = f(x)$ bends at x: $d^2 y / dx^2$ is proportional to the reciprocal of the radius of curvature of the curve $y = f(x)$ at x.

form of the diffusion equation:

$$\frac{\partial C}{\partial t} = D\left\{\frac{\partial^2 C}{\partial x^2} + \frac{\partial^2 C}{\partial y^2} + \frac{\partial^2 C}{\partial z^2}\right\}. \tag{2.5.6}$$

This expression will not be used further in this book.

2.5.3.2 Diffusion with concurrent mass production

We now consider the situation where the diffusion process takes place in a medium where the diffusing substance also is *produced*. The source of *mass production* is assumed to be distributed as a function of the coordinate x. The intensity of this source, which is denoted by $\rho(x)$, represents the amount of substance being produced per *unit volume* and *unit time*,

$$\rho(x) \overset{\Delta}{=} \text{(produced amount) m}^{-3}\,\text{s}^{-1}.$$

The principle of mass conservation applied to the volume element $\Delta V = A\,\Delta x$ in the time Δt takes the following form*

$$\left\{\begin{array}{l}\text{Accumulated}\\ \text{substance in } A\,\Delta x\\ \text{during } \Delta t\end{array}\right\} = \left\{\begin{array}{l}\text{Flux through}\\ \text{surface } A \text{ at } x\\ \text{during } \Delta t\end{array}\right\} + \left\{\begin{array}{l}\text{Produced}\\ \text{substance in } A\,\Delta x\\ \text{during } \Delta t\end{array}\right\}$$
$$- \left\{\begin{array}{l}\text{Flux through}\\ \text{surface } A \text{ at } x + \Delta x\\ \text{during } \Delta t\end{array}\right\}.$$

The equivalent mathematical formulation reads

$$\frac{\partial}{\partial t}\{C(x,t)A\,\Delta x\}\Delta t = J(x)A\,\Delta t + \{\rho(x)A\,\Delta x\}\,\Delta t - J(x+\Delta x)A\,\Delta t.$$

Calculations analogous to those in Section 2.5.3.1 lead to the following form of the diffusion equation

$$\frac{\partial C}{\partial t} = D\frac{\partial^2 C}{\partial x^2} + \rho(x), \tag{2.5.7}$$

where $\rho(x)$ would be a *negative* quantity if the diffusing substance had been removed (consumed) from the medium.

2.5.3.3 Classification of diffusion processes

The diffusion equation is a *partial differential equation*[†] of the second order, where the independent variables are the position x and the time t. Solutions

---

* This statement is equivalent to *Fick's principle*, which is formulated and used in almost any textbook of physiology.
[†] Mathematicians prefer to denote it as a partial differential equation of parabolic type.

that contain both time and position variables are called *time-dependent* or *non-stationary solutions*. Examples of solutions of the diffusion equation will be shown in the following sections. When t is sufficiently large (strictly $t \to \infty$) the solutions will describe one of the following situations.

(i) The substance in question has distributed itself uniformly throughout the available space. When all concentration gradients are therefore extinct everywhere and *no net flux* can be observed, the system is said to have attained a *state of equilibrium*. This state can be expressed mathematically as

$$J = 0, \qquad \textbf{Equilibrium state}, \qquad\qquad (2.5.8)$$

for *all* values.

(ii) By introducing certain external constraints, e.g. by the continuous addition and removal of the same amount at two different positions, it is possible to maintain a concentration gradient – and therefore also a flux through the system – with no change in the concentration profile of the substance. The system is in *a stationary state*, and the solutions of the diffusion equation corresponding to this state are called *stationary* or *time-independent* solutions. Mathematically the stationary solution is characterized by the conditions

$$\left.\begin{array}{l} \dfrac{\partial C}{\partial t} = 0 \\[2mm] J = \text{constant} \neq 0 \end{array}\right\} \textbf{Stationary state}, \qquad (2.5.9)$$

for all x.

(iii) Finally, the concentration profile may change with time, but at such a slow rate that the concentration profile for all practical purposes can be regarded as stationary. Such process is called a *quasi-stationary* process, which is often the condition preferred in experimental investigations, because the stationary situation is far simpler to handle than the corresponding time-dependent solution. Whether a process can be regarded and treated as belonging to class (ii) remains always a matter of judgement, the value of which depends upon the experience and tact of the investigator.

2.5.4 Stationary diffusion processes

Physical models of simple geometry such as a plate, a cylinder or a sphere are frequently used to simulate certain diffusion processes in biological systems.

We show below how to calculate the concentration profiles in such bodies when the *concentration* or *flux* of the diffusing substance is maintained at a constant value at certain well-defined positions. These fixed values, which generally form the end points of the concentration profile in the region, are attached *boundary values* belonging to the diffusion problem in question.

2.5.4.1 One-dimensional diffusion

To illustrate the usual practice to find the concentration profile corresponding to a given set of boundary values we examine two simple, basic examples.

(i) Diffusion through a plate

We consider diffusion through a homogeneous medium in the shape of a plate of thickness h. The plate separates two liquid phases (i) and (o) containing the diffusing solute, whose concentrations in the two phases are by some means maintained at constant values $C^{(i)}$ and $C^{(o)}$. The problems to be solved are: (a) how is the shape of the stationary concentration profile inside the plate related to the concentrations $C^{(i)}$ and $C^{(o)}$ in the two phases (i) and (o); and (b) how large is the diffusion flux through the plate, if the solute diffusion coefficient in the plate has the value D.

The coordinate system is placed as shown in Fig. 2.9. Inside the region $0 \leq x \leq h$ the concentration $C(x)$ must satisfy the diffusion equation

$$\frac{\partial C}{\partial t} = D\frac{\partial^2 C}{\partial x^2}, \quad \text{for} \quad 0 \leq x \leq h.$$

Since the diffusion process is assumed to be stationary the condition $\partial C/\partial t = 0$ holds everywhere for $0 \leq x \leq h$. Hence, the concentration profile in the plate is a solution of the ordinary second-order differential equation

$$\frac{d^2 C}{dx^2} = \frac{d}{dx}\left(\frac{dC}{dx}\right) = 0, \quad \text{for} \quad 0 \leq x \leq h. \tag{2.5.10}$$

Equation (2.5.10) is satisfied if*

$$\frac{dC}{dx} = A_1,$$

where A_1 is an arbitrary constant. Integration of this equation yields

$$C(x) = \int \left(\frac{dC}{dx}\right) dx = \int A_1 \, dx + B_1 = A_1 x + B_1, \tag{2.5.11}$$

---

* See Chapter 1, Section 1.2.3.1(a).

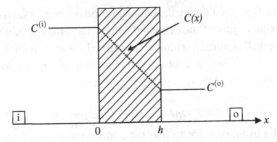

Fig. 2.9. Stationary one-dimensional diffusion through a homogeneous plate (membrane) of thickness h.

which is the *general solution* of Eq. (2.5.10). It contains two arbitrary constants A_1 and B_1 that arise from the result of integrating a second-order differential equation. It follows from Eq. (2.5.11) that the concentration profile of the stationary one-dimensional diffusion process is *linear*. The constants A_1 and B_1 are determined by the physical situation, as described by the boundary conditions. In the present case, the concentrations in the two surrounding solutions remain at the fixed values $C^{(i)}$ and $C^{(o)}$. To simplify matters we assume that solute solubility is the same in the membrane and in the solutions. The mathematical problem then consists of adjusting the two constants A_1 and B_1 in the general solution, Eq. (2.5.11), to describe correctly the physical situation. This process of adjustment (solving the *boundary value problem*) involves two steps.

(1) Write the conditions the function $C(x)$ must satisfy (the *boundary conditions*). In the present case the two set of numbers $(0, C^{(i)})$ and $(h, C^{(o)})$ belong to the function $C(x)$:

$$C(0) = C^{(i)}, \quad \text{for} \quad x = 0, \tag{B1}$$

and

$$C(h) = C^{(o)}, \quad \text{for} \quad x = h. \tag{B2}$$

(2) These boundary conditions are inserted successively in the general solution Eq. (2.5.11). This results in a set of equations where the integration constants are the unknowns to be determined. In the present case, applying the boundary condition (B1)* gives

$$C^{(i)} = A_1 \cdot 0 + B_1, \tag{I}$$

---

* In this text (B1), (B2) etc. are used to number the boundary conditions and Roman capitals (I), (II) etc. are used for the equations derived from the conditions.

or

$$B_1 = C^{(i)},$$

and applying the boundary condition (B2) yields

$$C^{(0)} = A_1 h + B_1 = A_1 h + C^{(i)}, \tag{II}$$

and replacing B_1 by $C^{(i)}$. Hence

$$A_1 = \frac{C^{(0)} - C^{(i)}}{h}.$$

The above values of A_1 and B_1 are inserted into Eq. (2.5.11) to give

$$C(x) = \frac{C^{(0)} - C^{(i)}}{h} x + C^{(i)}, \tag{2.5.12}$$

which describes the expression for the concentration profile in the plate under stationary conditions. To check the result, put $x = 0$ and $x = h$, which returns $C(0) = C^{(i)}$ and $C(h) = C^{(0)}$, as it should. In the present case, the solution of the *boundary value problem* is a fairly simple matter. More elaborate boundary conditions results in more complicated calculations. But, the technique to be used is fundamentally the same* no matter the degree of mathematical complexity an actual case may present.

Having obtained the equation for the concentration profile we calculate the flux by using Fick's law. We have

$$\frac{dC}{dx} = \frac{C^{(0)} - C^{(i)}}{h} \quad \text{for} \quad 0 \le x \le h,$$

and invoking Eq. (2.5.3) gives

$$J = -D\frac{C^{(0)} - C^{(i)}}{h} = \frac{D}{h}\left(C^{(i)} - C^{(0)}\right), \tag{2.5.13}$$

which is the expression for the flux through the plate if the solubility of the substance is the same in the two phases (i) and (o) and the membrane. It is often more convenient[†] to rewrite the flux expression Eq. (2.5.13) as

$$J = P\left(C^{(i)} - C^{(0)}\right), \tag{2.5.14}$$

where

$$P = \frac{D}{h}. \tag{2.5.15}$$

---

* As illustrated in greater detail in the next example.
[†] Since the positive direction of the flux is (i) → (o).

The coefficient P – i.e. the flux through the plate corresponding to a concentration difference $C^{(i)} - C^{(o)} = \Delta C = 1$ – is called the *permeability coefficient* of the membrane, or more simply the *permeability* for the substance in question. In Section 2.7.1 we adapt this concept to include the more common situation where the solute solubility in the membrane differs from that of the surrounding solutions.

Conversely, if the flux J and the concentration difference $\Delta C = C^{(i)} - C^{(o)}$ are known quantities, we use Eq. (2.5.15) to calculate the membrane permeability as

$$P = \frac{J}{C^{(i)} - C^{(o)}}. \tag{2.5.16}$$

(ii) Diffusion through two adjoining, different media

Next we examine a somewhat more complicated situation. The simple membrane in Fig. 2.9 is replaced by a composite membrane consisting of two contiguous, but different, layers. Layer (1) has thickness h_1 and a solute diffusion coefficient D_1, whereas layer (2) has thickness h_2 and a solute diffusion coefficient D_2. As in the above example, the concentrations of the two surrounding media (i) and (o) are maintained at $C^{(i)}$ and $C^{(o)}$. We wish to determine the concentration profiles $C(x)$ in the two media and how the flux depends on the concentration difference $C^{(i)} - C^{(o)} = \Delta C$.

We place the coordinate system as shown in Fig. 2.10. To simplify the writing we put $h_1 = a$ and $h_1 + h_2 = b$. When D_1 and D_2 differ, the diffusion equation must be solved separately for each of the two regions (1) and (2). We put

$$C(x) = \begin{cases} C_1(x) & \text{in layer (1),} \quad \text{for} \quad 0 \leq x \leq a, \\ C_2(x) & \text{in layer (2),} \quad \text{for} \quad a \leq x \leq b, \end{cases} \tag{2.5.17}$$

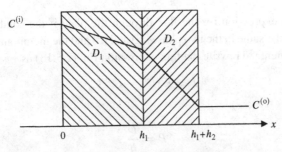

Fig. 2.10. Stationary diffusion from concentration $C^{(i)}$ to concentration $C^{(o)}$ through a compound medium composed of two contiguous layers of thicknesses h_1 and h_2 and diffusion coefficients D_1 and D_2 for the diffusing substance.

where $C_1(x)$ and $C_2(x)$ are functions to be determined. Because the process is stationary the concentration profiles in both media must be solutions of the equation

$$\frac{d^2C}{dx^2} = 0,$$

whose general solution, Eq. (2.5.11), is valid in both regions. Accordingly, we write

$$C(x) = \begin{cases} C_1(x) = A_1x + B_1, & \text{for} \quad 0 \le x \le a, \\ C_2(x) = A_2x + B_2, & \text{for} \quad a \le x \le b. \end{cases} \quad (2.5.18)$$

Thus, we have four constants A_1, B_1, A_2 and B_2, whose values must be adjusted so as to make the behavior of $C_1(x)$ and $C_2(x)$ conform to the physical constraints imposed on the system. To solve the problem we need *four* equations containing the four unknown constants $A_1, \ldots B_2$. To obtain these equations one has to set up four boundary conditions that fix the proper concentration profile through the two layers. The first two are similar to those of example (i), since the concentration must be continuous at the transition from the surrounding phases (i) and (o) to layer (1) and layer (2) respectively. This implies that

$$C_1(x) = C^{(i)}, \quad \text{for} \quad x = 0, \quad (B3)$$

and

$$C_2(x) = C^{(o)}, \quad \text{for} \quad x = b. \quad (B4)$$

The next two boundary conditions are determined by the physical situation at the transition between the two layers corresponding to the position $x = a$. The concentration profile is continuous at the transition between the two layers, since, by assumption, the distribution coefficient has the same value 1.0 in the two layers*, that is

$$C_1(x) = C_2(x), \quad \text{for} \quad x = a. \quad (B5)$$

Since the process is stationary, no accumulation occurs within the membrane. The fluxes must therefore have the *same* value at $x = a - \varepsilon$ and $x = a + \varepsilon$, where $\varepsilon \ll 1$. Invoking Fick's law yields

$$-D_1\left(\frac{dC_1}{dx}\right)_{a-\varepsilon} = -D_2\left(\frac{dC_2}{dx}\right)_{a+\varepsilon}. \quad (B6)$$

* Otherwise an infinite flux would result at $x = a$.

To make the general solutions for $C_1(x)$ and $C_2(x)$ in Eq. (2.5.14) conform to the above conditions (B3), ... (B6) we proceed as follows.

Condition (B6) applied to $C_1(x)$ and $C_2(x)$ of Eq. (2.5.18) gives

$$D_1 A_1 = D_2 A_2, \tag{III}$$

and condition (B5) applied to the same equations gives

$$A_1 a + B_1 = A_2 a + B_2. \tag{IV}$$

Condition (B3) applied to $C_1(x)$ of Eq. (2.5.18) gives

$$C^{(i)} = B_1, \tag{V}$$

and finally condition (B4) applied to $C_2(x)$ gives

$$A_2 b + B_2 = C^{(o)}. \tag{VI}$$

Thus, at our disposal we have *four simultaneous equations* to determine the *four* constants, and the problem is in principle solved, leaving only a more or less laborious computation.

These four equations are solved step by step. We insert (III) and (V) into (IV)

$$a\left[\frac{D_2}{D_1} - 1\right] A_2 - B_2 = -C^{(i)},$$

which, when added to (VI), gives

$$A_2 = \frac{C^{(o)} - C^{(i)}}{a(D_2/D_1 - 1) + b},$$

or

$$A_2 = -\frac{D_1\left(C^{(i)} - C^{(o)}\right)}{a(D_2 - D_1) + b D_1}.$$

Inserting A_2 into (III) yields

$$A_1 = -\frac{D_2\left(C^{(i)} - C^{(o)}\right)}{a(D_2 - D_1) + b D_1}.$$

Finally, inserting A_2 into (VI) gives

$$B_2 = -\frac{b D_1\left(C^{(i)} - C^{(o)}\right)}{a(D_2 - D_1) + b D_1} + C^{(o)}.$$

Inserting the values for A_1, A_2, A_3 and A_4 into Eq. (2.5.18) and replacing the values $x = a$ and $x = b$ by $a = h_1$ and $b = h_1 + h_2$ we obtain

$$C(x) = \begin{cases} C_1(x) = C^{(i)} - \dfrac{D_2\left(C^{(i)} - C^{(0)}\right)}{h_1 D_2 + h_2 D_1}\, x, & \text{for} \quad 0 \le x \le a, \\[3mm] C_2(x) = C^{(0)} + \dfrac{D_1\left(C^{(i)} - C^{(0)}\right)}{h_1 D_2 + h_2 D_1}\, (h_1 + h_2 - x), & \text{for} \quad a \le x \le b, \end{cases}$$

(2.5.19)

as the expression for the concentration profile through the composite membrane.

Fick's law, $J = -D_1(dC_1/dx)$, or $J = -D_2(dC_2/dx)$ applied to Eq. (2.5.19) gives the following expression for the flux through the membrane

$$J = \frac{D_1 D_2}{h_1 D_2 + h_2 D_1}\left(C^{(i)} - C^{(0)}\right).$$

(2.5.20)

A comparison with Eq. (2.5.14) shows that the factor on the right-hand side $(D_1 D_2)/(h_1 D_2 + h_2 D_1)$ has the character of the permeability of the compound membrane. Accordingly, we generalize Eq. (2.5.14) and define an *equivalent permeability* $\langle P \rangle$ for the system

$$J = \langle P \rangle \left(C^{(i)} - C^{(0)}\right),$$

where

$$\langle P \rangle = \frac{D_1 D_2}{h_1 D_2 + h_2 D_1} = \frac{1}{\dfrac{h_1}{D_1} + \dfrac{h_2}{D_2}} = \frac{1}{\dfrac{1}{P_1} + \dfrac{1}{P_2}}$$

since $P_1 = D_1/h_1$ and $P_2 = D_2/h_2$ are the permeabilities of each of the layers with diffusion coefficients D_1 and D_2 and thicknesses h_1 and h_2. Equating the left-hand side and the far right-hand term gives

$$\frac{1}{\langle P \rangle} = \frac{1}{P_1} + \frac{1}{P_2},$$

(2.5.21)

that is, addition of membrane permeabilities arranged in series obeys the same law as, for example, the addition law of conductances in series in electric circuits.

(iii) Unstirred layers

The last expression is more easily obtained as follows. Let J denote the stationary flux through the compound membrane in Fig. 2.10. The concentration

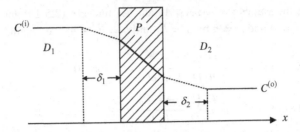

Fig. 2.11. Example of concentration profiles when a membrane is surrounded by two unstirred layers of thickness δ_1 and δ_2 and with solute diffusion coefficients D_1 and D_2. The dotted lines indicate a possible concentration profile near the membrane if the intensity of stirring decreases gradually.

changes through the two layers can then be written

$$\Delta C_1 = \frac{J}{P_1} \quad \text{and} \quad \Delta C_2 = \frac{J}{P_2}, \tag{2.5.22}$$

where ΔC_1 and ΔC_2 are the concentration drops in layer (1) and layer (2) with permeabilities P_1 and P_2, respectively. Further, let ΔC denote the total concentration drop across the compound membrane and $\langle P \rangle$ its equivalent permeability. Then we can write

$$\Delta C_1 + \Delta C_2 = \Delta C = \frac{J}{\langle P \rangle}.$$

Invoking Eq. (2.5.22) gives

$$\frac{J}{P_1} + \frac{J}{P_2} = \frac{J}{\langle P \rangle}$$

or

$$\frac{1}{\langle P \rangle} = \frac{1}{P_1} + \frac{1}{P_2}. \tag{2.5.23}$$

This result can be generalized to a compound membrane consisting of n different layers:

$$\frac{1}{\langle P \rangle} = \frac{1}{P_1} + \frac{1}{P_2} + \cdots \frac{1}{P_n}. \tag{2.5.24}$$

Usually, the membrane permeability of a substance is determined experimentally as the ratio between the stationary diffusion flux and the concentration difference across the membrane ΔC that causes the flux. Generally it is assumed that $\Delta C = C^{(i)} - C^{(o)}$, i.e. the two surrounding media are so well stirred that the macroscopic concentration of the solutions is maintained right up to each of the two membrane/solution interfaces. In most cases this assumption may be legitimate. Sometimes, however, the solute flux is so high that even the most vigorous stirring cannot bring about an interfacial solute concentration at the bulk solution value. This situation can be simulated by assuming that the membrane – whose permeability we wish to determine – is surrounded by two unstirred layers of the solution with thickness δ_1 and δ_2, respectively. The mass transfer between the bulk phases with concentrations $C^{(i)}$ and $C^{(0)}$ will have to take place by diffusion through these two unstirred layers as illustrated in Fig. 2.11. In this case the ratio between the stationary flux J and the macroscopic concentration difference $C^{(i)} - C^{(0)}$ will result in an equivalent permeability $\langle P \rangle$, which according to Eq. (2.34) consists of the following terms

$$\frac{1}{\langle P \rangle} = \frac{\delta_1}{D_1} + \frac{1}{P} + \frac{\delta_2}{D_2}, \tag{2.5.25}$$

where P is the "true" membrane permeability and D_1 and D_2 are their solute diffusion coefficients in the two unstirred layers with thicknesses δ_1 and δ_2.

In the model in Fig. 2.11 it is assumed that the stirring stops abruptly at distances δ_1 and δ_2 from the membrane. This assumption is hardly reasonable except in quite special cases. In more realistic situations, the intensity of the stirring will decrease gradually as the membrane is approached. This means that the unstirred layer thickness becomes somewhat indeterminate*.

(iv) Plate covered on one side by a membrane of permeability P

An important special case of section (ii) is when the thickness of one layer shrinks in such a way that a considerable barrier against diffusion remains even as the thickness ϵ becomes infinitely small relatively to the thickness h_2 of the other layer. The solution can be obtained directly from Eq. (2.5.20) by letting $h_1 = \varepsilon \to 0$ and assigning a finite value P to the ratio D_1/ε. We shall, however, abstain from taking such a short cut as this situation is of sufficient importance to be considered and formulated as a boundary value problem in its own right. We arrange the coordinate system as shown in Fig. 2.12(A). The membrane whose thickness we ignore is placed at $x = 0$ with the plate of thickness h placed to the right. Let the diffusing substance have the concentrations $C^{(i)}$ and $C^{(0)}$ in the surrounding media and a diffusion coefficient D in the plate. We seek the formula for the concentration profile through the plate and, in particular, the concentration in the region $0 < x \leq h$ inside the membrane.

---

* This should not be ignored.

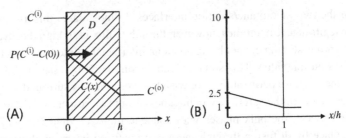

Fig. 2.12. Stationary concentration profile through a homogeneous plate of thickness h and diffusion coefficient D for the substance. On the left-hand side the plate is covered by an infinitely thin membrane with permeability P. (A) The boundary value problem. (B) Example: $C^{(i)}/C^{(o)} = 10$ and $P_i/P = 5$, where $P_i = D/h$.

We have shown* that the concentration profile in the plate must be of the form

$$C(x) = A_1 x + B_1, \quad \text{for} \quad \varepsilon \leq x \leq h, \tag{2.5.26}$$

where $\varepsilon \ll 1$. The conditions to be satisfied are as follows.

The concentration is maintained at $C^{(o)}$ on the right-hand side of the plate, i.e.

$$C(x) = C^{(o)}, \quad \text{for} \quad x = h. \tag{B7}$$

On the left-hand side the conditions are more complex. (a) At the membrane side facing the solution the concentration is $C^{(i)}$. (b) The flux of the substance is continuous in crossing the interface from the inner side of the membrane at $x = \varepsilon$ to the plate. This requires that

$$P \left(C^{(i)} - C(x) \right) = -D \frac{dC}{dx}, \quad \text{for} \quad x = \varepsilon. \tag{B8}$$

The left-hand side represents the flux through the membrane, the right-hand side is the flux entering the plate†. The general equation for $C(x)$, Eq. (2.5.26), is inserted into the two boundary conditions. This leads to two linear equations in A_1 and B_1

$$h A_1 + B_1 = C^{(o)}, \tag{VII}$$

and

$$-D A_1 + P B_1 = P C^{(i)}, \tag{VIII}$$

---

* See Section 2.5.4.1 (ii), Eq. (2.5.18).
† Compare Section 2.5.4.1 (i), Eq. (2.5.14) and Section 2.5.4.1 (ii), Eq. (B6).

where the term εA_1 has been disregarded in Eq. (VIII). Solving for A_1 and B_1, using Cramer's rule, gives

$$A_1 = \frac{\det \begin{vmatrix} C^{(0)} & 1 \\ PC^{(i)} & P \end{vmatrix}}{\det \begin{vmatrix} h & 1 \\ -D & P \end{vmatrix}} = \frac{PC^{(0)} - PC^{(i)}}{hP + D} = -\frac{P(C^{(i)} - C^{(0)})}{P_i + P}\frac{1}{h},$$

and

$$B_1 = \frac{\det \begin{vmatrix} h & C^{(0)} \\ -D & PC^{(i)} \end{vmatrix}}{hP + D} = \frac{hPC^{(i)} + DC^{(0)}}{hP + D} = \frac{PC^{(i)} + P_iC^{(0)}}{P_i + P},$$

where $P_i = D/h$ is the permeability coefficient of the substance in the plate. Inserting the expressions for A_1 and B_1 into Eq. (2.5.26) gives the desired expression for the concentration profile through the plate in the range $\varepsilon \leq x \leq h$

$$C(x) = -\frac{P(C^{(i)} - C^{(0)})}{P_i + P}\frac{x}{h} + \frac{PC^{(i)} + P_iC^{(0)}}{P_i + P}. \tag{2.5.27}$$

An example is shown in Fig. 2.12 for $C^{(i)} = 10$, $C^{(0)} = 1$ and $P_i/P = 5$.

(v) Diffusion with mass consumption

This slightly more complicated situation is of particular interest in biological systems in connection with the transport of O_2 and CO_2. Consider a rectangular plate, e.g. a tissue slice, immersed in a solution containing a substance that is kept at the constant value $C^{(0)}$. The substance enters the slice by diffusion – with a diffusion coefficient D – and is *consumed* at a constant intensity ϱ. Since the supply of substance to the slice is the same on the two sides one can expect a shape of the concentration profile that is symmetric with respect to the median plane of the slice. Accordingly, we place the origin of the x-axis half-way inside the plate whose thickness we set to $2h$. In general the concentration in the plate is governed by

$$\frac{\partial C}{\partial t} = D\frac{\partial^2 C}{\partial x^2} - \varrho,$$

where the constant ϱ is *negative* because the substance is consumed. Furthermore, since the process is *stationary*, we have $\partial C/\partial t = 0$. Thus, the equation for the concentration profile is governed by a second-order ordinary differential equation

$$D\frac{d^2 C}{dx^2} = \varrho, \tag{2.5.28}$$

or

$$\frac{d}{dx}\left(\frac{dC}{dx}\right) = \frac{\varrho}{D}. \tag{2.5.29}$$

The first integration of this equation gives*

$$\frac{dC}{dx} = \frac{\varrho}{D}\int dx + A_1 = \frac{\varrho}{D}x + A_1,$$

where A_1 is a constant. Integrating this equation gives the general equation for the stationary concentration profile in case the diffusing substance is also *consumed at a constant rate*

$$C(x) = \int \left(\frac{\varrho}{D}x + A_1\right)dx = \tfrac{1}{2}\frac{\varrho}{D}x^2 + A_1 x + B_1, \tag{2.5.30}$$

where A and B are constants to be adjusted to the boundary conditions. These are

$$C(x) = C^{(0)}, \quad \text{for} \quad x = +h, \tag{B9}$$

and

$$C(x) = C^{(0)}, \quad \text{for} \quad x = -h. \tag{B10}$$

For $x = h$, we obtain

$$\tfrac{1}{2}\frac{\varrho}{D}h^2 + A_1 h + B_1 = C^{(0)}, \tag{IX}$$

and for $x = -h$ we get

$$\tfrac{1}{2}\frac{\varrho}{D}h^2 - A_1 h + B_1 = C^{(0)}. \tag{X}$$

Subtracting (X) from (IX) gives

$$2A_1 h = 0,$$

from which it follows that $A_1 = 0$, and

$$B_1 = C^{(0)} - \tfrac{1}{2}\frac{\varrho}{D}h^2.$$

Inserting these value for A_1 and B_1 into Eq. (2.5.30) gives

$$C(x) = \tfrac{1}{2}\frac{\varrho}{D}x^2 - \tfrac{1}{2}\frac{\varrho}{D}h^2 + C^{(0)}, \tag{2.5.31}$$

---

* See Chapter 1, Section 1.2.3.1, (f) and (g).

the final solution for the boundary value problem. It appears that the concentration profile has the shape of a parabola with downward convexity. As expected the minimum concentration is in the middle of the layer and has the value

$$C_{min} = C^{(0)} - \frac{1}{2}\frac{\varrho}{D}h^2. \qquad (2.5.32)$$

Hence, if D and ϱ are known one can estimate the value of the external concentration $C^{(0)}$ that is needed to ensure that C_{min} is larger than a given value C'.

The flux into the layer at $x = h$ is

$$J = -D\left(\frac{dC}{dx}\right)_{x=h} = -D\left(\frac{1}{2}\frac{\varrho}{D}\right)2h = -\varrho h. \qquad (2.5.33)$$

The flux at $x = -h$ is $J(-h) = \varrho h$, i.e. of equal magnitude to the flux at $x = h$ but in the opposite direction in accordance with the loading of the slice from both sides. The value ϱh also could be obtained without using Fick's law, as ϱ is the consumption per unit volume and unit time. The consumption in time Δt inside a volume extending from $x = 0$ to $x = h$ and bounded by end-surfaces A is equal to $\varrho h A \Delta t$. This amount must be compensated by a supply of equal magnitude across the surfaces A, which is $J A \Delta t$. Hence setting $\varrho h A \Delta t = A J \Delta t$ yields the flux into the medium as in Eq. (2.5.35).

It follows from Eq. (2.5.31) that $dC/dx = 0$ for $x = 0$. This means that the *flux always is zero* at $x = 0$. Thus, Eq. (2.5.31) considered only in the region $0 \leq x \leq h$ represents the concentration profile in a layer of thickness h, which is supplied by material at one surface only (in the present case at $x = h$) and is *impermeable* to the substance at the other surface (at $x = 0$).

Negative concentrations are not possible physically. Thus, Eq. (2.5.31) is valid only for external concentrations $C^{(0)}$ that result in $C_{min} \geq 0$, or

$$C^{(0)} \geq \frac{1}{2}\frac{\varrho}{D}h^2.$$

For lower values of $C^{(0)}$ the concentration inside the layer will still decrease with a parabolic profile and become *zero* at some position $x_0 < h$ where no more of the diffusing material will be available for consumption. Hence, no material will diffuse into the remaining region $0 \leq x \leq x_0$ and therefore dC/dx is also zero at x_0. To evaluate x_0 we adjust the general solution Eq. (2.5.30) to the following boundary conditions

$$C(x) = 0, \quad \text{for} \quad x = x_0, \qquad (B11)$$

and

$$\frac{dC}{dx} = 0, \quad \text{for } x = x_0. \tag{B12}$$

Insertion into Eq. (2.5.30) gives

$$\tfrac{1}{2} \frac{\varrho}{D} x_0^2 + A_1 x_0 + B_1 = 0, \tag{XI}$$

and

$$\frac{\varrho}{D} x_0 + A_1 = 0, \tag{XII}$$

from which we find $A_1 = -\varrho x_0 / D$ and $B_1 = \varrho x_0^2 / 2D$. Inserting these values in Eq. (2.5.30) gives

$$C(x) = \tfrac{1}{2} \frac{\varrho}{D} x^2 - \frac{\varrho}{D} x_0 x + \tfrac{1}{2} \frac{\varrho}{D} x_0^2, \tag{2.5.34}$$

which describes the concentration profile in the region $x_0 \leq x \leq h$, whereas the concentration is zero in the remaining region $0 \leq x \leq x_0$. Examples of such profiles are shown in Fig. 2.13. In many applications the essential question is how far the substance moves into the tissue for a given value of $C^{(o)}$. Replacing

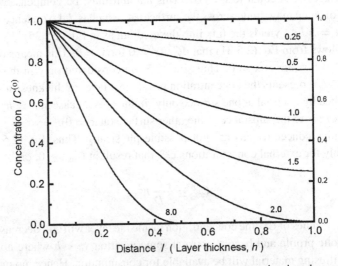

Fig. 2.13. Concurrent stationary diffusion and mass consumption in a homogeneous layer of thickness h. The substance diffuses from the surrounding medium at concentration $C^{(o)}$ into the plate through the surface plane at $x = 0$. In the layer the diffusion coefficient and the rate of consumption are D and $-\varepsilon$ respectively. The plane at $x = h$ is not penetrable to the substance. Ordinate: concentration. Abscissa: distance $X = x/h$ in units of the layer thickness h. The curves are calculated from Eq. (2.5.31) and Eq. (2.5.34) after displacing the x-axis to $x = -h$ and using the relative units $C(x)/C^{(o)}$ and x/h. The numbers on the curves are the values of $\varrho/2D$.

x with h in Eq. (2.5.34) yields

$$C^{(0)} = \tfrac{1}{2}\frac{\varrho}{D}h^2 - \frac{\varrho}{D}x_0 h + \tfrac{1}{2}\frac{\varrho}{D}x_0^2 = \tfrac{1}{2}\frac{\varrho}{D}(h - x_0)^2,$$

or

$$x_0 = h - \sqrt{\frac{2DC^{(0)}}{\varrho}}. \qquad (2.5.35)$$

It may be helpful to visualize geometrically the process of solving boundary conditions, e.g. those associated with Eq. (2.5.7). For a given set of values for D and ε, the shape of the concentration profile is given by the parabola $\tfrac{1}{2}\frac{\varrho}{D}x^2$ – without regard to the boundary values. The adjustment of the boundary conditions amounts to moving the parabola around in the plane until it contains the two points in space x_1, $C(x_1)$ and x_2, $C(x_2)$ that correspond to the specific set of boundary values.

EXAMPLE To illustrate the use of Eq. (2.5.31) or Eq. (2.5.34) we consider the diffusion of a gas, e.g. O_2, in a homogeneous tissue slice of thickness $2h$ having a constant oxygen consumption. Principally there is no difference between the diffusion in an aqueous solution of a salt or of a gas. In both cases the process is governed by Fick's law.

An aqueous solution of gas molecules is often obtained by bubbling a volume V of water with the gas – either in its pure form or in a mixture of gases – for a sufficient time to *equilibrate* the water with the gas molecules. Kinetically this equilibrium means that the number of gas molecules entering the gas phase from the liquid phase per unit time is the same as the number moving from the gas in to the water phase. Thermodynamically the criterion for this equilibrium is that the chemical potentials for the species have the same value in the gas phase and in the liquid phase.

No commonly accepted unit exists to describe the solubility of a gas in a liquid. A convenient measure of the concentration of the dissolved gas molecules of a given species, e.g. O_2, *is the volume of the gaseous species V_{gas} – reduced to the temperature $0\,^\circ C$ (273.16 K) and pressure of 1 atm* – which is dissolved per unit volume solvent at the current temperature T.* Closely connected to this definition as a measure for the solubility of a gas is the Bunsen absorption coefficient[†] α. This is defined as the gas volume (reduced to STP) which dissolves and equilibrates per unit volume solute at the temperature T at 1 atm

---

[*] This state is called "STP" or "Standard Temperature and Pressure": 1 atm = 760 mm Hg = 760 Torr = 101.3 kPa. See Section 2.81.

[†] Introduced in 1857 by R. W. Bunsen (1811–1899), Professor of chemistry at several universities in Germany, finally in Heidelberg (1852–1889). Bunsen had a very broad chemical repertory. Together with Kirchhoff he founded spectral analysis, and discovered rubidium and cesium.

partial pressure of the gas. When the volume V_{gas} of the gas in question at STP is dissolved in a solute volume V_{sol} at a partial pressure p_j (in atm) the absorption coefficient at the temperature T for the gas is computed as

$$\alpha = \frac{V_{gas}}{V_{sol}} \cdot \frac{1}{p_j}, \qquad (2.5.36)$$

where the two volumes must be expressed in the same units. The dimension of α is $(m^3 \text{ gas})/(m^3 \text{ solution}) \text{ atm}^{-1}$. Thus, when a gas at a partial pressure p_j is in contact with a given aqueous volume V_{sol}, then at equilibrium the amount of gas that enters in solution whose volume V_{gas} at STP is given by

$$V_{gas} = \alpha \, p_j \, V_{sol}. \qquad (2.5.37)$$

In keeping with the above concentration measure, the dissolved gas at a partial pressure p_j has the concentration

$$C_{gas} = \frac{V_{gas}}{V_{sol}}, \qquad (2.5.38)$$

expressed in the units employed for V_{gas} and V_{sol}. Putting the last two equations together we have also

$$C_{gas} = \alpha p_j. \qquad (2.5.39)$$

Table 2.1 gives the values for α for several gases at three different temperatures. Note the relatively large values for CO_2 and H_2S, which result from the fact that these molecules also undergo a *chemical reaction* with the H_2O molecules. In this connection one should remember that quite different relations exist between the dissolved gas volume and the equilibrium pressure for O_2 and CO_2 when the liquid phase is not water but *blood*.

Table 2.1. *Solubilities of gases in water*

The numerical values are Bunsen's absorption coefficients α, i.e. the gas volume reduced to STP that is absorbed in a volume unit of H_2O when the partial pressure of the gas is 1 atm

Gas	0 °C	25 °C	50 °C
N_2	0.023 54	0.014 34	0.010 88
O_2	0.048 89	0.028 31	0.020 90
H_2	0.021 48	0.017 54	0.016 08
CO	0.035 37	0.021 42	0.016 15
CO_2	1.713	0.759	0.436
H_2S	4.670	2.282	1.392

Equation (2.5.38) is a consequence of Henry's law[*]

$$p_j = K_j X_j, \qquad (2.5.40)$$

which connects the partial pressure of the gas p_j at equilibrium with the mole fraction $X_j = N_j/(N_j + N_w)$ of the dissolved gas where N_j and N_w are the number of moles of gas and water that make up the solution.

To handle solutions of one or more gases in biological liquids, respiratory physiologists and clinical physicians have found it useful to extend the considerations about the equilibrium between the gaseous and the liquid phase as follows. When a liquid contains a gas of species j with the concentration $C_j(x)$ at the position x, the amount can alternatively be characterized by an equivalent partial pressure p_j called the *tension*, T_j, *of the gas* in the liquid, which is equal to the equilibrium partial pressure p_j of the gas that produces the concentration $C_j(x)$ in the liquid. In the case of the simple situation of an aqueous solution of the gas j with the concentration C_j, we have from Eq. (2.5.38)

$$T_j = \frac{C_j}{\alpha_j}. \qquad (2.5.41)$$

Experimentally the tension of a gas T is determined by establishing at the required position inside the liquid a small cavity into which the dissolved gas molecules enter by diffusion. When equilibrium is established, the partial pressure p_j of the gaseous species in the cavity is determined experimentally and referred to as the value of the gas tension, since by definition $p_j = T_j$. This finishes our excursion into the solubility of gases in liquids.

We shall now describe the diffusion of O_2 into a living tissue (a frog muscle), where some of the data due to August Krogh[†] are used.

Krogh described the O_2-transport through a tissue slice by measuring the stationary O_2-flux when the partial pressures of O_2 on the two sides (i) and (o)

---

[*] Discovered in 1803 by W. Henry (1775–1836), a factory owner in Manchester. His original formulation was as follows. The amount of gas which at constant temperature and pressure is dissolved into a given liquid volume is proportional to the equilibrium pressure of the gas over the liquid. The form of Eq. (2.5.38) comes out as a result of formulating the equilibrium thermodynamically.

[†] August Krogh (1874–1947), was Professor of zoophysiology in Copenhagen. The combination of remarkable physical insight, with very wide interests in zoology and physiology, taken together with a unique talent for understanding and formulating the heart of the problem made him one of the greatest experimental physiologists of all times. He received the Nobel Prize in 1920 for his work on the function of the capillaries. He demonstrated that O_2 transfer in the lungs can be exclusively accounted for by a diffusion process. He also performed important studies in the human physiology of respiration and circulation and in osmoregulation and flight of insects. He developed a number of elegant methods for experimental investigations which could provide clear answers to well-formulated questions.

were $p^{(i)}$ and $p^{(o)}$. Accordingly he gave Fick's law the following form

$$J = -K\frac{dp}{dx} = -K\frac{dT}{dx}, \qquad (2.5.42)$$

where the units used for J were (cm^3 O$_2$ at STP) cm^{-2} min^{-1}. K is *Krogh's constant of diffusion*, which equals the amount of O$_2$ in cm^3 at STP that crosses 1 cm^2 min^{-1} under action of a pressure gradient of 1 atm cm^{-1}. Thus, the units of K are cm^2 atm^{-1} min^{-1}.

We write our usual form of Fick's law

$$J = -D\frac{dC}{dx}, \qquad (2.5.43)$$

where – considering Krogh's definition – we also specify the O$_2$-flux in units of cm^3 O$_2$[STP] cm^{-2} min^{-1}. Accordingly, we also express the distance x (in cm) and O$_2$-concentration (in cm^3) at STP per unit volume (in cm^3). The diffusion coefficient D in Eq. (2.5.43) is then in units of cm^2 min^{-1}. Since α is assumed not to depend upon the distance x we rewrite Eq. (2.5.42) as

$$J = -D\frac{dC}{dx} = -\alpha D\frac{d}{dx}\left(\frac{C}{\alpha}\right)$$

$$= -\alpha D\frac{dT}{dx} = -K\frac{dT}{dx},$$

from which it follows that

$$K = \alpha D. \qquad (2.5.44)$$

Equation (2.5.32) gives the minimum concentration $C_{\min}$ corresponding to an external concentration $C^{(o)}$. Dividing on both sides with α gives

$$\frac{C_{\min}}{\alpha} = \frac{C^{(o)}}{\alpha} - \frac{1}{2}\frac{\varrho}{\alpha D}h^2, \qquad (2.5.45)$$

or by invoking Eq. (2.5.43) and Eq. (2.5.40)

$$T_{\min} = T^{(o)} - \frac{1}{2}\frac{\varrho}{K}h^2. \qquad (2.5.46)$$

For a frog muscle at 20 °C Krogh found

$$K = 1.4 \times 10^{-5} \text{ cm}^2 \text{ min}^{-1} \text{ atm}^{-1}$$
$$\varrho = 5 \times 10^{-4} \text{ cm}^3 \text{ cm}^{-3} \text{ min}^{-1}.$$

Hence $\varrho/2K = 17.86$ atm cm^{-2}. Inserting this value in Eq. (2.5.45) gives

$$T_{\min} = T^{(o)} - 17.86 \times h^2. \qquad (2.5.47)$$

Let $T^{(o)}$ represent the partial pressure of O_2 of the atmosphere, i.e. $T^{(o)} = 0.21$ atm $= 160$ mm Hg. Further, we assume $T_{min} = 30$ mm Hg $= 0.04$ atm, corresponding to T_{O_2} in venous blood. Using these values we obtain

$$h = \sqrt{\frac{0.21 - 0.04}{17.86}} = 0.0976 \text{ cm} \approx 1 \text{ mm}$$

as the distance from the tissue surface to the position where T_{O_2} is down to 30 mm Hg.

This value of h may turn out to be unacceptably small. The only way out is to increase the value of $T^{(o)}$, e.g. to 1 atm by bubbling with pure oxygen. Using this value in the above equation for h gives instead

$$h = 2.32 \text{ mm}.$$

This example illustrates the fact that the supply of oxygen by diffusion to the oxygen-consuming living tissues is only effective over a rather short distance. With the longer distances of supply other mechanisms of transport are necessary, e.g. transport via the circulatory system where diffusion is replaced by transport by convection*.

2.5.4.2 Diffusion in a cylinder with radial symmetry

Many biological preparations, e.g. a nerve or a muscle fiber, are approximately cylindrical in shape. When describing the transfer of matter between the interior of these cells and their surrounding medium it is no longer possible to maintain the concept of a diffusion flux taking place only in *one direction* as we have done hitherto. One alternative is to introduce the three-dimensional Cartesian coordinates x, y, z using the diffusion equation expressed in three coordinates[†]. It frequently occurs that the concentration is maintained constant on the surface of the cylindrical cell. To express and apply this requirement in rectangular coordinates x, y, z would result in extremely clumsy algebra. By changing to the cylindrical coordinate reference system z, r, θ, the problem is greatly simplified, since the whole region comprising the surface of a given circular cylinder is now specified by the single coordinate r, namely the radius of the cylinder. The connection between these two coordinate systems is illustrated in Fig. 2.14.

In cylindrical coordinates the derivatives of $C(z, r, \theta)$ are

$$\frac{\partial C}{\partial z}, \quad \frac{\partial C}{\partial r}, \quad \frac{\partial C}{\partial \theta}.$$

---

* The example illustrates the disastrous effects a coronary infarction may have on the oxygen supply to the tissue around an occluded arterial branch.
[†] See Eq. (2.5.6).

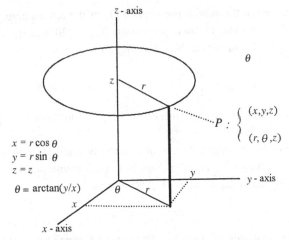

Fig. 2.14. Illustration of the localization of a point P in space using rectangular coordinates x, y, z and circular cylindrical coordinates z, r, θ. The point P is in cylindrical coordinates located relative to the x, y, z-axes by passing through P three surfaces. (1) A circular cylindrical surface around the z-axis and a radius $r = \sqrt{x^2 + y^2}$. (2) Cutting the cylinder by a plane parallel to the x, y-plane at a distance z, the point P is then contained in the circle with radius r. This circle represents the points $z = $ constant and $r = $ constant. (3) A plane containing the z-axis and making an angle θ with the x, z-plane of the Cartesian system then cuts the circle in the point P. The angle θ is related to the polar coordinates of the points x, y, where $x = r \cos\theta$, $y = r \sin\theta$, which gives $r^2 = x^2 + y^2$ and $\theta = \arctan(y/x)$.

However, in the examples considered in this section the concentrations $C(z, r, \theta)$ never vary with the distance along the z-axis or the angle of rotation around the z-axis, resulting in $\partial C/\partial z = \partial C/\partial\theta = 0$ everywhere in space. As a consequence, the only position coordinate required is the radius r from the axis of the cylinder.

Some relatively simple situations will now be considered. However, since the substance now spreads out in space *radially* from the z-axis it is necessary to adapt the diffusion equation to these new conditions.

(i) The diffusion equation

We derive the form of the diffusion equation according to the situation where the independent variable now is the radius r of the cylindrical surface on which the concentration is $C(r, t)$ at the time t. Because r is the only spatial coordinate, the diffusion flux will have the *same* value everywhere on the cylindrical surface with radius r and will at every point on the surface have a direction corresponding to the direction of the radius from the cylinder axis to the point in question. Thus

$$J(r, t) = -D\frac{\partial C(r, t)}{\partial r}. \tag{2.5.48}$$

Similar to the treatment in Section 2.5.3.2 we assume that a matter consumption $\varrho(r)$ also is present.

We consider a volume element ΔV in the shape of a cylindrical ring of height Δz and with inner radius r and outer radius $r + \Delta r$. We calculate the change in concentration in this volume element during an infinitely small time between t and $t + \Delta t$. In the time Δt an amount of matter ${}^{(i)}M_r^{(\Delta t)}$ crosses the inner surface of the ring of radius r and area $A_r = 2\pi r \Delta z$ that is

$$^{(i)}M_r^{(\Delta t)} = J(r, t)\bar{A}^{(r)} \Delta t = J(r, t)2\pi r \Delta z \Delta t,$$

where $J(r, t)$ is the radially directed flux at the distance r from the cylinder axis at the time t. In the same time interval an amount ${}^{(o)}M_{r+\Delta r}^{(\Delta t)}$ is removed from the volume element through the outer surface of the ring of radius $r + \Delta r$ with area $A_{r+\Delta r} = 2\pi(r + \Delta r)\Delta z$. This amount is

$$^{(o)}M_{r+\Delta r}^{(\Delta t)} = J(r + \Delta r, t)A_{r+\Delta r} \Delta t = J(r + \Delta r, t)2\pi(r + \Delta r)\delta x \Delta t.$$

Further, an amount ${}^{(pr)}M_{\Delta V}^{(\Delta t)}$ is produced in the ring of volume $\Delta V = 2\pi r \Delta r \Delta z$ in the time Δt that is

$$^{(pr)}M_{\Delta V}^{(\Delta t)} = \varrho\Delta V \Delta t = 2\pi\Delta r \Delta z\varrho\Delta t.$$

The amount ${}^{(ac)}M_{\Delta V}^{(\Delta t)}$ accumulated in time Δt inside the ring equals the amount received by diffusion ${}^{(i)}M_r^{(\Delta t)}$ plus the amount produced ${}^{(pr)}M_{\Delta V}^{(\Delta t)}$ in the volume element $\Delta V = 2\pi r \Delta r \Delta z$ minus the amount ${}^{(o)}M_{r+\Delta r}^{(\Delta t)}$ which has left the volume element by diffusion in the time Δt. Hence

$$^{(ac)}M_{\Delta V}^{(\Delta t)} = {}^{(i)}M_r^{(\Delta t)} + {}^{(pr)}M_{\Delta V}^{(\Delta t)} - {}^{(o)}M_{r+\Delta r}^{(\Delta t)}$$
$$= J(r, t)2\pi r \Delta z \Delta t + 2\pi r \Delta z \varrho\Delta t - J(r + \Delta r, t)2\pi(r + \Delta r)\Delta z \Delta t$$
$$= 2\pi h\left(J(r, t)r - J(r + \Delta r, t)(r + \Delta r)\right)\Delta t + 2\pi r \Delta r \Delta z\varrho(r) \Delta t,$$

where it is assumed that the concentration profile changes so slowly with time that $J(r, t) \approx J(r, t + \Delta t)$ and $J(r + \Delta r, t) \approx J(r + \Delta r, t + \Delta t)$. But the accumulated amount ${}^{(ac)}M_{\Delta V}^{(\Delta t)}$ also equals the increment in concentration $\Delta C^{(\Delta t)}$ in the volume element in the time Δt times the volume element

$\Delta V = 2\pi r\, \Delta r\, \Delta z$. Hence

$$^{(ac)}M_{\Delta V}^{(\Delta t)} = \Delta C^{(\Delta t)}\Delta v = (C(r', t+\Delta t) - C(r', t))2\pi r\, \Delta r\, \Delta z,$$

where the average concentrations in the volume element ΔV at the times $t + \Delta t$ and t respectively are $C(r', t + \Delta t)$ and $C(r', t)$ with $r \leq r' \leq r + \Delta r$, i.e. $r' \to r$ when $\Delta r \to 0$. Collecting together these two equations gives

$$-2\pi \Delta z\, \Delta t\, (J(r+\Delta r, t)(r+\Delta r) - J(r, t)r) + 2\pi r\, \Delta r \Delta z \varrho(r)\, \Delta t$$
$$= 2\pi r\, \Delta r \Delta z(C(r', t+\Delta t) - C(r', t))$$

or

$$-\Delta t\, (J(r+\Delta r, t)r - J(r, t)r + J(r+\Delta r, t)\, \Delta r) + r\, \Delta r\varrho(r)\, \Delta t$$
$$= r\, \Delta r(C(r', t+\Delta t) - C(r', t)).$$

Division on both sides with $r\, \Delta r\, \Delta t$ gives

$$-\left(\frac{J(r+\Delta r, t) - J(r, t)}{\Delta r} + \frac{J(r+\Delta r, t)}{r} \right) + \varrho(r)$$
$$= \frac{C(r', t+\Delta t) - C(r', t)}{\Delta t}.$$

When $\Delta t \to 0$ and $\Delta r \to 0$ we have $r' \to r$ and $J(r+\Delta r, t) \to J(r, t)$. Further, the fractions containing Δr and Δt in the denominators tend towards their respective partial derivatives*. Thus, the mass conservation formulated in this coordinate system has the form

$$\frac{\partial C(r, t)}{\partial t} = -\frac{\partial J(r, t)}{\partial r} - \frac{1}{r}J(r, t) + \varrho(r). \qquad (2.5.49)$$

Inserting the expression for Fick's law into the cylindrical coordinates $J = -D\partial C/\partial r$ gives

$$\frac{\partial C(r, t)}{\partial t} = D\left(\frac{\partial^2 C(r, t)}{\partial r^2} + \frac{1}{r}\frac{\partial C(r, t)}{\partial r} \right) + \varrho(r), \qquad (2.5.50)$$

provided the diffusion coefficient D is independent of r or $C(r, t)$, respectively. Equation (2.5.49) can also be written in a more compact way

$$\frac{\partial C(r, t)}{\partial t} = D\frac{1}{r}\frac{\partial}{\partial r}\left(r\frac{\partial C}{\partial r} \right) + \varrho(r). \qquad (2.5.51)$$

* See Chapter 1, Section (1.8.2), Eq. (1.8.5).

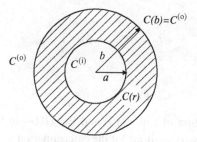

Fig. 2.15. Illustration of the stationary diffusion through a cylindrical shell.

When the process has become *stationary* we have $\partial C/\partial t = 0$, and the concentration profiles are now governed by the following equations

$$D\left(\frac{d^2C(r)}{dr^2} + \frac{1}{r}\frac{dC(r)}{dr}\right) = -\varrho(r) \qquad (2.5.52)$$

or

$$D\frac{1}{r}\frac{d}{dr}\left(r\frac{dC}{r}\right) = -\varrho(r). \qquad (2.5.53)$$

(ii) Diffusion through a cylindrical shell

By analogy with the diffusion through a plate in Section 2.5.4.1 (i), we consider the stationary diffusion in the space between the two concentric cylinders with radii a and b as shown in Fig. 2.15. The concentrations at the surfaces of the cylinder are maintained at the constant values $C^{(i)}$ for $r = a$ and at $C^{(o)}$ for $r = b$. No consumption/production of matter is present. Hence, the concentration profile in the region $a \le r \le b$ is found by solving Eq. (2.5.52) subject to $\varrho(r) = 0$, i.e.

$$\frac{d}{dr}\left(r\frac{dC}{dr}\right) = 0. \qquad (2.5.54)$$

The general solution results from two successive integrations. The first gives

$$r\frac{dC}{dr} = A, \quad \text{or} \quad \frac{dC}{dr} = \frac{A_1}{r},$$

where A_1 is a constant. A second integration gives

$$C(r) = A_1\int \frac{1}{r}\,dr + B_1 = A_1\ln r + B_1, \qquad (2.5.55)$$

where B_1 is the second constant of integration. Equation (2.5.55) is the *general solution* of the *stationary radial* diffusion process. The boundary conditions to be satisfied are

$$C(r) = C^{(i)}, \quad \text{for} \quad r = a, \tag{B13}$$

and

$$C(r) = C^{(0)}, \quad \text{for} \quad r = b. \tag{B14}$$

One-by-one application of Eq. (B13) and Eq. (B14) to the general equation gives the following two equations for the determination of the constants A_1 and B_1

$$A_1 \ln b + B_1 = C^{(0)} \tag{XIII}$$

and

$$A_1 \ln a + B_1 = C^{(i)}. \tag{XIV}$$

Subtracting (XIV) from (XIII) gives

$$A_1 = \frac{C^{(0)} - C^{(i)}}{\ln b - \ln a} = \frac{C^{(0)} - C^{(i)}}{\ln(b/a)} = \frac{C^{(i)} - C^{(0)}}{\ln(a/b)},$$

which inserted into (XIII) gives

$$B_1 = \frac{C^{(i)} \ln b - C^{(0)} \ln a}{\ln(b/a)} = \frac{C^{(0)} \ln a - C^{(i)} \ln b}{\ln(a/b)}.$$

Inserting these values for A_1 and B_1 in Eq. (2.5.55) we obtain

$$C(r) \ln \frac{a}{b} = \left(C^{(i)} - C^{(0)}\right) \ln r + C^{(0)} \ln a - C^{(i)} \ln b.$$

Addition of $C^{(0)} \ln b - C^{(0)} \ln b$ on the right-hand side results in the rearrangement

$$C(r) = \frac{\ln(r/b)}{\ln(a/b)} \left(C^{(i)} - C^{(0)}\right) + C^{(0)}, \quad \text{for } a \leq r \leq b. \tag{2.5.56}$$

To determine the radial flux $J(r) = -D(dC/dr)$, we note that $d(\ln r/b)/dr = [1/(r/b)] \cdot [d(r/b)/dr] = (b/r) \cdot (1/b) = 1/r$. Hence

$$J(r) = -\frac{1}{r} \frac{D}{\ln(a/b)} \left(C^{(i)} - C^{(0)}\right). \tag{2.5.57}$$

From this it follows that the flux through the cylindrical shell *decreases* with increasing values of r. This is not inconsistent with the existence of a stationary diffusion process, because the flux is only constant when it proceeds in *one* direction without accumulation or consumption of matter. The above expression for the radial symmetric flux has precisely the form which ensures that no accumulation of matter takes place, in agreement with the prescribed condition $\varrho(r) = 0$. To check that Eq. (2.5.53) is in keeping with $\varrho(r) = 0$ we calculate the total mass transport $M(r)$ through a cylindrical surface somewhere in the region $a \leq r \leq b$ of height Δz. Reckoned per unit time, the mass transport by diffusion is $M(r) = 2\pi r \, \Delta z \, J(r)$. Inserting the value for $J(r)$ from Eq. (2.5.56) gives

$$M(r) = -2\pi \, \Delta z \frac{D}{\ln(a/b)} \left(C^{(i)} - C^{(0)} \right),$$

which is independent of r.

If the thickness of the cylindrical shell $h = b - a$ is small compared with the outer radius b, the expression for the flux takes on a simpler form. We have

$$\ln \frac{b - h}{b} = \ln \left(1 - \frac{h}{b} \right) \approx -\frac{h}{b},$$

provided $h/b \ll 1^*$. The expression for the flux can then be written as

$$J(b) = \frac{D}{h} \left(C^{(i)} - C^{(0)} \right) = P \left(C^{(i)} - C^{(0)} \right). \tag{2.5.58}$$

Formally, this is identical to expression Eq. (2.5.13) for a one-dimensional stationary flux through a plate of thickness h. With a nerve of radius $b = 2 \, \mu\mathrm{m}$ and a membrane thickness of $h = 100 \, \mathring{\mathrm{A}}$, the discrepancy between Eq. (2.5.57) and Eq. (2.5.58) is less than 1%. Also in cases where the thickness of the cylinder shell is comparable with the radius of the cylinder we can employ a flux equation of the same form as Eq. (2.5.57). The flux through the outer surface of the cylinder in the direction of r is

$$J(b) = \frac{1}{b} \frac{D}{\ln(b/a)} \left(C^{(i)} - C^{(0)} \right).$$

Introducing an average permeability coefficient $\langle P \rangle$ by

$$\langle P \rangle = \frac{1}{b} \frac{D}{\ln(b/a)} \tag{2.5.59}$$

leads to an expression that is formally of the same form as Eq. (2.5.57).

(iii) Diffusion in a cylinder with mass consumption

We now consider a cylinder of radius r. The external concentration is maintained at a constant value $C(b) = C^{(0)}$. In the interior the inward diffusion and a matter consumption ϱ of constant intensity are taking place at the same time. The

---

* See Chapter 1, Section 1.6.3, Eq. (1.6.21).

stationary concentration profile is governed by

$$D \frac{1}{r} \frac{d}{dr} \left(r \frac{dC}{dr} \right) = -\varrho. \qquad (2.5.60)$$

The first integration gives

$$r \frac{dC}{dr} = \frac{\varrho}{D} \int r \, dr + A_1 = \frac{1}{2} \frac{\varrho}{D} r^2 + A_1,$$

where A_1 is a constant. Integration of this expression leads to

$$C(r) = \frac{1}{2} \frac{\varrho}{D} \int r \, dr + A_1 \int \frac{1}{r} \, dr + B_1 = \frac{1}{4} \frac{\varrho}{D} r^2 + A_1 \ln r + B_1, \quad (2.5.61)$$

which is the general solution containing two constants of integration A_1 and B_1. In the present situation the concentration is *finite* everywhere in the interior including the axis of the cylinder ($r = 0$). But $\ln r \to -\infty$, for $r \to 0$. Thus, the term $A_1 \ln r$ in Eq. (2.5.61) does not contribute to the problem with a physically realistic solution. Consequently, the constant A_1 must be zero. An alternative argumentation is that the concentration profile must display radial symmetry with respect to the cylinder axis and have its minimum there, i.e. $dC/dr = 0$ for $r = 0$. In the expression for the gradient

$$\frac{dC}{dr} = \frac{1}{2} \frac{\varrho}{D} r + \frac{A_1}{r},$$

this condition will only be satisfied if $A_1 = 0$. Thus, the concentration profile in the region $0 \le r \le b$ must be given by

$$C(r) = \frac{1}{4} \frac{\varrho}{D} r^2 + B_1, \qquad (2.5.62)$$

which implies that it is only the constant B_1 that needs adjustment to make Eq. (2.5.60) conform to the condition: $C(r) = C^{(0)}$ for $r = b$. Applying this to Eq. (2.5.62) gives the value $B_1 = C^{(0)} - \varrho b^2/4D$. The solution to our problem then takes the following form

$$C(r) = C^{(0)} - \frac{1}{4} \frac{\varrho}{D} (b^2 - r^2). \qquad (2.5.63)$$

The concentration on the cylinder axis becomes

$$C(0) = C^{(0)} - \frac{1}{4} \frac{\varrho}{D} b^2. \qquad (2.5.64)$$

Note the similarity between this expression and the identical problem corresponding to the one-dimensional process.

To control this result we note that when stationarity prevails in the interior of the cylinder the mass consumption per unit time and per unit length is

$$M = \pi b^2 \varrho.$$

The mass transfer must be supplied by diffusion at the surface of the cylinder. The flux in the direction of r at the surface, i.e. at radius b, is

$$J(b) = -D \left(\frac{dC}{dr}\right)_{r=b} = -D \left(\frac{1}{2}\frac{\varrho b}{D}\right) = -\tfrac{1}{2}\varrho b,$$

where the minus sign indicates that the direction of the flux is towards the axis (as it should be). The amount that crosses the surface by diffusion per unit time and per unit length of the cylinder is

$$M = 2\pi b \times \tfrac{1}{2}\varrho b = \pi b^2 \varrho,$$

which is identical to the total mass consumption in the interior.

We now consider the slightly more complicated situation where the external surface of the cylinder is covered by a membrane of thickness h, which – although being orders of magnitude less than the radius b of the cylinder – offers a considerable barrier to diffusion. Let D_m denote the diffusion coefficient in the membrane for the substance in question. As indicated in Fig. 2.16 the concentration in the surrounding medium is maintained at the constant value $C^{(o)}$. As the concentration in the interior must be finite (eventually zero in a certain region $0 \le r \le r_o < b$), the shape of the concentration profile must in general still be given by

$$C(r) = \tfrac{1}{4}\frac{\varrho}{D}r^2 + B, \tag{2.5.65}$$

in the region $0 \le r \le b - h \approx b$. However, since the membrane presents a barrier to diffusion the constant B cannot be determined by the condition $C(b) = C^{(o)}$ as in the above example. Here, the constant B must be adjusted to conform

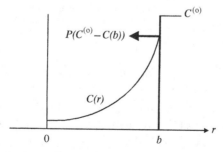

Fig. 2.16. Diffusion in a cylinder with the surface covered by a thin membrane.

to the situation where the flux through the membrane is equal to the flux into the interior of the cylinder at $r = b$ where the concentration is $C(b)$. The flux through the membrane is then (according to Eq. (2.5.57))

$$J(b) = \frac{D_m}{h} \left(C(b) - C^{(0)} \right) = P \left(C(b) - C^{(0)} \right), \qquad (2.5.66)$$

where $P = D_m/h$ is the permeability of the membrane. The flux in the interior of the cylinder from $r = b$ is

$$J(b) = -D \left(\frac{dC}{dr} \right)_{r=b}.$$

The flux at the transition between the two regions must be continuous. Hence

$$-D \left(\frac{dC}{dr} \right)_{r=b} = P \left(C(b) - C^{(0)} \right). \qquad (2.5.67)$$

Invoking Eq. (2.5.65) the term on the left becomes $-\varrho b/2$. Hence

$$C(b) = C^{(0)} - \tfrac{1}{2} \frac{\varrho b}{P}.$$

The value of the constant B in Eq. (2.5.65) gives the condition

$$C(r) = C(b) = C^{(0)} - \tfrac{1}{2} \frac{\varrho b}{P}, \quad \text{for} \quad r = b,$$

which applied to Eq. (2.5.65) gives

$$B_1 = C^{(0)} - \tfrac{1}{4} \frac{\varrho}{D} b^2 - \tfrac{1}{2} \frac{\varrho b}{P}.$$

The concentration profile in the interior of the cylinder is then given by

$$C(r) = C^{(0)} + \tfrac{1}{4} \frac{\varrho}{D} (r^2 - b^2) - \tfrac{1}{2} \frac{\varrho b}{P}. \qquad (2.5.68)$$

The concentration profile is again a parabola having the minimum value

$$C_{\min} = C^{(0)} - \tfrac{1}{4} \frac{\varrho}{D} b^2 - \tfrac{1}{2} \frac{\varrho b}{P} \qquad (2.5.69)$$

at the cylinder axis.

(iv) Diffusion from a cylinder into the surrounding medium with mass consumption (Krogh's cylinder)

In connection with his classical investigations on the O_2-supply from the capillaries to the tissues, August Krogh put forward a model in 1919 for this diffusion regime, now called Krogh's cylinder, consisting of two concentric cylinders: (1) an inner cylinder of radius r, representing the capillary, is covered by a thin membrane which does not act as a diffusion barrier to the substance $-O_2-$ contained in the interior and is in every direction maintained at the constant value $C^{(i)}$; and (2) an outer cylinder of radius $r = r_o > a$. The oxygen diffuses from the capillary wall with a diffusion coefficient D into the surrounding medium (tissue) where there is constant oxygen consumption ϱ caused by the metabolic processes. The oxygen flux is zero at the surface of the outer cylinder*. The problem consists of finding the concentration $C(r) = C^{(0)}$ at the position $r = r_o$.

The general solution for $C(r)$, where $a \leq r \leq r_o$, is given by Eq. (2.5.54)

$$C(r) = \tfrac{1}{4}\frac{\varrho}{D}r^2 + A_1 \ln r + B_1, \qquad (2.5.54)$$

where $\ln r$ is also a solution since $r \geq a > 0$. The solution of the problem therefore requires an adjustment of both constants A_1 and B_1. However, the solution must also conform to the following two boundary conditions

$$C(r) = C^{(i)} \quad \text{for} \quad r = a, \qquad (B15)$$

and

$$\frac{dC}{dx} = 0 \quad \text{for} \quad r = r_o. \qquad (B16)$$

Applying (B15) and (B16) to Eq. (2.5.54) results in two equations for the determination of A_1 and B_1

$$A_1 \ln a + B_1 = C^{(i)} - \tfrac{1}{4}\frac{\varrho}{D}a^2, \qquad (XV)$$

and

$$\frac{1}{r_o}A_1 = -\tfrac{1}{2}\frac{\varrho}{D}r_o, \qquad (XVI)$$

from which we obtain

$$A_1 = -\tfrac{1}{2}\frac{\varrho}{D}r_o^2 \quad \text{and} \quad B_1 = C^{(i)} + \tfrac{1}{2}\frac{\varrho}{D}r_o^2 \ln a - \tfrac{1}{4}\frac{\varrho}{D}a^2.$$

Inserting the values of A_1 and B_1 into Eq. (2.5.54) leads to the following expressions for the concentration profile in the region $a \leq r \leq r_o$

$$C(r) = C^{(i)} - \tfrac{1}{2}\frac{\varrho}{D}r_o^2 \ln\left(\frac{r}{a}\right) + \tfrac{1}{4}\frac{\varrho}{D}r^2 - \tfrac{1}{4}\frac{\varrho}{D}a^2, \qquad (2.5.70)$$

---

* This assumption is motivated from structural considerations. In the tissue in question (a striated muscle) there is a regular, e.g. hexagonal, arrangement of capillaries in parallel to the muscle fibers. The oxygen tension in the region around any single capillary will be influenced by the counter-diffusion from the surrounding, hexagonally arranged, capillaries leading to the state $dC/dr = 0$ at a certain distance r_0, which is the radius of the outer cylinder surface.

or

$$C(r) = C^{(i)} - \frac{1}{4}\frac{\varrho}{D}\left[2r_0^2 \ln\left(\frac{r}{a}\right) + a^2 - r^2\right]. \qquad (2.5.71)$$

In the following numerical calculations we shall consider $C(r)$ in units of cm^3 O_2 at STP per cm^3 tissue and D in units of cm^2/min^{-1}. Conforming with the procedure used in setting up Eq. (2.5.45), both sides of Eq. (2.5.71) are divided by the absorption coefficient α for O_2 in the tissue[*]. This gives

$$T(r) = T^{(i)} - \frac{1}{4}\frac{\varrho}{K}\left[2r_0^2 \ln\left(\frac{r}{a}\right) + a^2 - r^2\right], \qquad (2.5.72)$$

where $T(r) = C(r)/\alpha$ is the O_2-tension at the position r and $K = \alpha D$ is Krogh's diffusion coefficient. For the sake of convenience we write the equation as

$$T(r) = T^{(i)} - \frac{\varrho a^2}{4K}\left[2\left(\frac{r_0}{a}\right)^2 \ln\left(\frac{r}{a}\right) - \left(\frac{r}{a}\right)^2 + 1\right]. \qquad (2.5.73)$$

The oxygen tension $T^{(0)}$ on the cylindrical surface at $r = r_0$ is

$$T^{(0)} = T^{(i)} - \frac{\varrho a^2}{4K}\left[2\left(\frac{r_0}{a}\right)^2 \ln\left(\frac{r_0}{a}\right) - \left(\frac{r_0}{a}\right)^2 + 1\right],$$

$$= T^{(i)} - \frac{\varrho a^2}{4K}[\Phi^2(2\ln\Phi - 1)], \qquad (2.5.74)$$

where $\Phi = r_0/a$ and the value $+1$ in the upper bracket [ ] is left out since $(r_0/a)^2 \gg 1$ in the present case.

We now apply again the value determined by Krogh for frog muscle at $20\,^\circ C$.

$$K = 1.4 \times 10^{-5}\,cm^2\,min^{-1}\,atm^{-1} \quad \text{and} \quad \varrho = 5 \times 10^{-4}\,cm^3\,cm^{-3}\,min^{-1}.$$

This results in $\varrho/K = 35.7\,atm\,cm^{-2}$. We put the diameter of the internal cylinder (capillary) equal to $2a = 10\,\mu m$, or $a = 5 \times 10^{-4}\,cm$. At the arterial end of the capillary the O_2-tension is $\approx 110\,mm\,Hg = 0.144\,atm$, and $c.\,40\,mm$ $Hg = 0.05$ in the venous portion. This gives $0.097\,atm$ for the average tension in the capillary, which we use as the value for $T^{(i)}$.

In Table 2.2 are given the O_2-tensions at different values of the radius r_0 of Krogh's cylinder, i.e. corresponding to the distance where $dC/dr = 0$, calculated from Eq. (2.5.73), partly for resting muscle and partly for a muscle having maximum O_2-consumption ($\varrho_{max} = 10 \times \varrho$). Even though this model may appear to represent an idealization, and even if some guesswork is involved in the above consideration, the numbers do illustrate vividly the demands and the limitations of the oxygen supply in working muscles and also give a first-order

[*] For frog muscle at $20\,^\circ C$ we have $\alpha = 0.0254\,cm^3\,cm^{-3}\,atm^{-1}$.

Table 2.2. *Data for Krogh's cylinder*

O_2-tensions $T^{(o)}$ on the surface of Krogh's cylinders for different values r_0 in μm calculated from Eq. (2.5.73) partly for resting muscle and partly at maximum muscle work

r_0 (μm)	$T^{(o)}$ (mm Hg) Rest	$T^{(o)}$ (mm Hg) Max. work
50	73.1	67.6
100	70.3	39.9
110	69.5	31.2
120	68.5	21.4
130	67.4	10.5
150	64.9	—
200	56.4	—
250	44.8	—

estimate of the structural arrangement of the capillaries which is necessary to provide an effective O_2-supply to the working muscles.

2.5.4.3 Diffusion with radial symmetry in a sphere

In experimental biology the spherical cell is a favorite subject of experiments. The preference is chiefly because of the simple conditions of spatial symmetry which facilitate many considerations, among these mass transfer into and out of the cell. The situation often encountered is that the concentration is maintained constant at the surface of a sphere. In such cases calculations become much easier by changing from the Cartesian rectangular coordinates x, y, z to the spherical coordinate reference system r, θ, ϕ where a point in space is specified by its position on a spherical surface of radius r – sharing its origin with that of the Cartesian system – and the two angles θ and ϕ made by r with the x–z-plane and the x–y-plane of the Cartesian system. As in the previous sections we shall only treat processes which do not depend upon the coordinates θ and ϕ, i.e. the surfaces in space on which $C = constant$ and $J = constant$ are concentric spherical surfaces. Thus, only a *single* position coordinate, namely the radius r from the center of the sphere, is required to describe the diffusion processes in the sphere.

(i) The diffusion equation

We shall now derive the form that the diffusion equation must adopt when the independent variable is the radius r of a sphere on the surface of which the concentration $C(r, t)$ has the same value at a given time t. The shortest distance between a point where the concentration is $C(r)$ and that where it is $C(r + \Delta r)$

has the direction of r. Hence the diffusion flux will radiate from the spherical surface with the same value for a given value of r. Thus

$$J(r, t) = -D \frac{\partial C(r, t)}{\partial r}. \tag{2.5.75}$$

As before, we assume that a mass production is present in the region. We consider a volume element ΔV in the shape of a spherical shell with inner radius r and outer radius $r + \Delta r$. We calculate the change in concentration in this volume element during an infinitely small time between t and $t + \Delta t$. During the time Δt an amount of matter ${}^{(i)}M_r^{(\Delta t)}$ crosses the inner surface of the ring of radius r and area $A_r = 4\pi r^2$ that is

$${}^{(i)}M_r^{(\Delta t)} = J(r, t)A_r \, \Delta t = J(r, t)4\pi r^2 \, \Delta t,$$

where $J(r, t)$ is the radially directed flux at the distance r from the center of the sphere at the time t. In the same time interval an amount ${}^{(o)}M_{r+\Delta r}^{(\Delta t)}$ is removed from the volume element through the outer surface of the shell of radius $r + \Delta r$ with area $A_{r+\Delta r} = 4\pi(r + \Delta r)^2$. This amount is

$${}^{(o)}M_{r+\Delta r}^{(\Delta t)} = J(r + \Delta r, t)A_{r+\Delta r} \, \Delta t = J(r + \Delta r, t)4\pi(r + \Delta r)^2 \Delta t.$$

Further, an amount ${}^{(pr)}M_{\Delta V}^{(\Delta t)}$ is produced in the shell of volume $\Delta V = 4\pi r^2 \, \Delta r$ in the time Δt which is

$${}^{(pr)}M_{\Delta V}^{(\Delta t)} = \varrho \Delta V \, \Delta t = 4\pi r^2 \Delta r \, \varrho \Delta t.$$

The amount ${}^{(ac)}M_{\Delta V}^{(\Delta t)}$ accumulated during the time Δt inside the shell equals the amount received by diffusion ${}^{(i)}M_r^{(\Delta t)}$ plus the amount produced ${}^{(pr)}M_{\Delta V}^{(\Delta t)}$ in the volume element $\Delta V = 4\pi r^2 \, \Delta r$ minus the amount ${}^{(o)}M_{r+\Delta r}^{(\Delta t)}$ which has left the volume element by diffusion during the time Δt. Hence

$$\begin{aligned}
{}^{(ac)}M_{\Delta V}^{(\Delta t)} &= {}^{(i)}M_r^{(\Delta t)} + {}^{(pr)}M_{\Delta V}^{(\Delta t)} - {}^{(o)}M_{r+\Delta r}^{(\Delta t)} \\
&= J(r, t)4\pi r^2 \, \Delta t + 4\pi r^2 \, \Delta r \, \varrho \Delta t - J(r + \Delta r, t)4\pi(r + \Delta r)^2 \Delta t \\
&= 4\pi (J(r, t)r^2 - J(r + \Delta r, t)(r + \Delta r)^2)\Delta t + 4\pi r^2 \, \Delta r \varrho(r) \, \Delta t,
\end{aligned}$$

where it is assumed that the concentration profile changes so slowly with time that $J(r, t) \approx J(r, t + \Delta t)$ and $J(r + \Delta r, t) \approx J(r + \Delta r, t + \Delta t)$. But the accumulated amount ${}^{(ac)}M_{\Delta V}^{(\Delta t)}$ also equals the increment in concentration $\Delta C_{\Delta V}^{(\Delta t)}$ in the volume element during the time Δt times the volume element $\Delta V = 4\pi r^2 \Delta r$. Hence

$${}^{(ac)}M_{\Delta V}^{(\Delta t)} = \Delta C_{\Delta V}^{(\Delta t)} \Delta V = (C(r', t + \Delta t) - C(r', t))4\pi r^2 \, \Delta r,$$

where the average concentrations in the volume element ΔV during the time intervals $t + \Delta t$ and t respectively are $C(r', t + \Delta t)$ and $C(r', t)$ with $r \leq r' \leq r + \Delta r$, i.e. $r' \to r$ when $\Delta r \to 0$.

Combining these two equations gives

$$4\pi r^2 \, \Delta r \left(C(r', t + \Delta t) - C(r', t) \right)$$
$$= -4\pi r^2 \left(J(r + \Delta r, t)\frac{(r + \Delta r)^2}{r^2} - J(r, t) \right) \Delta t + 4\pi r^2 \Delta r \, \varrho(r) \, \Delta t,$$

or

$$4\pi r^2 \Delta r (C(r', t + \Delta t) - C(r', t))$$
$$= -4\pi r^2 \left(J(r + \Delta r, t)\left\{ 1 + 2\frac{\Delta r}{r} + \left(\frac{\Delta r}{r}\right)^2 \right\} - J(r, t) \right)\Delta t$$
$$+ 4\pi r^2 \Delta r \varrho(r) \, \Delta t,$$

or

$$\Delta r \, (C(r', t + \Delta t) - C(r', t))$$
$$= -\left(J(r + \Delta r, t)\left\{ 1 + 2\frac{\Delta r}{r} + \left(\frac{\Delta r}{r}\right)^2 \right\} - J(r, t) \right)\Delta t + \Delta r \varrho(r) \, \Delta t.$$

Division on both sides by $\Delta r \, \Delta t$ gives

$$\frac{C(r', t + \Delta t) - C(r', t)}{\Delta t}$$
$$= -\left(\frac{J(r + \Delta r, t) - J(r, t)}{\Delta r} + J(r + \Delta r, t)\left\{ \frac{2}{r} + \frac{\Delta r}{r^2} \right\} \right) + \varrho(r).$$

When $\Delta t \to 0$ and $\Delta r \to 0$ then $r' \to r$ and $J(r + \Delta r, t) \to J(r, t)$. The fractions with Δr and Δt in their respective denominators tend towards their partial derivatives*. The principle of mass conservation expressed in the spherical coordinate system with radial symmetry then takes the form

$$\frac{\partial C(r, t)}{\partial t} = -\frac{\partial J(r, t)}{\partial r} - \frac{2}{r}J(r, t) + \varrho(r). \tag{2.5.76}$$

Inserting the expression Eq. (2.5.74) for Fick's law gives

$$\frac{\partial C(r, t)}{\partial t} = D\left(\frac{\partial^2 C(r, t)}{\partial r^2} + \frac{2}{r}\frac{\partial C(r, t)}{\partial r} \right) + \varrho(r), \tag{2.5.77}$$

---

* See Chapter 1, Section 1.8.2, Eq. (1.8.2).

which can also be written in a more compact form as

$$\frac{\partial C(r, t)}{\partial t} = D \frac{1}{r^2} \frac{\partial}{\partial r} \left(r^2 \frac{\partial C(r, t)}{\partial r} \right) + \varrho(r). \tag{2.5.78}$$

When *stationary conditions* prevail the concentration profiles are governed by

$$D \left(\frac{d^2 C(r, t)}{dr^2} + \frac{2}{r} \frac{dC(r, t)}{dr} \right) + \varrho(r) = 0, \tag{2.5.79}$$

or

$$D \frac{1}{r^2} \frac{d}{dr} \left(r^2 \frac{dC(r, t)}{dr} \right) + \varrho(r) = 0. \tag{2.5.80}$$

(ii) Diffusion through a spherical shell

We now consider stationary diffusion through a homogeneous region contained between two concentric spheres – a spherical shell – of inner and outer radii a and b. The concentrations of the diffusing substance at the inner and outer surfaces are maintained at the constant values $C^{(i)}$ and $C^{(o)}$. It is assumed that there is no mass production in the shell ($\varrho(r) = 0$). Hence Eq. (2.5.80) reduces to

$$\frac{d}{dr} \left(r^2 \frac{dC(r)}{dr} \right) = 0. \tag{2.5.81}$$

Integration with respect to r gives

$$r^2 \frac{dC(r)}{dr} = A_1,$$

or

$$\frac{dC(r)}{dr} = \frac{A_1}{r^2},$$

where A_1 is a constant. A second integration results in

$$C(r) = A_1 \int \frac{1}{r^2} \, dr + B_1 = -\frac{A_1}{r} + B_1, \tag{2.5.82}$$

where B_1 is another constant. Equation (2.5.81) is the general form of the solution for spherical coordinates with radial symmetry of the stationary diffusion without mass consumption.

The boundary conditions that Eq. (2.5.81) satisfies are

$$C(r) = C^{(i)}, \quad \text{for} \quad r = a, \tag{B17}$$

and

$$C(r) = C^{(o)}, \quad \text{for} \quad r = b. \tag{B18}$$

Applying these two conditions one-by-one to Eq. (2.5.81) gives two simultaneous linear equations in A_1 and B_1. Solving these and inserting the value for A_1 and B_1 back into Eq. (2.5.81) gives the following expression for the concentration profile in the region $a \leq r \leq b$

$$C(r) = \frac{a(b-r)C^{(i)} + b(r-a)C^{(0)}}{(b-a)r}. \qquad (2.5.83)$$

The flux $J(b) = -D(dC/dr)_{r=b}$ through the outer surface of the shell in the direction r is

$$J(b) = D\frac{ab}{b^2(b-a)}\left(C^{(i)} - C^{(0)}\right). \qquad (2.5.84)$$

We consider again the situation where the thickness of the shell $h = b - a$ is small in proportion to the radius b. In that case $ab \approx b^2$, and Eq. (2.5.83) takes the form

$$J(b) = \frac{D}{h}\left(C^{(i)} - C^{(0)}\right) = P\left(C^{(i)} - C^{(0)}\right), \qquad (2.5.85)$$

i.e. the flux in the direction (i) → (o) is formally of the same form as the expression for the one-dimensional flux through a plate. In those cases where the thickness of the shell cannot be disregarded relatively to b, an expression of the form of Eq. (2.5.83) is still of practical use but now with an average permeability coefficient $\langle P \rangle$ of magnitude

$$\langle P \rangle = \frac{a}{b(b-a)}D. \qquad (2.5.86)$$

(iii) Sphere covered by a thin membrane, mass consumption in the interior
We consider a sphere of radius b covered by a thin membrane of thickness $h \ll b$ and with a permeability P for the diffusing substance. In the interior there is mass consumption of constant intensity ϱ. The concentration in the external medium of the substance in question is maintained at $C^{(0)}$. To determine the stationary concentration profile in the interior of the sphere we have to find the solution in the region $0 \leq r \leq b$ of

$$D\frac{1}{r^2}\frac{d}{dr}\left(r^2\frac{dC(r)}{dr}\right) = \varrho, \qquad (2.5.87)$$

where the mass consumption implies a negative value of ϱ.

Integration with respect to r gives

$$r^2\frac{dC(r)}{dr} = \frac{\varrho}{D}\int r^2\, dr + A_1 = \frac{1}{3}\frac{\varrho}{D}r^3 + A_1,$$

where A_1 is a constant. The next integration results in

$$C(r) = \tfrac{1}{3}\frac{\varrho}{D}\int r\,dr + A_1\int r^{-2}\,dr + B_1 = \tfrac{1}{6}\frac{\varrho}{D}r^2 - \frac{A_1}{r} + B_1, \quad (2.5.88)$$

where B_1 is another constant. The above equation represents the general solution of Eq. (2.5.85). But the present case requires that the concentration must be *finite* for $r = 0$, which implies the vanishing of A_1. Thus, the general expression for the concentration profile inside the sphere is

$$C(r) = \tfrac{1}{6}\frac{\varrho}{D}r^2 + B_1. \quad (2.5.89)$$

The constant B_1 is determined by the condition

$$C(r) = C(b), \quad \text{for} \quad r = b, \quad (B19)$$

which gives

$$B_1 = C(b) - \frac{\varrho b^2}{6D},$$

where $C(b)$ is the concentration in the sphere at the inner side of the membrane. To determine $C(b)$ we proceed as follows. The amount $^{(\text{tot})}M_b^{(\Delta t)}$ that in time Δt enters the total surface of area A of the sphere of radius b from the inner side of the covering membrane is

$$^{(\text{tot})}M_b^{(\Delta t)} = A J(b)\,\Delta t = 4\pi b^2 J(b)\,\Delta t,$$

where $J(b)$ is the flux through the membrane into the sphere. As the membrane thickness h is infinitely smaller than the radius b of the sphere we have from Eq. (2.5.84)

$$^{(\text{tot})}M_b^{(\Delta t)} = 4\pi b^2 P\left(C^{(0)} - C^{(\text{i})}\right)\Delta t = 4\pi b^2 \varrho\left(C^{(0)} - C(b)\right)\Delta t.$$

During stationary conditions this amount $^{(\text{tot})}M_b^{(\Delta t)}$ equals the mass consumption in the same time Δt inside the sphere*, i.e. $\tfrac{4}{3}\pi b^3 \varrho\,\Delta t$. Hence it follows that

$$4\pi b^2 P\left(C^{(0)} - C(b)\right) = \tfrac{4}{3}\pi b^3 \varrho.$$

Thus

$$C(b) = C^{(0)} - \frac{\varrho b}{3P}.$$

Inserting $C(b)$ in the expression for B_1 gives

$$B_1 = C^{(0)} - \frac{\varrho b}{3P} - \frac{\varrho b^2}{6D}.$$

---

* A sphere with radius r has the volume $\tfrac{4}{3}\pi r^3$.

Replacing this value for B_1 in Eq. (2.5.88) gives

$$C(r) = C^{(0)} - \frac{1}{6}\frac{\varrho}{D}(b^2 - r^2) - \frac{\varrho b}{3P}, \qquad (2.5.90)$$

which is the equation for the concentration profile in the interior of the sphere.

The profile is again a parabola with the minimum of concentration at $r = 0$, namely

$$C_{\min} = C^{(0)} - \frac{\varrho b^2}{6D} - \frac{\varrho b}{3P}. \qquad (2.5.91)$$

If we compare Eq. (2.5.88) with its counterparts corresponding to the cylinder and the slab, we can see that the differences between the profiles are only their curvatures.

2.5.5 Time-dependent diffusion processes

In this book we do not intend to present a systematic treatment of the various methods available to solve the time-variant diffusion equation. However, in consideration of the following chapters we shall treat two important situations. Moreover, they possess the advantage that they can be handled with the tools presented in Chapter 1.

2.5.5.1 An extended initial distribution (Boltzmann's trick)

The Austrian physicist Ludwig Boltzmann* showed in 1894 that the one-dimensional diffusion equation can be transformed into an ordinary second-order differential equation. Solving this equation, the two arising constants of integration can be adjusted so that the solutions have a strong resemblance to the graphs illustrated in Fig. 2.4.

Boltzmann's starting point was the diffusion equation

$$\frac{\partial C}{\partial t} = D\frac{\partial^2 C}{\partial x^2} \qquad (2.5.92)$$

covering the infinite range $-\infty \leq x \leq \infty$. He then introduced a new variable ξ, defined as

$$\xi = \frac{x}{\sqrt{t}} = x\,t^{-\frac{1}{2}}, \qquad (2.5.93)$$

---

* Ludwig Boltzmann (1844–1906) was Professor of theoretical physics in Vienna. He is particularly known for his work on kinetic gas theory and electricity together with his endeavors to account for the second law of thermodynamics from molecular kinetic considerations. He was the founder of statistical mechanics.

and then regarded $C(x, t)$ as a function of the new variable ξ, i.e.

$$C(x, t) = C(\xi), \quad \text{where} \quad \xi = \xi(x, t).$$

We shall now see how Eq. (2.5.92) transforms as a function of ξ. We have*

$$\frac{\partial C}{\partial t} = \frac{dC}{d\xi}\frac{\partial \xi}{\partial t}.$$

Likewise from Eq. (2.5.93)

$$\frac{\partial \xi}{\partial t} = x\frac{d}{dt}\left(t^{-\frac{1}{2}}\right) = -\tfrac{1}{2}xt^{-\frac{3}{2}} = -\tfrac{1}{2}\frac{1}{t}\frac{x}{\sqrt{t}} = -\tfrac{1}{2}\frac{\xi}{t},$$

which inserted into the above equation gives

$$\frac{\partial C}{\partial t} = -\frac{1}{2}\frac{\xi}{t}\frac{dC}{d\xi}. \qquad (2.5.94)$$

We also have from Eq. (2.5.93)

$$\frac{\partial C}{\partial x} = \frac{dC}{d\xi}\frac{\partial \xi}{\partial x} = \frac{1}{\sqrt{t}}\frac{dC}{d\xi}. \qquad (2.5.95)$$

Hence

$$\frac{\partial^2 C}{\partial x^2} = \frac{\partial}{\partial x}\left(\frac{\partial C}{\partial x}\right) = \frac{\partial}{\partial \xi}\left(\frac{\partial C}{\partial x}\right)\frac{\partial \xi}{\partial x} = \frac{\partial}{\partial \xi}\left(\frac{1}{\sqrt{t}}\frac{dC}{d\xi}\right)\frac{1}{\sqrt{t}} = \frac{1}{t}\frac{d^2 C}{d\xi^2}.$$

$$(2.5.96)$$

Inserting this equation and Eq. (2.5.94) into Eq. (2.5.92) gives

$$-\frac{1}{2}\frac{\xi}{t}\frac{dC}{d\xi} = D\frac{1}{t}\frac{d^2 C}{d\xi^2},$$

or

$$\frac{d^2 C}{d\xi^2} + \frac{1}{2D}\xi\frac{dC}{d\xi} = 0, \qquad (2.5.97)$$

whereby the original partial differential equation Eq. (2.5.91) is now rewritten in the form of an ordinary second-order differential equation, the independent variable $\xi = x/\sqrt{t}$ is of course a function of x and t. Equation (2.5.97) can be written in the form[†]

$$\frac{d}{d\xi}\left(\ln\frac{dC}{d\xi}\right) = -\frac{1}{2D}\xi.$$

---

* Compare with Chapter 1, Section 1.8.4, Eq. (1.8.17).
[†] $\frac{d\ln y}{dx} = \frac{d\ln y}{dy}\frac{dy}{dx} = \frac{1}{y}\frac{dy}{dx}$ and put $y = \frac{dC}{dx}$. See Chapter 1, Section 1.2.5.1, Eq. (1.2.15) and Section 1.4.2, Eq. (1.4.3).

The equation is multiplied on both sides with $d\xi$ and integrated with respect to ξ. This gives

$$\int \frac{d}{d\xi}\left(\ln \frac{dC}{d\xi}\right) d\xi = -\frac{1}{2D}\int \xi \, d\xi$$

$$\ln \frac{dC}{d\xi} = -\frac{1}{4D}\xi^2 + \text{const.},$$

or

$$\frac{dC}{d\xi} = A e^{-\xi^2/4D}.$$

The next integration gives

$$C(\xi) = A \int e^{-\xi^2/4D} \, d\xi + B_1,$$

where B_1 is a constant. It is convenient to write this indefinite integral in the following form with a dummy variable u of integration*

$$C(\xi) = A_1' \int_a^\xi e^{-u^2/4D} \, du + B_1'', \qquad (2.5.98)$$

where a is an arbitrary number and A_1' and B_1'' are the two constants of integration. We now introduce a new variable y of integration

$$y = \frac{u}{2\sqrt{D}}, \quad \text{and} \quad du = 2\sqrt{D}\,dy,$$

and put $a = 0$ in the lower limit. Equation (2.5.98) then takes the form†

$$C(\xi) = A_1 \int_0^{\xi/2\sqrt{D}} e^{-y^2} dy + B_1'.$$

Inserting $\xi = x/\sqrt{t}$ gives

$$C(x,t) = A_1 \int_0^{x/2\sqrt{Dt}} e^{-y^2} dy + B_1', \qquad (2.5.99)$$

which is a solution of Eq. (2.5.92). That applies also to

$$C(x,t) = A_1 \int_0^{(x-x_0)/2\sqrt{Dt}} e^{-y^2} dy + B_1, \qquad (2.5.100)$$

where x_0 is an arbitrary number.

---

* Compare Chapter 1, Section 1.3.2, Eq. (1.3.7).
† The modified symbols A and B' for the constants are introduced intending to reflect the changes of the variable of integration and of the lower limit of integration to $a = 0$.

As will appear below the two constants A_1 and B_1 can be adjusted in such a way that Eq. (2.5.99) satisfies the two boundary conditions

$$C(x, t) = 0, \quad \text{for} \quad x \to +\infty \quad \text{and} \quad t > 0, \tag{B20}$$

and

$$C(x, t) = C_0 = \text{const.}, \quad \text{for} \quad x \to -\infty \quad \text{and} \quad t > 0. \tag{B21}$$

To this end we shall use a classical result*

$$\int_0^\infty e^{-y^2} dy = \frac{1}{2}\sqrt{\pi} \, {}^\dagger. \tag{2.5.101}$$

Applying Eq. (B20) to Eq. (2.5.99) gives

$$0 = A_1 \int_0^\infty e^{-y^2} \, dy + B_1 = \frac{1}{2}\sqrt{\pi} A_1 + B_1, \tag{XVII}$$

and from Eq. (B21) we obtain

$$C_0 = A_1 \int_0^{-\infty} e^{-y^2} \, dy + B_1 = A_1(-1) \int_{-\infty}^0 e^{-y^2} \, dy + B_1 = -\frac{1}{2}\sqrt{\pi} A_1 + B_1. \tag{XVIII}$$

Addition of Eq. (XVII) and Eq. (XVIII) gives $B_1 = -\frac{1}{2}C_0$, while Eq. (XVIII) subtracted from Eq. (XVII) gives $A_1 = -C_0/\sqrt{\pi}$. Insertion of these value for A_1 and B_1 in Eq. (2.5.100) gives

$$C(x, t) = \frac{1}{2}C_0 \left\{ 1 - \frac{2}{\sqrt{\pi}} \int_0^{(x-x_0)/2\sqrt{Dt}} e^{-y^2} \, dy \right\}. \tag{2.5.102}$$

For small values of t the behavior of Eq. (2.5.102) is remarkable, since

(i) when $t \to 0$ for $x - x_0 > 0$ we have

$$C(x, 0) = \frac{1}{2}C_0 \left\{ 1 - \frac{2}{\sqrt{\pi}} \int_0^\infty e^{-y^2} \, dy \right\} = \frac{1}{2}C_0 \left\{ 1 - \frac{2}{\sqrt{\pi}} \frac{\sqrt{\pi}}{2} \right\} = 0,$$

(ii) and when $t \to 0$ for $x - x_0 < 0$ then

$$C(x, 0) = \frac{1}{2}C_0 \left\{ 1 - \frac{2}{\sqrt{\pi}} \int_0^{-\infty} e^{-y^2} \, dy \right\} = \frac{1}{2}C_0 \left\{ 1 - \frac{2}{\sqrt{\pi}} \left(-\frac{\sqrt{\pi}}{2} \right) \right\} = C_0.$$

---

* See Appendix A.

† A mathematician is one to whom *that* is as obvious as that twice two makes four is to you (Lord Kelvin, during a lecture).

Thus, Eq. (2.5.102) gives the concentration profile $C(x, t)$ in the range $-\infty < x < \infty$ at the time t, when the initial distribution of the diffusing substance is given by

$$C(x, 0) = \begin{cases} C_0, & \text{for } -\infty < x < x_0 \\ 0, & \text{for } x > x_0, \end{cases}$$

i.e. at the time $t = 0$ the total amount of substance is present only in the region $-\infty < x < x_0$ at a constant concentration C_0.

The integral $\int e^{-x^2} dx$ in Eq. (2.5.102) cannot be expressed in terms of elementary mathematical functions, and consequently numerical methods must be used to determine the values of the definite integral. Moreover, as the integral frequently occurs in mathematical physics, probability theory and statistics – in conformity with the definition of the natural logarithm, for example – it is convenient to *define* two *new* special functions by means of the integral.

The *error function* Erf(x) is defined by

$$\text{Erf}(x) = \frac{2}{\sqrt{\pi}} \int_0^x e^{-u^2} du. \tag{2.5.103}$$

It follows from Eq. (2.5.101) that

$$\text{Erf}(\infty) = 1. \tag{2.5.104}$$

Since e^{-t^2} is an even function we have*

$$\text{Erf}(-x) = \frac{2}{\sqrt{\pi}} \int_0^{-x} e^{-u^2} du = \frac{2}{\sqrt{\pi}} \int_0^x e^{-v^2} \frac{du}{dv} dv$$

$$= -\frac{2}{\sqrt{\pi}} \int_0^x e^{-v^2} dv = -\text{Erf}(x). \tag{2.5.105}$$

The *complementary error function* Erfc(x)[†] is defined by

$$\text{Erfc}(x) = \frac{2}{\sqrt{\pi}} \int_x^{\infty} e^{-u^2} du. \tag{2.5.106}$$

Since

$$\frac{2}{\sqrt{\pi}} \int_0^{\infty} e^{-u^2} du = \frac{2}{\sqrt{\pi}} \int_0^x e^{-u^2} du + \frac{2}{\sqrt{\pi}} \int_x^{\infty} e^{-u^2} du,$$

---

* Introduce a new variable of integration $v = -u$, i.e. $du/dv = -1$, and for $u = -x$, $v = x$. See Chapter 1, Section 1.7.1.

[†] Both functions were tabulated long ago and are now available in all standard computing programs.

2. Migration and diffusion

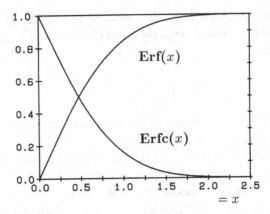

Fig. 2.17. The graphs of the functions Erf(x) and Erfc(x).

Invoking Eq. (2.5.104) and Eq. (2.5.105) we have

$$\text{Erfc}(x) = 1 - \text{Erf}(x), \qquad (2.5.107)$$

and

$$\text{Erfc}(-x) = 1 + \text{Erf}(x) = 2 - \text{Erfc}(x). \qquad (2.5.108)$$

The graphs of Erf(x) and Erfc(x) are shown in Fig. 2.17.

After this small excursion we can write Eq. (2.5.102) more compactly as

$$C(x, t) = \tfrac{1}{2} C_0 \, \text{Erfc} \left\{ \frac{x - x_0}{2\sqrt{Dt}} \right\}. \qquad (2.5.109)$$

The concentration profile of $C(x, t)$ depends upon the value of the single parameter $w = (x - x_0)/2\sqrt{Dt}$. A graph of Eq. (2.5.109) is shown in Fig. 2.18(A) for values of w in the range $0 \le w \le +2.5$. The function $C(w) = C(x, t)$ is an *odd* function with respect to the coordinate lines $w = 0$, and $y = \tfrac{1}{2}$. For $w = 0$ we have $C(0) = \tfrac{1}{2}$ irrespective of the value of $t > 0^*$. This is illustrated in further detail on Fig. 2.18(B) by assigning the value $D = 5.0 \times 10^{-10}\,\text{m}^2\,\text{s}^{-1}$ and calculating a family of concentration profiles $C(x, t)$ as a function of the distance x from the origin ($x_0 = 0$) at various times $t > 0$ after the start of the diffusion process.

---

* The function $C(x, t)$ *jumps* instantaneously from the value $C(0, t) = C_0$ for $t = -\epsilon$ to $C(0, t)$ $= \tfrac{1}{2} C_0$ for $t = +\epsilon$, where $0 < \epsilon \ll 1$.

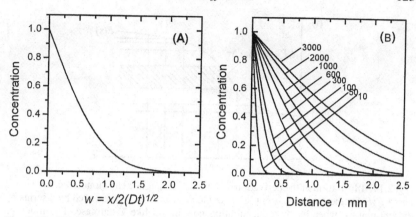

Fig. 2.18. (A) Concentration profiles resulting from an initially extended distribution: $C = C_0$ for $x < 0$ and $C = 0$ for $x > 0$ as a function of the parameter $w = x/2\sqrt{Dt}$ corresponding to Eq. (2.5.109). (B) Concentration profiles as a function of the distance in millimeters from the initial position $x = 0$ for a substance with a diffusion coefficient $D = 5.0 \times 10^{-10}$ m^2 s^{-1}. The numbers on the curves represent time in seconds.

2.5.5.2 Diffusion from a region with constant concentration

Figure 2.18(B) shows that the concentration profiles represented by Eq. (2.5.109) for *all* values of $t > 0$ take the value $\frac{1}{2}C_0$ at $x = x_0$. This implies that the equation

$$C(x, t) = C_0 \operatorname{Erfc}\left\{ \frac{x - x_0}{2\sqrt{D(t - t_0)}} \right\}, \quad x > x_0 \quad \text{and} \quad t > t_0, \quad (2.5.110)$$

describes the concentration $C(x, t)$ in the region $x > x_0$ when we imagine that at the time $t = t_0$ a solution is placed in the region $x < x_0$ and kept at the constant concentration C_0. The argument in Eq. (2.5.109) has $2(D(t - t_0))^{1/2}$ in the denominator as the appearance of a concentration $C(x, t)$ in the position $x > x_0$ depends upon the time $t - t_0 \geq 0$ after the establishment at the time t_0 of the concentration C_0 in the region $x < x_0$. To emphasize the discontinuity in concentration at $t = t_0$ it is considered practical to introduce the function

$$H(t - t_0) = \begin{cases} 0 & \text{for } t < t_0 \\ 1 & \text{for } t > t_0, \end{cases} \quad (2.5.111)$$

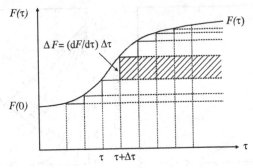

Fig. 2.19. Illustration of the derivation of Duhamel's integral. The monotonic increasing curve $F(t)$ for the concentration $C(0, t)$ at the position $x = 0$ is replaced by a series of discontinuities, whereby the concentration now grows like a staircase. The function starts with a discontinuity by jumping to the value $F(0)$ at $t = 0$.

denoted as *Heaviside's unit step function**. Applying Eq. (2.5.111) to Eq. (2.5.110) we have

$$C(x, t) = C_0 \operatorname{Erfc}\left\{ \frac{x - x_0}{2\sqrt{D(t - t_0)}} \right\} H(t - t_0), \quad \text{for } x > x_0, \quad (2.5.112)$$

which emphasizes that $C(x, t) = 0$ for $t < t_0$.

2.5.5.3 Duhamel's integral

In the previous section we considered the case of the instantaneous establishing of a fixed concentration C_0 at the time $t = t_0$ at the position $x = x_0$. We shall now treat the somewhat more complicated situation where the concentration C_0 impressed at the position x_0 *varies* with time, and we shall calculate the concentration $C(x, t)$ produced in the region $x_0 < x < \infty$ for all times $t > t_0$. To avoid complicating the algebra unnecessarily we put $x_0 = 0$ and $t_0 = 0$. The concentration in the region $x \geq 0$ is initially zero, but at the time $t = 0$ the concentration increases in the position $x = 0$ according to the time function

$$C(0, t) = F(t), \quad \text{for } x = 0 \quad \text{and} \quad t > 0. \quad (2.5.113)$$

An example of a time course of $F(t)$ is shown in Fig. 2.19. The problem consists of calculating the concentration $C(x, t)$ at the position $x > 0$ and at a given time $t > 0$, when the concentration in the position $x = 0$ has changed during

---

* In veneration of Oliver Heaviside (1850–1925), a man of exceptional – but also most controversial – intelligence and an innovator in applied mathematics, operator calculus, vector analysis and theory of electromagnetic fields.

the time $0 \le \tau \le t$ according to the function $C(0, \tau) = F(\tau)$. To this end we replace the continuous monotonic increasing function $F(t)$ in Fig. 2.19 with the discontinuous curve that results from dividing the time $\tau = t$ in n time intervals $\Delta\tau = t/N$ of equal size*. The staircase-shaped curve can be looked upon as the summation of n step functions succeeding each other, the amplitudes of which ΔF_i depend upon the shape of $F(\tau)$. According to Section 2.5.5.2 the function

$$A(x, t \mid 0, \tau) = \text{Erfc}\left\{\frac{x}{2\sqrt{D(t - \tau)}}\right\} H(t - \tau) \equiv A(x, t - \tau) \quad (2.5.114)$$

states the concentration at the position x to the time t when at the time $\tau < t$ the concentration $C = 1$ is established instantaneously at the position $x = 0$. We consider the staircase at time τ_i to the following jump at time $\tau_i + \Delta\tau$. The jump in concentration $\Delta F(\tau_i)$ will at the time $t > \tau_i$ transfer a concentration $\delta C_i(x, t)$ at the position x, which is equal to $\Delta F(\tau_i) A(x, t - \tau_i)$. We have[†]

$$\Delta F(\tau_i) = F(\tau_i + \Delta\tau) - F(\tau_i) = \left(\frac{dF(\tau)}{d\tau}\right)_{\hat{t}_i} \Delta\tau,$$

where $\tau_i \le \hat{t}_i \le \tau_i + \Delta\tau$. Hence

$$\delta C_i(x, t) = \left(\frac{dF(\tau)}{d\tau}\right)_{\hat{t}_i} \Delta\tau A(x, t - \tau_i).$$

The concentration in the position x resulting from the whole staircase contour of concentrations is now obtained by summing all the n contributions from the time $\tau = 0$ to $\tau = t$, i.e.

$$C(x, t) = F(0)A(x, t) + \sum_{i=1}^{i=n}\left(\frac{dF(\tau)}{d\tau}\right)_{\hat{t}_i} A(x, t - \tau_i)\Delta\tau,$$

where the first term comes from the jump $F(0)$ at the time $\tau = 0$. When the number of intervals grows beyond bounds, the staircase contour approaches the actual concentration curve $F(\tau)$, and the expression above takes the form

$$C(x, t) = F(0)A(x, t) + \lim_{\substack{n \to \infty \\ \Delta\tau_i \to 0}} \sum_{i=1}^{i=N}\left(\frac{dF(\tau)}{d\tau}\right)_{\hat{t}_i} A(x, t - \tau_i)\Delta\tau,$$

---

* The division in *equally large* intervals $\Delta\tau$ is not essential for the argument.
[†] Compare Chapter 1, Section 1.2.2, Eq. (1.2.14).

or*

$$C(x,t) = F(0)A(x,t) + \int_0^t \frac{dF(\tau)}{d\tau} A(x, t - \tau)\, d\tau, \qquad (2.5.115)$$

which is the concentration $C(x,t)$ at the position x at the time t resulting from the diffusion from the time-variable concentration $C(0,t) = F(t)$ located at the position $x = 0$. In general, numerical methods (e.g. Simpson's method) are required to integrate Eq. (2.5.115). This should not give rise to any difficulties if a program for $\mathrm{Erfc}(x)$ is at one's disposal. The first term in the integral $dF/d\tau$ presents no difficulties if $F(\tau)$ is given analytically. On the other hand, if $F(\tau)$ results as numerical data obtained experimentally a numerical computation of $dF/d\tau$ may be subject to some uncertainty that again turns up in the evaluation of Eq. (2.5.115). For that reason its may be desirable to transform the integral in Eq. (2.5.115) so that $F(\tau)$ appears instead of $dF/d\tau$. To this end we apply the rule for integration by parts[†] to the integral on the right-hand side. We have

$$C(x,t) = F(0)A(x,t) + [A(x,t-\tau)F(\tau)]_{\tau=0}^{\tau=t} - \int_0^t F(\tau)\frac{\partial A(x,t-\tau)}{\partial \tau}\, d\tau$$

$$= F(0)A(x,t) + A(x,0)F(t) - F(0)A(x,t) - \int_0^t F(\tau)\frac{\partial A(x,t-\tau)}{\partial \tau}\, d\tau.$$

The first and third terms cancel. Moreover, we have $A(x,0) = 0$. Thus, the solution of our problem can be written in the following compact form

$$C(x,t) = -\int_0^t F(\tau)\frac{\partial A(x,t-\tau)}{\partial \tau}\, d\tau. \qquad (2.5.116)$$

This important relation and Eq. (2.5.115) were first derived by J.M.C. Duhamel[‡] in 1833.

We shall now rewrite the equation in a form more suitable for numerical calculation by evaluating the last term in the integral. To simplify the typography we write

$$u = u(x,\tau) = \frac{x}{2\sqrt{D(t-\tau)}} = \frac{x}{2\sqrt{D}}(t-\tau)^{-\frac{1}{2}} = \frac{x}{2\sqrt{D}}w^{-\frac{1}{2}}. \qquad (2.5.117)$$

---

* Compare Chapter 1, Section 1.3.1, Eq. (1.3.2).
† See Chapter 1, Section 1.7.2, Eq. (1.7.5).
‡ J.M.C. Duhamel (1797–1872) was a French mathematician attached to École Polytechnique in 1830–1860, and from 1836 was professor of mathematical analysis.

We have from Eq. (2.5.114)

$$-\frac{\partial}{\partial \tau}(A(x,t-\tau)) = -\frac{\partial}{\partial \tau}\left(\mathrm{Erfc}\left\{\frac{x}{2\sqrt{D(t-\tau)}}\right\}\right) = -\frac{\partial}{\partial \tau}(\mathrm{Erfc}\{u\})$$

$$= -\frac{\partial}{\partial \tau}\left(\frac{2}{\sqrt{\pi}}\int_u^0 e^{-y^2}dy\right) = \frac{2}{\sqrt{\pi}}\frac{\partial}{\partial \tau}\left(\int_0^u e^{-y^2}dy\right)$$

$$= \frac{2}{\sqrt{\pi}}\frac{d}{du}\left(\int_0^u e^{-y^2}dy\right)\frac{\partial u}{\partial \tau} = \frac{2}{\sqrt{\pi}}e^{-u^2}\frac{\partial u}{\partial \tau}.$$

From Eq. (2.5.117) we have

$$\frac{\partial u}{\partial \tau} = \frac{x}{2\sqrt{D}}\frac{du}{dw}\frac{dw}{d\tau} = \frac{x}{2\sqrt{D}}\left(-\frac{1}{2}\right)(t-\tau)^{-\frac{3}{2}}(-1) = \frac{x}{4\sqrt{D(t-\tau)^3}},$$

so the expression for $-\partial A(x,t-\tau)/\partial \tau$ takes the form

$$-\frac{\partial}{\partial \tau}[A(x,t-\tau)] = \frac{2}{\sqrt{\pi}}\left[\frac{x}{4\sqrt{D(t-\tau)^3}}\right]e^{-x^2/4D(t-\tau)}.$$

Replacing this expression for $\partial A(x,t-\tau)/\partial \tau$ in Eq. (2.5.116) changes Duhamel's integral to the following form

$$C(x,t) = \frac{x}{2\sqrt{\pi D}}\int_0^t \frac{F(\tau)}{\sqrt{(t-\tau)^3}}e^{-x^2/4D(t-\tau)}\,d\tau, \qquad (2.5.118)$$

which in some instances may be more easy to handle, particularly in connection with numerical procedures.

2.5.5.4 An instantaneous surface distribution

In Section 2.5.5.1 we described the situation where the total amount of the diffusing material was located initially – e.g. at the time $t=0$ – only in the region to the left of the plane at $x=x_0$ at a constant concentration C^0 ($-\infty < x < x_0$). This result shall now be used to describe the other extreme situation where the material is initially distributed uniformly on the surface of a plane which is established instantaneously at the position $x=x_0$ at the time $t=t_0$. This event is referred to as the establishment of an *instantaneous source*.

We know that the function

$$C_1(x,t) = \tfrac{1}{2}C_0\left\{1 - \frac{2}{\sqrt{\pi}}\int_0^{(x-x_0)/2\sqrt{Dt}} e^{-y^2}\,dy\right\} \qquad (2.5.119)$$

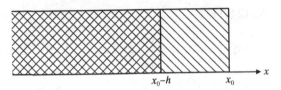

Fig. 2.20. An initially "box-shaped" distribution of width h is considered as the result of the difference between two extended distributions with their boundaries at x_0 and $x_0 - h$.

represents the distribution in space at the time t resulting from the initial distribution

$$C_1(x, 0) = \begin{cases} C_0, & \text{when } -\infty \leq x \leq x_0 \\ 0, & \text{when } x > x_0 \end{cases} \tag{2.5.120}$$

for $t \leq 0$ and the boundary conditions $C_1 = C_0$ for $x \to -\infty$, and $C_1 = 0$ for $x \to +\infty$ and $t > 0$. The initial distribution is shown in Fig. 2.20.

The solution

$$C_2(x, t) = \frac{1}{2}C_0 \left\{ 1 - \frac{2}{\sqrt{\pi}} \int_0^{(x+h-x_0)/2\sqrt{Dt}} e^{-y^2} \, dy \right\} \tag{2.5.121}$$

is in principle identical to that of Eq. (2.5.119) except that the initial jump in concentration *no longer* is located at $x = x_0$, but at $x = x_0 - h$. This initial distribution is likewise shown in Fig. 2.20. Thus, the solution

$$C(x, t) = C_1(x, t) - C_2(x, t) \tag{2.5.122}$$

corresponds to the distribution corresponding to the initial concentration profile

$$\left. \begin{array}{ll} C = 0 & \text{for } x < x_0 - h \\ C = C_0 & \text{for } x_0 - h < x < x_0 \\ C = 0 & \text{for } x > x_0 \end{array} \right\}, \quad \text{and for } \quad t = 0_+.$$

This "box-shaped" profile is shown in Fig. 2.20. For $t > 0$ the "box" will collapse gradually concurrently with the outward diffusion to both sides by the substance. The time course of this process is found by inserting Eq. (2.5.119) and

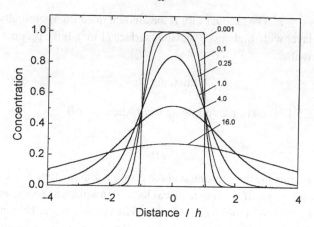

Fig. 2.21. Concentration profiles resulting from the collapse of the initial distribution shown in Fig. 2.20. Distance $X = x/h$ in units of h. The numbers attached to the curves are times in units of $(Dt/h^2)^{1/2}$.

Eq. (2.5.121) in Eq. (2.5.122). This gives

$$C(x, t) = C_1(x, t) - C_2(x, t)$$

$$= -\frac{C_0}{\sqrt{\pi}} \int_0^{(x-x_0)/2\sqrt{Dt}} e^{-y^2} \, dy + \frac{C_0}{\sqrt{\pi}} \int_0^{(x+h-x_0)/2\sqrt{Dt}} e^{-y^2} \, dy$$

$$= \frac{C_0}{\sqrt{\pi}} \left\{ \int_{(x-x_0)/2\sqrt{Dt}}^0 e^{-y^2} \, dy + \int_0^{(x+h-x_0)/2\sqrt{Dt}} e^{-y^2} \, dy \right\}$$

$$= \frac{C_0}{\sqrt{\pi}} \int_{(x-x_0)/2\sqrt{Dt}}^{(x+h-x_0)/2\sqrt{Dt}} e^{-y^2} \, dy. \tag{2.5.123}$$

Figure 2.21 describes how the "box-shaped" initial distribution in Fig. 2.20 collapses with time. We now apply the mean value theorem for the definite integral[*]

$$\int_a^b f(x) \, dx = (b - a) f(\eta), \quad \text{where } a \le \eta \le b,$$

on Eq. (2.5.123). This gives

$$C(x, t) = \frac{C_0}{\sqrt{\pi}} \left\{ \frac{x + h - x_0}{2\sqrt{Dt}} - \frac{x - x_0}{2\sqrt{Dt}} \right\} e^{-(x+\epsilon-x_0)^2/4Dt}$$

$$= \frac{C_0 h}{2\sqrt{\pi Dt}} e^{-(x+\epsilon-x_0)^2/4Dt}, \tag{2.5.124}$$

[*] See Chapter 1, Section 1.3.4, Eq. (1.3.16).

where $0 \leq \epsilon \leq h$. The product $C_0 h$ is the amount of substance initially located inside the layer with thickness h and unit surface (1 m^2). If the product $C_0 h$ has the property that

$$\lim_{h \to 0} (C_0 h) = \mathcal{N},$$

then Eq. (2.5.124) takes the following form when $h \to 0$

$$C(x, t) = \frac{\mathcal{N}}{2\sqrt{\pi D t}} e^{-(x-x_0)^2/4Dt}, \qquad (2.5.125)$$

which represents the concentration at the position x at the time $t > 0$ when an amount of substance M per unit area was located in a plane at $x = x_0$ at the time $t = 0$, while the rest of the space $x \neq x_0$ was free of substance. This situation is also described by saying that at the time $t = 0$ there is an *instantaneous source* of strength $\mathcal{N}^*$ in the position $x = x_0$. The response, Eq. (2.5.125), resulting from the establishment of the instantaneous source of strength $\mathcal{N} = 1$, is also known as Green's function for the one-dimensional diffusion process.

2.5.5.5 Green's function

Green's function[†] for diffusion is often written as

$$C(x, t) \equiv G(x, t \mid \xi, 0) = \frac{1}{2\sqrt{\pi D t}} e^{-(x-\xi)^2/4Dt}, \qquad (2.5.126)$$

where $G(x, t \mid \xi, 0)$ stands for the concentration $C(x, t)$ at the position x and at time t as the result of the diffusion from the instantaneous plane source of unit strength $N = 1$ established at the position ξ at the time $t = 0$. Figure 2.22 shows concentration profiles at various times from such an instantaneous source placed at the position $\xi = 0$. Green's function holds a prominent position in mathematical physics since it is possible by means of this primitive function to construct solutions of far greater complexity. In the following we shall illustrate this property with a few examples.

(i) A varying initial distribution in space

We assume that initially the concentration of the diffusing matter is distributed with the space variable ξ according to

$$C(\xi, 0) = f(\xi), \quad \text{for } -\infty < \xi < \infty \quad \text{and} \quad t = 0. \qquad (2.5.127)$$

---

[*] $\mathcal{N}$ is the surface concentration, e.g. in mol m^{-2}.
[†] Named after G. Green (1793–1841), a self-taught English mathematician who among other things introduced this function – although in a more general form – in his famous, but rather belatedly recognized, work *An Essay on the Application of Mathematical Analysis to the Theories of Electricity and Magnetism*, Nottingham, 1828, which he published privately.

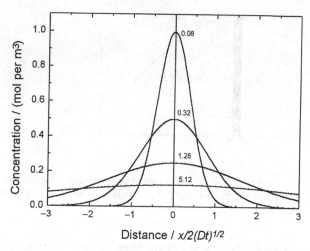

Fig. 2.22. Illustration of Green's function: the spread of matter by diffusion from an instantaneous plane source of strength $N = 1$ in the position $x_0 = 0$. Ordinate: concentration. Abscissa: distance from the source in units of $(Dt)^{1/2}$. The numbers on the single curves are the values of Dt.

Unless $f(\xi)$ is constant the matter will move by diffusion towards the equilibrium state. So we want to know the concentration $C(x, t)$ in the plane at the fixed position $\xi = x$ and at the time t. We consider a thin layer between the planes at the distances* ξ_j and $\xi_j + \Delta \xi_j$ from origin at $\xi = 0$. See Fig. 2.23. The amount that is confined inside this layer of unit area is

$$f(\xi_j)\,\Delta\xi_j,$$

and can be regarded as the strength of an instantaneous source provided that $\Delta\xi$ is infinitely small. The distance of this source to the fixed point x is equal to $x - \xi_j$. According to Eq. (2.5.116) this source results in the contribution $\Delta C_j(x, t)$ to the concentration at the position x and time t that is

$$\Delta C_j(x, t) = \frac{f(\xi_j)\,\Delta\xi_j}{2\sqrt{\pi Dt}}\,e^{-(x-\xi_j)^2/4Dt}.$$

We now consider the region delimited between the two planes at $\xi = a$ and $\xi = b$ and divide the whole region into n similar layers of thickness $\Delta\xi_j$. Their contribution to the concentration at x is.

$$\sum_{j=1}^{j=n} \Delta C_j(x, t) = \sum_{j=1}^{j=n} \frac{f(\xi_j)}{2\sqrt{\pi Dt}}\,e^{-(x-\xi_j)^2/4Dt}\,\Delta\xi_j.$$

* The running position variable is denoted ξ, and a *fixed* point on the ξ-axis by $\xi = x$.

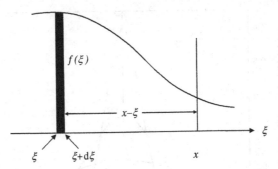

Fig. 2.23. Calculation of the concentration profile when the initial distribution $C(\xi, 0) = f(\xi)$ varies throughout the infinite space. The location of origin $\xi = 0$ (not shown) is somewhere to the left.

We now let n increase beyond any bound while at the same time $\xi_j \to 0$. The sum on the left-hand side tends towards the value of $C(x, t)$ while the right-hand side becomes the limit

$$C(x, t) = \lim_{\substack{n \to \infty \\ \Delta\xi_j \to 0}} \sum_{j=1}^{j=n} \frac{f(\xi_j)}{2\sqrt{\pi Dt}} \, e^{-(x-\xi_j)^2/4Dt} \, \Delta\xi_j.$$

or

$$C(x, t) = \int_a^b \frac{f(\xi)}{2\sqrt{\pi Dt}} \, e^{-(x-\xi)^2/4Dt} d\xi. \qquad (2.5.128)$$

When we let $a \to -\infty$ and $b \to +\infty$ the solution takes the final form which contains the contributions from the whole range $f(\xi)$ as defined by Eq. (2.5.127)*

$$C(x, t) = \int_{-\infty}^{\infty} \frac{f(\xi)}{2\sqrt{\pi Dt}} \, e^{-(x-\xi)^2/4Dt} d\xi. \qquad (2.5.129)$$

---

* From now on we omit writing the limiting procedure in detail and reduce the writing to show only the essential steps in the argument, namely: consider the layer of thickness $d\xi$ located at the position ξ. The quantity $f(\xi) \, d\xi$ (multiplied by the unit area) can be regarded as the strength of an instantaneous source at the position ξ whose distance to the point is $x - \xi$. At the time t the contribution from this source to the concentration in x is

$$\frac{f(\xi) \, d\xi}{2\sqrt{\pi Dt}} e^{-(x-\xi)^2/4Dt}.$$

Hence, the contribution from the total profile of concentration $C(\xi, 0) = f(\xi)$ to the concentration $C(x, t)$ in the position x is therefore

$$C(x, t) = \frac{1}{2\sqrt{\pi Dt}} \int_{-\infty}^{\infty} f(\xi) e^{-(x-\xi)^2/4Dt} \, d\xi.$$

(ii) Initial uniform distribution in the infinite half-space

We first consider the situation where the substance is distributed initially ($t = 0$) in the infinite half-space $-\infty < x \le x_0$ with the constant concentration C_0, i.e.

$$C(x, 0) = \begin{cases} C_0 & \text{for } -\infty < x \le x_0 \\ 0 & \text{for } x > x_0. \end{cases} \qquad (2.5.130)$$

Then we have from Eq. (2.5.129)

$$C(x, t) = \frac{1}{2\sqrt{\pi Dt}} \int_{-\infty}^{\infty} C(\xi, 0)\, e^{-(x-\xi)^2/4Dt}\, d\xi = \frac{C_0}{2\sqrt{\pi Dt}} \int_{-\infty}^{x_0} e^{-(x-\xi)^2/4Dt}\, d\xi.$$

We change the variable of integration from ξ to that of u by putting

$$u = \frac{x - \xi}{\sqrt{2Dt}}.$$

In accordance with the rule for integration by substitution*, $(d\xi/du)\, du$ replaces $d\xi$ in the integral together with the replacement in the lower limit of $\xi = -\infty$ with $u = (x - (-\infty))/2\sqrt{Dt} = \infty$ and in the upper limit $\xi = x_0$ by $u = (x - x_0)/2\sqrt{Dt}$. This gives

$$\begin{aligned} C(x, t) &= \frac{C_0}{2\sqrt{\pi Dt}} \int_{\infty}^{(x-x_0)/2\sqrt{Dt}} e^{-u^2} \left(\frac{d\xi}{du}\right) du \\ &= \frac{C_0}{2\sqrt{\pi Dt}} \int_{\infty}^{(x-x_0)/2\sqrt{Dt}} e^{-u^2} (-2\sqrt{Dt})\, du \\ &= \tfrac{1}{2} C_0 \frac{2}{\sqrt{\pi}} \int_{(x-x_0)/2\sqrt{Dt}}^{\infty} e^{-u^2}\, du^\dagger \\ &= \tfrac{1}{2} C_0 \operatorname{Erfc}\left\{ \frac{x - x_0}{2\sqrt{Dt}} \right\}. \end{aligned} \qquad (2.5.131)$$

This expression is – as it should be – identical to that of Eq. (2.5.109). Despite the undeniable beauty of Eq. (2.5.129) it is only integrable in terms of elementary functions for a limited number of analytical expressions for $f(\xi)$. So, in general, numerical methods of integration must be used.

(iii) The effect of an impermeable barrier

We imagine the instantaneous source of strength $\mathcal{N}$ placed at the position x_0 in the positive half-space $0 \le x < \infty$ whose plane of demarcation is to the left of an impermeable wall placed at the position $x = 0$. The problem is now to evaluate how the presence of this wall will modify the diffusion from the source

---

* See Chapter 1, Section 1.7.1, Eq. (1.72).
† See the definition of the complementary error function Section 2.5.5.1, Eq. (2.5.106).

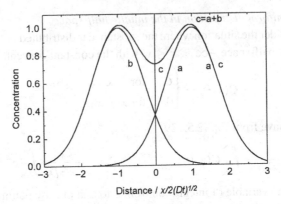

Distance / x/2(Dt)¹ᐟ²

Fig. 2.24. Superposition of the concentration profile $G(x, t \mid x_0, 0)$ from an instantaneous surface source placed in the position $x = x_0$ and the profile $G(x, t \mid -x_0, 0)$ from an identical surface distribution placed in the position $x = -x_0$ produce the concentration profile in the space $0 \leq x \leq \infty$, which is equivalent to the situation where the diffusion of matter into the space $-\infty \leq x \leq 0$ from the source at $x = x_0$ is prevented by a totally reflecting wall in the position $x = 0$.

at times $t > 0$. Since the flux at the wall is zero it follows that the concentration profile $C(x, t)$ at any $t > 0$ must be subject to the condition

$$\frac{\partial C}{\partial x} = 0, \quad \text{for} \quad x = 0 \quad \text{and} \quad t > 0. \tag{2.5.132}$$

This condition can be obtained by means of the following artifice*: we imagine the wall at $x = 0$ is removed and another source of the same strength is placed at the position $-x_0$. Taken together these two sources will generate a concentration profile that is

$$C(x, t) = \frac{\mathcal{N}}{2\sqrt{\pi D t}} e^{-(x-x_0)^2/4Dt} + \frac{\mathcal{N}}{2\sqrt{\pi D t}} e^{-(x+x_0)^2/4Dt}. \tag{2.5.133}$$

A graph of the two terms and their sums at time t is shown in Fig. 2.24. Since each of the two terms is symmetrically distributed with respect to $x = 0$ this also applies to their sum, which for that reason will have a local extremum at $x = 0$ and therefore also a horizontal slope at this position, i.e. $\partial C/\partial x = 0$. This judgment by eye may be re-examined by a direct calculation of the gradient. This gives

$$\frac{\partial C}{\partial x} = -\mathcal{N} \frac{x - x_0}{4\sqrt{\pi (Dt)^3}} e^{-(x-x_0)^2/4Dt} - \mathcal{N} \frac{x + x_0}{4\sqrt{\pi (Dt)^3}} e^{-(x+x_+)^2/4Dt}.$$

---

* Technically designated as "reflection of the source" into the inaccessible space.

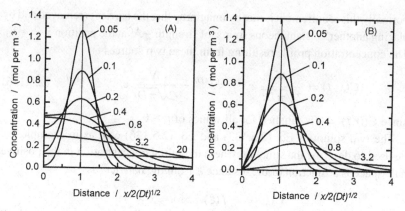

Fig. 2.25. Concentration profiles in the infinite positive half-space arising from the diffusion of matter from an instantaneous surface source placed at the position x_0 when (A) the diffusion into the space $-\infty \leq x \leq 0$ is prevented by a totally reflecting wall placed at the position $x = 0$, and (B) the wall is totally absorbing.

Then, putting $x = 0$ yields

$$\frac{\partial C}{\partial x} = \mathcal{N}\frac{x_0}{4\sqrt{\pi(Dt)^3}}e^{-x_0^2/4Dt} - \mathcal{N}\frac{x_0}{4\sqrt{\pi(Dt)^3}}e^{-x_0^2/4Dt} = 0.$$

Figure 2.25 illustrates the reflection at the impermeable wall of the diffusing substance liberated from the instantaneous source placed at the position $x = x_0$. The characteristic feature is a rise at the barrier of the amount of matter followed by a wave, just like a water wave hitting a wharf. When the position of the instantaneous source moves to the position $x = 0$ we have

$$C(x, t) = \frac{\mathcal{N}}{\sqrt{\pi Dt}}e^{-x^2/4Dt}, \quad x \geq 0. \tag{2.5.134}$$

The value of this expression is twice that given by Eq. (2.5.116.). This is because the total matter $\mathcal{N}$ initially present per unit area in the surface distribution is now forced to move in *one* direction only, namely into the positive half-space $0 \leq x \leq \infty$.

(iv) The effect of a matter-absorbing wall
Again we consider the instantaneous source of strength $\mathcal{N}$ placed at the position $X = 0$ in the infinite half-plane $0 \leq x < \infty$, but now we ascribe to the wall at $x = 0$ the property of absorbing *all* the matter with which it comes into contact by diffusion. The condition to be satisfied now is

$$C(0, t) = 0, \quad \text{for } x = 0 \quad \text{and for} \quad t \geq 0. \tag{2.5.135}$$

Formally this requirement is met by imagining that the barrier is removed and by placing another* instantaneous source of strength $-\mathcal{N}$ at the position $x = -x_0$. The concentration profile resulting from these two sources is

$$C(x,t) = \frac{\mathcal{N}}{2\sqrt{\pi Dt}}\, e^{-(x-x_0)^2/4Dt} - \frac{\mathcal{N}}{2\sqrt{\pi Dt}}\, e^{-(x+x_0)^2/4Dt}, \quad (2.5.136)$$

since $C(0,t) = 0$ is satisfied for all values of $t \geq 0$.

The two solutions Eq. (2.5.133) and Eq. (2.5.136) can, as in example (i), be used to describe the situation where the initial concentration in the positive half-space is a function of the distance ξ from the barrrier[†] as

$$C(\xi, 0) = f(\xi), \quad \text{for} \quad x \geq 0.$$

Let us consider the case of an absorbing wall: the layer at the distance ξ and of thickness $d\xi$ contributes to the concentration in the plane $\xi = x$ at time t by the amount

$$\frac{f(\xi)d\xi}{2\sqrt{\pi Dt}}\, e^{-(x-\xi)^2/4Dt}.$$

The contribution at x from the source with strength $-f(\xi)d\xi$ and reflected as regards $\xi = 0$ to $-\xi$ is

$$-\frac{f(\xi)d\xi}{2\sqrt{\pi Dt}}\, e^{-(x+\xi)^2/4Dt}[‡].$$

The resultant contribution

$$\left(\frac{1}{2\sqrt{\pi Dt}}\, e^{-(x-\xi)^2/4Dt} - \frac{1}{2\sqrt{\pi Dt}}\, e^{-(x+\xi)^2/4Dt} \right) f(\xi)\, d\xi,$$

causes the concentration to be zero at $x = 0$ for all values of t. That also holds for the summation of the contributions from all the layers in the range $0 \leq x \leq \infty$, thus

$$C(x,t) = \frac{1}{2\sqrt{\pi Dt}} \int_0^\infty f(\xi)\left[e^{-(x-\xi)^2/4Dt} - e^{-(x+\xi)^2/4Dt} \right] d\xi. \quad (2.5.137)$$

We shall apply this solution to the following example. The half-space $0 \leq x < \infty$ has an impermeable wall at $x = 0$. The concentration is everywhere C_0 at the time $t < t_0$. At $t = t_0$ the wall at $x = 0$ is replaced by an absorbing barrier, which maintains the concentration there at $C(0,t) = 0$. Putting $f(\xi) = C_0$

---

* Although this situation is difficult to realize physically.
† See footnote to example (i).
‡ Note that the reflection to $-\xi$ does not imply that ξ is negative in the formula.

we have

$$C(x,t) = \frac{C_0}{2\sqrt{\pi D(t-t_0)}} \int_0^\infty \left[e^{-(x-\xi)^2/4D(t-t_0)} - e^{-(x+\xi)^2/4D(t-t_0)} \right] d\xi,$$

for $t \geq t_0$, since $t - t_0$ is the time left for the wall to swallow the adjacent matter. As in example (ii) we put

$$y = \frac{\xi - x}{2\sqrt{D(t-t_0)}} \quad \text{and} \quad z = \frac{x + \xi}{2\sqrt{D(t-t_0)}},$$

after which the analogous calculations give

$$
\begin{aligned}
C(x,t) &= \frac{C_0}{\sqrt{\pi}} \left\{ \int_{-x/2\sqrt{D(t-t_0)}}^\infty e^{-y^2} dy - \int_{x/2\sqrt{D(t-t_0)}}^\infty e^{-z^2} dz \right\} \\
&= \frac{C_0}{\sqrt{\pi}} \left\{ \int_{-x/2\sqrt{D(t-t_0)}}^\infty e^{-y^2} dy + \int_\infty^{x/2\sqrt{D(t-t_0)}} e^{-z^2} dz \right\} \\
&= \frac{C_0}{\sqrt{\pi}} \left\{ \int_{-x/2\sqrt{D(t-t_0)}}^\infty e^{-u^2} du + \int_\infty^{x/2\sqrt{D(t-t_0)}} e^{-u^2} du \right\} \\
&= \frac{C_0}{\sqrt{\pi}} \int_{-x/2\sqrt{D(t-t_0)}}^{x/2\sqrt{D(t-t_0)}} e^{-u^2} du \\
&= C_0 \frac{2}{\sqrt{\pi}} \int_0^{x/2\sqrt{D(t-t_0)}} e^{-u^2} du,
\end{aligned}
$$

since the integrand $\exp[-u^2]$ is an even function, i.e. $\int_{-a}^0 \ldots du = \int_0^a \ldots du$. The above expression can also be written*

$$C(x,t) = C_0 \mathrm{Erf}\left\{ \frac{x}{2\sqrt{D(t-t_0)}} \right\}. \tag{2.5.138}$$

(v) A variable flux into one half-space

In biological systems we often encounter the situation where a substance is liberated from a surface either as a shortlasting spatter or as a spray of limited duration but declining with time. Such an event may be imitated by the following model. We consider the semi-infinite half-space $0 \leq x \leq \infty$, which initially is free of diffusing matter. At time $t = 0$ matter starts to flow into the half-space through the plane at $x = 0$ and a flux J of matter starts at time $t = 0$. Apart from this unidirectional flow the plane is assumed to be impermeable to matter. The flux $J(t)$ is assumed to vary with the time t. We want to know how the concentration $C(x,t)$ builds up by this matter flow at the position x in the

---

* See Eq. (2.5.103).

half-plane and at the time $t > 0$. To begin with we look at the plane at $x = 0$. In the time interval between $t = \tau$ and $t = \tau + d\tau$ an amount appears at this plane of magnitude $J(\tau)d\tau$ per unit area. This amount can be regarded as an instantaneous surface distribution appearing at the time τ in the position $x = 0$. According to Eq. (2.5.126) this source will at any later time $t > \tau$ contribute in the plane at x with the concentration

$$\frac{J(\tau)d\tau}{\sqrt{\pi D(t-\tau)}}e^{-x^2/4D(t-\tau)},$$

since the time elapsed after the appearance of the source is $t - \tau$. Summing all these contributions from the flux $J(t)$ from the time $\tau = 0$ to $\tau = t$ we obtain the concentration in x at the time t as

$$C(x,t) = \frac{1}{\sqrt{\pi D}}\int_0^t \frac{J(\tau)}{\sqrt{t-\tau}}e^{-x^2/4D(t-\tau)}\,d\tau. \qquad (2.5.139)$$

This integral can be integrated in terms of elementary functions for a limited number of functions of $J(\tau)$. Otherwise numerical methods must be used. However, it may be useful to bring the integral into a slightly different form. We can transform the integral by changing the variable of integration from τ to $u = t - \tau$. Hereby the lower limit changes from $\tau = 0$ to $u = t$ and the upper limit from $\tau = t$ to $u = 0$. Furthermore, $d\tau/du = -1$. We have then

$$C(x,t) = \frac{1}{\sqrt{\pi D}}\int_t^0 \frac{J(t-u)}{\sqrt{u}}e^{-x^2/4Du}\left(\frac{d\tau}{du}\right)du$$

$$= \frac{1}{\sqrt{\pi D}}\int_0^t \frac{J(t-u)}{\sqrt{u}}e^{-x^2/4Du}\,du, \qquad (2.5.140)$$

as an alternative expression for the concentration profile.

It may be instructive to visualize these formulas by means of a numeric example. As the simplest possibility we consider a constant "charging current" J_0. We then have, from Eq. (2.5.140)*,

$$C(x,t) = \frac{J_0}{\sqrt{\pi D}}\int_0^t \frac{1}{\sqrt{u}}e^{-x^2/4Du}\,du$$

$$= 2J_0\sqrt{\frac{t}{\pi D}}\,e^{-x^2/4Dt} - \frac{J_0}{D}\,x\,\mathrm{Erfc}\left\{\frac{x}{2\sqrt{Dt}}\right\}. \qquad (2.5.141)$$

Figure 2.26 shows concentration profiles calculated from Eq. (2.5.140) both as a function of the distance x (A) and time t (B).

---

* The details of the calculations leading to the final result are shown in Appendix B.

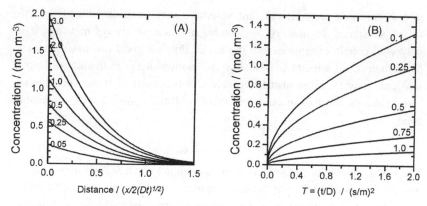

Fig. 2.26. Concentration profiles calculated from Eq. (2.5.140) with a "charging current" $J_0 = 1$. (A) Concentration profile into the half-space $0 \le x < \infty$ as a function of distance x given in dimensionless unit $X = x/2\sqrt{Dt}$. The numbers on the single curves are values of time $T = t/D$. (B) The growth of the concentration with time T. The numbers on the curves indicate the distance from the position $x = 0$ in units of the variable X.

2.5.6 Molecular description of diffusion

In the preceding sections we have presented several examples to illustrate the calculation of the diffusion in space and time of a substance in solution. The bases for these calculations were

(1) a pure empirical phenomenological physical law – Fick's law

$$J = -D\frac{\partial C}{\partial x},$$

stating how the gradient of concentration and the diffusion flux are interconnected at any point in space; and

(2) a general physical principle about mass conservation, which (mathematically formulated) is written as

$$\frac{\partial C}{\partial t} = -\frac{\partial J}{\partial x}.$$

Combining these two expressions results in the fundamental law – the diffusion equation – for the process of diffusion in one dimension

$$\frac{\partial C}{\partial t} = D\frac{\partial^2 C}{\partial x^2},$$

which makes the starting point in obtaining a solution of a diffusion problem that allows for numerical work up.

Fick's law gives a pure phenomenological – and macroscopic – description of the interdependence between diffusion flux and concentration gradient in a system that contains per unit volume a very large amount of the diffusing substance. In biology this condition is almost always met, and it is only with certain exceptions that the use of this law gives rise to difficulties. Nevertheless, an attempt to visualize the motion of the individual diffusing molecule may be of an instructive value, in particular as it may provide an insight into the mechanisms that underlie the diffusion process at the molecular level.

2.5.6.1 Brownian motion

The key to this understanding is the phenomenon known as *Brownian motion*. The English botanist Robert Brown* found that microscopic particles (pollen grain) in a water suspension executed incessant irregular movements of microscopic scale. In the years that followed, many explanations – either erroneous or imperfect – were put forward to account for the phenomenon until Albert Einstein (1905)[†] and M. von Smoluchowski (1906) independently developed the theory for Brownian motion that won acceptance almost immediately. The basic concept in this theory is the idea that all the molecules in a solution participate in a *chaotic, uninterrupted process of stirring* caused by the *thermal movements* executed by the water molecules and the suspended particles as well. Each particle is, therefore, in a state of receiving an irregular – but incessant – bombardment from its neighboring molecules. The totally random character of each molecular impact, taken together with the fact that the kinetic energy of each individual molecule in the solution perpetually *fluctuates* about a mean value

$$\left\langle \tfrac{1}{2}mv^2 \right\rangle = \tfrac{3}{2}kT,$$

where T is the temperature of the system in Kelvin and k is Boltzmann's constant[‡], will lead to the formation of a local anisotropy of momentum of the molecules that grows and falls continuously. If this excess momentum received by the single molecule is sufficiently large in a given direction, the impact will cause the molecule to undergo a finite (short) displacement δ_1 in one direction smearing out the excess momentum during the collision with the neighboring molecules. After a short time (less than $c.\ 10^{-13}$ s) the same "red" molecule will again receive an impact and jump another distance δ_2 of magnitude and direction which is totally independent of those of the previous

---

* Robert Brown (1828): A brief Account of Microscopical Observations in the Months of June, July, and August, 1827, on the Particles contained in the Pollen of Plants; and on the general Existence of active Molecules in Organic and Inorganic Bodies. *Phil. Mag. N.S.*, **4**, 161–173.
[†] Albert Einstein was born on 14 March 1879 in Ulm, and died on 18 April 1955 in Princeton.
[‡] $k = 1.380\,47 \times 10^{-23}$ J K^{-1} per molecule.

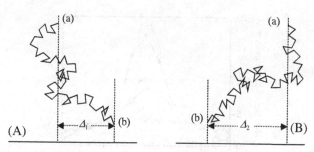

Fig. 2.27. Two examples of the reconstruction of the movements that might be executed by a suspended particle exhibiting Brownian motion in t seconds. In (A) the particle starts at (a) and ends at (b), corresponding to a net displacement to the right of Δ_1 along the x-axis. In (B) the net displacement is Δ_2 and to the left.

jump. This process will go on *ad infinitum* everywhere in the available space: all the molecules incessantly and chaotically bombard each other, and from time to time a certain fraction of the molecules will undergo small displacements (Platzwechslung) in a purely random manner because of their fluctuations in kinetic energy about the mean value $\langle mv^2/2 \rangle = 3kT/2$. It is these fluctuations in kinetic energy that also can be recorded experimentally by a study of the chaotic pattern of movement of small suspended microscopic particles.

Figure 2.27 shows an example of these zigzag movements where the attention has been focused on the motion of a single particle by taking photographs at fixed intervals. By holding together the sequence of pictures it is possible to make a crude reconstruction of the particle movement. In Fig. 2.27(A) the particle's starting point is (a), and after, say, 100 s it is found at (b). The points in between indicate the position observed on each photograph. The net displacement along the x-axis is Δ_1. An experiment repeated later (Fig. 2.27(B)) shows the same pattern in principle apart from the end-position (b) which now is to the left of the starting position (a) and the net displacement along the x-axis which is now Δ_2. When this type of experiment is repeated many times it turns out that the sums of positive and negative values of Δ_i tend to cancel out, i.e. the probability is the same of finding a particle at time t either in the position Δ or in $-\Delta$ from the starting point. Figure 2.28 illustrates the distributions of observing a net displacement Δ (positive and negative) after the same time interval t_1 together with the same kind of displacements observed at two later times t_2 and t_3. The curves can be interpreted as the probability that the particle is observed at time t at distance Δ from the starting position. The curves are symmetrically shaped around the starting position. This means that after a particle has executed Brownian motion in time t the probability for observing

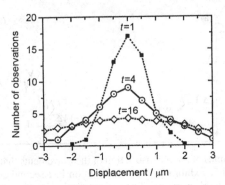

Fig. 2.28. Distribution of net displacements Δ at time t, when observations like those of Fig. 2.27 are repeated a large number of times. The three curves denote the displacements observed at three different times, where $t_1 < t_2 < t_3$.

a negative or positive net displacement is the same, and also that the *mean displacement* of n succeeding displacements of a single particle

$$\langle \Delta \rangle = \frac{1}{n} \sum_{i=1}^{i=n} \Delta_i = 0,$$

equals zero provided n is large. As a measure of wanderings of a single particle after a period of observation of t seconds one could either calculate the mean displacement of the particle in *one direction* or take the mean value of the square of each of the n displacements from the starting point

$$\langle \Delta^2 \rangle = \frac{1}{n} \sum_{i=1}^{i=n} \Delta_i^2,$$

which means that

$$\sqrt{\langle \Delta^2 \rangle}$$

is the *standard deviation* of the displacements.

At the later times $t_2 < t_3$ the curves shrink in height and grow wider on both sides. Thus, the longer the time of observation the less is the probability of finding the "red" particle again at the position first observed, whereas the standard deviation for the displacements increases. This means that the chance of finding the particle far away from the origin increases as time passes. We shall discuss this behavior in detail later.

2.5.6.2 Diffusion from a statistical point of view

Consider an instantaneous surface source of strength $\mathcal{N}$ molecules per m^2 established in the position $x = 0$ at time $t = 0$. The concentration produced at position x and at time t is*

$$C(x, t) = \frac{\mathcal{N}}{2\sqrt{\pi Dt}} e^{-x^2/4Dt}. \tag{2.5.142}$$

The number of particles dn present in the slab between x and $x + dx$ with surfaces A is

$$dn = C(x, t)\, dx \cdot A.$$

The fraction dP of this number dn and the number of particles, $\mathcal{N} \cdot A$, initially located on the source at the area A is

$$dP = \frac{dn}{\mathcal{N} \cdot A} = \frac{C(x, t)\, dx}{\mathcal{N}}. \tag{2.5.143}$$

This expression can be regarded as the probability of a particle – initially located at position $x = 0$ and time $t = 0$ – being found between x and $x + dx$ at time t. Combination with Eq. (2.5.142) gives

$$dP = \frac{1}{2\sqrt{\pi Dt}} e^{x^2/4Dt}\, dx = \varphi(x, t)\, dx, \tag{2.5.144}$$

where

$$\varphi(x, t) = \frac{1}{2\sqrt{\pi Dt}} e^{-x^2/4Dt} \tag{2.5.145}$$

is the *probability density* for the displacement x to time t. Note the similarity between this expression and the probability density for the *normal* law of distribution (the Gauss Distribution)

$$y = \frac{1}{\sigma\sqrt{2\pi}} e^{-x^2/2\sigma^2}, \tag{2.5.146}$$

where σ is the *standard deviation* from the *mean value* $\langle x \rangle$, where $\langle x \rangle$ is zero as Eq. (2.5.146) changes to Eq. (2.5.145) by substituting

$$\sigma^2 = \langle x^2 \rangle = 2Dt. \tag{2.5.147}$$

Naturally this leads to the concept that the displacements $x(t)$ of the single particles at time t during a diffusion process are characterized by a *spectrum*

* See Section 2.5.5.4.

of displacements from the initial position that are distributed normally with a mean value zero and a standard deviation $\sqrt{2Dt}$.

In this connection it is interesting to note that it is possible to derive Eq. (2.5.146) directly from the diffusion equation

$$\frac{\partial C}{\partial t} = D \frac{\partial^2 C}{\partial x^2},$$

where $C(x, t)$ represents the *number* of particles per unit volume (m^3). All particles are assumed to have started their diffusion at time $t = 0$ in a plane at $x = 0$ with a density of $\mathcal{N}$ particles per m^2. As the number of particles remains constant as time passes, we have

$$\int_{-\infty}^{\infty} C(x, t)\,dx = \mathcal{N} \quad \text{for} \quad t \geq 0,$$

and the fraction of particles that at time t is located between the layers at x and $x + dx$ is

$$\frac{C(x, t)\,dx}{\int_{-\infty}^{\infty} C(x, t)\,dx} = \frac{1}{\mathcal{N}} C(x, t)\,dx.$$

The dispersion $\langle x^2 \rangle$ of displacements x at time t is the initial position $x = 0$ is

$$\langle x^2 \rangle = \frac{1}{\mathcal{N}} \int_{-\infty}^{\infty} x^2 C(x, t)\,dx. \tag{2.5.148}$$

To determine how $\langle x^2 \rangle$ depends upon t we multiply the diffusion equation by x^2 and integrate over all values of x:

$$\int_{-\infty}^{\infty} x^2 \left(\frac{\partial C}{\partial t} \right) dx = D \int_{-\infty}^{\infty} x^2 \left(\frac{\partial^2 C}{\partial x^2} \right) dx.$$

On the left-hand side we reverse the order of integration and differentiation[*] and use Eq. (2.5.148) to obtain

$$\int_{-\infty}^{\infty} x^2 \left(\frac{\partial C}{\partial t} \right) dx = \frac{d}{dt} \left\{ \int_{-\infty}^{\infty} x^2 C(x, t)\,dx \right\} dx$$

$$= \mathcal{N} \frac{d}{dt} (\langle x^2 \rangle). \tag{2.5.149}$$

---

[*] This is permissable as long as the function $C(x, t)$ is continuous.

Integration by parts* of the right-hand side gives

$$D \int_{-\infty}^{\infty} x^2 \left(\frac{\partial^2 C}{\partial x^2} \right) dx = D \left[x^2 \left(\frac{\partial C}{\partial x} \right) \right]_{-\infty}^{\infty} - 2D \int_{-\infty}^{\infty} x \left(\frac{\partial C}{\partial x} \right) dx$$

$$= -2D \int_{-\infty}^{\infty} x \left(\frac{\partial C}{\partial x} \right) dx$$

$$= -2D[xC(x,t)]_{-\infty}^{\infty} + 2D \int_{-\infty}^{\infty} C(x,t) \, dx$$

$$= 2\mathcal{N}D, \tag{2.5.150}$$

as $\partial C / \partial x$ and C both tend to zero for $x \to \pm\infty$. Combining Eq. (2.5.149) and Eq. (2.5.150) gives

$$\frac{d}{dt}(\langle x^2 \rangle) = 2D,$$

or the identical equation

$$\langle x^2 \rangle = 2Dt + A,$$

where A is a constant that is zero because $\langle x^2 \rangle = 0$ for $t = 0$ since all particles initially started from the origin. Hence, at time t the mean value of the square of the displacements from the origin at $x = 0$ is

$$\langle x^2 \rangle = 2Dt,$$

which is identical to the expression in Eq. (2.5.147).

It is conceivable that such considerations led Einstein and Smoluchowski[†] to reverse the problem and advance a molecular model – by making certain simplifying assumptions about the movements of molecules – that on the basis of simple statistical considerations could account not only for the patterns of movement of the individual molecules but also for Fick's law, the diffusion equation and for the diffusion equation with a superimposed field of force.

2.5.6.3 Random walk

In his treatment of the diffusion process, Smoluchowski gives an idealized description of the dissolved molecules chaotically jumping around by solving

---

* See Chapter 1, Section 1.7.2, Eq. (1.7.5).

[†] Maryan Ritter von Smolan-Smoluchowski (1872–1917) was Professor of theoretical physics at the University in Lemberg 1900 and from 1913 at the Jagiellonian University in Kraków. He did pioneering work in his studies on Brownian motion, fluctuations in density of gases and realm of validity of the second law of thermodynamics along with studies of electrokinetic phenomena and coagulation of colloids.

the problem in the probability calculus that was first formulated explicitly by Karl Pearson (1905)* and is now called the *random walk* or *drunkard's walk*. The simplest version of the problem formulated for a walk in one direction only is as follows. A particle executes a *sequence of steps* of the *same length* λ. Each step is directed either *forwards* or *backwards* with the *same* probability, i.e. the probability for making a given step is $\frac{1}{2}$. Furthermore, the direction of each step is *independent* of the direction of the preceding step. After executing a total of N steps the expected location of the particle is on any of the points on the line

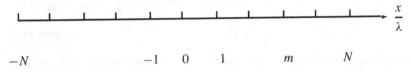

where the points $-N \leq m \leq N$ are regarded as coordinates on the x-axis with the step-length λ as the unit. This raises the following questions. (i) When the particle has executed N steps where will it most likely be found? (ii) What is the probability $W(N, m)$ that the particle – after executing a total of N steps – is located in a position x from the origin $x = 0$, i.e. corresponding to a *direct* walk from $x = 0$ to x that consists of m unidirectional steps , i.e. a displacement $x = \lambda m$, where $m \leq N$?

(i) The distribution function
To arrive at the position $x = \lambda m$ by walking a total of N steps the particle must necessarily – if $m < N$ – have made a certain number of steps N_+ along the positive direction of the x-axis and another number of steps N_- in the opposite direction in conformity with

$$N_+ - N_- = m, \qquad\qquad (2.5.151)$$

where the sequence of the mutually positive and negative steps can be established in a certain number of different ways depending upon the value of N and how large m is relatively to N. Moreover

$$N_+ + N_- = N. \qquad\qquad (2.5.152)$$

Adding the equations gives

$$N_+ = \frac{N + m}{2}, \qquad\qquad (2.5.153)$$

---

* Karl Pearson (1857–1936) was an English mathematician and statistician who, among other things, introduced the χ^2-test and wrote the still-readable *The Grammar of Science*.

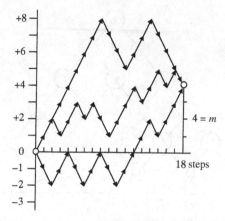

Fig. 2.29. Illustration of single steps in the progress of a "random walk". The figure shows three different ways one ends in the position corresponding to a direct walk of four unidirected successive steps by executing a total of 18 steps. This position can be reached by a total of 31 824 different ways.

and subtracting gives

$$N_- = \frac{N - m}{2}. \qquad (2.5.154)$$

As N_+ and N_- are both integers it follows that m is even/odd as N is even/odd. Further, m can only change in steps of 2, as

$$m = 2N_+ - N = \begin{cases} \dots -5, -3, -1, \quad 1, \ 3, \ 5 \dots \quad \text{for } N \text{ odd} \\ \dots -6, -4, -2, \ 0, \ 2, \ 4, \ 6 \dots \quad \text{for } N \text{ even.} \end{cases}$$

$$(2.5.155)$$

Before we attempt to work out a general expression for the probability $W(N, m)$, it may be instructive to consider the ways a random walk may be established from a given position when the values of N and m are small. In Fig. 2.29 are shown three different ways (out of a total of 31 824) in which the position $m = 4$ corresponding to four direct unidirected steps is reached by the execution of a total of 18 steps.

One way of managing the counting of the different patterns in a random is shown in Fig. 2.30. For $N = 1$ the particle can move to either left or right and thereby occupy the position $m = -1$ to $m = 1$. For $N = 2$ the particle manages to occupy the position $m = +2$ by making two steps pointing forwards (positive steps). Likewise the position $m = -2$ is occupied by two successive negative

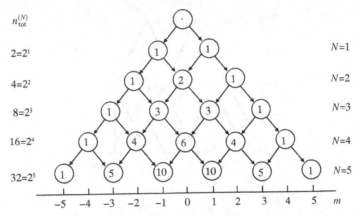

Fig. 2.30. Illustration of the feasibility of occupying the accessible positions by a "random walk" process. Abscissa: number of unidirected unidirectional steps m from origin. Right ordinate: total number of steps N. The numbers in the circles indicate the number of paths that lead to a given position m when executing in each path a total of N steps. Left ordinate: number of different paths $\mathcal{N}(n)$ that lead to all the accessible positions for given value of N. For further explanation see text.

steps $m = -2$. On the other hand the position $m = 0$ is attainable in two ways (as indicated by (2) in position $m = 0$): either one step *backwards* from position at $m = 1$ or by one step pointing forwards from the position at $m = -1$. The accessible positions for $N = 3$ are $-3, -1, 1, 3$. Here only one way, namely three successive steps pointing forwards, will lead to $m = 3$. By contrast the position $m = 1$ is accessible in three different ways (indicated by (3) in position $m = 1$): either by one step backwards from position $m = 2$, which was reached by two successive positive steps, or by one forward-pointing step from position $m = 0$. As shown above this position is accessible in two different ways. The general principle for counting the number of different paths that lead to a given position is as follows (see also Fig. 2.28): the number of paths to a position m that can be reached by N steps is equal to the sum of the ways by which the neighboring position $m - 1$ and $m + 1$ can be reached by a total of $N - 1$ steps. Inside the circles of the diagram in Fig. 2.30 are written the number of ways to reach the different positions for $N = 1, 2, 3, 4, 5$. For instance, for $N = 5$ the position at $m = +1$ is reached by six different paths from the preceding $(N = 4)$ neighboring position $m = 0$ and by four ways from the other neighbor. Thus, the total number of different paths leading to the position $+1$ in five steps is $6 + 4 = 10$. The numbers to the left in the diagram show the *total number* of different paths $\mathcal{N}(n)$ that will lead to the *totality* of accessible positions compatible with a given value of N. The probability for reaching a

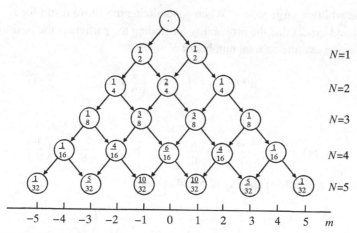

Fig. 2.31. Illustration of the occupation probabilities for the accessible positions m in a "random walk" process corresponding to the total number of steps $N = 1, 2, 3, 4, 5$. For further explanation see text.

given position m is therefore

$$W(N, m) = \frac{\text{Number of different paths to position } m}{\text{Number of different paths to all positions}}.$$

In Fig. 2.31 the occupation numbers (the values of the denominator in the above fraction) are replaced by the matching probabilities for reaching a given position when the total number of steps are $N = 1, 2, 3, 4, 5$. One observes that, notwithstanding the value of N, the probability is always largest for occupation positions located around the origin, whereas the more distant positions, which require a sequence of unidirectional steps, have the smaller probability. Finally, the probabilities are distributed symmetrically about the origin $x = 0$. Looking over the occupation probabilities corresponding to, for example, $N = 5$ we observe that each probability contains a constant factor $(\frac{1}{2})^N$. This factor is multiplied by a number that is identical to the coefficients of t in the binomial series* $(1 + t)^5$, their notation usually being

$$1, \quad \binom{5}{1}, \quad \binom{5}{2}, \quad \binom{5}{3}, \quad \binom{5}{4}, \quad 1,$$

where the terms in this sequence correspond to the positions at $m \equiv -5, -3, -1$, 1, 3, 5. The lower figure in each combination is equal to the number of *positive* steps, $N_+ = (N + m)/2$, that lead to the position in question. There is nothing to be said against an extension of this regularity to when N increases to

---

* Readers who remember Pascal's triangle from their college algebra might have smelled a rat when looking at the scheme in Fig. 2.30.

assume arbitrary large values. When generalizing the above result for $N = 5$, one should expect that the probability for finding the particle at the position m after having executed a total number of N steps is

$$W(N, m) = \binom{N}{N_+} \cdot \left(\frac{1}{2}\right)^N ,$$

or*

$$W(N, m) = \frac{N!}{N_+! \cdot (N - N+)!} \cdot \left(\frac{1}{2}\right)^N = \frac{N!}{N_+! \cdot (N_-)!} \cdot \left(\frac{1}{2}\right)^N .$$

Inserting Eq. (2.5.153) and Eq. (2.5.154) gives

$$W(N, m) = \frac{N!}{\left(\frac{N + m}{2}\right)! \cdot \left(\frac{N - m}{2}\right)!} \cdot \left(\frac{1}{2}\right)^N , \qquad (2.5.156)$$

which is the required expression for the probability of reaching the position m after performing a total of N steps by means of a mechanism based on "random walk" in one dimension.

Some readers may perhaps argue that this equation was obtained by cheating. But in a review article Smoluchowski (1916) himself used a similar argument based on induction. However, here we offer an alternative: the probability for a single step irrespective of direction is $\frac{1}{2}$. The probability of any possible sequence consisting of a total of N steps is then $(\frac{1}{2})^N$. But it is not the probability for *one* distinct sequence of steps leading to the position m that is required, but the probability that, after executing a total of N steps, the particle occupies the position m without regard to the particular sequence of steps that led to the position in question. This probability is equal to the probability for a distinct sequence of steps – *the a priori probability* – that is $(\frac{1}{2})^N$, times the *total number* of different paths that *all* lead to m. The last number is equal to the number of ways of mixing N_+ positive steps and N_- negative steps out of a total number of steps amounting to $N = N_+ + N_-$. This number is

$$\frac{N!}{N_+! \cdot N_-!} .$$

The underlying rationale is: the total ways of making N steps is $N!$. But this figure is too big as the given configuration results from a path consisting of N_+

---

* The number of different ways – the combination $_nC_r$ – to take r things from a greater sample n of objects is

$$C(n, r) \equiv {_nC_r} \equiv \binom{r}{n} = \frac{n(n - 1)(n - 2) \cdots (n - r + 1)}{r(r - 1) \cdots (1)} = \frac{n!}{r! \cdot (n - r)!} .$$

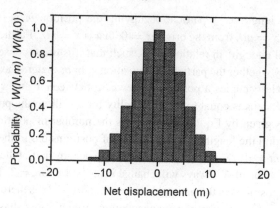

Fig. 2.32. The probability for a net displacement of m steps in a one-dimensional random walk consisting of a total of $N = 20$ steps.

positive steps and N_- negative steps, where each of these can be established in $N_+!$ and $N_-!$ different ways. And each may make a path leading to the position m. Thus, the number of different ways in which the particle will walk to the position m is then $N!/(N_+! \cdot N_-!)$, stated above. The probability of reaching the position m after a total of N steps is then equal to this number times the probability for a distinct sequence that has led to m (the *a priori* probability). Compare Eq. (2.5.156).

Figure 2.32 shows an example of probabilities calculated from Eq. (2.5.155). It appears again that around the *origin* the occupation probability is highest and that the probabilities decrease symmetrically to both sides. With a view to the treatment following later, it is convenient to replace the expression for $W(N, m)$ in Eq. (2.5.156), which gives discrete values corresponding to the integers N and m, with a continuous function. For large values of n, Stirling's approximation holds:

$$n! = \sqrt{2\pi n} \left(\frac{n}{e} \right)^n,$$

where $e = 2.71828\ldots$. Applying this formula to Eq. (2.5.156) gives, after some algebraic manipulation*,

$$W(N, m) = \sqrt{\frac{2}{\pi N}} \, e^{-m^2/2N}, \tag{2.5.157}$$

which is valid with great precision when $N \gg 1$ and the ratio m/N is small. It will now be shown that this solution of the one-dimensional "random walk" problem gives the same density of probability as that of Eq. (2.5.156): let each

* See Appendix C.

step have the length λ. A particle occupying the position m is subject to a displacement $x = m\lambda$ from the origin $x = 0$, that is $x = m\lambda$. Further, let $dm = dx/\lambda$ be small enough* in relation to N, such that it is of no consequence for the probability whether the particle is at position m or $m + dm$. The probability that the particle occupies a position somewhere between x and $x + dx$ after performing N steps is equal to the probability for a walk to the position $m = x/\lambda$, which is given by Eq. (2.5.157), times the number of positions that are accessible within the length dx. The number of positions dm in the interval of length dx is $dm = dx/\lambda$. But the particle can only occupy half of these places as m for a given value of N always will change in steps† of $\Delta m = 2$. The number of accessible positions in the interval between x and $x + dx$ is therefore $dx/2\lambda$. As all these positions have the same occupation probability we have

$$W(N, m) = \sqrt{\frac{2}{\pi N}}\, e^{-m^2/2N}.$$

Thus, the probability dP that the particle will be located in the region between x and $x + dx$ is

$$dP = W(N, m)\left(\frac{dx}{2\lambda}\right) = \sqrt{\frac{2}{\pi N}}\, e^{-m^2/2N} \cdot \frac{dx}{2\lambda}.$$

Let the time to execute the total of N steps be t. Hence, the time, τ, between two succeeding steps is $\tau = t/N$. Furthermore, the length of walking x, corresponding to the position m, is $x = m\lambda$. Thus, the above equation can be rewritten as follows

$$dP = \frac{1}{2\sqrt{\pi(\lambda^2/2\tau)t}}\, e^{-x^2/4(\lambda^2/2\tau)t}\, dx.$$

Putting

$$D = \frac{\lambda^2}{2\tau},$$

the expression assumes the form

$$dP = \varphi(x, t)\, dx = \frac{1}{2\sqrt{\pi Dt}}\, e^{-x^2/4Dt}\, dx, \qquad (2.5.158)$$

where $\varphi(x, t)$ is the probability density for the displacement x at the time t and the mechanism for the displacement is a one-dimensional "random walk".

---

* But still an integer.
† See Eq. (2.5.155).

It should be noted that this expression is identical to the probability density $\varphi(x, t)$ that we derived using the macroscopic theory for the diffusion process.

We now imagine that a large collection of particles of $\mathcal{N}$ per square meter *simultaneously* begin their "random walk" from the plane at $x = 0$ and at time $t = 0$. One expects then that inside a prism with end-surfaces of area A at the positions x and $x + dx$ the number of particles dn contained are

$$dn = \mathcal{N}\varphi(x, t)\, dx \cdot A,$$

and the particle concentration in the slab is

$$C(x, t) = \frac{dn}{A\, dx} = \frac{\mathcal{N}}{2\sqrt{\pi Dt}} e^{-x^2/4Dt}. \qquad (2.5.159)$$

Comparing this expression with Eq. (2.5.125) we see that the solution of the diffusion equation for the spread of matter in space and time from an instantaneous source is identical to the solution obtained by making use of considerations based on the "random walk", provided that one identifies the ratio $\lambda^2/2\tau$ between the square of length of the single step and twice the duration of the step with the coefficient of diffusion D belonging to the macroscopic problem. In both cases *mass conservation* was an absolute condition. But in the first the phenomenological mechanism behind the diffusion of matter is a flux that was described by Fick's law

$$J = -D\frac{\partial C}{\partial x}.$$

In the second case the same diffusion process is explained solely by the assumption that all the particles incessantly execute chaotic movements to and fro with the *same* probability for each step and where the probability for the direction of a given step is independent of the direction of the previous step. This suggests the line of thought that the mechanism underlying the flux by way of diffusion traces back to the idea that all the particles execute Brownian motion. From this point of view we lose the need to assume that the concentration gradient establishes a "driving force", as was originally assumed by Nernst and also later by Planck.

(ii) The mean displacement

The number of particles located in the slab between x and $x + dx$ with end-surfaces of unit area is $\mathcal{N}\varphi(x, t)\, dx$. The sum of the displacements of these particles is $x \cdot \mathcal{N}\varphi(x, t)\, dx$. The sum of the displacements that all the particles have executed at time t is

$$L = \mathcal{N} \int_{-\infty}^{\infty} x\, \varphi(x, t)\, dx.$$

From this we obtain the mean value $\langle x \rangle$ of all the displacements as $L/\mathcal{N}$, or

$$\langle x \rangle = \int_{-\infty}^{\infty} x\, \varphi(x, t)\, dx = 0, \qquad (2.5.160)$$

as $\varphi(x, t)$ is symmetrical with respect to $x = 0$. Naturally the above result is rather obvious, as a step pointing forwards and a step pointing backwards have the same probability, namely $\frac{1}{2}$. However, as the probability is *finite* and non-vanishing for finding the particle somewhere in the region $x \neq 0$ at time t it is useful to consider two different types of mean value.

(iii) The mean displacement in one direction
At any time t *half* of the particles will occupy the space $x \geq 0$ with the remainder in $x \leq 0$. The mean displacement $\bar{x}$ for each of the two swarms of particles, i.e. the mean displacement from the origin in *either* the positive *or* the negative direction, is

$$\bar{x} = \frac{\mathcal{N}}{\mathcal{N}/2} \int_{0}^{\infty} x\, \varphi(x, t)\, dx = \frac{1}{\sqrt{\pi D t}} \int_{0}^{\infty} x\, e^{-x^2/4Dt}\, dx.$$

Changing the variable of integration to $u = x^2/4Dt = x^2/\alpha$ (i.e. $du/dx = 2x/\alpha$ and $dx/du = \alpha/2x$) gives

$$\bar{x} = \frac{1}{\sqrt{\pi D t}} \int_{0}^{\infty} x\, e^{-u^2} \left(\frac{dx}{du} \right) du = \frac{\alpha}{2\sqrt{\pi D t}} \int_{0}^{\infty} e^{-u}\, du,$$

The integral on the right-hand side is $[-e^{-u}]_0^\infty = 1$, hence

$$\bar{x} = 2\sqrt{\frac{Dt}{\pi}}. \qquad (2.5.161)$$

(iv) The root mean square displacement (the Einstein–Smoluchowski relation)
The measure most frequently used to describe the displacement of a particle or a swarm of particles to the time t is the mean value of the *square* of the displacements. This is

$$\langle x^2 \rangle = \int_{-\infty}^{\infty} x^2\, \varphi(x, t)\, dx = 2 \int_{0}^{\infty} x^2\, \varphi(x, t)\, dx = \frac{1}{\sqrt{\pi D t}} \int_{0}^{\infty} x^2\, e^{-\beta x^2}\, dx,$$

where $\beta = 1/4Dt$. Since the value of the integral is a function of the parameter β, we make use of the following useful trick: the integrand can be written as

$$\int_0^\infty x^2 e^{-\beta x^2} \, dx = \int_0^\infty (-1) \frac{\partial}{\partial \beta} (e^{-\beta x^2}) \, dx = -\frac{d}{d\beta} \left(\int_0^\infty e^{-\beta x^2} \, dx \right),$$

by interchanging the order of differentiation and integration. We have*

$$\int_0^\infty e^{-\beta x^2} \, dx = \tfrac{1}{2} \sqrt{\pi/\beta} = \tfrac{1}{2} \sqrt{\pi} \beta^{-\frac{1}{2}},$$

and therefore

$$-\frac{d}{d\beta} \left(\int_0^\infty e^{-\beta x^2} \, dx \right) = -\tfrac{1}{2} \sqrt{\pi} \frac{d}{d\beta} (\beta^{-\frac{1}{2}}) = \tfrac{1}{4} \sqrt{\pi} \beta^{-\frac{3}{2}} = \tfrac{1}{4} \sqrt{\pi} (1/\beta)^{\frac{3}{2}}.$$

Inserting $\beta = 1/4Dt$, and collecting the pieces gives

$$\langle x^2 \rangle = \tfrac{1}{4} \sqrt{\pi} \sqrt{(4Dt)^3} / \sqrt{\pi Dt} = \tfrac{4}{4} \sqrt{4(Dt)^2} = 2Dt.$$

Alternatively this integral is evaluated by making use of the standard technique of integration by parts[†] by considering the integrand

$$I = \int_{-\infty}^\infty x^2 e^{-\beta x^2} dx = \int_{-\infty}^\infty x \cdot x \, e^{-\beta x^2} dx,$$

as the product of two functions

$$u = x \quad \text{and} \quad \frac{dv}{dx} = x e^{-\beta x^2}.$$

Consequently $du/dx = 1$ and $v = -e^{-\beta x^2}/(2\beta)$. Using the rule about integration by parts

$$I_1 = \int u \, dv = uv - \int v \, du = -\frac{x}{2\beta} e^{-\beta x^2} + \frac{1}{2\beta} \int e^{-\beta x^2} dx,$$

gives

$$I = \left[-\frac{x}{2\beta} e^{-\beta x^2} \right]_{-\infty}^\infty + \frac{1}{2\beta} \int_{-\infty}^\infty e^{-\beta x^2} \, dx = 0 + \frac{1}{2\beta} \sqrt{\frac{\pi}{\beta}} = 4\sqrt{\pi} (Dt)^{3/2}.$$

It follows then that

$$\langle x^2 \rangle = \frac{1}{2\sqrt{\pi Dt}} \times I = \frac{1}{2\sqrt{\pi Dt}} \times 4\sqrt{\pi} (Dt)^{3/2} = 2Dt.$$

---

* See Appendix A, Eq. (A.6).
[†] See Chapter 1, Section 1.72, Eq. (1.7.5).

This result is often written as

$$D = \frac{\langle x^2 \rangle}{2t}, \qquad (2.5.162)$$

and called the Einstein–Smoluchowski equation. It interconnects the mean value of the square of the displacements in the time t – that reflects the "random walk" performed by the particles – with the coefficient of diffusion, which is the basis parameter in the macroscopic description of diffusion*. Equation (2.5.161) or Eq. (2.5.162) are of practical use as they allow us, in a simple way, to estimate how far a swarm of particles is expected to move via a "random walk" in a given interval of time, provided that the value of the coefficient of diffusion D for the substance in question is known. The fraction Z of the particles which at time t is located inside the region $-\sqrt{2Dt} \le x \le \sqrt{2Dt}$ is

$$Z = \frac{1}{2\sqrt{\pi Dt}} \int_{-\sqrt{2Dt}}^{\sqrt{2Dt}} e^{-x^2/4Dt}\, dx = \frac{1}{\sqrt{\pi Dt}} \int_{0}^{\sqrt{2Dt}} e^{-x^2/4Dt}\, dx,$$

as the integrand is symmetrical around $x = 0$. We change the variable of integration to $u = x/2\sqrt{Dt}$ (i.e. $du/dx = 1/2\sqrt{Dt}$ and $dx/du = 2\sqrt{Dt}$) and to the limits $u = 0$ for $x = 0$ and for $x = \sqrt{2Dt}, u \to \sqrt{2Dt}/2\sqrt{Dt} = 1/\sqrt{2}$. This gives

$$Z = \frac{2\sqrt{Dt}}{\sqrt{\pi Dt}} \int_{0}^{1/\sqrt{2}} e^{-u^2}\, du = \frac{2}{\sqrt{\pi}} \int_{0}^{1/\sqrt{2}} e^{-u^2}\, du,$$

or rewritten as

$$Z = \text{Erf}\{1/\sqrt{2}\} = \text{Erf}\{0.7071\} = 0.683^{\dagger}.$$

Thus slightly more than *two thirds* of the particles will at time t occupy the space between the planes at

$$x = -\sqrt{2Dt} \quad \text{and} \quad x = \sqrt{2Dt},$$

while the remaining particles will have moved further away from their initial position.

To compare the reciprocal magnitudes of the mean displacement $\bar{x}$ in one direction (Eq. (2.5.161)) and the mean of the square of the displacements $\langle x^2 \rangle$ we take the ratio

$$\frac{(\bar{x})^2}{\langle x^2 \rangle} = \frac{4Dt/\pi}{2Dt} = \frac{2}{\pi},$$

---

* In Appendix D an example is given of the use of this equation in the verification of the theory of Brownian motion.
† Obtained either by looking up in a table or by using a PC-program for Erfc(x).

or

$$\bar{x} = 0.8 \sqrt{\langle x^2 \rangle}. \tag{2.5.163}$$

Thus, the two measures of displacement at time t are of almost equal size. More important, however, is the fact that these (net) displacements are *not* directly proportional to the time t but to the *square root* of t^*. The significance of this appears from the following example. A swarm of particles with a coefficient of diffusion $D = 10^{-6}$ cm^2 s^{-1}, is released from the position $x = 0$ at time $t = 0$. Putting $\sqrt{\langle x^2 \rangle} = 1$ cm we obtain

$$t = \frac{\langle x^2 \rangle}{2D} = \frac{1}{2} \frac{1}{10^{-6}} = 5 \times 10^5 \text{ s}$$

as the time required until 68.3% of the particles are in the range between $x = -1$ cm and $x = 1$ cm. However, putting $\sqrt{\langle x^2 \rangle} = 1$ μm, the analogous time is $t = 5$ ms. In other words, a mass transfer that is the result of Brownian motion – or diffusion, respectively – will only take place at a substantial speed if the distances concerned are small.

In this derivation it was assumed that the particles were executing their random walk in one dimension, i.e. along a straight line. We shall now extend the treatment to include random walks in two and three dimensions.

(α) Two-dimensional random walk We now consider a two-dimensional random walk process, i.e. the particle moves in a plane having as its system of reference the rectangular system with axes x and y. Consider first the situation where the particle is located at time $t = 0$ in the position $x = 0$ and moves solely by a one-dimensional process along the x-axis and is at time t located at a position between x and $x + dx$. For this the probability is, according to Eq. (2.5.157),

$$dP(x, t) = \frac{1}{(4\pi Dt)^{1/2}} e^{-x^2/4Dt} dx. \tag{2.5.164}$$

At the other extreme the particle could have executed a one-dimensional random walk from the position $x = 0$ in the y direction perpendicular to the x-axis and at time t become localized between y and $y + dy$. For this the probability is

$$dP(y, t) = \frac{1}{(4\pi Dt)^{1/2}} e^{-y^2/4Dt} dy. \tag{2.5.165}$$

---

* See also Appendix D.

2. *Migration and diffusion*

We now assume that the particle executes a two-dimensional random walk *both* in the x-direction and in the y-direction and at time t is located in the region between the positions x, y and $x + dx, y + dy$. The probability for this movement can be calculated from Eq. (2.5.161) and Eq. (2.5.162) since the individual displacement δr for position (x, y) has both an x-component δx and a y-component δy, which separately make up the elements in a random walk along the x- and y-axis. According to the law of compound probabilities for two independent probabilities* we have

$$dP((x, y), t) = dP(x, t) \times dP(y, t)$$

$$= \frac{1}{(4\pi Dt)} e^{-(x^2+y^2)/4Dt} \, dx \, dy. \qquad (2.5.166)$$

This expression gives the probability that at time t the particle is located within an infinitesimal area $dx \, dy$, around the position (x, y), as specified by the two rectangular coordinates. Moreover, it appears that all infinitesimal target areas $dx \, dy$ having the same distance $r = \sqrt{x^2 + y^2}$ from the point of origin $(0,0)$ have the *same* probability irrespective of the direction of the displacement. This implies that any final position that is located within the circular rings of radius r and $r + dr$ has the same probability. To calculate the probability according to this condition we transform the equation above to polar coordinates r, θ, where $r^2 = x^2 + y^2$, and $\theta = \arctan(y/x)$ is the angle between radius vector r and the x-axis, since $x = r \cos\theta$, $y = r \sin\theta$. The surface element in these coordinates is $r \, d\theta \, dr^\dagger$. Hence, Eq. (2.5.165) takes the form

$$P((r, \theta), t) = \frac{1}{(4\pi Dt)} e^{-r^2/4Dt} r \, d\theta \, dr. \qquad (2.5.167)$$

Integration with respect to θ from 0 to 2π yields

$$dP(r, t) = \frac{1}{2Dt} e^{-r^2/4Dt} r \, dr, \qquad (2.5.168)$$

which is the probability that after time t the particle is located somewhere in a circular band in the range between r and $r + dr$.

We search again for the mean value of the square of the displacements over the whole accessible area, that is

$$\langle r^2 \rangle = \int_0^\infty r^2 \, dP.$$

---

* The probability of the simultaneous occurrence of *both* an event A with probability $P(A)$ *and* an event B with probability $P(B)$ is equal to the product $P(A)P(B)$ of the two probabilities.
† The length of the arc between the coordinates θ and $\theta + d\theta$ along the circle of radius r is $r \, d\theta$.

Insertion of Eq. (2.5.167) gives

$$\langle r^2 \rangle = \frac{1}{2Dt} \int_0^\infty r^3 \, e^{-r^2/4Dt} dr. \tag{2.5.169}$$

Using the earlier procedure of putting $\beta = 1/4Dt$ gives

$$\langle r^2 \rangle = 2\beta \int_0^\infty r^3 \, e^{-\beta r^2} dr.$$

Making use of the identity

$$-r^2 e^{-\beta x^2} = \frac{\partial}{\partial \beta} \{ e^{-\beta r^2} \},$$

the above expression can be written as

$$\langle r^2 \rangle = -2\beta \int_0^\infty r \frac{\partial}{\partial \beta} \{ e^{-\beta r^2} \} dr$$

or, by changing the order of integration and differentiation,

$$\langle r^2 \rangle = -2\beta \frac{\partial}{\partial \beta} \left\{ \int_0^\infty r e^{-\beta r^2} dr \right\}.$$

Putting $u = \beta r^2$ (i.e. $du/dr = 2\beta r$ and $dr/du = 1/2\beta r$) gives

$$\langle r^2 \rangle = -2\beta \frac{\partial}{\partial \beta} \left\{ \int_0^\infty r e^{-u} \left(\frac{dr}{du} \right) du \right\} = -2\beta \frac{\partial}{\partial \beta} \left\{ \int_0^\infty \frac{1}{2\beta} e^{-u} du \right\}$$

$$= -2\beta \frac{\partial}{\partial \beta} \left\{ \frac{1}{2\beta} \right\} = -2\beta \left(-\frac{1}{2\beta^2} \right) = \frac{1}{\beta} = 4Dt. \tag{2.5.170}$$

Thus, the two-dimensional Einstein–Smoluchowski equation takes the form

$$D = \frac{\langle r^2 \rangle}{4t}. \tag{2.5.171}$$

(*β*) *Three-dimensional random walk* We consider next a particle executing a random walk in three dimensions. Let $(0,0,0)$ be the position of the particle at time $t = 0$. The probability that at time t the particle has made a displacement $r = (x^2 + y^2 + z^2)^{1/2}$ from the origin $(0,0,0)$ and is located somewhere in the region between x and $x + dx$, y and $y + dy$ and z and $z + dz$, i.e. in the volume element $dv = dx\,dy\,dz$, is accordingly

$$dP((x, y, z), t) = \frac{1}{(4\pi Dt)^{1/2}}\,e^{-x^2/4Dt}dx \times \frac{1}{(4\pi Dt)^{1/2}}\,e^{-y^2/4Dt}dy$$

$$\times \frac{1}{(4\pi Dt)^{1/2}}\,e^{-z^2/4Dt}dz$$

$$= \frac{1}{(4\pi Dt)^{3/2}}\,e^{-(x^2+y^2+z^2)/4Dt}dx\,dy\,dz. \qquad (2.5.172)$$

Replacing $x^2 + y^2 + z^2 = r^2$ in the exponent yields

$$P_{dv}(r, t) = \frac{1}{(4\pi Dt)^{3/2}}\,e^{-r^2/4Dt}dv, \qquad (2.5.173)$$

which denotes the probability that at time t the particle has moved by a random walk to the volume element dv that is located on any position on the surface on the circle of radius r. The volume element dv in the expression above is replaced by $dv = 4\pi r^2 dr$, which is the volume of the spherical shell with radius r and $r + dr$. Hence the probability that at time t the particle is located somewhere in the space between r and $r + dr$ but *irrespective* of the direction from $(0,0,0)$ is

$$dP(r, t) = \frac{1}{(4\pi Dt)^{3/2}}\,4\pi r^2\,e^{-r^2/4Dt}dr. \qquad (2.5.174)$$

We shall again calculate the mean value of the squared displacements $\langle r^2 \rangle$. This is

$$\langle r^2 \rangle = \int_0^\infty r^2\,dP$$

$$= \frac{4\pi}{(4\pi Dt)^{3/2}} \int_0^\infty r^4\,e^{-r^2/4Dt}dr. \qquad (2.5.175)$$

Putting $\beta = (4Dt)^{-1}$ again gives

$$\langle r^2 \rangle = \frac{4\beta^{3/2}}{\sqrt{\pi}} \int_0^\infty r^4\,e^{-\beta r^2}dr = \frac{4\beta^{3/2}}{\sqrt{\pi}}\,I. \qquad (2.5.176)$$

The identity

$$r^2\,e^{-\beta r^2} = -\frac{\partial}{\partial \beta}\{e^{-\beta r^2}\}$$

is again used for the calculation of the integral. We have

$$I = -\int_0^\infty r^2\,\frac{\partial}{\partial \beta}\{e^{-\beta r^2}\}\,dr = -\frac{\partial}{\partial \beta}\left\{\int_0^\infty r^2\,e^{-\beta r^2}dr\right\}, \qquad (2.5.177)$$

where the order of integration and differentiation is reversed. We already have

$$\int_0^\infty r^2 e^{-\beta r^2} dr = \tfrac{1}{4} \sqrt{\pi} \, \beta^{-3/2},$$

which when inserted into Eq. (2.5.176) gives

$$I = -\frac{\partial}{\partial \beta} \{ \tfrac{1}{4} \sqrt{\pi} \, \beta^{-3/2} \} = \tfrac{1}{4} \sqrt{\pi} \, \frac{3}{2} \beta^{-5/2}.$$

Thus the final form of Eq. (2.5.175) becomes

$$\langle r^2 \rangle = \frac{4\beta^{3/2}}{\sqrt{\pi}} \tfrac{1}{4} \sqrt{\pi} \, \frac{3}{2} \beta^{-5/2} = \tfrac{3}{2} \beta^{-1} = (3/2)4Dt = 6Dt, \quad (2.5.178)$$

since $\beta = (4Dt)^{-1}$.

Thus, the three-dimensional Einstein–Smoluchowski equation for the radial symmetric random walk becomes

$$D = \frac{\langle r^2 \rangle}{6t}. \qquad (2.5.179)$$

2.5.6.4 Random walk and Fick's law

We consider a swarm of particles in solution. Each particle is assumed to execute a "random walk" due to thermal agitation. We shall now demonstrate that in a non-uniform distribution the Brownian motion alone will cause a net mass transfer that follows Fick's law of diffusion. We assume that the elementary displacements δx_i of each particle – and with that also the resulting displacement x in the time t – comprise a *spectrum* of displacements $-\infty \le x \le \infty$ having the probability density $\Phi(x, t)$ for a displacement x in time t. Thus, $dP = \Phi(x, t)dx$ is the probability that a particle at time t is located between x and $x + dx$. Naturally, but not unreasonably, restrictions on the properties of $\Phi(x, t)$ will be required.

(i) Einstein's simplified treatment

Following a call from the Dutch chemist R. Lorentz in 1908, Einstein gave a simplified version of his theory of Brownian motion. We shall reproduce that part of the argument that led to Fick's law and to the connection between the coefficient of diffusion of the substance and the "random walk" of the particles in solution. Naturally Einstein's presentation was based upon the results he had obtained previously. But an unusual talent is required to find an argument that at the same time is so simple and convincing. Einstein's starting point was to simplify the behavior of the swarm of particles by reducing the total spectrum of displacements in a given time to *two single* displacements $\pm \bar{x}$, that are the

2. Migration and diffusion

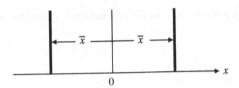

Fig. 2.33. Einstein's simplified replacement of a spectrum of molecular displacements with two single displacements $\pm\bar{x}$.

mean displacements at the time t to either side from the origin. In Section 2.5.6.3 (ii) this quantity was defined and calculated from the "random walk" theory. This simplification means that all the particles in the swarm have been assigned the *same* displacement $\bar{x}$ in the time t, as indicated in Fig. 2.33. We consider a swarm of particles with the concentration decreasing approximately linearly along the x-axis. A plane with surface area A is placed at right angles to the direction of the flux (the hatched surface in Fig. 2.34). This surface constitutes the common terminal surface of two adjacent prisms both having $\bar{x}$ – the displacement in time t in one direction – as the length of their sides. The volume element to the left for the surface A is named $V^{(1)}$, and as mean concentration we take the concentration $N(x)$ in the center of the prism at the position x. As the measure of concentration we use number of particles per unit volume. Similarly $V^{(2)}$ is the volume of the prism to the right of the surface A. Using the center position of the left prism as a reference this prism has its center at $x + \bar{x}$ and a mean concentration of $N(x + \bar{x})$. Owing to the "random walk" the traffic of particles at the hatched area A consists of a certain number of particles passing through the area A in the direction from $V^{(1)} \rightarrow V^{(2)}$ and of another number that in the same time pass through in the opposite direction* $V^{(2)} \rightarrow V^{(1)}$. By choosing $\bar{x}$ in time t as the length of each prism Einstein managed to establish that among the number of particles initially occupying the volume element $V^{(1)}$ *one half* of this number had the chance to pass the plane of the hatched area A within the time t. The number of particles that were initially present in the volume element $V^{(1)}$ is $N(x)A\bar{x}$. After the time t has elapsed one half of these particles have moved out of the volume element in the direction towards the left while the remaining half has moved towards the right and by this process have passed through the hatched area A. The transfer $M^{1\rightarrow2}$ through the plane in *time t* in the direction $1 \rightarrow 2$ is

$$M^{1\rightarrow2} = \tfrac{1}{2}N(x)A\bar{x}.$$

---

* For certain applications it is convenient to use the term **unidirectional fluxes** as the designation of the collection of particles that move simultaneously but in opposite directions through a given plane.

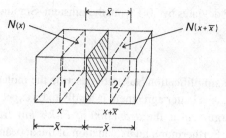

Fig. 2.34. Illustration of Einstein's simplified treatment of a process of diffusion based on a "random walk".

Similarly, the transfer $M^{2\to1}$ through the plane A in the direction $2 \to 1$ is

$$M^{2\to1} = \tfrac{1}{2}N(x+\bar{x})A\,\bar{x}.$$

Hence, the *net movement* $M^{(\text{tot})}$ in the direction $1 \to 2$ through the plane A is

$$M^{(\text{tot})} = M^{1\to2} - M^{2\to1}$$

$$= \tfrac{1}{2}N(x)A\,\bar{x} - \tfrac{1}{2}N(x+\bar{x})A\,\bar{x}$$

$$= -\tfrac{1}{2}\bar{x}^2\, A\, \frac{N(x+\bar{x}) - N(x)}{\bar{x}}.$$

The flux J in the direction $1 \to 2$ is $J = M^{(\text{tot})}/At$. Thus

$$J = -\frac{1}{2t}\bar{x}^2\, \frac{N(x+\bar{x}) - N(x)}{\bar{x}}.$$

If the concentration profile is approximately rectilinear through the two prisms and only short times t are considered, i.e. $\bar{x} \ll 1$, the above expression tends to*

$$J = -\frac{1}{2}\frac{\bar{x}^2}{t}\frac{dN}{dx}. \qquad (2.5.180)$$

The factor $\bar{x}^2/2t$ is a constant, which Einstein demonstrates directly by squaring the sums of n displacements of length Δ for the time t[†]. The above expression then becomes identical with Fick's law by putting

$$D = \frac{1}{2}\frac{\bar{x}^2}{t}. \qquad (2.5.181)$$

---

* Compare Chapter 1, Section 1.2.3, Eq. (1.2.4).
[†] See also Appendix D.

This value for D deviates by 30% from the Einstein–Smoluchowski equation

$$D = \frac{\langle x^2 \rangle}{2t},$$

but, in view of the simplifications taken to describe the molecular behavior of the diffusion process, a better agreement is hardly to be expected. The essence, however, of the argument is the *deduction of Fick's law* from the theory of Brownian motion. Furthermore, an explanation of the appearance of the diffusion flux has been advanced without the need for introducing specific undirected forces that drive the separate particles down along the declining concentration profile: all the particles are performing their Brownian motion with an equal probability for a step pointing forwards or backwards. In this way the particles will move chaotically to and fro, and net transport in *one direction* appears when the *density of particles* varies in a given direction. During a given time Δt a layer having the greater density of particles will *give off more* particles to the adjoining layer than it *receives* in the same time Δt, since the probabilities for the forward and backward steps are equal to $\frac{1}{2}$.

(ii) A more exact derivation of Fick's law

Some readers may argue that simplifying the spectrum of displacements to only two values $+\bar{x}$ and $-\bar{x}$ looks perhaps like a shot in the dark, and may therefore ask what kind of result could be expected if one – more realistically – assumed that there existed at each time a whole *spectrum* of displacements that were all accessible for the particle, although with a different probability. We shall now show what modifications will be the result of such an assumption[*].

We define the probability dP that in time τ a particle has taken a step of length lying between ξ and $\xi + d\xi$ as

$$dP = \Phi(\xi, \tau \,|\, 0, 0)\, d\xi, \qquad (2.5.182)$$

where

$$\Phi(\xi, \tau \,|\, 0, 0) \equiv \Phi(\xi, \tau) \qquad (2.5.183)$$

is the probability density that a particle at time $t = 0$ is in the position $x = 0$ to make a displacement of length ξ in the time τ. The function $\Phi(\xi, \tau)$ could be a Gaussian curve, but this is not essential for the argument that follows. Only the following properties for $\Phi(\xi, \tau)$ are required.

---

[*] The exposition that follows is based on: Le Claire, A.D. (1958): Random walk and drift in chemical diffusion. *Phil. Mag.*, **3**, 921.

(a) $\Phi(\xi, \tau)$ is symmetrical around $\xi = 0$, i.e.

$$\Phi(\xi, \tau) = \Phi(-\xi, \tau), \tag{2.5.184}$$

as the probability for a forward and a backward jump is the same.

(b) $\Phi(\xi, \tau) \to 0$, for $\xi \to \pm\infty$, the shortest displacements having probabilities that by far dominate over the larger ones. It is convenient to formulate this condition by stating that

$$\int_x^\infty \Phi(\xi, \tau)\, d\xi \tag{2.5.185}$$

decreases more strongly for $x \to \infty$ than the function x^{-n}, where n is a positive number, thus

$$\lim_{x \to \infty} \left[x^n \int_x^\infty \Phi(\xi, \tau)\, d\xi \right] = 0.$$

(c) The probability is 1 for finding the particle somewhere in $-\infty < \tau < \infty$, or

$$\int_{-\infty}^\infty \Phi(\xi, \tau)\, d\xi = 1, \tag{2.5.186}$$

as $\Phi(\xi, \tau)$ is normalized.

Figure 2.35 shows a one-dimensional concentration profile in units of the number of particles per unit volume. Let $C(x, t)$ be the concentration at time t. We consider a plane at the position x_0 and will calculate the net transfer of particles passing through a unit area in the time τ, that is assumed to be small enough so that the concentration profile remains practically unchanged, i.e.

$$C(x, t) \approx C(x, t + \tau).$$

One approach is the following.

Let x' represent a position coordinate *to the right* of the plane x_0, i.e. $x_0 \leq x' < \infty$. A particle that initially is located in the position $x \leq x_0$, i.e. to the *left* of the plane x_0, has the probability to have moved in time τ to the region between x' and $x' + dx'$, which is

$$dP = \Phi((x' - x), \tau)\, dx',$$

or

$$dP = \Phi(\xi, \tau)\, d\xi,$$

where $\xi = x' - x$ is the displacement in time τ. The probability that the particle at time τ will be found somewhere in the region $x_0 \leq x' < +\infty$, i.e. *to the right*

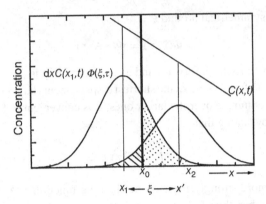

Fig. 2.35. Derivation of Fick's law assuming that the random walks of the dissolved particles are represented by a *spectrum* of displacements with probability density $\Phi(\xi, \tau)$ for a displacement ξ to time τ. Net flux through the plane at x_0 is the difference between the unidirectional flux in the direction left $\rightarrow$ right from the whole region to the left of x_0 and the unidirectional flux in the direction right $\rightarrow$ left from the whole region to the right of x_0. The figure illustrates the displacements to the time τ from the two layers of thicknesses dx_1 and dx_2, that are located in the positions x_1 and x_2 at either side of the plane at x_0 and initially ($\tau = 0$) containing the amounts $C(x_1)\,dx_1$ and $C(x_2)\,dx_2$. Net transfer through the plane at x_0 in the direction left $\rightarrow$ right is the difference between the amounts contained in the hatched and the dotted areas. For further explanation, see text.

of the plane x_0 is

$$P(\xi \geq x_0 - x) = \int_{x_0}^{\infty} \Phi((x' - x), \tau)\,dx' = \int_{x_0 - x}^{\infty} \Phi(\xi, \tau)\,d\xi, \quad (2.5.187)$$

as $\xi = x_0 - x$ for $x' = x_0$. $P(\xi \geq x_0 - x)$ is the probability that the particle, initially located at the position x, has passed the plane at x_0 in time $t \leq \tau$.

The amount of substance that initially (at time t) is located in the layer between x and $x + dx$ and surface area 1 m² is

$$C(x, t) \cdot 1 \cdot 1 \cdot dx.$$

We consider this as the strength of an instantaneous surface source at the position x at time t. Among these particles the number[*]

$$dx \cdot C(x, t)P(\xi \geq x_0 - x) = dx \cdot C(x, t) \cdot \int_{x_0 - x}^{\infty} \Phi(\xi, \tau)\,d\xi$$

will be located *to the right* of the place at x_0 at the time $t + \tau$. The total number of particles that in time τ have moved from the region $-\infty < x \leq x_0$ through

---

[*] As τ is assumed to be small enough that $C(x, t) \approx C(x, t + \tau)$.

an unit area at the plane x_0 to be found somewhere in the region $x \geq x_0$ is

$$\overrightarrow{M}_\tau = \int_{-\infty}^{x_0} dx \left(C(x, t) \int_{x_0 - x}^{\infty} \Phi(\xi, \tau) \, d\xi \right), \qquad (2.5.188)$$

where $\overrightarrow{M}_\tau$ is the unidirectional mass transfer in the direction *left* $\rightarrow$ *right* along the direction of the x-axis, passing through the plane at x_0 in time $t = \tau$.

We then apply the same reasoning to all the particles that initially occupy the region $x \geq x_0$, i.e. to the right of the plane at x_0. Let x' now represent a position in the region $-\infty < x' \leq 0$. The probability that a particle – initially located in the position $x \geq x_0$ – in the time t will be found between x' and $x' + dx'$ is

$$dP = \Phi((x' - x), \tau) \, dx' = \Phi(\xi, \tau) \, d\xi.$$

Hence, the probability that this particle will be found somewhere in the region $-\infty < x' \leq x$ is

$$P(\xi \leq x_0 - x) = \int_{-\infty}^{x_0 - x} \Phi(\xi, \tau) \, d\xi. \qquad (2.5.189)$$

Among the particles that are initially present between x and $x + dx$ with the concentration $C(x, t)$, the number

$$dx \cdot C(x, t) \int_{-\infty}^{x_0 - x} \Phi(\xi, \tau) \, d\xi$$

are located somewhere in the region $x \leq x_0$ at the time $t + \tau$. Among all the particles initially present in the region $x_0 \leq x < +\infty$, the number

$$\overleftarrow{M}_\tau = \int_{x_0}^{\infty} dx \left(C(x, t) \cdot \int_{-\infty}^{x_0 - x} \Phi(\xi, \tau) \, d\xi \right) \qquad (2.5.190)$$

have passed the plane at x_0 in the time τ in the direction *right* $\rightarrow$ *left* and end up somewhere in the region $-\infty < x \leq x_0$.

The net transfer M_τ of particles along the direction of the x-axis through a unit area at the plane at x_0 in the time τ is

$$M_\tau = \overrightarrow{M}_\tau - \overleftarrow{M}_\tau$$

$$= \int_{x_0}^{\infty} \left(C(x, t) \int_{x_0 - x}^{\infty} \Phi(\xi, \tau) \, d\xi \right) dx - \int_{x_0}^{\infty} \left(C(x, t) \int_{-\infty}^{x_0 - x} \Phi(\xi, \tau) \, d\xi \right) dx.$$

$$(2.5.191)$$

Thus, principally the problem is solved, since the net flux through the plane at x_0 is described by means of the concentration profile $C(x, t)$ at the time t and by the probability density $\Phi(\xi, \tau)$ for the displacement ξ after lapse of time of τ.

But the actual problem is to examine if this expression may assume the shape of Fick's law, i.e. whether M_τ can be described by means of the concentration gradient $\partial C/\partial x$ in the position $x = x_0$. To this end it is natural to expand $C(x, t)$ as a Taylor series around $x = x_0$

$$C(x, t) = C(x_0, t) + (x - x_0)\left(\frac{\partial C}{\partial x}\right)_{x_0} + \frac{(x - x_0)^2}{2}\left(\frac{\partial^2 C}{\partial x^2}\right)_{x_0}$$

$$+ \cdots \frac{(x - x_0)^n}{n!}\left(\frac{\partial^n C}{\partial x^n}\right)_{x_0}, \tag{2.5.192}$$

and insert this expression for $C(x, t)$ in Eq. (2.5.174). This gives*

$$M_\tau(x_0) = -\frac{1}{2}\left(\frac{\partial C}{\partial x}\right)_{x_0}\int_{-\infty}^{\infty}\xi^2\Phi(\xi, \tau)\,d\xi$$

$$-\frac{1}{24}\left(\frac{\partial^3 C}{\partial x^3}\right)_{x_0}\int_{-\infty}^{\infty}\xi^4\Phi(\xi, \tau)\,d\xi - \frac{1}{720}\left(\frac{\partial^5 C}{\partial x^5}\right)_{x_0}\int_{-\infty}^{\infty}\xi^6\Phi(\xi, \tau)\,d\xi$$

$$-\cdots\frac{1}{(n+1)n!}\left(\frac{\partial^n C}{\partial x^n}\right)_{x_0}\int_{-\infty}^{\infty}\xi^{n+1}\Phi(\xi, \tau)\,d\xi. \tag{2.5.193}$$

Thus, the mass transfer $M_\tau(x_0)$ through the plane at x_0 depends on the values of the derivatives

$$\left(\frac{\partial C}{\partial x}\right)_{x=x_0}, \quad \left(\frac{\partial^3 C}{\partial x^3}\right)_{x=x_0}, \quad \left(\frac{\partial^5 C}{\partial x^5}\right)_{x=x_0}, \quad \left(\frac{\partial^7 C}{\partial x^7}\right)_{x?=x_0}, \quad \text{etc.}$$

In the case where the concentration profile is rectilinear around x_0 or is in the shape of a parabolic segment

$$C(x) = \begin{cases} A_1(x - x_0) + C(x_0), & \text{or} \\ A_2(x - x_0)^2 + A_1(x - x_0) + C(x_0). \end{cases}$$

Equation (2.5.176) reduces to

$$M_\tau(x_0) = -\frac{1}{2}\left(\frac{\partial C}{\partial x}\right)_{x_0}\int_{-\infty}^{\infty}\xi^2\Phi(\xi, \tau)\,d\xi. \tag{2.5.194}$$

If the concentration profile $C(x, t)$ in the region around x_0, which is an essential contribution to the two unidirectional currents $\overrightarrow{M}_\tau$ and $\overleftarrow{M}_\tau$, does not deviate markedly from a parabolic segment, the numerical values of the derivatives $\partial^n C/\partial x^n$ for $n = 3, 5, 7\ldots$ are much less that $\partial C/\partial x$ and attenuate strongly

---

* The details of the calculations are given in Appendix E.

as n increases. Furthermore, the integrals

$$\int_{-\infty}^{\infty} \xi^2 \Phi(\xi, \tau)\, d\xi, \qquad \int_{-\infty}^{\infty} \xi^4 \Phi(\xi, \tau)\, d\xi, \qquad \int_{-\infty}^{\infty} \xi^6 \Phi(\xi, \tau)\, d\xi, \quad \text{etc.}$$

are all of the same order of magnitude. For all the concentration profiles that are well-behaved functions as regards their derivatives, it will do to include only the first term on the right-hand side of Eq. (2.5.176). Therefore the flux through the plane at x_0

$$J(x_0) = \frac{M_\tau(x_0)}{\tau},$$

can be written as

$$J(x) = -\frac{1}{2\tau}\left(\frac{\partial C}{\partial x}\right)\int_{-\infty}^{\infty} \xi^2 \Phi(\xi, \tau)\, d\xi, \qquad (2.5.195)$$

where the subscript x_0 is left out. The integral

$$\int_{-\infty}^{\infty} \xi^2 \Phi(\xi, \tau)\, d\xi = \langle \xi^2 \rangle \qquad (2.5.196)$$

is the mean value of the square of displacements. Thus, Eq. (2.5.195) can also be written

$$J = -\frac{\langle \xi^2 \rangle}{2\tau}\frac{\partial C}{\partial x}, \qquad (2.5.197)$$

or

$$J = -D\frac{\partial C}{\partial x}, \qquad (2.5.198)$$

corresponding to Fick's law. Thus, taking into account a whole spectrum of displacements at the time τ with the probability density $\Phi(\xi, \tau)$ has resulted in the following expression for the coefficient of diffusion

$$D = \frac{\langle \xi^2 \rangle}{2\tau}, \qquad (2.5.199)$$

which is identical to the Einstein–Smoluchowski equation (Eq. (2.5.162)), which we previously derived by determining the density of probability $\Phi(\xi, \tau)$ for the displacements in the one-dimensional random walk process (Section 2.5.6.3 (i)) and thereby also the distribution of matter at the time t following the establishment of an instantaneous surface source (see Section 2.5.6.3 (iv)).

2.5.6.5 Random walk and the diffusion equation

In his first paper on Brownian motion, Einstein (1905) showed that the "random walk" assumption made the swarm of particles spread out in space and

            2. Migration and diffusion

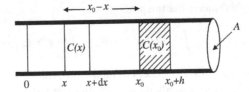

Fig. 2.36. Derivation of the diffusion equation. For further explanation see text.

time in agreement with an equation that formally is identical to the diffusion equation Eq. (2.5.5). The argument will now be reproduced although part of the presentation and the terminology is based on a paper by M. Planck (1917)[*]. The argument does not present any real difficulties apart from those of keeping track of the position coordinates.

We consider a solution of particles contained in a tube of infinite length and cross-sectional area $A = 1$ m^2, as shown in Fig. 2.36. The solute particles are assumed to be uniformly distributed across the cross section of the tube, i.e. the only concentration gradient is along the length of the tube. Let x be a position coordinate along the cylinder axis reckoned from an arbitrarily placed origin. The particle concentration $C(x, t)$ is the number of particles per unit volume. We consider a volume element[†] ΔV_0 located between the planes at x_0 and $x_0 + h$, as shown in Fig. 2.36, where $h \ll 1$. At time t the number of particles in the volume element ΔV_0 is[‡]

$$C(x_0, t)h\, A = C(x_0, t)h \,,$$

where $C(x_0, t)$ is the concentration between x_0 and $x_0 + h$ at time t. The number of particles contained in the same volume element at time $t + \tau$ is

$$C(x_0, t + \tau)h.$$

The increment in particles ΔN_τ in ΔV_0 at time τ is[§]

$$\Delta N_\tau = (C(x_0, t + \tau) - C(x_0, t))h = h\left(\frac{\partial C(x_0, t)}{\partial t}\right)\tau, \quad (2.5.200)$$

provided $\tau \ll 1$. This increment in the number of particles ΔV_0 is the result of the random displacements to and fro that all the particles in the swarm execute in the time τ. Two oppositely directed tendencies operate here. (i) A number of particles that at time t were *outside* the layer between x_0 and $x_0 + h$, have

---

[*] Planck, M: *Sitz der preuss. Akad.*, p. 324 (1917).
[†] The subscript $_0$ refers to a volume element ΔV at the position x_0.
[‡] From now on the writing of the area A will be suppressed.
[§] See Chapter 1, Section 1.2.4, Eq. (1.2.11).

during the time τ moved into the volume element ΔV_0 and are now located somewhere in the region between x_0 and $x_0 + h$. (ii) Within the same time τ a number of particles that at time t were *inside* the volume element ΔV_0 likewise move *out* of ΔV_0 by Brownian motion, and are found somewhere *outside* ΔV_0 in the regions $x \leq x_0$ and $x \geq x_0 + h$. Thus, the increment ΔN_τ of the number of particles in ΔV_0 at the time τ can also be written

$$\Delta N_\tau = N^{(in)} - N^{(out)}, \tag{2.5.201}$$

where $N^{(in)}$ and $N^{(out)}$ are the number of particles that in time τ have moved *in* to and *out* of, respectively, the element of volume ΔV_0. What remains is to calculate $N^{(in)}$ and $N^{(out)}$.

Again we assume that the spectrum of displacements executed by the particles in the time τ as a result of their Brownian motion is described by a stochastic process similar to that used previously, i.e. the probability that a particle – at time t occupying the position x – will at time $t + \tau$ be located in the region between x' and $x' + dx'$ is

$$dP = \Phi(x' - x, \tau)\,dx' = \Phi(\xi, \tau)\,d\xi,$$

where $\Phi(x' - x, \tau) = \Phi(\xi, \tau)$ is the probability density* for the displacements ξ in the time τ.

Accumulation in ΔV_0: a particle occupying the position x at the time t (see Fig. 2.36) will have the probability

$$\Phi(x_0 - x, \tau)h$$

of occupying at the time $t + \tau$ the region between x_0 and $x_0 + h$, i.e. *inside* the volume element ΔV_0 at the position x_0 of height h and with end surfaces of area $1\ m^2$. The number of particles contained at time t in the volume element ΔV_0 between the planes at x and $x + dx$ is

$$C(x, t)\,dx.$$

Among these particles the number

$$dN^{(in)} = C(x, t)\,dx\,\Phi(x_0 - x, \tau)h$$

will have moved in time τ into the volume element ΔV_0 between the planes at x and $x + h$. The total number of particles that have moved in time τ into the volume element ΔV_0 from the regions outside is

$$N^{(in)} = \int_{-\infty}^{x_0} [C(x, t)\,\Phi(x_0 - x, \tau)h]\,dx + \int_{x_0+h}^{\infty} [C(x, t)\,\Phi(x_0 - x, \tau)h]\,dx,$$

* See Eq. (2.5.183) and Eq. (2.5.186).

or as $h \ll 1$

$$N^{(\text{in})} = h \int_{-\infty}^{\infty} C(x, t)\, \Phi(x_0 - x, \tau)\, dx. \tag{2.5.202}$$

The number of particles contained in the volume element ΔV_0 at the time t is $C(x_0, t)\, h$. Let x' represent a position coordinate $x_0 \le x' \le x_0 + h$ in the volume element ΔV_0. The probability that a particle – located at the position x' at time t – makes a displacement $x - x'$ to be found between the positions x and $x + dx$ in time τ is

$$dP = \Phi(x - x', \tau)\, dx.$$

The number of particles that at time t are in the layer between x' and $x' + dx'$ equals $C(x_0, t)\, dx'$. The number of particles from this layer that at time τ have moved *out* of the volume element ΔV_0, i.e. to the regions $x \le x_0$ and $x \ge x_0 + h$, is then

$$dN^{(\text{out})} = dx'\, C(x_0, t) \left\{ \int_{-\infty}^{x_0} \Phi(x - x', \tau)\, dx + \int_{x_0+h}^{\infty} \Phi(x - x', \tau)\, dx \right\}.$$

Among the totality of particles $C(x_0, t)\, h$ contained in the volume element at time t, the number

$$N^{(\text{out})} = \int_{x_0}^{x_0+h} C(x_0, t) \left\{ \int_{-\infty}^{x_0} \Phi(x - x', \tau)\, dx + \int_{x_0+h}^{\infty} \Phi(x - x', \tau)\, dx \right\} dx', \tag{2.5.203}$$

will at time $t + \tau$ have moved out of the volume element ΔV_0. The integral inside the curly bracket $\{\ \}$ is rewritten like this

$$\begin{aligned}
I &= \int_{-\infty}^{x_0} \Phi(x - x', \tau)\, dx + \int_{x_0}^{\infty} \Phi(x - x', \tau)\, dx \\
&\quad + \int_{x_0+h}^{\infty} \Phi(x - x', \tau)\, dx - \int_{x_0}^{\infty} \Phi(x - x', \tau)\, dx \\
&= \int_{-\infty}^{\infty} \Phi(x - x', \tau)\, dx + \int_{x_0+h}^{\infty} \Phi(x - x', \tau)\, dx - \int_{x_0}^{\infty} \Phi(x - x', \tau)\, dx \\
&= 1 - \int_{x_0}^{x_0+h} \Phi(x - x', \tau)\, dx.
\end{aligned}$$

Inserting this into Eq. (2.5.203) gives

$$N^{(\text{out})} = C(x_0, t) \int_{x_0}^{x_0+h} dx' - C(x_0, t) \int_{x_0}^{x_0+h} dx' \left(\int_{x_0}^{x_0+h} \Phi(x - x', \tau)\, dx \right).$$

As $h \ll 1$ we put the integral in the parenthesis equal to $h\, \Phi(x_0 + h/2 - x', \tau)$. Hence

$$N^{(\text{out})} = C(x_0, t)\, h - C(x_0, t) \int_{x_0}^{x_0+h} h\Phi(x_0 + h/2 - x', \tau)\, dx'$$

$$= C(x_0, t)\, h - C(x_0, t) \int_{-h/2}^{h/2} \Phi(\xi, \tau)\, d\xi, \qquad (2.5.204)$$

as the variable of integration is changed from x' to $\xi = x' - x_0 - h/2$. The magnitude of the definite integral is determined by the thickness h of the volume element ΔV_0 relative to the time τ it takes the particles to move out of the volume element. If τ is large enough to make the standard deviation of the displacements $\sqrt{\xi^2}$ much larger than $h/2$ we have

$$\int_{-h/2}^{h/2} \Phi(\xi, \tau)\, d\xi \ll 1.$$

With this condition of constraint on τ, Eq. (2.5.204) tends approximately towards

$$N^{(\text{out})} = C(x_0, t)\, h, \qquad (2.5.205)$$

which implies that during the lapse of time τ the majority of particles, that at the time t were contained inside the volume element ΔV_0, have moved out by the exchange process. In his treatment of the problem, Planck also made this assumption[*]. Inserting Eq. (2.5.204) and Eq. (2.5.205) in Eq. (2.5.201) gives

$$\Delta N_\tau = h \int_{-\infty}^{\infty} C(x, t)\Phi(x_0 - x, \tau)\, dx - hC(x_0, t). \qquad (2.5.206)$$

It is an advantage to measure the displacements from the position x_0, where the accumulation is calculated. Hence we change the variable of integration x to $\xi = x - x_0$. This results in

$$\Delta N_\tau = h \int_{-\infty}^{\infty} C(x_0 + \xi, t)\Phi(\xi, \tau)\, d\xi - hC(x_0, t), \qquad (2.5.207)$$

as $dx/d\xi = 1$; $\Phi(-\xi, \tau) = \Phi(\xi, \tau)$ and $x = x_0 + \xi$. Invoking Eq. (2.5.183) gives the expression for the accumulation

$$\tau \left(\frac{\partial C}{\partial t}\right)_{x=x_0} = \int_{-\infty}^{\infty} C(x_0 + \xi, t)\, \Phi(\xi, \tau)\, d\xi - C(x_0, t). \qquad (2.5.208)$$

The result obtained so far to describe the accumulation in time τ is known as an integro-differential equation. The usual procedure to transform this equation

[*] In this way he avoided having to write one extra page of rather tedious calculations.

to a partial differential equation is to expand the function $C(x_0 + \xi)$ around $C(x_0, t)$ as a Taylor series

$$C(x_0 + \xi, t) = C(x_0, t) + \xi \left(\frac{\partial C}{\partial x}\right)_{x_0} + \frac{1}{2!}\xi^2 \left(\frac{\partial^2 C}{\partial x^2}\right)_{x_0} + \frac{1}{3!}\xi^3 \left(\frac{\partial^3 C}{\partial x^3}\right)_{x_0} + \cdots$$

$$(2.5.209)$$

Insertion in Eq. (2.5.190) gives

$$\tau \left(\frac{\partial C}{\partial t}\right)_{x_0} = \int_{-\infty}^{\infty} \left\{ C(x_0, t) + \xi \left(\frac{\partial C}{\partial x}\right)_{x_0} + \frac{1}{2}\xi^2 \left(\frac{\partial^2 C}{\partial x^2}\right)_{x_0} \right.$$
$$\left. + \frac{1}{6}\xi^3 \left(\frac{\partial^3 C}{\partial x^3}\right)_{x_0} + \frac{1}{24}\xi^4 \left(\frac{\partial^4 C}{\partial x^4}\right)_{x_0} + \cdots \right\} \Phi(\xi, \tau)\,d\xi - C(x_0, t)$$

or by rearranging

$$\tau \left(\frac{\partial C}{\partial t}\right)_{x_0} = C(x_0, t) \int_{-\infty}^{\infty} \Phi(\xi, \tau)\,d\xi - C(x_0, t)$$

$$+ \left(\frac{\partial C}{\partial x}\right)_{x_0} \int_{-\infty}^{\infty} \xi \Phi(\xi, \tau)\,d\xi + \frac{1}{6}\left(\frac{\partial^3 C}{\partial x^3}\right)_{x_0} \int_{-\infty}^{\infty} \xi^3 \Phi(\xi, \tau)\,d\xi$$

$$+ \cdots + \frac{1}{2}\left(\frac{\partial^2 C}{\partial x^2}\right)_{x_0} \int_{-\infty}^{\infty} \xi^2 \Phi(\xi, \tau)\,d\xi$$

$$+ \frac{1}{24}\left(\frac{\partial^4 C}{\partial x^4}\right)_{x_0} \int_{-\infty}^{\infty} \xi^4 \Phi(\xi, \tau)\,d\xi + \cdots.$$

The upper row on the right-hand side vanishes. So does the second row as the integrand is an odd function because $\Phi(\xi, \tau) = \Phi(-\xi, \tau)$ is an even function that is multiplied by an odd function $\xi^{(2n+1)}$. In the last row we can disregard all the terms

$$\frac{1}{4!}\left(\frac{\partial^4 C}{\partial x^4}\right)_{x_0} \int_{-\infty}^{\infty} \xi^4 \Phi(\xi, \tau)\,d\xi, \quad \frac{1}{6!}\left(\frac{\partial^6 C}{\partial x^6}\right)_{x_0} \int_{-\infty}^{\infty} \xi^6 \Phi(\xi, \tau)\,d\xi, \text{etc.,}$$

either because they are zero or because quantitatively they are much smaller than the first term. Then, dividing on both sides by τ we have the final result

$$\frac{\partial C}{\partial t} = \frac{1}{2\tau}\frac{\partial^2 C}{\partial x^2} \cdot \int_{-\infty}^{\infty} \xi^2 \Phi(\xi, \tau)\,d\xi, \qquad (2.5.210)$$

where the index x_0 is omitted as Eq. (2.5.191) must be valid for any volume element ΔV_0 irrespective of position on the x-axis. The mean value of the squares of the displacements at the time t is

$$\langle x^2 \rangle = \int_{-\infty}^{\infty} \xi^2 \Phi(\xi, \tau) \, d\xi,$$

which inserted into Eq. (2.5.191) gives

$$\frac{\partial C}{\partial t} = \frac{\langle x^2 \rangle}{2\tau} \frac{\partial^2 C}{\partial x^2}. \tag{2.5.211}$$

This is identical to the diffusion equation

$$\frac{\partial C}{\partial t} = D \frac{\partial^2 C}{\partial x^2},$$

if we put

$$D = \frac{\langle x^2 \rangle}{2\tau}. \tag{2.5.212}$$

This is the expression for the diffusion coefficient that we previously have obtained in Section 2.5.6.3 (iv), Eq. (2.5.162) and Section 2.5.6.4, Eq. (2.5.182).

2.5.6.6 Random walk over an energy barrier

The nature of the diffusion process has so far been examined by using pure statistical considerations. As an extension to this approach we will now present a slightly different angle that also considers the physics about the "Platzwechslung" steps in more detail. This approach has the merit of providing an insight into the influence of *temperature* on the diffusion process.

It has been known for a long time that the dependence of the coefficient of diffusion of temperature T is expressed by the relation

$$D = Z \, e^{-A/\mathcal{R}T},$$

which formally is identical with Arrhenius's law for the temperature dependence of the rate constant k for a chemical reaction

$$X \xrightarrow{k} Y,$$

in which a substance X by a chemical reaction transforms into another substance Y. In Arrhenius's law[*], the quantity A – with the dimension J mol^{-1} and denoted the *energy of activation of the process* – is interpreted as an energy barrier that separates the substance X from Y. Among the molecules of X only those

---

[*] Svante Arrhenius (1859–1927) was made Professor of physics in 1895. The directorship of the Nobel Institute for Physical Chemistry, Stockholm, was created for him in 1905. His main contribution was the proposal of the theory of the ionic dissociation of electrolytes for which he was awarded the Nobel Prize for chemistry in 1903.

possessing an extra energy A may react in the conversion to substance Y. The factor Z – with the dimension Hz – determines the frequency by which a molecule having an energy greater than A – the activated molecule – transforms to molecule Y. Ideas of a similar kind have been very fertile by supplementing the statistical description of the dissolved molecule's random walk with models that include properties of the molecular structure in which the diffusion takes place.

The point of departure was the experimental demonstration by Hevesy* that when a piece of lead – initially not containing radioactive lead molecules – was brought into close contact with radioactive lead, one would observe that as time passed radioactive lead penetrated into the lead bar by a process having great similarity with an ordinary diffusion process. This observation could not be reconciled with the idea of lead bar as a solid, unalterable crystal lattice where the individual lead molecules executed thermal oscillations around their positions of equilibrium. In contrast, it became necessary to assume that the crystal had single *vacant* molecular positions (vacancies or "holes"). Now, the vacant position – the "hole" – has two possibilities to change its state. (1) A neighboring atom may – provided its vibration energy is by a fluctuation raised above a critical level – detach itself from the equilibrium position and ramble about until it eventually ends up in an vacant "hole" and, at the same time, leave a vacant position – a "hole" – behind. (2) The vibrating atoms may transiently build up such a pattern that the "hole" is forcibly pushed into a new vacant position in the crystal lattice. For both patterns of movement it holds that the initial and final states are separated by an energy barrier of a certain height that restricts the number of atoms – or "holes" – participating in the "Platzwechslung" only to those having an excess of kinetic energy. Physical models based on these ideas and submitted to a quantitative analysis, using methods of statistical mechanics, were able to account for the diffusion in crystals and, furthermore, to elucidate some of the physical factors that determine the magnitude of the self-diffusion coefficient for the diffusing substance. One of the pioneers in this development was the Russian physicist J. Frenkel[†], who also put forward several arguments in

---

* George von Hevesy (1885–1966) was a Hungarian physicist and chemist. He was Professor in Freiburg (1926–34) and collaborated with Niels Bohr in Copenhagen (1934–1943). He was professor in Stockholm from 1943 and became a Swedish citizen. He discovered, together with D. Coster, the element Hafnium. After the discovery of artificial radioactivity Hevesy developed methods for using these substances as tracer elements, among other things to elucidate the movements and fate of various substances in living organisms. He was awarded the Nobel Prize for chemistry in 1943 for this contribution.

[†] During World War II; Frenkel summarized in 1942 these and related subjects in: J. Frenkel: *Kinetic Theory of Liquids*, Oxford, 1946. Also in Dover Publications Inc., New York, 1955.

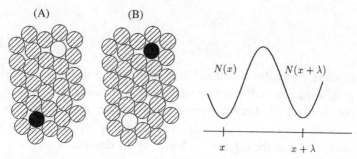

Fig. 2.37. Illustrations (left) of a single jump in the random walk of a particle. (A) The black particle increases its kinetic energy with respect to the surrounding water molecules. In a distance of a few molecular diameters there is a vacant position (a "hole"). (B) The "hot" molecule has now pushed the neighboring molecules away and made a jump to occupy the vacant "hole", leaving a vacant space in the molecular structure. The right-hand curve shows how the initial position is separated from the final position by a potential barrier of height U_0 and width λ. The situation is generalized to that where the diffusing molecules are located in the "valleys" in the positions x and $x + \lambda$ with concentrations $N(x)$ and $N(x + \lambda)$.

support of the view that the similarity between a liquid and a solid is far greater than between a liquid and a compressed gas, at least near the melting point for the liquid. It did not present greater difficulties – not conceptually at least – to transfer these ideas to an analogous description of the "Platzwechslung" of dissolved molecule. Even if the liquid is less structured than the crystal, there are well-defined structural arrangements of molecules that appear *locally* but in a transient manner as they move around in space. Owing to the fluctuations in kinetic energy a solute molecule may gain an excess of energy and momentum in a given direction and jump *out* of a given position in the liquid structure and move *into* a neighboring vacant position (a "hole"). The environment corresponding to a single jump is illustrated in Fig. 2.37. The situation that only molecules possessing excess energy enter into the "Platzwechslung" is brought into being by assuming that the initial and final positions are separated *spatially* by a distance λ as well as *energetically* by means of a "mountain", namely a potential barrier of height U_0 only being passable to molecules possessing a kinetic energy exceeding U_0. We then have to imagine the cooperative action of the traffic over a very large number of such potential barriers. We can illustrate this by considering a single potential barrier enclosed by two "valleys" that contain the diffusing molecules in a finite concentration. Let this be $N(x)$ molecules per unit volume in the position x (the "valley" to the left in Fig. 2.37). A certain fraction of these molecules – called *activated* molecules – have a thermal energy with a velocity component v_{min} in the direction x,

where

$$\tfrac{1}{2}m\, v_{\min}^2 \ge U_0. \tag{2.5.213}$$

Each of these molecules will then move over the top of the barrier with a velocity $v_x \ge v_{\min}$ and fall down into the position towards $x + \lambda$ (the "valley" to the right in Fig. 2.37). Let $N^*(x)$ represent the number of activated molecules in the position x and let $\langle \overrightarrow{v^*} \rangle$ be the average thermal velocity with which these molecules move over the ridge of the barrier in the direction $x \to x + \lambda$. The flux of these particles is*

$$\overrightarrow{J} = \tfrac{1}{2}N^*(x)\,\langle \overrightarrow{v^*} \rangle, \tag{2.5.214}$$

where the factor $\tfrac{1}{2}$ makes allowance for the remaining half to move in the opposite direction over the adjoining barrier† of particles that among a population of N particles have an energy in the range between U and $U + dU$, given by Boltzmann's law of distribution‡

$$dN^* = n^*(U)dU = \frac{N}{kT}\,e^{-U/kT}dU. \tag{2.5.215}$$

As a consequence, the number $N^*(U)$ of particles having an energy *equal to or larger than* U_0 is given by

$$N^*(U) = \int_{U_0}^{\infty} n^*(U)\,dU = \frac{N}{kT}[-kTe^{-U/kT}]_{U_0}^{\infty} = N\,e^{-U_0/kT}, \tag{2.5.216}$$

which inserted into Eq. (2. 5.214) gives

$$\overrightarrow{J} = \tfrac{1}{2}N(x)\,\langle \overrightarrow{v^*} \rangle\,e^{-U_0/kT}. \tag{2.5.217}$$

Similarly, the flux in the *opposite* direction from the valley at $x + \lambda$ – where the diffusing particles have the concentration $N(x + \lambda)$ – to the valley at the position x is given by

$$\overleftarrow{J} = N(x + \lambda)\,\langle \overleftarrow{v^*} \rangle\,e^{-U_0/kT}. \tag{2.5.218}$$

Both $\overleftarrow{J}$ and $\langle \overleftarrow{v^*} \rangle$ are positive quantities. We put $\langle \overrightarrow{v^*} \rangle = \langle \overleftarrow{v^*} \rangle = \langle v^* \rangle$. The resulting flux in the direction of x is then the difference between the two unidirectional

---

* See Section 2.4.2, Eq. (2.4.6.).
† Compare also the arguments in Section 2.5.6.4.
‡ See, for example, G.S. Rushbroke: *Introduction to Statistical Mechanics*, Oxford, 1951.

fluxes $\overrightarrow{J}$ and $\overleftarrow{J}$

$$J = \overrightarrow{J} - \overleftarrow{J}$$
$$= \tfrac{1}{2}N(x)\,\langle v^*\rangle\,e^{-U_0/kT} - \tfrac{1}{2}N(x+\lambda)\,\langle v^*\rangle\,e^{-U_0/kT}$$
$$= -\tfrac{1}{2}\langle v^*\rangle\,e^{-U_0/kT}\,(N(x+\lambda) - N(x))$$
$$= -\tfrac{1}{2}\langle v^*\rangle\,\lambda\,e^{-U_0/kT}\,\frac{N(x+\lambda) - N(x)}{\lambda}.$$

If the structure is fine-grained enough that it is meaningful to identify the difference quotient above with the derivative of $N(x)$ with respect to x we can also write *

$$J = -\tfrac{1}{2}\langle v^*\rangle\,\lambda\,e^{-U_0/kT}\,\frac{\partial N}{\partial x}. \qquad (2.5.219)$$

This expression passes into Fick's law if we identify the coefficient of diffusion D with

$$D = \tfrac{1}{2}\langle v^*\rangle\,\lambda\,e^{-U_0/kT}.$$

If $\langle \tau_0 \rangle$ represents the mean passage time over the potential barrier we have: $\langle v^*\rangle = \lambda/\langle \tau_0\rangle$. Insertion in the above expression gives

$$D = \tfrac{1}{2}\frac{\lambda^2}{\langle \tau_0\rangle}e^{-U_0/kT} = D_0\,e^{-U_0/kT}, \qquad (2.5.220)$$

where the factor $D_0 = \lambda^2/2\langle\tau_0\rangle$ to the exponential now contains the same elements as the expression that we developed on the basis of statistical arguments. Thus, the concept of an energy barrier over which the diffusing molecules must jump to achieve their "Platzwechslung" has resulted in introducing the *temperature* into the diffusion process and in accounting for the experimentally observed dependence of the diffusion coefficient upon the temperature. Furthermore, it appears natural to characterize the properties of the diffusing substance by the magnitude of the *activation energy* for the diffusion process. This quantity is determined experimentally by measuring the diffusion coefficient at a series of different temperatures $T_1, T_2, \ldots, T_n$ and plotting the values of D thus obtained logarithmically as a function of the reciprocal temperatures $1/T$. This results in a straight line – provided that no phase transitions occur in the system at the chosen range of temperatures – with the slope

$$\alpha = -\frac{U_0}{k} = -\frac{A}{\mathcal{R}}, \qquad (2.5.221)$$

---

* See also Section 2.5.6.4, Eq. (2.5.180).

where A is the energy of activation usually given in units of J mol^{-1}.

The energy of activation is a participating factor both for the magnitude of the diffusion coefficient and also for its temperature dependence, as

$$\frac{d \ln D}{dT} = \frac{1}{D} \frac{dD}{dT} = \frac{1}{D} \frac{d}{dT} \left(D_0 \, e^{-U_0/kT} \right)$$

$$= \frac{1}{D} D_0(-1) e^{-U_0/kT} \frac{d}{dT} \left(\frac{U_0}{kT} \right) = \frac{-1}{D} D_0 \, e^{-U_0/kT} \left(-\frac{U_0}{kT^2} \right),$$

whence

$$\frac{1}{D} \frac{dD}{dT} = \frac{U_0}{kT^2} = \frac{A}{\mathcal{R}T^2}, \qquad (2.5.222)$$

i.e. the larger the energy of activation, the larger is the temperature coefficient for the substance in question.

2.6 Diffusion and migration superimposed

2.6.1 The Smoluchowski equation

In biological systems we often encounter the situation where passive mass transfer takes place not only by diffusion – caused by a non-uniform distribution of the dissolved particles – but also because of the presence of an external field of force – most frequently an electric field acting on an ion – that drives each particle in the direction of the force. Thus, the mass transfer results from the superposition of these two essentially different transport mechanisms. From a macroscopic point of view, the expression for the flux is based on the following argument.

With a non-uniform distribution and in the absence of an external driving force X, the net flux is *purely diffusive* and is given by Fick's law:

$$J_{\text{diff}} = -D \frac{\partial C}{\partial x}, \quad \text{for} \quad X = 0,$$

where D is the diffusion coefficient of the substance.

With a uniform distribution in space but in the presence of an external driving force X acting on each dissolved particle, the net flux is now a *migration flux* of magnitude

$$J_{\text{migr}} = vC, \quad \text{for} \quad \frac{\partial C}{\partial x} = 0,$$

where v is the stationary migration velocity resulting from the action of the force X^*.

* Compare Eq. (2.4.4): $v = BX$, where B is the mechanical mobility of the particle.

If both tendencies to cause mass transfer act *simultaneously* it seems most reasonable to assume that the net result is in the first instance the *linear addition* of the two separate contributions, each of which taken at the time produces a net flux as indicated above, thus

$$J = -D\frac{\partial C}{\partial x} + vC. \tag{2.6.1}$$

Introducing instead the driving force X, the expression for the flux is

$$J = -D\frac{\partial C}{\partial x} + BCX. \tag{2.6.2}$$

If *mass conservation* characterizes the system then

$$\frac{\partial C}{\partial t} = -\frac{\partial J}{\partial x}, \quad \text{mass conservation,} \tag{2.6.3}$$

is still valid.

Insertion of Eq. (2.6.2) gives

$$\frac{\partial C}{\partial t} = D\frac{\partial^2 C}{\partial x^2} - BX\frac{\partial C}{\partial x}, \tag{2.6.4}$$

or

$$\frac{\partial C}{\partial t} = D\frac{\partial^2 C}{\partial x^2} - v\frac{\partial C}{\partial x}, \tag{2.6.5}$$

provided that D and B are independent of the distance or concentration. Equation (2.6.2) and the following derived equations are often called Smoluchowski equations. As shown by Smoluchowski[*], Eq. (2.6.5) is simplified by introducing a new variable U defined as

$$C = U\exp\left\{\frac{v(x-x_0)}{2D} - \frac{v^2 t}{4D}\right\} = U\mathcal{E}, \tag{2.6.6}$$

where $\mathcal{E}$ stands for the exponential term. We have

$$\frac{\partial C}{\partial t} = \frac{\partial U}{\partial t}\mathcal{E} + U\mathcal{E}(-1)\frac{v^2}{4D}.$$

Furthermore

$$\frac{\partial C}{\partial x} = \frac{\partial U}{\partial x}\mathcal{E} + U\mathcal{E}\frac{v}{2D},$$

and

$$\frac{\partial^2 C}{\partial x^2} = \frac{\partial^2 U}{\partial x^2}\mathcal{E} + \frac{\partial U}{\partial x}\mathcal{E}\frac{v}{D} + U\mathcal{E}\left(\frac{v}{2D}\right)^2.$$

[*] Smoluchowski, M. v. (1915): *Ann. d. Physick*, **48**, 1103.

Insertion of these expressions into Eq. (2.6.6) and dividing on both sides by $\mathcal{E}$ gives

$$\frac{\partial U}{\partial t} - U\frac{v^2}{4D} = D\frac{\partial^2 U}{\partial x^2} + D\frac{\partial U}{\partial x}\frac{v}{D} + DU\left(\frac{v}{2D}\right)^2 - v\frac{\partial U}{\partial x} - U\frac{v^2}{2D},$$

or

$$\frac{\partial U}{\partial t} = D\frac{\partial^2 U}{\partial x^2}, \tag{2.6.7}$$

whereby Eq. (2.6.5) is transformed into a form that is identical with the one-dimensional diffusion equation. The solutions of this equation – of which we already have a few – are in general far easier to obtain than trying to solve the Smoluchowski equation (Eq. (2.6.5)) directly. The facilitation, however, is only up to a point since the initial and boundary conditions connected to the solution of Eq. (2.6.5) for $C(x, t)$ must naturally be translated via the transforming equation Eq. (2.6.6) to the corresponding initial and boundary conditions for the function $U(x, t)$. We shall illustrate the procedure with a few examples.

2.6.1.1 An instantaneous plane source in infinite space

The solution of the diffusion equation

$$C(x, t) = \frac{\mathcal{N}}{2\sqrt{\pi Dt}}e^{-(x-x_0)^2/4Dt} \tag{2.6.8}$$

represents one-dimensional diffusion following the release at time $t = 0$ of particles from an instantaneous plane source* of strength $\mathcal{N}$ (mol m^{-2}) placed at the position $x = x_0$. We shall now find the counterpart satisfying the Smoluchowski equation (Eq. (2.6.5)), i.e. every particle released from the plane source now also moves with the velocity v as the result of the force $X^\dagger$. Putting $x = 0$ and $t = 0$ in the Smoluchowski transformation

$$C(x, t) = U(x, t)\exp\left\{\frac{v(x - x_0)}{2D} - \frac{v^2 t}{4D}\right\} \tag{2.6.9}$$

we have $C = U$, i.e. the two functions have the same initial conditions, namely the instantaneous source $\mathcal{N}$ at $x = x_0$. The solution of the transformed equation

$$\frac{\partial U}{\partial t} = D\frac{\partial^2 U}{\partial x^2}$$

that satisfies the initial condition

$$U = \begin{cases} \mathcal{N}, & \text{for} \quad x = x_0 \quad \text{and} \quad t = 0 \\ 0, & \text{for} \quad x \neq x_0 \quad \text{and} \quad t = 0 \end{cases}$$

---

* See Section 2.5.4.4, Eq. (2.5.125).
$\dagger$ The sign of v is positive if the force X acts in the direction of the x-axis.

and the boundary conditions $U = 0$ for $x \to \pm\infty$ when $t > 0$ is then obtained from Eq. (2.6.8) as

$$U(x, t) = \frac{\mathcal{N}}{2\sqrt{\pi Dt}}\, e^{-(x-x_0)^2/4Dt}.$$

Insertion in Eq. (2.6.9) gives then

$$
\begin{aligned}
C(x, t) &= \frac{\mathcal{N}}{2\sqrt{\pi Dt}}\exp\left\{-\frac{(x-x_0)^2}{4Dt}\right\} \cdot \exp\left\{\frac{v(x-x_0)}{2D} - \frac{v^2 t}{4D}\right\} \\
&= \frac{\mathcal{N}}{2\sqrt{\pi Dt}}\exp\left\{-\frac{(x-x_0)^2 - 2(x-x_0)vt + (vt)^2}{4Dt}\right\} \\
&= \frac{\mathcal{N}}{2\sqrt{\pi Dt}}\exp\left\{\frac{(x-x_0-vt)^2}{4Dt}\right\}.
\end{aligned}
\tag{2.6.10}
$$

This expression differs from Eq. (2.6.8) by containing the extra term $-vt$ instead of $x - x_0$, where x_0 was a *fixed* position. Thus, the concentration profile $C(x, t)$ is displaced as a whole with a constant velocity $v = BX$ in the direction of the force X, but the distribution of symmetry around the vertex, which at the time t is localized at the position $x_0 = vt$, remains *unchanged*. Figure 2.38 shows the concentration profile at time $T = Dt0.08$ after the establishment of the instantaneous plane source (A). At this time the force producing the

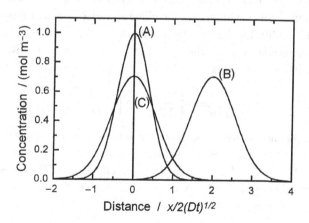

Fig. 2.38. The spreading by diffusion at time t from an instantaneous matter source of strength 1 that is placed in the position $x = 0$. Ordinate: concentration. Abscissa: distance X in units of $2\sqrt{Dt}$. (A) Diffusion only. The concentration profile at time t_0 corresponds to $Dt = 0.08$. (B) Diffusion with superimposed migration. At time $Dt = 0.08$ each solute particle is acted upon by an external driving force giving rise to a migration velocity v in the direction of the x-axis, causing a displacement of the peak of the profile to the position $X = 2$ at the time $Dt = 0.16$. (C) By way of comparison is shown the profile at the same time $Dt = 0.16$, but without any driving force acting.

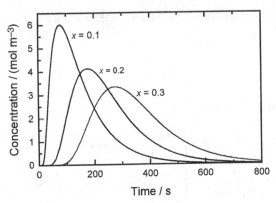

Fig. 2.39. The time course of the concentrations from a swarm of particles moving by action of diffusion and migration past the positions at $x = 0.1$ cm, $x = 0.2$ cm and $x = 0.3$ cm. The swarm was initially liberated instantaneously at $x = 0$.

migration velocity v is impressed and the profile is shown at the time $2T$ in (B). By way of comparison we also show the profile (C) at the same time $2T$ for the pure diffusion ($v = 0$).

(i) *The concentration change with time at a fixed position in space*

Consider now the situation where the position coordinate x is kept at a constant value $x = x'$. $C(x', t)$ is then a function of t alone and indicates how the concentration changes with time at the plane at $x = x'$. For that reason it may be useful to modify Eq. (2.6.10) to contain only t. Let t' be the time for the vertex of the profile to move from $x = 0$ to $x = x'$, i.e. $x' = vt'$. Replacing x in Eq. (2.6.10) with vt' gives

$$C(x', t) = \frac{\mathcal{N}}{2\sqrt{\pi D t}} \exp\left\{\frac{v^2 t'}{2D}\right\} \exp\left\{-\frac{v^2 t'}{4D}\left[\frac{t'}{t} - \frac{t}{t'}\right]\right\}. \qquad (2.6.11)$$

Figure 2.39 shows three time courses corresponding to the positions $x' = 0.1$ cm, $x' = 0.2$ cm and $x' = 0.3$ cm for a substance with a diffusion coefficient $D = 2.5 \times 10^{-5}$ cm^2 s^{-1} and a migration velocity of $v = 10^{-3}$ cm s^{-1} in the positive direction of the x-axis. Note the resemblance of the curves to that of an indicator dilution curve.

(ii) *Driving the swarm towards a reflecting barrier: a case of sedimentation*

We now consider the somewhat more complicated situation where the particles released at time $t = 0$ from the instantaneous plane source of strength $\mathcal{N}$ at the position $x = x_0$ *are driven* by the external force *towards* the origin where the totally reflecting plane is placed at $x = 0$, i.e. the flux there is zero. The situation is similar to a process of sedimentation in a jar the bottom of which is placed at $x = 0$.

The particles of the swarm are moving with the flux

$$J = -D\frac{\partial c}{\partial x} - vC, \tag{2.6.12}$$

as v in Eq. (2.6.3) is replaced by $-v$. Correspondingly the concentration profile of the released particles is governed by

$$\frac{\partial C}{\partial t} = D\frac{\partial^2 C}{\partial x^2} + v\frac{\partial C}{\partial x}, \tag{2.6.13}$$

with the initial value

$$C(x, t) = \begin{cases} \frac{N}{2\sqrt{\pi Dt}} e^{-(x-x_0)^2/4Dt}, & \text{for} \quad t = 0 \quad \text{and} \quad x = x_0 \\ 0, & \text{for} \quad t = 0 \quad \text{and} \quad x \neq x_0 \end{cases} \tag{2.6.14}$$

and the boundary conditions

$$C(x, t) \to 0, \quad \text{for} \quad x \to \infty \quad \text{and} \quad t < 0, \tag{B22}$$

and

$$D\frac{\partial C}{\partial x} + vC = 0, \quad \text{for} \quad x = 0 \quad \text{and} \quad t > 0. \tag{B23}$$

The transformation (Eq. (2.6.6))

$$C(x, t) = U \exp\left\{\frac{-v(x - x_0)}{2D} - \frac{v^2 t}{4D}\right\} = U\mathcal{E} \tag{2.6.15}$$

applied to Eq. (2.6.13) leads to the equation in $U(x, t)$

$$\frac{\partial U}{\partial t} = D\frac{\partial^2 U}{\partial x^2}, \tag{2.6.7}$$

which has the same initial conditions as $C(x, t)$ and the boundary condition $U \to 0$ for $x \to \infty$. Applying Eq. (2.6.15) to the boundary condition Eq. (B23) gives

$$\frac{\partial C}{\partial x} + \frac{v}{D}U = \frac{\partial U}{\partial x}\mathcal{E} - U\mathcal{E}\frac{v}{2D} + \frac{v}{D}\mathcal{E}U = \frac{\partial C}{\partial x} + \frac{v}{2D}U.$$

Thus the boundary condition Eq. (B23) for $C(x, t)$ to be satisfied for $U(x, t)$ at $x = 0$ is

$$\frac{\partial U}{\partial x} + \frac{v}{2D}U = 0$$

or

$$\frac{\partial U}{\partial x} + mU = 0, \tag{B24}$$

by putting

$$m = \frac{v}{2D}. \tag{2.6.16}$$

To solve Eq. (2.6.7) for $U(x, t)$ with the initial and boundary conditions we draw on solutions obtained previously that to some extent resemble the present problem. We put

$$U(x, t) = u(x, t) + w(x, t) \tag{2.6.17}$$

where $u(x, t)$ and $w(x, t)$ taken together shall satisfy the initial and boundary conditions for $U(x, t)$. For $u(x, t)$ we choose the function*

$$u = \frac{\mathcal{N}}{2\sqrt{\pi Dt}} e^{-(x-x_0)^2/4Dt} + \frac{\mathcal{N}}{2\sqrt{\pi Dt}} e^{-(x+x_0)^2/4Dt} \tag{2.6.18}$$

or to reduce redundancy we write

$$u = A\left\{ e^{-(x-x_0)^2/B} + e^{-(x+x_0)^2/B} \right\} \tag{2.6.19}$$

by putting

$$A = \frac{\mathcal{N}}{2\sqrt{\pi Dt}}, \quad \text{and} \quad B = 4Dt. \tag{2.6.20}$$

As for $u(x, t)$ it represents initially an instantaneous source at $x = x_0$ and, furthermore, $\partial u/\partial x = 0$ for $x = 0$ and $t > 0$. Thus, the boundary condition Eq. (B24) is satisfied for the special case of $v = 0$.

We have from Eq. (2.6.17)

$$\frac{\partial U}{\partial x} + mU = \frac{\partial u}{\partial x} + mu + \frac{\partial w}{\partial x} + mw, \tag{2.6.21}$$

and because of the boundary condition Eq. (B24) this leads to

$$\left(\frac{\partial w}{\partial x} + mw \right)_{x=0} = -\left(\frac{\partial u}{\partial x} \right)_{x=0} - (mu)_{x=0}$$

for $x = 0$. Invoking Eq. (2.6.19) gives for the right-hand side[†]

$$\left(\frac{\partial u}{\partial x} \right)_{x=0} \quad 0, \quad \text{and} \quad (mu)_{x=0} = 2mAe^{-x_0^2/B}$$

---

* See Section 2.5.5.5 (iii), Eq. (2.5.133).
† Compare also Section 2.5.4.3 (iii).

and consequently

$$\left(\frac{\partial w}{\partial x} + mw\right)_{x=0} = -2mAe^{-x_0^2/B}.$$

As it stands the equation is of little use for determining the function w unless its range of validity can be extended to values other than $x = 0$. A reasonable guess, however, is to put

$$2mAe^{-x_0^2/B} = \lim_{x \to 0} 2mAe^{-(x+x_0)^2/B}$$

and consider the right-hand side as a particular value of the function of x

$$\phi(x) = -2mAe^{-(x+x_0)^2/B},$$

which represents an instantaneous plane source placed at the position $x = -x_0{}^*$. To find the function w we have to solve

$$\frac{dw}{dx} + mw = \phi(x, t) = -2mAe^{-(x+x_0)^2/B}, \qquad (2.6.22)$$

which is an ordinary first-order linear equation. The general solution is[†]

$$w\,e^{mx} = -2mA \int \exp\{mx\} \exp\{-(x+x_0)^2/B\} \, dx + K_1. \qquad (2.6.23)$$

As $w(x, t) = 0$ for $x \to \infty$ the right-hand side can also be written as the definite integral

$$w(x, t)\,e^{mx} = -2mA \int_{\infty}^{x} \exp\{mX\} \exp\{-(X+x_0)^2/B\} \, dX, \qquad (2.6.24)$$

where X is the formal variable of integration. Evaluation of the integral[‡] leads to

$$w(x, t) = \mathcal{N}\frac{v}{2D} \exp\left\{\frac{v^2 t}{4D} - \frac{v}{2D}x_0 - \frac{vx}{2D}\right\} \text{Erfc}\left\{\frac{x + x_0 - vt}{2\sqrt{Dt}}\right\}.$$

---

[*] Formally the function $2mAe^{-(x-x_0)^2/B}$ would have done equally well. The source is placed at $x = -x_0$ for two reasons: (1) it is required that w initially is zero for $0 \le x < \infty$, and (2) the reflection of particles at $x = 0$ implies that $(\partial C/\partial x)_{x=0}$ is *negative*. Since $(\partial u/\partial x)_{x=0} = 0$ this requires sources placed in the negative half plane to produce the proper amount of flow across the plane at $x = 0$ into the positive half plane.
[†] See Chapter 1, Section 1.9.1.3, Eq. (1.9.17).
[‡] See Appendix F.

Returning to $U = u + w$ we have

$$U = \frac{\mathcal{N}}{2\sqrt{\pi Dt}} e^{-(x-x_0)^2/4Dt} + \frac{\mathcal{N}}{2\sqrt{\pi Dt}} e^{-(x+x_0)^2/4Dt}$$
$$+ \mathcal{N}\frac{v}{2D} \exp\left\{\frac{v^2 t}{4D} - \frac{v}{2D}x_0 - \frac{vx}{2D}\right\} \text{Erfc}\left\{\frac{x + x_0 - vt}{2\sqrt{Dt}}\right\}$$

and – using Eq. (2.6.15) – this expression in $U(x, t)$ is then transformed backwards to $C(x, t)$

$$C(x, t) = \frac{\mathcal{N}}{2\sqrt{\pi Dt}}\left[e^{-(x-x_0)^2/4Dt} + e^{-(x+x_0)^2/4Dt}\right] \exp\left\{-\frac{v(x - x_0)}{2D} - \frac{v^2 t}{4D}\right\}$$
$$+ \mathcal{N}\frac{v}{2D}\exp\left\{-\frac{v(x - x_0)}{2D} - \frac{v^2 t}{4D}\right\}$$
$$\times \mathcal{N}\frac{v}{D\sqrt{\pi}}\exp\left\{\frac{v^2 t}{4D} - \frac{v}{2D}x_0 - \frac{vx}{2D}\right\} \text{Erfc}\left\{\frac{x + x_0 - vt}{2\sqrt{Dt}}\right\}$$

or finally

$$C(x, t) = \frac{\mathcal{N}}{2\sqrt{\pi Dt}}\left[e^{-(x-x_0)^2/4Dt} + e^{-(x+x_0)^2/4Dt}\right] \exp\left\{-\frac{v(x - x_0)}{2D} - \frac{v^2 t}{4D}\right\}$$
$$+ \mathcal{N}\frac{v}{2D}\exp\left\{-\frac{vx}{D}\right\}\text{Erfc}\left\{\frac{x + x_0 - vt}{2\sqrt{Dt}}\right\}, \qquad (2.6.25)$$

which was first derived by Smoluchowski (1915)[*]. The patterns displayed by this formula are not easy to see except for large values of t where in the limit $t \to \infty$ the right-hand side will represent the final, i.e. *time independent*, distribution of the particles that were initially liberated from the instantaneous plane source at $x = x_0$. This distribution is also the *equilibrium distribution* of the sedimentation process as the profile is not maintained by means of a constant supply and removal of matter for the system.

When $t \to \infty$ the first term in Eq. (2.6.25) becomes zero and the argument to Erfc tends to ∞. As Erfc$-\infty = 2$ we have

$$C(x) = \mathcal{N}\frac{v}{D} e^{-vx/D} = \frac{\mathcal{N}}{\lambda}e^{-x/\lambda}, \qquad (2.6.26)$$

[*] Smoluchowski, M.v. (1915): *Ann. d. Phys.*, IV Folge, **48**, 1103. Smoluchowski did not give any details of his derivation apart from stating that the derivation of $U(x, t)$ was a problem solved in heat conduction. Perhaps he had in mind the papers of Hobson, E.W. (1887): *Proc. Lond. Math. Soc. (1)*, **19**, 279, and Bryan, G.H. (1891): *Proc. Camb. Phil. Soc.*, **7**, 246; Bryan, G.H. (1891): *Proc. Lond. Math. Soc. (1)*, **22**, 424; or Byerly, W.E. (1893): *An Elementary Tretise on Fourier's Series and Spherical, Cylindrical and Ellipsoidal Harmonics*. Ginn and Company, Boston, pp. 123 and 126. Although differing in details, the tactic used in the present treatment derives mainly from Byerly and from Fürth, R. (1917): *Ann. d. Phys.*, IV Folge, **53**, 177.

where

$$\lambda = D/v$$

is the distance where $C(\lambda) = e^{-1} = 0.37$ and is often referred to as the *thickness of the sedimentation layer**. The characteristic of this exponentially decreasing concentration profile is that at every position in space there exists a balance – that results in a zero net flux – between the tendency to transfer by diffusion $-D dC/dx$ *away from* the reflecting barrier and the tendency to transfer vC *towards* the barrier due to the force X. The equilibrium profile can also be determined from the condition $J = 0$. From Eq. (2.6.12) we have

$$J = -D\frac{dC}{dx} - vC = 0.$$

Rearranging

$$\frac{1}{C}\frac{dC}{dx} = -\frac{v}{D} = \frac{d\ln C}{dx}$$

and integrating gives

$$C = A_1 e^{-vx/D},$$

where A_1 is a constant of integration that depends on the magnitude of the liberated $\mathcal{N}$ (mol m^{-2}) that still remains in the space $0 \le x < \infty$. We therefore have

$$\mathcal{N} = A_1 \int_0^\infty e^{-vx/D}\, dx = A_1\left[-\frac{D}{v}e^{-vx/D}\right]_0^\infty = A_1\frac{D}{v}.$$

Hence it follows that

$$C(x) = \mathcal{N}\frac{v}{D}e^{-vx/D}.$$

This distribution may also be obtained from the Smoluchowski equation by putting $\partial C/\partial t = 0$ and integrating the stationary form

$$\frac{d^2C}{dx^2} + \frac{v}{D}\frac{dC}{dx} = 0.$$

Naturally this procedure is more troublesome.

A series of concentration profiles calculated from Eq. (2.6.25) is shown in Fig. 2.40. The release of the swarm results first in a purely diffusive profile (curve 1), but the effect of the force comes to the fore very soon and the maximum of the profiles moves towards the origin while gradually decreasing and becoming flatter on account of the random movements experienced

* The name is due to Smoluchowski (1915).

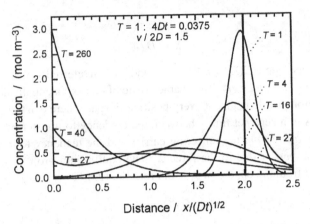

Distance / $x/(Dt)^{1/2}$

Fig. 2.40. Illustration of a sedimentation process. Particles released from an instanta-neous plane source placed at $x = 2.0$ are driven by an external force X with the velocity v towards a reflecting barrier at $x = 0$. Abscissa: distance in length, units as those of the diffusion coefficient D. Ordinate: concentration. The value of $v/2D = 1.5$. The curves are concentration profiles calculated corresponding to the values of $4Dt$. Curve 1: 0.0375, relative time $T = 1$. Curve 2: $T = 4$. Curve 3: $T = 16$. Curve 4: $T = 27$. Curve 5: $T = 40$. Curve 6: $T = 260$. The equilibrium profile is curve 6. The profiles do not differ substantially from those shown by Smoluchowski (1915).

by the particles (curves 2 and 3). Once the particles approach the reflect-ing barrier in an appreciable number the curves begin again to rise because of the increasing numbers of reflections (curves 4 and 5) until finally for $t \to \infty$ the profile has reached its exponentially decreasing equilibrium shape (curve 6).

2.6.2 *"Random walk" considerations*

Some readers may perhaps argue that the setting up of the flux equation (Eq. (2.6.1)) and the Smoluchowski equation (Eq. (2.6.4)) in Section 2.6.1 requires a better foundation than the justification of belonging to the class of "educated guesses". To meet this objection it will now be shown that the same result is obtained from considerations based upon random walk statis-tics as used in Section 2.5.6.4. A molecular model of the migration process will also be presented, aiming at throwing some light on the basic mechanism involved.

2.6.2.1 *The flux equation*

We start off with simplifications similar to those that Einstein used in his ele-mentary derivation of Fick's law (see Section 2.5.6.4.). However, the present

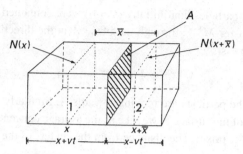

Fig. 2.41. A simplified model to derive the net flux in a swarm of particles moving under the combined effect of a concentration gradient and an external driving force. The spectrum of displacements is replaced by one mean forward displacement $\bar{x}_+$ and one mean backward displacement $\bar{x}_-$.

situation differs since the effect of the external force X is to impress a *unidirectional* component of velocity on the irregular random movements of each dissolved particle. The average value of this component is the migration velocity v. Therefore, the displacements to time t are no longer distributed symmetrically forwards and backwards as in the force-free case. Figure 2.41 illustrates the difference between the two situations. In the force-free case the mean displacement $\bar{x}$ at time t is symmetric around the initial position $x = 0$. However, when an external force X produces a migration velocity v in the direction of the force, the result is that the plane of symmetry of the swarm at the time t moves a distance vt from the initial position $x = 0$. This implies that the mean displacement from the initial position becomes *asymmetric* as the mean displacement $\bar{x}_+$ in the direction of the force now becomes

$$\bar{x}_+ = \bar{x} + vt,$$

where $\bar{x}$ is the mean displacement in the force-free case. Similarly, the mean displacement $\bar{x}_-$ in the opposite direction is

$$\bar{x}_- = \bar{x} - vt.$$

With $\bar{x}_+$ and $\bar{x}_-$ at our disposal we construct (Fig. 2.41) the analogy to "Einstein's box"[*] with the surface of area A constituting the common terminal surface of the two adjacent prisms having $\bar{x}_+$ and $\bar{x}_-$ as sides and volumes $V^{(1)} = A\bar{x}_+$ and $V^{(2)} = A\bar{x}_-$. The direction of the force is from left to right. Let N denote the number of particles dissolved per unit volume. As the mean concentration in the left prism we take the value $N(x)$ at the center with position x, i.e. at a distance $vt/2$ to the left of the plane A. Then, $N(x)V^{(1)} = N(x)A\bar{x}_+$

---

[*] Section 2.5.6.4 (i).

is the number of particles that initially ($t = 0$) were contained in the prism to the left. The transfer $M^{1\to2}$ through the plane A in the direction $1 \to 2$ after the lapse of time t is

$$M^{1\to2} = \tfrac{1}{2}N(x)(\bar{x} + vt)\,A,$$

as only half of the particles are moving towards A. Similarly, for the volume $V^{(2)}$ to the right of the surface A we take as the average concentration the value at the center of the prism. The distance from the centers of the two volumes is

$$(\bar{x} + vt)/2 + (\bar{x} - vt)/2 = \bar{x}.$$

Using the center position x of the left prism as reference, the prism on the right has its center at $x + \bar{x}$, where the concentration $N(x + \bar{x})$ is taken as the average concentration in the volume $V^{(2)}$. Note that the distance from center to center is the mean displacement $\bar{x}$ in time t when the particles are undergoing Brownian motion without the presence of a driving force. The transfer $M^{2\to1}$ through the plane A in the direction $2 \to 1$ after time t is

$$M^{1\leftarrow2} = \tfrac{1}{2}N(x + \bar{x})(\bar{x} - vt)A.$$

The net mass transfer $M^{(\text{tot})}$ in the direction $1 \to 2$ is $M^{1\to2} - M^{1\leftarrow2}$. Therefore

$$\begin{aligned}
J &= \frac{M^{1\to2} - M^{1\leftarrow2}}{At} \\
&= \tfrac{1}{2}\frac{A\,[N(x)(\bar{x} + vt) - N(x + \bar{x})(\bar{x} - vt)]}{At} \\
&= -\tfrac{1}{2}\frac{\bar{x}^2}{t}\frac{N(x + \bar{x}) - N(x)}{\bar{x}} + \tfrac{1}{2}N(x)v + \tfrac{1}{2}N(x + \bar{x})v \\
&= -\tfrac{1}{2}\frac{\bar{x}^2}{t}\frac{N(x + \bar{x}) - N(x)}{\bar{x}} + \tfrac{1}{2}[N(x) + N(x + \bar{x})]v.
\end{aligned}$$

We replace the difference quotient $[N(x + \bar{x}) - N(x)]/\bar{x}$ with the derivative dN/dx as in Section 2.5.6.4. Furthermore, we assume only small values of t giving $\bar{x} \ll 1$, i.e. $\tfrac{1}{2}[N(x) + N(x + \bar{x})] \approx N(x)$. Then, the above expression for the flux becomes

$$J = -\frac{1}{2}\frac{\bar{x}^2}{t}\frac{dN}{dx} + N\,v, \tag{2.6.27}$$

or

$$J = -D\frac{dN}{dx} + N\,v, \tag{2.6.28}$$

by replacing $\bar{x}^2/2t = D$ as in Section 2.5.6.4, Eq. (2.5.181). Finally, dividing by Avogadro's constant $\mathcal{N}_A$ on both sides we obtain the Smoluchowski flux equation

$$J = -D\frac{dC}{dx} + C v. \tag{2.6.29}$$

2.6.2.2 Random walk and the diffusion–migration equation

Following the same lines of treatment as in Section 2.5.6.5 we shall now derive the Smoluchowski equation by considering a swarm of particles executing Brownian motion with superimposed *unidirectional* velocity component having the mean value

$$v = BX,$$

where X is the force that acts on each particle in the solution. If v is much smaller than the thermal velocities of the particles, the shape of the probability density function $\Phi(\xi, \tau)$ for a displacement ξ in the time τ will differ only insignificantly from the probability density $\phi(\xi, \tau)$ corresponding to the force-free case. The manner in which the two probabilities differ is that in the force-free case $\phi(\xi, \tau)$ is distributed symmetrically with respect to initial position $\xi = 0$ for all values of τ, whereas the plane of symmetry for $\Phi(\xi, \tau)$ has moved in time τ the distance $v\tau$ in the direction of the force*. The probability dP that a particle, which at time t occupies the position x, will be located in the region between x_0 and $x_0 + h$ at the time $t + \tau$ is

$$dP = \Phi(x_0 - x, \tau)h = \phi(x_0 + (x - v\tau), \tau)h,$$

as the plane of symmetry $\phi(\xi, \tau)$ is now pushed to the position $x + v\tau$ by the force. The accumulation in the layer of thickness h between x_0 and $x_0 + h$ is

$$\Delta N_\tau = h\tau \left(\frac{\partial C}{\partial t}\right)_{x_0} = h \int_{-\infty}^{\infty} C(x, t)\phi(x_0 - (x + v\tau))\,dx - hC(x_0, t),$$

in analogy to Eq. (2.5.191). Introducing the displacement $\xi = x + v\tau - x_0$ results in

$$\tau \left(\frac{\partial C}{\partial t}\right)_{x_0} = \int_{-\infty}^{\infty} C(x_0 + \xi - v\tau, t)\phi(\xi, \tau))\,d\xi - C(x_0, t).$$

The function $C(x_0 + \xi - v\tau, t)$ in the integrand is expanded in a Taylor's series about x_0, i.e. in powers of $\xi - v\tau$. Thus

$$\tau \left(\frac{\partial C}{\partial t}\right)_{x_0} = \sum_{n=1}^{n=\infty} \frac{1}{n!} \left(\frac{\partial^n C}{\partial x^n}\right)_{x_0} \int_{-\infty}^{\infty} (\xi - v\tau)^n \phi(\xi, \tau)\,d\xi.$$

* A possible configuration of $\phi(\xi, \tau)$o and $\phi^*(\xi, \tau)$ is illustrated by Eq. (2.5.145) and Eq. (2.6.10).

All integrals in this expansion containing *odd* powers of n in the integrand $\xi^n \phi(\xi, \tau)$ will vanish, as $\phi(-\xi, \tau) = \phi(\xi, \tau)$ and $(-\xi)^n = -\xi^n$. When τ is macroscopically small it can be shown that all the higher terms of the integral containing $\xi^n \phi(\xi, \tau)$ for n even can be disregarded in comparison with the contribution from $\xi^2 \phi(\xi, \tau)$. With this approximation and after omission of the subscripts, the above expression becomes

$$\frac{\partial C}{\partial t} \tau = \frac{\partial C}{\partial x} \int_{-\infty}^{\infty} (\xi - v\tau) \phi(\xi, \tau) \, d\xi + \frac{1}{2} \frac{\partial^2 C}{\partial x^2} \int_{-\infty}^{\infty} (\xi - v\tau)^2 \phi(\xi, \tau) \, d\xi$$

$$= \frac{\partial C}{\partial x} \left\{ \int_{-\infty}^{\infty} \xi \phi(\xi, \tau) \, d\xi - v\tau \int_{-\infty}^{\infty} \phi(\xi, \tau) \, d\xi \right\}$$

$$+ \frac{1}{2} \frac{\partial^2 C}{\partial x^2} \left\{ \int_{-\infty}^{\infty} \xi^2 \phi(\xi, \tau) \, d\xi - 2v\tau \int_{-\infty}^{\infty} \xi \phi(\xi, \tau) \, d\xi \right.$$

$$\left. + (v\tau)^2 \int_{-\infty}^{\infty} \phi(\xi, \tau) \, d\xi \right\}.$$

Here the integrals containing $\xi \phi(\xi, \tau)$ will vanish. Furthermore

$$\int_{-\infty}^{\infty} \phi(\xi, \tau) \, d\xi = 1$$

as the particle must be somewhere in the range $-\infty \leq \xi \leq \infty$. We have then

$$\frac{\partial C}{\partial t} = \frac{1}{2\tau} \int_{-\infty}^{\infty} \xi^2 \phi(\xi, \tau) \, d\xi \cdot \frac{\partial^2 C}{\partial x^2} - v \frac{\partial C}{\partial x} + \frac{(v\tau)^2}{2\tau} \frac{\partial^2 C}{\partial x^2}.$$

The integral

$$\langle \xi^2 \rangle = \int_{-\infty}^{\infty} \xi^2 \phi(\xi, \tau) \, d\xi$$

represents the average value of the square of the displacements. Hence, we also have

$$\frac{\partial C}{\partial t} = \frac{\langle \xi \rangle^2}{2\tau} \frac{\partial^2 C}{\partial x^2} - v \frac{\partial C}{\partial x} + \frac{(v\tau)^2}{2\tau} \frac{\partial^2 C}{\partial x^2}, \tag{2.6.30}$$

as an expression for the accumulation provided the influence of the perturbing force results only in a translocation of $\phi(\xi, \tau)$ along the x-axis by the amount vt but not to a change in the shape of $\phi(\xi, \tau)$ compared with the force-free case. This expression can be simplified provided

$$\langle \xi^2 \rangle \gg (v\tau)^2.$$

This implies that the external field of force is so small that the translocation of $v\tau$ along the x-axis of the probability density ϕ in time τ is much less than the square root of the average value of the square of the Brownian displacements at the time τ. Hereby Eq. (2.6.30) becomes

$$\frac{\partial C}{\partial t} = \frac{1}{2\tau} \int_{-\infty}^{\infty} \xi^2 \phi(\xi, \tau) \, d\xi \cdot \frac{\partial^2 C}{\partial x^2} - v \frac{\partial C}{\partial x}, \qquad (2.6.31)$$

or

$$\frac{\partial C}{\partial t} = \frac{1}{2\tau} \int_{-\infty}^{\infty} \xi^2 \phi(\xi, \tau) \, d\xi \cdot \frac{\partial^2 C}{\partial x^2} - BX \frac{\partial C}{\partial x}. \qquad (2.6.32)$$

Again we put

$$D = \frac{\langle \xi^2 \rangle}{2\tau}, \qquad (2.6.33)$$

which is the Einstein–Smoluchowski equation, and finally we get

$$\frac{\partial C}{\partial t} = D \frac{\partial^2 C}{\partial x^2} - BX \frac{\partial C}{\partial x}, \qquad (2.6.34)$$

as derived by Smoluchowski (1913) and used in his studies of Brownian motion under the influence of an external field of force*. Naturally Eq. (2.6.34) could have been derived in Section 2.5.6.5 by extending the treatment that covered only the force-free case to also cover the effect of the force. But this would have complicated that section further and, therefore, we have chosen this sequence of presentation.

2.6.2.3 Migration over an energy barrier

In Section 2.5.6.6 we incorporated in the "Platzwechslung" of the diffusion process an energy barrier with height U_0 which limited the exchange of positions only to those molecules that had the necessary excess energy. A net transfer of molecules – diffusion – was brought about if the number of molecules in the two "valleys" adjacent to the energy barrier were different. Spatially, such an energy structure can be visualized as a periodic landscape of mountain ridges $U_1(x)$ with the same height U_0, where all the "valleys" are at the same horizontal level, i.e. have the same energy level. Figure 2.42 shows such a range of potential profiles. If the particle concentration $N(x)$ is uniformly distributed spatially, i.e. $N(x)$ does not change from valley to valley, the traffic across the top of each potential mountain consists of two streams of particles of equal size but oppositely directed. We now imagine that an external field of force X is set up

---

* The Smoluchowski equation is a special case of a more general equation that describes the behavior of the swarm with regard to both velocities and positions. This equation was put forward somewhat later than Smoluchowski's by the Dutch physicist A.D. Fokker (1915) and derived by Max Planck (1917).

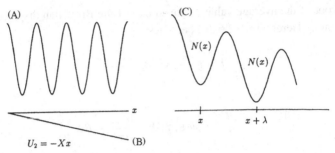

$U_2 = -Xx$ (B)

Fig. 2.42. (A) Illustration of the horizontal rows of potential profiles, each of height U_0, which the solute molecules pass when no external driving force X is present. (B) The potential energy curve $(-X x)$ for the external driving force. (C) The effect of application of the driving force X is to deform the original potential profiles into an asymmetric shape.

that acts perpendicularly to the direction of the ridges. Let the x-axis coincide with the direction of X. The force produces a migration flux

$$J = Nv = NBX,$$

where v is the migration velocity of each particle X and $B = x/X$ is its mechanical mobility. The questions to consider are: how does the force field affect the traffic across the potential barrier in such a way that the flux J above results? And, how is the mobility B related to the height U_0 of the energy barrier?

Since the height U_0 of the energy barrier represents the amount of work required to enter the top it is expedient to consider the presence of the driving force in the same way, i.e. by linking the force to the presence of an energy potential* U_2 in space that we define as

$$U_2 = -X x,$$

which represents the work we have to do in moving a distance x *against* the force reckoned from an arbitrary initial position x_0. A particle at position x is then in the total potential field that is the superposition of the periodic potential linked to the barrier and that due to the force, thus

$$U(x) = U_1(x) + U_2 = U_1(x) - X x.$$

The two potentials and their superpositions are illustrated in Fig. 2.42, which shows that the resultant potential profile still preserves the periodic character, but falls towards the direction of the x-axis, from left to right. We now consider a single potential barrier deformed in this way and with width λ. In the "valley"

* This is, for example, analogous to the potential energy associated with gravity.

at position x the particles with concentration $N(x)$ particles per unit volume moving in the direction $x \to x + \lambda$ are confronted with a potential barrier of height

$$U_0 - \tfrac{1}{2}\lambda X,$$

i.e. *reduced* by $\tfrac{1}{2}\lambda X$ as compared with the force-free case ($X = 0$), while the height in the opposite direction is

$$U_0 + \tfrac{1}{2}\lambda X,$$

i.e. *increased* by the amount $\tfrac{1}{2}\lambda X$. The same potential barrier confronts the particles in the "valley" at $x + \lambda$ with concentration $N(x)$ that are jumping in the direction $x + \lambda \to x$. As we will see below, it is just this difference in height with which the potential barrier faces the two "valleys" that set up a net transfer of particles in the direction of the force irrespective of a macroscopically uniform concentration.

To avoid unnecessary complications we assume that the two oppositely directed particle streams – unidirectional fluxes – have the same mean velocity across the potential barrier, thus

$$\langle \overrightarrow{v^*} \rangle = \langle \overleftarrow{v^*} \rangle = \langle v^* \rangle.$$

Then, in analogy to Section 2.5.6.4, Eq. (2.5.197), the unidirectional flux $\overrightarrow{J}$ in the direction $x \to x + \lambda$ is

$$\overrightarrow{J} = \tfrac{1}{2} N(x) \langle \overrightarrow{v^*} \rangle e^{-(U_0 - \lambda X/2)/kT}, \qquad (2.6.35)$$

and the unidirectional flux $\overleftarrow{J}$ in the direction $x \leftarrow x + \lambda$ is

$$\overleftarrow{J} = \tfrac{1}{2} N(x) \langle \overleftarrow{v^*} \rangle e^{-(U_0 + \lambda X/2)/kT}. \qquad (2.6.36)$$

The resultant flux J in the direction of x is then given by the difference

$$
\begin{aligned}
J &= \overrightarrow{J} - \overleftarrow{J} \\
&= \tfrac{1}{2} N(x) \langle \overrightarrow{v^*} \rangle e^{-(U_0 - \lambda X/2)/kT} - \tfrac{1}{2} N(x) \langle \overleftarrow{v^*} \rangle e^{-(U_0 + \lambda X/2)/kT} \\
&= \tfrac{1}{2} N(x) \langle \overrightarrow{v^*} \rangle e^{-U_0/kT} e^{(\lambda X/2)/kT} - \tfrac{1}{2} N(x) \langle \overleftarrow{v^*} \rangle e^{-U_0/kT} e^{(-\lambda X/2)/kT} \\
&= N(x) \langle v^* \rangle e^{-U_0/kT} \times \tfrac{1}{2} \big[e^{(\lambda X/2)/kT} - e^{(-\lambda X/2)kT} \big] \\
&= N(x) \langle v^* \rangle e^{-U_0/kT} \sinh\left(\frac{\lambda X}{2kT} \right).
\end{aligned}
\qquad (2.6.37)
$$

Series expanding the hyperbolic sinus* gives

$$J = N(x)\langle v^* \rangle e^{-U_0/kT} \left(\frac{\lambda X}{2kT} + \frac{1}{3!} \left[\frac{\lambda X}{2kT} \right]^3 + \frac{1}{5!} \left[\frac{\lambda X}{2kT} \right]^5 + \cdots \right). \quad (2.6.38)$$

For forces X that are sufficiently small to make $\lambda X \ll kT$ the first term in the series suffices. We have then

$$J = N(x)\langle v^* \rangle e^{-U_0/kT} \frac{\lambda X}{2kT}.$$

Let the mean passage time across the barrier be $\langle \tau_0 \rangle$. Then we have $\langle v^* \rangle = \lambda/\langle \tau_0 \rangle$, which, when inserted in the above expression, gives

$$J = N(x) \frac{\lambda^2}{2\langle \tau_0 \rangle} \frac{1}{2} kT \, e^{-U_0/kT} X. \quad (2.6.39)$$

In Section 2.5.6.6 the quantity $\lambda^2/2\langle \tau_0 \rangle$ was identified with a diffusion coefficient D_0. Similarly, it appears natural to regard the quantity $(\lambda^2/2\langle \tau_0 \rangle)/kT$ as the analogous mobility B_0 and write

$$B = B_0 \, e^{-U_0/kT}, \quad (2.6.40)$$

which in principle leaves B and D with the same type of temperature dependence. Inserting Eq. (2.6.40) into Eq. (2.6.38) gives

$$J = N(x) B X, \quad (2.6.41)$$

which is identical to Eq. (2.4.7), which was derived on the basis of the simplifying assumption that the dissolved molecules were driven through the solvent with a constant velocity and, directly to the point, without involving them jumping over a periodic row of potential barriers.

When the driving force is large enough to make the product λX quantitatively comparable to kT, the higher terms in the series expansion of the hyperbolic sinus in Eq. (2.6.38) can no longer be neglected, which leads to a nonlinear relation between the migration flux J and the driving force X. This important feature of the model is also observed experimentally.

We shall complete this line of argument by relaxing the idea that the particle concentrations $N(x)$ were the same in all the "valleys". Instead we assume that the particle concentrations change from "valley" to "valley", but that $N(x)$ changes in such a well-behaved manner that it makes sense to replace the difference $\Delta N = N(x + \lambda) - N(x)$ with its differential $(\partial N/\partial x)\lambda$. This implies that the concentrations $N(x)$, $N(x + \lambda)$, etc., are discrete points on a continuous, differentiable curve that reproduces correctly the concentrations at each "valley". On this basis the unidirectional flux $\vec{J}$ across the barrier in Fig. 2.42

---

* Since $\sinh y = y + \frac{y^3}{3!} + \frac{y^5}{5!} + \frac{y^7}{7!} \cdots$.

in the direction $x \rightarrow x + \lambda$ is

$$\overrightarrow{J} = \tfrac{1}{2} N(x) \langle v^* \rangle e^{-(U_0 - \lambda X/2)/kT}, \tag{2.6.42}$$

and the unidirectional flux $\overleftarrow{J}$ in the direction $x \leftarrow x + \lambda$ is

$$\overleftarrow{J} = \tfrac{1}{2} N(x + \lambda) \langle v^* \rangle e^{-(U_0 + \lambda X/2)/kT}$$
$$= \tfrac{1}{2} N(x) \langle v^* \rangle e^{-(U_0 + \lambda X/2)/kT} + \left(\tfrac{1}{2} \frac{\partial N}{\partial x} \lambda \right) \langle v^* \rangle e^{-(U_0 + \lambda X/2)/kT}. \tag{2.6.43}$$

From these two equations we again calculate the net flux $J = \overrightarrow{J} - \overleftarrow{J}$. A glance at the calculations leading to Eq. (2.6.37) shows that the expression for J now considered will only differ by containing the second term in Eq. (2.6.43). Thus we can immediately write

$$J = N(x) \langle v^* \rangle e^{-U_0/kT} \sinh \left(\frac{\lambda X}{2kT} \right) - \tfrac{1}{2} \frac{dN}{dx} \lambda \langle v^* \rangle e^{-U_0/kT} e^{(-\lambda X/2)/kT} \tag{2.6.44}$$

as the expression for the net flux of a swarm of particles moving over a potential barrier of height U_0 under the common effect of an external driving force X and varying concentration, the rate of change with distance being $\partial N/\partial x$.

When $\lambda X \ll kT$ the first term on the right-hand side is $N(x) B X$. For the second term we have as $e^\alpha \rightarrow 1$ for $\alpha \rightarrow 0$

$$\tfrac{1}{2} \frac{\partial N}{\partial x} \lambda \langle v^* \rangle e^{-U_0/kT} e^{(-\lambda X/2)/kT} \rightarrow \tfrac{1}{2} \lambda \langle v^* \rangle e^{-U_0/kT} \frac{\partial N}{\partial x}.$$

In Section 2.5.6.6 we made the identifications

$$\tfrac{1}{2} \lambda \langle v^* \rangle e^{-U_0/kT} = \tfrac{1}{2} \frac{\lambda^2}{\langle \tau_0 \rangle} e^{-U_0/kT} = D.$$

Thus, for $\lambda X \ll kT$ we have

$$J = -D \frac{\partial N}{\partial x} + B N X \tag{2.6.45}$$

in accordance with the results obtained previously.

2.6.3 Kramers' equation

We now consider the case where the external driving force X is a *conservative force*. This class of forces has the property that the work performed during the movement from one position (i) – the initial position – to another (f) – the final position – *depends only* on the two positions (i) and (f) and not upon the details of the movement between (i) and (f)*. Gravity and the electrostatic field strength are typical examples of conservative fields of force. Frictional forces are typical examples of *non-conservative forces*: the longer the displacement path, the larger is the performed work. To bodies acted upon by conservative forces we ascribe a *potential energy*. The potential energy of a body at a given point in space is equal to the work done *against* the conservative forces in moving the body to the position in question from an initial position where the potential energy is arbitrarily put equal to zero. Therefore, the work *done by* the conservative forces when displacing the body a given distance equals the *decrease* in the body's potential energy. Thus, the potential energy of a body is a function of its position in space. We denote the potential $U(x)$ as long as one-dimensional cases are considered. The potential energy $U(x)$ and the conservative force are interconnected by the relation[†]

$$X = -\frac{\partial U}{\partial x}. \qquad (2.6.46)$$

When X is a driving force in a migration process the flux in Eq. (2.6.2)

$$J = -D\frac{\partial C}{\partial x} + BXC\ k \qquad (2.6.47)$$

can be written in a more compact form that serves as a convenient starting point for many calculations of stationary fluxes, in particular when the particles have to move across one or more potential barriers. We divide Eq. (2.6.28) on both sides by D and insert Eq. (2.6.27). This gives

$$-\frac{J}{D} = \frac{\partial C}{\partial x} + \frac{B}{D}C\frac{\partial U}{\partial x}. \qquad (2.6.48)$$

Consider the identity

$$\frac{\partial}{\partial x}\left(C\,e^{U(x)B/D}\right) = \frac{\partial C}{\partial x}e^{U(x)B/D} + Ce^{U(x)B/D}\frac{B}{D}\frac{\partial U}{\partial x},$$

and rearrange in the form

$$\frac{\partial C}{\partial x} + \frac{B}{D}C\frac{\partial U}{\partial x} = e^{-U(x)B/D}\frac{\partial}{\partial x}\left(C\,e^{U(x)B/D}\right).$$

It follows that Eq. (2.6.28) can be written as

$$J\,e^{U(x)B/D} = -D\frac{\partial}{\partial x}\left(C\,e^{U(x)B/D}\right), \qquad (2.6.49)$$

---

* In thermodynamics the phrase *state function* refers to systems in which the changes depend only upon the initial and final state of the system and not on the way in which the changes are brought about.
† See Chapter 3, Section 3.2.

as an alternative form of the Smoluchowski equation when the driving force can be derived from a potential function. Multiplying on both sides by dx results in

$$J e^{U(x)B/D} dx = -D d(C e^{U(x)B/D}).$$

(2.6.50)

Integrating this equation between the two positions x_0 and x_1 in space where the concentrations and potentials are $C(x_0)$, $U(x_0)$ and $C(x_1)$, $U(x_1)$, respectively, gives

$$\int_{x_0}^{x_1} J e^{U(x)B/D} dx = -\int_{\substack{U(x_0) \\ C(x_0)}}^{\substack{U(x_1) \\ C(x_1)}} D d(C e^{U(x)B/D}).$$

If the situation is *stationary*, i.e. $\partial J / \partial x = 0$, we can move J outside the integration sign on the left-hand side. If, in addition, D is independent of concentration we have

$$J \int_{x_0}^{x_1} e^{U(x)B/D} dx = -D \int_{\substack{U(x_0) \\ C(x_0)}}^{\substack{U(x_1) \\ C(x_1)}} x d(C e^{U(x)B/D})$$

$$= -D\left[C(x_1) e^{U(x_1)B/D} - C(x_0) e^{U(x_0)B/D}\right].$$

Thus, the flux between the points x_0 and x_1 is given by

$$J = D \frac{C(x_0) e^{U(x_0)B/D} - C(x_1) e^{U(x_1)B/D}}{\int_{x_0}^{x_1} e^{U(x)B/D} dx},$$

(2.6.51)

which describes the stationary flux in a swarm of particles executing Brownian motion in the presence of a conservative field of force to which the potential energy $U(x)$ can be assigned. This important equation was first derived by H.A. Kramers (1940)*.

2.6.4 Diffusion coefficient and mobility

In the expression of Fick's law

$$J = -D \frac{\partial C}{\partial x},$$

the diffusion coefficient D is the parameter that connects the concentration gradient with the flux. In Section 2.5.6.5 it was shown that the diffusion coefficient D and mean of the squared displacements $\langle \xi^2 \rangle$ of the individual particles

---

* Kramers, H.A. (1940): *Physica*, **7**, 240. The Dutch physicist H.A. Kramers (1894–1952) collaborated closely with Niels Bohr in Copenhagen during the time 1916–1926. Later he became Professor of physics in Utrecht.

executed in time τ are connected by the Einstein–Smoluchowski relation

$$D = \frac{\langle \xi^2 \rangle}{2\tau}.$$

The magnitude of $\langle \xi^2 \rangle$ will again depend on how far a particle is displaced after receiving a shortlasting impact on account of an asymmetric bombardment from the surrounding molecules. For a given impulse on a particle the resistance to movement that the surrounding molecules offer – as reflected in the coefficient of friction β or the mobility B – will determine the length of the displacement. Therefore, one may expect that two coefficients D and B that enter in the Smoluchowski equation are interconnected uniquely. Such a relation was first derived by Einstein (1905). In the same paper he used this relation to derive an important formula that allowed a determination of Avogadro's constant from the data of the Brownian motion executed by a suspension of microscopic particles of uniform size. This formula has also turned out to be useful in the determination of the size of dissolved molecules.

2.6.4.1 The Einstein relation

We again consider a swarm of particles driven towards a totally reflecting barrier by an external force X as already dealt with in Section 2.6.1.1(ii). The field of force is assumed to be conservative so we put $X = -dU/dx$, where U is the potential energy of the particles at the distance x from the reflecting plane at $x = 0$. The force X tends to drive all the particles down to the plane at $x = 0$, which is the position of lowest potential energy, whereas the thermal movements tends to distribute the particles uniformly over the accessible space. After a certain time these two opposing tendencies will develop into a stationary state that also corresponds to an *equilibrium state*, where the tendency of the swarm to move in one direction because of the sloping concentration profile $C(x)$ is counterbalanced by the tendency towards motion in the opposite direction due to the presence of the force X. The equilibrium state is characterized by the condition

$$J = 0, \tag{2.6.52}$$

everywhere in space. We consider the two positions x and $x = 0$ with the concentrations (number of particles per m^3) $C(x)$ and $C(0)$. Furthermore we arbitrarily put the potential energy of the particles $U(x) = 0$ for $x = 0$. The equilibrium condition $J = 0$ means that the numerator in Kramers' equation (Eq. 2.6.51) must be zero. Hence

$$C(x)\,e^{U(x)B/D} = C(0)\,e^0,$$

or

$$C(x) = C(0)\, e^{-U(x)B/D}.$$

The correlation statistical thermodynamics to this dynamic equilibrium state is the *Maxwell–Boltzmann energy distribution*

$$C(x) = C(0)\, e^{-U(x)/kT}, \tag{2.6.53}$$

where k is Boltzmann's constant*. As the two expressions above are identical, we must have

$$B/D = 1/kT,$$

or

$$D = kT\, B. \tag{2.6.54}$$

This is the famous Einstein equation[†], which links the diffusion coefficient D of the particles with their mechanical mobility B. The relation implies – among other things – that only *one parameter* enters into the Smoluchowski equation; according to convenience, this may be either be D or B. Likewise, Kramers' equation takes the more general, and simpler, form

$$J = D\, \frac{C(x_0)\, e^{U(x_0)/kT} + C(x_1)\, e^{U(x_1)/kT}}{\displaystyle\int_{x_0}^{x_1} e^{U(x)/kT}\, \mathrm{d}x}. \tag{2.6.55}$$

2.6.4.2 Einstein–Stokes relation

On the basis of the macroscopic hydrodynamic laws, Sir G.G. Stokes showed (1856)[‡] that when a spherical body of radius r immersed in a fluid of viscosity η is driven with a stationary velocity v under the action of a force X, these quantities are interconnected by the formula

$$X = 6\pi r \eta v, \tag{2.6.56}$$

---

* $k = 1.380\,47 \times 10^{-23}\,\mathrm{J\,K^{-1}\,molecule^{-1}}$.

[†] Einstein derived this equation by considering the equilibrium between a diffusion flux and an oppositely directed migration whose virtual driving force was identified with an osmotic pressure gradient of the solvent.

[‡] G.G. Stokes (1819–1903), became professor of mathematics at Cambridge in 1849. He made many important contributions to mathematics and mathematical physics, in particular hydrodynamics, optics, electricity and acoustics. He became President of the Royal Society, London, 1885–90, and was raised to the peerage in 1889.

which is called *Stokes' equation*. Using the relation $v = BX$ to eliminate v in Stokes' equation, we obtain for a spherical particle the expression

$$B = \frac{1}{6\pi r \eta},$$ (2.6.57)

provided that the radius r is large enough relative to the radius of the water molecules that are still within the range of validity of the hydrodynamical laws. Inserting the Einstein equation, Eq. (2.6.54), gives

$$D = \frac{kT}{6\pi r \eta},$$

or by multiplying the numerator and denominator by Avogadro's constant $\mathcal{N}_A$

$$D = \frac{\mathcal{R}T}{6\pi r \eta \mathcal{N}_A},$$ (2.6.58)

making use of the relation $\mathcal{R} = k\mathcal{N}_A$, where $\mathcal{R}$ is the *gas constant* having the numerical value*

$$\mathcal{R} = 8.314 \text{ J mol}^{-1} \text{ K}^{-1}.$$

Equation (2.6.58) is called the Einstein–Stokes relation, even though it is due solely to Einstein. This relation is important as – among other things – it makes determination of molecular dimensions possible by measurements of the diffusion coefficient of the substance in question when diffusing in a medium with a known viscosity η. In Appendix G we show that application of the above equations derived from the theory of Brownian motion to the spectrum of displacements of a Brownian particle in the time t discloses a fundamental constant of nature, Avagadro's constant $\mathcal{N}_A$.

The Einstein–Stokes relation also throws light upon the fact that the magnitude of the diffusion coefficient depends only slightly upon the molecular weight of the substance. We have

$$Dr = \frac{\mathcal{R}T}{6\pi \eta \mathcal{N}_A} = constant,$$ (2.6.59)

for the diffusion in a particular medium at constant temperature. The mass m of a sphere of radius r and mass density ρ is

$$m = \tfrac{4}{3}\pi r^3 \rho,$$

* In other units: $\mathcal{R} = 1.986\ 46 \text{ cal}_{15} \text{ mol}^{-1}\text{K}^{-1} = 8.205\ 44 \times 10^{-2} \text{ dm}^3 \text{ atmmol}^{-1} \text{ K}^{-1}$.

Table 2.3. *Relation between diffusion coefficient and molecular weight*

Molecule	M (Dalton)	D (m^2 s^{-1} × 10^{12})	$D\sqrt[3]{M}$ × 10^9
Glycine	76	95	4.02
Arginine	174	58	3.24
Cytochrome C	13 000	10.1	2.73
Pepsin	36 000	9.0	2.94
CO-hemoglobin	68 000	6.2	2.53
Urease	480 000	3.5	2.74
TMV	40 000 000	0.53	1.81

From Setlow R.B. & Pollard E.C. (1962): *Molecular Biophysics*, p. 83, Addison-Wesley Publishing Company Inc., Reading, Massachusetts, USA.

as the volume of a sphere is $\frac{4}{3}\pi r^3$. If this mass represents a single spherical molecule the molecular weight is

$$M = \mathcal{N}_A\, m = \tfrac{4}{3}\pi r^3 \rho \mathcal{N}_A,$$

as Avogadro's constant $\mathcal{N}_A$ equals the number of molecules per mol. From this it follows that

$$r \propto \sqrt[3]{M}.$$

Inserting this in Eq. (2.6.58) it follows that*

$$D\sqrt[3]{M} = \text{constant.} \qquad (2.6.60)$$

Table 2.3 presents a collection of substances of widely different molecular weights together with their values of diffusion coefficients in water (°C) and the product of $D\sqrt[3]{M}$. The table shows that even with a very large spectrum of molecular weights the product $D\sqrt[3]{M}$ remains reasonably constant.

There is no inconsistency between Eq. (2.6.61) and the law of diffusion of T. Graham (1829)[†] states that the rate of diffusion – and of effusion – of a *gas* is inversely proportional to the square root of the density or, alternatively, to the square root of the molecular weight M of the diffusing gas molecules. Graham obtained his law from careful quantitative experiments without any reference

---

* The impatient reader may object to the redundancy of precision presented here by saying that from Eq. (2.6.58) all that is needed is the next almost obvious statement that $r \propto \sqrt[3]{M}$.

[†] Graham, T. (1829): On the law of diffusion of gases. *Quart. J. Sci.*, **2**, 74. Reprinted in *Chemical and Physical Researches*, Edinburgh University Press, Edinburgh, 1876.

to diffusion coefficients. Later theoretical studies started with Maxwell (1860)[*] who, considering the gas molecules as elastic spheres that move with a mean free path much larger than the radius of the molecule and with an average speed of $\langle v \rangle = \sqrt{3kT/M}$, calculated from the occurrence of incessant collisions between the gas molecules *themselves* both the diffusion flux and the diffusion coefficient of the gas molecules and found both quantities to be inversely proportional to $\sqrt{M}$.

In a dense fluid system, however, the collisions between the diffusing molecules themselves are a rare event and the vacant space accessible is very sparse, so the Brownian particle ploughs its way through a viscous fluid exposed to a viscous drag acting in opposition to the direction of motion, and loses its excess momentum on the way until it ends up in a vacant position (a "hole"). This implies that a characteristic coefficient of friction β and a mobility $B = 1/\beta$ are assigned to the particle, both of which depend upon the size and shape. For particles above a certain size the relation between the driving force X and the particle velocity v holds approximately well by Stoke's law, i.e. $B \propto 1/r$. Invoking the Einstein relation $D = kTB$ it follows that $D \propto 1/r$ and hence $D \propto \sqrt[3]{M}$.

2.6.4.3 The "driving force" behind the diffusion process

Mass transfer by migration appears as a result of an external driving force X that causes the particles to move through the solvent with an average velocity v in the direction of the force that superimposes upon irregular random movements of the particles due to thermal agitation. The migration flux is

$$J = BCX_{\text{migr}} \tag{2.6.61}$$

(see also Eq. (2.4.9)). In the diffusion process, however, there is no question of a finite diffusion velocity in a *certain direction*, e.g. in the direction of the concentration drop. Neither does the presence of a diffusion gradient establish a special "diffusions force" that drives each diffusing molecule in the direction of the concentration drop.

In connection with certain formal calculations it is sometimes convenient to overlook the processes that on a molecular level are the causes of the transport process by diffusion, and instead regard the diffusion process as the cause of a force that drives each solvent particle in the direction of the decreasing concentration. We shall now determine the magnitude of this equivalent force.

Consider a swarm of diffusing particles having concentration C and concentration gradient dC/dx at the position x. The diffusion flux J_{dif} is, according to

---

[*] Maxwell, J.C. (1860): Illustrations of the dynamical Theory of Gases. *Phil. Mag.*, **20**, 21.

Fick's law,

$$J_{dif} = -D\frac{dC}{dx}.$$

We multiply the right-hand side of this equation by $1 = (B/B) \cdot (C/C)$, and rearrange the factors in this way

$$J_{dif} = BC\left[-\frac{D}{B}\frac{1}{C}\frac{dC}{dx}\right].$$

This disguised expression for Fick's law is of the same form as the expression for the migration flux $J_{migr} = BCX_{migr}$ given previously. Thus, a diffusion flux in a position with concentration gradient dC/dx and concentrations C, can *formally* be considered as a migration process that is caused by an equivalent "force of diffusion" of magnitude

$$X_{dif} = -\frac{D}{B}\frac{1}{C}\frac{dC}{dx} = -\frac{D}{B}\frac{d\ln C}{dx}, \qquad (2.6.62)$$

i.e. that driving force on each solute particle in a region with an uniform concentration of concentration C gives rise to a mass transfer of just the same magnitude as that taking place by diffusion if the gradient were dC/dx. Invoking the Einstein relation gives the following alternative expression for the force of diffusion

$$X_{dif} = -kT\frac{d\ln C}{dx}. \qquad (2.6.63)$$

Introducing the chemical potential for the solute particles at the position x

$$\mu(x) = \mu^0 + \mathcal{R}T\ln C(x),$$

leads to the following expression for the force of diffusion

$$X_{dif} = -\frac{1}{\mathcal{N}_A}\frac{d\mu}{dx}. \qquad (2.6.64)$$

The expression on the right-hand side is sometimes called the "thermodynamic force" for the diffusion process, although it remains unclear how a gradient of a chemical potential can give rise to a force that acts on each single solute particle. However, giving this force an imposing name does not change the situation that the "force of diffusion" results from a few purely formal manipulations of Fick's law, which only gives this force the character of a *pseudoforce*, just like, for example, a centrifugal force, being without any real physical content and whose introduction has not contributed to any further unsight into the underlying mechanism or the diffusion process.

Also geht die Diffusion einfach als Resultat der ungestörten BROWNschen Bewegungen der einzelnen Teilschen hervor, und es ist ganz falsch, wenn manche Forscher meinen, das dabei noch ein spezieller, der BROWNschen Bewegung eine Richtung gebender Einfluss, – etwa das osmotische Druckgefälle – tätig sein. Der fiktive, der makroskopischen Auffassungsweise der Diffusion entsprechende Begriff des osmotischen Druckes vertritt die Betrachtung der "verborgenen" Molekularbewegungen und is mit derselbender – soweit die klassischen Diffusionserscheinungen in Betracht kommen – vollkommen äkvivalent, darf aber mit ihr nicht verquickt werden. Entweder denke man sich die Substanzteilschen als passiv durch den osmotischen Druck getrieben, ohne die BROWNschen Bewegungen zu berücksichtigen, oder man ziehe die letztere in Rechnung, ohne den fiktiven osmotischen Druck einzuführen.

(M.v. Smoluchowski: Drei Vorträge über Diffusion, Brownsche Molekularbewegung
und Koagulation von Kolloidteilschen, 1918).

2.7 Diffusion through membranes

In comparison with the cell's cytoplasm and its surrounding fluid, the cell membrane presents a considerably larger barrier to the transfer of ions and hydrophilic molecules. Thus, when considering the passive transfer between the cytoplasm and the exterior we may safely assume that the total concentration change between the cytoplasm and the exterior is located across the cell membrane.

As an introduction to the treatment of mass transfer through biological membranes that follows, we consider an artificial, homogeneous membrane of thickness h, surrounded by two solutions of neutral molecules, e.g. sucrose, of concentrations $C^{(i)}$ and $C^{(o)}$ in the two surrounding phases, which we denote by (i) and (o). The transfer of the dissolved substance through the membrane takes place by diffusion. Let the direction (i) $\rightarrow$ (o) be the positive direction of the x-axis. The flux through the membrane in the direction (i) $\rightarrow$ (o) is, according to Fick's law*,

$$J = -D\frac{dC}{dx},$$ (2.7.1)

where dC/dx is the concentration gradient at position x inside the membrane and D is the diffusion coefficient for the substance in question *in the membrane*. We assume that the concentration profile in the membrane is linear. This assumption will be correct for a stationary flux of uncharged molecules[†], Eq. (2.5.11.), but will also be valid for ions when there is no electric field in the membrane. Let $C_m^{(i)}$ and $C_m^{(o)}$ represent the membrane concentrations bordering the cytoplasm and the external fluid. If the membrane thickness is h we have

---

* Section 2.5.2, Eq. (2.5.3).
† See Section 2.5.4.

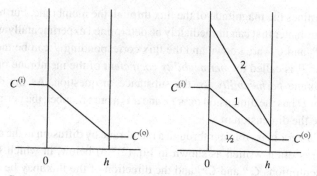

Fig. 2.43. Concentration profiles through a homogeneous membrane of thickness h. (A) The diffusing substance has the same solubility as the surrounding media. (B) Different solubility in membrane and external medium. The figures on the curves denote the values of the distribution coefficients.

(Fig. 2.43(A))

$$\frac{dC}{dx} = \frac{C_m^{(0)} - C_m^{(i)}}{h},$$

which inserted in Eq. (2.7.1) gives

$$J = -D\frac{C_m^{(0)} - C_m^{(i)}}{h} = -\frac{D}{h}\left[C_m^{(0)} - C_m^{(i)}\right]. \tag{2.7.2}$$

2.7.1 Permeability coefficient

The practical applicability of the above expression is rather limited. The thickness of the membrane is generally not known with greater precision. A further complication enters if the solubility of the diffusing substance is not the same in the membrane as in the surrounding medium. This effect is illustrated in Fig. 2.42(B). A formal remedy – or rather a cover of our ignorance – is to write

$$C_m^{(0)} = \alpha C^{(0)} \quad \text{and} \quad C_m^{(i)} = \alpha C^{(i)}, \tag{2.7.3}$$

where α is the distribution coefficient for the substance. We can then write Eq. (2.7.3) in the form

$$J = -\frac{\alpha D}{h}\left[C^{(0)} - C^{(i)}\right]. \tag{2.7.4}$$

Thus, with a given set of values for $C^{(0)}$ and $C^{(i)}$ it is the factor

$$P = \frac{\alpha D}{h} = \frac{\alpha kT B}{h}, \tag{2.7.5}$$

that determines the magnitude of the flux through the membrane. Furthermore, it is the parameter that can immediately be determined experimentally when the values $C^{(o)}$ and $C^{(i)}$ are known and the flux corresponding J can be measured. The factor P is called the *permeability coefficient* of the membrane or simply the membrane *permeability* for the substance in question. As the diffusion coefficient D has the dimension $\mathrm{m^2\,s^{-1}}$, and α is a pure number, the permeability must have the dimension $\mathrm{m\,s^{-1}}$.

Fick's law for mass transfer through a membrane by diffusion in the direction (i) → (o) is often written as shown in Eq. (2.7.6) below, in which form the two concentrations $C^{(i)}$ and $C^{(o)}$ and the direction of the flux may be easier to remember:

$$J = P\left[C^{(i)} - C^{(o)}\right]. \tag{2.7.6}$$

Still, it is worthwhile to bear in mind the inverse relationship that exists between D and α. A large value of α will counteract the tendency to reduce the flux if the diffusion coefficient D in the membrane is small in proportion to the diffusion coefficient D' in the solution. For example, as seen from Eq. (2.7.5) the membrane does not present any barrier whatsoever in relation to the surrounding medium provided

$$\alpha D = D'.$$

Conversely, the transfer with the same diffusion coefficient in the solution and in the membrane can nevertheless be restricted by the membrane if the solubility in the membrane is considerably less than in aqueous solution. A membrane thus presents the *greatest transfer barrier* to those substances whose *mobility and solubility in the membrane* are much less than in the surrounding solutions.

2.7.2 Kinetics of exchange

To determine the membrane permeability P, the most frequent method is – having Eq. (2.7.6) in mind – to establish a convenient experimental arrangement such as that illustrated in Fig. 2.5, where the two phases (i) and (o) surrounding the membrane are, by and large, kept constant during the determination of the mass transfer through the membrane. By determining permeabilities of biological membranes this can only rarely be realized, as the small cell volume necessarily involves a change in the cytoplasmic concentration during the exchange to the external medium. However, there are many experimental situations where this restriction will not cause any serious complication in the experimental design. In the simplest case the volume of the external phase is made very large in proportion to the total cellular volume, so that its concentration for all practical purposes can be considered as constant. We shall now consider two cases.

2.7.2.1 Outer concentration kept at zero

Let C^0 be the molar concentration in phase (i) at time $t = 0$ and $C(t)$ at time t. The flux in the direction (i) $\rightarrow$ (o) at the time t is, according to Eq. (2.7.6),

$$J = P \left[C(t) - C^{(0)} \right] = PC(t), \tag{2.7.7}$$

as $C^{(0)} = 0$ for all values of t. The amount $\Delta M^{(\Delta t)}$ that in time Δt is removed from phase (i) through the membrane is

$$\Delta M^{(\Delta t)} = J A \, \Delta t,$$

where A is the area of the membrane. The concentration in phase (i) is, within the same time interval, changed from $C(t)$ to $C(t + \Delta t)$. Hence, we have

$$\Delta M^{(\Delta t)} = [C(t) - C(t + \Delta t)] \, V = J A \, \Delta t,$$

where V is the volume of phase (i). Dividing on both sides by $V \, \Delta t$ and invoking Eq. (2.7.6) gives

$$-\frac{C(t + \Delta t) - C(t)}{\Delta t} = \frac{A P}{V} C(t'),$$

where $C(t')$ is the mean concentration in the time lapse of $t \leq t' \leq t + \Delta t$. When $\Delta t \rightarrow 0$, the left-hand side approaches $-dC/dt$ and $C(t') \rightarrow C(t)$, thus

$$\frac{dC}{dt} = -\frac{A P}{V} C = -kC, \tag{2.7.8}$$

where the functional dependence $C(t)$ of C is now omitted. The quantity

$$k = \frac{A P}{V} \triangleq s^{-1}, \tag{2.7.9}$$

is called the *rate constant* for the exchange process.

The movement of the substance from phase (i) has a time course that is governed by the equation

$$\frac{dC}{dt} + kC = 0, \tag{2.7.10}$$

the general solution of which can be written as*

$$C(t) = K_1 e^{-kt}, \tag{2.7.11}$$

---

* See Chapter 1, Section 1.9.1.1, Eq. (1.9.5).

where K_1 is an arbitrary constant that depends upon the character of the problem under consideration, in this case the initial condition

$$C(t) = C^0 \quad \text{for} \quad t = 0.$$

Inserting this condition in Eq. (2.7.11) gives

$$K_1 = C^0,$$

leading to the final solution of Eq. (2.7.8)

$$C(t) = C^0 e^{-kt}, \qquad (2.7.12)$$

or

$$\ln C(t) = \ln C^0 - kt, \qquad (2.7.13)$$

from which it appears that the concentration in phase (i) decreases exponentially with time, and the rate of change depends upon the magnitude of the rate constant k

$$k = \frac{AP}{V}. \qquad (2.7.14)$$

The greater the rate constant, the faster the substance will disappear with time from phase (i). The factors that determine this rate are the ratio between the membrane A and volume V of phase (i) together with the membrane permeability P.

A convenient way of determining the rate constant for a number of values of $C(t)$ and t belonging together is either to plot $\ln C$ versus t or to use a program for linear regression analysis. The resulting slope $d\ln C/dt$ is, according to Eq. (2.7.14), equal to $-k$. If, in addition, the values of the membrane area A and the volume of phase (i) are known then by using Eq. (2.7.5) the membrane permeability P for the substance in question can be calculated.

2.7.2.2 *Outer concentration kept constant: cell concentration initially zero*

We now consider the reverse situation where the inner phase (i) is kept initially at zero concentration ($C(0) = 0$) and is loaded from the outer phase (o) that is maintained at a constant concentration C^0. Let $C(t)$ be the concentration in phase (i) at the time t. The flux in the direction (o) → (i) at time t is

$$J = P[C^0 - C(t)]. \qquad (2.7.15)$$

In the time Δt the amount (moles)

$$\Delta M^{(\Delta t)} = J A \Delta t$$

is transferred into phase (i) with area A and volume V. This amount can also be written as

$$\Delta M^{(\Delta t)} = [C(t + \Delta t) - C(t)] \, V = J A \, \Delta t = P[C^0 - C(t)] A \Delta t.$$

Dividing both sides by $V \Delta t$ we have

$$\frac{C(t + \Delta t) - C(t)}{\Delta t} = \frac{AP}{V} [C^0 - C(t)],$$

and letting $\Delta t \to 0$ gives

$$\frac{dC(t)}{dt} = k[C^0 - C(t)],$$

or

$$\frac{dC(t)}{dt} + kC(t) = kC^0. \tag{2.7.16}$$

This is the differential equation for the mass accumulation in phase (i). The solution for the initial condition $C(0) = 0$ for $t = 0$ is*

$$C(t) = C^0 [1 - e^{-kt}], \tag{2.7.17}$$

showing that the concentration $C(t)$ in phase (i) grows exponentially asymptotically towards the value C^0 and with the *same* rate constant as in the previous situation. It may also be observed that the relation between the time t and $\ln[1 - C(t)/C^0]$ is linear and again with the slope $-k$.

2.7.2.3 Both phases comparable in size

Sometimes it is not practicable to establish the experimental situation described above where the surrounding phase is regarded as infinitely large as compared with the object that is being studied. When both phases are comparable in size the kinetics of exchange may be altered as it is no longer possible to maintain the concentration in the outer phase at a constant level. This situation that involves an account of the state in both phases is treated below.

In phase (i) with volume $V^{(i)}$ an amount of substance m_0 is in solution at time $t = 0$, while phase (o) with volume $V^{(o)}$ initially contains no dissolved substance. We consider the state at time t where the concentrations are $C^{(i)}$ and $C^{(o)}$ in phases (i) and (o) respectively. The flux through the membrane in the direction (i) $\to$ (o) is

$$J(t) = P\left(C^{(i)} - C^{(o)}\right). \tag{2.7.18}$$

---

* See Chapter 1, Section 1.9.1.2, Eq. (1.9.7). Put $\alpha = k$ and $K = kC^0$. It may be instructive to solve the equation directly by putting $U(t) = C^0 - C(t)$, i.e. $dU/dt = -dC/dt$ and first solve for $U(t)$.

Thus, phase (o) has in the time between t and $t + \mathrm{d}t$ received an amount of substance $\mathrm{d}m^{(\mathrm{o})}$ that is

$$\mathrm{d}m^{(\mathrm{o})} = AJ(t)\,\mathrm{d}t = AP\left(C^{(\mathrm{i})} - C^{(\mathrm{o})}\right)\mathrm{d}t. \tag{2.7.19}$$

As the total dissolved substance remains constant we have

$$m_0 = m^{(\mathrm{i})} + m^{(\mathrm{o})},$$

where $m^{(\mathrm{i})}$ and $m^{(\mathrm{o})}$ are the amounts in phases (i) and (o) at time t. The concentrations in the two phases are then written as

$$C^{(\mathrm{i})} = \frac{m^{(\mathrm{i})}}{V^{(\mathrm{i})}} = \frac{m_0 - m^{(\mathrm{o})}}{V^{(\mathrm{i})}} \quad \text{and} \quad C^{(\mathrm{o})} = \frac{m^{(\mathrm{o})}}{V^{(\mathrm{o})}}.$$

Insertion of these values for $C^{(\mathrm{i})}$ and $C^{(\mathrm{o})}$ in Eq. (2.7.17) gives

$$\begin{aligned}
\frac{\mathrm{d}m^{(\mathrm{o})}}{\mathrm{d}t} &= AP\left\{\frac{m_0 - m^{(\mathrm{o})}}{V^{(\mathrm{i})}} - \frac{m^{(\mathrm{o})}}{V^{(\mathrm{o})}}\right\} \\
&= AP\left\{\frac{m_0}{V^{(\mathrm{i})}} - \left(\frac{1}{V^{(\mathrm{i})}} + \frac{1}{V^{(\mathrm{o})}}\right)m^{(\mathrm{o})}\right\} \\
&= AP\frac{m_0}{V^{(\mathrm{i})}} - AP\frac{V^{(\mathrm{i})} + V^{(\mathrm{o})}}{V^{(\mathrm{i})}V^{(\mathrm{o})}}m^{(\mathrm{o})} \\
&= b - k\,m^{(\mathrm{o})}, \tag{2.7.20}
\end{aligned}$$

where

$$k = AP\frac{V^{(\mathrm{i})} + V^{(\mathrm{o})}}{V^{(\mathrm{i})}V^{(\mathrm{o})}} \quad \text{and} \quad b = AP\frac{m_0}{V^{(\mathrm{i})}}.$$

Dividing by $V^{(\mathrm{o})}$ on both sides of Eq. (2.7.20) gives the following equation for $C^{(\mathrm{o})}$

$$\frac{\mathrm{d}C^{(\mathrm{o})}}{\mathrm{d}t} + kC^{(\mathrm{o})} = \frac{b}{V^{(\mathrm{o})}}. \tag{2.7.21}$$

As the phase (o) contains no substance at time zero, Eq. (2.7.21) must satisfy the initial condition $C^{(\mathrm{o})} = 0$ for $t = 0$. The appropriate solution is*

$$C^{(\mathrm{o})} = \frac{m_0}{V^{(\mathrm{i})} + V^{(\mathrm{o})}}[1 - \mathrm{e}^{-kt}] = C_\infty[1 - \mathrm{e}^{-kt}], \tag{2.7.22}$$

where $C_\infty = m_0/(V^{(\mathrm{i})} + V^{(\mathrm{o})})$ is the final concentration $C^{(\mathrm{o})}$ in phase (o) when equilibrium $(\mathrm{d}C^{(\mathrm{o})}/\mathrm{d}t = 0)$ is established. As expected, this concentration appears to be equal to the total amount initially present divided by the total volume

---

* See Chapter 1, Section 1.9.1.2, Eq. (1.9.8). Put $\alpha = k$ and $K = b/V^{(\mathrm{o})}$.

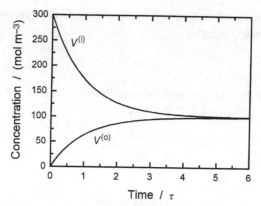

Fig. 2.44. The time course of the exchange by diffusion between two compartments of finite sizes: ($V^{(i)} : V^{(o)} = 2$). At time $t = 0$ we have $C^{(i)} = 300$ and $C^{(o)} = 0$. Abscissa: time with the time constant $\tau = 1/k$ as unit. Ordinate: concentration.

of the system. Thus, according to Eq. (2.7.22), the concentration in phase (o) rises exponentially and asymptotically towards C_∞ with a rate constant

$$k = AP\frac{V^{(o)}}{V^{(i)}V^{(o)}}, \tag{2.7.23}$$

that contains both volumes between which the exchange takes place*.

To determine the time course of $C^{(i)}$ in phase (i) we note that

$$m_0 = C^{(i)}V^{(i)} + C^{(o)}V^{(o)},$$

is valid for any time. Solving for $C^{(i)}$

$$C^{(i)} = \frac{m_0}{V^{(i)}} - \frac{V^{(o)}}{V^{(i)}}C^{(o)},$$

and replacing $C^{(o)}$ from Eq. (2.7.23) gives

$$C^{(i)} = \frac{m_0}{V^{(i)}} - \frac{V^{(o)}}{V^{(i)}}C_\infty[1 - e^{-kt}],$$

or by substituting $m_0 = C_\infty V^{(i)} + C_\infty V^{(o)}$

$$C^{(i)} = C_\infty\left(1 + \frac{V^{(o)}}{V^{(i)}}e^{-kt}\right). \tag{2.7.24}$$

The time course towards the state of equilibrium in the system is illustrated in Fig. 2.44. The two concentrations $C^{(i)}$ and $C^{(o)}$ exponentially approach the

---

* Equation (2.7.22) could also be interpreted as the loading of a fictitious, equivalent volume $\overline{V}$ from an external phase of constant concentration $C_\infty = m_0/(V^{(i)} + V^{(o)})$, where $1/\overline{V} = 1/V^{(o)} + 1/V^{(i)}$.

2. *Migration and diffusion*

equilibrium concentration C_∞, and the time courses are governed by the rate constant k or time constant τ

$$\tau = \frac{1}{k}. \tag{2.7.25}$$

To determine the rate constant k from a collection of data of $C^{(0)}$ and t obtained experimentally we write Eq. (2.7.22) as

$$Y(t) = \frac{C_\infty - C^{(0)}(t)}{C_\infty} = e^{-kt}, \tag{2.7.26}$$

or

$$\ln Y(t) = -kt.$$

Thus, the plot of $\ln Y(t)$ versus T represents a straight line with a slope $(-k)$, from which the permeability P is calculated by using Eq. (2.7.23), provided the three parameters A, $V^{(i)}$ and $V^{(0)}$ are known*. To determine the rate constant k from the time course of $C^{(i)}$ we plot instead

$$\ln Z(t) = \frac{C^{(i)}(t) - C_\infty}{C_\infty} = \frac{V^{(0)}}{V^{(i)}} e^{-kt},$$

or

$$\ln Z(t) = \ln\left(\frac{V^{(0)}}{V^{(i)}}\right) - kt,$$

which also yields a linearly decreasing function with slope $-k$. The value of $\ln Z$ extrapolated to $t = 0$ gives in addition the value of $\ln(V^{(0)}/V^{(i)})$.

It may happen that the conditions of the experiment result in a set of data corresponding to two times, e.g. t_1, $C_1^{(0)}$ and t_2, $C_2^{(0)}$. We insert the data in the integrated equation Eq. (2.7.26) for $C^{(0)}$

$$\ln Y(t_1) = -kt_1,$$
$$\ln Y(t_2) = -kt_2.$$

Subtraction of the lower equation from the upper one yields

$$k(t_2 - t_1) = \ln Y(t_1) - \ln Y(t_2) = \ln\left(\frac{Y(t_1)}{Y(t_2)}\right)$$

$$= \ln\left(\frac{C_\infty - C_1^{(0)}}{C_\infty} \times \frac{C_\infty}{C_\infty - C_2^{(0)}}\right) = \ln\left(\frac{C_\infty - C_1^{(0)}}{C_\infty - C_2^{(0)}}\right).$$

---

* This procedure only works if a reliable value of C_∞ is known. If such a value is not available one has to apply a nonlinear regression analysis instead.

Inserting $C_\infty = m_0/(V^{(i)}V^{(o)})$ and the expression for the rate constant (Eq. (2.7.23)) gives

$$AP\frac{V^{(i)} + V^{(o)}}{V^{(i)}V^{(o)}}(t_2 - t_1) = \ln\frac{m_0 - C_1^{(o)}\left(V^{(i)} + V^{(o)}\right)}{m_0 - C_2^{(o)}\left(V^{(i)} + V^{(o)}\right)}.$$

Hence

$$P = \frac{1}{t_2 - t_1}\frac{V^{(i)}V^{(o)}}{A\left(V^{(i)} + V^{(o)}\right)}\ln\frac{m_0 - C_1^{(o)}\left(V^{(i)} + V^{(o)}\right)}{m_0 - C_2^{(o)}\left(V^{(i)} + V^{(o)}\right)}, \qquad (2.7.27)$$

for calculation of the permeability coefficient*, provided the total amount m_0 that initially was present in phase (o) is known. But as the total amount in the system is constant we also have

$$m_0 = C_1^{(i)}V^{(i)} + C_1^{(o)}V^{(o)},$$

which allows Eq. (2.7.26) to be written alternatively[†] as

$$P = \frac{1}{t_2 - t_1}\frac{V^{(i)}V^{(o)}}{A\left(V^{(i)} + V^{(o)}\right)}\ln\frac{\left(C_1^{(i)} - C_1^{(o)}\right)V^{(i)}}{C_1^{(i)}V^{(i)} + C_1^{(o)}V^{(o)} - C_2^{(o)}\left(V^{(i)} + V^{(o)}\right)}.$$

$$(2.7.28)$$

Sometimes the situation emerges that substance is initially present in both phases, i.e.

$$C^{(i)} = C_0^{(i)} \quad \text{and} \quad C^{(o)} = C_0^{(o)} \quad \text{for} \quad t = 0.$$

As $m_0 = C_0^{(i)}V^{(i)} + C_0^{(o)}V^{(o)}$ still applies, insertion in Eq. (2.7.26) gives[‡]

$$P = \frac{1}{t_2 - t_1}\frac{V^{(i)}V^{(o)}}{A\left(V^{(i)} + V^{(o)}\right)}\ln\frac{C_0^{(i)}V^{(i)} + C_0^{(o)}V^{(o)} - C_1^{(o)}\left(V^{(i)} + V^{(o)}\right)}{C_0^{(i)}V^{(i)} + C_0^{(o)}V^{(o)} - C_2^{(o)}\left(V^{(i)} + V^{(o)}\right)}.$$

$$(2.7.29)$$

The last three formulae represent a kind of relief work to be used only if two pairs of data are available. It should be remembered that in such cases we are prevented from making any control of the validity of the basic assumptions involved. This undesirable situation might inspire us to seek another experimental procedure that provides a larger number of data to use in the determination of the rate constant for the exchange or the permeability coefficient P.

---

* Northrop, J.H. & Anson, M.V. (1929): *J. Gen. Physiol.* **12**, 543.
[†] Robbins, M. & Mauro, A. (1960): *J. Gen. Physiol.* **43**, 523.
[‡] Dainty, J. & House, C.R. (1966): *J. Physiol.* **185**, 172.

2.7.3 Compartment analysis

In this section we shall briefly refer to some of the elements of a transport model that is very often used in biology to simulate mass transfer – in particular of tracer molecules – through a single region or several adjacent regions. Such a region is denoted a *compartment*. In the simplest form of the model, the transport of the tracer molecule takes place into or through a single layer of cells. The model used here has the form of a box with volume V and end surfaces $A^{(1)}$ and $A^{(2)}$. The transport through these two areas takes place from the two surrounding phases having concentrations C_1 and C_2, that may either be constant or time variant. The interior of the compartment is considered to be effectively stirred so that the concentration in the interior is constant and equal to $C(t)$. We consider now two simple situations, where the initial concentration of the tracer in the compartment is zero, i.e. $C(t) = 0$ for $t = 0$.

2.7.3.1 Transport with passive membrane permeabilities

Let the concentration $C_1(t)$ in phase$^{(1)}$ change with time t. Further, we assume that the volume of phase$^{(2)}$ is sufficiently large that for all practical purposes the concentration $C_2(t)$ of tracer can be disregarded. Thus, the flux $J^{(2)}$ from the compartment through the surface $A^{(2)}$ into phase$^{(2)}$ depends only on the concentration $C(t)$ in the compartment. We shall now calculate how the concentration inside the compartment $C(t)$ changes when the concentration difference $\Delta C = C_1(t) - C(t)$ exists across the surface $A^{(1)}$. Let $J^{(1)}$ be the flux. The amount $\Delta M^{(1)}(\Delta t)$ that in the time Δt is transported across $A^{(1)}$ is

$$\Delta M^{(1)}(\Delta t) = J^{(1)} A^{(1)} \Delta t = P^{(1)}(C_1(t) - C(t)) A^{(1)} \Delta t,$$

where $P^{(1)}$ is the permeability of the surface $A^{(1)}$. If no efflux were present across the other surface of the compartment the concentration inside would increase by the amount

$$\Delta C^{(1)} = \frac{\Delta M^{(1)}(\Delta t)}{V} = k^{(1)}(C_1(t) - C(t))A^{(1)}\Delta t, \qquad (2.7.30)$$

where $k^{(1)} = P^{(1)}A^{(1)}/V$ is a *rate constant* for the loading with tracer to the compartment from phase$^{(1)}$.

Considering then the opposite situation with an impermeable surface $A^{(1)}$ and a permeable surface $A^{(2)}$. In the time Δt an amount $\Delta M^{(2)}(\Delta t) = J^{(2)} A^{(2)} \Delta t$ will leave the compartment and cause a *decrease* $-\Delta C^{(2)}$ in concentration $C(t)$ of the compartment that amounts to

$$-\Delta C^{(2)} = -\frac{\Delta M^{(\Delta 2)}(\Delta t)}{V} = -k^{(2)}(C(t) - 0)\Delta t, \qquad (2.7.31)$$

where $k^{(2)} = P^{(2)}A^{(2)}/V$ is the rate constant for the removal of tracer from the compartment across $A^{(2)}$. When both surfaces are open the resultant inflow and outflow lead to an increment $\Delta C = C(t + \Delta t) - C(t)$ in concentration in the compartment during the time Δt. As a first approximation we write

$$\Delta C = \Delta C^{(1)} - \Delta C^{(2)} = \frac{dC}{dt} \Delta t,$$

i.e. the two contributions $\Delta C^{(1)}$ and $\Delta C^{(2)}$ are considered as purely additive, which holds as long as Δt is small. Inserting Eq. (2.7.30) and Eq. (2.7.31) and dividing on both sides by Δt gives

$$\frac{dC}{dt} = k^{(1)}C_1(t) - \left(k^{(1)} + k^{(2)}\right) C(t),$$

or

$$\frac{dC}{dt} + \left(k^{(1)} + k^{(2)}\right) C(t) = k^{(1)}C_1(t), \qquad (2.7.32)$$

which is the differential equation that governs the time course of the concentration $C(t)$ in the compartment. When the *stationary* situation is established, i.e. $dC/dt = 0$, the outer concentration $C_1(t)$ and the inside concentration $C(t)$ have their stationary values $C_1(\infty)$ and $C(\infty)$, which then are linked as

$$\left(k^{(1)} + k^{(2)}\right) C(\infty) = k^{(1)}C_1(\infty). \qquad (2.7.33)$$

In other words: *inflow to* the compartment $k^{(1)} [C_1(\infty) - C(\infty)]$ and *outflow from* the compartment $k^{(2)}C(\infty)$ are *equal*.

We shall now set up two solutions of Eq. (2.7.33), both of which have as their initial value $C(0) = 0$, i.e. the compartment contains no tracer at $t = 0$.

(i) A step change in outer concentration
We seek a solution of Eq. (2.7.32) for which

$$C_1(t) = \begin{cases} 0, & \text{for } t < 0, \\ C_1^0, & \text{for } t \geq 0. \end{cases}$$

The solution is already given in Chapter 1, Section 1.9.1.2, Eq. (1.9.9). Replacing in this equation $\alpha = k^{(1)} + k^{(2)}$ and $K = k^{(1)}C_1^{(0)}$ we obtain

$$C(t) = \frac{k^{(1)}}{k^{(1)} + k^{(2)}} C_1^{(0)}\left[1 - e^{-(k^{(1)}+k^{(2)})t}\right]. \qquad (2.7.34)$$

Thus, the concentration in the compartment rises exponentially and asymptotically towards the stationary value $k^{(1)}C_1^{(0)}/(k^{(1)} + k^{(2)})$. As shown above, this value is obtained more easily by putting $dC/dt = 0$ in Eq. (2.7.32).

(ii) Outer concentration grows asymptotically

Let $C_1(t)$ rise according to

$$C_1(t) = C_1^{(0)}[1 - e^{-\beta t}],$$

where β is the rate constant for the rise of $C_1(t)$. Initially no tracer is present inside the compartment, i.e. $C(0) = 0$. The time course of $C(t)$ is then governed by

$$\frac{dC}{dt} + \left(k^{(1)} + k^{(2)}\right) C(t) = k^{(1)} C_1^{(0)}[1 - e^{-\beta t}]. \tag{2.7.35}$$

The general solution is given in Chapter 1, Section 1.9.1.3, Eq. (1.9.14), from which it follows

$$
\begin{aligned}
C(t) e^{\alpha t} &= \int e^{\alpha t} k^{(1)} C_1^{(0)}[1 - e^{-\beta t}]\, dt + A_1 \\
&= k^{(1)} C_1^{(0)} \int \left[e^{\alpha t} - e^{(\alpha - \beta)t} \right] dt + A_1 \\
&= k^{(1)} C_1^{(0)} \left(\frac{e^{\alpha t}}{\alpha} - \frac{e^{(\alpha - \beta)t}}{\alpha - \beta} \right) + A_1, \tag{2.7.36}
\end{aligned}
$$

where $\alpha = k^{(1)} + k^{(2)}$ and A_1 is a constant of integration whose magnitude is determined by the initial condition $C(t) = 0$ for $t = 0$. This implies that $A_1 = 1/(\alpha - \beta) - 1/\alpha k^{(1)}$. Inserting this value in Eq. (2.7.36) we have

$$\frac{C(t)}{k^{(1)} C_1^{(0)}} e^{\alpha t} = \left[\frac{e^{\alpha t}}{\alpha} - \frac{1}{\alpha} \right] + \frac{1}{\alpha - \beta}\left[1 - e^{(\alpha - \beta)t}\right],$$

or

$$\frac{C(t)}{k^{(1)} C_1^{(0)}} = \frac{1}{\alpha}[1 - e^{-\alpha t}] + \frac{e^{-\alpha t} - e^{-\beta t}}{\alpha - \beta}. \tag{2.7.37}$$

The last term on the right-hand side presents a minor difficulty as both numerator and denominator become zero for $\alpha = \beta$. To examine the consequences we assume that α and β only differ slightly from each other and put $\alpha = \beta + h$. This gives

$$\frac{e^{-\alpha t} - e^{-\beta t}}{\alpha - \beta} = \frac{e^{-t(\beta + h)} - e^{-t\beta}}{h} = \frac{f(\beta + h) - f(\beta)}{h},$$

where $f(\beta) = e^{-t\beta}$. Letting $\alpha \to \beta$, i.e. $h \to 0$, we obtain

$$\lim_{\alpha \to \beta}\left[\frac{e^{-\alpha t} - e^{-\beta t}}{\alpha - \beta} \right] = \lim_{h \to 0}\left[\frac{f(\beta + h) - f(\beta)}{h} \right] = \frac{d}{d\beta}(e^{-t\beta}) = -t\, e^{-\beta t}. \tag{2.7.38}$$

Keeping this result* in mind we rearrange Eq. (2.7.21) and insert $\alpha = k^{(1)} + k^{(2)}$ to obtain

$$
\frac{C(t)}{C_1^0} = \frac{k^{(1)}}{k^{(1)} + k^{(2)}} \left[1 - e^{-(k^{(1)}+k^{(2)})t} \right]
$$

$$
+ \frac{k^{(1)}}{k^{(1)} + k^{(2)} - \beta} \left[e^{-(k^{(1)}+k^{(2)})t} - e^{-\beta t} \right] \quad \text{for} \quad \beta \neq k^{(1)} + k^{(2)}
$$

$$(2.7.39)$$

and

$$
\frac{C(t)}{C_1^0} = \frac{k^{(1)}}{k^{(1)} + k^{(2)}} \left[1 - e^{-(k^{(1)}+k^{(2)})t} \right]
$$

$$
- k^{(1)} t\, e^{-(k^{(1)}+k^{(2)})t}, \quad \text{for} \quad \beta = k^{(1)} + k^{(2)}. \qquad (2.7.40)
$$

The outflow $M^{(\text{out})}$ of tracer from the compartment per unit time is then

$$
M^{(\text{out})} = k^{(2)} C(t). \qquad (2.7.41)
$$

2.7.3.2 One-way transport

(i) Unidirectional flux

In Section 2.7.2.1 we considered the experimental situation where the concentration in phase (o) was maintained at zero at all times. This implies that once a molecule has moved through the membrane no possibility is left for it to return by a random walk. To single out this special transfer situation, the flux is designated as a *unidirectional flux*[†].

Some find it convenient to apply this concept of unidirectional flux also to situations where the concentration of substance is finite on both sides of the membrane. The rationale may be as follows. We rewrite Eq. (2.7.6)

$$
J = P \left(C^{(\text{i})} - C^{(\text{o})} \right)
$$
$$
= PC^{(\text{i})} - PC^{(\text{o})}
$$

and regard the (net) flux J in the direction (i) $\rightarrow$ (o) as the result of two opposite unidirectional fluxes. The first

$$
\overset{(\text{i})\rightarrow(\text{o})}{J} = PC^{(\text{i})}, \qquad (2.7.42)
$$

---

* This result could also be obtained by considering the function $F(\alpha) = f(\alpha)/g(\alpha)$, where $f(\alpha) = e^{-\alpha t} - e^{-\beta t}$ and $g(\alpha) = \alpha - \beta$. Now $F(\beta) = f(\beta)/g(\beta)$ is of the form 0/0. Using the rule of L'Hospital, $\lim_{\alpha \to \beta} F(\alpha) = f'(\beta)/g'(\beta) = -t\, e^{-\beta t}/1 = -t\, e^{-\beta t}$.

† In Section 2.5.6.4. we used the term **unidirectional fluxes** in a different context when considering the random walk movement across a *plane in space* of a collection of particles that move simultaneously but in *opposite* directions through the plane.

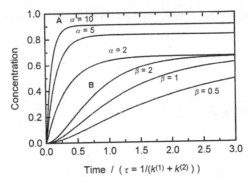

Fig. 2.45. Loading of a single compartment by a tracer with rate constants $k^{(1)}$ and $k^{(2)}$ when: (A) stepwise change of outer concentration with $k^{(2)} = 1$ and $k^{(1)}/k^{(2)} = a = 2, 5, 10$; and (B) outer concentration grows exponentially as $1 - e^{-\beta t}$ with rate constant $\beta = 2, 1, 0.5$ and $\alpha = 2$. For further explanation see text.

is called the "outflux", or "efflux", and the other

$$\overset{(0)\to(i)}{J} = PC^{(0)}, \tag{2.7.43}$$

the "influx". Naturally, this purely formal manipulation of Fick's law does not give an additional physical insight into the nature of the mass transfer through the membrane. However, by certain calculations it offers an advantage – provided it is done with tact* – formally to operate with the unidirectional fluxes, just as in those situations where the concept of a "force of diffusion" is useful.

(ii) Unidirectional transfer

As mentioned in Section 2.1 the transport through biological membranes does not take place exclusively by way of diffusion and migration processes. In the cell membrane there are localized specific molecules – transport molecules – that catalyze a *unidirectional transport* through the membrane of certain species, e.g. Na^+, K^+ and Ca^{2+}, with a rate that is largely independent of the concentration in the receiving phase. In the following we shall illustrate with a few examples how the presence of such one-way traffic influences the exchange kinetics in a compartment (see also Fig. 2.45).

We imagine that the unidirectional transport is directed towards the interior of the compartment and takes place at the surface $A^{(1)}$ where molecular aggregates are provided with specialized elements whereby a certain fraction of the

---

* For example, by avoiding provoking situations that directly are contradicting the second law of thermodynamics.

impinging molecules from phase$^{(1)}$, e.g. Na$^+$, are captured and transported with the velocity $v^{(1)}$ through the membrane to the *inside* of the compartment. This unidirectional flux $J^{(1)\to(i)}$ we denote $J^{(in)}$. When $C_1(t)$ is the concentration of tracer in phase$^{(1)}$ we can write the unidirectional flux formally* as

$$J^{(in)} = \gamma^{(1)}C_1(t)v^{(1)}, \tag{2.7.44}$$

where $\gamma^{(1)}$ is the fraction of the impinging molecules that are caught by the transport mechanism. The amount of tracer $\Delta M^{(in)}$ that in time Δt is transported unidirectionally through the membrane is

$$\Delta M^{(in)} = J^{(in)}A^{(1)}\Delta t = \gamma^{(1)}C_1(t)v^{(1)}A^{(1)}\Delta t.$$

The increment of the tracer concentration in the interior of the compartment is

$$\Delta C^{(1)} = \frac{\Delta M^{(in)}}{V} = k^{(in)}C_1(t)\Delta t, \tag{2.7.45}$$

where

$$k^{(in)} = \frac{\gamma^{(1)}v^{(1)}A^{(1)}}{V} \overset{\Delta}{\equiv} s^{-1} \tag{2.7.46}$$

is the *rate constant* for the *unidirectional influx* of the tracer through the surface $A^{(1)}$.

We imagine now that a similar unidirectional transport mechanism is located at the compartment's other surface $A^{(2)}$ and causes another unidirectional flux – a unidirectional efflux – $J^{(out)}$ of tracer *out* of the compartment. In the time Δt this mechanism will *decrease* the tracer concentration $C(t)$ inside by the amount

$$-\Delta C^{(2)} = -k^{(out)}C(t)\,\Delta t, \tag{2.7.47}$$

where $k^{(out)}$ is the *rate constant* for the *unidirectional efflux* from the compartment.

We assume – as in Section 2.7.3.1 – that the two increments $\Delta C^{(1)}$ and $\Delta C^{(2)}$ contribute additively in the time Δt to the resultant increment in concentration $\Delta C = (dC/dt)\Delta t$ in the compartment. We then have

$$\frac{dC}{dt} = k^{(in)}C_1(t) - k^{(out)}C(t), \tag{2.7.48}$$

or

$$\frac{dC}{dt} + k^{(out)}C(t) = k^{(in)}C_1(t), \tag{2.7.49}$$

---

* This formula should not be taken too literally. Its only purpose is to write the relevant elements that must enter in an expression for $J^{(in)}$.

which formally is identical to Eq. (2.7.32) but differs from a physical point of view on account of the significance to be attributed to the two rate constants $k^{(in)}$ and $k^{(out)}$.

(iii) Passive influx and unidirected efflux

In biology one very often encounters the situation where one surface of the compartment (the cell) is passively permeable to, for example, Na^+ ions, whereas at the other surface the outward transport of the same species is unidirectional (active transport). Let $C_1(t)$ and $C(t)$ represent the concentrations in phase$^{(1)}$ and in the compartment respectively. Bringing together Eq. (2.7.32) and Eq. (2.7.49), the traffic through the compartment will be governed by the following expression

$$\frac{dC}{dt} + \left(k^{(in)} + k^{(out)}\right) C = k^{(1)}C_1(t), \qquad (2.7.50)$$

which only differs from Eq. (2.7.32) by replacing the rate constant $k^{(2)}$ for the passive efflux with the rate constant $k^{(out)}$ for the unidirectional efflux. Thus, the solution is already given by Eq. (2.7.34).

2.7.4 Stationary diffusion with superimposed migration

Passive mass transfer through membranes is very often the result of diffusion overlaid by migration. In biological systems ions are transferred across cell membranes by this mechanism, the driving force being electric. A more detailed description of this subject is given in Chapter 3. As far as the transfer of uncharged molecules is concerned, an overlaid migration sometimes enters as the result of a *pressure gradient* or as a *convection* in cell layers through which a substantial amount of water transport takes place.

As an example we consider the transfer through a membrane of thickness h of an uncharged molecule moving with the migration/convection velocity v in the direction at right angles to the membrane. The substance has a diffusion coefficient D in the membrane. To determine the stationary concentration profile the point of departure is the Smoluchowski equation, Eq. (2.6.5), in its stationary form

$$\frac{d^2C}{dx^2} - \frac{v}{D}\frac{dC}{dx} = 0, \qquad (2.7.51)$$

and find the solution that satisfies a given set of boundary conditions, e.g. the concentrations on the two sides of the membrane are kept constant. To solve

Eq. (2.7.51) we put

$$\frac{dc}{dx} = p,$$

and obtain

$$\frac{dp}{dx} + \frac{v}{D}p = 0.$$

Solving for p gives

$$p = A_1 e^{-vx/D} = \frac{dC}{dx},$$

from which we obtain the general solution of Eq. (2.7.51)

$$C(x) = A_1 e^{vx/D} + B_1, \qquad (2.7.52)$$

where A_1 and B_1 are two constants to be adjusted, e.g. so that the solution reproduces correctly the concentrations at the two sides of the membrane. When $C(x)$ is determined the flux is calculated from

$$J = -D\frac{dC}{dx} + vC. \qquad (2.7.53)$$

The intervening computations are unproblematic but rather extensive – and a little boring. As an alternative – and perhaps also simpler – strategy we shall solve the problem by considering first the flux.

2.7.4.1 Determination of the flux

Let $C^{(i)}$ and $C^{(o)}$ be the concentrations in the surrounding phases: phase$^{(i)}$ and phase$^{(o)}$. The membrane thickness is h. The direction of the x-axis is (i) $\rightarrow$ (o) and the crossing from phase$^{(i)}$ to the membrane is at $x = 0$, while the other side of the membrane is at $x = h$. The concentrations $C(0)$ and $C(h)$ in the membrane at these two positions are assumed to be equal to the concentrations $C^{(i)}$ and $C^{(o)}$ in the two phases, i.e. the distribution coefficient* α is 1.0. We write Eq. (2.7.52) in the form

$$\frac{dC}{dx} - \frac{v}{D}C = -\frac{J}{D}. \qquad (2.7.54)$$

This equation is of the form (see Chapter 1, Section 1.9.1.2)

$$\frac{dy}{dx} + \alpha y = K,$$

---

* See Section 2.7.1, Eq. (2.7.3).

which has the general solution

$$y = \frac{K}{\alpha} + A_1 e^{-\alpha x},$$

where A_1 is an integration constant. Substituting $K = -J/D$ and $\alpha = -v/D$ and replacing y by $C(x)$ we obtain

$$C(x) = -\frac{J}{D}\left(-\frac{D}{v}\right) + A_1 e^{vx/D} = \frac{J}{v} + A_1 e^{vx/D}. \qquad (2.7.55)$$

To adjust A_1 we note that $C(x) = C(0)$ for $x = 0$. Inserting this pair in the above equation and solving for A_1 gives

$$A_1 = C(0) - \frac{J}{v}.$$

Hence, the equation for the concentration profile in the membrane originating from $(C(0), 0)$ is

$$C(x) = \frac{J}{v}\left(1 - e^{vx/D}\right) + C(0)\,e^{vx/D}, \quad \text{for } 0 \le x \le h. \qquad (2.7.56)$$

For $x = h$, $C(x)$ assumes the value

$$C(h) = \frac{J}{v}\left(1 - e^{vh/D}\right) + C(0)\,e^{vh/D}. \qquad (2.7.57)$$

Replacing $C(0)$ and $C(h)$ respectively by $C^{(i)}$ and $C^{(o)}$ and solving for J we obtain the expression for the stationary flux

$$J = v\frac{C^{(i)}e^{vh/D} - C^{(o)}}{e^{vh/D} - 1}, \qquad (2.7.58)$$

usually called Hertz's equation*. It appears that the flux comes out as an odd mixture of migration and diffusion with the concentration $C^{(i)}$ weighted by the exponential $e^{vh/D}$ relatively to $C^{(o)}$. Thus, for a given value $\Delta C = C^{(i)} - C^{(o)}$ the diffusion flux is assisted strongly – and nonlinearly – when the convective velocity v acts in the direction of the concentration drop. Conversely, a convective velocity v in the contrary direction will oppose the diffusive flux and eventually reduce it to zero or in the end drive the substance *against* the concentration fall. Figure 2.46 illustrates the dependence of the flux on the magnitude and direction of the convection velocity when $C(0) = 10$ and $C(h) = 1$. It is also seen that there is a certain velocity $v^{(eq)}$ at which the flux is zero,

---

* The equation was derived in 1923 by the German physicist Gustav Hertz (1887–1975). He was awarded the Nobel Prize for physics 1925, together with J. Franck.

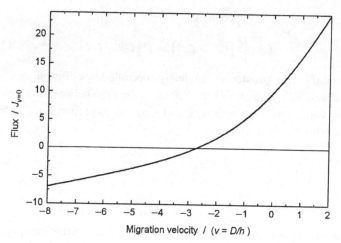

Fig. 2.46. Flux through a membrane of thickness h under combined influence of diffusion and convection. Abscissa: migration velocity v in units of D/h. Ordinate: flux in units of $J_{v=0} = D(C(0) - C(h))/h$. Concentrations are arbitrarily put equal to $C(0) = 10$ and $C(h) = 1$.

i.e. a state of equilibrium is established of the mass transfer between the two phases. Putting $J = 0$ and $v = v^{(eq)}$ in Eq. (2.7.57) yields

$$C^{(i)} \exp\left\{\frac{hv^{(eq)}}{D}\right\} - C^{(0)} = 0.$$

Hence[*]

$$v^{(eq)} = \frac{D}{h} \ln\left(\frac{C^{(0)}}{C^{(i)}}\right) = \frac{kTB}{h} \ln\left(\frac{C^{(0)}}{C^{(i)}}\right). \qquad (2.7.59)$$

An analogous expression is shown in Chapter 3 for the condition of equilibrium for *ions* that move across a membrane under the influence of an electric field.

If $hv \gg D$ the exponential term dominates both in the numerator and denominator of Eq. (2.7.57) with the result that

$$J \to vC^{(i)}, \quad \text{for } hv \gg D, \qquad (2.7.60)$$

corresponding to a transfer dominated by convection. This is also the case for $|hv| \gg D$, when $v < 0$. The result is instead

$$J \to vC^{(0)}, \quad \text{for} \quad |hv| \gg D \quad \text{but} \quad v < 0. \qquad (2.7.61)$$

[*] See Section 2.6.4.1, Eq. (2.6.35).

When $v = 0$, Eq. (2.7.57) becomes*

$$J = \frac{D}{h}\left[C^{(i)} - C^{(0)}\right] = P\left[C^{(i)} - C^{(0)}\right],\qquad (2.7.62)$$

as it should, i.e. the transfer is completely controlled by diffusion.

The *degree of rectification* Y is defined as the ratio between the two fluxes resulting when v tends towards $+\infty$ and $-\infty$. We have from Eq. (2.7.59) and Eq. (2.7.60)

$$Y = \frac{J_{v=+\infty}}{J_{v=-\infty}} = \frac{C(0)}{C(h)}.\qquad (2.7.63)$$

2.7.4.2 *Unidirectional fluxes and flux ratio*

A useful tool in membrane biology to clarify whether an active transport mechanism participates in the transfer of a given species is flux ratio analysis[†]. The procedure consists of determining experimentally the two unidirectional fluxes[‡] across the membrane $J^{(in)}$ and $J^{(out)}$ and comparing the value of the ratio $J^{(in)}/J^{(out)}$ with the ratio of the unidirectional fluxes that is evaluated theoretically, assuming that the transfer process in question is purely passive, i.e.

$$\left(\frac{J^{(i)\to(0)}}{J^{(i)\leftarrow(0)}}\right)^{(exp)}\quad \text{and}\quad \left(\frac{J^{(i)\to(0)}}{J^{(i)\leftarrow(0)}}\right)^{(calc)}.\qquad (2.7.64)$$

The procedure is applied almost exclusively to the study of the transfer of ionic species where the migration velocity is due to an electric field. However, in the evaluation of the so-called "solvent drag" effect, the flux ratio is applied to the transfer of non-electrolytes[§].

The unidirectional fluxes follow immediately from Eq. (2.7.57). Putting $C^{(0)} = 0$ the unidirectional flux in the direction (i) $\to$ (o) becomes

$$J^{(0)\to(h)} = v\frac{C^{(i)}\,e^{hv/D}}{e^{hv/D} - 1}.\qquad (2.7.65)$$

When $C^{(i)} = 0$ we obtain instead the opposite unidirectional flux

$$J^{(0)\leftarrow(h)} = -v\frac{C^{(0)}}{e^{hv/D} - 1},\qquad (2.7.66)$$

---

* When $hv/D \ll 0$ we have $\exp\{hv/D\} \approx 1 + hv/D$ (Chapter 1, Section 1.6). Hence: $v/(\exp\{hv/D\} - 1) \approx v/(hv/D) = D/h$.

[†] The systematic use of flux ratio analysis in biology is due to H.H. Ussing.

[‡] See Section 2.7.3.2(i).

[§] Ussing, H.H. (1978): Interpretation of tracer fluxes. In: *Membrane Transport in Biology*, Giebisch, G., Tosteson, D. C. & Ussing, H.H. (eds), Vol. 1. Concepts and Models, chapter 3, pp. 115–140, Springer-Verlag, New York.

where the minus sign indicates that the direction for $J^{(h0)}$ is opposite to the direction of the x-axis. However, the convention in flux ratio analysis is to consider only the numerical values. Putting $J^{(i)\to(h)} = \overset{(i)\to(o)}{J}$ and $\overset{(i)\leftarrow(o)}{J} = -J^{(0)\leftarrow(h)}$, we have accordingly

$$\frac{\overset{(i)\to(o)}{J}}{\overset{(i)\leftarrow(o)}{J}} = \frac{C^{(i)}}{C^{(0)}} e^{hv/D}, \tag{2.7.67}$$

as a special case of a more general relation for the flux ratio due to Ussing*.

2.7.4.3 Concentration profile

As mentioned before the standard procedure to determine the concentration profile is to solve Eq. (2.7.50) directly and adjust the two constants of integration corresponding to the boundary conditions for $C(x) = C(0)$ for $x = 0$ and $C(x) = C(h)$ for $x = h$. However, in Section 2.7.4.1 we arrived at an expression involving the flux J (Eq. (2.7.55)), which we now rewrite as

$$J = v\frac{C^{(i)}e^{xh/D} - C(x)}{e^{vh/D} - 1}, \tag{2.7.68}$$

where $C(0)$ is replaced by $C^{(i)}$. As this value for J is equal to that of Eq. (2.7.57) we have

$$v\frac{C^{(i)}e^{hv/D} - C^{(0)}}{e^{hv/D} - 1} = v\frac{C^{(i)}e^{xv/D} - C(x)}{e^{xv/D} - 1}.$$

Solving this equation with respect to $C(x)$ we obtain the expression for the concentration profile

$$C(x) = \frac{C^{(i)}e^{hv/D} - C^{(0)} - \left[C^{(i)} - C^{(0)}\right]e^{xv/D}}{e^{hv/D} - 1}. \tag{2.7.69}$$

It appears that the concentration profile now is nonlinear except for $v = 0$. In this case we have†

$$C(x) = C^{(i)} - \left[C^{(i)} - C^{(0)}\right]\frac{x}{h},$$

which is the profile corresponding to a pure diffusion controlled transfer process (compare Eq. (2.5.12)). Figure 2.47 shows a family of concentration profiles corresponding to different values of $\alpha = hv/D$. It may be noted that the shape of the profile is like a sail suspended between $C^{(i)}$ and $C^{(0)}$ and formed by the

---

* Ussing, H.H. (1949): *Acta Physiol. Scand.,* **19**, 43. Ussing, H.H. (1952): *Adv. Enzymol.,* **13**, 21.
† Compare the limit used in connection with the derivation of Eq. (2.7.61).

2. *Migration and diffusion*

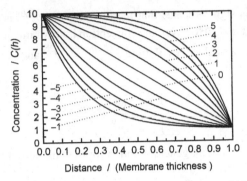

Fig. 2.47. Stationary concentration profiles through a membrane of thickness h by superimposed diffusion and convection. Ordinate: concentration in units of $C(h)$. Abscissa: distance through the membrane in units $X = x/h$ of membrane thickness h. The numbers associated to the curves are different values of the parameter $\alpha = hv/D$.

direction v of the wind blowing. This form appears as the flux

$$J = -D(\mathrm{d}C/\mathrm{d}x) + vC$$

and must be constant through the membrane. Thus, a large value of the contribution vC requires a small value of the other term $-D(\mathrm{d}C/\mathrm{d}x)$, and vice versa for a small value of vC. For $v > 0$ the value of vC is smallest at $x = h$, and, consequently, $|\mathrm{d}C/\mathrm{d}x|$ is at a maximum. For $v < 0$ the two contributions through the membrane counteract each other. Thus the demand of a constant flux through the membrane requires a balanced increase in $|vC|$ and $|\mathrm{d}C/\mathrm{d}x|$. The contribution of vC – now in the direction $h \to 0$ – increases as the membrane side at $x = 0$ is approached. To compensate for this $|\mathrm{d}C/\mathrm{d}x|$ increases concurrently. Hence, the slope of the concentration profile has now its greatest value at $x = 0$.

2.8 Convective and osmotic water movement through membranes

> Few phenomena are so well understood thermodynamically, and so ill understood kinetically, as osmotic flow of solvent through a semipermeable membrane.
>
> *(Longuet-Higgins & Austin, 1966).*

So far we have described passive mass transfer through a membrane of uncharged molecules tacitly assumed to be dissolved in water. All biological membranes, however, are also more or less permeable to *water*. In this section we shall give an account of those two mechanisms that are of particular relevance in this context, namely water transport by *convection* and water transport by *osmosis*.

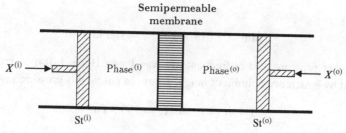

Fig. 2.48. Set up to illustrate water transport through a semipermeable membrane by convection (filtration) and osmosis. See text for further explanation.

2.8.1 Convective water movement

The two types of water transport are clearly demonstrated by examining water movement through a special type of membrane, namely an *ideal semipermeable membrane*. Such a membrane is permeable *exclusively for the solvent* – water in this case – and does not allow the passage of any other species. Figure 2.48 illustrates a measuring device for studying the water movements through such a membrane. The semipermeable membrane is placed inside a cylindrical tube and surrounded on both sides by a fine-meshed grid to carry a substantial pressure difference without bulging or, eventually, bursting. The semipermeable membrane separates two chambers (i) and (o), the end-walls of which are two tightly fitting, frictionless pistons $St^{(i)}$ and $St^{(o)}$. Both chambers are filled with water containing no air bubbles. We imagine now that a force $X^{(i)}$ acts on the piston $St^{(i)}$. If the other piston is locked in, piston $St^{(i)}$ transmits a *pressure $p^{(i)}$* to the liquids in the chambers of magnitude

$$p^{(i)} = \frac{X^{(i)}}{A^{(i)}},$$

where $A^{(i)}$ is the area of $St^{(i)}$. Pressure in S.I. units has the dimension newton per square meter, and is designated the special unit pascal (Pa)*. The pressure unit most frequently used earlier was 1 atmosphere (1 atm), which is the pressure required to carry a mercury column of height 760 mm. This column (mass density 13.6×10^3 kg m^{-3}) exerts a pressure p in newton per square meter that is:

$$p = 0.76\,(\mathrm{m})\; 13.6 \times 10^3\;(\mathrm{kg\ m^{-3}}) \cdot 9.81\;(\mathrm{m\ s^{-2}})$$
$$= 1.013 \times 10^5\;(\mathrm{N\ m^{-2}}) = 1.013 \times 10^5 \quad \mathrm{Pa},$$

---

* This unit is named in memory of Blaise Pascal (1623–1662), a French mathematician and theologian, who made important contributions to the theory of numbers and conic sections and who founded the theory of probability. He performed fundamental investigations on the pressure in gases and liquids.

or

$$1 \text{ atm} = 1.013 \times 10^5 \text{ Pa} \approx 100 \text{ kPa.}$$

Another pressure unit in frequent use was 1 mm Hg = 1 Torr = pressure exerted by a mercury column of height 1 mm. In pascal this pressure in

$$1 \text{ Torr} = 1 \text{ mm Hg} = 133.3 \text{ Pa.}$$

Similarly, with piston $St^{(i)}$ locked in, a force $X^{(o)}$ applied to piston $St^{(o)}$ produces a pressure $P^{(o)} = X^{(o)}/A^{(o)}$ where $A^{(o)}$ is the area of $St^{(o)}$. If neither of the pistons is locked and $p^{(i)} > p^{(o)}$, water molecules are forced through the porous structure of the semipermeable membrane. Thus the pressure difference between the chambers – and across the membrane – will cause a net transport of water from chamber (i) to chamber (o). To the displacement Δx, e.g. of piston $St^{(o)}$, recorded with a suitable device, these corresponds an increment $\Delta V^{(o)} = \Delta x A^{(o)}$ in volume. To the transfer of the volume ΔV in the time Δt there corresponds a *water flux* of magnitude

$$J_v = \frac{\Delta V^{(o)}}{A \, \Delta t},$$

where A is the area of the semipermeable membrane. In this case the dimensions of the water flux J_v are $\text{m}^3 \text{ m}^{-2} \text{ s}^{-1} = \text{m s}^{-1}$, and is denoted for this reason the *volume flux*. If, instead, it appears preferable to express the amount of transferred water in mol, we divide the volume flux J_v by the *molar volume* $\overline{V}_v$ of water, i.e. the water volume that contains 1 mol H_2O at the temperature in question, i.e. very near to 18 dm^3 mol^{-1} at room temperature. Figure 2.43, line (A), shows an example of the dependence of the water flux J_v on the pressure difference $p^{(i)} - p^{(o)} = \Delta p$ between the two chambers. The resistance that the two chambers offer against the movement of the water is of no significance compared with that presented by the membrane, and the pressure drop across the membrane is considered to be identical to the pressure difference Δp. It appears from the figure that the water flux is proportional to pressure difference across the membrane. We write

$$J_v = L_P \left[p^{(i)} - p^{(o)} \right], \tag{2.8.1}$$

where the coefficient L_P with dimensions (m s^{-1} Pa^{-1}) is designated the *hydraulic conductance* of the membrane. This simple expression for J_v is only valid in a certain range of pressures $|p^{(i)} - p^{(o)}|$ that depends on the particular properties of the semipermeable membrane. Outside this range the relation becomes nonlinear and L_P is no longer a constant. In this case we maintain Eq. (2.8.1) as the *definition* of the hydraulic conductance and for the prescription for calculating its value from the experimental data of J_v and Δp belonging together

$$L_P = \frac{J_v}{p^{(i)} - p^{(o)}},$$

(2.8.2)

where $L_P = F(J_v)$ is now regarded as a function of the volume flux J_v of water through the semipermeable membrane.

2.8.2 Osmotic water movement

However, it is also possible to establish a water flow through a semipermeable membrane without using an external device to establish a pressure difference across the membrane. We replace the pure water phase in chamber (i), for example, with a solution of molecules of so large a size that they cannot penetrate the semipermeable membrane, e.g. sucrose molecules. When the pressures in both chambers are equal ($p^{(i)} = p^{(o)}$) one will observe a spontaneous flow of water from the chamber (o) – containing the *pure* solvent – through the semipermeable membrane into chamber (i). In due time the volume of chamber (i) increases as that of chamber (o) decreases and – when the system is left alone – the final result is that all the water left in chamber (o) has moved into chamber (i) containing the sucrose solution. On the face of it one could imagine that from chamber (i) – that contains the dissolved but impermeable molecules – *a suction* is exerted through the semipermeable membrane upon the water molecules in the pure water phase in chamber (o). If the content in chamber (o) is replaced by a sucrose solution similar to that in chamber (i), but at a *lower* concentration, a water flow in the direction (o) → (i) is still set up. During this process the solution in chamber (i) is diluted, while that in chamber (o) becomes more concentrated. The spontaneous flow will come to an end when *equal concentrations* in the two chambers are established.

This spontaneous movement of water from the dilute solution to the more concentrated one when the solutions are separated from each other by a semipermeable membrane is called *osmosis* (R. Dutrochet, 1827) – using the Greek word for "push" – or an *osmotic water movement*. Originally the designations "endosmosis" and "exosmosis" were used as a specification for the direction of the water movement, but these terms are now rarely used.

2.8.2.1 Osmotic pressure

Using the two pistons $St^{(i)}$ and $St^{(o)}$ in Fig. 2.48 we apply a pressure difference

$$\Delta p = p^{(i)} - p^{(o)},$$

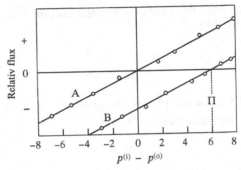

Fig. 2.49. Illustration of the connection between the water flux through a semipermeable membrane and the hydraulic pressure difference between the two sides of the membrane. The water flux in the direction (i) → (o) is assumed to be positive. (A) The two chambers (i) and (o) both contain pure water. (B) Chamber (i) contains a solution of sucrose, chamber (o) contains pure water. At the pressure difference $p^{(i)} - p^{(o)} = \Pi$ the water flux is zero in the direction (o)→(i). The excess pressure Π designates as the *osmotic pressure* of the sucrose solution.

across the membrane and measure simultaneously the water flux J_V, taking care that each volume displacement ΔV does not influence significantly the sucrose concentration in chamber (i). It appears from Fig. 2.49, line (B) that the relation between Δp and J_V is still linear and with the same slope L_P but displaced parallel to line (A) – passing through (0,0) – and intersects the x–axis, i.e. the pressure axis, in the point $p = \Pi$. The relation between Δp and J_V can therefore be written as

$$J_V = L_P \left[p^{(i)} - p^{(o)} - \Pi \right] = L_P [\Delta p - \Pi]. \tag{2.8.3}$$

Thus, for pressure differences $p^{(i)} - p^{(o)} < \Pi$ water will move in the direction chamber (o) → chamber (i), which contains the sucrose solution, whereas for all pressure differences $p^{(i)} - p^{(o)} > \Pi$ water will be driven out of the sucrose solution through the membrane into chamber (o). The dividing line corresponds to a situation with *zero net flux* of water through the membrane and a pressure difference between the phase containing the solution and the pure water phase equal to Π, or

$$J_V = 0, \quad \text{for} \quad p^{(i)} - p^{(o)} = \Pi. \tag{2.8.4}$$

This characteristic pressure Π is designated the *osmotic pressure*, or rather *the solution's osmotic pressure* because of its causal relation to the presence of the solution facing the semipermeable membrane. This unfortunate terminology is somewhat misleading as it supports illusions about a pressure being generated by the solution and in this way producing the osmotic water movement. For this reason it is important to remember that an osmotic pressure appears *only if* a solution is in contact with the pure solvent by way of an ideal semipermeable membrane. This results in water transport *into* the solution from the pure water phase (osmosis), and the osmotic pressure is the excess pressure to be *impressed on the solution* to prevent a water transport into the solution by osmosis.

2.8.2.2 Colligative properties

Every solution has certain common properties that, to a large degree, are unaffected by the special physical or chemical properties of the dissolved molecules. As far as the so-called ideal solutions* are concerned these properties depend exclusively on the *number of dissolved molecules* in proportion to the *total number* of molecules contained in the solution, i.e. of the mole fraction X_s for the dissolved substance

$$X_s = \frac{N_s}{N_s + N_w},\qquad (2.8.5)$$

where N_s is the number of dissolved molecules and N_w is the number of solvent molecules (water in this book). These properties are therefore denoted the *colligative properties* (colligare: tie together), used as a collective designation for the following properties of the solution which, as referred to the pure solvent, implies the following.

(1) A depression of the vapor pressure.
(2) An elevation of the boiling point.
(3) A depression of the freezing point.
(4) An osmotic pressure.

One can – for example, by using thermodynamic arguments – calculate how a given colligative property is related to the mole fraction X_s of the dissolved substance. As each of the colligative properties depends only on X_s there will also exist a precise, well-defined relation between one colligative property and the remaining three[†]. Conversely, by measuring a colligative property, e.g. the freezing point depression, one can determine the mole fraction X_s of the dissolved substance.

---

* In an ideal solution the forces of interaction between solvent/solute molecules do not differ from those acting between the solvent molecules themselves.
[†] For example, it can be shown that the ratio between the osmotic pressure of the solution and its freezing point depression is equal to $\Delta H_{fus}/T_w \overline{V}_w$, where ΔH_{fus} is the molar heat of melting of water and T_w the freezing point. $\overline{V}_w$ is the molar volume of water. Compare Eq. (2.6.4).

2.8.2.3 The underlying mechanism of osmotic water movement

Using thermodynamic arguments the formulation of the osmotic equilibrium
and calculation in precise terms of the quantities involved that determine the
magnitude of the osmotic pressure have been fully worked out for many years,
beginning with the Dutch physical chemist J. van't Hoff in 1887. By comparison, the detailed description of a kinetic, molecular theory for water movements
by osmosis is still rather incomplete. For that reason we shall only consider the
more general points of view.

Osmosis represents a flow of water through a semipermeable membrane
from a region of high water concentration to the adjacent region with lower
concentration. Hereby osmosis does not diverge in principle from the process
of diffusion already described, that is a current of dissolved particles flows from
the region of high concentration to that of low concentration. Both processes
represent a spontaneous tendency to attain a mixing of the whole available
space, a process that on the molecular level is accounted for by Brownian
motion caused by the thermal agitation in which all the molecules participate.

A diffusion process in a free unbounded medium differs – even in the simplest
case – from osmosis in including *two* concurrent, oppositely directed transfer
processes. Consider, for example, a solution of sucrose in water. The mole
fractions for sucrose X_s and for water x_w are given by

$$X_s = \frac{N_s}{N_s + N_w} \quad \text{and} \quad X_w = \frac{N_v}{N_s + N_w}.$$

It follows that

$$X_s + X_v = 1$$

is valid all over the solution whether the sucrose concentration X_s is constant
or varies in space. In the latter case we have, in any direction x, that

$$\frac{dX_s}{dx} = -\frac{dX_w}{dx},$$

which means that to a given value of the concentration gradient dX_s/dx for
the dissolved substance there corresponds an equally large but oppositely directed gradient dX_w/dx for water. One should therefore expect that even in the
simplest situation as described in the example in Section 2.5.2.2, two diffusion
fluxes should be considered and therefore also two diffusion equations: one for
the dissolved substance and one for the water. However, this is not necessary
because with the free diffusion process the flux of the solute J_s and that of the
solvent J_v are subject to the additional condition of constraint that the process
must take place without volume changes, which implies that at every position

in space equal volumes of solute and of solvent are moved simultaneously but in opposite directions. For this reason it suffices to consider only the diffusion flux of the solute*.

The situation mentioned above, with two equally large but oppositely directed volume transfers of water and solute, is rarely met when the two concentrations of solutes are separated by a membrane with widely different permeability to water and the dissolved substance. When the permeability to water is much larger than the permeability to the dissolved substance, the flux of water by diffusion into the more concentrated solution exceeds the flux by diffusion of solute in the opposite direction. The result is that this phase may undergo a transient increase in volume, that may end dramatically with the bursting of the cell. Naturally, the most extreme situation is that when the membrane for all practical purposes behaves as an ideal semipermeable membrane.

With equal pressures on both sides of the membrane the transfer of water through the membrane has the character of a diffusion flux, the strength being determined by the difference in water concentration between the membrane sides and the water permeability in the membrane. When the water concentrations are equal on both sides, no net transfer of water will take place unless the hydrostatic pressures on the two phases are unequal. When different pressures act on the two solutions, which also differ in concentration, the hydraulic component of the water movement will add to the diffusive component, i.e. will *increase water movement* when directions are the same as the drops in pressure and concentration, and if opposite *counteract* the water movement by diffusion. To a given difference in water concentration across the membrane there is a definite pressure difference that will just compensate for the water movement by diffusion and make the net movement of water equal to zero. This pressure difference is equal to the difference between the osmotic pressure of the two solutions (compare Eq. (2.8.3)).

2.8.2.4 *The equation for the osmotic pressure*

The first well-sustained and systematic investigations of the osmotic pressure of solutions were done by Wilhelm Pfeffer[†]. He used membranes of cupro-ferrocyanide, which are permeable to water but not to sucrose molecules, and observed that the "osmotic pressure", as he denoted the equilibrium pressure, was proportionate to the sucrose concentration and to the absolute temperature. Pfeffer's investigations were instrumental in prompting the following

---

* For a very careful treatment of this problem and of the various way of defining a diffusion coefficient see Hartley, G.S. & Crank, J. (1949): *Trans. Faraday Soc.*, **45**, 801.

[†] Wilhelm Pfeffer (1845–1920) was a German plant physiologist, Professor in Basel (1877), in Tübingen (1878) and in Leipzig (1887); he published *Osmotische Untersuchungen* in 1877.

extensive experimental inquiries about the osmotic pressure and, in particular, to the thermodynamic description that the Dutch physical chemist J.H. van 't Hoff published in 1885, containing among other things the explicit expression for the manner in which the osmotic pressure depends on concentration and temperature*.

The excess pressure Π on an ideal solution in contact with the pure solvent through a semipermeable membrane that prevents the water molecules from moving into the solution is nowadays calculated by using the criteria for a heterogeneous equilibrium[†] for a chemical species j that is distributed between two phases (a) and (b), e.g. across a semipermeable membrane. The two phases are assumed to be in thermal equilibrium, so $T^{(a)} = T^{(b)} = T$. In the equilibrium condition one needs only to consider the solvent (water), as any spontaneous passage between the phases of the solute is prevented by the semipermeable membrane. Thus, the only way the free energy of the system can decrease, or its entropy increase, is by transfer of water between the phases. The chemical potential of water is represented by μ_w. The Gibbs criterion for equilibrium of water between the two phases (a) and (b) is[‡]

$$\mu_w^{(a)} = \mu_w^{(b)}. \tag{2.8.6}$$

Let X_w and X_s be the mole fractions for the solvent (water) and the solute in the solution. The chemical potential $\mu_w^{(b)}$ for water in the solution at temperature T and pressure $p^{(b)}$ is written as

$$\mu_w^{(b)} = \mu_w^\circ \left(T, p^{(b)} \right) + \mathcal{R}T \ln X_w, \tag{2.8.7}$$

provided the solution behaves as ideal. The phase (a) that contains the pure solvent has the mole fraction $X_w = 1.0$. Let $p^{(a)}$ be the pressure in this phase. We have then

$$\mu_w^{(a)} = \mu_w^\circ \left(T, p^{(a)} \right). \tag{2.8.8}$$

Thus, for a given value of X_w in phase (b), the two pressures $p^{(a)}$ and $p^{(b)}$ must have values that satisfy the equilibrium condition

$$\mu_w^\circ \left(T, p^{(a)} \right) = \mu_w^\circ \left(T, p^{(b)} \right) + \mathcal{R}T \ln X_w. \tag{2.8.9}$$

---

* J.H. van 't Hoff (1852–1911) was Professor of chemistry in Amsterdam 1876 and Berlin 1894. He did fundamental work on chemical equilibrium, reaction rates and osmotic pressure. He was awarded the Nobel Prize for chemistry in 1901.
† Developed by J. Willard Gibbs (1839–1903), Professor of mathematical physics at Yale University. He introduced the chemical potential and was founder of the modern formulation of chemical thermodynamics in his work *The Equilibrium of Heterogeneous Substances*. He was also the founder – together with Maxwell and Boltzmann – of statistical mechanics.
‡ See, for example, Denbigh, K. (1957): *The Principles of Chemical Equilibrium*. Cambridge University Press, p. 84.

The mole fraction X_w of the solvent in phase (b) is always less than one and, therefore, the term $\mathcal{R}T \ln X$ is always negative. Thus, the condition of the above equation is never fulfilled – and equilibrium is never attained – if the two pressures $p^{(a)}$ and $p^{(b)}$ are equal. To obtain equilibrium the pressure $p^{(b)}$ must be changed relative to $p^{(a)}$ so that the chemical potential in its pure form for water in phase (b) $\mu_w^o(T, p^{(b)})$ increases by an amount that numerically is equal to $\mathcal{R}T \ln X$. We shall now calculate the pressure difference $p^{(b)} - p^{(a)}$ at equilibrium. We rewrite Eq. (2.8.9) as

$$\mathcal{R}T \ln X_v = \mu_v^o\left(T, p^{(a)}\right) - \mu_v^o\left(T, p^{(b)}\right).$$

At constant temperature T the chemical potential of a pure substance depends only upon the pressure p^*. Formally, the right-hand side of the above expression can be written

$$\mathcal{R}T \ln X_w = \int_{p^{(b)}}^{p^{(a)}} \left(\frac{d\mu_w^o}{dp}\right) dp, \qquad (2.8.10)$$

where $d\mu_w^o/dp$ is the rate of change of the chemical potential with pressure p. The change with pressure in free energy of a pure substance at constant temperature T is given by

$$dG = V\,dp.$$

The chemical potential of a pure substance – in this case water – is identical to the molar free energy of the substance. We therefore have

$$d\mu_w^o = \overline{V}_w\,dp,$$

where $\overline{V}_w$ is the molar volume of the substance (for water $c.$ 18 cm^3) [†]. Hence

$$\frac{d\mu_w^o}{dp} = \overline{V}_w.$$

Insertion into Eq. (2.8.10) gives

$$\mathcal{R}T \ln X = \int_{p^{(b)}}^{p^{(a)}} \left(\frac{d\mu^o}{dp}\right) dp = \langle \overline{V}_w\rangle \left(p^{(a)} - p^{(b)}\right) = -\langle \overline{V}_w\rangle\, \Pi,$$

where $\langle V_w\rangle$ is the mean value of the molar volume $\overline{V}_w$ in the pressure range $p^{(a)}$ to $p^{(b)}$ and

$$\Pi = p^{(b)} - p^{(a)}$$

---

[*] The Gibbs function for a pure substance is only a function $G(T, p)$ of temperature T and pressure p, and its change is given by $dG = -S dT + V dp = (\partial G/\partial T)_p\, dT + (\partial G/\partial p)_T\, dp$.
[†] $\overline{V}_w = 1.801\,83 \times 10^{-5}$ m^3 mol^{-1} at 0 °C and $1.806\,89 \times 10^{-5}$ m^3 mol^{-1} at 25 °C.

is the osmotic pressure. Then, we have

$$\Pi = -\frac{\mathcal{R}T}{\langle \overline{V}_w \rangle} \ln X_w = -\frac{\mathcal{R}T}{\langle \overline{V}_w \rangle} \ln(1 - X_s), \qquad (2.8.11)$$

as $X_w + X_s = 1$. This expression is thermodynamically *exact*. The usual practice is to introduce a few approximations. We have * $\ln(1 \pm Y) \approx \pm Y$, if $|Y|$ is much less than 1.0. For a *dilute* solution we have $X_s \ll 1$. Therefore, for this solution the expression holds approximately

$$\Pi = \frac{\mathcal{R}T}{\langle \overline{V}_w \rangle} X_s.$$

This expression is simplified further since the solution is dilute. We have $X_s = n_s/(n_w + n_s) \approx n_s/n_w$, where n_s and n_w are the number in gram molecules of dissolved substance and of water in the solution. Inserting this approximated expression for X_s in the above equation yields

$$\Pi = \frac{\mathcal{R}T}{n_w \langle \overline{V}_w \rangle} n_s.$$

The product $n_w \langle \overline{V}_w \rangle = [V_w]$ represents the volume that the water molecules constitute of the total volume V of the solution. The quantity $m' = n_s/[V_w]$ represents the amount of dissolved substance in moles per unit volume of water and states the *volume molality*[†]. The equation above is then written as

$$\Pi = \mathcal{R}T \frac{n_s}{[V_w]} = \mathcal{R}T \, m'. \qquad (2.8.12)$$

If the solution is very dilute we can write $n_w \langle \overline{V}_w \rangle \approx V$, which gives

$$\Pi = \mathcal{R}T \frac{n_s}{V} = \mathcal{R}T \, C \qquad (2.8.13)$$

(since $C = n_s/V$ is the *molar volume concentration*, e.g. in mol dm^{-3}).

Equation (2.8.13) is the celebrated equation first derived by van't Hoff in 1885, and is referred to as *van't Hoff's equation*.

The osmotic water movements are capable of producing or resisting quite appreciable pressure differences. As an example, we consider a sucrose solution containing 1 mol sucrose dissolved in 1 kg H_2O (1 molal sucrose). The molecular weight M_w of water is 18.016 g mol^{-1}. Its mass density at 0 °C is $\rho(0) = 0.999\,87$ g cm^{-3}, and thus its molar volume is $\overline{V}_w = 18.0160/0.999\,87 =$

---

* See Chapter 1, Section 1.6, Eq. (1.6.17).
† For dilute solutions this measure deviates insignificantly from the more frequently used weight molality m, that is the amount of dissolved substance reckoned in *moles per kilogram of water*.

Table 2.4. Osmotic pressures of sucrose solutions in water at 25 °C

Molality m (mol kg^{-1})	Molarity C (mol dm^{-3})	$\Pi^{(obs)}$ (atm)	$\Pi^{(calc)}$ (atm)		
			Eq. (2.8.13)	Eq. (2.8.12)	Eq. (2.8.11)
0.1	0.098	2.59	2.36	2.40	2.44
0.2	0.193	5.05	4.63	4.81	5.46
0.3	0.282	7.61	6.80	7.21	7.82
0.4	0.370	10.14	8.90	9.62	10.22
0.5	0.453	12.75	10.90	12.00	12.62
0.6	0.533	15.39	12.80	14.40	15.00
0.7	0.610	18.13	14.70	16.80	17.40
0.8	0.685	20.91	16.50	19.20	19.77
0.9	0.757	23.72	16.50	21.60	22.15
1.0	0.825	26.64	19.80	24.99	24.49

Partly from Moore, Walter J. (1972): *Physical Chemistry*. Longman Group Ltd. Fifth Edition, London.

18.01 83 cm^3/mol^{-1} = $1.801\,83 \times 10^{-5}$ m^3 mol^{-1}. As 1 kg H$_2$O contains $1000/18.0160 = 55.505$ mol H$_2$O, in 1 molal solution the water has the mole fraction $X_w = 55.505/(55.505 + 1) = 0.982\,30$. Inserting these values in Eq. (2.8.12) yields

$$\Pi = -\frac{\mathcal{R}T}{V_w}\ln X_w = -\frac{8.314\,(\text{joule mol}^{-1}\,\text{K}^{-1}) \times 273.16\,(\text{K})}{1.801\,83 \times 10^{-5}(\text{m}^3\,\text{mol}^{-1})}\ln(0.9823)$$

$$= 2.251 \times 10^6\,(\text{N m}^{-2}) = \frac{2.251 \times 10^6\,(\text{Pa})}{1.013 \times 10^5(\text{Pa atm}^{-1})} = 22.2\,\text{atm}.$$

The dependence of the osmotic pressure on the concentration is further illustrated in Table 2.4, which contains a number of experimentally determined osmotic pressures $\Pi^{(obs)}$ of a sucrose solution. This solution displays ideal behavior reasonably well in the concentration range $0 \leq m \leq 0.5$ mol kg^{-1}, and the data may serve as illustrative of the range of validity of the above formulas.

There is a good agreement between the values of $\Pi^{(obs)}$ and those calculated from Eq. (2.8.12) that is correct for the ideal solution. As one could expect, the van't Hoff equation agrees reasonably well with the experimental values in the range of low concentration, where $C \leq 0.2$, whereas Eq. (2.8.13) – corresponding to a less forceful approximation – manages with somewhat higher concentrations than the van't Hoff equation, whose built-in approximations imply a validity for very dilute solutions. The osmotic pressures met in biology rarely exceed 7–8 atm. For this reason this equation is widely used, disregarding its approximations.

We consider now the situation where a solution in chamber (i) in Fig. 2.48 contains a number of different non-electrolytes with concentrations $C_1, C_2, C_3, \ldots$ As the osmotic pressure is a colligative property of the solution, i.e. Π depends only on the total *number* of molecules dissolved and not on the individual properties, the osmotic pressure of the solution calculated from the van't Hoff equation is

$$\Pi = \mathcal{R}T \sum_{i=1}^{i=n} C_i. \qquad (2.8.14)$$

Next we consider a solution of an electrolyte of molar concentration C, where each molecule in the solution is dissociated into v_+ cations and v_- anions. In keeping with Eq. (2.8.14) the osmotic pressure of the solution should be

$$\Pi = \mathcal{R}T[v_+ + v_-]C, \qquad (2.8.15)$$

provided the electrolyte solution at the concentration concerned behaves as an ideal solution*. Thus a solution of NaCl should have twice the osmotic pressure as a sucrose solution at the same concentration. It is important to bear this fact in mind although the exact value of the osmotic pressure may be difficult to assess at a glance.

Finally, let both chambers (i) and (o) in Fig. 2.48 contain solutions with osmotic pressures $\Pi^{(i)}$ and $\Pi^{(o)}$. If the pressures in the two chambers are $p^{(i)}$ and $p^{(o)}$, respectively, the volume flux of water through the semipermeable membrane is

$$J_v = L_P\left[p^{(i)} - p^{(o)} - \left(\Pi^{(i)} - \Pi^{(o)}\right)\right], \qquad (2.8.16)$$

from which it follows that the excess pressure Δp on the solution in chamber (i) that prevents a water flow through the membrane is

$$\Delta p = p^{(i)} - p^{(o)} = \Pi^{(i)} - \Pi^{(o)}, \quad \text{for} \quad J_v = 0. \qquad (2.8.17)$$

2.8.2.5 Osmotic coefficient

The van't Hoff equation and the expressions derived from it are only valid for ideal solutions that are rather diluted. In the ideal solution the forces of interactions between the water molecules do not differ from those that act between molecules of water and dissolved molecules. Thus, the properties of the ideal solution that may differ from those of pure water are solely attributable

---

* For electrolyte solutions this condition is only fulfilled in extremely dilute solutions.

to the *smaller number of water molecules* per unit volume in the solution than in pure water. For the ideal solution the chemical potential of each of the species j involved is written in the simple form

$$\mu_j = \mu_j^o + \mathcal{R}T \ln X_j, \tag{2.8.18}$$

where X_j is the mole fraction of the species j in the solution. This simple behavior, however, holds only in the lower range of concentration. The higher the concentration is of the solution, i.e. in practice $C > 0.2 \text{ mol dm}^{-3}$ for a non-electrolyte, the larger are the deviations from Eq. (2.8.18). In the case of electrolyte solutions the deviations from ideality occur at much greater dilutions. To retain the form of the chemical potential above under all conditions, in 1909 the Danish physical chemist N. Bjerrum* introduced the following modification of the chemical potential for the solvent (water)

$$\mu_j = \mu_j^o + g\,\mathcal{R}T \ln X_j,$$

where the factor g, that must be determined by experiment, is called the *osmotic coefficient*. Using this form of the chemical potential to carry out the analogous calculations of the osmotic pressure $\Pi^{(\text{real})}$ leads to

$$\Pi^{(\text{real})} = g\,\Pi^{(\text{ideal})}.$$

The osmotic coefficient g is the proper correction factor to be used when mole fraction is the concentration measure. Later "practical" osmotic coefficients have also been introduced when other units of concentration are in use[†]. To preserve the form of van't Hoff's equation in cases of substantial departures from ideality the *molar osmotic coefficient* ϕ is introduced. This coefficient is defined as

$$\Pi = \phi\,\mathcal{R}TC, \tag{2.8.19}$$

and is determined experimentally by measuring Π at a various concentrations. Examples of such determinations are shown in Table 2.5.

2.8.2.6 A simple dynamic model of osmotic equilibrium

Osmotic equilibrium is usually described using the methods of equilibrium thermodynamics, just as we did in Section 2.8.2.4. However, it may be of interest to see that considerations based on the random walk will lead to the same result, also for the reason that this point of view provides a more vivid picture of the

---

[*] Niels Bjerrum (1879–1958) was Professor of chemistry at the Royal Veterinary and Agricultural College, Copenhagen.
[†] For example, a molal or a molar osmotic coefficient.

Table 2.5. *Values of molar osmotic*
coefficients (0.15 M) at 25 °C

Solution	ϕ
K_2SO_4	0.76
NaCl	0.93
Na_2SO_4	0.77
KCl	0.92
Sucrose	1.01

molecular movements at equilibrium. To remove as many additional compli-cating factors as possible we reduce the model to contain only the essentials: a semipermeable membrane $\mathcal{M}$ of *no* thickness, represented as a rigid math-ematical plane, which is impenetrable to matter except at the holes that just allow the passage of water molecules. The membrane $\mathcal{M}$ separates two liquids. Phase (o) to the right of $\mathcal{M}$ consists of pure water, containing N molecules per unit volume. In phase (i) to the left of $\mathcal{M}$ is the solution having the water con-centration of $N - \Delta N$ molecules per unit volume. The solute molecules – not being able to move across the membrane – do not participate in the osmotic flow or in creating an osmotic pressure, except by representing a volume that otherwise had been at the disposal of the water molecules*. The macroscopic pressure in phase (i) is $P + \Delta P$ and P in phase (o). In a small region around $\mathcal{M}$ the pressure decreases from the value $P + \Delta P$ to P.

To describe the kinetics of exchange between phase (i) and phase (o) of the water molecules at the molecular level, we use the same type of argument as that used in Section 2.5.6.6 and Section 2.6.2.2 with the exception that we now focus our attention on the movements of the *actual water molecules* and not on the dissolved molecules. This point of view does not require the introduction of new physical principles as both the water and solute molecules execute the same pattern of random walk. Thus, the jumping of a water molecule ("Platzwechslung") from an initial position to a vacant "hole" and from this position to the next vacant position can also be visualized by the concept of a landscape of periodic energy barriers of height U_0 over which the water molecules have to jump when moving from one position to the next. Figure 2.50 shows a sequence of such potential ridges of height U_0 with the same distance (let us call this λ) between the valleys. We imagine the membrane $\mathcal{M}$ being placed at the top of a potential barrier, and thus achieving the situation where the pure water of concentration N_w and the solution with the water concentration

---

* This assertion is flagrantly at variance with that favored by those who consider the osmotic pressure to originate as a result of the bombardment of the solute molecules.

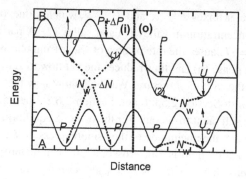

Fig. 2.50. Illustration of the flow of water molecules in the establishment of an osmotic equilibrium, when the water molecules have to jump across a sequence of potential barriers of height U_0. The two phases containing the solution (phase (i)) and pure water (phase (o)) are separated by an infinitely thin semipermeable membrane. (A) Equal pressure in both phases. (B) The pressure difference ΔP establishes a pressure gradient in the transition zone between (1) and (2), and adds to the water molecules in phase (i) an extra potential energy that leads to a deformation of the potential profile, that facilitates the jumping of the molecules over the barrier in the direction (1)→(2).

$N_w - \Delta N$ are located in the two adjacent valleys (1) and (2), i.e. separated by just one potential barrier of height U_0. Figure 2.50(A) illustrates the state of things where the pressures are *equal* in both phases. As the particle number in valley (2) differs from that in valley (1) by the amount ΔN, the traffic over the top of the barrier will be unsymmetrical, giving rise to a *net transfer* of water molecules – i.e. an *osmotic water flux* – in the direction (2)→(1). The deficit in the number of water molecules in valley (2) that tends to develop in this way is made up by a compensating net flux from the next valley, and so on. In a similar way the excess water molecules that are filling valley (1) move to the adjacent valleys to the left. Thus, although water molecules will move in both directions over the potential crest where the membrane is placed, there will be a more rapid flow of water from phase (o) containing the pure water into the solution than from the solution into the pure water. The quantitative treatment of this situation using Boltzmann statistics has already been given in Section 2.5.6.6.

When the osmotic equilibrium is established there is in phase (i) an excess pressure of magnitude ΔP over the pressure P in phase (o). This excess pressure is generated either by means of pistons as illustrated in Fig. 2.48 or as result of the spontaneous movement of water into phase (i) when its volume is kept constant. The effect of this pressure is to increase the tendency of the water molecules in phase (i) to flow into the pure water solution contained in phase (o). Thus, the pressure difference counteracts the difference between the two unidirectional water fluxes $J^{(1)\to(2)}$ and $J^{(2)\to(1)}$, which is due to the difference ΔN in water concentration in valley (1) and valley (2). A particular pressure difference $\Delta P = \Pi$ causes the rate of flow of water from phase (i) to be equal to the flow from phase (o) and thus reduces the *net* flow to zero. This pressure difference Π is the *osmotic pressure*. We shall show now that a combination of the arguments in Section 2.5.6.6 and Section 2.6.2.2 based on Boltzmann statistics will be sufficient to account for the osmotic equilibrium as described above and to predict the magnitude of the osmotic pressure Π.

The pressure P in a fluid is a macroscopic physical quantity* – in units of force per unit area – that on a microscopic scale is a reflection of the mechanical tensions between the individual water molecules[†]. The evaluation of the pressure from the modern theories of liquids[‡] is a far more involved matter than that of calculating the pressure of a ideal gas. But these details need not concern us. However, of immediate importance in this context is the significance that should be assigned to the notion P in the small regions under consideration. In any molecular system, irrespective of size, all the significant quantities like energy, volume, number of particles included, pressure, etc., undergo *fluctuations*, i.e. spontaneous *deviations* from their *average value*. As the volume considered becomes smaller the fluctuations grow in size but, in spite of this inherent uncertainty, when the quantity in question is averaged over a sufficient time it attains the macroscopic value that applies to thermodynamic considerations. Thus, when we come to apply the term pressure and particle number to regions of molecular dimensions we have to bear in mind terms representing the *averages* at the actual positions and refrain from considering events in so small a time interval that the fluctuations cannot be ignored safely.

In our case of osmotic equilibrium the two phases (i) and (o) are under the homogeneous pressures $P + \Delta P$ and P *except* at the region of transition between phase (i) and phase (o). For physical reasons an instantaneous drop in pressure is hardly possible, as a discontinuity in pressure would call for the

* Actually this is a special case of a stress tensor
$$\begin{pmatrix} P_{11} & P_{12} & P_{13} \\ P_{21} & P_{22} & P_{23} \\ P_{31} & P_{32} & P_{33} \end{pmatrix} = \begin{pmatrix} P & 0 & 0 \\ 0 & P & 0 \\ 0 & 0 & P \end{pmatrix} = P \begin{pmatrix} 1 & 0 & 0 \\ 0 & 1 & 0 \\ 0 & 0 & 1 \end{pmatrix}$$
with the tangential stresses P_{21}, P_{12}, etc., being zero in the absence of viscous drags and at equilibrium having identical normal stresses P_{11}, P_{22}, P_{33} equal to P.

[†] The dynamical theory of a gas pressure was first developed by Daniel Bernoulli (1738) with further developments largely due to Clausius (1857) and especially to Clerk Maxwell (1859) and Ludwig Boltzmann (1868).

[‡] See, for example Frenkel, J. (1946): *Kinetic Theory of Liquids.* Oxford University Press, Oxford; Cole, G.H.A. (1967): *The Statistical Theory of Classical Simple Dense Fluids.* Pergamon Press, Oxford; McQuarrie, D.A. (1973): *Statistical Mechanics.* Harper & Row, New York.

appearance of a discontinuity in the acting force, which is physically unrealistic. We therefore assume that in the transition region there is a gradual change in pressure from one value $P + \Delta P$ to the other value P. What the actual width of the transition region may be is difficult to assess. As a limit we assume that the transition region extends from the valley (1) over a distance λ to the bottom of valley (2).

To apply the methods of Boltzmann statistics to the osmotic equilibrium we must first consider the effect that the pressure difference may have on the energy of the water molecules in the two phases. The change in energy must be associated with the pressure change in the transition zone, as the pressures otherwise remain at two constant levels. To evaluate the energy change of a water molecule associated with the pressure change we need a relation derived from hydrodynamics: if the pressure in a fluid is changing, e.g. in the direction of the x-axis, it will exert a force X on a body of volume V that is*.

$$X = -\frac{dP}{dx} V. \tag{2.8.20}$$

The pressure gradient in the transition zone in Fig. 2.50(B) between valley (1) and valley (2) where the pressures are $P + \Delta P$ and P is

$$\frac{dP}{dx} = \frac{P - (P + \Delta P)}{\lambda} = -\frac{\Delta P}{\lambda}. \tag{2.8.21}$$

Let $\bar{v}_w$ be the volume of a water molecule. Then, a molecule that is at present in the transition zone is acted upon by a force in the direction (1)→(2) that is

$$X_w = -\frac{dP}{dx} \bar{v}_w = \frac{\Delta P}{\lambda} \bar{v}_w.$$

The work W_w required to move a water molecule solely against the hydrostatic pressure difference from valley (2) to valley (1) is then

$$W_w = X_w \lambda = \frac{\Delta P}{\lambda} \bar{v}_w \lambda = \Delta P \bar{v}_w.$$

This extra energy has to be added to the energy represented by the profile of height U_0 in the transition zone. As in Section 2.6.2.3 we divide this extra energy into two halves so that the particles in valley (1) are confronted in the direction (1)→(2) with a potential barrier of height

$$U_0 - \tfrac{1}{2} \Delta P \bar{v}_w,$$

* See Appendix H.

i.e. *decreased* by the amount $\frac{1}{2}\Delta P \bar{v}_w$, while the height in the opposite direction (2)→(1) as seen from valley (2) is

$$U_0 + \tfrac{1}{2}\Delta P \bar{v}_w,$$

i.e. *increased* by the amount $\frac{1}{2}\Delta P \bar{v}_w$. This deformation of the potential profile in the transition zone is illustrated in Fig. 2.50(B). In analogy to Section 2.5.6.6 we can now write the unidirectional flux in the direction (1)→(2) as

$$\overrightarrow{J} = \tfrac{1}{2}(N - \Delta N)\,\langle \overrightarrow{v^*}\rangle\, e^{-(U_0 - \Delta P \bar{v}_w/2)/kT} \qquad (2.8.22)$$

where $\langle \overrightarrow{v^*}\rangle$ is the average thermal velocity with which the water molecules move over the ridge of the barrier in the direction (1)→(2), and the factor $\frac{1}{2}$ makes allowance for the remaining half moving in the opposite direction.

Similarly, the unidirectional flux in the direction (2)→(1) from valley (2) to valley (1) is

$$\overleftarrow{J} = \tfrac{1}{2}N\,\langle \overleftarrow{v^*}\rangle\, e^{-(U_0 + \Delta P \bar{v}_w/2)/kT}. \qquad (2.8.23)$$

At osmotic equilibrium we have

$$\overrightarrow{J} = \overleftarrow{J} \qquad (2.8.24)$$

and

$$\Delta P = \Pi. \qquad (2.8.25)$$

Furthermore, there are no grounds for not assuming that

$$\langle \overrightarrow{v^*}\rangle = \langle \overleftarrow{v^*}\rangle = \langle v^*\rangle.$$

Thus, we have from Eq. (2.8.24)

$$\tfrac{1}{2}(N - \Delta N)\,\langle v^*\rangle\, e^{-(U_0 - \Pi \bar{v}_w/2)/kT} = \tfrac{1}{2}N\,\langle v^*\rangle\, e^{-(U_0 + \Pi \bar{v}_w/2)/kT},$$

or

$$1 - \frac{\Delta N}{N} = e^{-\Pi \bar{v}_w/kT}.$$

Taking logarithms on both sides gives

$$\ln\left(1 - \frac{\Delta N}{N}\right) = -\frac{\Pi \bar{v}_w}{kT}. \qquad (2.8.26)$$

If the solution is ideal and the solute molecules differ greatly in size from the water molecules, the deficit in water molecules per unit volume in phase (i) as compared with the number N in the pure water phase (o) is not different from the number of solute molecules per unit volume in phase (i), i.e. $\Delta N \approx N_s$, here N_s is the number of solute molecules per unit volume. Thus, we have

$$\frac{\Delta N}{N} = X_s,$$

where X_s is the molar fraction of the solute in phase (i). Hence, we then arrive at the following expression for the pressure Π that ensures osmotic equilibrium between the two phases

$$\Pi = -\frac{kT}{\bar{v}_w} \ln(1 - X_s) \qquad (2.8.27)$$

or, as $\mathcal{N}_A k = \mathcal{R}$ and $\mathcal{N}_A \bar{v}_w = \bar{V}_w$ is the molar volume of water, we can also write

$$\Pi = -\frac{\mathcal{R}T}{\bar{V}_w} \ln(1 - X_s). \qquad (2.8.28)$$

This expression is identical to the result obtained in Section 2.8.2.4, Eq. (2.8.11), by using considerations based on equilibrium thermodynamics. Then, carrying out the same approximations as in Section 2.8.2.4 we arrive at the van't Hoff formula

$$\Pi = \mathcal{R}T C, \qquad (2.8.29)$$

where C is the molar concentration of the solute in phase (i).

To some readers this agreement between Eq. (2.8.13) and Eq. (2.8.29) may appear trivial, since we have not delved into detailed molecular considerations and have only used the average values of the variables in the application of Boltzmann statistics. But the advantage of applying, even in the simplest manner, Boltzmann statistics to describe the equilibrium situation over that of applying equilibrium thermodynamics is that the first-mentioned treatment presents the picture of water molecules at equilibrium moving in *both directions* through the membrane in equal numbers, whereas equilibrium thermodynamics has nothing to say about the movement of water between the two phases at equilibrium.

The arguments presented here and the equilibrium result obtained will not be different if the infinitely thin membrane is replaced by one of finite thickness h and the hole by a cylindrical pore.

2.8.3 The freezing-point depression

All solutions have a lower freezing point than the pure solvent. As mentioned earlier, this lowering of the freezing point, the *freezing-point depression*, belongs (just as the osmotic pressure) to the colligative properties of the solution. The freezing-point depression ΔT_f for the solution is defined as

$$\Delta T_f = T_w - T_s,$$

where T_w is the freezing point for the pure solvent (water) and $T_s (< T_w)$ is the freezing point for the solution. A thermodynamic treatment of the temperature equilibrium between pure ice and water in an ideal solution, following the same lines as those in Section 2.8.2.4 but with the temperature T being the variable instead of the pressure p, leads to the following formula for the freezing-point depression:

$$\Delta T_f = -\frac{\mathcal{R} T_v^2}{\Delta H_{fus}} \ln X_v = -\frac{\mathcal{R} T_v^2}{\Delta H_{fus}} \ln(1 - X_s) \approx \frac{\mathcal{R} T_v^2}{\Delta H_{fus}} X_s, \quad (2.8.30)$$

where $\Delta H_{fus} = 6.01$ kJ mol^{-1} is the heat of melting of the reaction $H_2O^{(ice)} \rightarrow H_2O^{(liq)}$ and $T_v = 273.16$ K is the freezing point of water. We replace in this expression the mole fraction X_s for the solute with the *molality* (m_s), i.e. the number of gram moles contained in 1 kg of the solvent (water)*. For a dilute solution we have

$$X_s = \frac{M_v}{1000} m_s,$$

where $M_v = 18.016$ g mol^{-1} is the molecular weight of water. We then have

$$\Delta T_f = \frac{\mathcal{R} T_v^2 M_v}{1000\, \Delta H_{fus}} m_s = K_f m_s,$$

and from this it appears that the freezing-point depression is directly proportional to the molality m_s of the dissolved substance as the constant

$$K_f = \frac{\mathcal{R} T_v^2 M_v}{1000\, \Delta H_{fus}},$$

only depends on properties belonging to the pure water, i.e.

$$K_f = \frac{8.314 \times (273.16)^2 \times 18.016}{1000 \times 6010} = 1.860.$$

Hence

$$\Delta T_f = 1.860\, m_s. \qquad (2.8.31)$$

---

* See Appendix R, Eq. (R.4).

Thus, the freezing-point depression ΔT_f^* of an ideal solution of *one molal* (1 mol per kg H_2O) concentration is

$$\Delta T_f^* = 1.860\ K. \qquad (2.8.32)$$

2.8.3.1 The freezing-point depression and osmotic pressure

As mentioned earlier, the four colligative properties are mutually dependent. Connecting together Eq. (2.8.12) and Eq. (2.8.20) by eliminating $\ln X_v$ we obtain the following relation between the freezing-point depression ΔT_f of a solutions and its osmotic pressure:

$$\Pi = \frac{\Delta H_{fus}}{T_v \overline{V_v}} \Delta T_f. \qquad (2.8.33)$$

The numerical value of the fraction is

$$\frac{6010\ (\text{j mol}^{-1})}{273.16\ (\text{K})\ 1.8018 \times 10^{-5}\ (\text{m}^3\ \text{mol}^{-1})} = 1.2211 \times 10^6 \quad (\text{Pa K}^{-1}),$$

and so Eq. (2.8.23) takes the form

$$\Pi = 1.2211 \times 10^6\ \Delta T_f \quad (\text{Pa}).$$

Putting $\Delta T_f = 1.860$ gives

$$\begin{aligned}
\Pi &= 1.2211 \times 10^6 \times 1.86\ (\text{Pa}) = 2.2712 \times 10^6\ (\text{Pa}) \\
&= \frac{2.2712 \times 10^6\ (\text{Pa})}{1.013 \times 10^5\ (\text{Pa atm}^{-1})} = 22.4\ \text{atm}.
\end{aligned}$$

In other words: a solution having a freezing-point depression $\Delta T_f = 1.860\ \text{K}$ has an osmotic pressure of 2.2712×10^3 kPa or 22.4 atm at 0 °C. A determination of the freezing–point depression of a solution very often replaces the direct measuring of the osmotic pressure of the solution, partly because a freezing-point depression is easier to perform experimentally and partly because even very small volumes – that do not allow a direct pressure measurement – are accessible to freezing-point determinations.

The osmotic pressure of a solution and the three other colligative properties indicate that the addition of the dissolved molecules changes the properties of solvent molecules relative to those of the pure solvent. Principally, these changes are only determined by the *number* of molecules dissolved per unit volume. For non-electrolytes this number is equal to the number of moles dissolved multiplied by Avogadro's number $\mathcal{N}_A$ (6.023×10^{23} molecules per mole of substance). For an electrolyte solution this number has to be multiplied by the number

$$\nu = \nu_+ + \nu_-$$

of particles that results from the dissociation of the salt molecule, ν_+ and ν_- being the number of cations and anions of which the salt molecules are composed. Thus, $\nu = 3$ for Na_2SO_4. Highly diluted solutions of NaCl and Na_2SO_4 will have osmotic pressures equal to twice and three times the pressure of a sucrose solution with the same molar concentration (compare Eq. (2.5.218)). Of secondary importance are various factors such as high concentration, the shape of the dissolved molecules, electrostatic interaction between the ionic species in the solution and between ions and water, etc., that are pooled together in the osmotic coefficient. It appears from Table 2.5 that a 0.15 M solution of Na_2SO_4 has an osmotic pressure found in an ideal solution of a non-electrolyte of concentration

$$C = 0.77 \times 3 \times 0.15 = 0.3465 \, \text{M},$$

and at 25 °C both solutions will have an osmotic pressure

$$\Pi = 0.3465 \times 27.4 \, \text{atm} = 9.494 \, \text{atm} = 9.494 \times 1.013 \times 10^5 \, \text{Pa}$$
$$= 9.618 \times 10^5 \, \text{Pa} = 961.8 \, \text{kPa}.$$

Thus, two solutions having the same molar concentration may have widely different osmotic pressures, just as two solutions having the same osmotic pressure may differ as regard to concentration. Finally two solutions that contain the same number of dissolved particles per unit volume may have different osmotic pressures due to divergences from ideality.

2.8.3.2 Osmolarity

Biological cell membranes are compliant and are not capable of resisting great pressure differences between their two sides. The preservation of the various factors that determine the cell volume is therefore essential for the survival of the cell. One important factor is to maintain a constant difference in the water activity between the cell interior and its surrounding medium. For historical reasons it is the practice instead to talk and argue with differences in osmotic pressure. Biological fluids such as blood plasma, and extra- and intracellular liquids, etc., contain many different substances in different concentrations. An estimation of the osmotic pressure on the basis of the concentrations of the constituents and the magnitudes of their osmotic coefficients would be a rather laborious and inconvenient procedure. Instead, direct measurement of the osmotic pressure or the freezing-point depression is preferred, in particular the latter as it allows us to work with a very small volume. To have a measure for the number of particles that are dissolved, e.g. per kilogram of water, in a given biological fluid, it has been convenient to introduce a *solution of reference*, which gives a precise measure of the water activity of the biological fluid in question and at the same time reflects the number of free particles in the solution. A reference is considered an ideal solution containing a non-specified non-electrolyte. The amount dissolved is denoted by the suggestive unit *osmol*, to emphasize the ideality of the solution and its relation to its osmotic pressure*. A solution containing 1 *osmol per kilogram of water* is called a 1 *osmolal* solution. A 1 *osmolal solution* (containing 1 osmol per kilogram of H_2O) is characterized by having *a freezing-point depression* of 1.860 K and thus also an *osmotic pressure* of 22.4 atm at 0 °C, as well as containing 6.023×10^{23} dissolved particles of non-specified constitution per kilogram of $H_2O^†$ displaying ideal behavior.

Consider a solution of unknown composition but having a freezing-point depression of 0.93 K. This value corresponds in our reference solution to an osmolality of $0.93/1.86 = 0.50$ osmol per kg $H_2O = 500$ mosmol per kg H_2O. It can then be said that, as far as *water activity* is concerned, the test solution behaves as a 0.5 osmolal = 500 mosmolal solution and has an osmotic pressure of 11.2 atm at 0 °C.

The freezing-point depression of normal human blood plasma is $\Delta T_f^{(plasma)}$ 0.52–0.54 K. Hence the osmolality of the plasma is

$$0.53/1.86 = 285 \text{ mosmol kg}^{-1}$$

or, as often expressed in common physiological or clinical jargon, 285 mosmol dm^{-3}, as one disregards the minor differences that exist at these concentrations between the measures molality and molarity.

It is also possible to make an estimate of the osmolality of a non-ideal solution if the osmotic coefficient ϕ is known, since a non-ideal solution of a single salt of molality m and the osmotic coefficient is

$$\text{osmolarity} = \phi \, v \, m,$$

where v is the number of dissociating particles. Thus, a 0.154 M solution of

---

* The concept *osmolar concentration* was introduced E.J. Warburg (1892–1969), Professor of internal medicine, Copenhagen University. Warburg, E.J. (1922): Studies on carbonic acid components and hydrogen ions activities in blood and salt solutions, *Biochem. J.* **16**, 11–340, is an exemplary work on the distribution of small ions between blood plasma and red cells. Warburg was a great exponent for the clinician trained in the natural sciences.
† Thus equal to Avogadro's number $\mathcal{N}_A$.

NaCl* contains

$$0.93 \times 2 \times 0.154 = 0.286 \, \text{osmol kg}^{-1} \quad \text{or} \quad 286 \, \text{mosmol kg}^{-1}.$$

In this section we have often changed between the different measures of concentration. An overview is presented in Appendix N.

2.8.3.3 Reflection coefficient

Many membranes, especially biological membranes, are in varying degrees permeable also to the dissolved molecules. To put a stop to osmotic water movement through this kind of membrane requires less excess pressure on the solution side than that required if the membrane were ideally semipermeable. In the limiting case, where the membrane is equally permeable to all the species in the solution, an interdiffusion with participation of all the molecules present takes place between the two phases. And using this membrane to measure the osmotic pressure of the solution would result in zero pressure. At the other end of the scale is the pressure $\Pi^{(\text{ideal})}$ that would be obtained if the membrane were ideally semipermeable. Somewhere between these extremes

$$0 \leq \Pi^{(\text{eff})} \leq \Pi^{(\text{ideal})},$$

the effective pressures $\Pi^{(\text{eff})}$ are actually encountered.

The Dutch physical chemist A.J. Staverman[†] showed in 1951 – using arguments based on irreversible thermodynamics – that a non-ideal semipermeable membrane could be characterized by means of a single parameter, which he called the *reflection coefficient* of the membrane. We consider the set up of Fig. 2.48, but the membrane is no longer an ideal semipermeable membrane. Let phase (i) contain a solution with concentration C, whereas phase (o) initially contains pure water. Using the two pistons we impress a pressure difference $\Delta p = p^{(i)} - p^{(o)} \geq \Pi^{(\text{ideal})}$ on the two phases (i) and (o) and a *filtrate*, i.e. water + solutes, is squeezed through the membrane into phase (o). The concentration of the filtrate C_f depends on the ease with which solvent molecules permeate the membrane, i.e. $C_f = C^{(i)}$ when there is no discrimination between solvent and solute, and $C_f = 0$ for the ideal semipermeable membrane. As a measure for the relative ease by which a dissolved component j is transferred by the filtration process, Staverman introduced the parameter

$$\sigma_j = 1 - \frac{C_f}{C^{(i)}}, \tag{2.8.34}$$

---

* This is the molality of the *physiological salt solution*, containing 0.9% NaCl, i.e. $9/58.44 = 0.154$ mol per kg H_2O.

[†] Staverman, A.J. (1951): The theory of measurement of osmotic pressure. *Rec. trav. chim.* **70**, 344.

where σ_j is designated the *reflection coefficient* for the molecular species j. Equation (2.8.34) serves as a measure for how effectively the membrane acts as a molecular "sieve". When $\sigma = 0$ the membrane does not distinguish between solvent and solute; whereas $\sigma = 1$ corresponds to a total impermeability to the dissolved molecules. Furthermore, Staverman showed that the excess pressure on the solution with concentration C that is required to stop the water flow through a membrane having a reflection coefficient σ_j is given by the following modification of van't Hoff's equation

$$\Pi^{(\text{eff})} = \sigma_j \mathcal{R} T C^{(i)}, \qquad (2.8.35)$$

where $\Pi^{(\text{eff})}$ is called the *effective osmotic pressure* of the substance j in the solution, and the osmotic water flow through the membrane takes the form

$$J_v = L_P \left[\Delta P - \sigma_j \mathcal{R} T C^{(i)} \right]. \qquad (2.8.36)$$

But the solution will still have an *osmotic pressure*, Π, that is

$$\Pi = \mathcal{R} T C^{(i)},$$

and a *molal freezing-point depression* of 1.86 K, provided it behaves as an ideal solution.

The reflection coefficient can be determined as

$$\sigma_j = \frac{\Pi^{(\text{eff})}}{\Pi^{(\text{ideal})}}, \qquad (2.8.37)$$

if the quantities on the right-hand side are known. $\Pi^{(\text{eff})}$ results from measuring the osmotic pressure difference across the semipermeable membrane in question and $\Pi^{(\text{ideal})}$ results from measuring the freezing-point depression of the same solution.

2.8.4 *Water movement across cell membranes*

In living cells water forms by far the greatest constituent of the protoplasm. The cell water interchanges with the external surrounding fluid through the cell membrane. The water permeability is determined either by measuring the influx or efflux of tagged water (D_2O) between the cell interior and the external fluid of normal tonicity, or by measuring the volume changes caused by the bulk flow of water through the membrane, when the cell is exposed to slightly hypertonic or hypotonic solutions. The ease of penetration depends on the cell type and species of animal*. A typical value of water permeability in

---

* An extensive list of water permeabilities from animal cells ranging from 5×10^{-5} cm s^{-1} to 4×10^{-2} cm s^{-1} is given in Dick, D.A.T (1966): *Cell Water*, Butterworth, London, pp. 90–92.

mammalian cells is 5×10^{-3} cm s^{-1}. In the unmodified lipid bilayer membrane, values have been determined ranging from 2×10^{-5} to 1×10^{-3} cm s^{-1}*, i.e. values whose sizes are comparable to those found among the animal species. By way of comparison, a water layer of thickness $h = 1000$ nm $= 10^{-4}$ cm (about the same thickness as a cell membrane) with a self-diffusion coefficient[†] equal to $D = 2 \times 10^{-5}$ cm^2 s^{-1} presents a permeability to water that is $P = 2 \times 10^{-5}$ cm^2 s$^{-1}/10^{-4}$ cm $= 0.2$ cm s^{-1}. When the cell membrane is impermeable to all the surrounding solute molecules, the cell membrane behaves as an ideal semipermeable membrane where the water exchange across the cell membrane takes place exclusively by osmosis, and the cell volume shrinks or swells to adjust the tonicity of the cell interior according to the changes in water concentration in the external solution. When one or more of the solutes in the external solution has a permeability of the same order as the water the situation becomes more complex. As two extreme examples we consider the following. (1) Human erythrocytes that are suddenly exposed to an *isosmotic* solution of urea, a substance to which the erythrocyte membrane is permeable. Initially, water is in equilibrium, but urea, which is only sparingly present inside the cell, diffuses into the cell interior and in so doing reduces the cell water concentration, which in turn leads to a bulk flow of water by osmosis into the cell tending to restore the equilibrium state of water across the membrane. This state, however, will never be established as urea, which still is far from equilibrium, continues to diffuse into the cell together with the concomitant water inflow, resulting in a gradual increase in the cell volume. This process continues until the cell membrane swells beyond the limit of its extensibility and bursts. (2) A physiological salt solution, i.e. a 0.154 M NaCl solution ($= 0.9\%$ NaCl solution) is not only isosmotic with the human erythrocytes and other body cells but is also an *isotonic* solution, i.e. in contrast to example (1) the water equilibrium between the cell water and solution is retained. The reason for this different osmotic response arises because in example (1) permeabilities of solvent and solute are almost of the same order of magnitude whereas in example (2) the permeabilities of Na$^+$ and Cl$^-$ are several orders of magnitude smaller than the water permeability. Thus, in spite of a considerable concentration difference of sodium ions between the interior and exterior of the cell, the membrane will for practical purposes behave as a semipermeable membrane.

---

* Huang, C. & Thompson, T.E. (1966): Properties of lipid bilayer membranes separating two aqueous phases: Water permeability. *J. Mol. Biol.*, **15**, 539–554. Finkelstein, A. (1976): Water and nonelectrolyte permeability of lipid bilayer membranes. *J. Gen. Physiol.*, **68**, 127–135.

† Kohn, P.G. (1965): Tables of some Physical and Chemical Properties of Water. In *The State and Movements of Water in Living Organism. Symposia of the Society for Experimental Biology*, no. XIX, Cambridge, pp. 1–16.

Both in single cells and in whole organisms it is essential for survival that the water content is maintained within rather narrow limits. Many investigations have tried to understand the mechanisms of water transport and their control both across the single cell membrane and layers of epithelia such as those of the kidney and the gut, where *paracellular* water transport also plays an important part. Water flow is often found to take place in a direction that is contrary to the direction of flow that should take place according to the known magnitudes and directions of the gradients of concentration and pressure. Therefore, although being a prerequisite for describing biological water transport, the laws of passive water transport form only part of the explanation. Much experimental and theoretical work has been done in search of the additional mechanisms that underlie active water transport across the single cell membrane and across the epithelial cell layer where the isotonic water transport across leaky epithelia presents a problem of particular interest. Comprehensive lists of references to the works on biological water transport are found in Finkelstein (1987)[*] and House (1974)[†].

An important step towards the understanding of the nature of the water permeability of the cell membrane was the finding (Preston & Agre, 1991)[‡] of an integral membrane protein in the erythrocyte membrane that functions as a selective water permeable channel and is now known as *aquaporin-1*(AQP1). The clue to this search was the old observation that the diffusion of water through lipid bilayer membranes occurred with a high activation energy, whereas the diffusion through red blood cell membranes had a much lower activation energy and, in addition, was of the same size as that of the self-diffusion of H_2O in water. A subsequent search has demonstrated a widespread presence of specific water channels in many cell types and in that connection also a whole family of aquaporins, thereby once and for all solving the question about the presence of water-filled pores in the cell membrane. However, other types of membrane proteins (i.e. co-transporters) also contribute significantly to the transport of water across cell membranes (Zeuthen 2000)[§].

Some epithelia transport fluid that is in osmotic equilibrium with the bathing solution. A model for this *isotonic transport* that operates unassisted by an active water transport system has been established[‖]. The basic feature of this

---

[*] Finkelstein, A. (1987): *Water Movement Through Lipid Bilayers, Pores, and Plasma Membranes. Theory and Reality.* Wiley-Interscience publication: Distinguished Lecture of the Society of General Physiology, Volume 4, p. 228, John Wiley and Sons, New York.

[†] House, C.R. (1974): *Water Transport in Cells and Tissues*, Edward Arnold, London, p. 562.

[‡] Preston, G.M. & Agre, P. (1991): Isolation of the cDNA for erythrocyte integral membrane protein of 28 kilodalton: member of an ancient channel family. *Proc. Nat. Acad. Sci. USA*, **88**, 11 110–11 114.

[§] Zeuthen, T. (2000): Molecular water pumps. *Rev. Physiol. Biochem. Pharmacol.*, **141**, 97–151.

[‖] Larsen, E.H., Sørensen, J.N. & Sørensen, J.B. (2002): Analysis of the sodium recirculation theory of solute coupled water transport in small intestine. 10-1013/jphysiol.2001.013248

"sodium recirculation theory" is a *closed loop* of ionic flow that is driven by sodium pumps in the lateral membranes transporting to the lateral interspace, generating there a slight hyperosmotic fluid that causes excess fluid inflow by osmosis from the serosal side, mainly through the tight junctions. A slightly excess hydrostatic pressure drives the lateral space fluid into the inside bath, where ions secreted in excess are absorbed by the mucosal part of the epithelial cell and returned (recirculated) to the main transporter source, leaving the rest of the transported fluid by and large isosmotic.

Chapter 3

Membrane potentials

Although far different from the model diffusion set up with a parchment or ion exchange membranes, nevertheless I believe that the elementary parts at the molecular level of these biological events, complicated as they are, do contain just the same fundamental elements as we can reveal by the model study in the simpler, well defined systems.

(Torsten Teorell, 1956)

3.1 Introduction

So far we have only considered the passive transport of uncharged molecules, e.g. sucrose. In this chapter we shall discuss some of the characteristic features of the passive transfer of *electrolytes* in a free, unlimited medium, and also describe some of the basic mechanisms that lead to the presence of an electric potential difference (*membrane potential*) across the membrane.

An electrolyte solution contains ions having a positive charge (*cations*) and ions with negative charge (*anions*). However, with only a few exceptions is it possible to describe the transfer of an electrolyte by the movement of its salt alone. In general it is necessary to describe separately the movements for the ionic constituents of the salt. The dissociation to ions with negative and positive charges results in an additional restriction on the mutual movements among the ionic species. Because of the strong electrostatic forces of interaction between anions and cations an excess charge due to a deviation in the number of positive and negative charges will not appear anywhere in the solution in amounts that can be detected by simple chemical analysis. To characterize this situation we say that *macroscopic electroneutrality* is preserved in an electrolyte solution.

It very often happens that the ions in the electrolyte solution, in addition to their microscopically electric forces of interaction, are also under the influence of a large-scale electric field or potential difference that may extend over the total volume of the solution.

This complicates the description of the passive transfer of the ions as the field of force associated with the potential gradient acts on the ions according to their respective charges. When no concentration gradient exists in the solution this electric field of force will cause the ions to move along – according to the sign of their charge – solely by a process of *migration*. Thus, the point of departure for describing a passive transfer process of ions is not Fick's law (Eq. (2.2.3)) but the *Smoluchowski equation* (Eq. (2.6.2) or Eq. (2.6.4)), which describes the passive transfer process resulting from the presence of both a concentration gradient and an external field of force.

When two electrolyte solutions are separated by a membrane that is permeable to one or all of the ions in solution, one will often observe an electric potential difference between the two solutions and – on closer inspection – find that the potential jumps across the membrane that separates the two solutions. The potential difference appears if the ions, in diffusing across the membrane, also carry a net charge across the membrane with the result that one solution is allocated an excess of e.g. *positive* charges, while at the same time an equivalent charge deficit – i.e. excess of *negative* charges – is left in the other solution. The electric potential difference across the membrane that results from this *charge separation* is called a *membrane potential*. The following three condition are required for the establishment of a membrane potential.

(1) The membrane must be permeable to one or all of the ions contained in solutions surrounding the membrane.
(2) If more than one ionic species can penetrate the membrane their rates of penetration must be different.
(3) The two solutions must be different (either in concentration or composition or in both).

In this chapter we shall give in more detail an account of the factors that determine the magnitude and sign of the electric potential differences that may arise in free solutions as well as across membranes separating two electrolyte solutions.

3.2 Electric field and potential

I shall give a brief outline of the basic concepts of electric field and electric potential for those readers unfamiliar with the subjects. All electric phenomena are due to the presence of *electric charges*, of which there are two types: positive and negative charges. The natural, elementary charge is the *negative* charge carried by the *electron*. Assigning the negative charge to the electron has no deeper significance. It turns out to be the consequence of an arbitrary choice made in 1781 – many years before the discovery of the electron – when Benjamin Franklin* demonstrated the presence of both positive and negative charges and arbitrarily defined them.

Around each charged particle is a larger or smaller region, where other electrically charged particles come under the action of a force. When the charges do not move relative to each other we talk about an *electrostatic force*. The basic experiment for describing this kind of interaction was performed in 1785 by C.A. Coulomb[†], and the result is expressed in Coulomb's law, which states that the force X between two charges of magnitudes Q_1 and Q_2 is proportional to the product of the charges and inversely proportional to the square of the distance r_{12} between the point charges, thus

$$X = k\frac{Q_1 Q_2}{r_{12}^2},$$ (3.2.1)

where k is a constant of proportionality whose magnitude depends upon the units that are chosen for X, Q and r, and on the properties of the *medium* in which the charges are located. Charges of the *same sign repel* each other, while an *attraction* occurs between *opposite* charges. In SI units the electric charge is the coulomb (C), taken over from the already existing practical system of units. When the units for length r and force X are in meters (m) and newtons (N) the constant k in Coulomb's law takes the value

$$K_c = 8.987\,42 \times 10^9 \approx 9 \times 10^9 \triangleq \frac{\text{N m}^2}{\text{C}^2},$$ (3.2.2)

provided the charges are located in a vacuum. K_c is called the *Coulomb constant*. When developing the mathematical theory for electrostatics one finds that the factor $4\pi k$ appears very frequently in the equations. When developing the SI system of units it was found to be a convenient time to refurnish the positions in the equations where the factor 4π appeared. For that reason a new constant ε_0 was defined

$$K_c = \frac{1}{4\pi\varepsilon_0},$$

---

* Benjamin Franklin (1706–1790) was a North American statesman, physicist and man of letters with interest in all kinds of natural science. He was the committee member who wrote "The American Declaration of Liberty", and was the inventor of the lightning rod.
† Charles Augustin Coulomb (1736–1806) was a French engineer and physicist. He made important investigations on the resistance of friction and on electric and magnetic forces, and was the inventor of the torsion balance.

which implies that the factor 4π now appears instead in Coulomb's law, which now takes the form

$$X = \frac{1}{4\pi\varepsilon_0}\frac{Q_1 Q_2}{r^2}. \qquad (3.2.3)$$

It follows from the numerical value of K_c that

$$\varepsilon_0 = \frac{1}{4\pi K_c} = 8.854 \times 10^{-12} \triangleq \frac{C^2}{N\,m^2} \triangleq \text{farad per meter}. \qquad (3.2.4)$$

The constant ε_0 is called the *vacuum permittivity*.

When an electric charge is influenced by an electrostatic force – naturally arising from the presence of other charges either moving or at rest – it is an advantage in many instances to disregard the presence of the charges and their interactions and consider the situation as that of the force being caused by the presence of an *electric field* at the region in question. The magnitude of the electric field E in a given position x in space is defined in this way. We consider in the position x a charge q so small that its presence does not disturb the original field. We then measure the electrostatic force $\delta X(x)$ that acts on the charge δq. Then the electric field $E(x)$ – or just E – in the position x is

$$E(x) = \frac{\delta X(x)}{\delta q} \triangleq \text{newton per coulomb}. \qquad (3.2.5)$$

This vector quantity E that characterizes the field at every point has many names of which we choose a few at random: *electric field, electric field strength, electric field intensity* and *electric intensity*. In this book there is no need to specify explicitly the field as a vector, as we shall always consider its direction as coincident with that of the x-axis.

A charge of magnitude q at a position in space with the field strength E is acted upon by a force X

$$X = qE. \qquad (3.2.6)$$

In the case of an ion, say H^+, the force is

$$X = eE,$$

where

$$e = 1.602 \times 10^{-19}\,\text{coulomb}$$

is the *positive elementary charge*, that is, the *proton charge*. An alternative to the script e is to write q_e. For a cation K^{z+} carrying z_+ positive elementary charges the force is

$$X_{z_+} = z_+ e E,\tag{3.2.7}$$

where X and E have the same directions. For an anion A^{z-} with z_- negative elementary charges the force is

$$X_{z_-} = -|z_-| e E,\tag{3.2.8}$$

that is, *oppositely* directed to the force acting on the cation. We have chosen the notation $-|z_-|$ to avoid any misunderstanding that might be associated with the notation z_- that is often in use.

Closely related to the electric field is the *electric potential*[*]. This quantity, which also characterizes the presence of the electric field, is very useful in connection with electrical measurements and calculations.

When a charge in a region containing an electric field is moved around, work W (N m = J) is performed since all the time the charge is acted upon by a force. If we are moving the charge *we* are doing work *against* the electric field. Conversely we say that the field is performing work by moving it without external interference. The work required to move a charge from a point (1) to a point (2) in space depends not only on the magnitude of the charge q in question but also on the strength of the electric field between the two positions. The larger the field, the larger the work. Therefore, the *work* required to move the charge is also a measure of the *field strength* in the space. We shall now demonstrate the connection between these two quantities. To characterize the work in moving a charge, a new concept – a space function – is introduced, called the *electrostatic potential* or just simply the *potential*. As a symbol for this quantity we use in this book the Greek letter ψ (*psi*).

The physical meaning of the potential at a point is illustrated in this way. For the sake of simplicity we consider the field that is set up by a point charge. This field displays *spherical symmetry*, i.e. the field around the point charge has the same value for the same distance r from the charge and is everywhere directed as the direction of the radius r drawn from the charge to the point in question. The field lines, that is curves drawn in the electric field, such that the tangent at any point is parallel to the electric intensity at that point, are thus lines radiating from the point charge. Furthermore, the field strength – decreasing as the square of the distance from the point charge (Coulomb's law) – vanishes as the distance approaches infinity. Supposing that we move a small, positive charge δq from "infinity" to a point (1) in space. In doing so we perform work W_1 against the

---

[*] The concept *electric potential* and its relation to the electric field strength was introduced by the English mathematician George Green (1793–1841).

field. Then the potential at this point is defined as

$$\psi_1 = \frac{W_1}{\delta q}, \quad \text{joule per coulomb, (J C}^{-1}\text{)}. \qquad (3.2.9)$$

The potential at point (1) represents the work required to move a charge unit (1 coulomb) from infinity to point (1)*. The electric potential has the dimension of *work per coulomb*, and is designated by a special unit, volt:

$$1 \text{ joule per coulomb} \overset{\text{def}}{=} 1 \text{ volt (V)}.$$

We repeat the procedure but move the charge to another position (2)† and perform work W_2. The potential at point (2) is

$$\psi_2 = \frac{W_2}{\delta q} \text{ (volt)}.$$

Thus, the work required to move the charge from point (1) to point (2) is

$$W_2 - W_1 = \delta q(\psi_2 - \psi_1), \qquad (3.2.10)$$

and the *potential difference* between point (2) and point (1) is

$$\psi_2 - \psi_1 = \frac{W_2 - W_1}{\delta q}. \qquad (3.2.11)$$

In everyday speech the names *voltage* or *tension* between the points (2) and (1) are used as synonyms for the potential difference. In both cases the *volt* is the preferred unit.

A simple relation exists between the potential ψ and the field strength E.

$$\psi(x) \qquad \psi(x + h)$$

$$x \qquad x + h$$

To simplify matters we let the x-axis be a radius from the point charge to ensure a merging with the field direction. We move the points (1) and (2) sufficiently near to each other to the positions x and $x + h$ to ensure that the force

---

* This definition is due to the British physicist William Thomson (1824–1907), Professor of physics at Glasgow University 1848–1898. Thomson was one of the greatest physicists of his time and has made important contributions to almost all fields of physics. In 1892 he was raised to the peerage as Lord Kelvin of Largs.
† The two points need not to lie on the same radius, as no work is required to move the charge on a spherical surface of any radius r.

required in moving the charge from (1) to (2) along the distance h is approximately constant, with $\langle X \rangle$ representing the mean value. Let the potentials at points (1) and (2) be $\psi(x)$ and $\psi(x + h)$. The work we have to do in order to move the charge q over the distance h is, according to Eq. (3.2.10),

$$\Delta W = \langle X \rangle h = \delta q[\psi(x + h) - \psi(x)].$$

Thus,

$$\langle X \rangle = \delta q \frac{\psi(x + h) - \psi(x)}{h}.$$

When $h \to 0$, $\langle X \rangle \to X$, which is the force acting in the position x. At the same time the right-hand side tends to $(d\psi/dx)^*$. Hence

$$X = \delta q \frac{d\psi}{dx}.$$

The force X that we use to move the charge is numerically equal to the force acted upon the charge by the field. This force is $E\delta q$, and *oppositely directed* to the force X that we impress on the charge. Thus, we have

$$X = -E\delta q.$$

This gives

$$-E\delta q = \delta q \frac{d\psi}{dx},$$

or

$$E = -\frac{d\psi}{dx}. \tag{3.2.12}$$

In other words, the *field strength*, E, at a given point x is equal to the rate at which the potential declines per unit length $(-d\psi/dx)$ in the direction of the field.

The primary unit for the field is newton per coulomb ($N\,C^{-1}$), but Eq. (3.2.12) shows that the field is also in units of volt per meter ($V\,m^{-1}$). This derived unit is much more convenient in practical use, as it is far easier to measure volts and meters than newtons and coulombs.

3.3 Transport of ions in solutions

In this section we shall summarize some of the concepts and laws that especially apply to the description of the transfer of ions in solution.

---

* Compare Chapter 1, Section 1.2, Eq. (1.2.4).

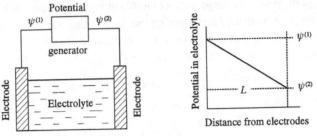

Fig. 3.1. An example of the course of the potential profile through an electrolyte of length L.

3.3.1 Migration

We consider a solution of an electrolyte of uniform concentration. Two plate-shaped electrodes are immersed in the solution and connected to a voltage generator (e.g. a battery, see Fig. 3.1). Let V be the potential difference (volt) between the plates which are at a distance L (m) apart. Owing to this there is a field in the solution in the direction of the voltage drop of magnitude*

$$E = \frac{V}{L}. \qquad (3.3.1)$$

This field impresses a force on each of the ions, having the same direction as the field for cations and the opposite direction for anions. Let N_j represent the concentration (number of ions per cubic meter) of the ion of type j, e.g. Ca^{2+}, with the charge number $z_j = z_{j+} = z_+ = z_{Ca^{2+}} = 2$, where a bunch of alternative symbols for the charge number are written. The ions of type j carry the charge $Q = z_j e$, where e is the positive elementary charge. The force X_j that acts on these ions is

$$X_j = QE = z_j eE, \qquad (3.3.2)$$

and where the direction of the force is given by the sign of the charge number z_j. The force X_j results in a migration velocity v_j, that is

$$v_j = B_j X_j, \qquad (3.3.3)$$

where B_j is the mechanical mobility[†] of the ion. Inserting the expression for X_j results in

$$v_j = z_j e B_j E = u_j E, \qquad (3.3.4)$$

---

* In this section the presence of a potential difference in the electrolyte is taken for granted. Section 3.4.3 contains an account of the function of the electrodes used to measure potentials in an electrolyte.

[†] Compare Chapter 2, Section 2.4.1, Eq. (2.4.5).

where

$$u_j = z_j e B_j \qquad (3.3.5)$$

is called the *electric mobility* of the ion, that is, the stationary velocity of the ion when exposed to the action of a field of 1 V m^{-1}. In aqueous solutions, field strengths of the order of 10 V m^{-1} produce velocities of about 5×10^{-7} m s^{-1}. Thus, their electric mobilities are of magnitude 5×10^{-8} m^2 s^{-1} V^{-1}.

The migration flux J_j for an ion of type j due to a force X_j is $v_j N_j{}^*$. Insertion in the above expression (Eq. (3.3.3)) for v_j gives

$$J_j = u_j N_j E. \qquad (3.3.6)$$

The charge carried by an ion of type j is $Q = z_j e$. Then, the flux above J_j carries an electric *current* that is

$$I_j = z_j e J_j. \qquad (3.3.7)$$

As the flux represents the number of ions that in unit time crosses a unit area placed at right angles to the flux, it follows that I_j is the current in ampere (coulomb per second) that flows through the unit area. Thus, I_j is the *current density* – i.e. ampere per square meter – of the ion of type j. Insertion of Eq. (3.3.6) in Eq. (3.3.7) gives

$$I_j = z_j e u_j N_j E. \qquad (3.3.8)$$

We multiply this expression by $1 = N_j / \mathcal{N}_A$, where $\mathcal{N}_A$ is Avogadro's constant, and rewrite it in the form

$$I_j = z_j \left(e \mathcal{N}_A\right) u_j \left(N_j / \mathcal{N}_A\right) E.$$

Now $N_j / \mathcal{N}_A$ is the concentration C_j in mol m^{-3}, and

$$e \mathcal{N}_A = \mathcal{F} \overset{\text{def}}{\equiv} \text{Faraday's constant} = 96\,492 \text{ coulomb per equivalent.} \qquad (3.3.9)$$

Hence, we can also write Eq. (3.3.8) as

$$I_j = z_j \mathcal{F} u_j C_j E. \qquad (3.3.10)$$

The connection between a current flow i in a conducting medium – a wire – and the voltage difference V applied at the ends of the conductor to cause the current is given by Ohm's law[†]

$$V = iR, \qquad (3.3.11)$$

---

[*] Compare Chapter 2, Section 2.4.2, Eq. (2.4.6).
[†] Formulated in 1827 by the German physicist Georg Simon Ohm (1789–1854).

where R is the *resistance* of the conductor. If V and i are given in volt and ampere, respectively, the SI unit of resistance is *ohm* (Ω), that is $R = V/i =$ volt/ampere. Ohm's law is often written in the form

$$i = \frac{V}{R} = GV, \qquad (3.3.12)$$

where

$$G = \frac{1}{R} \qquad (3.3.13)$$

is called the *conductance* of the conductor. The unit for conductance used to be reciprocal ohm (mho $\overset{\text{def}}{=} \Omega^{-1}$). The SI unit for conductance is siemens* (S). The conductance of a conductor of sectional area A (m^2) and length L (m) can be written

$$G = \kappa \frac{A}{L}, \qquad (3.3.14)$$

where the constant κ is called the *specific conductivity* or just the *conductivity* of the conductor κ has the dimension S m^{-1}. It appears from Eq. (3.3.14) that κ represents the conductance of a conductor of sectional area 1 m^2 and length 1 m. We multiply the voltage drop V over the conductor by the sectional area A and length L. This gives (see Eq. (3.2.12) and Eq. (3.3.11))

$$GV = \kappa A \frac{V}{L} = \kappa A E = i, \qquad (3.3.15)$$

where i is the current in ampere flowing through the area A. Dividing on both sides by A yields the current density $I = i/A$ on the right-hand side. Thus

$$I = \kappa E, \qquad (3.3.16)$$

is an alternative expression for Ohm's law written in terms of current density and field strength of the conductor. Taking together Eq. (3.3.10) and Eq. (3.3.16) gives

$$\kappa_j = z_j \mathscr{F} u_j C_j, \qquad (3.3.17)$$

where κ_j is the *partial conductivity*, that is the contribution from the conductivity of the electrolyte that arises from the ion of type j.

---

* In veneration of E.W.v. Siemens (1816–1892), German engineer and manufacturer, who built up one of the largest factories for the construction of electrotechnical machinery and apparatus, where he contributed personally with several new inventions. He also reinvented the dynamo.

The *molar conductivity* for the ion j is

$$\lambda_j = \frac{\kappa_j}{C_j} = z_j \mathscr{F} u_j. \tag{3.3.18}$$

We can build up a relation between λ_j and the diffusion coefficient D_j in the following way

$$\lambda_j = z_j \mathscr{F}(z_j e B_j) = z_j^2 \mathscr{F} e \frac{D_j}{kT} = z_j^2 \mathscr{F}\left(\frac{\mathscr{N}_A e}{\mathscr{N}_A k}\right) \frac{D_j}{T} = z_j^2 \mathscr{F} \frac{\mathscr{F}}{\mathscr{R}T} D_j = \frac{(z_j \mathscr{F})^2}{\mathscr{R}T} D_j, \tag{3.3.19}$$

where use is made of Eq. (3.3.5), the Einstein relation (Eq. (2.6.35)) together with the relation $e/k = \mathscr{F}/\mathscr{R}$. Originally this relation was derived by Nernst.

Tables of physical–chemical data often give the molar conductivity λ_j° corresponding to the infinite dilution of the electrolyte, that is when the electric interactions between the ions are minimized. The corresponding electric mobility of the ions is

$$u_j^\circ = \frac{\lambda_j^\circ}{z_j \mathscr{F}}.$$

It follows from Eq. (3.3.10) that the total current density $I = \sum I_j$ resulting from the migration of all the n different ion types in the solution is

$$I = \sum_{j=1}^{j=n} I_j = \left(\sum_{j=1}^{j=n} z_j \mathscr{F} u_j C_j\right) E, \tag{3.3.20}$$

and the total conductivity $\kappa = \sum_{j=1}^{j=n} \kappa_j = I/E$ is

$$\kappa = \sum_{j=1}^{j=n} z_j \mathscr{F} u_j C_j. \tag{3.3.21}$$

Correspondingly the total *resistivity* of the solution is

$$\rho = \frac{1}{\kappa}, \tag{3.3.22}$$

that is, the resistance of the electrolyte in the shape of a cube with the sides 1 m. The SI unit of ρ is $\Omega \cdot$ m. If the length (in meters) in the direction of the current flow is instead h, the corresponding *resistance* is

$$R = \rho h = \frac{h}{\kappa}, \tag{3.3.23}$$

which is the resistance of a prism of the electrolyte of sectional area in square meters, but of length h. The conductance of the prism is $G = 1/R$, i.e.

$$G = \frac{\kappa}{h}.$$ (3.3.24)

As κ has dimensions S/m, the conductance above has the dimension S/m^2. Similarly, any ion of type j has a resistance R_j and a conductance G_j that is given as

$$R_j = \frac{h}{z_j \mathscr{F} u_j C_j} \quad \text{and} \quad G_j = \frac{z_j \mathscr{F} u_j C_j}{h}.$$ (3.3.25)

When the electrolyte concentration varies throughout the solution we can still calculate the conductance if the concentration profiles $C_j(x)$ for the individual ions are know. We consider a layer of infinite thickness dx. The resistance is according to Eq. (3.3.25)

$$dR = \frac{dx}{\kappa} = \frac{dx}{\sum z_j \mathscr{F} u_j C_j(x)}.$$

Hence, the resistance of the whole layer of thickness h is

$$R = \frac{1}{G} = \int_0^h \left(\sum z_j \mathscr{F} u_j C_j(x) \right)^{-1} dx.$$ (3.3.26)

The resistance R represented by Eq. (3.3.26) is sometimes referred to as the *integral resistance* and the conductance $G = 1/R$ as the *integral conductance*.

In biological membranes we very often have to consider the situation where the individual ionic currents each flow through specialized structures (channels) of their own. In that case we have to consider the individual, or *partial resistances* and *partial conductances*, respectively, that are associated with each current. A layer of thickness dx has, according to Eq. (3.3.25), a resistance per unit area that is

$$dR_j = \frac{dx}{z_j \mathscr{F} u_j C_j(x)},$$

wherefore the layer of thickness h has per unit area a partial resistance R_j or a partial conductance G_j that is

$$R_j = \frac{1}{G_j} = \int_0^h \frac{1}{z_j \mathscr{F} u_j C_j(x)} dx,$$ (3.3.27)

and the *total* conductance now takes the form

$$G = \sum_{j=1}^{j=n} G_j. \tag{3.3.28}$$

It should be noted that in biological membranes this total conductance will in general differ from the total conductance that is calculated from Eq. (3.3.26). In the case of a *homogeneous* regime the total conductance is given by Eq. (3.3.26) and the individual conductances as calculated above are only of a formal significance. If, on the other hand, each ion moves separately through a channel of its own, as often is the case in biological membranes, it is only the individual (partial) conductances G_j as calculated from Eq. (3.3.27) that are of real physical significance, and the total conductance is $G = \sum G_j$.

We shall later use these special conductances, that is the conductance of a material of unit sectional area 1 m^2 and length or thickness h.

3.3.2 Electrodiffusion (Nernst–Planck equations)

We consider now the situation where both the electrolyte concentration and the electric potential $\psi(x)$ vary one-dimensionally in the direction of the x-axis*. This transfer process, where the ions in the solution move under the influence of the presence of both a concentration gradient dC/dx for the different ions species and a potential gradient dψ/dx, is called *electrodiffusion*. In writing the explicit expressions for individual ionic fluxes the starting point is the Smoluchowski equation[†]

$$J = -D\frac{dC}{dx} + BCX, \tag{3.3.29}$$

but the situation is now more complex since the electrolyte solution is dissociated into anions and cations, and this requires a flux equation to be written for each ionic type j. In the case of certain solution, e.g. NaCl, we can use the chemical symbols Na$^+$ and Cl$^-$ as indices (e.g. J_{Na^+}, D_{Cl^-}, etc.), or to facilitate the typography use the more convenient way of writing J_{Na}, D_{Cl}, etc., with the implication that designations refer to the ion and *not* to the chemical element. When the content of the solution is unspecified we shall use the same notation as in Section 3.3.1, though when the need arises we shall amplify the sub-index to j+, when referring to a cation and to j− for an anion. Thus, reference to a cation, e.g. Na$^+$ is $j+$, and the concentration of this ion at the position x is written as $C_{j+}(x)$. Furthermore, the electrical valence is written as z_{j+} and the

---

* See footnote on page.
[†] See Chapter 2, Section 2.6.1, Eq. (2.6.1) and Eqs. (2.6.4–2.6.5).

diffusion coefficient and mechanical mobility as D_{j+} and B_{j+}, respectively. The analogous symbols for an anion, e.g. Cl$^-$, are j$-$, $C_{j-}(x)$, D_{j-} and B_{j-} etc.

Because of the interconnection of the diffusion coefficient D_j and the mobility B_j in the Einstein relation we are at liberty to decide whether to write the flux equations in terms of either D_j or B_j. In this way we arrive at a number of equivalent expressions for the ionic flux that all are designated the *Nernst–Planck equations*.

3.3.2.1 Equivalent forms

We consider a point at the position x where the electric potential is ψ. An ion of the type j, which for the moment is a *cation*, with the electric valency z_j, is (owing to the electric field) under the influence of a force $X_j^{(el)}$ that – according to Eq. (3.3.7) and Eq. (3.3.12) – is

$$X_j^{(el)} = z_j e E = -z_j e \frac{d\psi}{dx}. \tag{3.3.30}$$

Replacing the driving force X in Eq. (3.3.29) with the above expression for the electric force we obtain

$$J_j = -D_j \frac{dC_j}{dx} - z_j e B_j C_j \frac{d\psi}{dx}, \tag{3.3.31}$$

where J_j is the flux at the position x of the cation j with the diffusion coefficient D_j and mobility B_j and concentration $C_j(x)$ under the influence of the concentration gradient $dC_j(x)/dx$ and the potential gradient $d\psi/dx$.

The Einstein relation* is valid for both uncharged and charged particles. Then, for the ion j we have

$$D_j = kTB_j. \tag{3.3.32}$$

Elimination of D_j from Eq. (3.3.31) gives

$$J_j = -kTB_j \frac{dC_j}{dx} - z_j e B_j C_j \frac{d\psi}{dx}, \tag{3.3.33}$$

or

$$J_j = -z_j e B_j \frac{kT}{z_j e} \frac{dC_j}{dx} - z_j e B_j C_1 \frac{d\psi}{dx}.$$

Inserting the expression Eq. (3.3.5) for the electrical mobility $u_j = z_j e B_j$ of the ion in the above expression gives

$$J_j = -u_j \frac{kT}{z_j e} \frac{dC_j}{dx} - u_j C_1 \frac{d\psi}{dx}. \tag{3.3.34}$$

---

* See Chapter 2, Section 2.7.4.1, Eq. (2.6.35).

or using $k/e = \mathcal{R}/\mathcal{F}$

$$J_j = -u_j \frac{\mathcal{R}T}{z_j\mathcal{F}} \frac{dC_j}{dx} - u_j C_1 \frac{d\psi}{dx}. \qquad (3.3.35)$$

Alternatively we write

$$J_j = -u_j C_j \left[\frac{\mathcal{R}T}{z_j\mathcal{F}} \frac{1}{C_j} \frac{dC_j}{dx} + \frac{d\psi}{dx} \right], \qquad (3.3.36)$$

where the term in square brackets [ ] contains an "equivalent potential gradient". Furthermore, taking the term $1/z_j\mathcal{F}$ outside the brackets yields

$$J_j = -\frac{u_j}{z_j\mathcal{F}} C_j \left[\mathcal{R}T \frac{d\ln C_j}{dx} + \frac{d}{dx}(z_j\mathcal{F}\psi(x)) \right],$$

as $d\ln C_j/dx = (1/C_j) \cdot dC_j/dx$. Now, $u_j/(z_j\mathcal{F}) = z_j e B_j/(z_j e \mathcal{N}_A) = B_j \mathcal{N}_A = b_j$, where b_j is the *molar mechanical mobility*. Moreover, we define

$$\bar{\mu}_j = \mu_j^\circ + \mathcal{R}T \ln C_j(x) + z_j\mathcal{F}\psi(x),$$

as the *electrochemical potential* of the ion at the position x. Thus, the above equation is also written

$$J_j = b_j C_j \left(-\frac{d\bar{\mu}}{dx} \right), \qquad (3.3.37)$$

as an analogue to Eq. (2.6.45) with the "driving force per mole" given by the negative gradient of the electrochemical potential of the ion in the position x^*.

We can also eliminate B_j instead of D_j in Eq. (3.3.31). This gives

$$-\frac{J_j}{D_j} = \frac{dC_j}{dx} + z_j \frac{e}{k} C_j \frac{d\psi}{dx}. \qquad (3.3.38)$$

In calculations it is often inconvenient to write

$$\frac{e\psi}{kT} \quad \text{or} \quad \frac{\psi}{kT/e}.$$

To facilitate the writing a new variable Ψ is introduced that is defined as

$$\Psi = \frac{\psi}{kT/e}, \quad \text{and thus} \quad \frac{d\Psi}{dx} = \frac{e}{kT} \frac{d\psi}{dx}, \qquad (3.3.39)$$

which allows Eq. (3.3.38) to be written in terms of the dimensionless potential ψ

$$-\frac{J_j}{D_j} = \frac{dC_j}{dx} + z_j C_j \frac{d\Psi}{dx}. \qquad (3.3.40)$$

---

* Compare this with often-used expressions "the ions move down along their electrochemical gradient" or "... move *against* its electrochemical gradient."

The dimension of kT/e is

$$\frac{kT}{e} = \frac{((\text{J per molecule}) \, \text{K}^{-1}) \, \text{K}}{\text{coulomb per molecule}} = \text{joule per coulomb} \overset{\text{def}}{=} \text{volt},$$

that is, Ψ gives the value of the potential ψ in units of kT/e^{*}.

Applying Kramers' transformation[†] to Eq. (3.3.40) gives

$$J_j \, e^{z_j \psi} = -D_j \frac{d}{dx}\left(C_j e^{z_j \psi}\right), \tag{3.3.41}$$

where the right-hand side is directly integrable. This equation is very useful in the calculation of the ion fluxes between two well-defined limits as, for example, in the case of flux through membranes. The term $C_j \, e^{z_j \psi}$ is sometimes referred to as the *electrochemical concentration/activity* of the ion.

To arrive at the expressions for the flux of an anion J_{j-} we replace the charge numbers z_{j+} with $-|z_{j-}|$ and the electrical mobilities u_{j+} with $-|u_{j-}|$, e.g.

$$J_{j-} = -kT \, B_{j-} \frac{dC_{j-}}{dx} + |z_{j-}| e B_{j-} \, C_{j-} \frac{d\psi}{dx}, \tag{3.3.42}$$

$$J_{j-} = -|u_{j-}| \frac{\mathcal{R}T}{|z_{j-}|\mathcal{F}} \frac{dC_{j-}}{dx} + |u_{j-}| \, C_{j-} \frac{d\psi}{dx}, \tag{3.3.43}$$

and

$$J_{j-} \, e^{-|z_{j-}|\psi} = -D_{j-} \frac{d}{dx}\left(C_{j-} e^{-|z_{j-}|\psi}\right). \tag{3.3.44}$$

Equations (3.3.35), (3.3.36) and (3.3.40) along with (3.3.42), (3.3.43) and (3.3.44) are called the *electrodiffusion equations* or the *Nernst–Planck equations*[‡]. These equations form the basic tools in the description of the passive transfer of ions through membranes.

3.3.2.2 Poisson's equation

A problem in electrodiffusion comprising n different ionic species is in principle solved when the *concentration profiles*, $C_j(x)$, for each of the ions participating in the electrodiffusion and the *potential profile* $\psi(x)$ in the region $a \leq x \leq b$ under consideration are determined from the boundary conditions that each

---

[*] $kT/e = \mathcal{R}T/\mathcal{F} = 25.7$ mV at 25 °C. See Section 3.4.2.

[†] See Chapter 2, Section 2.6.3, Eq. (2.6.51).

[‡] The German physicist Max Planck (1858–1947), who was awarded the Nobel Prize in 1918 for his epoch-making contributions to the foundation of the quantum theory, gave the fundamental theory for electrodiffusion in 1890 (Planck, M. (1890): *Ann. Physik u. Chem. N.F.* **35**, 561, Planck, M. (1890): *Ann. Physik u. Chem. N.F.* **39**, 161) that still is of current interest.

ionic species must satisfy. To calculate the individual ionic fluxes we must set up a flux equation

$$J_j = -u_j \frac{\mathscr{R}T}{z_j\mathscr{F}} \frac{dC_j}{dx} - u_j C_j \frac{d\psi}{dx}, \quad j = 1, 2, 3, \ldots n \qquad (3.3.45)$$

for each of the n ionic species present.

However, this set of equations is insufficient to solve even the stationary problem, because we have only $\sum j_+ + \sum j_- = n$ and there are $n + 1$ unknowns, i.e. the n concentrations C_j and the potential $\psi(x)$ of which the latter does *not* enter into the boundary conditions. It follows from the theory of electrostatics that the potential at every point in space must in addition agree with *Poisson's equation**

$$\frac{\partial^2 \psi}{\partial x^2} + \frac{\partial^2 \psi}{\partial y^2} + \frac{\partial^2 \psi}{\partial z^2} = -\frac{\rho}{\varepsilon_0 K}, \qquad (3.3.46)$$

where ρ is the spatial *charge density* (C m^{-3}), $\varepsilon_0 = 8.85 \times 10^{-12}$ F m^{-1} is the permittivity of vacuum, and $K = \varepsilon/\varepsilon_0$ is the relative dielectric constant. Poisson's equation expresses mathematically the fact that electric field lines of force originate *only* at positive charges $(+Q)$ and end up in negative charges $(-Q)$. In those cases where the potential can be described one-dimensionally as $\psi(x)$, Poisson's equation is written as

$$\frac{d^2 \psi}{dx^2} = -\frac{\rho}{\varepsilon_0 K}. \qquad (3.3.47)$$

When the charges arise from ions – as in our case – the charge density is

$$\rho = \mathscr{F} \sum_{}^{n} (z_{j+} C_{j+} - |z_{j-}| C_{j-}), \qquad (3.3.48)$$

and Poisson's equation takes the form

$$\frac{d^2 \psi}{dx^2} = -\mathscr{F} \sum_{}^{n} (z_{j+} C_{j+} - |z_{j-}| C_{zj-})/\varepsilon_0 K. \qquad (3.3.49)$$

This equation combines with the n flux equations in Eq. (3.3.45) to constitute a complete set of $n + 1$ equations, from which the n concentrations and the potential $\psi(x)$ can now be determined – at least in principle. The mathematical problems encountered in practice are usually quite enormous. Fortunately,

---

* Siméon Denis Poisson (1781–1840) was a French mathematician. In 1802 he became Professor of mathematics and mechanics at École Polytechnique, and in 1809 was made Professor of mechanics at Faculté des Sciences. Poisson did fundamental work on the theory of electricity and magnetism and celestial mechanics.

in many situations it is permissible to make use of one of the following approximations that both simplify the mathematical procedures and facilitate the computational work.

3.3.2.3 Electroneutrality

We consider the electrodiffusion of a monovalent, binary single salt, e.g. HCl. To solve a given problem involves the two Nernst–Planck equations

$$J_H = -u_H \frac{\mathcal{R} T}{\mathcal{F}} \frac{dC_H}{dx} - u_H C_H \frac{d\psi}{dx},$$

$$J_{Cl} = -u_{Cl} \frac{\mathcal{R} T}{\mathcal{F}} \frac{dC_{Cl}}{dx} + u_{Cl} C_{Cl} \frac{d\psi}{dx},$$

combined with Poisson's equation in one dimension

$$\frac{d^2\psi}{dx^2} = -\frac{\mathcal{F}}{\varepsilon_0 K}(C_H(x) - C_{Cl}(x)).$$

In an aqueous solution we have $K = 80$. Inserting the numerical values of $\mathcal{F}$, ε_0 and K gives

$$\frac{d^2\psi}{dx^2} = -\frac{dE}{dx} = -1.37 \times 10^{14} (C_H - C_{Cl}).$$

Thus, a departure from electroneutrality amounting to $\Delta C = 1 \text{ mol m}^{-3} = 1 \text{ mmol dm}^{-3}$ implies that the field will change by the amount equal to 10^{14} V m^{-1}. Under the conditions that are usual in a biology laboratory it is rare to encounter potential differences between two phases that surpass 200–300 mV. And in by far the majority of situations this potential difference is established over distances large enough to make the field E and its spatial rate of change dE/dx many orders of magnitude less than 10^{14}. This state of affairs was recognized by W. Nernst (1888)* and later by M. Planck (1890) who assumed that as for ordinary electrolyte solutions the difference $C_H - C_{Cl}$ can safely be disregarded as immeasurable for all practical purposes. Hence we can write

$$C_H(x) = C_{Cl}(x) = C(x) \tag{3.3.50}$$

in the two flux equations for H^+ and Cl^- without introducing any significant error in the succeeding calculations. In this way the solution of the above set of two Nernst–Planck equations is greatly simplified as they now only contain

---

* Walther Nernst (1864–1942) was a German physicist and physical chemist. In 1891 he became Professor of chemistry at the University of Göttingen and in 1905 became Professor of physical chemistry in Berlin. His work on the theory of galvanic elements and thermodynamics was fundamental to the development of physical chemistry. In 1920 he was awarded the Nobel Prize in chemistry.

two unknowns to be determined, namely $C(x)$ and $d\psi/dx$, where the latter can vary only slightly with the distance x because of the assumption of a vanishing value of $\rho = \mathscr{F}(C_H(x) - C_{Cl}(x)) \propto d\,(d\psi/dx)dx \approx 0$.

Equation (3.3.50) and its generalization

$$\sum^{n+} z_{j+}C_{j+} = \sum^{n-} |z_{j-}|C_{j-} \qquad (3.3.51)$$

are called the *Nernst–Planck electroneutrality condition*. This assumption has the result that Eq. (3.3.49) now takes the form

$$\frac{d^2\psi}{dx^2} = 0, \quad \text{or} \quad \frac{d\psi}{dx} = -E = \text{constant}. \qquad (3.3.52)$$

Poisson's equation is now automatically included in the Nernst–Planck equations, which can be solved separately if $\psi(x)$ is known. This assumption will hold in most of the following examples. In Section 3.5.3 we shall consider a situation in which the condition of electroneutrality breaks down and where the inclusion of the Poisson equation is required to calculate the profiles of concentration and potential.

3.3.2.4 The constant field

When evaluating the magnitudes of cellular membrane potentials some authors have taken more drastic measures by assuming *a priori* that the electric field inside the membrane is *constant*, and have then used this assumption as the starting point for the calculations. As we shall see later, the flux equation can now be integrated separately. This assumption, which for obvious reasons is in favor in many applications, was introduced in 1943 by D.E. Goldman*.

The constant field in a membrane is not just an *a priori* assumption. Under certain experimental conditions[†], the electric field through the membrane is always constant. We shall consider this case later in this chapter.

3.4 The equilibrium potential

In this section we shall give an account of two important concepts, namely the *electrochemical equilibrium* and the *equilibrium potential* for an ion. These concepts are crucial in the description of the mutual exchange of ions between two electrolyte solutions that are separated by a membrane, which is permeable to one or all the ions present. A knowledge of the magnitude of the equilibrium potential for the ion is important both for deciding in which direction an ion

---

* Goldman, D.E. (1943): Potential, impedance, and rectifications in membranes. *J. Gen. Physiol.*, **27**, 37.

[†] Same total concentration on both sides of the membrane.

will move passively through the membrane and for an understanding of those factors that determine the magnitude of the electric potential difference (the *membrane potential*) that may eventually build up across the membrane.

3.4.1 A qualitative description of the origin of the membrane potential across an ion-selective permeable membrane

To begin we consider the extremely simple case of a membrane that is *selectively permeable* to either cations or anions; that is, only one or the other ionic species is allowed to pass through. Such membranes – called ion-exchange membranes – are widely used in technology. Such a membrane consists of a matrix of carbohydrates to which anions, e.g. $-SO_3^-$, or cations, e.g. $-NH_3^+$, are bound covalently. The presence of these *fixed* charges will cause an attraction of the ions of opposite charge – also called *counter-ions* – and will prevent ions of the same charge – also called *co-ions* – from invading the membrane matrix because of the restraint implied by the principle of electroneutrality. Thus, if the concentration of the fixed charges is high the membrane will behave as if it were impermeable to the co-ions. (A more detailed account of this situation is given in Section 3.5.2.2 leading to Eq. (3.5.26) and Eq. (3.5.27).) In the example of Fig. 3.2, we consider a cationic permeable membrane that separates the two phases (i) and (o) containing KCl solutions of concentrations $C^{(i)}$ and $C^{(o)}$. If $C^{(i)} > C^{(o)}$ the permeable K^+ ion tends to diffuse from phase (i) to phase (o) according to the concentration difference $\Delta C = C^{(i)} - C^{(o)}$, between the two phases, whereas the impermeable Cl^- will remain in phase (i). The result is that the electroneutrality in the two phases is preserved as an excess of the positively charged K^+ ions will accumulate in phase (o) while an equivalent amount constitutes a charge deficit in phase (i). As a result, a *potential difference* across the membrane builds up, where phase (o) will be at an electric potential $\psi^{(o)}$ that is higher than the potential $\psi^{(i)}$ in phase (i). In other words: if the membrane is permeable *only* for K^+ then

$$\psi^{(i)} < \psi^{(o)} \quad \text{where} \quad C^{(i)} > C^{(o)}. \tag{3.4.1}$$

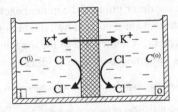

Fig. 3.2. The ion-selective membrane.

And in the membrane there is now an electric field

$$E = -\frac{d\psi}{dx},\tag{3.4.2}$$

having the direction from phase (o) → phase (i)*. This field will act upon a K^+ ion in the membrane with a force

$$X_{K^+} = eE\tag{3.4.3}$$

in the direction of the field and tend to move the K^+ ion in the direction from (o) → (i), that is *contrary* to the direction of ion transfer due to the presence of the concentration gradient in the membrane. In terms of the random walk performed by the K^+ ions in the membrane, this implies an asymmetry of the jumping pattern such that the probability for a jump in the direction (o) ↔ (i) is larger than for a jump in the (i) → (o) direction, thereby making the transfer by diffusion due to the concentration gradient less effective. Thus, the potential difference $\psi^{(o)} - \psi^{(i)}$ generated between the two phases as a result of the diffusion of free K^+ – without concomitant charge compensation from the equivalent number of Cl^- ions – will *counteract* the diffusion of K^+ ions from phase (i) to phase (o) and result in a reduction of the flux of K^+ ions from (i) to (o). In the end a sufficient number of free K^+ ions will have moved into phase (o) where the potential difference – the field in the membrane – has built up to the value that prevents a *net transport* of K^+ ions from phase (i) to phase (o) owing to the concentration difference $C^{(i)} - C^{(o)}$ between the two phases. (Compare Fig. 3.2.) Thus, a state of *equilibrium* is now established where the same number of K^+ ions moving per unit time from phase (i) to phase (o) equals the number moving in the same time from phase (o) to phase (i), resulting in a zero *net flux* of K^+ ions through the membrane, i.e.

$$J_{K^+} = 0, \quad \text{at equilibrium.}\tag{3.4.4}$$

The state of equilibrium is also called an *electrochemical equilibrium* to emphasize that the movement of the K^+ ions takes place under the influence of both a gradient of concentration and electric potential[†].

When the equilibrium is established no more macroscopic changes in the system can be detected. The potential difference across the membrane that corresponds to the equilibrium for the diffusible ion between the two phases is called the *equilibrium potential for the ions* at the existing concentration difference $\Delta C = C_K^{(i)} - C_K^{(o)}$ for the K^+ ions. In the present case it is the equilibrium

---

[*] If the potential profile through the membrane is linear we have $E = -(\psi^{(o)} - \psi^{(i)})/L$, where L is the thickness of the membrane.

[†] In equilibrium thermodynamics, the equilibrium formulation is that the electrochemical potential for K^+ – no matter what it may mean – has the same value in the two phases, i.e. $\overline{\mu}_{K^+}^{(i)} = \overline{\mu}_{K^+}^{(o)}$.

potential for the K^+ ion and accordingly we denote it as

$$V_K^{(eq)} \equiv \text{equilibrium potential for the } K^+ \text{ ion.} \qquad (3.4.5)$$

Consider now the Cl^- ions in the two phases. Does the potential difference $\psi^{(o)} - \psi^{(i)}$ across the membrane exert any influence on the pattern of movements of these ions? As long as the Cl^- ions are unable to enter and penetrate the membrane, the potential difference is of no consequence as the potentials $\psi^{(o)}$ and $\psi^{(i)}$ are constant throughout their respective phases and, therefore, no potential gradient or field ($E = -(\partial \psi / \partial x)$) will act upon the ions in the two phases, which may be lifted for that matter to any value of ψ. But the situation becomes different if we allow the Cl^- ions even a tiny permeability in the membrane. Each Cl^- ion that enters the membrane phase will now be subjected to the action of the field in the membrane, which acts on the negatively charged Cl^- ion in the direction phase (i) $\rightarrow$ phase (o). This is the *same* direction as the ions would move in consequence of the concentration difference. Thus, with polarity of the membrane potential, where $\psi^{(o)} > \psi^{(i)}$, the Cl^- ions are further removed from their equilibrium than prescribed by their concentration difference $\Delta C = C_{Cl}^{(i)} - C_{Cl}^{(o)}$.

We imagine now that the membrane is of an exceptional design, and that by way of some design we can suddenly change the membrane permeability from the state of being selectively permeable to K^+ ions to that of Cl^- ions instead. What happens? The crossing of the membrane by the K^+ ions will suddenly come to a stop. But the free charges that are accumulated on both sides of the membrane will remain at the instant of the blockade of the movements of the K^+ and sustain the potential difference across the membrane and the field $-d\psi/dx$ belonging to the equilibrium potential, $V_K^{(eq)}$ for K^+ ions with the direction (i) $\leftarrow$ (o). As the Cl^- ions carry a negative charge this field will drive the Cl^- ions in the direction (i) $\rightarrow$ (o) and thereby support the movement of the Cl^- ions through the membrane by way of diffusion. In other words: the Cl^- ions will now be under the influence of *both* factors that contribute to the flux of Cl^- in the direction (i) $\rightarrow$ (o). As the membrane is now selectively permeable to Cl^- ions they will move from phase (i) to phase (o) without an escort of an equivalent number of K^+ ions. The result is that phase (o) is being charged by an excess of negative charges whereas phase (i) contains the deficit of negative ions in an equivalent amount, that is positive charges. Thus, the potential in phase (o) will now be *lower* than that of phase (i). When equilibrium is established again there is a jump in potential across the membrane of the same magnitude as in the previously considered situation, where the membrane was permeable to K^+. But the *polarity* of the membrane potential – the equilibrium potential – will be that of the *opposite*, thus

$$\psi^{(i)} > \psi^{(o)} \quad \text{with} \quad C^{(i)} > C^{(o)}, \tag{3.4.6}$$

if the membrane is permeable for Cl^- ions.

3.4.2 The Nernst equation

The equilibrium potential was introduced and defined operationally by considering the ionic movements through a selective permeable membrane. However, these considerations are also applicable to the situation where the two phases are separated by a membrane that is permeable to several different ionic species having different concentrations in the two phases. Each permeable ionic species tends to move by the process of diffusion from the phase containing a high concentration to that of the lower one. This net movement of a particular ion can – irrespective of the presence of all the other ionic species and their concentrations – be counteracted by the application of an electric field that is set up by impressing a potential difference of the correct polarity between the two phases. A field of the proper direction and magnitude will lead to a *zero net flux* for the ion in question, without regard to the movements across the membrane of the remaining ions. The potential difference between the phases that establishes this particular field is the *equilibrium potential* for the ion concerned. The greater the tendency for the ions to move by diffusion, the greater is the potential difference required to cause a zero net flux. Therefore, one should expect the existence of a quantitative connection between the concentrations in the two phases of the ion referred to and its equilibrium potential. The German physical chemist Wilhelm Ostwald showed in 1890 that this relation had the form*

$$\psi^{(i)} - \psi^{(o)} = V_j^{(eq)} = \frac{\mathcal{R}T}{z_j \mathcal{F}} \ln \frac{C_j^{(o)}}{C_j^{(i)}}, \tag{3.4.7}$$

usually denoted Nernst's equation[†]. The quantities in the fraction are: the gas constant $\mathcal{R}$ (8.314 J mol^{-1} K^{-1}), the Faraday constant $\mathcal{F}$ (96 492 C mol^{-1}),

---

[*] Wilhelm Ostwald (1853–1932) was a German physicist, chemist and philosopher. He was Professor of chemistry in Riga (1881–1887) and of physical chemistry in Leipzig (1887–1906). Ostwald made a fundamental contribution to physics, chemistry, energetics, natural philosophy and theory of colors. He is considered – together with van't Hoff, Arrhenius and Nernst – as the founder of physical chemistry. He was awarded the Nobel Prize for chemistry in 1909.

[†] Probably because of the close resemblance to the equation describing the electromotive force of the galvanic cell that was derived in 1888 by the German physical chemist Walther Nernst.

Table 3.1. $\mathcal{R}T/\mathcal{F}$ at different temperatures

°C	$\mathcal{R}T/\mathcal{F}$ (mV)	$2.3026 \cdot \mathcal{R}T/\mathcal{F}$ (mV)	K
0	23.5	54.3	273.15
25	25.7	59.2	298.15
37	26.7	61.5	310.16

T the temperatures in K, and z_j the charge number for an ion of type j, that is

$$z_j = \begin{cases} 1, 2, 3, \ldots & \text{for cations;} \\ -1, -2, -3, \ldots & \text{for anions.} \end{cases}$$

The dimension of quantity $\mathcal{R}T/\mathcal{F}$ must be in volts as will appear from the dimension analysis below

$$\mathcal{R}T/\mathcal{F} \triangleq \text{J mol}^{-1}\text{K}^{-1}\text{K}/(\text{C mol}^{-1}) = \text{J C}^{-1} \triangleq \text{volt.}$$

According to usual practice the natural logarithm (base $e = 2.718\,281\ldots$) in Eq. (3.4.7) above is replaced by the logarithm to the base of 10*. The relation between the two types of logarithms is $\ln X = 2.3026 \cdot \log X$. The Nernst equation is often written as

$$\psi^{(i)} - \psi^{(0)} = V_j^{(eq)} = 2.3026 \frac{\mathcal{R}T}{z_j\mathcal{F}} \log \frac{C_j^{(0)}}{C_j^{(i)}}. \tag{3.4.8}$$

Table 3.1 shows the value of $\mathcal{R}T/\mathcal{F}$ at various temperatures of frequent occurrence. In the table the values of $2.3026 \cdot \mathcal{R}T/\mathcal{F}$ are likewise shown, since the log-scale is usually used in the practical handling of the Nernst equation.

Thus, at 25 °C one has

$$\psi^{(i)} - \psi^{(0)} = V_j^{(eq)} = \frac{59.2}{z_j} \log \frac{C_j^{(0)}}{C_j^{(i)}} \text{ (mV).} \tag{3.4.9}$$

For the K^+ ions ($z_j = +1$) the equilibrium potential corresponding to a concentration ratio of $C^{(0)}/C^{(i)} = 1/10$ at this temperature $V_K^{(eq)} = -59.2$ mV, and for the Cl^- ions ($z_j = -1$) at the same concentration ratio we have $V_{Cl}^{(eq)} = +59.2$ mV.

---

* Although the widespread use of pocket calculators has made this kind of facilitation less urgent, it is still of value to remember that $\ln_{10} 10 = \log 10 = 1$.

3.4.2.1 The charge density of the excess charges on the two membrane sides

It may be instructive to estimate the magnitude of excess charges – e.g. K^+ ions – that have accumulated on the two faces of the membrane in order to establish an equilibrium potential of the same order of magnitude as that above. Simple electrostatic considerations will do. Any membrane capable of charge separation acts as a capacitor. Biological membranes all have an electric capacitance C_m of the order

$$C_m \approx 1 \ \mu F \ cm^{-2}$$

($1 \ \mu F = 10^{-6} \ F$). Assume then that the potential difference V_m (membrane potential) across the membrane is 60 mV. To this value of V_m there corresponds a *charge density* q on the membrane side at the high potential that is

$$q = C_m V_m = 10^{-6} \ F \ cm^{-2} \times 60 \times 10^{-3} \ V = 6 \times 10^{-8} \ C \ cm^{-2},$$

whereas a negative charge density of the same amount is located at the opposite membrane side. The charge carried by one mol of K^+ ions is $\mathscr{F} = 96\ 492 \approx 10^5 \ C \ mol^{-1}$. Thus the above excess positive charge corresponds to an excess molar K^+ concentration ΔC_{K^+} of magnitude

$$\Delta C_{K^+} = \frac{6 \times 10^{-8} \ C \ cm^{-2}}{10^5 \ C \ mol^{-1}} = 6 \times 10^{-13} \ mol \ cm^{-2}.$$

Multiplication of ΔC_{K^+} with Avogadro's number $\mathscr{N}_A$ yields

$$6 \times 10^{-13} \ (mol \ cm^{-2}) \times 6.03 \times 10^{23} \ (ions \ mol^{-1}) = 3.3 \times 10^{11} \ ions \ cm^{-2},$$

which represents the number of unneutralized K^+ ions per cm^2 on the positive face of the membrane, which – in combination with the same number of excess negative charges (Cl^- ions) on the opposite membrane side – produces the potential difference across the membrane of 60 mV. This number corresponds to a single layer of tightly packed hydrated K^+ ions. Thus, the number of uncompensated ions required to produce a membrane potential of the order of 100 mV is quite negligible and far too small to be measured by ordinary chemical means. Artificial membranes have in general a much smaller capacitance, and accordingly the above number of free charges will diminish concurrently with a decrease of the capacitance of the ionic exchange membrane.

3.4.2.2 Derivation of Nernst's equation

There are several ways to derive this equation. Many chemists would presumably prefer equilibrium thermodynamics and base their argument on the

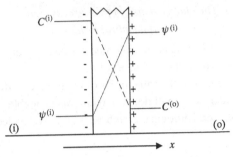

Fig. 3.3. Illustration of the ionic equilibrium across the ion-selective membrane.

electrochemical potentials of ions. Our starting point – more in keeping with the rest of the approach in this book – is the Nernst–Planck equations*.

Figure 3.3 illustrates a membrane of selective ionic permeability of thickness h that is surrounded by two electrolyte solutions, e.g. KCl, at different concentrations, where $C^{(i)} > C^{(o)}$. The membrane is assumed to be permeable to K^+ ions, which will diffuse through the membrane in the direction (i) $\rightarrow$ (o) generating the potential difference $\psi^{(i)} - \psi^{(o)}$ that builds up as described above. The flux of the K^+ ions is described by the Nernst–Planck equations. The form most suited for this case is

$$J_j = -u_j C_j \left[\frac{\mathscr{R}T}{z_j \mathscr{F}} \frac{1}{C_j} \frac{dC_j}{dx} + \frac{d\psi}{dx} \right], \tag{3.4.10}$$

with the x-axis placed perpendicularly to the membrane and with origin $x = 0$ at the membrane side facing phase (i) and with direction (i) $\rightarrow$ (o). We call dC_j/dx the concentration gradient of the ionic species j (cation or anion) and $d\psi/dx$ is the potential gradient at any position $0 \leq x \leq h$ on the x-axis. When equilibrium is established in the system the potential difference across the membrane has grown to the characteristic value, the equilibrium potential $V_j^{(eq)}$ for the K^+ ion, which makes the net flux through the membrane zero, i.e.

$$\psi^{(i)} - \psi^{(o)} = V_j^{(eq)} \quad \text{and} \quad J_j = 0; \quad \text{at equilibrium.} \tag{3.4.11}$$

In Eq. (3.5.10) $u_j C_j \neq 0$. The condition $J_j = 0$ in Eq. (3.5.11) then requires that the term inside the brackets is zero for all values of x in the region $0 \leq x \leq h$. Thus, at membrane equilibrium the potential gradient $d\psi/dx$ is adjusted exactly

* In both cases the usual practice is to consider the extremely simplified situation where the membrane is impermeable to water, thus disregarding the need to consider a concurrent water movement. This type of equilibrium is denoted as a *non-osmotic* equilibrium.

relative to the concentration gradient dC_j/dx so that

$$\frac{d\psi}{dx} = -\frac{\mathcal{R}T}{z_j\mathcal{F}}\frac{1}{C_j}\frac{dC_j}{dx}.$$

We multiply both sides by dx. This gives *

$$d\psi = -\frac{\mathcal{R}T}{z_j\mathcal{F}}\frac{dC_j}{C_j}, \qquad (3.4.12)$$

which means that the potential change $d\psi$ in the membrane between two closely situated points x and $x+dx$ equals the relative decrease in concentration $-dC_j/C_j$ between the same points multiplied by $\mathcal{R}T/z_j\mathcal{F}$. The total potential difference across the membrane is obtained by integration of Eq. (3.5.12) between the proper limits. As long as we remain in a macroscopic distance from the two sides of the membrane the potentials and the concentrations are constant everywhere in the two phases, i.e.

$$C_j = C_j^{(i)} \quad \text{and} \quad \psi = \psi^{(i)} = \text{constant in phase (i)},$$

and

$$C_j = C_j^{(o)} \quad \text{and} \quad \psi = \psi^{(o)} = \text{constant in phase (o)}.$$

Thus, it is of no consequence whether the integration path just includes the membrane or extends from an arbitrary point in phase (i) with concentration $C_j^{(i)}$ and potential $\psi^{(i)}$ through the membrane $0 \le x \le h$ to another point in phase (o) with concentration $C_j^{(o)}$ and potential $\psi^{(o)}$. Hence

$$\int_{\psi^{(i)}}^{\psi^{(o)}} d\psi = -\frac{\mathcal{R}T}{z_j\mathcal{F}}\int_{C_j^{(i)}}^{C_j^{(o)}}\frac{dC_j}{C_j}.$$

We then have[†]

$$[\psi]_{\psi^{(i)}}^{\psi^{(o)}} = -\frac{\mathcal{R}T}{z_j\mathcal{F}}\left[\ln C_j\right]_{C_j^{(i)}}^{C_j^{(o)}},$$

$$\psi^{(o)} - \psi^{(i)} = -\frac{\mathcal{R}T}{z_j\mathcal{F}}\left[\ln C_j^{(o)} - \ln C_j^{(i)}\right] = -\frac{\mathcal{R}T}{z_j\mathcal{F}}\ln\frac{C_j^{(o)}}{C_j^{(i)}},$$

or, to conform to Eq. (3.5.7),

$$\psi^{(i)} - \psi^{(o)} = -\frac{\mathcal{R}T}{z_j\mathcal{F}}\ln\frac{C_j^{(i)}}{C_j^{(o)}}, \qquad (3.4.13)$$

which is Nernst's equation.

* See Chapter 1, Section 1.2.5, Eq. (1.2.13).
[†] See Chapter 1, Section 1.3.2, Eq. (1.3.9), Section 1.4.1, Eq. (1.4.2) and Section 1.4.2, Eq. (1.4.10).

3.4.3 Establishing the electric contact to the electrolyte solution: electrodes

In the preceding pages we have described the accumulation of excess charges of opposite signs on the two sides of a boundary structure like a membrane giving rise to a potential difference across the boundary. However, to measure this potential difference directly one runs into the basic difficulty that the object to be measured resides in an electrolyte solution, whereas the appropriate electronic device for measuring an electric voltage or current requires metallic wire connections. This means that in any electric circuit that is set up to measure a potential difference inside an electrolyte solution at least two metal–electrolyte interfaces are involved. Interactions across the boundary between the metal ions in the wire and ions in the solution create an electric double layer at each interface. Thus, in the circuit designed to measure the target potential difference in the solution there are inserted at least two extra potential jumps that may be as large as or even larger than the potential difference of the object and may thereby jeopardize the measurement. To be able to eliminate the influence on the actual measurement of these extra potential jumps, it is imperative that they remain stable and constant during the measurement. To this end the two metallic wires from the electronic measuring device are separately connected to specialized structures – a *galvanic half-cell* of some kind – that are dipping into the electrolyte and provide well-defined stable transitions from the metallic wires to the electrolyte solution.

3.4.3.1 The galvanic cell

When metallic Zn powder is sprinkled into a $CuSO_4$ solution the metal dissolves according to the scheme

$$Zn + Cu^{2+} \rightarrow Zn^{2+} + Cu + heat. \qquad (3.4.14)$$

The free energy change that accompanies this reaction can be stored as electric energy by letting the reaction occur in a *galvanic cell*, of which a familiar example is the *Daniell cell**, consisting of a Zn plate immersed in an aqueous solution of $ZnSO_4$ and a Cu plate dipping into a solution of $CuSO_4$. The two solutions are separated by a partition – usually a porous wall or a U-tube filled with an electrolyte – that provides electric contact but prevents mixing of the solutions by diffusion/migration. When thus connected a potential difference of about 1.10 V between the Cu electrode and the Zn electrode will be found when the current flow is insignificant[†]. When the two electrodes are connected by a conducting wire, a spontaneous electric current of positive charges flows

---

* Invented by J.F. Daniell (1790–1845), Professor of chemistry at King's College, London, and before the advent of electromagnetic current generators used widely in physics.
† This potential difference is often called the *electromotive force* of the cell, also denoted the e.m.f.

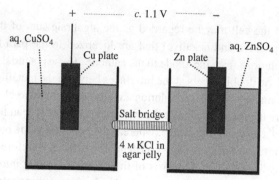

Fig. 3.4. A galvanic battery consisting of a copper plate dipping into a jar containing a solution of $CuSO_4$ and a zinc plate dipping into a $ZnSO_4$ solution (a Daniell cell). The two solutions are electrically connected via a salt bridge consisting of a strong KCl solution in agar jelly that prevents mixing of the solutions by diffusion/convection.

from the Cu electrode, the *anode*, to the Zn electrode, the *cathode* (or electrons flow in the opposite direction), whereas in the interior of the cell the same current is carried by cations in the direction from the cathode to the anode. The above reaction now takes place as the net result of the reactions that occur at the two electrodes, namely at the interface $Zn/ZnSO_4$ solution a Zn atom goes into the solution according to the scheme:

$$Zn \rightarrow Zn^{2+} + 2\mathcal{E}, \tag{3.4.15}$$

where $\mathcal{E}$ denotes an electron, and at the $Cu/CuSO_4$ interface a Cu ion is deposited according to

$$Cu^{2+} + 2\mathcal{E} \rightarrow Cu. \tag{3.4.16}$$

The constituents that enter into the Daniell cell – or any similar galvanic cell – are usually written

$$Zn(metal)|ZnSO_4(aqueous)\|CuSO_4(aqueous)|Cu(metal), \tag{3.4.17}$$

where | symbolizes the metal–electrolyte interface and ‖ represents the conducting partition between the two electrolyte solutions. Furthermore, the left-to-right sequence of the entering elements is written to conform to the convention that a positive electric current passing through the cell in the direction left-to-right results in the overall chemical reaction (Eq. (3.3.1)). Often the scheme for the galvanic cell is written as

$$Zn(metal)|Zn^{2+}(aqueous)\|Cu^{2+}(aqueous)|Cu(metal), \tag{3.4.18}$$

to show only the elements that take part in the electrochemical reactions.

(i) Half-cells

The e.m.f. of the cell may be regarded as the algebraic sum of the potential differences (positive and negative) that are localized in the path through the electrolyte solutions from the anode to the cathode. These potential jumps arise from the electric double layers at the interfaces between the metallic electrodes and the surrounding electrolyte solutions and between the two electrolyte solutions, where the interdiffusion of the two electrolytes may lead to a potential jump denoted a *diffusion potential**, whereas the potential in the bulk solutions away from these interfaces remains constant. The cell can therefore be split into two *half-cells*. One half-cell consists of the Cu plate dipping into the $CuSO_4$ solution, while the other half-cell contains the Zn plate dipping into the $ZnSO_4$ solution. It is convenient in general to consider any galvanic cell as being composed of two half-cells – which may be called A and B – with the respective electrode reactions

$$A \rightleftharpoons A^{z_A+} + z_A \mathcal{E} \quad \text{and} \quad B \rightleftharpoons B^{z_B+} + z_B \mathcal{E}. \qquad (3.4.19)$$

The electrodes that enter into a half-cell may be divided into three categories according to their mode of operation.

(1) The first type of electrode is a soluble metal such as Zn or Cu in contact with a solution of its own ions as in the Daniell cell above, i.e. it is ions of the electrode material that enter in the electrode reaction in question. This type of electrode is called a *cation electrode*.

(2) The second type involves a metal and a sparingly soluble salt of the metal in contact with a soluble salt of the same anion. An example that is important in biological applications is the electrode consisting of metallic silver, and solid AgCl in contact with a solution containing soluble chloride ions. The electrode reaction in this case may be written as

$$Ag(s) + Cl^- \rightleftharpoons AgCl(s) + \mathcal{E}, \qquad (3.4.20)$$

the chloride ions being those in the solution of the soluble chloride, such as HCl or NaCl. Similarly, this electrode is called an *anion electrode*.

(3) The third type is an insoluble metal such as gold or platinum dipping into a solution containing both oxidized and reduced states of an *oxidation–reduction* system. The unassailable metal operates as a donor/acceptor of electrons in the oxidation–reduction reaction such as

$$Fe^{2+} \rightleftharpoons Fe^{3+} + \mathcal{E}, \qquad (3.4.21)$$

---

* Also denoted a *liquid junction potential*. See Section 3.6.

or when hydrogen gas is oxidized to hydrogen ions in the hydrogen electrode, thus

$$\tfrac{1}{2}H_2 \rightleftharpoons H^+ + \mathcal{E}. \qquad (3.4.22)$$

It will be noted that, irrespective of differences in construction of the three types of electrode, their fundamental principles of operation are the same, since every half-cell reaction can in general be written as

$$\text{Reduced State} \rightleftharpoons \text{Oxidated State} + \mathcal{E} \qquad (3.4.23)$$

if we decide to use the terms "oxidation" and "reduction" in their broadest sense; that is, oxidation implies the *liberation* of electrons while reduction refers to the *taking up* of electrons. Accordingly, in case (1) the metal Cu is the reduced state while Cu^{2+} represents the oxidated state. In case (2) the metallic silver and the chloride ion together make up the reduced state while the AgCl forms the oxidized state.

(ii) Electrode potentials

In many applications of galvanic cells a detailed knowledge of the seat of the e.m.f. or of the mechanism of its origin is not required. Thus, classical thermodynamics relates precisely the cell's e.m.f. to the chemical reaction that occurs in the cell by identifying the electric work W performed by passing an electric charge through the cell with the resultant change in free energy of the cell reaction. When the process takes place reversibly the electric work $W^{(\text{rev})} = -z_j \mathcal{F} \times$ (e.m.f.) per unit extent of reaction* ξ involving a transfer of $z_j \mathcal{F}$ (in coulomb) across the e.m.f. is equal to the change in reaction free energy ΔG of the resultant electrode processes, i.e. $\pm \Delta G$ according to the direction of transfer, and thereby interconnecting a measureable quantity – the potential difference measured between two metal electrodes – to the concentrations (activities) in the bulk solutions of the reacting components.

An exhaustive treatment of the kinetics of the electrode processes in a galvanic cell is a project on its own. Nevertheless, even a crude picture of the electrode kinetics in the form of a simplified, semi-quantitative treatment may be of value for us, i.e. the description of how to *establish a proper electric connection to an electrolyte solution.*

---

* This is equal to the change in free energy of the chemical reaction

$$\alpha A + \beta D \rightarrow \gamma C + \delta D$$

after α moles of A and β moles of B have been consumed to form γ moles of C and δ moles of D, i.e. the extent of reaction ξ has changed by one unit.

First of all, the transfer of charge across the electrode interface consists essentially of the exchange of electrons between the electrode and particles (neutral molecules, complex ions or simple ions) in the solution side of the interface. An ion's valence changes if it donates or accepts an electron from the electrode, thus

$$M^+ \xrightarrow[\text{donation to electrode}]{\text{electron}} M^{2+} \quad \text{or} \quad M^{2+} \xrightarrow[\text{received from electrode}]{\text{electron}} M^+, \quad (3.4.24)$$

where the electrode in the first case behaves as an *electron sink* and in the second case as an *electron source*. The direction of the electron flow depends on the nature of the electrode and on that of the surrounding electrolyte solution, but for any interface one direction of the electron flow is *spontaneous*. Consider again the Daniell cell at comparable concentrations of Cu^{2+} and Zn^{2+}. When the two terminals are connected by a conducting wire to allow for the electron flow from the Zn electrode to the Cu electrode, the directions of the spontaneous processes at the two electrode interfaces are

$$Zn \xrightarrow[\text{to electrode (sink)}]{\text{minus 2 electrons}} Zn^{2+} \quad \text{and} \quad Cu^{2+} \xrightarrow[\text{from electrode (source)}]{\text{plus 2 electrons}} Cu, \quad (3.4.25)$$

where, at the Zn electrode (the electron sink), Zn atoms are spontaneously de-electronated and dissolved in return as Zn^{2+} ions, whereas at the Cu electrode (the electron source) Cu^{2+} is electronated and deposited. The rate at which these processes proceed can be controlled by connecting to the terminals of the cell a variable voltage supply whose output $V^{(var)}$ opposes the e.m.f. of the cell. When $V^{(var)}$ approaches the value V^0 of the e.m.f. the process slows down and comes to a stop when $V^{(var)} = V^0$ with no current flow between the terminals of the cell and no net current flow across the two metal interfaces in the solution. Raising $V^{(var)}$ above V^0 reverses the direction of the current flow and of the electron exchange at the interfaces, thus

$$Zn^{2+} \xrightarrow[\text{from electrode (source)}]{\text{plus 2 electrons}} Zn \quad \text{and} \quad Cu \xrightarrow[\text{to electrode (sink)}]{\text{minus 2 electrons}} Cu^{2+}. \quad (3.4.26)$$

This indicates that electric fields are set up at the two metal–electrolyte interfaces and that their values control the rate and direction of the processes that occur at the electrode. It is reasonable to assume that at $V^{(var)} = V^0$ (which also corresponds to the state when the cell is left to itself), where the processes at the interfaces have reached a state of equilibrium, *characteristic potential differences* are built up across the interfaces from the metallic parts of the electrodes to the bulk of the solutions at which the processes of Eq. (3.3.13) and those taking the opposite direction (Eq. (3.3.14)) proceed at the *same rate*, i.e. the *net current is zero* across each interface, thus

$$\text{Rate of } \left\{ \text{Zn} \xrightarrow[\text{to electrode (sink)}]{\text{minus 2 electrons}} \text{Zn}^{2+} \right\} = \text{Rate of } \left\{ \text{Zn}^{2+} \xrightarrow[\text{from electrode (source)}]{\text{plus 2 electrons}} \text{Zn} \right\},$$

(3.4.27)

and similarly at the Cu electrode.

To account for the characteristic potential difference we consider a piece of zinc that is dipped into a solution of $ZnSO_4$. Initially there are no excess charges either on the metal or in solution. However, excess charges of opposite polarity are built up very quickly on the two sides of the interface as the result of two opposing processes, namely the passage of Zn^{2+} ions from the metal into the solution leaving the two valence electrons behind and the reverse, i.e. the passage of Zn^{2+} from the solution to the Zn electrode. At the $Zn|ZnSO_4$ interface the former process is initially the more rapid, and cations will move into the solution faster than they return. Thus, free electrons will be left in the metal and an excess amount of Zn^{2+} ions will accumulate in the solution in the interface region, and thereby build up an *electric double layer*, causing the electric potential in the solution to rise above that in the Zn electrode. The electric field – directed from the solution into the metal – restrains the tendency of the Zn^{2+} ions to leave the negatively charged Zn electrode and enter the solution at the higher potential, whereas the tendency of the Zn^{2+} ions in the solution to return to the Zn electrode is facilitated. The potential difference will continue to increase until the two processes are occurring at the *same rate*, and the system has attained its equilibrium and no more net changes will occur. The potential difference $\psi^{(Zn)} - \psi^{(soln)} = V_{Zn}^{(el)}$ between the Zn electrode and the bulk of the solution that is associated with the state of Eq. (3.4.27) is called the *electrode potential* of the Zn electrode dipping into the $ZnSO_4$ solution.

A kinetic treatment of the origin of the electrode potential $V^{(el)}$ based on Boltzmann statistics was first given by Butler (1924)[*]. In outline it runs as follows. We consider the potential energy configuration that surrounds a metal atom that is located in the surface of the metal electrode. Displacing the atom from its equilibrium position requires work against the interacting forces from the neighboring atoms, giving rise to a change in the atom's potential energy. Figure 3.5 shows how the potential energy will depend on the position perpendicular to the surface. Valley (a) to the left represents the equilibrium position of the atom in the crystal lattice. If one attempts to move the atom inwards an intense repulsion is encountered as illustrated by the steepness of the left leg L. When an ion of valence z_j moves from the metal phase (m) across the interface towards the solution phase (s) the forces of attraction fall off rapidly

---

[*] Butler, J.A.V. (1924): *Trans. Faraday Soc.*, **19**, 729.

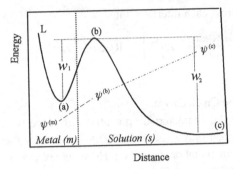

Distance

Fig. 3.5. Potential energy–distance profile for the displacement of a metal atom from the position in the immediate vicinity to the metal–electrolyte interface (dotted line). The equilibrium position in the three-dimensional crystal lattice is at (a). The steep part to the left is due to strong forces of repulsion encountered by moving the atom further into the metal. The forces of attraction decrease rapidly with distance along the path (a) → (b), requiring work W_1 to reach the peak at (b). The decline in energy W_2 along (b) → (c) is due to the reduction of the electrostatic energy of the ion due to shielding by dipolar water molecules. The electric potentials in the bulk of the metal (m) and solution (s) are $\psi^{(m)}$ and $\psi^{(m)}$, respectively; $\psi^{(b)}$ is the potential at the position (b) corresponding to the peak.

and the energy curve reaches a peak at position (b), where the height from the valley represents the work W_1 required to remove the metal atom as a metal ion M^{z_j+} from the lattice leaving z_j electrons behind. When moving the metal ion M^{z_j+} further into the interface the polar water molecules will be attracted and orient themselves with the negative sides towards the M^{z_j+} ion making an aggregate of an M^{z_j+} ion in the middle of a sheath of water. The result of this shielding is a reduction in the electrostatic energy of the ion and the energy curve descends to the flatland (c) by the amount W_2 that is the energy of solvation. The draft of the energy diagram does not include the additional work that is required to move the electric charge $z_j e$ of the ion M^{z_j+} across an electric potential profile $\psi(x)$ from the metal of the electrode to the bulk of the solution. Let a distribution of excess charges across the interface give rise to the electric potentials $\psi^{(m)}$, $\psi^{(b)}$ and $\psi^{(soln)}$ at the positions (a), (b) and (c). Hence, the transition in the direction $m \to s$ across the barrier requires the work

$$U_1 = W_1 + z_j e\left(\psi^{(b)} - \psi^{(m)}\right) \quad \text{for} \quad m \to s, \tag{3.4.28}$$

and correspondingly for the transition in the opposite direction $s \to m$

$$U_2 = W_2 + z_j e\left(\psi^{(b)} - \psi^{(s)}\right) \quad \text{for} \quad s \to m, \tag{3.4.29}$$

where W_2 is the work required to strip the M^{z_j+} ion of its attached sheath of water molecules. To describe the traffic of ions across the barrier at (b) we apply Boltzmann statistics in the same way as in Section 2.5.6.6. Thus, the unidirectional flux in the direction $m \rightarrow s$ is

$$\overset{m \rightarrow s}{J} = N_M \langle \overset{\rightarrow}{v^*} \rangle e^{-U_1/kT}, \qquad (3.4.30)$$

where N_M is the number of metal ions per unit surface and $\langle \overset{\rightarrow}{v^*} \rangle$ is the average thermal velocity with which the M^{z_j+} ions move across the ridge at (b). Similarly the unidirectional flux $\overset{m \leftarrow s}{J}$ in the opposite direction $s \rightarrow m$ is

$$\overset{m \leftarrow s}{J} = N_s \langle \overset{\leftarrow}{v^*} \rangle e^{-U_2/kT}, \qquad (3.4.31)$$

where N_s is the number of ions per unit volume in the solution and $\langle \overset{\leftarrow}{v^*} \rangle$ is the average velocity across the peak in the direction $s \rightarrow m$. Hence, the net flux J in the direction $m \rightarrow s$ is

$$\overset{m \rightarrow s}{J} - \overset{m \leftarrow s}{J} = J = N_M \langle \overset{\rightarrow}{v^*} \rangle e^{-U_1/kT} - N_s \langle \overset{\leftarrow}{v^*} \rangle e^{-U_2/kT}. \qquad (3.4.32)$$

At equilibrium the potential difference across the interface has attained the characteristic value $\psi^{(m)} - \psi^{(s)} = V_M^{(eq)}$ where the two unidirectional fluxes are equal. Assuming that $\langle \overset{\rightarrow}{v^*} \rangle = \langle \overset{\leftarrow}{v^*} \rangle = \langle v^* \rangle$, Eq. (3.4.32) then reads

$$N_M \langle v^* \rangle e^{-U_1/kT} = N_s \langle v^* \rangle e^{-U_2/kT}$$

or

$$U_1 - U_2 + kT \ln \frac{N_s}{N_M} = 0.$$

Insertion of Eq. (3.4.28) and Eq. (3.4.29) and solving for $\psi^{(m)} - \psi^{(s)}$ gives

$$\psi^{(m)} - \psi^{(s)} = \frac{W_1 - W_2}{z_j e} + \frac{kT}{z_j e} \ln \frac{N_s}{N_m}$$

$$= \frac{W_1 - W_2}{z_j e} - \frac{kT}{z_j e} \ln N_m + \frac{kT}{z_j e} \ln N_s, \qquad (3.4.33)$$

which may also be written as

$$V_M^{(eq)} = V_m^{\otimes} + \frac{\mathcal{R}T}{z_j \mathcal{F}} \ln C_m, \qquad (3.4.34)$$

where C_m is the molar/molal concentration and the first two terms are collected in the constant $V_m^{\otimes}$, which represents the value of the electrode potential when $C_m = 1$. This expression for the electrode potential derived by means of a kinetic argument is of the correct form in so far as it is formally identical to the usual expression

$$V_{\mathrm{M}}^{(\mathrm{eq})} = \mathcal{V}_{\mathrm{m}}^{(0)} + \frac{\mathcal{R}T}{z_j \mathcal{F}} \ln a_+, \qquad (3.4.35)$$

which is derived by applying chemical thermodynamics. Both expressions share the shortcoming that they are of *no use* in calculating the *absolute value of the electrode potential*. Neither is it possible by any experimental design to measure the value of *a single electrode potential*. To measure the potential difference between, say, the Zn electrode and the bulk ZnSO$_4$ solution would require another metal wire that had to be electrically connected in some way to the electrolyte solution. But this would result in the creation of a new interface layer and a new electrode potential in the circuit, and what we would measure is the *difference* between the electrode potentials of two half-cells. This is the unavoidable fact that is inseparably bound to any potential measurement in an electrolyte solution.

Although the values of the relevant factors in Eq. (3.4.34) are unknown, the present kinetic treatment – however crude – may be of a complementary value in providing a more vivid insight into the establishment of the electrode potential.

(iii) Standard electrode potentials
The potential difference between the electrodes in any galvanic cell, i.e. the difference between the values of the electrode potentials of two half-cells, is a well-defined quantity* and always measurable. The ambiguity that is associated with fixing the value of the electrode potential could be removed if an electrode with an absolute *zero* electrode potential were available. Thus, any e.m.f. measured against that cell would represent the electrode potential of the other half-cell. Unfortunately, no such zero reference electrodes have been found. Therefore, as the second-best move, electrochemists have chosen an *arbitrary zero of potential* and refer all other cell potentials to it. The arbitrary electrochemical zero of potential is that of the *standard hydrogen electrode*

$$(\mathrm{Pt})\mathrm{H}_2\,(p = 1\ \mathrm{atm})|\mathrm{H}^+\,(a_+ = 1),$$

which is defined as having a zero electrode potential *at any temperature*, when the partial pressure of hydrogen around the platinum electrode is 1 atm[†] and the mean activity of the hydrogen ion constituent is unity[‡]. Thus, according to this

---

* If two different electrolyte solutions are connected electrically by means of a *salt bridge*, additional liquid junction considerations are required. This holds also if the electric wiring cannot be arranged in a symmetric way.
† 1 atm = 101.325 kPa.
‡ This convention is a consequence of the basic convention $G_{\mathrm{H}_2}^0 = G_{\mathrm{H}^+}^\ominus = 0$.

convention, the e.m.f. of the cell

$$(Pt)H_2 \, (p = 1 \text{ atm})|H^+ \, (a_+ = 1) \, \| \, M^{z_j+} \, |M$$

is the value of the electrode potential of the half-cell: $M \mid M^{z_j+}$ in relation to the *standard hydrogen electrode*, and the thermodynamic equation for the electrode potential Eq. (3.4.35) can be read without any ambiguity by using a script $\mathcal{V}$ to imply a potential difference relative to the standard hydrogen electrode

$$\mathcal{V}_M^{(eq)} = \mathcal{V}_M^0 + \frac{RT}{z_j \mathcal{F}} \ln a_{M^+}. \tag{3.4.36}$$

$\mathcal{V}_M^0$ is the *standard electrode potential*, i.e. the e.m.f. of the half-cell $M \mid M^{z_j+}$ measured against the standard hydrogen electrode with unit activity of M^{z_j+} in the electrolyte solution.

Standard electrode potentials play an important role in many branches of electrochemistry, and their values for numerous half-cells have been determined and tabulated. We shall not pursue the subject of standard electrode potentials further, since none of the following topics in this book requires use of their absolute values. However, it is important to keep in mind the different physical significance that is attached to the electrode potential $V_M^{(eq)}$ of Eq. (3.4.35) and that of $\mathcal{V}_M^{(eq)}$ of Eq. (3.4.36).

(iv) Non-equilibrium electrode current

When the potential difference between the metal electrode and the electrolyte solution $\psi^{(m)} - \psi^{(s)}$ differs from the electrode potential $V_M^{(eq)}$, a net current flows across the interface, whose density I in the direction $m \to s$ is

$$I = z_j eJ = z_j e \left(\overset{m \to s}{J} - \overset{m \leftarrow s}{J} \right) = \overset{m \to s}{I} - \overset{m \leftarrow s}{I}, \tag{3.4.37}$$

where J is the net flux and $\overset{m \to s}{I}$ and $\overset{m \leftarrow s}{I}$ are the unidirectional currents. Invoking Eq. (3.4.31) yields

$$I = z_j e \left[N_M \langle \vec{v} \rangle \, e^{-U_1/kT} - N_s \langle \overleftarrow{v} \rangle \, e^{-U_2/kT} \right]. \tag{3.4.38}$$

At equilibrium the net current through the interface is zero. Hence, the two unidirectional currents $\overset{m \to s}{I}$ and $\overset{m \leftarrow s}{I}$ flowing across the barrier can be designated by the same term of magnitude I_0 – sometimes denoted the *equilibrium exchange-current density*[*] – that according to Eq. (3.4.30) and Eq. (3.4.31) is

$$I_0 = z_j e N_M \langle v^* \rangle e^{-U_1^*/kT} = z_j e N_s \langle v^* \rangle e^{-U_2^*/kT}, \tag{3.4.39}$$

[*] The existence of this current was first suggested by J.A.V. Butler.

where

$$U_1^* = W_1 + z_j e \left(\psi_e^{(b)} - \psi_e^{(m)} \right) \Bigg\}$$
$$U_2^* = W_2 + z_j e \left(\psi_e^{(b)} - \psi_e^{(s)} \right) \Bigg\}$$

$$(3.4.40)$$

are the values of U_1 and U_2 at equilibrium corresponding to the potentials $\psi_e^{(m)}$, $\psi_e^{(b)}$ and $\psi_e^{(s)}$.

We wish to express the above non-equilibrium current I in terms of I_0 and the *excess potential* displacement $\eta = (\psi^{(m)} - \psi^{(s)}) - V_M^{(eq)}$ from the electrode potential, also denoted the *overvoltage*. To this end we split each of the energy terms U_1 and U_2 in this way

$$U_1 = W_1 + z_j e \left(\psi^{(b)} - \psi^{(m)} \right) = U_1^* + \Delta U_1 \Bigg\}$$
$$U_2 = W_2 + z_j e \left(\psi^{(b)} - \psi^{(s)} \right) = U_2^* + \Delta U_2 \Bigg\}$$

$$(3.4.41)$$

by means of which Eq. (3.4.38) is written

$$I = z_j e \left[N_M \langle \vec{v} \rangle \, e^{-(U_1^* + \Delta U_1)/kT} - N_s \langle \overleftarrow{v} \rangle \, e^{-(U_2^* + \Delta U_2)/kT} \right]$$
$$= z_j e \left[N_M \langle \vec{v} \rangle \, e^{-U_1^*/kT} e^{-\Delta U_1/kT} - N_s \langle \overleftarrow{v} \rangle \, e^{-U_2^*/kT} e^{-\Delta U_2/kT} \right]$$
$$= z_j e N_M \langle v^* \rangle \, e^{-U_1^*/kT} \times e^{-\Delta U_1/kT} - z_j e N_s \langle v^* \rangle \, e^{-U_2^*/kT} \times e^{-\Delta U_2/kT},$$

and introduce in the last line the assumption that the average velocities $\langle \vec{v} \rangle$ and $\langle \overleftarrow{v} \rangle$ do not differ significantly from the equilibrium value $\langle v^* \rangle$. Invoking Eq. (3.4.39) yields

$$I = I_0 \left[e^{-\Delta U_1/kT} - e^{-\Delta U_2/kT} \right].$$

$$(3.4.42)$$

We have from Eq. (3.4.41)

$$\Delta U_1 = U_1 - U_1^*$$
$$= W_1 + z_j e \left(\psi^{(b)} - \psi^{(m)} \right) - W_1 - z_j e \left(\psi_e^{(b)} - \psi_e^{(m)} \right)$$
$$= z_j e \left(\left(\psi^{(b)} - \psi_e^{(b)} \right) + \left(\psi_e^{(m)} - \psi^{(m)} \right) \right)$$
$$= z_j e \left(\Delta \psi^{(b)} - \Delta \psi^{(m)} \right)$$

and similarly

$$\Delta U_2 = z_j e \left(\Delta \psi^{(b)} - \Delta \psi^{(s)} \right).$$

Inserting ΔU_1 and ΔU_2 in Eq. (3.4.42) gives

$$I = I_0 \left[e^{-z_j e(\Delta \psi^{(b)} - \Delta \psi^{(m)})/kT} - e^{-z_j e(\Delta \psi^{(b)} - \Delta \psi^{(s)})/kT} \right]$$

$$= I_0 \left[e^{-z_j e \Delta \psi^{(b)}/kT} e^{-z_j e \Delta \psi^{(m)}/kT} - e^{-z_j e \Delta \psi^{(b)}/kT} e^{-z_j e \Delta \psi^{(s)}/kT} \right]$$

$$= I_0 e^{-z_j e \Delta \psi^{(b)}/kT} \left[e^{-z_j e \Delta \psi^{(m)}/kT} - e^{-z_j e \Delta \psi^{(s)}/kT} \right]. \qquad (3.4.43)$$

Consider then the simple situation of a *symmetric distribution* of the overvoltage η across the barrier; that is, the additional electric work due to η is distributed in equal amounts on the two sides of the potential barrier. Accordingly, the displacements of the equilibrium values of $\psi_e^{(m)}$ and $\psi_e^{(s)}$ are

$$\Delta \psi^{(m)} = -\tfrac{1}{2}\eta, \quad \text{and} \quad \Delta \psi^{(s)} = \tfrac{1}{2}\eta.$$

Insertion into Eq. (3.4.43) gives

$$I = I_0 e^{-z_j e \Delta \psi^{(b)}/kT} \left[e^{-z_j e \eta/2kT} - e^{-z_j e \eta/2kT} \right]. \qquad (3.4.44)$$

This formula can be expressed in a more compact form by means of a hyperbolic function*, since

$$\sinh y = \frac{e^y - e^{-y}}{2}.$$

Furthermore, in the case of a symmetric displacement of η we have in addition $\Delta \psi^{(b)} = 0$. Hence, instead of Eq. (3.4.44) we can write

$$I = 2 I_0 \sinh \left[\frac{z_j e}{2kT} \eta \right] = 2 I_0 \sinh \left[\frac{z_j \mathcal{F}}{2\mathcal{R} T} \eta \right]. \qquad (3.4.45)$$

This is the basic *electrodic equation* in its simplest form[†]. Since $\sinh y$ is an odd function of y it appears from Eq. (3.4.45) that equal displacements of η that are applied on either sides of $V_M^{(eq)}$, produce a positive and negative current of the same size. Thus, in this case, application of an alternating voltage does not lead to a *voltage rectification* across the interface; that is, a constant voltage displacement to one side of the barrier that lasts until the alternating voltage is turned off.

---

* See Appendix O.

[†] Also denoted the Butler–Volmer equation (Bockris, J.O.M. & Reddy, A.K. (1970): *Modern Electrochemistry*, vol. 2, chapter 8, p. 883. Macdonald, London.). Note also the formal identity to the formula for the migration flux that is derived by similar methods in Chapter 2, Section 2.6.2.2.

Since $\sinh y \approx y$ for $y \ll 1$ we can develop an approximate expression for Eq. (3.4.45). If $z_j \mathscr{F} \eta / 2 \mathscr{R} T \ll 1$ we have accordingly

$$\sinh \left[\frac{z_j \mathscr{F}}{2 \mathscr{R} T} \eta \right] \approx \frac{z_j \mathscr{F}}{2 \mathscr{R} T} \eta,$$

and Eq. (3.4.45) assumes the form

$$I = I_0 \frac{z_j \mathscr{F}}{\mathscr{R} T} \eta, \tag{3.4.46}$$

where the electrode current I and the applied overvoltage η now are linearly related as in Ohm's law, with the quantity $I_0 \mathscr{F} / \mathscr{R} T$ serving as the *conductance* at the electrode–solution interface. What are the ranges for η where Eq. (3.4.46) can be considered valid? Letting $y = 0.2$ we have $\sinh 0.2 = 0.201\,33$, and the error of using the linear approximation is less than 0.1 %. Accepting then this range $y = \pm 0.2$ we have $(z_j \mathscr{F} / 2 \mathscr{R} T) \eta = 0.2$, and therefore

$$\eta = 0.2 \frac{2 \mathscr{R} T}{z_j \mathscr{F}}.$$

At room temperature $\mathscr{R} T / \mathscr{F}$ is near to 25 mV. Thus the linear range of Eq. (3.4.46) is $\eta = \pm 10 / z_j$ mV.

(v) Reversibility

An important property of the galvanic half-cells is their capability to operate *reversibly*. We consider a galvanic cell with an e.m.f. of value V^0 that is connected to a variable voltage supply with the output $V^{(\text{var})}$ opposing the e.m.f. When $V^{(\text{var})}$ is adjusted exactly to V^0 there is no current flow through the cell and no chemical reactions should take place in the cell. If the external voltage $V^{(\text{var})}$ is increased by a very small amount ΔV, current will pass into the cell, and half-cell reactions like those of Eq. (3.4.19) occur. On the other hand, if the external e.m.f. is decreased to the value $V^0 - \Delta V$ the current will flow in the opposite direction and the processes taking place at the two electrodes should be exactly reversed. Thus, with these small currents the departure from equilibrium is small enough so that the system is always virtually in equilibrium. With larger currents, however, concentration gradients arise within the cell because the relatively slow diffusion processes cannot cope with the rate of matter conversion at the electrodes, and the cell cannot be regarded as existing in a state of equilibrium, as the changes produced will not be reversed by passing the current in the opposite direction. This also holds for an irreversible process such as electrolysis that produces H_2 and O_2 at the electrodes. In such situations one will observe that after disconnecting $V^{(\text{var})}$ the voltage of the cell remains for some time at $V^{(\text{var})}$, to be followed by a more or less slow return to the initial e.m.f. The terms *polarization* and *overpotential* are used rather indiscriminately to refer to some irreversible electrode process that is associated with the current flow through the half-cells. Polarization processes cause undesirable sources of error in potential recordings and should be avoided. This objective is achieved by employing a proper galvanic half-cell and operating it reversibly, that is by drawing only an infinitely small current during the measurement. In practice this works by using a voltmeter with an input resistance that is several orders of magnitude higher than the resistance of the galvanic half-cell.

3.4.3.2 Two recording electrodes

Two electrodes that are used extensively in biological applications are the silver–silver chloride electrode and the calomel electrode.

(i) The silver–silver chloride electrode

The AgCl electrode consists of a silver rod that is coated with a layer of crystalline AgCl. This coat of solid AgCl is usually deposited by operating the silver rod as an anode for a short time in a strong Cl^- solution, such as 0.1–1.0 M HCl. This electrode will work properly in an electrolyte containing Cl^- ions in sufficient amounts to exceed the solubility product of AgCl, and thus ensures that the solid AgCl remains present at the silver rod. This electrode functions reversibly with respect to chloride ions and, therefore, belongs to the class of *anion electrode*. The function may be described as follows. Essentially an equilibrium will be established between the Ag atoms in the silver rod/AgCl coat and Ag^+ ions in solutions near the interface. The electrode reaction can be written as

$$Ag^+ + Cl^- \longleftrightarrow AgCl + \mathcal{E},$$

or, to give more weight to the Ag/Ag^+ reaction,

$$-\mathcal{E} \quad \begin{matrix} Cl^- + Ag^+ \rightleftharpoons AgCl \\ \uparrow\downarrow \\ Ag \end{matrix} \quad + \mathcal{E}.$$

The equilibrium electrode potential corresponding to the electrode reaction $Ag \rightleftharpoons Ag^+ + \mathcal{E}$ can, in accordance with Eq. (3.4.35), be written as

$$V_{Ag,Ag^+}^{(\text{eq})} = V_{Ag,Ag^+}^0 + \frac{\mathcal{R}T}{\mathcal{F}} \ln a_{Ag^+}. \tag{3.4.47}$$

However, silver is sparingly soluble in an environment consisting of chloride ions, the concentration being determined by the *solubility product*

$$a_{Ag}a_{Cl} = K_s. \tag{3.4.48}$$

Therefore

$$V_{Ag,Ag^+}^{(eq)} = V_{Ag,Ag^+}^0 + \frac{\mathscr{R}T}{\mathscr{G}} \ln K_s - \frac{\mathscr{R}T}{\mathscr{G}} \ln a_{Cl^-}.$$

In this expression we replace a_{Cl^-} by the molal concentration [Cl] – after all individual ionic activities cannot be obtained by independent methods – and collect the first two terms plus a term containing an activity coefficient into a single constant. This gives

$$V_{Ag,Cl^-}^{(eq)} = V_{Ag,Cl^-}^0 - \frac{\mathscr{R}T}{\mathscr{G}} \ln [Cl], \tag{3.4.49}$$

showing that the electrode potential varies linearly with the logarithm of the chloride ion concentration. When the silver–silver chloride electrode is connected to a half-cell whose electrode potential V_1 is not susceptible to changes in concentrations of Cl^- ions, the potential difference V between the AgCl electrode and the other half-cell takes the form

$$V = V_{Ag,Cl^-}^{(eq)} - V_1 = V^* - \frac{\mathscr{R}T}{\mathscr{G}} \ln [Cl], \tag{3.4.50}$$

where V^* is the potential difference for [Cl] = 1. Thus, V is a measure of the chloride concentration in the solution. In this case the electrode operates as an *indicator electrode* for Cl^- ions. Alternatively, if the chloride concentration around the AgCl electrode is held constant – e.g. in glass tube of a suitable shape where escape of Cl^- ions by diffusion to the bulk of electrolyte solution is minimized – the electrode potential will remain constant in spite of varying the electrolyte concentration or composition. In this case the AgCl electrode now functions as a *reference electrode*, and is a basis for comparing the values of other potential sources in the electrolyte. When two such identical reference AgCl electrodes are placed in the same electrolyte solutions, both silver rods will be elevated by the *same electrode potential* with respect to the bulk of the solution, and *no* potential difference will be observed between the electrodes. For some reason – e.g. by the action of external electrodes in the electrolyte – let a potential drop of magnitude ΔV exist between the two electrodes. Hence, the electrode potential of one reference electrode is elevated by the amount ΔV with respect to that of the other electrode, but both electrodes will remain in equilibrium. A conducting wire – or a low-resistance voltmeter – connected to the electrodes will result in a perturbation of the electrode potentials away from their equilibrium values with the result that a current flows in the external circuit. The higher the resistance of the voltmeters, the smaller the current load and, ultimately, a voltmeter reading may be obtained that differs from ΔV by a vanishingly small amount. In this case the AgCl electrodes have operated as perfect *recording electrodes*, and the influences of the two electrode potentials have exactly cancelled, leaving only ΔV in the measurement. AgCl electrodes are used extensively in biology because of their ease of preparation and stability. Their drawbacks are a relatively high resistance and low capacity to draw current without deterioration.

(ii) The calomel electrode

Other sparingly soluble salts reacting with the corresponding metal are used similarly as a reversible electrode with respect to the anion. The *calomel electrode* consists of a small quantity of pure mercury that is mixed with a paste of Hg_2Cl_2 that is in contact with a standard solution of KCl, usually a saturated one with excess KCl crystals. The electrode process may be written as

$$2Hg + 2Cl^- \longleftrightarrow Hg_2Cl_2 + 2\mathcal{E},$$

or alternatively

$$-2\mathcal{E} \quad \begin{array}{c} Hg_2Cl_2 \rightleftharpoons Hg_2^{2+} + 2Cl^- \\ \uparrow\downarrow \\ 2Hg \end{array} \quad + 2\mathcal{E},$$

and by analogy with the Ag–AgCl electrode, the electrode potential for the calomel electrode $V_{cal}^{(eq)}$ is

$$V_{cal}^{(eq)} = V_{cal}^0 + \frac{\mathcal{R}T}{2\mathcal{F}} \ln a_{Hg_2^{2+}}. \tag{3.4.51}$$

Like silver, the solubility of Hg_2Cl_2 is low. Utilizing the solubility product to replace $a_{Hg_2^{2+}}$ in the above expression we finally end up with

$$V_{cal}^{(eq)} = V_{cal}^\oplus - \frac{\mathcal{R}T}{2\mathcal{F}} \ln [Cl], \tag{3.4.52}$$

showing that the calomel electrode behaves like the Ag–AgCl electrode to changes in surrounding content of chloride ions. The employment of the saturated KCl solution around the calomel mercury paste makes the electrode very stable and also well suited as an all-round *reference electrode* because the diffusion potential between two electrolytes is generally very small when the one solution consists of saturated KCl*.

---

* See Section 3.6.3.3.

3.4.4 The equivalent electric circuit for the ionic-selective membrane

The flux of an ionic species is – by virtue of the charge-carrying capacity – equivalent to the flow of an electric current. These two quantities are interrelated as follows. Any ions with charge number z_j carry the charge $z_j \mathscr{F}$ coulomb per mol (C mol^{-1}). If the flux J_j of type j is specified in mol m^{-2} s^{-1} the *current density* in C m^{-2} s^{-1} = A m^{-2} belonging to the flux is

$$I_j = z_j \mathscr{F} J_j. \tag{3.4.53}$$

According to convention, the flow direction of the positive charges is reckoned as the *positive direction* of the *current flow*. Thus, as regards cations – carrying positive charges – the directions of the flux J_j and the corresponding current density I_j are the *same*. Anions, on the other hand, carry negative charges. Thus, the anionic flux and its corresponding current density are *oppositely* directed. These circumstances are correctly reflected by way of the *sign* of the charge number z_j, that is

$$I_{j+} = z_{j+} \mathscr{F} J_{j+} \quad \text{and} \quad I_{j-} = -|z_{j-}| \mathscr{F} J_{j-}.$$

The transfer of ions through membranes is very often studied by electric methods of measurement, that is measuring electric current density instead of ionic flux, membrane resistance or conductance as a measure for the ionic permeability. Electric methods are very sensitive and of great precision and their ability to follow time courses of high speed are superior to all other methods used in the study of the transfer of ions. Thus, it is possible to follow processes that proceed within microseconds (1 μs = 10^{-6} s). For that reason electric methods of measurement have been of particular value in the study of the properties of the excitable membranes.

3.4.4.1 Measurement of the current–voltage characteristic

Figure 3.6 illustrates the principal elements in an experimental arrangement designed to measure simultaneously the values of the potential difference across

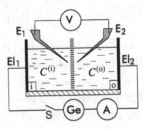

Fig. 3.6. Diagram of the elements involved in the simultaneous measuring of membrane potential and membrane current.

a membrane (membrane potential) and the membrane current. The membrane under study separates the chamber into two halves (i) and (o), which contain an electrolyte with concentrations $C^{(i)}$ and $C^{(o)}$. To measure the potential difference across the membrane $V = \psi^{(i)} - \psi^{(o)}$, a pair of electrodes, E_1 and E_2, in the shape of glass pipettes filled with a suitable electrolyte, are placed as close as possible to each membrane side and connected – by means of appropriate metal electrodes that dip into the pipettes – to the voltmeter V. The external control of the voltage across the membrane is brought about via two plate electrodes, El_1 and El_2, that are connected to the voltage generator, Ge, capable of producing an electromotive force of variable strength and polarity. The switch, S, turns the generator on and off. The current I that flows in the external circuit – and through the membrane – is measured with the ammeter A. In the following we assume that the membrane is selectively permeable to cations and that the solutions in the two chambers are NaCl.

Which processes are involved in the current flow in the whole circuit when we are studying the relationship between the current density I and voltage drop V across the ionic-selective membrane? Naturally, the membrane current (I_{Na}) is carried by Na^+ ions having one positive elementary charge. In the metallic wires connecting the voltage generator (Ge) to the current producing electrodes El_1 and El_2 the movable metallic electrons constitute the current flow. At the electrodes there occurs an exchange of the electrons in the metal with those of the ions in the solution. At the cathode – being connected to the negative pole of the voltage generator – the water molecules (the H^+ in the hydroxonium molecule H_3O^+) nearest to the cathode will capture an electron and undergo a reduction according to the scheme

$$H^+ + \mathcal{E} \rightarrow \tfrac{1}{2}H_2. \tag{3.4.54}$$

At the *anode* an electron is donated from the OH^- ions to the metallic electrode according to the scheme

$$OH^- \rightarrow \tfrac{1}{2}H_2O + \tfrac{1}{4}O_2 + \mathcal{E}. \tag{3.4.55}$$

The net result is that an electron is carried from the cathode to the anode. Furthermore, electroneutrality in the two chambers is conserved as one Na^+ ion is transferred through the membrane from the chamber containing the anode to that of the cathode. With sufficiently strong currents, O_2 bubbles from the anode and H_2 from the cathode.

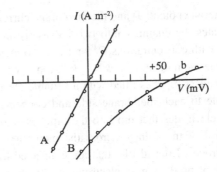

Fig. 3.7. Examples of current–voltage characteristics for an ion selective membrane for (A) equal salt concentrations in both chambers and (B) for different concentrations $C_{NaCl}^{(i)} < C_{NaCl}^{(o)}$. Abscissa: membrane potential $\psi^{(i)} - \psi^{(o)}$. Ordinate: membrane current per unit membrane area.

3.4.4.2 Membrane current and membrane conductance

Let the two chambers contain the same electrolyte and at identical concentrations, e.g. 100 mM NaCl. With the voltage generator turned off, no potential difference exists across the membrane and, naturally, no current will traverse the membrane. The voltage generator is then turned on to obtain a series of data pairs of current density I and potential difference V across the membrane*. Figure 3.7, curve A, shows an example of such a series of measurements. In the diagram, the direction chamber (i) → chamber (o) is reckoned as the positive direction of current flow. It appears from the figure that the current density I is directly proportional to the membrane potential V, thus

$$I_{Na} = G_{Na} V, \qquad (3.4.56)$$

which corresponds to Ohm's law (compare Eq. (3.3.11)). The constant G_{Na} has the dimension $ohm^{-1} \, m^{-2} = S \, m^{-2}$ and represents the conductance per unit membrane area. G_{Na} is called the *sodium conductance* of the membrane and indicates the current – carried by the Na^+ ions – that flows through one unit area of the membrane owing to a potential difference of 1 V. In the present case the membrane is permeable only to Na^+ ions and the surrounding solution contains only NaCl. Thus, no difficulty is encountered in identifying the conductance measured as a sodium conductance G_{Na}. The magnitude of the conductance G_{Na} depends on two factors.

(1) The drift velocity v_{Na} with which a field of unit strength drives the individual Na^+ ion through the membrane, that is on the ion's electric

---

* To find the current density I from the reading of the current i on the ammeter requires naturally a knowledge of the membrane area through which the current flows.

mobility u_{Na}. This quantity is again proportional to the ion's mechanical mobility* B_j.

(2) The concentration C_{Na} of the sodium ions in the membrane. The greater the concentration the greater the number of charge carriers and the greater the current per unit area for a given velocity of migration v for the ions in question.

In the present case ($z_{Na} = 1$) we express the sodium conductance of the membrane analogous to Eq. (3.3.25)

$$G_{Na} = \frac{\mathscr{F} u_{Na} C_{Na}}{h}, \qquad (3.4.57)$$

where h now represents the thickness of the ion-exchange membrane.

If the membrane is surrounded by identical solutions, the concentration of sodium C_{Na} in the membrane will not change irrespective of the magnitudes of the current I_{Na}. As long as the mobility of sodium B_{Na} does not change with the strength of the impressed field, a direct proportionality between I_{Na} and V should be expected as indicated in Fig. 3.7, curve A.

We consider next the situation with different concentrations in the two chambers, e.g. $C_{Na}^{(i)} = 10$ mM and $C_{Na}^{(0)} = 100$ mM. With the voltage generator turned off, the system is left to itself to arrive at the equilibrium, where the membrane potential V has grown – owing to the initial, solitary movements of the Na$^+$ ions – from the value zero to that of the equilibrium potential $V_{Na}^{(eq)}$ for sodium, which can be calculated from the Nernst equation. At 25 °C we have

$$V = \psi^{(i)} - \psi^{(0)} = V_{Na}^{(eq)} = 59.2 \cdot \log \frac{100}{10} = 59.2 \, \text{mV}.$$

For $V = V_{Na}^{(eq)}$ we have $I_{Na} = 0$. These values constitute the basic point on the current–voltage diagram for the new situation where the concentrations on the two sides of the membrane *differ* from each other. We now switch the voltage generator on and push the membrane potential to some other value than $V_{Na}^{(eq)}$, e.g. to $V = +65$ mV. What will be the effect on the movements of the Na$^+$ ions? Raising the membrane potential above $V_{Na}^{(eq)} = 59.2$ mV has the effect of increasing the field in the direction (i) → (o) *above* the field at $V = V_{Na}^{(eq)}$ that prevents a net flux of Na$^+$ ions in the direction (o) → (i) owing to concentration difference $\Delta C = C_{Na}^{(0)} - C_{Na}^{(i)} = 90$ mM. This additional field will cause the Na$^+$ ions to move through the membrane in the direction (i) → (o) and give rise to a current flow in the *same* direction. It is the value of this current that is read off on the ammeter. On the other hand, by clamping the membrane potential to a value *lower* than +59.2 mV, that is $V < V_{Na}^{(eq)}$, the field is now too small to

---

* Compare Eqs. (3.3.3),(3.3.4) and (3.3.5).

counteract the diffusion of the Na^+ ions in the direction (o) $\rightarrow$ (i), and the result is now a current in this direction carried by the Na^+ ions. Thus, a perturbation of the membrane potential about $V_{Na}^{(eq)}$ results in a current in the direction (i) $\rightarrow$ (o) when $V > V_{Na}^{(eq)}$, and a current in the opposite direction for $V < V_{Na}^{(eq)}$. Figure 3.7, curve B, shows a series of pairs of I_{Na} and V belonging together. For small displacements ($\Delta V < 10$ mV) of the membrane potential V about the equilibrium potential $V_{Na}^{(eq)}$ one will observe a linear relation between I_{Na} and V (as shown by the line element a–b on curve B), which in this range can be written by the equation

$$I_{Na} = G_{Na}\left[V - V_{Na}^{(eq)}\right], \tag{3.4.58}$$

where G_{Na} is the sodium conductance of the membrane for $V \approx V_{Na}^{(eq)}$.

In most cases one will find – as illustrated in Fig. 3.5, curve B – that the I–V diagram deviates from the rectilinear course for greater displacements from the equilibrium potential. In biological membranes such deviations are often quite substantial. In these cases it has been of practical value to retain the formula above to describe the connection between membrane potential and ionic current (the *current–voltage characteristic*) for the membrane, i.e.

$$I_j = G_j\left[V - V_j^{(eq)}\right], \tag{3.4.59}$$

but the ion conductance in the membrane G_j for the ion of species j is no longer constant, but varies with the membrane potential. A closer functional connection $G_j = G_j(V)$ must be determined by experiments to measure pairs of values of I_j and V and then use Eq. (3.4.59) as the *equation of definition* for the determination of G_j from the values of I_j and V, that is

$$G_j = \frac{I_j}{V - V_j^{(eq)}}, \tag{3.4.60}$$

where the equilibrium potential $V_j^{(eq)}$ is the value of the membrane potential V where the membrane current I is zero.

Geometrically G_j equals the *slope* of the line drawn from the origin $(0, V_j^{(eq)})$ on the V-axis to a given point (I, V) on the I–V curve, that is, the slope of a *secant* to the I–V curve from the point $(0, V_j^{(eq)})$. For that reason it is the practice to designate G_j as the *secant-conductance* or *chord-conductance*. Another way of obtaining a measure for the membrane conductance at a given point (I, V) on the I–V curve is to impress a small perturbation of the membrane potential (or membrane current) ΔV and measure the change in membrane current (or membrane potential) ΔI brought about. For small values of ΔI and ΔV the ratio

$$\frac{\Delta I}{\Delta V} \approx \left(\frac{dI}{dV}\right)_V = g_j, \tag{3.4.61}$$

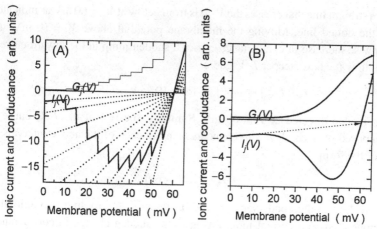

Fig. 3.8. Illustration of the formation of an I–V characteristic that displays a negative slope conductance. (A) A cation-selective permeable membrane surrounded by NaCl solutions providing a $V_{Na}^{(eq)} = 60$ mV. The Na^+ conductance is supposed to increase with V in a stepwise manner (upper curve). The dotted lines all intersecting at $V = 60$ mV represent the I–V characteristics of Eq. (3.4.61) associated with each of the above values of G_{Na}. The saw-tooth-shaped solid line is the I–V characteristic that results from the stepwise change in G_{Na} with V. (B) The I–V characteristic when the G_{Na} increases continuously in the range of V where the current is inward directed. Note the transient phase of a *negative* slope conductance.

also having the dimensions of S m^{-2}, equals the slope of the *tangent* to the I–V curve at the point (I, V). This ratio g_j is also designated as the *slope conductance* for the ion of species j. The more nonlinear the I–V curve, the greater the differences between the values of the two conductances. It should also be noted that the chord conductance G_j is *always* a positive quantity irrespective of the shape of the I–V curve, whereas there may appear examples where the slope conductance assumes a *negative* value within a limited range of values of membrane potentials.

The formation of this apparent paradoxical behavior is illustrated in Fig. 3.8. In (A) a cation-selective membrane is surrounded by, for example, NaCl solutions with concentration ratio $C_{Na}^{(i)}/C_{Na}^{(o)} = 1/10$ giving an equilibrium potential $V_{Na}^{(eq)} = 60$ mV as before. The ionic conductance is assumed to be voltage dependent in a stepwise manner as indicated in the upper part of (A), i.e. for membrane potentials less than $+10$ mV the Na^+ conductance has a low, constant value $G_{Na}^{(1)}$, whereas at the values $V = 10, 15, 20, 25, \ldots, 55$ mV the conductance increases stepwise. As the membrane potential is displaced from $V = 0$ towards more positive values the inward-directed current follows the course

$$I = G_{Na}^{(1)}[V - 60], \quad \text{for} \quad V < 10 \text{ mV},$$

i.e. a straight line that crosses the V-axis from below at $V = 60\,\text{mV}$ as indicated by the dotted line. Moving the membrane potential above $V = 10$ mV, the stepwise increase in conductance causes a sudden jump in the inward current to follow the new steeper I–V curve

$$I = G_{\text{Na}}^{(2)}[V - 60], \quad \text{for} \quad 10 < V < 15\,\text{mV},$$

that is also directed at $V = 60$ mV. Similarly, the rest of the I–V curve is composed of linear I–V sub-elements being in each of the above intervals of V of the form

$$I = G_{\text{Na}}^{(i)}[V - 60],$$

where $G_{\text{Na}}^{(i)}$ is the conductance in the ith interval. In crossing two adjoining regions there is a current jump owing to the sudden change in conductance. The solid line in Fig. 3.8(A) shows the resultant I–V curve in the range $0 < V < +65$ mV. This saw-tooth-shaped curve results from the interplay between a stepwise increase in G_{Na} – making bigger the inward current and the slope of the I–V sub-elements – and a declining driving force tending to reduce the current until its direction reverses that at $V = 60$ mV. The overall effect is an inward, decreasing Na^+ current that transiently increases to a peak in the range 35–40 mV from which it declines and crosses the V-axis at $V = 60$ mV. In (B) the voltage dependence of the conductance has been changed to a continuous curve by letting the number of steps increase. The I–V curve changes into the corresponding continuous curve that exhibits a region where the slope conductance is *negative* in spite of the steady increase in core conductance.

One reason for the voltage dependence of the membrane conductance for an ion is that the strength and direction of the field may determine the concentration in the membrane of the ion in question. Such models will be considered later. However, in many biological membranes – especially the excitable membranes* – such simple model considerations are insufficient to account for the much more pronounced voltage dependence of some of the ions involved. The explanation on the molecular level of the underlying mechanisms still represents one of the great challenges in membrane biophysics.

3.4.4.3 The equivalent circuit diagram

In this section we shall demonstrate that the empirically established connection between the membrane potential V and the single ionic current can be derived from the Nernst–Planck equation. This leads to an explicit expression that involves the different factors that determine the magnitude of the partial

---

* See Chapter 4.

conductance for an ion. In this context it also becomes convenient to represent electrodiffusion for a single ion through the membrane in pure electric terms as an equivalent electric circuit diagram, which sometimes may be advantageous. We still consider the situation of Fig. 3.6 of a membrane that is selectively permeable to an ion of type j. On the two phases we impress a potential difference $V = \psi^{(i)} - \psi^{(o)}$, which causes a current flow through the membrane with a current density I_j – the positive direction being chosen as the direction (i) $\rightarrow$ (o). We search after the connection between V and I_j when the concentrations in the surrounding phases are $C_j^{(i)}$ and $C_j^{(o)}$. The ionic current carried by the flux J_j is

$$I_j = z_j \mathscr{F} J_j. \tag{3.4.62}$$

We eliminate J_j on the right-hand side by using Eq. (3.3.36). This gives

$$I_j = -z_j \mathscr{F} u_j \, C_j \left[\frac{\mathscr{R}T}{z_j \mathscr{F}} \frac{1}{C_j} \frac{dC_j}{dx} + \frac{d\psi}{dx} \right],$$

or

$$\frac{I_j}{z_j \mathscr{F} u_j \, C_j} = -\frac{\mathscr{R}T}{z_j \mathscr{F}} \frac{d \ln C_j}{dx} - \frac{d\psi}{dx}.$$

This expression is multiplied by dx and integrated through the membrane from phase (i) at $x = 0$ with concentration $C_j^{(i)}$ and potential $\psi^{(i)}$ to phase (o) at $x = h$ with concentration $C_j^{(o)}$ and potential $\psi^{(o)}$. As the system is assumed to be stationary we have $dI_j/dx = 0$, and consequently I_j is put outside the integration sign. We have then

$$I_j \int_0^h \frac{1}{z_j \mathscr{F} u_j \, C_j(x)} \, dx = -\frac{\mathscr{R}T}{z_j \mathscr{F}} \int_{C_j^{(i)}}^{C_j^{(o)}} d \ln C_j - \int_{\psi^{(i)}}^{\psi^{(o)}} d\psi,$$

or

$$I_j \int_0^h \frac{1}{z_j \mathscr{F} u_j \, C_j(x)} \, dx = \frac{\mathscr{R}T}{z_j \mathscr{F}} \ln \left(\frac{C_j^{(i)}}{C_j^{(o)}} \right) + \psi^{(i)} - \psi^{(o)}.$$

According to Eq. (3.4.13) the first term on the right-hand side is equal to $-V_j^{(eq)}$, that is, the equilibrium potential with reversed sign for the ion of species j, while the second term is the impressed potential difference $V = \psi^{(i)} - \psi^{(o)}$ over the membrane. The integral on the left-hand side is, according to Eq. (3.3.27), equal to the resistance R_j of the membrane for the ion of type j. We may therefore write the expression above as

$$I_j = \frac{V - V_j^{(eq)}}{R_j},$$

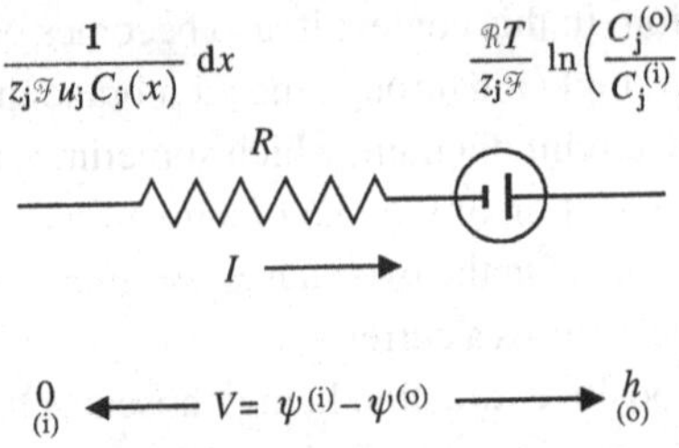

$$\int_0^h \frac{1}{z_j \mathcal{F} u_j C_j(x)}\, dx \qquad\qquad \frac{\mathcal{R}T}{z_j \mathcal{F}} \ln\!\left(\frac{C_j^{(o)}}{C_j^{(i)}}\right)$$

Fig. 3.9. The equivalent circuit for an ion-selective membrane.

or, as the membrane conductance G_j for the ion of type j is equal to $1/R_j$,

$$I_j = G_j\left[V - V_j^{(eq)}\right]. \qquad (3.4.63)$$

The form of this equation is the same as that of Eq. (3.4.59), but the membrane conductance is now expressed explicitly as

$$\frac{1}{G_j} = \int_0^h \frac{1}{z_j \mathcal{F} u_j C_j(x)}\, dx, \qquad (3.4.64)$$

from which G_j can be calculated if the concentration profile $C_j(x)$ and the electric mobility in the membrane are known.

The relationship between the current I_j and the voltage V by which it is generated can be visualized by an *electric equivalent circuit* that is illustrated in Fig. 3.9. It consists of a resistance $R_j = 1/G_j$ in series with an *electromotive force* equal to the equilibrium potential for the ion in question both as for magnitude and sign. The voltage across the terminals (i) and (o) is the impressed membrane potential V. Such an equivalent circuit may sometimes be of value when treating more complex situations, where several different ionic species move across the membrane.

It should be noted that the measuring arrangement of Fig. 3.9 only allows determination of a partial ionic conductance G_j if the appropriate actions have been taken to ensure that it is only the ions j in question that are driven through the membrane by the field. If such measures cannot be taken and the membrane current is carried by several different ions – both anions and cations – we can of course still record a current–voltage characteristic for the membrane. And the relation between membrane current I and membrane potential V can still be written as

$$I = G(V - V_m), \qquad (3.4.65)$$

but here I is the *sum* of the single ion currents, that is, the *total* ionic current, and

G is the *total* membrane conductance. Furthermore, V_m is not an equilibrium potential but a more or less *complex diffusion* potential, that – as described in Section 3.6 – is established when the system is left on its own, so that the sum of all the ion fluxes $\sum J_j$ adjusts itself to the situation where the total ion current $\sum I_j$ through the membrane (the membrane current) is zero.

In case of a biological membrane, V_m is in general designated as the *resting potential* of the cell. In other situations experimental practice calls the term the *reversal potential* for V_m, as it is at this value of the membrane potential that the membrane current changes its direction as the membrane potential changes from $V_{rev} - \Delta V$ to $V_{rev} + \Delta V$.

3.4.4.4 Membrane conductance and membrane permeability

We have met the parameters, ionic conductance and ion permeability, that both reflect the ion's ability to penetrate the membrane. As both quantities are basically derived from the diffusion coefficient D_j or mechanical mobility B_j respectively, one may put the question: what is the connection between these two quantities? To attempt an answer to this question is not an easy matter apart from stating that one parameter is directly proportional to the other, but that their ratio depends on the particular environment to which the membrane is exposed. Therefore, we shall only consider a few simple examples. For a homogeneous membrane we defined in Chapter 2, Section 2.7.1, Eq. (2.7.5), the permeability as

$$P_j = \frac{\alpha D_j}{h} = \frac{\alpha kT B_j}{h}, \tag{3.4.66}$$

from which it appears that the permeability is proportional to the diffusion coefficient D_j or mechanical mobility B_j, that is, with quantities that indicate how far the single ion moves per unit time under the action of a driving force X_j. The electric mobility u_j of the ion is defined as

$$u_j = z_j e B_j, \tag{3.4.67}$$

by means of which B_j in Eq. (3.4.61) is replaced by $u_j{}^*$.

$$P_j = \frac{\alpha_j kT}{z_j eh} u_j = \frac{\alpha_j \mathcal{R} T}{z_j \mathcal{F} h} u_j, \tag{3.4.68}$$

which shows that the permeability P_j is proportional to the electric mobility u_j of the ion. The conductance G_j, however, is proportional to the product of the mobility u_j and the concentration C_j in the membrane. The explicit

---

* See Section 2.7.4.2, $\mathcal{N}_A \cdot k = \mathcal{R}$, and Eq. (3.3.8), $\mathcal{N}_A \cdot e = \mathcal{F}$.

relation between P_j and G_j is therefore only relatively simple in the situation where the membrane is surrounded on both sides by the same salt and in equal concentrations. In this case the ion concentration C_j in the membrane is constant and the conductance can be written as*

$$G_j = \frac{z_j \mathscr{F} u_j\, C_j}{h}. \tag{3.4.69}$$

Combining these expressions to eliminate u_j gives

$$G_j = \frac{z_j^2 \mathscr{F}^2}{\mathscr{R}T}\frac{C_j}{\alpha_j}\,P_j = \frac{z_j^2 \mathscr{F}^2}{\mathscr{R}T}\,C_j^{(0)}P_j, \tag{3.4.70}$$

as $C_j = \alpha C_j^{(i)} = C_j^{(0)\dagger}$.

When $C_j^{(i)} \neq C_j^{(0)}$ the corresponding relation between P_j and G_j can be derived with little effort provided G_j is constant or the value of G_j is known for $V = 0$. Consider the equation

$$I_j = G_j\bigl(V - V_j^{(eq)}\bigr) = z_j \mathscr{F} J_j.$$

Then making use of a voltage generator we imagine that the membrane potential is forced to assume the value $V = 0$ (the membrane is said to be short circuited). The ion flux J_j then occurs exclusively by diffusion. We have accordingly

$$(I_j)_0 = z_j \mathscr{F}(P_j)_0\bigl(C_j^{(i)} - C_j^{(0)}\bigr)$$

and

$$(I_j)_0 = -(G_j)_0 V_j^{(eq)},$$

where the subscript zero refers to the values for the short-circuited membrane. Eliminating $(I_j)_0$ and replacing the value of $V_j^{(eq)}$ we obtain

$$z_j \mathscr{F}(P_j)_0\bigl(C_j^{(i)} - C_j^{(0)}\bigr) = -(G_j)_0\left[-\frac{\mathscr{R}T}{z_j \mathscr{F}}\ln\left(\frac{C_j^{(i)}}{C_j^{(0)}}\right)\right],$$

and by rearrangement

$$(G_j)_0 = \frac{(z_j \mathscr{F})^2}{\mathscr{R}T}\frac{C_j^{(i)} - C_j^{(0)}}{\ln\bigl(C_j^{(i)}/C_j^{(0)}\bigr)}(P_j)_0, \tag{3.4.71}$$

which sometimes is a useful expression.

With less restrictive conditions the relationship between P_j and G_j becomes rather involved. In Section 3.7.3 we shall consider the special case where the

---

* See Section 3.3.1, Eq. (3.3.25).
† Compare Chapter 2, Section 2.7.1, Eq. (2.7.3).

membrane is homogeneous and the field in the membrane is constant – the so-called Goldman regime.

3.5 The Donnan potential

Until now we have considered two types of membrane equilibrium: (1) the *ideally selective permeable membrane*, which is permeable only to either positively or negatively charged ions, and (2) the *ideally semipermeable membrane*, which is permeable only to the solvent and totally impermeable to the dissolved substances, the study of which resulted in concepts such as the equilibrium potential and the osmotic pressure to characterize an electrochemical and an osmotic equilibrium. A third group of membrane phenomena appears when the membrane is permeable to water, ions and neutral molecules of *small size*. We call this type of membrane a *semipermeable membrane*. To this group belong also the biological membranes.

In this section we shall consider a *special type of equilibrium* over the semipermeable membrane that appears when both the surrounding phases are electrolytes but one of the phases contains dissolved ionized macromolecules of a molecular dimension large enough to prevent their penetration through the membrane. Originally this type of membrane equilibrium became interesting in connection with the determination of the molecular weight of charged macromolecules in a colloid solution, by measuring the osmotic pressure of the solution and considerations of stability required an addition of a certain amount of salt. As we shall see below, the presence of the impermeable ionized component influences the distribution of the permeable ions so that at equilibrium the system possesses the following features:

(1) an asymmetrical distribution of the diffusible ions between the two phases,
(2) an electric potential difference (*the Donnan potential*) across the membrane, and
(3) an osmotic pressure difference between the phases.

Originally this type of membrane equilibrium was described by the American physicist Willard Gibbs (1875) and later studied in detail – theoretically as well as experimentally – by the Irish physical chemist F.G. Donnan. The equilibrium is generally called the *Gibbs–Donnan equilibrium*. In keeping with the practice used so far we shall first describe in qualitative terms the way this equilibrium comes about. Next follows a macroscopic quantitative description using equilibrium thermodynamics. Finally, we shall calculate the courses of

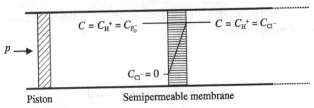

Piston   Semipermeable membrane

Fig. 3.10. Initial distribution in connection with the development of a Donnan equilibrium between two phases that are separated by a membrane that is impermeable for the macroion Po^-. Initially the phase (i) contains in solution HPo (H^+ and Po^- in concentrations C, while phase (o) contains HCl (H^+ and Cl^-) with the same concentration. The piston can exert a pressure p on phase (i) to prevent water movements from phase (o) to phase (i).

the profiles of potential and concentration that connect the equilibrium values belonging together in the two phases.

3.5.1 Qualitative description of the Donnan distribution

As the simplest imaginable situation we consider the situation that is shown in Fig. 3.10, where the two phases (i) and (o) are separated by the semipermeable membrane. Initially, phase (i) contains a solution of a macromolecule HPo with concentration C, and which we assume to be completely dissociated into H^+ and Po^-. The membrane is permeable to water and small ions, but not to the macroion Po^-. In the other phase (o) we put an HCl solution with the same concentration C. Furthermore, we assume that volume of phase (o) is much larger than that of phase (i), which implies that any consecutive exchange of water, H^+ and Cl^- ions between the two phases does not have any influence on the concentration of HCl in phase (o). Initially, no equilibrium exists between the two phases as the concentration of Cl^- in phase (i) is zero and is $C^{(o)}$ in phase (o). This difference in concentration causes Cl^- ions to diffuse into phase (i), with the result that an infinitely small amount of uncompensated negatively charged Cl^- ions assign to the phase a lower (negative) potential than that of phase (o). This potential difference tends – as described in Section 3.4.1 – to oppose any further in-flow of Cl^- ions. But the membrane is permeable to the H^+ ions in phase (o), at a higher potential than in phase (i). Thus, although $C_H^{(i)} = C_H^{(o)}$ the ions are no longer in equilibrium, and the H^+ ions will move according to the electric field from phase (o) to phase (i), and in doing so will tend to put out the electric imbalance between the two phases. In consequence, more Cl^- ions will diffuse from phase (o) to phase (i), which is followed by a further movement of H^+ ions, and so on. The two ions will rapidly adjust their velocities of movement such that H^+ and Cl^- ions move in (almost) equivalent amounts from phase (o) to phase (i) until equilibrium is established. At this stage phase (i), which contains the diffusible anion, will also contain a certain Cl^- concentration, ΔC_{Cl}, and an H^+ concentration that exceeds that of phase (o) by $\Delta C_H = \Delta C_{Cl}$. At equilibrium the concentrations of the diffusible ions have adjusted themselves in relation to the potential difference

$$V_D = \psi^{(i)} - \psi^{(o)},$$

so that the tendency of each ionic species to cross the membrane in one direction by way of diffusion is balanced by the electric potential difference between the phases that tends to move the ion in the opposite direction. The membrane potential – that is attached to this rather peculiar situation – owes its presence solely to the indiffusible macroion Po^-, and is usually called the *Donnan potential*. It relates to the class of equilibrium potentials where the relationship between the magnitude and sign of the Donnan potential and the concentration of the diffusible ions is given by the Nernst equation, Eq. (3.4.7). In the present case where $z_{H^+} = 1$ and $z_{Cl^-} = -1$, we have*

$$\psi^{(i)} - \psi^{(o)} = V_D^{(eq)} = \frac{\mathcal{R}T}{\mathcal{F}} \ln \frac{C_H^{(o)}}{C_H^{(i)}} = \frac{\mathcal{R}T}{\mathcal{F}} \ln \frac{C_{Cl}^{(i)}}{C_{Cl}^{(o)}}. \tag{3.5.1}$$

It follows that at equilibrium the ionic distribution of ions is given by

$$r_D = \frac{C_H^{(o)}}{C_H^{(i)}} = \frac{C_{Cl}^{(i)}}{C_{Cl}^{(o)}}, \tag{3.5.2}$$

where r_D is called the *Donnan ratio*. At equilibrium the H^+ and the Cl^- ions have distributed themselves between the phases (i) and (o), where the ratio between the H^+ concentrations is the inverse of the ratio of the Cl^- concentrations. Thus, the above equation can also be written as

$$C_H^{(i)} C_{Cl}^{(i)} = C_H^{(o)} C_{Cl}^{(o)}, \tag{3.5.3}$$

which also can be accounted for by the following heuristic[†] argument. The number of H^+ ions which strike the membrane per second from phase (i) is proportional to the H^+ concentration $C_H^{(i)}$ in phase (i). The number of Cl^- ions which in the same time also strike the membrane is likewise proportional to the Cl^- concentration in phase (i). The number of H^+ and Cl^- ions that strike the membrane per second *simultaneously* is – according to the law of

---

* See Chapter 1, Section 1.4.2, Eq. (1.4.10): $-\ln x = \ln(1/x)$.
[†] Heuristic argument ≡ reasoning not regarded as final and strict but provisional and plausible only, whose purpose is to discover the solution of the present problem. (Polya, G. (1957): *How to solve it*, Doubleday & Company Inc., New York.)

composite probabilities – proportional to the product $C_H^{(i)} \times C_{Cl}^{(i)}$ of the two concentrations. Hence, the part of the H^+ and Cl^- ions that *cross* the membrane per second from phase (i) to phase (o) is proportional to the same product, thus

$$\text{passage of HCl from phase (i) to phase (o)} \propto K \, C_H^{(i)} \times C_{Cl}^{(i)},$$

where K is a constant of proportionality. An analogous argument applied to the H^+ and Cl^- ions present in phase (o) gives similarly

$$\text{passage of HCl from phase (o) to phase (i)} \propto K \, C_H^{(o)} \times C_{Cl}^{(o)}.$$

At equilibrium the passage of HCl is the same in the two directions. As the constant K must be the same for the two unidirectional fluxes above, we can write the condition of equilibrium for the H^+ and Cl^- ions in the two phases as

$$C_H^{(i)} \, C_{Cl}^{(i)} = C_H^{(o)} \, C_{Cl}^{(o)},$$

in agreement with Eq. (3.5.3).

3.5.2 *Quantitative treatment of the Donnan system*

In this section we shall use equilibrium thermodynamics to calculate the relationship between the concentration of the impermeable macroion and the distribution ratio for the diffusible ions when equilibrium is established in the system. To avoid complicating the treatment unnecessarily we let the two phases contain a solution of a single salt, e.g. NaCl, that dissociates in *monovalent* cations and anions.

The thermodynamic criterion for the equilibrium in the Donnan system is that the chemical potential of water and the electrochemical potentials for their permeable ions separately have the *same value* in the two phases. The chemical potential of macroion Po^- is not considered, as there is no possibility for exchange between the two phases. For our purpose it is convenient to write the electrochemical potential $\bar{\mu}_j$ for the ionic species j in the form

$$\bar{\mu}_j^{(l)} = \mu_j^0 + p^{(l)} \bar{v}_j + \mathcal{R} T \ln a_j^{(l)} + z_j \mathcal{F} \psi^{(l)},$$

where the superscript (l) refers to the value of $\bar{\mu}_j$ in phase (l), $p^{(l)}$ and $\psi^{(l)}$ are pressure and potential in the phase, $a_j^{(l)}$ is the activity of the ion of type j in the solution, and $\bar{v}_j$ is the partial molar volume at the pressure $p_j^{(l)}$ of the ion with the charge number z_j. The above condition written out for each component

$$\left.\begin{array}{l} \overline{\mu}_{j+}^{(i)} = \overline{\mu}_{j+}^{(o)} \\ \overline{\mu}_{j-}^{(i)} = \overline{\mu}_{j-}^{(o)} \\ \mu_{w}^{(i)} = \mu_{w}^{(o)} \end{array}\right\} \tag{3.5.4}$$

leads to the following expressions for the state of equilibrium

$$p^{(i)}\overline{v}_{H+} + \mathscr{R}T \ln a_{j+}^{(i)} + \mathscr{F}\psi^{(i)} = p^{(o)}\overline{v}_{j+} + \mathscr{R}T \ln a_{j+}^{(o)} + \mathscr{F}\psi^{(o)}$$

$$p^{(i)}\overline{v}_{j-} + \mathscr{R}T \ln a_{j-}^{(i)} - \mathscr{F}\psi^{(i)} = p^{(o)}\overline{v}_{j-} + \mathscr{R}T \ln a_{j-}^{(o)} - \mathscr{F}\psi^{(o)}$$

$$p^{(i)}\overline{v}_{w} + \mathscr{R}T \ln a_{w}^{(i)} = p^{(o)}\overline{v}_{w} + \mathscr{R}T \ln a_{w}^{(o)},$$

which we rearrange as

$$\left(p^{(i)} - p^{(o)}\right)\overline{v}_{j+} + \mathscr{F}\left(\psi^{(i)} - \psi^{(o)}\right) = \mathscr{R}T\left(\ln a_{j+}^{(o)} - \ln a_{j+}^{(i)}\right)$$

$$\left(p^{(i)} - p^{(o)}\right)\overline{v}_{j-} - \mathscr{F}\left(\psi^{(i)} - \psi^{(o)}\right) = \mathscr{R}T\left(\ln a_{j-}^{(o)} - \ln a_{j-}^{(i)}\right)$$

$$\left(p^{(i)} - p^{(o)}\right)\overline{v}_{w} = \mathscr{R}T\left(\ln a_{w}^{(o)} - \ln a_{w}^{(i)}\right).$$

Now

$$\Pi = p^{(i)} - p^{(o)} \tag{3.5.5}$$

is the *pressure difference* (osmotic pressure) between the phases (i) and (o) at equilibrium, and

$$V_{D} = \psi^{(i)} - \psi^{(o)} \tag{3.5.6}$$

is the potential difference (the *Donnan potential*) between the phases. Inserting these values in the above set of equations gives

$$\Pi \, \overline{v}_{j+} + \mathscr{F}V_{D} = \mathscr{R}T \ln \left(\frac{a_{j+}^{(o)}}{a_{j+}^{(i)}}\right) \tag{3.5.7}$$

$$\Pi \, \overline{v}_{j-} - \mathscr{F}V_{D} = \mathscr{R}T \ln \left(\frac{a_{j-}^{(o)}}{a_{j-}^{(i)}}\right) \tag{3.5.8}$$

$$\Pi \, \overline{v}_{w} = \mathscr{R}T \ln \left(\frac{a_{w}^{(o)}}{a_{w}^{(i)}}\right). \tag{3.5.9}$$

Adding Eq. (3.5.7) and Eq. (3.5.8) yields

$$\frac{a_{j+}^{(0)}}{a_{j+}^{(i)}} \cdot \frac{a_{j-}^{(0)}}{a_{j-}^{(i)}} = \mathrm{Exp}\left\{\frac{\Pi(\bar{v}_{j+} + \bar{v}_{j-})}{\mathcal{R}T}\right\}.$$

In all biological situations of practical importance the osmotic pressure difference Π is so small that $\Pi(\bar{v}_{j+} + \bar{v}_{j-}) \ll \mathcal{R}T^*$. The right-hand side in the expression above then takes the value 1.0, and the expression is reduced to

$$\frac{a_{j+}^{(0)}}{a_{j+}^{(i)}} = \frac{a_{j-}^{(i)}}{a_{j-}^{(0)}}, \qquad (3.5.10)$$

which is called the *Gibbs–Donnan distribution condition*.

We ignore for the same reason the pressure contributions on the left-hand side of Eq. (3.5.7) and Eq. (3.5.8). This gives

$$V_{\mathrm{D}} = \frac{\mathcal{R}T}{\mathcal{F}} \ln\left(\frac{a_{j+}^{(0)}}{a_{j+}^{(i)}}\right) = -\frac{\mathcal{R}T}{\mathcal{F}} \ln\left(\frac{a_{j-}^{(0)}}{a_{j-}^{(i)}}\right), \qquad (3.5.11)$$

which – according to the thermodynamical procedure used – involves the ratio between the individual ionic activities $a_j^{(i)}/a_j^{(0)}$ and *not* the ratio between concentrations of the chemical substances as used in Eq. (3.5.1). However, this matter should be of little concern in this book. In the first place we are always dealing with fairly dilute solutions. Secondly, we are always considering the ratio between ions that, by and large, are at the *same ionic strength* in both phases. Thus, no serious error will be committed by putting

$$\frac{a_j^{(i)}}{a_j^{(0)}} \approx \frac{C_j^{(i)}}{C_j^{(0)}},$$

and, accordingly, we shall continue using concentration, that is, putting everywhere the activity coefficients equal to $1.0^\dagger$. Thus, we shall now write the expression for the *Donnan potential* as

$$V_{\mathrm{D}} = \frac{\mathcal{R}T}{\mathcal{F}} \ln\left(\frac{C_{j+}^{(0)}}{C_{j+}^{(i)}}\right) = -\frac{\mathcal{R}T}{\mathcal{F}} \ln\left(\frac{C_{j-}^{(0)}}{C_{j-}^{(i)}}\right), \qquad (3.5.12)$$

---

* Typical value for a monovalent ion at room temperature is $\bar{v}_j = 20$ cm^3 mol^{-1}. This gives for $\Pi = 1$ atm $= 10^5$ Pa: $\Pi\bar{v}_j = 10^5 \times 20 \times 10^{-6} = 2$ J, to be compared with $\mathcal{R}T = 8.314 \times 298 = 2438$ J.

$\dagger$ The scrupulous reader may readily make the desired corrections by replacing the concentration C_1 with the activity $a_j = f_j C_j$ according to requirement.

and the *Gibbs–Donnan distribution ratio* r_D as

$$r_D = \frac{C_{j+}^{(o)}}{C_{j+}^{(i)}} = \frac{C_{j-}^{(i)}}{C_{j-}^{(o)}}. \qquad (3.5.13)$$

These relations are of limited value even if the concentration in phase (o) is known, since the value of r_D must also depend on the concentration of the non-permeant macroion in phase (i) – the Donnan phase. However, if the concentration of the macroion is known the relationship between all the quantities involved in the Donnan equilibrium can be calculated. Let the macroion be a polyelectrolyte of concentration C_{Po} with a charge number z that may be either a positive or negative number. For both phases we impose the principle of macroscopic electroneutrality. This implies for phase (o) that*

$$C_+^{(o)} = C_-^{(o)} = C_o \qquad (3.5.14)$$

while for the Donnan phase (i) we have

$$C_+^{(i)} - C_-^{(i)} + z\,C_{Po} = 0. \qquad (3.5.15)$$

Using Eq. (3.5.12) the relationship between the concentrations of the diffusible anions and cations in the phases (i) and (o) can be written as

$$C_+^{(i)} = C_o \exp\left\{-\frac{\mathscr{F} V_D}{\mathscr{R} T}\right\} = C_o\, e^v, \qquad (3.5.16)$$

and

$$C_-^{(i)} = C_o \exp\left\{\frac{\mathscr{F} V_D}{\mathscr{R} T}\right\} = C_o\, e^v, \qquad (3.5.17)$$

where we for reasons of simplifying the typography have expressed the Donnan potential in units of $\mathscr{R} T / \mathscr{F}^\dagger$ as

$$v = \frac{V_D}{\mathscr{R} T / \mathscr{F}}. \qquad (3.5.18)$$

Insertion of Eq. (3.5.16) and Eq. (3.5.17) in Eq. (3.5.14) gives

$$C_o\,[e^v - e^{-v}] = zC_{Po}.$$

The term inside the brackets equals $2 \sinh v^\ddagger$. We have then

$$\sinh v = \sinh\left(\frac{\mathscr{F} V_D}{\mathscr{R} T}\right) = \frac{z C_{Po}}{2C_o}, \qquad (3.5.19)$$

---

* From now on we write C_+ and C_- instead of C_{j+} and C_{j-}.
† See Section 3.3.2.1, Eq. (3.3.39).
‡ See Appendix I.

3. *Membrane potentials*

and

$$V_D = \frac{\mathscr{R}T}{\mathscr{F}}v = \frac{\mathscr{R}T}{\mathscr{F}}\text{arsinh}\left(\frac{zC_{Po}}{2C_o}\right), \qquad (3.5.20)$$

which are convenient* relations to determine one of the three quantities V, zC_{Po} and C_o provided two of them are known. It appears[†] from Eq. (3.5.20) that the polarity of the Donnan potential $V_D = \psi^{(i)} - \psi^{(o)}$ is determined by the sign of the charge number z of the macroion. If the macroion ionizes positively the Donnan phase will be at a positive potential with respect to the potential in the phase (o). Similarly a negative value of z corresponds to a lower potential than that of phase (o).

We will now consider two situations that represent rather extreme cases.

3.5.2.1 Low polyelectrolyte concentration

If $|z|C_{Po} \ll C_o$ in Eq. (3.5.19) then $\sinh v \ll 1$. For $y \ll 1$ we have[‡]

$$\sinh y = \tfrac{1}{2}[e^y - e^{-y}] \approx \tfrac{1}{2}[1 + y - (1 - y)] = y.$$

Hence

$$\sinh v \approx v = \frac{zC_{Po}}{2C_o} \quad \text{or} \quad V = \frac{\mathscr{R}T}{\mathscr{F}}\frac{zC_{Po}}{2C_o}. \qquad (3.5.21)$$

We use the same approximations $e^{-y} \approx 1 - y$ and $e^y \approx 1 + y$ on Eq. (3.5.16) and Eq. (3.5.17). This yields

$$\begin{aligned}
C_+^{(i)} = C_-^{(i)} - zC_{Po} &= C_o e^v - zC_{Po} \\
&= C_o[1 + v] - zC_{Po} = C_o + C_o\frac{zC_{Po}}{2C_o} - zC_{Po} \\
&= C_o - \tfrac{1}{2}zC_{Po}
\end{aligned} \qquad (3.5.22)$$

and correspondingly

$$\begin{aligned}
C_-^{(i)} = C_+^{(i)} + zC_{Po} &= C_o e^{-v} + zC_{Po} \\
&= C_o[1 - v] + zC_{Po} = C_o - C_o\frac{zC_{Po}}{2C_o} + zC_{Po} \\
&= C_o + \tfrac{1}{2}zC_{Po},
\end{aligned} \qquad (3.5.23)$$

---

* We have decided to make use of the hyperbolic functions and not, as is customary, second-degree polynomials, since $\sinh x$ and its inverse $\text{arsinh}\, x$ are programmed into any pocket calculator.

† Because $\sinh(-x) = -\sinh x$. Besides, about the properties of the hyperbolic functions see Appendix O.

‡ See Chapter 1, Section 1.6.

which means that half of the charge of the non-permeant macroions is compensated by an excess of ions – counter-ions – having the sign opposite to that of the charge number z, and the other half by a deficit of ions – co-ions – having the same sign as z. The osmotic pressure difference between the phases is

$$\Pi = \Re T\left[C_+^{(i)} + C_-^{(i)} + C_{Po} - 2C_o\right] = \Re T C_{Po}, \qquad (3.5.24)$$

using the van't Hoff equation, Eq. (2.8.13). It appears that the osmotic pressure in this case is determined exclusively by the concentration of the non-permeable macroions.

3.5.2.2 High polyelectrolyte concentration

In this case we have $|z|C_{Po} \gg C_o$. It is in general valid that

$$\text{arsinh}\, y = \begin{cases} \ln\left(y + \sqrt{y^2 + 1}\right) \approx \ln 2y & \text{for } y \gg 1 \\ -\ln\left(\sqrt{y^2 + 1} - y\right) \approx -\ln|2y| & \text{for } y \ll -1. \end{cases}$$

Equation (3.5.26) can then be written as

$$V = \pm \frac{\Re T}{\mathscr{F}} \ln\left(\frac{|z|C_{Po}}{C_o}\right), \qquad (3.5.25)$$

where the sign of V corresponds to that of z. If $|V| \gg \Re T/\mathscr{F}$, Eq. (3.5.16) and Eq. (3.5.17) taken together with the earlier condition of electroneutrality give

$$\left.\begin{array}{l} C_+^{(i)} \to 0 \\ C_-^{(i)} \to zC_{Po} \end{array}\right\} \quad \begin{array}{l} \text{for } V \gg \Re T/\mathscr{F} \\ \text{and } z \text{ positive} \end{array} \qquad (3.5.26)$$

and

$$\left.\begin{array}{l} C_+^{(i)} \to |z|C_{Po} \\ C_-^{(i)} \to 0 \end{array}\right\} \quad \begin{array}{l} \text{for } |V| \gg \Re T/\mathscr{F} \\ \text{and } z \text{ negative.} \end{array} \qquad (3.5.27)$$

In other words: the charge on the macroion is now completely compensated by diffusible ions of *opposite* charge sign, and the system is apparently completely impermeable to the ions of the *same* sign – co-ions – as that of the macroion. It is this situation that is at work in the ion-exchange membranes. The osmotic pressure difference between the phases is now

$$\Pi = \Re T[|z| + 1]C_{Po}, \qquad (3.5.28)$$

since the condition $|z|C_{Po} \gg C_o$ implies that the co-ions contribute only insignificantly to the magnitude of the osmotic pressure.

In general the osmotic pressure is given by the equation

$$\Pi = \Re T\left[C_+^{(i)} + C_-^{(i)} + C_{Po} - 2C_o\right]. \qquad (3.5.29)$$

Insertion of Eq. (3.5.16) and Eq. (3.5.17) gives

$$\Pi = \Re T[C_o e^{-v} + C_o e^{v} - C_{Po} - 2C_o] = \Re T[C_o(e^{v} + e^{-v} - 2) + C_{Po}].$$

However, $2\cosh y = e^{y} + e^{-y}$. Whence

$$\Pi = \Re T[2C_o(\cosh v - 1) + C_{Po}], \qquad (3.5.30)$$

which is valid in all cases – also the two extreme situations just considered. When $|v| \ll 1$ we have $\cosh v \to 1$, and Eq. (3.5.30) tends to Eq. (3.5.24). Similarly, when $|v| \gg 1$, Eq. (3.5.30) tends to Eq. (3.5.28) as $2C_o(\cosh v - 1) \approx 2C_o(\frac{1}{2}e^{v} - 1) \approx C_o e^{v} \approx zC_{Po}$.

3.5.3 Concentration and potential profiles

In the previous section we applied the standard procedures of thermodynamics to obtain equations for the equilibrium ratios of the diffusible anion and cation in the two phases (i) and (o), together with the corresponding values of pressure and potential differences. By introducing the value of the (non-thermodynamic variable) concentration of the macroion C_{Po} and the (also non-thermodynamic) restriction of macroscopic neutrality in each of the two phases we could calculate the particular values belonging to the Donnan equilibrium. However, these values refer to those of the two macroscopic phases (i) and (o). But in the transition from one phase to the other, one must cross an intermediate phase where the separation between the concentration of anions and cations occurs leading to a breakdown of the electroneutrality and thus to the emergence of the *free charges* that generate the potential difference between the phases. Hence, the potential profile in this region can no longer be obtained as a solution of the Nernst–Planck equations only, but requires in addition inclusion of the Poisson equation. Presenting a solution of this problem* may – apart from the inherent didactic merits – be of value by throwing light on which factors determine the width of the transition zone.

---

* This was first carried out by Bartlett & Kromhout (1952). The work is not easy reading and the solutions are given in terms of elliptic functions, which are hardly at the finger tips of the majority of biologists. For that reason the treatment given here (Sten–Knudsen, O.: Passive Transport Processes. In *Membrane Transport in Biology, Volume I, Concepts and Models*, Chapter 2, p. 73, Springer Verlag, Berlin, Heidelberg, New York, 1978) uses only elementary mathematical functions. The price is a longer text.

The membrane – which we assume to be infinitely thin – is placed as a plane at $x = 0$. The phase (i) – containing the Donnan phase – is confined to the region $x < 0$, and phase (o) – containing the pure electrolyte – is the other half-space for $x > 0$. We use the following symbols.

x The distance from the membrane with the positive direction from Donnan phase (i) to outer phase (o).

$\psi(x)$ The potential profile; $\psi(x)$ assumes the constant values $\psi^{(o)}$ for $x \to +\infty$ and $\psi^{(i)}$ for $x \to -\infty$. However, in practice this variation of potential will only occur within the range $-x_0 \leq x \leq x_0$, so we can put $\psi = \psi^{(o)}$ for $x > x_0$ and $\psi = \psi^{(i)}$ for $x < -x_0$. Arbitrarily we put $\psi^{(o)} = 0$. The potential difference $\psi^{(i)} - \psi^{(o)} = \psi^{(i)}$ between the two macroscopic phases is the Donnan potential, V_D.

$\Psi(x)$ $\psi/(kT/e) = \psi/(\mathcal{R}T/\mathcal{F})$ is the normalized potential in units of $\mathcal{R}T/\mathcal{F}$.

v The Donnan potential V_D in units of $\mathcal{R}T/\mathcal{F}$.

$M(x)$ The concentration of the impermeable macroion with charge number z. For $x < -x_0$, $M(x)$ is constant and equal to M_v.

$C_+(x)$ The concentration (mol m^{-3}) of the permeable monovalent cation in the position x. For $x > x_0$ $C_+(x)$ takes the constant value $C_+^{(0)} = C_0$.

$C_-(x)$ The concentration (mol m^{-3}) of the permeable monovalent anion in the position x. For $x > x_0$ $C_-(x)$ takes the constant value $C_-^{(0)} = C_0$.

When the Donnan equilibrium is established the fluxes J_+, J_- and J_M are zero everywhere in the system. In the transition zone $-x_0 < x < x_0$, however, their counterparts $\partial C_+/\partial x$, $\partial C/\partial x$ and $\partial M/\partial x$ are not zero as the region contains exactly the distribution of free charges $\rho(x)$ that generates the equilibrium profile $\psi(x)$ for the potential. These two quantities are interconnected by way of the Poisson equation*

$$\frac{d^2\psi}{dx^2} = -\frac{\rho(x)}{\varepsilon_0 K},$$
(3.5.31)

or by using the normalized potential $\Psi = \psi/(\mathcal{R}T/\mathcal{F})$

$$\frac{\mathcal{R}T\varepsilon_0 K}{\mathcal{F}}\frac{d^2\Psi}{dx^2} = -\rho(x),$$
(3.5.32)

where $\rho(x)$ is the charge density.

---

* See Section 3.3.2.2, Eq. (3.3.47).

In the region $x \geq 0$ the charge density is *

$$\rho(x) = \mathscr{F}[C_+(x) - C_-(x)] \quad \text{for} \quad x \geq 0, \tag{3.5.33}$$

whereas for $x \leq 0$ we have

$$\rho(x) = \mathscr{F}[C_+(x) - C_-(x) + zM(x)] \quad \text{for} \quad x \leq 0. \tag{3.5.34}$$

But in this case an extra condition of restraint is imposed on the form of the Poisson equation since the equilibrium state continues throughout the transition zone for the diffusible ions. We now begin the determination of this form.

3.5.3.1 The Poisson–Boltzmann equations

To establish the relationship between the profiles of potential $\psi(x)$ and the concentration of e.g. C_+ in the range $-\infty < x < \infty$ we take the Nernst–Planck equation in the form[†]

$$J_j e^{z_j \psi} = -D_j \frac{d}{dx} \left(C_j e^{z_j \psi} \right). \tag{3.5.35}$$

As equilibrium prevails we have $J_+ = 0$. Hence, the general course of concentration profile must follow

$$\int \frac{d}{dx} \left(C_+ e^{\psi(x)} \right) dx = A \quad \text{or} \quad C_+(x) e^{\psi(x)} = A, \tag{3.5.36}$$

where the value of the constant A can be determined from the conditions where $x \to +\infty$. We have here $\Psi(x) = 0$ when $C_+ = C_+^{(0)} = C_0$. This gives

$$A = C_0 e^0 = C_0.$$

Hence the concentration of the cation follows

$$C_+(x) = C_0 e^{-\Psi(x)}. \tag{3.5.37}$$

Similarly we have

$$C_-(x) = C_0 e^{\Psi(x)}, \tag{3.5.38}$$

for the anions. These two equations could also have been obtained either by stating that the electrochemical potential $\bar{\mu}_+$ is continuous and constant throughout the whole region $-x_0 < x < x_0$, or by expressing the equilibrium condition by using Boltzmann's law of distribution. Inserting the values for C_+ and C_- in the expression for the space charge above we have for the domain $x \geq 0$ that

$$\rho(x) = -\mathscr{F}C_0 \left[e^{\Psi(x)} - e^{-\Psi(x)} \right] = -2\mathscr{F}C_0 \sinh \Psi(x), \tag{3.5.39}$$

---

* See Section 3.3.2.2, Eq. (3.3.48).
† See Section 3.3.2.1, Eq. (3.3.41).

as $\sinh y = (e^y - e^{-y})/2$. We insert this expression for $\rho(x)$ in the Poisson equation, Eq. (3.5.32), and obtain the following equation to determine the potential $\Psi(x)$ in the range $x \geq 0$

$$\frac{\mathscr{R}T\varepsilon_0 K}{2\mathscr{F}^2 C_0} \frac{d^2\Psi}{dx^2} = \lambda^2 \frac{d^2\Psi}{dx^2} = \sinh \Psi(x) \quad \text{for} \quad x \geq 0, \qquad (3.5.40)$$

where the constant λ is given by

$$\lambda^2 = \frac{\mathscr{R}T\varepsilon_0 K}{2\mathscr{F}^2 C_0} \triangleq m^2. \qquad (3.5.41)$$

The constant λ has the dimension length and is called the *Debye length**. As we will see later it is this quantity that determines the extent of the region containing the net free charges.

Similarly, in the region $x \leq 0$ we have, for the macroion, that

$$M(x)e^{z\Psi(x)} = A.$$

To determine the constant A we note that the final value of $\psi(x)$ is the Donnan potential V_D and that of the macroions is M_v. So we have: $M(x) = M_v$ when $\Psi(x) = v$. Inserting these boundary values in the above equation gives

$$M_v\, e^{zv} = A,$$

from which it follows

$$M(x)\, e^{z\Psi(x)} = M_v\, e^{zv},$$

or

$$M(x) = M_v\, e^{z(v - \psi(x))} \quad \text{for} \quad x \leq 0. \qquad (3.5.42)$$

Thus, for $x \leq 0$ the space charge is

$$\rho(x) = -\mathscr{F}C_0\left[e^{\Psi(x)} - e^{-\Psi(x)}\right] + zM_v\mathscr{F}e^{z(v-\psi(x))}$$

$$= -2\mathscr{F}C_0\left[\sinh \Psi(x) - \frac{zM_v}{2C_0}e^{z(v-\psi(x))}\right]. \qquad (3.5.43)$$

Insertion of Eq. (3.5.32) and Eq. (3.5.41) in this expression yields the following equation for the determination of the potential profile $\Psi(x)$ for $x \leq 0$

$$\lambda^2\frac{d^2\Psi}{dx^2} = \sinh \Psi(x) - \frac{zM_v}{2C_0}e^{z(v-\Psi(x))} \quad \text{for} \quad x \leq 0. \qquad (3.5.44)$$

---

* Peter J.W. Debye (1884–1966), was a Dutch physicist. He was Professor in Zürich 1911, thereafter in Göttingen, Leipzig, and subsequently (1935) in Berlin. From 1942 he was Professor at Cornell University, USA. He contributed greatly to several branches bordering between physics and chemistry, and was awarded the Nobel Prize for chemistry in 1938.

Equations like Eq. (3.5.440) and Eq. (3.5.44) are called Poisson–Boltzmann equations, since the electric state – which is determined by Poisson's equation – conforms at the same time to a state of equilibrium that is described by Boltzmann's law of distribution.

3.5.3.2 Solving the Poisson–Boltzmann equations

We shall now find the course of the potential profile $\psi(x)$, including the transition zone containing the membrane at $x = 0$. As the regions $x \geq 0$ and $x \leq 0$ are characterized by their respective Poisson–Boltzmann equations we must find a solution for each region, after which one must tie the two solutions together to give a smooth transition at $x = 0$. This implies that both the potential profile $\Psi(x)$ and the electric field $d\Psi/dx$ have to be continuous functions at the position $x = 0$. When $\Psi(x)$ is known as a function of x in the whole space $-\infty < x < \infty$, the concentration profiles in the transition zone are calculated using Eq. (3.5.37), Eq. (3.5.38) and Eq. (3.5.42).

(i) The solution for $x \geq 0$
When one is confronted with a nonlinear differential equation such as the Poisson–Boltzmann equation

$$\lambda^2 \frac{d^2\psi}{dx^2} = \sinh \Psi(x),$$

it is often instructive first to examine how the linear version – if it exists – behaves. If we restrict the computation to values $\Psi(x)$ that are *sufficiently small* to validate the approximation $\sinh \Psi \approx \Psi$, the equation above can be written as

$$\lambda^2 \frac{d^2\psi}{dx^2} = \Psi(x). \tag{3.5.45}$$

This equation is called the *linearized* Poisson–Boltzmann equation. We rearrange the equation

$$\frac{d^2\psi}{dx^2} - \left(\frac{1}{\lambda}\right)^2 \Psi = 0,$$

and look for the solution in which $\Psi \to 0$ for $x \to \infty$. This is*

$$\Psi(x) = \Psi(0)\, e^{-x/\lambda}. \tag{3.5.46}$$

Thus, for small values of Ψ the potential decreases mono-exponentially with the distance x from the membrane, where the potential is $\Psi(0)$. The physical meaning of the Debye length λ also appears from the equation above as a

---

* See Chapter 1, Section 1.9.2.2, Eq. (1.9.22).

measure for the extent of the space charges as the potential for $x = \lambda$ has fallen to $\Psi(0) \cdot e^{-1} = 0.37\Psi(0)$. Alternative names for λ are the *characteristic length* or the *length constant*, also referring to the extent of a space variable.

We now proceed to the solution of the nonlinear Poisson–Boltzmann equation. The search for a solution of a second-order differential equation such as Eq. (3.5.40) is often facilitated by transforming the equation to one of first order. We multiply Eq. (3.5.40) on both sides by $d\Psi/dx$. This gives

$$\lambda^2 \frac{d^2\Psi}{dx^2} \cdot \frac{d\Psi}{dx} = \sinh\Psi \cdot \frac{d\Psi}{dx}.$$

Making use of the identity

$$\frac{1}{2}\frac{d}{dx}\left(\frac{d\Psi}{dx}\right)^2 = \frac{d\Psi}{dx} \cdot \frac{d^2\Psi}{dx^2},$$

we can write the equation above as

$$\frac{1}{2}\lambda^2 \frac{d}{dx}\left(\frac{d\Psi}{dx}\right)^2 = \sinh\Psi \frac{d\Psi}{dx}.$$

Multiplication on both sides by the differential dx makes the conversion[*]

$$\frac{1}{2}\lambda^2 d\left(\frac{d\Psi}{dx}\right)^2 = \sinh\Psi \, d\Psi.$$

Integration of this equation yields[†]

$$\frac{1}{2}\lambda^2 \int d\left(\frac{d\Psi}{dx}\right)^2 = \int \sinh\Psi \, d\Psi$$

$$\frac{1}{2}\lambda^2 \left(\frac{d\Psi}{dx}\right)^2 = \cosh\Psi + A_1. \qquad (3.5.47)$$

The constant A_1 is determined by the boundary condition

$$\frac{d\Psi}{dx} \to 0 \quad \text{for} \quad x \to +\infty.$$

But $\Psi \to 0$ for $x \to +\infty$ (in practice for $x > x_0$), so the boundary condition can be written alternatively as

$$\frac{d\Psi}{dx} \to 0 \quad \text{for} \quad \Psi \to 0.$$

---

[*] See Chapter 1, Section 1.2.2, Eq. (1.2.13).
[†] Since $\int \sinh y \, dy = \frac{1}{2}\int(e^y - e^{-y})\,dy = \frac{1}{2}(e^y + e^{-y}) = \cosh y.$

Inserting this into Eq. (3.5.47) gives $A_1 = -1$. Hence, the first integral of Eq. (3.3.47) takes the form*

$$\frac{1}{2}\lambda^2 \left(\frac{d\Psi}{dx}\right)^2 = \cosh\Psi - 1 = 2\sinh^2\left(\frac{\Psi}{2}\right)$$

or

$$\lambda\frac{d\Psi}{dx} = \pm 2\sinh(\Psi/2), \qquad\qquad (3.5.48)$$

where the decision as to which sign to use depends on the polarity of $\psi^{(i)} - \psi^{(o)} = V_D$. When V_D is positive the potential $\Psi(x)$ is a monotonic increasing function as we move from $x = +\infty$ in phase (o) to $x = -\infty$ in the Donnan phase (i); that is, $d\Psi/dx \leq 0$ in the whole range $-\infty < x < +\infty$. Thus, a positive value of V_D corresponds to the minus sign in Eq. (3.5.48). In the following we assume that V_D is positive, i.e. the charge number z is also positive. We multiply Eq. (3.5.48) by the differential dx and rearrange to give

$$\frac{1}{2\sinh(\Psi/2)}d\Psi = -\frac{1}{\lambda}dx. \qquad\qquad (3.5.49)$$

We integrate this expression from the membrane, i.e. from $x = 0$ and $\Psi(0) = \Psi_{0+}$, to the position x with the potential $\Psi(x)$. This gives[†]

$$\int_{\Psi_{0+}}^{\Psi}\frac{d\Psi}{2\sinh(\Psi/2)} = -\frac{1}{\lambda}\int_0^x dx$$

or, by changing the variable of integration from Ψ to $u = \Psi/2$

$$\int_{\Psi_{0+}/2}^{\Psi/2}\frac{d(\Psi/2)}{\sinh(\Psi/2)} = \int_{\Psi_{0+}/2}^{\Psi/2}\frac{du}{\sinh u} = -\frac{x}{\lambda}.$$

The indefinite integral of $1/\sinh u$ equals $\ln\tanh(u/2)$. Thus, we have

$$\ln\tanh(\Psi/4) - \ln\tanh(\Psi_{0+}/4) = \ln\frac{\tanh(\Psi/4)}{\tanh(\Psi_{0+}/4)} = -\frac{x}{\lambda},$$

or

$$\Psi(x) = 4\,\text{artanh}\left\{\tanh\left(\frac{\Psi_{0+}}{4}\right)\cdot e^{-x/\lambda}\right\}. \qquad\qquad (3.5.50)$$

This expression contains the integration constant – so far unknown – that is the value of the potential Ψ_{0+} on the membrane at $x = 0+$. Later this constant

---

* About the relations: $\cosh y - 1 = 2\sinh^2(y/2)$, and $\int(\sinh u)^{-1}du = \ln\tanh(u/2)$ see Appendix I.

[†] Ψ_{0+} holds a position as an integration constant, whose value must be determined by an independent procedure.

will be determined by adjusting the above solution with the solution of the differential equation for $\Psi(x)$ in the range $x \leq 0$. This we shall now find.

(ii) The solution for $x \leq 0$

The potential profile $\Psi(x)$ in the range $x \leq 0$ is governed by Eq. (3.5.44). Unfortunately, a solution of Eq. (3.5.44) in a closed form is only attainable when $z = \pm 1$, otherwise we have to fall back on numerical integration. As our primary purpose is to illustrate the general principles of solution we put $z = +1$. We have from Eq. (3.5.21)

$$\sinh v = \sinh \left(\frac{\mathscr{F} V_D}{\mathscr{R} T} \right) = \frac{M_v}{2C_o}.$$

Equation (3.5.44) contains a term that is identical to the right-hand side above. We replace this term by $\sinh v$ and multiply by $d\Psi/dx$. This gives

$$\lambda^2 \frac{d^2\Psi}{dx^2} \cdot \frac{d\Psi}{dx} = \sinh \Psi \frac{d\Psi}{dx} - \sinh v \cdot e^{v-\Psi} \cdot \frac{d\Psi}{dx},$$

or

$$\tfrac{1}{2}\lambda^2 \frac{d}{dx} \left(\frac{d\Psi}{dx} \right)^2 = \sinh \Psi \frac{d\Psi}{dx} - \sinh v \cdot e^{v-\Psi} \cdot \frac{d\Psi}{dx},$$

which, after multiplication with the differential dx, can be rewritten as

$$\tfrac{1}{2}\lambda^2 d \left(\frac{d\Psi}{dx} \right)^2 = \sinh \Psi \, d\Psi - \sinh v \cdot e^{v-\Psi} \, d\Psi.$$

Integration yields

$$\tfrac{1}{2}\lambda^2 \int d \left(\frac{d\Psi}{dx} \right)^2 = \int \sinh \Psi \, d\Psi - \sinh v \cdot e^v \int e^{-\Psi} \, d\Psi$$

$$\tfrac{1}{2}\lambda^2 \left(\frac{d\Psi}{dx} \right)^2 = \cosh \Psi + \sinh v \cdot e^{v-\Psi} + A_2.$$

To determine the constant of integration A_2 we use the condition that $d\Psi/dx \to 0$ for $x \to -\infty$, and $\Psi(-\infty)$ equals the Donnan potentials v (in practice for $x < -x_o$). Inserting these boundary values gives

$$0 = \cosh v + \sinh v e^{v-v} + A_2 \quad \text{or} \quad A_2 = -(\sinh v + \cosh v) = -e^v.$$

Hence, the Poisson–Boltzmann equation in the region $x < 0$ is

$$\tfrac{1}{2}\lambda^2 \left(\frac{d\Psi}{dx} \right)^2 = \cosh \Psi + \sinh v \cdot e^{v-\Psi} - e^v, \tag{3.5.51}$$

or, when all the exponential terms contained in the right-hand side are written out in full

$$\lambda^2 \left(\frac{d\Psi}{dx}\right)^2 = e^{\Psi} + e^{2v} e^{-\Psi} - 2e^{v}.$$

The integration of this expression requires some algebraic manipulations. Here is one procedure*: a new independent variable u is introduced

$$e^{\Psi} = u^2.$$

Taking the derivative with respect to x on both sides gives

$$e^{\Psi}\frac{d\Psi}{dx} = 2u\frac{du}{dx} \quad \text{or} \quad u^2\frac{d\Psi}{dx} = 2u\frac{du}{dx},$$

which leads to

$$\left(\frac{d\Psi}{dx}\right) = \frac{2}{u}\left(\frac{du}{dx}\right).$$

Equation (3.5.51) can then be written as

$$\frac{4\lambda^2}{u^2}\left(\frac{du}{dx}\right)^2 = u^2 + e^{2v}u^{-2} - 2e^{v},$$

or

$$4\lambda^2\left(\frac{du}{dx}\right)^2 = u^4 - 2u^2 e^{v} + e^{2v}.$$

Putting $e^{v} = u_v^2$, the right-hand side takes the form

$$4\lambda^2\left(\frac{du}{dx}\right)^2 = u^4 - 2u^2 u_v^2 + u_v^4 = \left[u^2 - u_v^2\right]^2,$$

from which we have

$$2\lambda\left(\frac{du}{dx}\right) = \pm\left[u^2 - u_v^2\right].$$

In this expression we must use the positive sign as $u^2 \leq u_v^2$ and du/dx is negative because $d\Psi/dx < 0$ for $x \leq 0$. Thus, we have[†]

$$\int\frac{du}{u_v^2 - u^2} = -\frac{1}{2\lambda}\int dx$$

$$\frac{1}{u_v}\text{artanh}\left(\frac{u}{u_v}\right) = -\frac{x}{2\lambda} + B, \tag{3.5.52}$$

---

* The idea is to resolve the problem into an integration of simple power functions.
† About $\int (a^2 - y^2)^{-1}\, dy = (1/a)\, \text{ar tanh}(y/a)$, see Appendix I.

where B is a constant of integration that we find by prescribing the condition that u takes the value u_{0-} when $x \to 0-$. This gives

$$\frac{1}{u_v}\text{artanh}\left(\frac{u_{0-}}{u_v}\right) = B$$

Equation (3.5.52) then becomes

$$\text{artanh}\left(\frac{u}{u_v}\right) = \text{artanh}\left(\frac{u_{0-}}{u_v}\right) - \frac{u_v}{2\lambda}x.$$

Since $\tanh(\text{artanh}\,y) = y$, the equation can be written

$$\frac{u}{u_v} = \tanh\left[\text{artanh}\left(\frac{u_{0-}}{u_v}\right) - \frac{u_v}{2\lambda}\cdot x\right]. \tag{3.5.53}$$

Using the relations

$$u^2 = e^{\psi}; \quad u_v^2 = e^v; \quad u_{0-} = e^{\psi_{0-}},$$

we obtain

$$\frac{u}{u_v} = e^{(\psi-v)/2}; \quad \frac{u_{0-}}{u_v} = e^{(\psi-v)/2}; \quad u_v = e^{v/2}.$$

Inserting these expressions in Eq. (3.5.53) gives

$$e^{(\psi-v)/2} = \tanh\left[\text{artanh}\left(e^{(\psi_{0-}-v)/2}\right) - \frac{e^{v/2}}{2\lambda}x\right].$$

Taking the logarithm on both sides leads to

$$\frac{\psi - v}{2} = \ln\left\{\tanh\left[\text{artanh}\left(e^{(\psi_{0-}-v)/2}\right) - \frac{e^{v/2}}{2\lambda}x\right]\right\},$$

or, alternatively

$$\psi(x) = v + 2\ln\left\{\tanh\left[\text{artanh}\left(e^{(\psi_{0-}-v)/2}\right) - \frac{e^{v/2}}{2\lambda}x\right]\right\}, \tag{3.5.54}$$

which – with the inclusion of the arbitrary constant ψ_{0-} – represents the general solution for the potential profile in the region $x \le 0$.

(iii) Binding the two solutions together

The two equations that we have obtained so far represent the values of the potential corresponding to each side of the membrane at $x = 0$. They each contain constants of integration, ψ_{0+} and ψ_{0-}, of which we shall make use in binding together the two equations.

It holds for the potential that both $\Psi(x)$ and $d\Psi/dx$ are *continuous* for $x = 0$. Hence, the boundary conditions for $\Psi(x)$ at $x = 0$ are

$$\Psi_{0+} = \Psi_{0-} = \Psi(0)$$

and

$$\left(\frac{d\Psi}{dx}\right)_{x=0-} = \left(\frac{d\Psi}{dx}\right)_{x=0+},$$

which permits the determination of $\Psi(0)$. The form of the last condition requires that the dielectric constant K is the same in both phases. Applying the above conditions to Eq. (3.5.47) and Eq. (3.5.51) gives

$$\cosh \Psi(0) - 1 = \cosh \Psi(0) + \sinh v \cdot e^{v - \Psi(0)} - e^{v},$$

or, by rearranging and writing the hyperbolic functions in their exponential forms

$$e^{v} - 1 = \tfrac{1}{2}(e^{v} - e^{-v})e^{v}e^{-\Psi(0)} = \tfrac{1}{2}(e^{2v} - 1)e^{-\Psi(0)}$$

$$e^{v} - 1 = \tfrac{1}{2}(e^{v} + 1)(e^{v} - 1)e^{-\Psi(0)}$$

$$1 = \tfrac{1}{2}(e^{v} + 1)e^{-\Psi(0)},$$

which leads to the following relation

$$\Psi(0) = \ln \frac{e^{v} + 1}{2}, \tag{3.5.55}$$

which allows the determination – using the value of the Donnan potential v – of that value $\Psi(0)$ at which both Ψ and $d\Psi/dx$ are continuous functions at the transitions for $x = 0$. Therefore, we put $\Psi(0) = \Psi_{0+} = \Psi_{0-}$ in Eq. (3.5.50) and Eq. (3.5.54). Hence, the final solution of the potential problem is

$$\Psi(x) = 4\,\text{artanh}\left\{\tanh\left(\frac{\Psi(0)}{4}\right) \cdot e^{-x/\lambda}\right\} \quad \text{for} \quad x \geq 0, \tag{3.5.56}$$

and

$$\Psi(x) = v + 2\ln\left\{\tanh\left[\text{artanh}\left(e^{(\Psi(0)-v)/2}\right) - \frac{e^{v/2}}{2\lambda}x\right]\right\} \quad \text{for} \quad x \leq 0. \tag{3.5.57}$$

Once the course of $\Psi(x)$ is known, the values of $C_+(x)$, $C_+(x)$ and $M_v(x)$ can be calculated from Eq. (3.5.37), Eq. (3.5.38) and Eq. (3.5.39).

3.5.3.3 A numerical example

The two equations above, from which we can calculate the potential profile, contain two parameters, i.e. the Debye length λ, and the Donnan potential $V_D = (\mathscr{R}T/\mathscr{F})\,v$, as parameters whose values depend on the ionic environment that makes up a given Donnan system. The third parameter $\Psi(0)$ depends only on the Donnan potential. The Debye length is given by the relation

$$\lambda = \left(\frac{\mathscr{R}T\varepsilon_0 K}{2\mathscr{F}^2 C_0} \right)^{1/2}. \qquad (3.5.58)$$

Insertion of the numeric values of the physical constants* gives

$$\lambda = 10.85 \times 10^{-10}\,\sqrt{K/|C|} \stackrel{\Delta}{=} m = 10.85 \times \sqrt{K/|C|} \stackrel{\Delta}{=} \mathring{A},$$

where $|C|$ the numerical value of the concentration (in mol m^{-3} or mmol dm^{-3}). In aqueous solutions the relative dielectric constant is $K = 80$. In this case

$$\lambda = \frac{97.1}{\sqrt{|C|}} \stackrel{\Delta}{=} \mathring{A}.$$

We consider now a Donnan system, where the concentration of the macroions is $M^+ = 235$ mm dm^{-3}, and the outside concentration is $C_0 = 100$ mm dm^{-3}. For this system the Debye length is $\lambda = 97.1/\sqrt{100} = 9.71\,\mathring{A}$. From Eq. (3.5.19) the Donnan potential – in normalized units – is

$$v = \mathrm{arsinh}\left(\frac{235}{2 \times 100} \right) = \mathrm{arsinh}\,1.175 = 1.0,$$

or $V_D = 25.7 \times 1.0 = 25.7$ mV at 25 °C. Equation (3.5.55) gives

$$\Psi(0) = \ln \frac{e^1 + 1}{2} = 0.620,$$

so that $\Psi(0) = 0.620 \times 25.7 = 15.94$ mV. Using these values for λ, v and $\Psi(0)$ we calculate $\Psi(x)$ from Eq. (3.5.56) and Eq. (3.5.57). From $\Psi(x)$ thus obtained, the values of $C_+(x)$, $C_-(x)$ and $M_+(x)$ are calculated from Eq. (3.5.37), Eq. (3.5.38) and Eq. (3.5.42). Finally, the space charge $\rho(x)$, that causes the potential profile, is calculated as $\rho = \mathscr{F}(C_+(x) - C_-(x))$ for $x \geq 0$ and as $\rho = \mathscr{F}(C_+(x) + M(x) - C_-(x))$ for $x \leq 0$. It appears from Fig. 3.11 that there is a transition zone of width 80 Å extending on both sides of $x = 0$. And it is initially outside this region that the concentrations and potential take the

---

* $\mathscr{R} = 8.314$ J mol^{-1} K^{-1}, $T = 273.16 + 25 = 298.16$, $\mathscr{F} = 96492$ C mol^{-1} and $\varepsilon_0 = 8.85 \times 10^{-12}$ F m^{-1}.

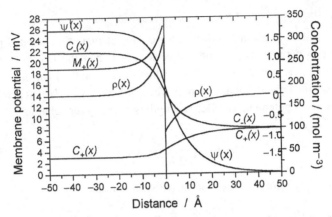

Fig. 3.11. Profiles through the transition zone of a Donnan system. Curves show: potential $\Psi(x)$; concentration $C_-(x)$ of negative ion; concentration $C_+(x)$ of positive ion; space charge density $\rho(x)$; and concentration of the positive macroion $M(x)$. Left ordinate: concentration, mol m^{-3}. External right ordinate: membrane potential (mV). Inner right ordinate: space charge density (in units of Faraday's constant per m^3). Abscissa: distance (Å).

macroscopic values that we calculate using a thermodynamic argument, and which also are those corresponding to conditions of electroneutrality. The deviations from electroneutrality for $x = 0$ may be considerable and the charge density may be correspondingly large to give rise to a local electric field of sufficient strength to balance the tendency of the ions to move down their concentration profiles by diffusion. At the position $x = 0$ the potential gradient is $d\psi/dx = -1.667 \times 10^7$ V m^{-1}. This corresponds to an electric force acting on the positive ion $X_{el} = e\,E$ or $\mathscr{F}E = 16 \times 10^{12}$ N mol^{-1}, which is the force required in this position to sustain the concentration gradients at the values $dC_+/dx = 3.49 \times 10^{10}$ and $dC_-/dx = 1.21 \times 10^{11}$ mol m^{-3} per m. To these gradients there corresponds – formulated alternatively – an "equivalent force of diffusion" of magnitude $-kT\,d\ln C/dx^*$ that balances the oppositely directed electric force X_{el}.

---

* See Eq. (2.6.63). From Eq. (3.5.37) we have: $C_+(x) = C_0\,e^{-\Psi(x)}$, or $\ln C_+(x) = \ln C_0 - \Psi(x) = \ln C_0 - (e/kT)\psi(x)$, by Eq. (3.3.39). Hence

$$X_{dif} = -kT\frac{d\ln C}{dx} = -kT\left(-\frac{e}{kT}\frac{d\psi}{dx}\right) = e\frac{d\psi}{dx}.$$

This kind of argument may be ill-chosen, if the introduction of an equal and opposite "diffusion force" leads to the concept of the ions being held fixed in the transition zone at equilibrium.

3.5.3.4 The total space charge

To calculate the total space charge in the zone of transition, the simplest point of departure is Poisson's equation

$$\frac{d^2\Psi}{dx^2} = -\frac{\rho(x)}{\varepsilon_0 K}. \tag{3.5.59}$$

The free charge dQ localized between the planes at x and $x + dx$ and reckoned per unit area of 1 m^2 is $dQ = \rho(x)\,dx$. We have then in the region $x > 0$

$$dQ = -\varepsilon_0 K \frac{d^2\psi}{dx^2}\,dx.$$

Hence it follows that

$$Q = -\varepsilon_0 K \int_0^\infty \left(\frac{d^2\psi}{dx^2}\right)dx$$

$$= -\varepsilon_0 K \left[\frac{d\psi}{dx}\right]_0^\infty = -\varepsilon_0 K \left[\left(\frac{d\psi}{dx}\right)_{x=\infty} - \left(\frac{d\psi}{dx}\right)_{x=0}\right],$$

or alternatively, as $d\psi/dx \to 0$ for $x \to \infty$

$$Q = \varepsilon_0 K \left(\frac{d\psi}{dx}\right)_{x=0}. \tag{3.5.60}$$

According to Eq. (3.5.48) we have

$$\left(\frac{d\psi}{dx}\right)_{x=0} = \frac{\mathscr{R}T}{\mathscr{F}}\left(\frac{d\Psi}{dx}\right)_{x=0} = -\frac{2\mathscr{R}T}{\lambda\mathscr{F}}\sinh\{\Psi(0)/2\}.$$

Inserting this expression together with that for λ (Eq. (3.5.41)) in Eq. (3.5.60) gives

$$Q = -2\sqrt{2\mathscr{R}T\varepsilon_0 K C_0}\,\sinh\left(\frac{\Psi(0)}{2}\right), \tag{3.5.61}$$

which is the total free charge in the region $x > 0$. A similar amount – but positive – is located in the range $x < 0$. In this connection it should be noted that the potential profile that is described by Eq. (3.5.50) for $x = 0$ is identical to the profile that would be found by removing the Donnan phase and placing instead at $x = 0$ free fixed charges of a surface density (in C m^{-2}) equal to the charge in the region $x < 0$. We have from the example

$$Q = 2\sqrt{2 \times 8.314 \times 298 \times 8.85 \times 10^{-12} \times 80 \times 200}\,\sinh(0.31)$$

$$= 2.21 \times \text{C m}^{-2} = 2.21 \times 10^{-8}\,\text{C cm}^{-2},$$

which should be compared with the other estimate of the space charge density from Section 3.4.2.1.

3.6 Diffusion potentials

The membrane potentials considered in Section 3.5 all belong to the class of *equilibrium potentials* that specifically refers to the particular condition of a *zero net flux* for the ion species involved. If the equilibrium potential refers to a package of different ions, the condition of zero net flux holds separately for each type of ion.

We shall now consider a type of membrane potential of more common occurrence – *diffusion potentials* – that differs markedly from an equilibrium potential as the underlying charge separation results from a sustained process of diffusion involving *both* anions and cations having *different* mobilities. The electric field thus generated forces the ions – in spite of their different mobilities – to move in equivalent amounts down their sloping concentration profile so that they carry a *zero net current*. Thus, a diffusion potential is – in contrast to the equilibrium potential – the reflection of ions moving towards a final equilibrium state. In the following sections we shall account in more detail for the underlying mechanisms that are of immediate interest in biology.

3.6.1 Qualitative description of the diffusion potential

As an introduction we start from an equilibrium situation already known that involves both cations and anions.

3.6.1.1 Collapse of the Donnan regime

In the Donnan system, equilibrium is sustained by the presence of the impermeable macroions in one of the phases, e.g. the inner phase (i). If the membrane – for some reason – suddenly becomes permeable also for the macroion, it will move into the outer phase (o) driven by both the electric field and fall of concentration in the direction (i) $\rightarrow$ (o), i.e. by a process of electrodiffusion. We imagine now – although rather unrealistically – that all the ions, including the macroion, have the same mobility. The electric double layer, consisting of a minute excess of negative and positive charges, will disappear in the time it takes to attain macroscopic electroneutrality by interdiffusion in the transition region less than 100 Å wide. When the electric restraints no longer exist, the system will then move purely by diffusion towards a *new state of equilibrium* with a uniform distribution in both phases. As all the ions were assumed to have the same mobility this process will take place without any additional electric manifestations.

Next we consider the process where a diffusion potential appears as a result of the diffusing ions have different mobilities.

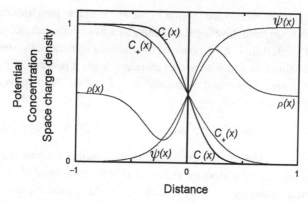

Fig. 3.12. Illustration of the origin of positive and negative space charges that underlie the appearance of a diffusion potential. Abscissa: potential. Ordinate: concentration of positive + and negative − ions, space charge ρ and potential ψ.

3.6.1.2 A binary electrolyte

The mechanism behind the establishment of such a regime is perhaps best illustrated by considering a situation similar to that in Chapter 2.5, Fig. 2.3 and Fig. 2.4, but where the diffusion cell now initially contains a solution of a binary electrolyte, e.g. HCl in the − infinitely large − chamber phase (i), whereas the other − also infinitely large − chamber phase (o) contains only water. When the two phases are put into contact the H^+ ions and the Cl^- ions will diffuse into phase (o). But as the H^+ ion is smaller that the Cl^- ion then according to the Einstein–Stokes relation* the H^+ ion diffuses faster[†] than the Cl^- ion. Therefore the advancing front line of the ions in phase (o) will be occupied by a small excess of H^+ ions owing to their attempt to diffuse ahead of the Cl^- ions, which in turn are left behind in the supply of diffusing ions from phase (i).

The lacking overlap of the two concentration profiles involves the emergence of a zone containing an excess of positive charges in phase (o) and an equivalent excess of negative charges left behind. An example of such a distribution of positive and negative space charges is shown in Fig. 3.12. The space charges will create an electric field through the transition zone, which − in the present case − is directed from phase (o) $\rightarrow$ phase (i), and will establish an electric potential profile that falls off through the transition zone in the direction phase (o) $\rightarrow$ phase (i), i.e. in the direction of the increasing HCl concentration. The electric field in the front line of diffusion acts on the H^+ ions in the direction

---

[*] See Chapter 2, Section 2.7.4.2, Eq. (2.6.58).
[†] For H^+ this is not the complete explanation for the great mobility of this ion. See, for example, K.J. Laidler (1978): *Physical chemistry with biological applications*, Benjamin/Cummings Publishing Co., Inc., California, USA.

(o) → (i) and thus retards the motion of the H^+ ions and increases that of the Cl^- ions to a corresponding degree. In the end a steady state sets in, where the H^+ and Cl^- ions diffuse from phase (i) to phase (o) with the *same speed* – that is with the same flux and with no net current – while a potential difference, a *diffusion potential*, is overlaid on the region of diffusion, the polarity of which is in the present case

$$\psi^{(o)} > \psi^{(i)} \quad \text{if} \quad B_+ > B_- \quad \text{and} \quad C^{(i)} > C^{(o)}.$$

As time passes the steepness of the concentration gradient decreases as phase (o) is successively loaded with more and more HCl until the point of equalization of concentration throughout the available space. As a consequence, the space charges gradually decrease and, accordingly, so do the field and the diffusion potential until extinction, when the net diffusion flux has ceased.

3.6.1.3 The salt bridge to eliminate the diffusion potential

It appears from the above-mentioned considerations that the larger the mobility of the cation B_+ is in proportion to that of the anion B_-, the larger will be the diffusion potential at a given difference in concentration $C^{(i)} - C^{(o)}$ between the phases. When, in the limit, we put $B_+/B_- \to \infty$ the value of the diffusion potential tends to that of the corresponding equilibrium potential. If, on the other hand, the cation and anion have the *same mobility* the two ions will also initially move with the same diffusion flux. Thus no spatial separation will occur among the positive and negative charges and electroneutrality is preserved throughout the whole system. Thus, in this case we have

$$\psi^{(i)} = \psi^{(o)} \quad \text{when} \quad B_+ = B_- \quad \text{irrespective of} \quad C^{(i)} = \begin{cases} > C^{(o)}, \\ = C^{(o)}, \\ < C^{(o)}. \end{cases}$$

The mobilities of the K^+ and Cl^- ions have almost the same values. Thus, the above statement would apply with great approximation to the situation where the diffusion cell had been filled with a KCl solution of any concentration instead of HCl. This would also hold if phase (o) initially contains another electrolyte solution, e.g. NaBr, having different mobilities for the cation and anion, provided that the concentration $C^{(o)}$ of NaBr is much less than the concentration $C^{(i)}$ of KCl in phase (i). During the interdiffusion of KCl in the direction (i) → (o) and of NaBr in the direction (o) → (i) both electrolytes attempt to establish the diffusion potentials belonging to their fluxes. The competition will enter into a resulting diffusion potential whose magnitude is dominated by the diffusion potential that is associated with the electrolyte that diffuses under the influence of the greatest concentration gradient. In the present case where the C_{KCl} in phase (i) is much larger than C_{NaBr} in phase (o) the diffusion potential across the transition zone will be dominated by the diffusion potential for KCl. This will amount at most to 1–2 mV because of the almost identical mobilities for K^+ and Cl^-. This particular property of a solution of KCl is exploited with advantage in all the situations where one wishes to establish an electrically conducting contact between two different electrolyte solutions without encountering an undesirable potential difference in consequence of the two electrolytes' interdiffusion. Consider as an illustration two solutions of 0.1 M HCl and 0.1 M NaCl. When put into direct contact a diffusion potential of about 31 mV will be produced. On the other hand, if the electric contact between the two solutions is brought about by a tube – *a salt bridge* – filled with a solution of saturated KCl (4.2 M) there will exist a fairly small diffusion potential between the HCl solution and the strong KCl solution (perhaps 4 mV) and an even smaller diffusion potential between the NaCl and KCl solutions (perhaps 1–1.5 mV). Thus, introducing the salt bridge has the effect of reducing – or eliminating – the otherwise existing diffusion potential by at least a factor of 10*.

3.6.2 Calculation of the diffusion potential for a binary monovalent electrolyte

The calculation of the diffusion potential under quite general conditions of diffusion for an arbitrary mixture of electrolytes containing several types of anions and cations having more than one valency presents almost insurmountable mathematical difficulties. On the other hand, it is a fairly tractable matter to calculate diffusion potentials corresponding to simple regimes of diffusion, e.g. a single binary electrolyte such as NaCl. In what follows we shall give a few examples of such regimes of diffusion that are also of relevance for certain estimations in biological systems. We consider the diffusion of a binary, monovalent electrolyte, e.g. NaCl, where the two phases are separated by "porous plug" that allows free passage for the ions by diffusion but acts to prevent a mechanical mixing between the two phases. The concentrations of NaCl in the chambers (i) and (o) that surround the plug are $C^{(i)}$ and $C^{(o)}$. After a short interval of contact a quantitative diffusion of Na^+ and Cl^- proceeds in the direction of the falling gradient. The fluxes in the porous plug for the two types of ions j_+ and j_- having charge numbers $z_+ = 1$ and $z_- = -1$ follow separately the

---

* A more detailed analysis of the effect of the salt bridge is given in Section 3.6.3.3.

respective Nernst–Planck equation, e.g. Eq. (3.3.35) and Eq. (3.3.43)

$$J_+ = -u_+ \frac{\mathcal{R}T}{\mathcal{F}} \frac{\mathrm{d}C_+}{\mathrm{d}x} - u_+ C_+ \frac{\mathrm{d}\psi}{\mathrm{d}x}, \tag{3.6.1}$$

and

$$J_- = -|u_-| \frac{\mathcal{R}T}{\mathcal{F}} \frac{\mathrm{d}C_-}{\mathrm{d}x} + |u_-| C_- \frac{\mathrm{d}\psi}{\mathrm{d}x}, \tag{3.6.2}$$

where the values refer to those inside and surrounding the plug*. When the diffusion has become stationary an electric field $E = -\mathrm{d}\psi/\mathrm{d}x$ is built up in the plug of such a magnitude that the fluxes of cations and anions are exactly equal or – in other words – the net ion current is zero. Hence

$$J_+ = J_-. \tag{3.6.3}$$

In addition, we assume that the condition of electroneutrality[†] prevails all over the system, that is

$$C_+(x) = C_-(x) = C(x), \tag{3.6.4}$$

which implies that the concentrations $C_+(x)$ and $C_-(x)$ in the two Nernst–Planck equations above can be replaced by the same concentration $C(x)$. Equation (3.6.3) then reads

$$-u_+ \frac{\mathcal{R}T}{\mathcal{F}} \frac{\mathrm{d}C}{\mathrm{d}x} - u_+ C \frac{\mathrm{d}\psi}{\mathrm{d}x} = -|u_-| \frac{\mathcal{R}T}{\mathcal{F}} \frac{\mathrm{d}C}{\mathrm{d}x} + |u_-| C \frac{\mathrm{d}\psi}{\mathrm{d}x},$$

or by rearrangement as

$$\mathcal{F} C \frac{\mathrm{d}\psi}{\mathrm{d}x} [u_+ + |u_-|] = -\mathcal{R}T \frac{\mathrm{d}C}{\mathrm{d}x} [u_+ - |u_-|],$$

which leads to

$$\frac{\mathrm{d}\psi}{\mathrm{d}x} = -\frac{\mathcal{R}T}{\mathcal{F}} \frac{u_+ - |u_-|}{u_+ + |u_-|} \frac{1}{C} \frac{\mathrm{d}C}{\mathrm{d}x}.$$

We multiply this equation on both sides by the differential $\mathrm{d}x$ to give

$$\frac{\mathrm{d}\psi}{\mathrm{d}x} \mathrm{d}x = \mathrm{d}\psi = -\frac{\mathcal{R}T}{\mathcal{F}} \frac{u_+ - |u_-|}{u_+ + |u_-|} \frac{\mathrm{d}C}{C}.$$

Apart from the factor $(u_+ - |u_-|)/(u_+ + |u_-|)$ this equation is identical to Eq. (3.4.12) in Section 3.4.2.2, and the integration follows the same procedure

$$\int_{\psi^{(\mathrm{i})}}^{\psi^{(\mathrm{o})}} \mathrm{d}\psi = -\frac{\mathcal{R}T}{\mathcal{F}} \frac{u_+ - |u_-|}{u_+ + |u_-|} \int_{C^{(\mathrm{i})}}^{C^{(\mathrm{o})}} \frac{\mathrm{d}C}{C},$$

---

* We write again $u_- = -|u_-|$ to obviate any misunderstandings as to sign.
[†] See Section 3.3.2.3, Eq. (3.3.50).

or

$$[\psi]_{\psi^{(i)}}^{\psi^{(o)}} = -\frac{\mathcal{R}T}{\mathcal{F}}\frac{u_+ - |u_-|}{u_+ + |u_-|}[\ln C]_{C^{(i)}}^{C^{(o)}},$$

from which follows the expression for the diffusion potential

$$\psi^{(o)} - \psi^{(i)} = V^{(\text{dif})} = \frac{\mathcal{R}T}{\mathcal{F}}\frac{u_+ - u_-}{u_+ + u_-}\ln\left(\frac{C^{(i)}}{C^{(o)}}\right). \qquad (3.6.5)$$

This equation was first derived* by W. Nernst in 1889. As the electric mobilities for cations and anions are in proportion both to their mechanical mobilities B_+ and B_- and to their diffusion coefficients D_+ and D_- the expression above can also be written in terms of these quantities, that is

$$V^{(\text{dif})} = \frac{\mathcal{R}T}{\mathcal{F}}\frac{B_+ - B_-}{B_+ + B_-}\ln\left(\frac{C^{(i)}}{C^{(o)}}\right) = \frac{\mathcal{R}T}{\mathcal{F}}\frac{D_+ - D_-}{D_+ + D_-}\ln\left(\frac{C^{(i)}}{C^{(o)}}\right). \qquad (3.6.6)$$

We shall now examine how large a diffusion potential one should expect from three binary electrolyte solutions (HCl, NaCl and KCl), where the concentration ratio is 10:1.

At infinite dilution we have the following values of electric mobilities: for H^+, N^+, K^+ and Cl^- ions in ($m^2\ s^{-1}\ V^{-1}$):

$$u_H = 36 \times 10^{-8}$$
$$u_{Na} = 5.17 \times 10^{-8}$$
$$u_K = 7.58 \times 10^{-8}$$
$$u_{Cl} = 7.88 \times 10^{-8}.$$

If $C^{(i)}/C^{(o)} = 10$, Eq. (3.6.5) predicts the following diffusion potentials[†] between each of the three binary solutions:

$$V_{\text{HCl}}^{(\text{dif})} = 25.7 \times (36.0 - 7.88)/(36.0 + 7.88) \times 2.3026 = 37.9 \text{ mV},$$

$$V_{\text{NaCl}}^{(\text{dif})} = 25.7 \times (5.17 - 7.88)/(5.17 + 7.88) \times 2.3026 = -12.3 \text{ mV},$$

$$V_{\text{KCl}}^{(\text{dif})} = 25.7 \times (7.58 - 7.88)/(7.58 + 7.88) \times 2.3026 = -1.1 \text{ mV},$$

which illustrates some of the characteristics that were described previously in Section 3.6.1.3.

3.6.3 Diffusion potential between solutions of different composition

The potential that results from the diffusion of a binary electrolyte represents the simplest case of the occurrence of a diffusion potential. In biology, however, the

---

* Nernst, W. (1889): Die electromotorische Wirksamkeit des Ionen. *Zeit. Phys. Chemie*, **4**, 129–181.
[†] $\mathcal{R}T/\mathcal{F} = 25.7$ mV at 25 °C and ln 10 = 2.3026.

origin of the majority of the diffusion potentials results from the interdiffusion between two electrolytes of *different* composition, e.g. between solutions of NaCl and KCl. To estimate the magnitude of diffusion potentials in such cases, Eq. (3.6.5) is of little use as it contains only the data for two monovalent types of ions. But, despite its shortcomings, Eq. (3.6.5) reflects the basic mechanism that causes the diffusion potential, namely the different mobilities of the ions taking part in the process of interdiffusion. This is evident in Eq. (3.6.5), where the contributions from differences in mobility and in concentration appear in two separate factors. The mathematics required in the calculation of a diffusion potential becomes increasingly complicated the more varied the ion composition becomes in the interdiffusing zone. The complexity is also revealed by the electric properties of the plug. Thus, there is in general a nonlinear relation between voltage V impressed across the plug and the resultant current I. Furthermore, in response to a step of voltage change ΔV the current I *lags* in reaching the new stationary value. The underlying cause is that the concentration profiles through the plug depend – contrary to the behavior of the single salt – on the magnitude and direction of the field in the membrane. When a voltage change ΔV is applied across the plug the various ionic species have to adjust their positions in the plug. But this change has to take place by electrodiffusion, and this requires a certain time known as *the redistribution time*, which among other things depends on the thickness of the transition zone. A given set of concentration profiles $C_j(x)$ in the plug represents a total resistance of magnitude[*]

$$R = \int_0^h \left(\sum z_j \mathscr{F} u_j C_j(x) \right)^{-1} dx. \tag{3.6.7}$$

An interchange of various ionic species in the plug with which a preponderance of ions of, for example, low mobility is moved to that of a high mobility will reduce the membrane resistance R and result in a change in the membrane current from its initial value to a new stationary value.

3.6.3.1 The Planck regime

We shall now summarize the essential results of the work of Planck on electrodiffusion and diffusion potentials[†], which has served since 1890 as the background for almost every following theoretical work on the subject. Planck considered the following situation: a homogeneous membrane (a "porous plug")

---

[*] See Section 3.3.1, Eq. (3.3.26).
[†] Planck, M. (1890a): Über die Erregung von Elektricität und Wärme in Elektrolyten. *Ann. Physik u. Chem., Neue Folge*, **39**, 161. Planck, M. (1890b): Über die Potentialdifferenz zwischen zwei verdünnten Lösungen binären Elektrolyten. *Ann. Physik u. Chem., Neue Folge*, **40**, 561.

of thickness h separates two electrolyte solutions (i) and (o) with the concentrations $C_j^{(i)}$ and $C_j^{(o)}$. Both phases are well stirred, making the concentrations uniformly distributed in each of the phases. All ions are assumed to be monovalent with charge numbers $z_+ = 1$ and $z_- = -1$. As the concentrations are kept constant in each of the phases the regime of interdiffusion in the plug will attain a stationary state, where the flux for each ion is given by the Nernst–Planck equation, that is for the cations

$$J_{j_+} = -u_{j_+}\frac{\mathcal{R}T}{\mathcal{F}}\frac{dC_{j_+}}{dx} - u_{j_+}C_{j_+}\frac{d\psi}{dx}, \tag{3.6.8}$$

and for the anions

$$J_{j_-} = -u_{j_-}\frac{\mathcal{R}T}{\mathcal{F}}\frac{dC_{j_-}}{dx} + u_{j_-}C_{j_-}\frac{d\psi}{dx}, \tag{3.6.9}$$

where neither the potential profile $\psi(x)$ nor the concentration profiles $C_j(x)$ are by necessity linear through the membrane. If the system is left alone, that is, no current is forced through the membrane by means of external electrodes, the stationary diffusion regime is characterized by a *zero total ion current* through the membrane. The potential difference across the membrane (*Planck's diffusion potential*) corresponding to this condition was then calculated by Planck by integration of the flux equations assuming the condition of electroneutrality to hold. We reproduce Planck's result in Section 3.6.3.1 (iii), but without going into detail as to the actual derivation.

(i) Planck's general relations
Planck derived two fundamental relations, which characterize his stationary diffusion regime. To facilitate the typography we replace in Eq. (3.6.8) the potential $\psi(x)$ with the normalized potential $\Psi = \psi/(\mathcal{R}T/\mathcal{F})$ and divide both sides by $\mathcal{R}Tu_{j_+}/\mathcal{F}$. This gives

$$-\frac{\mathcal{F}}{\mathcal{R}Tu_{j_+}}J_{j_+} = \frac{dC_{j_+}}{dx} + C_{j_+}\frac{d\Psi}{dx}.$$

When stationarity is attained the left-hand side will be a constant A_{j_+}. Summation over all positive ions gives

$$\sum A_{j_+} = A = \frac{d}{dx}\left(\sum C_{j_+}\right) + \frac{d\Psi}{dx}\sum C_{j_+}, \tag{3.6.10}$$

where $A_{j_+} = -\mathcal{F}J_{j_+}/\mathcal{R}Tu_{j_+}$. In a corresponding manner, summation over all the negative ions gives

$$\sum B_{j_-} = B = \frac{d}{dx}\left(\sum C_{j_-}\right) - \frac{d\Psi}{dx}\sum C_{j_-}, \tag{3.6.11}$$

where $B_{j_-} = -\mathscr{F}J_{j_-}/\mathscr{R}T|u_{j_-}|$. If Eq. (3.6.10) and Eq. (3.6.11) are added, we obtain

$$A + B = \frac{d}{dx}\left\{ \sum C_{j_+} + \sum C_{j_-} \right\} + \frac{d\Psi}{dx}\left\{ \sum C_{j_+} - \sum C_{j_-} \right\}. \qquad (3.6.12)$$

Since

$$\sum C_{j_+} + \sum C_{j_-} = 2C(x), \qquad (3.6.13)$$

where $C(x)$ is the *total concentration* of the ions in the position x, and since

$$\sum C_j^+ - \sum C_j^- = 0, \qquad (3.6.14)$$

because of the electroneutrality condition, we then obtain *Planck's first relation*

$$\frac{dC(x)}{dx} = \frac{1}{2}(A + B) = \text{constant}, \qquad (3.6.15)$$

that is, the *total concentration $C(x)$ varies linearly through the membrane*. Thus, when $C(x) = C^{(i)}$ for $x = 0$ and $C(x) = C^{(0)}$ for $x = h$, we have

$$\frac{dC}{dx} = \frac{C^{(0)} - C^{(i)}}{h},$$

from which it follows that

$$C(x) = \left[C^{(0)} - C^{(i)} \right] \frac{x}{h} + C^{(i)}. \qquad (3.6.16)$$

However, this does not imply that the individual concentrations also vary linearly through the membrane in the range $0 \leq x \leq h$.

We subtract Eq. (3.6.11) from Eq. (3.6.12) to obtain

$$A - B = \frac{d}{dx}\left\{ \sum C_{j_+} - \sum C_{j_-} \right\} + \frac{d\Psi}{dx}\left\{ \sum C_{j_+} + \sum C_{j_-} \right\}.$$

We insert Eq. (3.6.13) into this and obtain – because of the electroneutrality condition – *Planck's second relation*

$$\frac{d\Psi}{dx} = \frac{A - B}{2\,C(x)}, \qquad (3.6.17)$$

that is, the *field $E = -(d\Psi/dx)\cdot(\mathscr{R}T/\mathscr{F})$ is inversely proportional to the total concentration $C(x)$ at the position in question*. Inserting Eq. (3.6.16) in Eq. (3.6.17) gives

$$\frac{d\Psi}{dx} = \frac{1}{2}\frac{A - B}{\left[C^{(0)} - C^{(i)} \right](x/h) + C^{(i)}}. \qquad (3.6.18)$$

Thus, the rectilinear variation of the total concentration through the membrane, Eq. (3.6.16), implies that the potential in general will vary with the logarithm* of the position x in the membrane. Here comes the exception.

If the total concentration is the same on both sides of the membrane, that is, if $C^{(i)} = C^{(0)} = C_0$, we have

$$\frac{d\Psi}{dx} = \frac{A - B}{2\,C_0}, \tag{3.6.19}$$

i.e. under these conditions the field in the membrane is *always* constant. We shall consider this situation in greater detail in Section 3.7.3.

(ii) The electrical equivalent for the Planck regime

We shall now consider the more general situation where a variable potential difference V is applied across the membrane by means of external pairs of electrodes dipping into the two solutions. The total current through the membrane is no longer necessarily zero. We write again the flux for a positive ion j_+ as

$$J_{j_+} = -u_{j_+}\frac{RT}{\mathscr{F}}\frac{dC_{j_+}}{dx} - u_{j_+}C_{j_+}\frac{d\Psi}{dx}. \tag{3.6.20}$$

The current carried by this ion is $I_{j_+} = \mathscr{F}J_{j_+}$, or

$$I_j^+ = -RT\frac{d}{dx}(u_{j_+}C_{j_+}) - \mathscr{F}(u_{j_+}C_{j_+})\frac{d\Psi}{dx}, \tag{3.6.21}$$

as u_{j_+} is assumed to be constant through the membrane. Summation of all the currents carried by the positive ions gives

$$\sum I_{j_+} = I_+ = -RT\frac{d}{dx}\left\{\sum u_{j_+}C_{j_+}\right\} - \mathscr{F}\frac{d\Psi}{dx}\left\{\sum u_{j_+}C_{j_+}\right\}. \tag{3.6.22}$$

Following Planck we put

$$\mathscr{U} = \sum u_{j_+}C_{j_+}, \tag{3.6.23}$$

which, corresponding to the above equation, gives

$$I_+ = -RT\frac{d\mathscr{U}}{dx} - \mathscr{F}\mathscr{U}\frac{d\Psi}{dx}. \tag{3.6.24}$$

The flux for a negative ion is

$$J_j^- = -|u_j^-|\frac{RT}{\mathscr{F}}\frac{dC_j^-}{dx} + |u_j^-|C_j^-\frac{d\Psi}{dx}, \tag{3.6.25}$$

---

* Since $\int dx/(a + x) = \ln(a + x)$.

and the current carried by the ion is $I_{j_-} = -\mathcal{F}J_{j_-}$. The expression for the total current carried by the negative ions can therefore be written as

$$I_- = \mathcal{R}T\frac{d\mathcal{W}}{dx} - \mathcal{F}\mathcal{W}\frac{d\psi}{dx}, \tag{3.6.26}$$

where

$$\mathcal{W} = \sum |u_{j_-} \, C_{j_-}|. \tag{3.6.27}$$

The total current is

$$I = I_+ + I_-.$$

If we insert Eq. (3.6.24) and Eq. (3.6.26) in this equation we obtain the following expression for the total current through the membrane as

$$I = -\mathcal{R}T\frac{d(\mathcal{U} - \mathcal{W})}{dx} - \mathcal{F}(\mathcal{U} + \mathcal{W})\frac{d\psi}{dx}. \tag{3.6.28}$$

For $I = 0$ the field in the membrane has taken just that particular value that forces the ions in the membrane to move with such mutual velocities that *no net charge* flows through the membrane. Putting $I = 0$ in the above equation we get

$$\left(\frac{d\psi}{dx}\right)_{I=0} = -\frac{\mathcal{R}T}{\mathcal{F}}\frac{1}{\mathcal{U} + \mathcal{W}} \cdot \frac{d(\mathcal{U} - \mathcal{W})}{dx}, \tag{3.6.29}$$

which once again is Planck's expression for the stationary local field in the membranes based on the space charges resulting from the diffusion of ions through the membrane.

Equation (3.6.28) allows the construction of an electric equivalent diagram for Planck's regime to be set up. We rearrange the equation as

$$\frac{1}{\mathcal{F}(\mathcal{U} + \mathcal{W})}I = -\frac{\mathcal{R}T}{\mathcal{F}}\frac{1}{\mathcal{U} + \mathcal{W}}\frac{d(\mathcal{U} - \mathcal{W})}{dx} - \frac{d\psi}{dx}.$$

This equation is multiplied by the differential dx and integrated through the membrane from $x = 0$ to $x = h$. If a stationary state prevails, $dI/dx = 0$. We have then

$$I\int_0^h \frac{dx}{\mathcal{F}(\mathcal{U} + \mathcal{W})} = -\frac{\mathcal{R}T}{\mathcal{F}}\int_0^h \frac{d(\mathcal{U} - \mathcal{W})}{\mathcal{U} + \mathcal{W}} - \int_0^h d\psi, \tag{3.6.30}$$

where

$$-\int_0^h d\psi = \psi(0) - \psi(h) = V$$

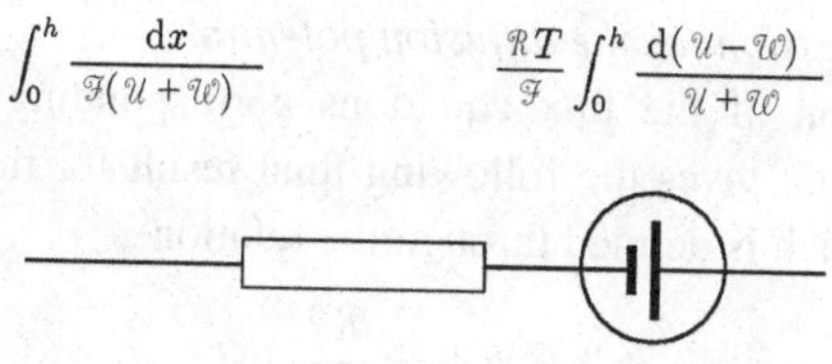

$$\int_0^h \frac{dx}{\mathcal{F}(\mathcal{U} + \mathcal{W})} \qquad \frac{\mathcal{R}T}{\mathcal{F}} \int_0^h \frac{d(\mathcal{U} - \mathcal{W})}{\mathcal{U} + \mathcal{W}}$$

Fig. 3.13. Elements of the electric equivalent diagram of Planck's diffusion regime. An external potential difference V acting across the layer causes a current I to flow. At zero current V equals the diffusion potential $V^{(\text{dif})}$.

is the potential difference across the membrane with a current flow I. Figure 3.13 shows the equivalent diagram corresponding to Eq. (3.6.30).

Switching off the current – and thereby putting the system on its own – we have for the immediately following time

$$V_{t=0+} = \frac{\mathcal{R}T}{\mathcal{F}} \int_0^h \frac{d(\mathcal{U} - \mathcal{W})}{\mathcal{U} + \mathcal{W}},$$

which Planck identified with the instantaneous diffusion potential*, that manifests itself immediately after switching off the external driving current/voltage. Similarly, he identified the integral on the left-hand side of Eq. (3.6.30) with the instantaneous resistance produced by the ions inside the membrane. After a sufficient lapse of time without current flow – i.e. for times longer than the redistribution time for the ions in the membrane – the system is again in a new stationary state, but for $I = 0$. The diffusion potential corresponding to this situation is then defined as

$$V^{(\text{dif})} = V_{I=0} = \frac{\mathcal{R}T}{\mathcal{F}} \int_0^h \frac{d(\mathcal{U} - \mathcal{W})}{\mathcal{U} + \mathcal{W}}, \qquad (3.6.31)$$

where the quantities $\mathcal{U}$ and $\mathcal{W}$ have now changed from the instantaneous values according to Eq. (3.6.28) to their new stationary values. The equation is Planck's precept for calculating the diffusion potential if the concentration profiles through the membrane are known for each of the participating ions when no current flows through the membrane.

Similarly, the total membrane resistance R is calculated by

$$R = \int_0^h \frac{1}{\mathcal{F}(\mathcal{U} + \mathcal{W})} dx. \qquad (3.6.32)$$

As they stand these equations are of limited practical value unless the concentration profiles through the membrane are known from another source.

* Compare Section 3.4.3.3 about the equivalent circuit for a single ion.

(iii) Planck's expression for the diffusion potential

Planck's integration of the flux equations corresponding to the stationary state of the diffusion gives the following final result for the diffusion potential $\psi^{(i)} - \psi^{(o)}$, which is defined through the relation

$$\psi^{(i)} - \psi^{(o)} = \frac{\mathcal{R}T}{\mathcal{F}} \ln \xi, \qquad (3.6.33)$$

where ξ is a parameter that is determined by solving the following transcendental equation

$$\frac{\xi \mathcal{U}^{(i)} - \mathcal{U}^{(o)}}{\mathcal{W}^{(i)} - \xi \mathcal{W}^{(o)}} = \frac{\ln\left(C^{(i)}/C^{(o)}\right) - \ln \xi}{\ln\left(C^{(i)}/C^{(o)}\right) + \ln \xi} \cdot \frac{\xi C^{(i)} - C^{(o)}}{C^{(i)} - \xi C^{(o)}}, \qquad (3.6.34)$$

where $C^{(i)}$ and $C^{(o)}$ are the total concentrations on the two sides of the membrane, and $\mathcal{U}$ and $\mathcal{W}$ are defined by Eq. (3.6.23) and Eq. (3.6.27), with the superscripts (i) and (o) corresponding to the ion concentrations in the two surrounding phases. We leave out the details as the long and rather tedious derivation is not essential in this context. The field in the range $0 \le x \le h$ is given by

$$-\frac{d\psi}{dx} = \frac{1}{h} \frac{\psi^{(i)} - \psi^{(o)}}{\ln\left(C^{(o)}/C^{(i)}\right)} \cdot \frac{C^{(o)} - C^{(i)}}{\left(C^{(o)} - C^{(i)}\right)(x/h) + C^{(i)}}, \qquad (3.6.35)$$

from which the potential profile $\psi(x)$ follows

$$\psi^{(i)} - \psi(x) = \frac{\psi^{(i)} - \psi^{(o)}}{\ln\left(C^{(o)}/C^{(i)}\right)} \cdot \left[\ln\left\{\left(\frac{C^{(o)}}{C^{(i)}} - 1\right) \cdot \frac{x}{h}\right\} + 1\right]. \qquad (3.6.36)$$

It appears that the potential profile $\psi(x)$ varies with the logarithm of the position x in the membrane as long as $C^{(o)} \neq C^{(i)}$. For $C^{(i)} \to C^{(o)}$ we have, in agreement with Eq. (3.6.19),

$$\psi^{(i)} - \psi(x) = \frac{\psi^{(i)} - \psi^{(o)}}{h} x,$$

that is, a linear potential profile and therefore a *constant* electric field. To obtain the complete solution of the Planck regime one has to determine the ξ from Eq. (3.6.34), in former times a rather discouraging enterprise to be used as a routine procedure. The diffusion potential is then obtained from Eq. (3.6.33). As the potential, or the field, profile is known, the concentration profile $C_j(x)$ can in principle be determined from the flux equations. We shall later illustrate this procedure.

3.6.3.2 The Henderson regime

The American physical chemist P. Henderson assumed in 1907 that the zone of transition consisted of a continuous mixture of the electrolytes in question, and that at any point in the transitional zone the concentrations are *linear mixtures* of the external concentrations $C_j^{(i)}$ and $C_j^{(o)}$ themselves. The concentration profiles therefore becomes

$$C_j(x) = C_j^{(i)} - \left(C_j^{(o)} - C_j^{(i)}\right)\frac{x}{h}. \tag{3.6.37}$$

By using a semi-thermodynamic argument similar to that used by W. Thomson (Lord Kelvin) in his treatment of the thermo-electric effect, Henderson could calculate the magnitude of the diffusion potential between two arbitrary electrolyte solutions if concentrations and conductances were known.

In biology we consider almost exclusively diffusion potentials arising between solutions that contain only monovalent ions. For that reason we shall only reproduce Henderson's equation that applies to this restriction. We consider two electrolyte solutions (i) and (o) containing a mixture of different electrolytes. Let the concentration in the two phases of a cation of type j_+ with an ionic mobility u_{j_+} be $C_{j_+}^{(i)}$ and $C_{j_+}^{(o)}$. Similarly the quantities for the anions of type j_- are designated u_{j_-}, $C_{j_-}^{(i)}$ and $C_{j_-}^{(o)}$. The diffusion potential between the two solution takes the form

$$\psi^{(i)} - \psi^{(o)} = \frac{\mathscr{R}T}{\mathscr{F}} \frac{\left(\mathscr{U}^{(o)} - \mathscr{W}^{(o)}\right) - \left(\mathscr{U}^{(i)} - \mathscr{W}^{(i)}\right)}{\left(\mathscr{U}^{(o)} + \mathscr{W}^{(o)}\right) - \left(\mathscr{U}^{(i)} + \mathscr{W}^{(i)}\right)} \ln\left(\frac{\mathscr{U}^{(o)} + \mathscr{W}^{(o)}}{\mathscr{U}^{(i)} + \mathscr{W}^{(i)}}\right),$$

$$\tag{3.6.38}$$

where the quantities $\mathscr{U}^{(i)}$, $\mathscr{U}^{(o)}$, $\mathscr{W}^{(i)}$ and $\mathscr{W}^{(o)}$ have the same meaning as defined previously, and when written in full read

$$\left.\begin{array}{l}
\mathscr{U}^{(i)} = u_{1+}C_{1+}^{(i)} + u_{2+}C_{2+}^{(i)} + \cdots = \sum u_{j+}C_{j+}^{(i)} \\[6pt]
\mathscr{U}^{(o)} = u_{1+}C_{1+}^{(o)} + u_{2+}C_{2+}^{(o)} + \cdots = \sum u_{j+}C_{j+}^{(o)} \\[6pt]
\mathscr{W}^{(i)} = u_{1-}C_{1-}^{(i)} + u_{2-}C_{2-}^{(i)} + \cdots = \sum u_{j-}C_{j-}^{(i)} \\[6pt]
\mathscr{W}^{(o)} = u_{1-}C_{1-}^{(o)} + u_{2-}C_{2-}^{(o)} + \cdots = \sum u_{j-}C_{j-}^{(o)}
\end{array}\right\}. \tag{3.6.39}$$

Equations (3.6.38) and (3.6.39) are Henderson's equations for the diffusion potential between two solutions of monovalent ions. The expressions may look rather difficult, but – as will be seen later – are in fact easy to handle. Henderson's equation can also be obtained by inserting the expressions for the ion profiles Eq. (3.6.39) in Eq. (3.6.38) and carrying out the integration.

We consider the simple situation where phase (i) and phase (o) consist of two binary electrolytes that have one ion in common, e.g. HCl in phase (i) and NaCl in phase (o). If the solutions also have the *same* concentrations C we have from Eq. (3.6.39)

$$\mathscr{U}^{(o)} = u_{Na}C \quad \text{and} \quad \mathscr{U}^{(i)} = u_H C,$$

Table 3.2. *Measured and calculated diffusion potentials between two binary electrolytes with same concentration (0.1 M) and with Cl^- the ion in common, at 25 °C*

	Diffusion potential in mV	
Solutions	Measured	Calculated
HCl:KCl	+26.78	+28.52
KCl:NaCl	+6.42	+4.86
KCl:LiCl	+8.76	+7.62
NaCl:LiCl	+2.62	+2.76
LiCl:NH$_4$Cl	−6.93	−7.57

Data from D.A. MacInnes & Y.L. Yeh (1921).

and

$$\mathcal{W}^{(o)} = u_{Cl}C \quad \text{and} \quad \mathcal{W}^{(i)} = u_{Cl}C,$$

which inserted in Eq. (3.6.38) give*

$$\psi^{(i)} - \psi^{(o)} = \frac{\mathcal{R}T}{\mathcal{F}} \ln\left(\frac{u_{Na} + u_{Cl}}{u_H + u_{Cl}}\right). \tag{3.6.40}$$

In the present situation with two solutions of the same concentration and with Cl^- as the common ion, the potential difference between the two solutions (i) and (o) (e.g. between HCl and NaCl) is easy to measure by using two Ag–AgCl electrodes, that is corresponding to the electrochemical cell

(i)   Ag|AgCl,   HCl : NaCl,   AgCl|Ag   (o) .

As the two solutions have the same Cl^- concentration, the potential difference between the Ag-wire and the solution into which the Ag–AgCl electrode is dipping will be the same for both electrodes. Any potential difference that may be measured between the silver wires must arise from the presence of a diffusion potential between the two solutions. In Table 3.2 we show a number of directly measured diffusion potentials generated between a number of different solutions and a comparison with the diffusion potentials calculated from Eq. (3.6.40) using the known values of the electric mobilities u_j of the ions.

As can be seen, there is a reasonably good agreement between the values of the diffusion potentials determined by experiment and by calculation.

* This equation is special case of an equation that was derived by the American chemists G.N. Lewis & L.W. Sargent in 1909.

3.6.3.3 The salt bridge once again

To measure a potential difference from any object, two metal wires must in the end be connected to the measuring device. The manner in which the other ends of the wires are connected depends entirely on the character of the object. In the case of an electrolyte it will not do to dip the copper wires directly into the solution. The connection has to be made to two *electrodes* of some kind, that are dipping into the solution. The principal element of an electrode is the *galvanic half-cell*, that is a metal immersed in an electrolyte containing one ion that can exchange reversibly with the metal to establish an electrochemical equilibrium between the metal and the metal ion in the solution. Typical half-cell reactions are: $Cu \rightleftharpoons Cu^{2+} + 2e$, $Zn \rightleftharpoons Zn^{2+} + 2e$, and $Ag \rightleftharpoons Ag^+ + e$, where the equilibria are controlled by concentrations of the metal ion. In the Ag–AgCl electrode the latter equilibrium is controlled by the Cl^- concentration via the solubility product $L_{AgCl} = [Ag^+] \times [Cl^-]$. The galvanic half-cell can be immersed directly in the electrolyte solution in question if it also contains the ion that belongs to the galvanic cell reaction. If this is not the case, the conducting connection from the galvanic cell to the electrolyte has to be established via a salt bridge of some kind, which in this way forms part of the total circuitry of the measuring system. These additional liquid junctions may generate diffusion potentials – also designated *liquid junction potentials* – that contribute as a source of error superimposed on a potential difference that is the ultimate objective of the measurement. Thus, if the presence of liquid junction potentials is unavoidable it is important to be able to assess their magnitude. In this connection the Henderson equation enters as a useful aid to estimate the magnitude of the liquid junction potentials formed at the transition zones between the different electrolytes that make up the system. However, the alternative to be preferred is to modify the liquid junctions in such a way that the liquid junction potentials are reduced as much as possible by means of a so-called *salt bridge*. Very often it consists of a glass tube – frequently in the form of a U-tube – that is filled with an electrolyte fixed in an agar-agar gel. A suitable electrolyte to reduce the liquid junction potential between the two electrolyte solutions is a saturated solution of KCl as the mobilities of K^+ and Cl^- ions are almost equal. In Section 3.6.1.3 we described qualitatively the use of a KCl salt bridge to eliminate the diffusion potential between the electrolyte solutions. We shall now support this argument with a quantitative example.

We consider the two electrolyte solutions that are in electric contact via interdiffusion through a porous plug:

(i) HCl (0.1 M) : NaCl (0.1 M) (o).

The electric contact to the solutions (i) and (o) is made by dipping an Ag–AgCl electrode into each solution. Since the Cl⁻ concentration is the same in both phases, the electrode potentials have the same – although unknown – value.

Using Eq. (3.6.40) we calculate the expected value of the liquid junction potential between the two solutions taking the values for u_H, u_{Na} and u_{Cl} from Section 3.6.2 . This gives

$$\psi^{(i)} - \psi^{(o)} = 25.7 \times \ln \frac{5.17 + 7.88}{36.0 + 7.88} = 25.7 \times \ln 0.2974 = -31.2 \text{ mV}.$$

We shall now compare this situation with the system

(i)　　HCl (0.1 M) : KCl (C) : NaCl (0.1 M) (o),

where the porous plug is removed and the two solutions are now in contact via a salt bridge that contains a solution of KCl of concentration C. We shall examine how the resultant liquid-junction potential between the two solutions

$$\psi^{(HCl)} - \psi^{(NaCl)} = \psi^{(HCl)} - \psi^{(KCl)} + \left(\psi^{(KCl)} - \psi^{(NaCl)} \right)$$

depends on the concentration C of KCl in the salt bridge. We have from Eq. (3.6.38) and Eq. (3.6.39), since $\psi^{(HCl)} \equiv \psi^{(i)}$ and $\psi^{(KCl)} \equiv \psi^{(o)}$,

$$\psi^{(HCl)} - \psi^{(KCl)}$$

$$= \frac{\mathscr{R}T}{\mathscr{F}} \frac{C(u_K - u_{Cl}) - 0.1(u_H - u_{Cl})}{C(u_K + u_{Cl}) - 0.1(u_H + u_{Cl})} \ln \left(\frac{C(u_K + u_{Cl})}{0.1(u_H + u_{Cl})} \right)$$

$$= \frac{\mathscr{R}T}{\mathscr{F}} \frac{C(7.58 - 7.88) - 0.1(36.0 - 7.88)}{C(7.58 + 7.88) - 0.1(36.0 + 7.88)} \ln \left(\frac{C(7.56 + 7.88)}{0.1(36.0 + 7.88)} \right)$$

$$= -\frac{\mathscr{R}T}{\mathscr{F}} \frac{0.3C + 2.81}{15.46C - 4.39} \ln(3.52\, C),$$

and, since $\psi^{(KCl)} \equiv \psi^{(i)}$ and $\psi^{(NaCl)} \equiv \psi^{(o)}$,

$$\psi^{(KCl)} - \psi^{(NaCl)}$$

$$= \frac{\mathscr{R}T}{\mathscr{F}} \frac{0.1(u_{Na} - u_{Cl}) - C(u_K - u_{Cl})}{0.1(u_{Na} + u_{Cl}) - C(u_H + u_{Cl})} \ln \left(\frac{C(u_K + u_{Cl})}{0.1(u_K + u_{Cl})} \right)$$

$$= \frac{\mathscr{R}T}{\mathscr{F}} \frac{0.1(5.17 - 7.88) - C(7.58 - 7.88)}{0.1(5.17 + 7.88) - C(7.58 + 7.88)} \ln \left(\frac{0.1(5.17 + 7.88)}{C(7.58 + 7.88)} \right)$$

$$= -\frac{\mathscr{R}T}{\mathscr{F}} \frac{0.3C - 0.27}{1.31 - 15.46C} \ln(11.80\, C).$$

Table 3.3. *KCl salt bridge to eliminate a liquid junction potential between an HCl and an NaCl solution (0.1 M): illustration of the effect of the KCl concentration*

KCl conc. (M)	$\psi^{(HCl)} - \psi^{(KCl)}$ (mV)	$\psi^{(KCl)} - \psi^{(NaCl)}$ (mV)	$\psi^{(HCl)} - \psi^{(NaCl)}$ (mV)
0.1	−26.65	−4.30	−30.95
1	−9.04	0.13	−8.91
2	−6.41	0.90	−5.51
3	−5.31	1.27	−4.05
4.2	−4.63	1.55	−3.08

Table 3.3 shows values for $\psi^{(HCl)} - \psi^{(KCl)}$, $\psi^{(KCl)} - \psi^{(NaCl)}$ and for $\psi^{(HCl)} - \psi^{(NaCl)}$ for different values of the KCl concentration C in the salt bridge. It appears that by using a saturated KCl solution in the salt bridge an electric connection between the KCl and NaCl solutions has now been attained with the result that the liquid junction potential has been *reduced* by a factor of 10 compared with that of the direct contact between the solutions. The example that we have considered represents a rather extreme situation where one ion, namely the H^+ ion, has a much greater mobility than the rest. The effectiveness of the KCl salt bridge will be even better when we are dealing instead with liquid junction potentials of the order of 5–10 mV, which will be reduced to less than 1 mV by the salt bridge.

3.7 Electrodiffusion through membranes

In Section 3.4 we described the rather special situation of a membrane being selectively permeable to either a cation or an anion. The diffusion regime in this membrane will never lead to a concentration equalization for the permeable ion since the net diffusion is very soon stopped by the build-up of the equilibrium potential across the membrane. In this section we shall discuss some simple stationary diffusion regimes across a membrane, which can be solved by Planck's theory for the diffusion potential, though without having to involve the complete, general solutions that we outlined in Section 3.6.3.

3.7.1 A single salt

It may be instructive first to consider the diffusion of a single salt – e.g. HCl – as it illustrates the elements in their simplest form. Let the membrane of thickness h separate the surrounding solutions that are kept at the constant values of $C = C^{(i)}$ for $x \leq 0$ and $C = C^{(o)}$ for $x \geq h$. Both ions can penetrate the membrane in which they have the constant mobilities u_+ and u_-.

3.7.1.1 The diffusion potential

We assume that the condition of electroneutrality holds inside the membrane; that is, in the region $0 \leq x \leq h$ we can put $C_+(x) = C_-(x) = C(x)$. When the diffusion process in the membrane becomes stationary, an electric field is established in the membrane of magnitude*

$$\frac{d\psi}{dx} = -\frac{\mathscr{R}T}{\mathscr{F}} \frac{1}{\mathscr{U} + \mathscr{W}} \cdot \frac{d(\mathscr{U} - \mathscr{W})}{dx}. \tag{3.7.1}$$

In the present case we have

$$\mathscr{U}(x) = u_+ C(x) \quad \text{and} \quad \mathscr{W}(x) = u_- C(x).$$

We have then

$$\mathscr{U}(x) + \mathscr{W}(x) = (u_+ + u_-)C(x) \quad \text{and} \quad \mathscr{U}(x) - \mathscr{W}(x) = (u_+ - u_-)C(x),$$

which, inserted in Eq. (3.7.1), gives

$$\frac{d\psi}{dx} = -\frac{\mathscr{R}T}{\mathscr{F}} \frac{u_+ - u_-}{u_+ + u_-} \cdot \frac{1}{C} \frac{dC}{dx}. \tag{3.7.2}$$

This expression is identical to that obtained directly from the Nernst–Planck equations in Section 3.6.2 and, naturally, the solution procedure is the same. We multiply Eq. (3.7.2) by the differential dx and integrate from $x = 0$, where $\psi(x) = \psi(0) = \psi^{(i)}$ and $C(x) = C(0) = C^{(i)}$, to $x = h$ where $\psi(x) = \psi(h) = \psi^{(o)}$ and $C(x) = C(h) = C^{(o)}$. Thus, we have again

$$\psi^{(o)} - \psi^{(i)} = V^{(dif)} = \frac{\mathscr{R}T}{\mathscr{F}} \frac{u_+ - u_-}{u_+ + u_-} \ln\left(\frac{C^{(i)}}{C^{(o)}}\right), \tag{3.7.3}$$

in agreement with Eq. (3.6.5). In the flux equation Eq. (3.6.5) we replace the potential gradients $d\psi/dx$ with the expression of Eq. (3.7.2) to give

$$J_+ = -u_+ \frac{\mathscr{R}T}{\mathscr{F}} \frac{dC}{dx} + u_+ C \frac{\mathscr{R}T}{\mathscr{F}} \frac{u_+ - u_-}{u_+ + u_-} \cdot \frac{1}{C} \frac{dC}{dx}$$

$$= -2 \frac{\mathscr{R}T}{\mathscr{F}} \frac{u_+ \cdot u_-}{u_+ + u_-} \cdot \frac{dC}{dx},$$

and we also have the corresponding expression for the flux of the negative ion

$$J_- = -2 \frac{\mathscr{R}T}{\mathscr{F}} \frac{u_+ \cdot u_-}{u_+ + u_-} \cdot \frac{dC}{dx}$$

---

* See Section 3.6.3.1 (ii), Eq. (3.6.29).

that is, the field $-d\psi/dx$ given by Eq. (3.7.2) retards the motion of the ion that has the greatest mobility and enhances the one lagging behind in such a way that both ions move in their stationary states through the membrane with the *same* flux, and therefore with no membrane current. Both ions move under the influence of the concentration gradient dC/dx with a *common* diffusion coefficient of magnitude*

$$D_\pm = 2\frac{\mathcal{R}T}{\mathcal{F}}\frac{u_+ \cdot u_-}{u_+ + u_-} = 2\frac{D_+ \cdot D_-}{D_+ + D_-}. \tag{3.7.4}$$

Thus, the flux of the salt $J_\pm$ through the membrane is described by a single equation like a simple process of diffusion

$$J_\pm = -D_\pm \frac{dC}{dx},$$

in complete agreement with Fick's law for the diffusion of a non-electrolyte[†].

For the sake of completeness we shall show how Eq. (3.7.3) derives from Planck's transcendental equation for ξ. As electroneutrality, that is $C^+ = C^- = C$, holds everywhere we have

$$\mathcal{U}^{(i)} = u_+ C^{(i)} \quad \text{and} \quad \mathcal{U}^{(o)} = u_+ C^{(o)},$$
$$\mathcal{W}^{(i)} = u_- C^{(i)} \quad \text{and} \quad \mathcal{W}^{(o)} = u_- C^{(o)}.$$

Inserting these values in Eq. (3.6.34) gives

$$\frac{u_+ \left(\xi C^{(i)} - C^{(o)}\right)}{u_- \left(\xi C^{(o)} - C^{(i)}\right)} = \frac{\xi C^{(i)} - C^{(o)}}{\xi C^{(o)} - C^{(i)}} \cdot \frac{\ln\left(C^{(i)}/\ln C^{(o)}\right) - \ln \xi}{\ln\left(C^{(i)}/\ln C^{(o)}\right) + \ln \xi},$$

which reduces to

$$[u_+ + u_-]\ln \xi = [u_+ - u_-]\ln\left(\frac{C^{(o)}}{C^{(i)}}\right).$$

Solving for $\ln \xi$ we have

$$\ln \xi = \frac{u_+ - u_-}{u_+ + u_-}\ln\left(\frac{C^{(o)}}{C^{(i)}}\right),$$

which applied to Eq. (3.6.33) yields

$$\psi^{(i)} - \psi^{(o)} = \frac{\mathcal{R}T}{\mathcal{F}}\ln \xi = \frac{\mathcal{R}T}{\mathcal{F}}\frac{u_+ - u_-}{u_+ + u_-}\ln\left(\frac{C^{(o)}}{C^{(i)}}\right),$$

in accordance with two other derivations.

---

* We have: $(\mathcal{R}T/\mathcal{F})u_+ = (kT/e)z_+ eB_+ = z_+ kT B_+ = z_+ D_+$.
† Equations (3.7.3) and (3.7.4) were first derived by W. Nernst (1888).

3.7.1.2 The membrane resistance

If the concentration profiles of both ions through the membrane are known the membrane resistance can be calculated from Eq. (3.6.32)

$$R = \frac{1}{\mathcal{F}} \int_0^h \frac{1}{\mathcal{U} + \mathcal{W}} \, dx.$$

The dependence of the total concentration on distance x in the membrane is

$$C(x) = \left[C^{(o)} - C^{(i)} \right] \frac{x}{h} + C^{(i)}. \tag{3.7.5}$$

Since electroneutrality $(C^+(x) = C^-(x) = C(x))$ holds, we have

$$\mathcal{U}(x) = u_+ C(x) \quad \text{and} \quad \mathcal{W}(x) = u_- C(x),$$

and

$$\mathcal{U}(x) + \mathcal{W}(x) = (u_+ + u_-)C(x)$$
$$= (u_+ + u_-) \left\{ \left[C^{(o)} - C^{(i)} \right] \frac{x}{h} + C^{(i)} \right\}.$$

If this is inserted in the above expression for the membrane resistance we obtain

$$R = \frac{1}{\mathcal{F}(u_+ + u_-)} \int_0^h \frac{1}{\left[C^{(o)} - C^{(i)} \right](x/h) + C^{(i)}} \, dx.$$

The integrand is of the form $1/(a + bx)$, with $a = C^{(i)}$ and $b = (C^{(o)} - C^{(i)})/h$, the indefinite integral of which is $\ln|a + bx|/b$. We have therefore

$$\int_0^h \frac{1}{a + bx} \, dx = \frac{1}{b} \ln \left(\frac{bh + a}{a} \right) = \frac{1}{b} \ln \left(\frac{bh}{a} + 1 \right),$$

from which it follows that

$$R = \frac{1}{\mathcal{F}(u_+ + u_-)} \cdot \frac{h}{C^{(o)} - C^{(i)}} \cdot \ln \left\{ \frac{(C^{(o)} - C^{(i)})/h}{C^{(i)}} h + 1 \right\},$$

or

$$R = \frac{h}{\mathcal{F}} \cdot \frac{\ln \left(C^{(o)} / C^{(i)} \right)}{(u_+ + u_-)(C^{(o)} - C^{(o)})}. \tag{3.7.6}$$

In this calculation of the resistance of the pore we have not taken the possibility of an increase in the concentration of the permeable ions into consideration as, for example, in a gramicidin A channel in a bimolecular lipid membrane.

For a HCl solution with $C^{(i)} = 100$ mol m^{-3} and $C^{(o)} = 10$ mol m^{-3} we obtain

$$R = \frac{h}{96942} \frac{2.3026}{(36 + 7.9) \times 10^{-8} \times (100 - 10)} = 0.615 \times h \stackrel{\triangle}{=} \Omega \text{ m}^2.$$

Thus, a "pore" radius of 5 Å $= 5 \times 10^{-10}$ m in a membrane of thickness $c.100$ Å $= 10^{-8}$ m has the resistance

$$R^{(\text{pore})} = \frac{R}{A^{(\text{pore})}} = 0.615 \frac{10^{-8}}{\pi (5 \times 10^{-10})^2} = 7.6 \times 10^9 \ \Omega.$$

3.7.1.3 The potential profile

We obtain the potential profile through the membrane by integration of Eq. (3.7.2) from $x = 0$ where the potential is $\psi^{(i)}$ and concentration $C^{(i)}$ to the position x with concentration $C(x)$ and potential $\psi(x)$. This gives

$$\psi^{(i)} - \psi(x) = \frac{\mathcal{R}T}{\mathcal{F}} \frac{u_+ - u_-}{u_+ + u_-} \ln \left(\frac{C(x)}{C^{(i)}} \right).$$

We insert the expression for $C(x)$ from Eq. (3.6.16) and eliminate the factor $\mathcal{R}T(u_+ - u_-)/\mathcal{F}(u_+ + u_-)$ using Eq. (3.7.3). This gives

$$\psi^{(i)} - \psi(x) = \frac{\psi^{(i)} - \psi^{(o)}}{\ln(C^{(o)}/C^{(i)})} \ln \left[\left(\frac{C^{(o)}}{C^{(i)}} - 1 \right) \frac{x}{h} + 1 \right], \qquad (3.7.7)$$

in accordance with the general expression by Planck (Eq. (3.3.36)). The potential varies logarithmically with the position x, whereas the concentration profile in the membrane is linear. These points are illustrated in Fig. 3.14, showing the courses of the concentration $C(x)$, the potential $\psi(x)$ and the electric field $-\mathrm{d}\psi/\mathrm{d}x$ presenting the electrodiffusion of a single salt, in this case an HCl solution ($u_\text{H} = 36 \times 10^{-8}$ V^{-1} s^{-1} m^2, $u_\text{Cl} = 7.8 \times 10^{-8}$ V^{-1} s^{-1} m^2) $C^{(i)} = 100$ mmol dm^{-3} and $C^{(o)} = 10$ mmol dm^{-3}.

3.7.1.4 The equivalent electric circuit

In the general I–V relation for the electrodiffusion regime, Eq. (3.6.30),

$$I \cdot \int_0^h \frac{\mathrm{d}x}{\mathcal{F}(\mathcal{U} + \mathcal{W})} = V - \frac{\mathcal{R}T}{\mathcal{F}} \int_0^h \frac{\mathrm{d}(\mathcal{U} - \mathcal{W})}{\mathcal{U} + \mathcal{W}} \qquad (3.7.8)$$

we have already calculated the integrals on the left side (Eq. (3.7.2)) and on the right side (Eq. (3.7.3)). Inserting these expressions in Eq. (3.7.8) above we

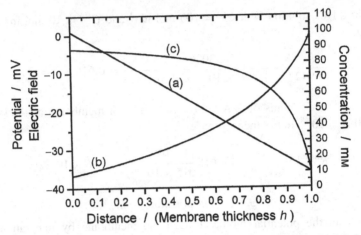

Fig. 3.14. Profiles of concentration (a), potential (b) and electric field (c) through a membrane of thickness h, at the stationary diffusion of HCl, when $C(0) = 100$ mol m^{-3} and $C(h) = 10$ mol m^{-3}. Left ordinate: potential (mV). Right ordinate: concentration (mol m^{-3} or mmol dm^{-3}). Abscissa: distance through the membrane, in units of membrane thickness h.

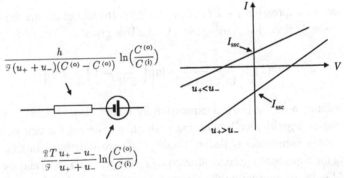

Fig. 3.15. The equivalent diagram for the diffusion regime of a single salt and the two I–V characteristics corresponding to the conditions $u_+ > u_-$ and $u_+ < u_-$ respectively.

obtain

$$I \frac{h}{\mathscr{F}(u_+ + u_-)(C^{(o)} - C^{(o)})} \ln\left(\frac{C^{(o)}}{C^{(i)}}\right) = V - \frac{\mathscr{R}T}{\mathscr{F}} \frac{u_+ - u_-}{u_+ + u_-} \ln\left(\frac{C^{(o)}}{C^{(i)}}\right).$$

(3.7.9)

The terms of this I–V relation are indicated as components in the equivalent electric circuit diagram in Fig. 3.15 together with two illustrations of I–V characteristics. The distinctive feature of the regime of electrodiffusion for a single salt is that the concentration profile is linear for both cation and anion, and depends neither on the magnitude of the voltage V applied across the membrane and, therefore, nor on the current that is forced through the membrane. The diffusion regime behaves like a *pure* ohmic resistance with a linear relation between current and voltage. The short circuit current I_{ssc} is the current flowing through the membrane that causes a voltage drop across the membrane (the membrane resistance), which is just compensated by the local electromotive force in the membrane, i.e. the diffusion potential. Putting $V = 0$ in Eq. (3.7.9) the following expression for the short circuit current is obtained*

$$I_{ssc} = -\Re T \frac{(u_+ - u_-)\left(C^{(o)} - C^{(i)}\right)}{h}.$$

3.7.1.5 Electroneutrality

It appears from Eq. (3.7.7) that the potential profile $\psi(x)$ is a function of the logarithm to the x in the membrane, i.e. $d^2\psi/dx^2 \neq 0$ in the region $0 \leq x \leq h$. Thus, the right-hand side of Poisson's equation is not zero, which implies that – in contradiction to our *a priori* assumption – the assumption of electroneutrality is continuously violated. Therefore, it may be of interest to examine whether this assumption may lead to obviously absurd predictions. We have from Poisson's equation

$$\rho(x) = \mathscr{F}[C^+(x) - C^-(x)] = -K\varepsilon_o \frac{d\psi^2}{dx^2} = -K\varepsilon_o \frac{d}{dx}\left(\frac{d\psi}{dx}\right).$$

Inserting the expression for $d\psi/dx$ (Eq. (3.7.2)) gives

$$\rho(x) = -K\varepsilon_o \frac{d}{dx}\left(-\frac{\Re T}{\mathscr{F}}\frac{u_+ - u_-}{u_+ + u_-}\cdot\frac{1}{C}\frac{dC}{dx}\right)$$

$$= K\varepsilon_o \frac{\Re T}{\mathscr{F}}\frac{u_+ - u_-}{u_+ + u_-}\frac{d}{dx}\left(\frac{1}{C}\frac{dC}{dx}\right)$$

$$= K\varepsilon_o \frac{\Re T}{\mathscr{F}}\frac{u_+ - u_-}{u_+ + u_-}\left[\frac{d}{dC}\left(\frac{1}{C}\right)\frac{dC}{dx}\frac{dC}{dx} + \frac{1}{C}\frac{d^2C}{dx^2}\right]$$

$$= -K\varepsilon_o \frac{\Re T}{\mathscr{F}}\frac{u_+ - u_-}{u_+ + u_-}\left(\frac{1}{C}\frac{dC}{dx}\right)^2,$$

as $d^2C/dx^2 = 0$ according to Eq. (3.7.5). Inserting the expression for CYx and dC/dx in this equation we obtain

$$C^+(x) - C^-(x) = -\frac{K\varepsilon_o}{h^2\mathscr{F}}\frac{\Re T}{\mathscr{F}}\frac{u_+ - u_-}{u_+ + u_-}\left[\frac{C^{(o)} - C^{(i)}}{\left(C^{(o)} - C^{(i)}\right)(x/h) + C^{(i)}}\right]^2.$$

$$(3.7.10)$$

---

* It might be a useful exercise to check that the right-hand side has the dimension A m^{-2}.

For an HCl solution with $C^{(i)} = 100$ mol m^{-3} and $C^{(i)} = 10$ mol m^{-3} we obtain for $h = 1$ cm $= 10^{-2}$ m a maximum value for $C^+(h) - C^-(h) = 9.7 \times 10^{-15}$ mol/m$^3 = 9.7 \times 10^{-15}$ mmol dm^{-3}, which is far beyond the limits for any macroscopic measurement. Thus, in this case Eq. (3.7.3) works well for calculating the diffusion potential. On the other hand, if $h = 10^{-8}$ m, Eq. (3.7.3) predicts a deviation from electroneutrality that is of the same order of magnitude as $C^{(o)}$, which obviously is absurd. But this prediction does not mean the presence of that amount of free space charges in practice. What it amounts to is that the basic assumptions $(C_+(x) = C_-(x) = C(x))$ for calculating the diffusion potential for such short distances are insufficient and a more thorough mathematical starting point must be formulated $(C_+(x) \neq C_-(x))$ that also incorporates the Poisson equation. An analysis along these lines* shows that the field is no longer that given by Planck's formula, Eq. (3.6.18), but is a solution of a second-order differential equation of *Airy's* type[†]. This solution contains a parameter x/λ, where λ is a Debye length[‡] in the membrane layer $0 < x < h$. For membranes where $(x/\lambda)^2 \gg 1$ the rigorous solution for the diffusion potential and the formula in Eq. (3.7.3) of Nernst and Planck becomes undistinguishable. When $(h/\lambda)^2 \to 1$ the formula of Eq. (3.7.3) overestimates the value of the diffusion potential. With $\lambda = 10$ Å – corresponding to $C^{(o)} = 100$ mol m^{-3} – and for $h = 100$ Å, Eq. (3.7.3) is still a reasonably good approximation.

3.7.2 Ion-selective membranes

We consider a membrane for which membrane permeabilities for cations and anions differ from each other by several orders of magnitudes. Thus, for all practical purposes the membrane can be regarded as permeable for ions of *only one charge type* – either positive or negative. Moreover, let the surrounding solutions contain several species of the ion type to which the membrane is permeable. In this situation a membrane equilibrium cannot be established unless the equilibrium potentials of each ion species are equal, that is

$$V_1^{(eq)} = V_j^{(eq)}$$

---

* See, for example, Bass, L. (1964): Potential of liquid junctions, *Trans. Faraday Soc.*, **60**, 1914.
† That is, equations of the rather innocent looking form: $d^2y/dx^2 + xy = 0$. The solutions can be written in terms of *Bessel functions* of the order $\frac{1}{3}$. As these functions are not loaded in our mathematical knapsack in Chapter 1 we shall not pursue the subject.
‡ See Section 3.5.3.1, Eq. (3.5.4).

for all the $j = 1 \ldots = n$ permeable ions, which implies that the concentration ratios are subject to the constraint

$$\left(\frac{C_1^{(i)}}{C_j^{(o)}} \right)^{1/z_1} = \left(\frac{C_1^{(j)}}{C_j^{(o)}} \right)^{1/z_j}. \tag{3.7.11}$$

When this condition is not satisfied we are dealing with a non-equilibrium diffusion regime such as those discussed previously, but with the difference that only ions of one charge sign participate in the process. So we shall end up with a system equalization of concentrations and no membrane potential. This process, however, may take some time and when inspected during times of shorter intervals the system's behavior may be regarded as stationary. According to usual practice such a system is called a *quasi-stationary system.* We consider now a membrane of thickness H that is surrounded exclusively by monovalent ions. According to Eq. (3.6.21) and Eq. (3.6.26) the ionic current is

$$I_+ = -\mathcal{R}T \frac{d\mathcal{U}}{dx} - \mathcal{F}\mathcal{U}\frac{d\psi}{dx}, \tag{3.7.12}$$

for the positive ions, and

$$I_- = \mathcal{R}T \frac{d\mathcal{W}}{dx} - \mathcal{F}\mathcal{W}\frac{d\psi}{dx}, \tag{3.7.13}$$

for the negative ions. If the membrane is only permeable for the *positive* we have

$$I_- = 0, \quad \text{for all values of } \psi^{(i)} - \psi^{(o)}.$$

When the system is left on its own the fluxes of the participating positive ions will adjust themselves in such a way that in the quasi-stationary state the net current flow is zero, that is $I_+ = 0$. We then have from Eq. (3.7.12)

$$-\frac{d\psi}{dx} = \frac{\mathcal{R}T}{\mathcal{F}} \frac{1}{\mathcal{U}} \frac{d\mathcal{U}}{dx} = \frac{\mathcal{R}T}{\mathcal{F}} \frac{d\ln \mathcal{U}}{dx}.$$

We multiply this expression by the differential dx and integrate from $x = 0$, where $\psi(0) = \psi^{(i)}$ and $\mathcal{U}(0) = \mathcal{U}^{(i)}$ to $x = h$, where $\psi(h) = \psi^{(o)}$ and $\mathcal{U}(h) = \mathcal{U}^{(o)}$. This gives

$$\psi^{(i)} - \psi^{(o)} = V^{(\text{dif})} = \frac{\mathcal{R}T}{\mathcal{F}} \ln \left(\frac{\mathcal{U}^{(o)}}{\mathcal{U}^{(i)}} \right). \tag{3.7.14}$$

We write the expressions for $\mathcal{U}^{(i)}$ and $\mathcal{U}^{(o)}$ out in full

$$\mathcal{U}^{(i)} = \sum_{j=1}^{j=n} u_{j+} C_{j+}^{(i)} \quad \text{and} \quad \mathcal{U}^{(o)} = \sum_{j=1}^{j=n} u_{j+} C_{j+}^{(o)},$$

and insert in the above to obtain

$$V^{(\text{dif})} = \frac{\mathscr{R}T}{\mathscr{F}} \ln \left(\frac{u_{1+}C_{1+}^{(\text{o})} + u_{2+}C_{2+}^{(\text{o})} + \cdots + u_{n+}C_{n+}^{(\text{o})}}{u_{1+}C_{1+}^{(\text{i})} + u_{2+}C_{2+}^{(\text{i})} + \cdots + u_{n+}C_{n+}^{(\text{i})}} \right). \qquad (3.7.15)$$

Since the mobility u_j is proportional to the diffusion coefficient D_j and therefore also to the membrane permeability $P_j = D_j / h$, we can also write Eq. (3.7.15) as

$$V^{(\text{dif})} = \frac{\mathscr{R}T}{\mathscr{F}} \ln \left(\frac{P_{1+}C_{1+}^{(\text{o})} + P_{2+}C_{2+}^{(\text{o})} + \cdots + P_{n+}C_{n+}^{(\text{o})}}{P_{1+}C_{1+}^{(\text{i})} + P_{2+}C_{2+}^{(\text{i})} + \cdots + P_{n+}C_{n+}^{(\text{i})}} \right). \qquad (3.7.16)$$

If, on the other hand, the membrane is permeable only to *negative* ions the field in the membrane is given by

$$\frac{d\psi}{dx} = \frac{\mathscr{R}T}{\mathscr{F}} \frac{1}{\mathscr{W}} \frac{d\mathscr{W}}{dx},$$

and the expression for the diffusion potential that corresponds to Eq. (3.7.16) becomes

$$V^{(\text{dif})} = \frac{\mathscr{R}T}{\mathscr{F}} \ln \left(\frac{P_{1-}C_{1-}^{(\text{i})} + P_{2-}C_{2-}^{(\text{i})} + \cdots + P_{n-}C_{n-}^{(\text{i})}}{P_{1-}C_{1-}^{(\text{o})} + P_{2-}C_{2-}^{(\text{o})} + \cdots + P_{n-}C_{n-}^{(\text{o})}} \right). \qquad (3.7.17)$$

Imagine now that the surrounding solutions contain NaCl and KCl respectively, and, further, let the membrane be permeable only to Na^+ and K^+ ions. The membrane potential is then, according to Eq. (3.7.16),

$$\psi^{(\text{i})} - \psi^{(\text{o})} = \frac{\mathscr{R}T}{\mathscr{F}} \ln \left(\frac{P_{\text{K}}C_{\text{K}}^{(\text{o})} + P_{\text{Na}}C_{\text{Na}}^{(\text{o})}}{P_{\text{K}}C_{\text{K}}^{(\text{i})} + P_{\text{Na}}C_{\text{Na}}^{(\text{i})}} \right). \qquad (3.7.18)$$

The same expression would also be valid if the membrane were permeable to Cl^- ions, *provided* the concentrations $C_{\text{Cl}}^{(\text{i})}$ and $C_{\text{Cl}}^{(\text{o})}$ have such a value that the membrane potential – the diffusion potential $(\psi^{(\text{i})} - \psi^{(\text{o})})$ in Eq. (3.7.17) – is equal to the equilibrium potential $V_{\text{Cl}}^{(\text{eq})}$ for the Cl^- ions. In this case $I_{\text{Cl}} = 0$ and the potential across the membrane is found by solving Eq. (3.7.16).

We consider the special case where the membrane is surrounded by KCl and NaCl solutions respectively, by the *same concentrations*, that is

$$C_{\text{K}}^{(\text{o})} = C_{\text{Na}}^{(\text{i})} \quad \text{and} \quad C_{\text{K}}^{(\text{i})} = C_{\text{Na}}^{(\text{o})} = C.$$

Equation (3.7.18) then takes the form

$$\psi^{(\text{i})} - \psi^{(\text{o})} = \frac{\mathscr{R}T}{\mathscr{F}} \ln \left(\frac{P_{\text{K}} \cdot 0 + P_{\text{Na}}C}{P_{\text{K}}C + P_{\text{Na}} \cdot 0} \right) = \frac{\mathscr{R}T}{\mathscr{F}} \ln \left(\frac{P_{\text{Na}}}{P_{\text{K}}} \right). \qquad (3.7.19)$$

Thus, under these special conditions a measurement of the membrane potential also gives the magnitude of the ratio P_{Na}/P_K, which is also designated as the membrane's *selectivity ratio* for the ions in question.

3.7.3 The Goldman regime

We shall now consider the somewhat more complicated situation, that is more frequently encountered in biology, where the membrane is permeable – though to a varying degree – to both anions and cations. We leave the treatment to the simplest situation where the surrounding media contain only monovalent ions, e.g. Na^+, K^+ and Cl^-, the concentrations of which are

In phase (i) : $C_{Na}^{(i)}, C_K^{(i)}, C_{Cl}^{(i)},$ and $C_{Na}^{(o)}, C_K^{(o)}, C_{Cl}^{(o)},$ in phase (o).

When the system has reached a stationary state* there is a potential difference $\psi^{(i)} - \psi^{(o)} = V_m$ across the membrane, and the three ion species will move across the membrane in such a way that the total ionic current is zero, provided that no external potential difference $V \neq V_m$ is applied. This diffusion regime is solved once we have calculated the stationary diffusion potential V_m, the *membrane potential*, the potential profile $\psi(x)$ through the membrane and the concentration profiles $C_j(x)$ for the three ions. From this knowledge the individual fluxes J_j and other relevant quantities can be obtained. Unfortunately – as pointed out previously – this path is easier prescribed than actually followed in practice. The character of the missing joker – even in this simple system – is perhaps illustrated most simply by writing the equation for the flux through the membrane of an individual ion in the form according to that of Kramers [†]

$$J_j\, e^{z_j \Psi(x)} = -D_j\, \frac{d}{dx}\left(C_j(x)\, e^{z_j \Psi(x)}\right),\qquad (3.7.20)$$

where $\Psi(x)$ is the normalized potential ($\psi/(kT/e)$) and the $C_j(x)$ are the concentrations at the position x in the membrane of thickness h. We multiply this equation by dx and integrate from $x = 0$, where $\Psi(0) = \Psi^{(i)}$ and $C_j(x) = C_j(0)$. to $x = h$, where $\Psi(h) = \Psi^{(o)}$ and $C_j(x) = C_j(h)$. This gives

$$\int_0^h J_j\, e^{z_j \Psi(x)} dx = \int_{\Psi^{(i)},\, C_j(0)}^{\Psi^{(o)},\, C_j(h)} D_j\, d\left(C_j(x)\, e^{z_j \Psi(x)}\right).$$

---

[*] In living biological cells stationarity is not an idealized hypothetical assumption, but a real condition on account of the active transport components that maintain a constant interior environment in the cell.

[†] See Section 3.3.2.1, Eq. (3.3.41).

Because of the prevailing steady state we have $\partial J_j/\partial x = 0$. In addition we assume that D_j has the same value inside the membrane, i.e. in the region $0 \leq x \leq h$. Thus, we can put both J_j and D_j outside the integration sign. Hence we have

$$J_j \int_0^h J_j \, e^{z_j \Psi(x)} dx = -D_j \big[C_j(x) \, e^{z_j \Psi(x)} \big]_{\Psi^{(i)}, C_j(0)}^{\Psi^{(o)}, C_j(h)},$$

or

$$J_j = D_j \frac{C_j(0) e^{z_j \Psi^{(i)}} - C_j(h) e^{z_j \Psi^{(o)}}}{\displaystyle\int_0^h e^{z_j \Psi(x)} dx}. \tag{3.7.21}$$

From this it appears that the flux J_j cannot be calculated from the procedure above *unless* we know explicitly the functional dependence of the potential $\Psi(x)$ of the distance x through the membrane. When the total concentrations on the two sides are widely different the only way ahead is to revert to Planck's general procedure to determine the diffusion potential and the potential profile $\psi(x)$, outlined in Section 3.6.3.1 (iii).

However, under certain conditions it is reasonable to assume that the electric field $(-d\psi/dx)$ is constant inside the membrane, i.e. the potential profile varies linearly all through the membrane, thus making the term $d\psi/dx$ in the Nernst–Planck equations degenerate to a constant factor. This assumption also entails that $d^2\psi/dx^2 = 0$, which, according to Poisson's equation, is equivalent to the presence of macroscopic electroneutrality throughout the membrane*. Then it becomes a relatively simple matter to integrate the Nernst–Planck equations separately for each ion – provided that we only are dealing with monovalent ions – to obtain the individual ion fluxes and ionic currents. The stationary value of the membrane potential V_m is then evaluated by putting the total ionic current equal to zero. In the case where the membrane is surrounded on both sides by Na^+, K^+ and Cl^- ions which penetrate the membrane with the respective permeabilities P_{Na}, P_K and P_{Cl}, the formula reads

$$V_m = \frac{\mathscr{R} T}{\mathscr{F}} \ln \left(\frac{P_K C_K^{(o)} + P_{Na} C_{Na}^{(o)} + P_{Cl} C_{Cl}^{(i)}}{P_K C_K^{(i)} + P_{Na} C_{Na}^{(i)} + P_{Cl} C_{Cl}^{(o)}} \right). \tag{3.7.22}$$

This formula was first derived by M. Planck (1890) corresponding to the physical situation where the electric field in the membrane is constant[†]. Later it was rediscovered by D.E. Goldman (1943)[‡], who introduced the presence of

---

* See Section 3.3.2.2, Eq. (3.3.51).
[†] That is: same total concentration on the both sides of the membrane. See Section 3.6.3.1, Eq. (3.6.19).
[‡] Goldman, D.E. (1942): Potential, impedance, and rectifications in membranes. *J. Gen. Physiol.*, **27**, 37.

a constant field as a postulate*. The form written above is due to A.L. Hodgkin & B. Katz (1949)[†], and it has become one of the tools most used in biological research to estimate the magnitude of membrane potentials in living cells. For that reason the equation is generally designated the Goldman–Hodgkin–Katz equation or just the Goldman equation.

It may appear tempting to regard the Goldman equation as the form of a generalized Nernst equation of some kind. But as it stands the Goldman equation represents a *diffusion potential* and *not* an equilibrium potential. However, the Goldman equation will transform itself to represent a Nernst equation in cases where the permeability of the membrane is dominated by a single ion. If, for example, $P_K \gg P_{Na} \approx P_{Cl}$ and the three concentrations are mutually so constituted that

$$P_K C_K^{(o)} \gg P_{Na} C_{Na}^{(o)} \approx P_{Cl} C_{Cl}^{(i)}, \quad \text{and} \quad P_K C_K^{(i)} \gg P_{Na} C_{Na}^{(i)} \approx P_{Cl} C_{Cl}^{(o)},$$

Eq. (3.7.22) degenerates into

$$V_m \approx \frac{\mathscr{R}T}{\mathscr{F}} \ln\left(\frac{P_K C_K^{(o)}}{P_K C_K^{(i)}}\right) = \frac{\mathscr{R}T}{\mathscr{F}} \ln\left(\frac{C_K^{(o)}}{C_K^{(i)}}\right) = V_K^{(eq)},$$

which is Nernst's equation.

Eq. (3.7.22) will also represent the membrane potential for the non-stationary situations provided that the membrane potential changes so slowly that the condition $I^{(tot)} = 0$ still remains a realistic approximation[‡].

3.7.3.1 Derivation of the Goldman equation

In view of the widespread use of the Goldman equation in biological situations it seems worth giving an account of the derivation of Eq. (3.7.22). This may also be justified because of the many additional useful concepts that are introduced.

Goldman's starting point was the set of assumptions – partly illustrated in Fig. 3.16.

(a) The membrane has the thickness h.
(b) All the ions in question are monovalent.
(c) The concentrations of an ion of type j inside the membrane are at the boundaries at $x = 0$ and $x = h$ equal to $C_j(0)$ and $C_j(h)$. Their connections

---

* By so doing total concentration no longer needs to be the same on both sides of the membrane.
[†] Hodgkin, A.L. & Katz, B. (1949): The effect of sodium ions on the electrical activity of the giant axon of the squid. *J. Physiol.*, **108**, 37.
[‡] This also implies that the component of the membrane current that charges the capacity of the membrane – of magnitude $I_c = C_m(dV/dt)$, where C_m is the membrane capacitance – can be disregarded safely.

3. Membrane potentials

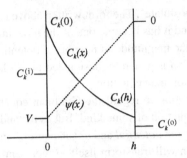

Fig. 3.16. The Goldman regime: specification of the quantities that enter into the calculation of the membrane potential. The positive direction of the x-axis is (i) → (o).

to the outside concentrations $C_j^{(i)}$ and $C_o^{(i)}$ are given by*

$$C_j(0) = \alpha_j C_j^{(i)}, \quad C_j(h) = \alpha_j C_j^{(o)}, \tag{3.7.23}$$

where α_j is the partition coefficient between membrane and the aqueous solution for the ion j.

(d) The potentials in the membrane at the surfaces at $x = 0$ and $x = h$ are equal to the potentials in the adjoining media, that is, $\psi(0) = \psi^{(i)}$ and $\psi(h) = \psi^{(o)}$. The electric field in the membrane is *assumed* to be constant. Hence, we have

$$\frac{d\psi}{dx} = \frac{\psi(h) - \psi(0)}{h} = \frac{\psi^{(o)} - \psi^{(i)}}{h} = -\frac{\psi^{(i)} - \psi^{(o)}}{h} = -\frac{V}{h},$$

$$\tag{3.7.24}$$

as $V = \psi^{(i)} - \psi^{(o)}$. In the following we shall use the normalized potential Ψ and put

$$v = \Psi^{(i)} - \Psi^{(o)} = \Psi(0) - \Psi(h).$$

Furthermore, for reasons of convenience we lock the potential in the outer phase at the value zero, i.e. $\Psi^{(o)} = \Psi^{(o)} = 0$.

(e) The state in the membrane is considered *stationary*, that is neither the membrane potential V, the concentration profiles $C_j(x)$ nor the ion fluxes J_j change with time.

We shall now find the fluxes for each ion by using as point of departure the integrated form of Kramers' equation (Eq. (3.7.21)). The integral in the denominator in this equation is evaluated from $x = 0$, where $\Psi(0) = v$, to $x = h$,

* See Chapter 2, Section 2.7.1, Eq. (2.7.3).

where $\Psi(h) = 0$, making use of $d\psi/dx = -V/h$. The result

$$\int_0^h e^{z_j\Psi(x)}\,dx = \int_v^0 e^{z_j\Psi(x)}\left(\frac{dx}{d\Psi}\right)d\Psi = -\frac{h}{v}\int_v^0 e^{z_j\Psi(x)}\,d\Psi$$

$$= -\frac{h}{z_j v}[1 - e^{z_j v}]$$

is inserted into Eq. (3.7.21) to give the following expression for the flux of the ion of the type j

$$J_j = z_j v\left(\frac{D_j}{h}\right)\frac{C_j(h) - C_j(0)\,e^{z_j v}}{1 - e^{z_j v}}.$$

Next by using Eq. (3.7.23) we substitute the membrane concentrations $C_j(0)$ and $C_j(h)$ with the concentrations $C_j^{(i)}$ and $C_j^{(o)}$ in the surrounding media . This gives

$$J_j = z_j v\left(\frac{\alpha_j D_j}{h}\right)\frac{C_j^{(o)} - C_j^{(i)}\,e^{z_j v}}{1 - e^{z_j v}}$$

$$= z_j P_j v\frac{C_j^{(o)} - C_j^{(i)}\,e^{z_j v}}{1 - e^{z_j v}}, \qquad (3.7.25)$$

where the permeability P_j is defined as

$$P_j = \frac{\alpha_j D_j}{h}, \qquad (3.7.26)$$

in agreement with Eq. (2.7.5).

We now assume that the membrane is permeable only to the three ions K^+, Na^+ and Cl^-. The fluxes for these ions* at the membrane potential $V = (\mathcal{R}T/\mathcal{F})v$ are

$$\left.\begin{aligned}
J_K &= P_K\, v\frac{C_K^{(o)} - C_K^{(i)}\,e^v}{1 - e^v}\\[2mm]
J_{Na} &= P_{Na}\, v\frac{C_{Na}^{(o)} - C_{Na}^{(i)}\,e^v}{1 - e^v}\\[2mm]
J_{Cl} &= -P_{Cl}\, v\frac{C_{Cl}^{(i)} - C_{Cl}^{(o)}\,e^v}{1 - e^v}
\end{aligned}\right\} \qquad (3.7.27)$$

---

* We have $z_K = z_{Na} = 1$ and $z_{Cl} = -1$.

together with the corresponding ionic currents $I_j = z_j J_j{}^*$

$$
\left.
\begin{aligned}
I_K &= \mathcal{F} J_K = \mathcal{F} P_K\, v\, \frac{C_K^{(o)} - C_K^{(i)}\, e^v}{1 - e^v} \\[2mm]
I_{Na} &= \mathcal{F} J_{Na} = \mathcal{F} P_{Na}\, v\, \frac{C_{Na}^{(o)} - C_{Na}^{(i)}\, e^v}{1 - e^v} \\[2mm]
I_{Cl} &= -\mathcal{F} J_{Cl} = \mathcal{F} P_{Cl}\, v\, \frac{C_{Cl}^{(i)} - C_{Cl}^{(o)}\, e^v}{1 - e^v}
\end{aligned}
\right\}.
\qquad (3.7.28)
$$

The total membrane current is

$$
I = I_K + I_{Na} + I_{Cl}.
$$

Inserting the above values for I_{Na}, I_K and I_{Cl} gives

$$
I = \mathcal{F} v\, \frac{P_K C_K^{(o)} + P_{Na} C_{Na}^{(o)} + P_{Cl} C_{Cl}^{(i)} - \left[P_K C_K^{(i)} + P_{Na} C_{Na}^{(i)} + P_{Cl} C_{Cl}^{(o)} \right] e^v}{1 - e^v}.
$$

$$(3.7.29)$$

It appears that the total current I is related to the membrane potential V in a nonlinear manner. But this is not a matter of concern just now, since our only interest is to find that particular value V_m of the membrane potential – the diffusion potential for the Goldman regime – that corresponds to the current free state to which the system adapts when not impressed by external electric forces. The condition

$$
I = 0, \quad \text{when} \quad V = V_m, \quad \text{and} \quad v_m = V_m/(\mathcal{R}T/\mathcal{F}),
$$

applied to Eq. (3.7.29) gives

$$
\mathcal{F} v_m\, \frac{P_K C_K^{(o)} + P_{Na} C_{Na}^{(o)} + P_{Cl} C_{Cl}^{(i)} - \left[P_K C_K^{(i)} + P_{Na} C_{Na}^{(i)} + P_{Cl} C_{Cl}^{(o)} \right] e^{v_m}}{1 - e^{v_m}} = 0,
$$

which requires that the numerator in the above fraction is zero[†]. Hence we have

$$
P_K C_K^{(o)} + P_{Na} C_{Na}^{(o)} + P_{Cl} C_{Cl}^{(i)} - \left[P_K C_K^{(i)} + P_{Na} C_{Na}^{(i)} + P_{Cl} C_{Cl}^{(o)} \right] e^{v_m} = 0,
$$

or by rearranging

$$
e^{v_m} = \frac{P_K C_K^{(o)} + P_{Na} C_{Na}^{(o)} + P_{Cl} C_{Cl}^{(i)}}{P_K C_K^{(i)} + P_{Na} C_{Na}^{(i)} + P_{Cl} C_{Cl}^{(o)}}.
$$

---

[*] In membrane biology these equations are often called the "constant field equations".
[†] The left-hand side of the equation does not become zero for $v = 0$ as $v/(1 - \exp v) \to -1$, when $v \to 0$.

Taking the logarithm on both sides gives*

$$v_m = \ln\left(\frac{P_K C_K^{(o)} + P_{Na} C_{Na}^{(o)} + P_{Cl} C_{Cl}^{(i)}}{P_K C_K^{(i)} + P_{Na} C_{Na}^{(i)} + P_{Cl} C_{Cl}^{(o)}}\right).$$

Using $\psi^{(i)} - \psi^{(o)} = V_m = v_m \Re T/\mathscr{F}$ we finally obtain

$$V_m = \frac{\Re T}{\mathscr{F}} \ln\left(\frac{P_K C_K^{(o)} + P_{Na} C_{Na}^{(o)} + P_{Cl} C_{Cl}^{(i)}}{P_K C_K^{(i)} + P_{Na} C_{Na}^{(i)} + P_{Cl} C_{Cl}^{(o)}}\right), \tag{3.7.30}$$

which is the celebrated form of the Goldman equation given by Hodgkin & Katz in 1949.

A few controls of the Goldman regime may perhaps be appropriate. We take the flux equation as a starting point

$$J_j = z_j P_j v \frac{C_j^{(o)} - C_j^{(i)} e^{z_j v}}{1 - e^{z_j v}}. \tag{3.7.31}$$

The membrane potential $V_j^{(eq)}$ corresponding to $J_j = 0$ – that is, where electrochemical equilibrium across the membrane holds for the ion – is determined by

$$C_j^{(o)} - C_j^{(i)} e^{z v_j^{(eq)}} = 0,$$

or

$$v_j^{(eq)} = \frac{1}{z_j} \ln\left(\frac{C_j^{(o)}}{C_j^{(i)}}\right), \tag{3.7.32}$$

which is identical to the Nernst equation for the equilibrium potential.

Short circuiting the membranes by forcing the membrane potential to the value $v = 0$ should result in a purely diffusive flux. We have

$$\lim_{v \to 0}\left(\frac{z_j v}{1 - e^{z_j v}}\right) \to -1,$$

and Eq. (3.7.31) becomes

$$J_j = P_j\big(C_j^{(i)} - C_j^{(o)}\big),$$

corresponding to the stationary flux of an uncharged particle†.

* See Chapter 1, Section 1.5.1, Eq. (1.5.10) and Eq. (1.5.11).
† See Chapter 2, Section 2.7.1, Eq. (2.7.6).

Putting $C_j^{(i)} = C_j^{(o)}$ in Eq. (3.7.31) gives

$$J_j = z_j P_j v C_j^{(i)} = z_j \frac{\alpha_j D_j}{h} v \frac{C_j(0)}{\alpha_j} = z k T B_j \frac{V_m}{kT/e} \frac{1}{h} C_j(0)$$

$$= B_j C_j(0) z_j e \frac{V_m}{h} = B_j C_j(0) z_j e \left(-\frac{d\psi}{dx}\right) = B_j C_j(0) X_j,$$

where $X_j = z_j e(-d\psi/dx)$ is the force acting on the ion j due to the field. Thus, the mechanism of transfer is – as one should expect – a pure migration*.

3.7.3.2 Concentration profiles

As previously mentioned, the shape of the concentration profiles $C_j(x)$ through the membrane will in general depend on the strength as well as the polarity of a voltage difference impressed by some external means across the membrane. We shall now examine how the profiles of the Goldman regime behave in this respect. We use the same tactic as in Chapter 2, Section 2.7.4.3, and express the flux partly by the values $C_j(0)$, $\psi(0)$ and $C_j(h)$, $\psi(h)$ at the boundary positions $x = 0$ and $x = h$, and partly by the value $C_j(0)$, $\psi(0)$ and $C_j(x)$, $\psi(x)$ corresponding to an arbitrarily chosen position $0 \le x \le h$ inside the membrane. As the flux is the same for all values of x, we have two identical expressions that, when combined, make an equation to be solved with respect to $C_j(x)$.

We start from the Nernst–Planck equation in the form of Eq. (3.7.20)

$$-\frac{J_j}{D_j} e^{z_j \psi(x)} dx = -\frac{J_j}{D_j} e^{z_j \psi(x)} \left(\frac{dx}{d\Psi}\right) d\Psi = d\left(C_j(x) e^{z_j \psi(x)}\right).$$

According to the assumption the potential profile is $\Psi(x) = v(1 - x/h)$. Hence, $dx/d\Psi = -h/v$. We have therefore

$$\frac{h J_j}{D_j v} e^{z_j \psi(x)} d\Psi = d\left(C_j(x) e^{z_j \psi(x)}\right). \tag{3.7.33}$$

We integrate this expression from the position $x = 0$, where $\Psi(0) = v$ and $C_j(x) = C_j(0) = \alpha_j C_j^{(i)}$ to $x = h$, where $\Psi(h) = 0$ and $C_j(x) = C_j(h) = \alpha_j C_j^{(o)}$. This gives

$$\frac{h J_j}{D_j v} \int_v^0 e^{z_j \psi(x)} d\Psi = \int_{C_j(0)}^{C_j(h)} d\left(C_j(x) e^{z_j \psi(x)}\right),$$

$$\frac{h J_j}{z_j D_j v}(1 - e^{z_j v}) = \alpha_j \left(C_j^{(o)} - C_j^{(i)} e^{z_j v}\right). \tag{3.7.34}$$

---

* In the substitutions enter in succession the results from Chapter 2, Section 2.7.4.1, Eq. (2.7.5), Chapter 2, Section 2.6.4, Eq. (2.6.54), Chapter 2, Section 2.4.2, Eq. (2.4.9).

We now integrate Eq. (3.7.33) from $x = 0$ to a position x inside the membrane $0 \leq x \leq h$, where the potential is $\Psi(x)$ and concentration $C_j(x)$. This gives

$$\frac{hJ_j}{z_j D_j v}\left(1 - e^{z_j\Psi(x)}\right) = \alpha_j C_j^{(0)} - C_j(x)\,e^{z_j\Psi(x)}. \qquad (3.7.35)$$

Dividing Eq. (3.7.35) by Eq. (3.7.34) yields

$$\frac{1 - e^{z_j\Psi(x)}}{1 - e^{z_j v}} = \frac{\alpha_j C_j^{(0)} - C_j(x)\,e^{z_j\Psi(x)}}{\alpha_j\left(C_j^{(0)} - C_j^{(i)}(x)\,e^{z_j v}\right)}.$$

Solving with respect to $C(x)$ gives the equation for the concentration profile as

$$C_j(x) = \alpha_j \frac{C_j^{(i)}e^{z_j v} - C_j^{(0)} - \left[C_j^{(i)} - C_j^{(0)}\right]e^{z_j(v-\Psi(x))}}{e^{z_j v} - 1}.$$

Now $\Psi(x) = v(1 - x/h)$, whence $v - \Psi(x) = vx/h$. Replacing then v by $\mathscr{F}V/\mathscr{R}T$ gives finally

$$C_j(x) = \alpha_j \frac{C_j^{(i)}\exp\{z_j\mathscr{F}V/\mathscr{R}T\} - C_j^{(0)} - \left[C_j^{(i)} - C_j^{(0)}\right]\exp\{(z_j\mathscr{F}V/\mathscr{R}T)(x/h)\}}{\exp\{z_j\mathscr{F}V/\mathscr{R}T\} - 1}.$$

$$(3.7.36)$$

Note the similarity between this formula and Eq. (2.7.68), which describes the concentration profile in the situation where a diffusion process is superimposed by convective flow.

Figure 3.17 shows the stationary concentration profiles through the membranes when the membrane potential V is held respectively at a negative

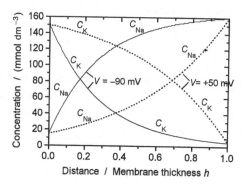

Fig. 3.17. Concentration profiles of Na^+, K^+ and Cl^- ions through a membrane of thickness h for two values of membrane potentials. Abscissa: position x in membrane in units of thickness h. Ordinate: concentration (mmol dm^{-3}). NB: The field in the membrane is constant because the total concentration is the same on both sides of the membrane.

(-90 mV) value and a positive ($+50$ mV) value. The concentrations in the surrounding media are $C_K^{(i)} = 150$, $C_K^{(o)} = 5$, $C_{Na}^{(i)} = 15$, $C_{Na}^{(o)} = 160$, $C_{Cl}^{(o)} = 165$, and $C_{Cl}^{(i)} = 165$. As the total concentrations $C^{(i)}$ and $C^{(o)}$ on either side are the same, the field in the membrane is constant. Experimentally it is possible to generate the change in the membrane potential almost instantaneously. The concentration profiles, however, do not change concurrently with the imposed potential change, since the rearrangement in space of the material involved will have to take a certain time until the new stationary state is established. To calculate the transition from the one stationary state to the other implies solving the *time-dependent* Nernst–Planck equations for each of the $1, 2, \ldots, n$ ion species that are involved in the process

$$\frac{\partial C_j}{\partial t} = -\frac{\partial J_j}{\partial x} \qquad j = 1, 2, \ldots, n,$$

where J_j is the flux*, together with Poisson's equation

$$\frac{\partial^2 \psi}{\partial x^2} = -\frac{\rho}{\varepsilon_o K}.$$

The result of such a calculation showing the transition of the concentration profiles between two stationary states is illustrated in Fig. 3.18. The calculations and numerical computations are due to Cohen, J. & Cooley, J.W. (1965).

3.7.3.3 Membrane conductance and membrane permeability

In principle, the partial ionic conductances G_j are calculated by inserting the analytic expression Eq. (3.7.35) for the concentration profile $C_j(x)$ through the membrane into the general formula (Eq. (3.3.27))

$$\frac{1}{G_j} = \int_0^h \frac{1}{z_j \mathscr{F} u_j C_j^{(x)}} dx \qquad (3.7.37)$$

for the conductance G_j. The evaluation is lengthy and rather tedious. But in the present case we can make a short cut, as we already know the expressions for the individual ion currents corresponding to the Goldman regime (Eq. (3.7.28)). These currents enter formally into the defining equation for the ion conductance (Eq. (3.4.21))

$$I_j = G_j [V - V_j^{(eq)}], \qquad (3.7.38)$$

---

* Using one of the forms written in Section 3.3.2.1.

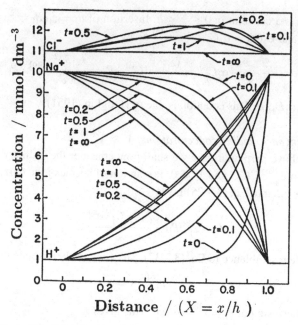

Fig. 3.18. Illustration of the time development of the transition of the concentration profile from one steady state to the other in the system 0.1 M NaCl + 0.01 M HCl |*membrane*| 0.1 M NaCl + 0.01 M HCl. In the inital steady state a current of $I = 0.5$ A cm^{-2} flows through the system in the direction left → right. At time $t = 0$ the current is cut off. Abscissa: distance through the membrane (in units of 10^{-2} cm). Ordinate: concentration ($C_j = 1.0 \equiv 0.1$ mol dm^{-3}). The numbers attached to the curves represent the time(s) after the current switches to $I = 0$. The curves are obtained by solving the time-dependent Nernst–Planck equations. (After Cohen & Cooley, 1965.*)

thus making it possible to write down the expression for G_j for the Goldman regime.

(i) Concentrations equal on both sides

We begin by considering the simple situation where the ions of species j in the two phases are – or very nearly are – in equilibrium. This requires a zero potential difference across the membrane. As the concentration profiles for the ions considered are horizontal, i.e. $C_j(x) = C_j 0 = C_j h$, we have from Eq. (3.7.37)

$$G_j = \frac{z_j \mathscr{F} u_j C_j(x)}{h}$$

* Cohen, R. & Cooley, J.W. (1965): The numerical solution of the time-dependent Nernst–Planck equations. *Biophys. J.*, **5**, 145.

3. Membrane potentials

since $C_j(x)$ is a constant in $0 \leq x \leq h$. Insertion of $u_j = z_j e B_j = z_j e D_j / kT = z_j \mathscr{F} D_j / \mathscr{R} T$ and $C_j(x) = \alpha_j C_j^{(i)} = \alpha_j C_j^{(o)}$ gives

$$G_j = \frac{z_j \mathscr{F}}{h} \frac{z_j \mathscr{F} D_j}{\mathscr{R} T} \alpha_j C_j^{(i)} = \frac{(z_j \mathscr{F})^2}{\mathscr{R} T} C_j^{(i)} \cdot P_j = \frac{(z_j \mathscr{F})^2}{\mathscr{R} T} C_j^{(o)} \cdot P_j, \quad (3.7.39)$$

since $P_j = \alpha_j D_j / h$. This equation is due to Hodgkin (1951).

(ii) Different surrounding concentrations: $V \approx V_j^{(eq)}$

From the above particular case we shall now consider the ordinary Goldman regime with $C_j^{(i)} \neq C_j^{(o)}$ and $V \neq 0$. For an ion of species j the current is $I_j = z_j \mathscr{F} J_j$. Insertion into Eq. (3.7.25) gives

$$I_j = z_j^2 \mathscr{F} P_j v \frac{C_j^{(o)} - C_j^{(i)} e^{z_j v}}{1 - e^{z_j v}}, \quad (3.7.40)$$

having little resemblance to Eq. (3.7.37). Since we wish to isolate the term

$$v_j - v_j^{(eq)} = \Delta v \quad (3.7.41)$$

in Eq. (3.7.40) we put $v_j = v_j^{(eq)} + \Delta v$. This gives

$$I_j = z_j^2 \mathscr{F} P_j v \frac{C_j^{(o)} - C_j^{(i)} e^{z_j v_j^{(eq)}} e^{z_j \Delta v}}{1 - e^{z_j v}} = z_j^2 \mathscr{F} P_j v C_j^{(o)} \frac{1 - e^{z_j (v - v_j^{(eq)})}}{1 - e^{z_j v}},$$

where the right-hand term is obtained by using

$$C_j^{(o)} = C_j^{(i)} e^{z_j v_j^{(eq)}},$$

according to Eq. (3.7.32). We then expand the exponential* in the numerator. Since we are nearly at equilibrium ($v_j^{(eq)} \approx v$) the first two terms $e^{z_j(v - v_j^{(eq)})} \approx 1 + z_j(v - v_j^{(eq)})$ will do. This linearization gives

$$I_j = -z_j^3 \mathscr{F} P_j C_j^{(o)} \frac{v}{1 - e^{z_j v}} [v - v_j^{(eq)}].$$

Insertion of $v = \mathscr{F} V / \mathscr{R} T$ and $v_j^{(eq)} = \mathscr{F} V_j^{(eq)} / \mathscr{R} T$, gives

$$I_j = -\frac{(z_j \mathscr{F})^3 P_j C_j^{(o)} V}{(\mathscr{R} T)^2 [1 - e^{z_j \mathscr{F} V / \mathscr{R} T}]} [V - V_j^{(eq)}]. \quad (3.7.42)$$

This expression is identical to Eq. (3.7.38), if we put

$$G_j = -\frac{(z_j \mathscr{F})^3 C_j^{(o)} V}{(\mathscr{R} T)^2 [1 - e^{z_j \mathscr{F} V / \mathscr{R} T}]} \cdot P_j, \quad (3.7.43)$$

---

* See Chapter 1, Section 1.6.

which is the expression for the conductance G_j for the ion of species j provided that the membrane potential V does not deviate more than a few millivolts from the equilibrium potential $V_j^{(eq)}$ for the ion in question. To estimate the accuracy, consider the approximation

$$1 - e^{z_j \left(v - v_j^{(eq)}\right)} \approx -z_j \left(v - v_j^{(eq)}\right), \qquad (3.7.44)$$

and $z_j(v - v_j^{(eq)}) = 0.1$, which corresponds to a deviation from equilibrium amounting to $V - V_j^{(eq)} \leq 0.1 \times 25.7 = 2.5$ mV. Then the left-hand side of Eq. (3.7.44) equals -0.1051. Thus, Eq. (3.7.43) is correct with a precision of 5% for deviations from equilibrium of ± 2.5 mV.

If the ion is in equilibrium then $V = V_j^{(eq)}$, that is

$$V = \frac{\mathcal{R}T}{z_j \mathcal{F}} \ln \left(\frac{C_j^{(o)}}{C_j^{(i)}}\right) \quad \text{and} \quad e^{z_j \mathcal{F} V / \mathcal{R}T} = \frac{C_j^{(o)}}{C_j^{(i)}},$$

which inserted in Eq. (3.7.43) gives

$$G_j = -\frac{(z_j \mathcal{F})^2 C_j^{(i)} C_j^{(o)}}{\mathcal{R}T \left(C_j^{(i)} - C_j^{(o)}\right)} \ln \left(\frac{C_j^{(o)}}{C_j^{(i)}}\right) \cdot P_j. \qquad (3.7.45)$$

This expression could also be obtained by calculating G_j from the equation of definition Eq. (3.7.37) if the concentration profile $C_j(x)$ for the ion at equilibrium is known. To determine $C_j(x)$ we apply the equilibrium condition $J_j = 0$ to Eq. (3.7.20). This gives

$$d \left(C_j(x) e^{z_j \Psi(x)}\right) = 0, \quad \text{for } 0 \leq x \leq h,$$

or

$$C_j(x) e^{z_j \Psi(x)} = A,$$

where A is a constant, the value of which is obtained from the boundary condition: $C_j(x) = C_j(0) = \alpha_j C_j^{(i)}$ and $\Psi(0) = v_j^{(eq)}$ for $x = 0$, which leads to

$$A = \alpha_j C_j^{(i)} e^{z_j v_j^{(eq)}},$$

and therefore

$$C_j(x) = \alpha_j C_j^{(i)} e^{z_j v_j^{(eq)}} \cdot e^{-z_j \Psi(x)} = \alpha_j C_j^{(i)} e^{z_j \left(v_j^{(eq)} - \Psi(x)\right)}.$$

Since the potential profile in the membrane is linear

$$\Psi(x) = v_j^{(eq)} - \left(\frac{v_j^{(eq)}}{h}\right) x,$$

the concentration profile at equilibrium has to obey the relation

$$C_j(x) = \alpha_j C_j^{(i)} e^{z_j(v_j^{(eq)}/h)x}.$$

Inserting this expression in Eq. (3.7.37) gives

$$R_j = \int_0^h \left(z_j \mathcal{F} u_j \alpha_j C_j^{(i)} e^{z_j(v_j^{(eq)}/h)x}\right)^{-1} dx = \left(z_j \mathcal{F} u_j \alpha_j C_j^{(i)}\right)^{-1} \int_0^h e^{-z_j(v_j^{(eq)}/h)x} dx$$

$$= \frac{1}{z_j \mathcal{F} u_j \alpha_j C_j^{(i)}} \cdot \frac{-1}{z_j v_j^{(eq)}/h} \left[e^{-z_j(v_j^{(eq)}/h)x}\right]_0^h = \frac{-h}{z_j^2 \mathcal{F} \alpha_j u_j v_j^{(eq)} C_j^{(i)}} \left[e^{-z_j v_j^{(eq)}} - 1\right].$$

Hence

$$G_j = -\frac{z_j^2 \mathcal{F} \alpha_j u_j v_j^{(eq)} C_j^{(i)}}{h\left[e^{-z_j v_j^{(eq)}} - 1\right]} = -\frac{z_j^2 \mathcal{F} \alpha_j u_j v_j^{(eq)} C_j^{(i)} e^{z_j v_j^{(eq)}}}{h\left[1 - e^{z_j v_j^{(eq)}}\right]}.$$

Substituting $\exp[z_j v_j^{(eq)}]C_j^{(i)} = C_j^{(o)}$; $\alpha_j u_j/h = \alpha_j z_j \mathcal{F} D_j/h \mathcal{R} T = z_j \mathcal{F} P_j/\mathcal{R} T$ and $v_j^{(eq)} = \mathcal{F} V_j^{(eq)}/\mathcal{R} T$ gives

$$G_j = -\frac{(z_j \mathcal{F})^3 C_j^{(o)} V_j^{(eq)}}{(\mathcal{R} T)^2 \left[1 - e^{z_j \mathcal{F} V_j^{(eq)}/\mathcal{R} T}\right]} \cdot P_j, \qquad (3.7.46)$$

which is identical to Eq. (3.7.42)*, provided we replace V by $V_j^{(eq)}$.

(iii) $V \neq V_j^{(eq)}$

The more frequent occurrence is that where the membrane potential differs from each of the equilibrium potentials $V_j^{(eq)}$ of the permeable ions. We take the expression for the ionic current

$$I_j = \frac{(z_j \mathcal{F})^2 P_j V\left[C_j^{(o)} - C_j^{(i)} e^{z_j \mathcal{F} V/\mathcal{R} T}\right]}{\mathcal{R} T(1 - e^{z_j \mathcal{F} V/\mathcal{R} T})},$$

where the normalized membrane potential $v = \mathcal{F} V/\mathcal{R} T$ is replaced by V. The right-hand side is multiplied by unity given by $(V - V_j^{(eq)})/(V - V_j^{(eq)})$. This gives

$$I_j = \frac{(z_j \mathcal{F})^2 P_j V\left[C_j^{(o)} - C_j^{(i)} e^{z_j \mathcal{F} V/\mathcal{R} T}\right]}{\mathcal{R} T(1 - e^{z_j \mathcal{F} V/\mathcal{R} T})(V - V_j^{(eq)})}(V - V_j^{(eq)}),$$

---

* The former method has the advantage of providing some insight of the value of a statement like "this equation is valid when the ion is near equilibrium" or like "for very small currents we have the following relation".

which is now put into the form $I_j = G_j(V - V_j^{(eq)})$. It then follows by inspection that

$$G_j = \frac{(z_j \mathscr{F})^2 \, V \left[C_j^{(o)} - C_j^{(i)} e^{z_j \mathscr{F} V / \mathscr{R} T} \right]}{\mathscr{R} T (1 - e^{z_j \mathscr{F} V / \mathscr{R} T})(V - V_j^{(eq)})} \cdot P_j, \qquad (3.7.47)$$

which is the expression for the conductance for the ion j at any value of the membrane potential V.

Insertion of the expression for the equilibrium potential

$$V_j^{(eq)} = -\frac{\mathscr{R} T}{z_j \mathscr{F}} \ln \left(\frac{C_j^{(i)}}{C_j^{(o)}} \right)$$

gives

$$G_j = \frac{(z_j \mathscr{F})^3 \left[C_j^{(o)} - C_j^{(i)} e^{z_j \mathscr{F} V / \mathscr{R} T} \right]}{(\mathscr{R} T)^2 (1 - e^{z_j \mathscr{F} V / \mathscr{R} T}) \left(z_j \mathscr{F} V / \mathscr{R} T + \ln \left(C_j^{(i)} / C_j^{(i)} \right) \right)} \cdot P_j, \qquad (3.7.48)$$

where G_j is now given by $C_j^{(o)}$ $C_j^{(i)}$ and V. This result could also be obtained by integration of Eq. (3.7.37), but the method used above is faster and less laborious. Equation (3.7.47) includes the previous, more restricted, formulas for G_j. When $V \to V_j^{(eq)}$ we obtain Eq. (3.7.43), which was derived assuming $V \approx V_j^{(eq)}$, and letting $V \to 0$ together with $C_j^{(i)} \to C_j^{(o)}$ gives Eq. (3.7.39).

3.7.3.4 Total current and membrane potential

The connection between the total membrane current $I^{(tot)} = I_{Na} + I_K + I_{Cl}$ and the membrane potential V was derived in Section 3.7.3.1. In this relation we introduced the surrounding concentrations C_{Na}, C_K and C_{Cl} and the separate ion permeabilities P_{Na}, P_K and P_{Cl}. We used this expression to bring to light one particular value of the membrane potential, that is the current free diffusion potential or the membrane potential V_m of the Goldman regime given by Eq. (3.7.30). We shall now derive an expression for the I–V relation when $V \neq V_m$ and corresponding to the situation where the field in the membrane is constant because the total concentrations are the same on both sides of the membrane. In our case this implies that $C_{Cl}^{(i)} = C_{Cl}^{(o)}$. It will also be more instructive to use ion conductances u_j instead of ion permeabilities $P_j = \alpha_j D_j / h$, which we shall rewrite as

$$P_j = \frac{\alpha_j}{h} kT \, B_j = \frac{\alpha_j kT}{he} e B_j = \frac{\alpha_j \mathscr{R} T}{h \mathscr{F}} u_j,$$

as $D = \mathcal{k}T\,B_j$ and $u_j = eB_j$ [*]. Hence it follows that

$$v P_j = \frac{\mathcal{F} V}{\mathcal{R} T} P_j = \left(\frac{V}{h}\right)\alpha_j u_j,$$

which inserted in Eq. (3.7.30) gives

$$I = \mathcal{F}\left(\frac{V}{h}\right)\left\{\frac{u_K\alpha_K C_K^{(o)} + u_{Na}\alpha_{Na} C_{na}^{(o)} + u_{Cl}\alpha_{Cl} C_{Cl}^{(i)}}{1 - e^{z_j \mathcal{F} V/\mathcal{R} T}}\right.$$
$$\left. - \frac{u_K\alpha_K C_K^{(i)} + u_{Na}\alpha_{Na} C_{Na}^{(i)} + u_{Cl}\alpha_{Cl} C_{Cl}^{(o)}}{1 - e^{z_j \mathcal{F} V/\mathcal{R} T}} e^{\mathcal{F} V/\mathcal{R} T}\right\}. \qquad (3.7.49)$$

A constant field requires in our case – a Planck regime – that $C_{Cl}^{(i)} = C_{Cl}^{(o)}$. The first numerator in Eq. (3.7.48) can then be written as

$$\kappa^{(o)} = \mathcal{F}\left(u_K\alpha_K C_K^{(o)} + u_{Na}\alpha_{Na} C_{Na}^{(o)} + u_{Cl}\alpha_{Cl} C_{Cl}^{(o)}\right),$$

which represents the total membrane conductance if the membrane were surrounded on both sides by the solution in phase (o). In the same manner

$$\kappa^{(i)} = \mathcal{F}\left(u_K\alpha_K C_K^{(i)} + u_{Na}\alpha_{Na} C_{Na}^{(i)} + u_{Cl}\alpha_{Cl} C_{Cl}^{(i)}\right)e^{\mathcal{F} V/\mathcal{R} T}$$

is the total membrane conductance corresponding to the ions in phase (i). Thus, we have

$$I = \frac{\kappa^{(o)} - \kappa^{(i)} e^{\mathcal{F} V/\mathcal{R} T}}{1 - e^{\mathcal{F} V/\mathcal{R} T}}\left(\frac{V}{h}\right) = \frac{\kappa^{(o)} e^{-\mathcal{F} V/\mathcal{R} T} - \kappa^{(i)}}{e^{-\mathcal{F} V/\mathcal{R} T} - 1}\left(\frac{V}{h}\right).$$
$$(3.7.50)$$

For large positive values of V we have $\exp\left(-\mathcal{F} V/\mathcal{R} T\right) \approx 0$ and

$$I = \kappa^{(i)}\left(\frac{V}{h}\right) = \kappa^{(i)} E,$$

in agreement with Eq. (3.3.16), as the field in the membrane is $V/h = E$.[†] The total conductance of the membrane is thus determined by the ions in phase (i). The electric field $E = V/h$ is directed (i) → (o), and will drive Na^+ and K^+ ions from phase (i) through the membrane. If the field is sufficiently large the ions arising from phase (i) will accumulate in the membrane and the concentration profiles $C_K(x) = \alpha_K C_K^{(i)}$ and $C_{Na}(x) = \alpha_{Na} C_{Na}^{(i)}$ will undergo no further changes with the field.

---

[*] See Chapter 2, Section 2.7.4.1, Eq. (2.6.54), and Section 3.3.1, Eq. (3.3.5).
[†] See Section 3.3.1, Eq. (3.3.1).

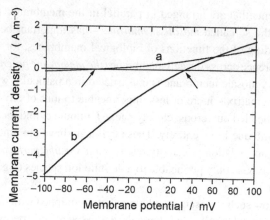

Fig. 3.19. Two examples of the relation between the total current through the membrane and the potential difference across the membrane calculated from Eq. (3.7.47). The concentrations in the surrounding media are (in mol dm^{-3}): $C_{Cl}^{(o)} = C_{Cl}^{(o)} = 165$, $C_{Na}^{(i)} = 15$, $C_{Na}^{(o)} = 160$, $C_{K}^{(i)} = 150$, $C_{K}^{(o)} = 5$. $P_K = 1.5 \times 10^{-8}$ m s^{-1}. Abscissa: membrane potential (mV). Ordinate: membrane current (A m^{-2}). Curve (a): $P_{Na} = 0.01\, P_K$, $P_{Cl} = 0.1 P_K$. Curve (b): $P_{Na} = 5.0\, P_K$, $P_{Cl} = 0.1 P_K$.

For large negative values of V we have: $\exp(\mathscr{F}V/\mathscr{R}T) \approx 0$ for $V \ll 0$. This gives

$$I = \kappa^{(o)} \left(\frac{V}{h}\right) = \kappa^{(o)} E,$$

where the conductance of the membrane is now dominated by the concentrations of Na$^+$ and K$^+$ in phase (o) and of their conductances in the membrane. The rectification ratio Y between these two states is

$$Y = \frac{I_{V \gg 0}}{I_{V \ll 0}} = \frac{\kappa^{(i)}}{\kappa^{(o)}}. \tag{3.7.51}$$

A substantial deviation from this ratio indicates that some additional mechanisms may be involved apart from those that can be attributed to a pure passive diffusion through a homogeneous membrane according to the Goldman regime. Figure 3.19 shows two examples of $I - V$ relations calculated from Eq. (3.7.47).

3.7.4 The mosaic membrane (the Millman equation)

All the transfer processes described up to now have taken place in a homogeneous medium or through layers arranged in series. We shall now consider a membrane with a heterogeneous structure consisting of a variety of distinct elements that are arranged in parallel in the membrane. This kind of membrane – the so-called *mosaic membrane* – has important bearings upon the understanding of the functions of biological membranes, for example the potential difference across the cell membrane (*the membrane potential*), among other things. A mosaic membrane is constructed in a particular way such that it consists of a matrix – more or less impermeable to ions of any kind. In this matrix are imbedded numerous, closely packed, minute cylinders, that all penetrate the membrane in its entirety. These cylinders have the unique property of displaying not only ion selectivity but also *ion-specific permeability*, that is each cylinder type is only permeable to a definite ion species (e.g. specific K^+ permeability).

Consider now such a mosaic membrane, that is equipped with three different types of cylinders (also called *ion channels*) being specifically permeable to Na^+ ions, K^+ ions and Cl^- ions. We imagine that the two chambers surrounding the membrane are filled with solutions containing Na^+ ions, K^+ ions and Cl^- ions but of such a composition that the concentrations for each ion species differ. This membrane will be permeable for all three kinds of ions, and the end result will therefore be equal concentrations on both sides for each ion type. But before this situation arises the interdiffusion process will have established a diffusion potential across the membrane and also across all the three types of ion channels. The magnitude of this diffusion potential, the *membrane potential* – written as V_m – shall now be evaluated. We consider first the totality of the Na^+-permeable cylinders (Na channels). If the two other types of ion channel had been absent the membrane would have behaved not only as a selective – but also an Na^+-specific – permeable membrane. The *I–V* characteristic for the membrane would have been described by Eq. (3.4.20) and by an equivalent circuit as shown in Fig. 3.6. The same would apply if the membrane had been equipped with either K^+ channels or Cl^- channels. The membrane that contains all three types of channel will then be represented by an equivalent circuit such as that shown in Fig. 3.20. Let the potential across the membrane be $V_m = \psi^{(i)} - \psi^{(o)}$. The currents that each type of ion carry are given by

$$\left. \begin{array}{l} I_{Na} = G_{Na}\left[V - V_{Na}^{(eq)} \right] \\[2mm] I_K = G_K\left[V - V_K^{(eq)} \right] \\[2mm] I_{Cl} = G_{Cl}\left[V - V_{Cl}^{(eq)} \right] \end{array} \right\}, \qquad (3.7.52)$$

where G_{Na}, G_K and G_{Cl} are the membrane conductances for the Na^+, K^+ and Cl^- ions, that is the conductance represented by number of ion channels per unit

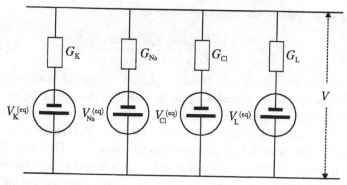

Fig. 3.20. The equivalent circuit for a mosaic membrane that is specifically permeable to Na^+, K^+ and Cl^- ions. The battery $V_L^{(eq)}$ represents the contribution to the membrane current from the additional collection of ionic sources. Owing to their relative smallness, their minor significance is usually neglected.

area for each ionic species. The three equilibrium potentials $V_j^{(eq)}$ are calculated from the Nernst equation, if the concentrations in the surrounding media are known, since

$$V_{Na}^{(eq)} = \frac{\mathcal{R}T}{\mathcal{F}} \ln\left(\frac{C_{Na}^{(o)}}{C_{Na}^{(i)}}\right), \quad V_K^{(eq)} = \frac{\mathcal{R}T}{\mathcal{F}} \ln\left(\frac{C_K^{(o)}}{C_K^{(i)}}\right),$$

$$V_{Cl}^{(eq)} = -\frac{\mathcal{R}T}{\mathcal{F}} \ln\left(\frac{C_{Cl}^{(o)}}{C_{Cl}^{(i)}}\right).$$

If the membrane conductances for the three ions are much smaller than those in free solution, the equalization of concentrations by diffusion would take place very slowly. Because of that the membrane potential V_m would also change slowly with the time. As long as we consider time intervals with which V_m does not decrease noticeably, the net membrane current must be negligibly small, so for all practical purposes we put it equal to zero as we did in the derivation of the Goldman equation (Eq. (3.7.30)). The condition, which is Kirchoff's second law*

$$I_{Na} + I_K + I_{Cl} = 0,$$

combined with Eq. (3.7.52) results in

$$G_{Na}\left[V_m - V_{Na}^{(eq)}\right] + G_K\left[V_m - V_K^{(eq)}\right] + G_{Cl}\left[V_m - V_{Cl}^{(eq)}\right] = 0.$$

* G. Kirchoff (1824–1887) was a German physicist. He was Professor from 1850 in Breslau, 1854 in Heidelberg and 1874 in Berlin. He discovered and formulated the laws for the distribution of the electric current that bear his name, and made important contributions to the elucidation of the physics of radiation of light.

Solving this equation with respect to V_m gives

$$V_m = \frac{G_{Na}\, V_{Na}^{(eq)} + G_{Ka}\, V_K^{(eq)} + G_{Cl}\, V_{Cl}^{(eq)}}{G_{Na} + G_K + G_{Cl}}, \qquad (3.7.53)$$

which we choose to denote the *Millman equation**. It appears from this equation that the membrane potential can be described as the sum of the products of the equilibrium potential and membrane conductance for each participating ion divided by the sum of the membrane conductances, that is the total conductance of the membrane for the ions that cross the membrane. Thus, Eq. (3.7.53) differs from the Goldman equation in the respect that it is now the individual conductances G_K, G_{Na} and G_{Cl} that enter as parameters and not the ionic permeabilities P_K, P_{Na} and P_{Cl}. Furthermore, the concentrations of the surrounding ions K^+, Na^+ and Cl^- enter only implicitly by way of the respective equilibrium potentials $V_K^{(eq)}$, $V_{Na}^{(eq)}$ and $V_{Cl}^{(eq)}$. But both equations express each in their own way the same thing: the magnitude of the membrane potential is determined by that ion whose flux through the membrane is dominant. Beyond this the two procedures for derivation are principally different. The expression for V_m in the Goldman equation derives from assuming from the outset that the system conforms to certain physical conditions, e.g. homogeneity of the membrane, that ion currents obey the Nernst–Planck equations, a constant field prevails in the membrane, etc. With those premises – which makes the merit of the theory – one can then calculate the manner in which the concentration profile and conductances depend upon the membrane potential provided that the ionic permeabilities are known. The Millman equation, however, is based only upon Kirchoff's second law, and is therefore more general and also more flexible than the Goldman equation, but naturally it gives no information about the deeper nature of the conductances. But with a known set of conductances it is a simple matter to calculate the membrane potential V_m. Therefore, several practical methods have been developed to determine the ion conductances by using macroscopic membrane areas of a finite size. During the past 10–20 years experimental techniques have also been developed that use a microelectrode technique to allow the localization of single ion channels in the membrane, and a measuring technique to determine the I–V characteristic of a single ion channel and hence the conductance for the ion in question. Thus, the structure of Eq. (3.7.53) reflects in a reasonably realistic way the situation in the living cell membrane as regards the membrane potential at rest.

---

* An equation of this form was first formulated by the American electroengineer J.M. Millman (*Proc. IRE*, **28** (1940)) in a paper dealing with electronic circuitry, but with the difference that $V_j^{(eq)}$ represented electromotive forces in the form of ordinary batteries and G_j were conductances of usual Ohm resistances.

It may be instructive to see that the Millman equation transforms into the Goldman equation if the expressions for the conductances of the Goldman regime are inserted into Eq. (3.7.53). The current through an Na$^+$ channel, for example, is

$$I_{Na} = G_{Na}\big[V - V_{Na}^{(eq)}\big].$$

Insertion of the general expression for G_{Na} from Eq. (3.7.52) gives

$$I_{Na} = \mathscr{F}\big(C_{Na}^{(o)} - C_{Na}^{(i)} e^{v}\big) \frac{v}{1 - e^{v}} P_{Na},$$

where the term $z_j \mathscr{F} V / \mathscr{R} T$ is replaced by the normalized potential $z_j v = z_j V / (\mathscr{R} T / \mathscr{F})$. The analogous expressions for I_K and I_{Cl} become

$$I_K = \mathscr{F}\big(C_K^{(o)} - C_K^{(i)} e^{v}\big) \frac{v}{1 - e^{v}} P_K,$$

and

$$I_{Cl} = \mathscr{F}\big(C_{Cl}^{(o)} - C_{Cl}^{(i)} e^{-v}\big) \frac{v}{1 - e^{-v}} P_{Cl}$$

$$= -\mathscr{F}\big(C_{Cl}^{(o)} - C_{Cl}^{(i)} e^{-v}\big) \frac{v\, e^{v}}{1 - e^{v}} P_{Cl}.$$

Since $I_{Na} + I_K + I_{Cl} = 0$ we have

$$\mathscr{F}\big(C_{Na}^{(o)} - C_{Na}^{(i)} e^{v}\big) \frac{v}{1 - e^{v}} P_{Na} + \mathscr{F}\big(C_K^{(o)} - C_K^{(i)} e^{v}\big) \frac{v}{1 - e^{v}} P_K$$

$$- \mathscr{F}\big(C_{Cl}^{(o)} - C_{Cl}^{(i)} e^{-v}\big) \frac{v\, e^{v}}{1 - e^{v}} P_{Cl} = 0,$$

from which we obtain

$$\big(C_{Na}^{(o)} - C_{Na}^{(i)} e^{v}\big) P_{Na} + \big(C_K^{(o)} - C_K^{(i)} e^{v}\big) P_K - \big(C_{Cl}^{(o)} - C_{Cl}^{(i)} e^{-v}\big) e^{v} P_{Cl} = 0k$$

or

$$\big(C_{Na}^{(o)} - C_{Na}^{(i)} e^{v}\big) P_{Na} + \big(C_K^{(o)} - C_K^{(i)} e^{v}\big) P_K - \big(C_{Cl}^{(o)} e^{v} - C_{Cl}^{(i)}\big) P_{Cl} = 0.$$

Rearrangement gives

$$P_{Na} C_{Na}^{(o)} + P_K C_K^{(o)} + P_{Cl} C_{Cl}^{(i)} = \big(P_{Na} C_{Na}^{(i)} + P_K C_K^{(i)} + P_{Cl} C_{Cl}^{(o)}\big) e^{v},$$

from which the Goldman equation is recovered after taking the logarithm on both sides and replacing v by $\mathscr{F} V_m / (\mathscr{R} T)$.

Like the Goldman equation*, the Millman equation correctly represents the sign and magnitude of the membrane potential when the membrane

---

* See the end of Section 3.7.3.

permeability – and therefore also the membrane conductance – is chiefly due to one particular ion species. In the extreme case where, for example,

$$G_K \gg G_{Na} \approx G_{Cl},$$

we have the approximation that

$$V_m \approx \frac{G_K V_K^{(eq)}}{G_K} = \frac{\mathcal{R}T}{\mathcal{F}} \ln\left(\frac{C_K^{(o)}}{C_K^{(i)}}\right) = V_K^{(eq)}.$$

It is sometimes useful to apply these extremes in the evaluation of the character of a membrane potential.

The Millman equation is – in contrast to the Goldman equation – also valid for other than monovalent ions. This is due to the fact that the defining equation for the membrane conductance

$$I_j = G_j[V - V_j^{(eq)}], \tag{3.7.54}$$

is valid for any species of ion. Summating over all the currents $1, 2, \ldots, n$ gives

$$I = \sum_{j=1}^{j=n} I_j = V \sum_{j=1}^{j=n} G_j - \sum_{j=1}^{j=n} G_j V_j^{(eq)}$$

or

$$V = \frac{I}{\sum G_j} + \frac{\sum G_j V_j^{(eq)}}{\sum G_j}, \quad j = 1, 2, \ldots, n, \tag{3.7.55}$$

in keeping with the equivalent diagram Fig. 3.20. Putting $I = 0$ in the above equation we obtain the Millman equation in its more general form

$$V = \frac{\sum G_j V_j^{(eq)}}{\sum G_j}. \tag{3.7.56}$$

3.8 The membrane potential of a biological cell

In the preceding sections of this chapter we have described some of the concepts and physical mechanisms that are involved in the passive transfer of ions through a homogenenous medium and through various types of artificial membrane with known properties. One objective was to account for the processes that generate an electric potential difference across these membranes and for the factors that determine the magnitude as well as polarity. In this section we shall illustrate the application of these concepts to living cell membranes. In all living cells a potential difference exists between the cytoplasm of the cell and its extracellular ionic fluid. As this potential difference turns out to be localized entirely across

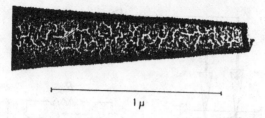

I μ

Fig. 3.21. Electron microscope picture of the tip of a glass microelectrode (before filling with 3 M KCl). To obtain a clear picture the surface of the pipette tip was made electrically conducting by applying a gold spray to the surface, hence the crackled surface. (After Lassen & Rasmussen 1978.)

the cell membrane we call it the *membrane potential of the cell*. According to well-established practice we define the membrane potential as the difference between the potential of the cytoplasm $\psi^{(i)}$ and that of the extracellular phase $\psi^{(o)}$, that is

$$V_m = \psi^{(i)} - \psi^{(o)}. \tag{3.8.1}$$

A negative value of the membrane potential means that potential level in the cell interior is *lower* than that of the surrounding extracellular fluid. It is often profitable to simplify Eq. (3.8.1) by putting $\psi^{(o)} = 0$, but if doubts arise as to sign of polarity, it is useful to return to Eq. (3.8.1).

3.8.1 Measuring the membrane potential

A direct measurement of a membrane potential requires that an electric contact of some kind can be established to the cytoplasm of the cell. Such a contact is nearly always provided by a salt bridge*, worked out as a so-called *microelectrode*. This device is prepared by heating and pulling a thin glass tube in a controlled manner that aims at obtaining the smallest possible tip diameter. In practice diameters down to 0.1–0.2 μm are obtainable. Such a tip is shown in Fig. 3.21. The interior of the micropipette is made electrically conducting by filling it with an electrolyte solution, most frequently a concentrated (3 M) KCl solution that serves a twofold purpose[†]. The K^+ and Cl^- ions have almost the same mobilities. Thus, the diffusion potential at the electrode tip arising from diffusion of KCl from the microelectrode into the cytoplasm can be disregarded. Using a concentrated KCl solution has the effect of making the diffusion potential from KCl the dominating component in the overall diffusion potential at the tip, and thus suppressing the influence of the diffusion into the microelectrode of the ions in the cytoplasm. Furthermore, the use of a concentrated electrolyte

---

* See Section 3.6.1.3 and Section 3.6.2.1.
[†] See the discussion in Section 3.6.3.3.

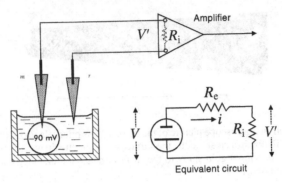

Equivalent circuit

Fig. 3.22. Sketch of set up to measure membrane potential in cells using microelectrodes. For more details see the text.

is advantageous in minimizing the electric resistance of the microelectrode. However, using even a 3M KCl solution, a microelectrode with a tip diameter of $0.1 - 0.2\ \mu\text{m}$ has a considerable resistance of 20–$50\ \text{M}\Omega$, which may necessitate special demands of the recording apparatus. The metallic contact to the micro-electrode is made through an Ag/AgCl electrode that is pushed into the shaft of the electrode[*]. As the reference electrode one can use another microelec-trode – with its tip cut off to reduce the resistance in the circuit – that is placed somewhere in the extracellular medium. In this way the measuring arrangement becomes almost symmetric, which may be preferable in some cases. Of course other types of reference electrode can be used. This may give rise to an initial potential difference at the input of the recording system, which then has to be outbalanced before starting the actual measurements. A sketch of the main ele-ments entering into the experimental set up to measure the membrane potential of a living biological cell is shown in Fig. 3.22. The object, which could be a cross-striated muscle or an isolated single muscle fiber from a frog, is suspended in a chamber that contains an artificial extracellular solution. For frog muscles we use a solution – the so-called *Ringer's solution*[†] – containing NaCl (110 mM), KCl (2.5 mM) and CaCl$_2$ (1.5 mM). The figure shows the chamber with a cross section of a single muscle fiber $\mathcal{M}$ (the actual physical diameter about 100 μm). At the start of the experiment both the measuring electrode (m) and the refer-ence electrode (r) are placed in the extracellular medium (the Ringer's solution). Each Ag/AgCl electrode is connected to the input stage of the recording system.

---

[*] The design of the glass capillary micropipette used to penetrate the cell membrane for recording transmembrane potentials that has gained wide applicability is due to Ling, G. & Gerard, R.W. (1949): The normal resting membrane potential of frog sartorius fibers. *J. Cell. Comp. Physiol.*, **34**, 383.

[†] In memory of the work of the English physiologist Sydney Ringer who in the years 1880–1883 developed the first satisfactory perfusion liquid.

The very large value of the microelectrode resistance places some demand on the design of the input pre-amplifier as to the value of the input resistance R_i – that is the electric resistance (impedance) between the two input sockets of the preamplifier stage. To avoid errors in the potential measurement with the microelectrode the input resistance has to be 200–1000 times larger than the resistance of the recording electrode, i.e. $R_i \approx 10^{10}\Omega$. We want to record a potential difference V between the two electrode tips. If R_e is the total resistance of the two electrodes, the current i that flows in the input circuit is then (according to Ohm's law) $i = V/(R_e + R_i)$. Likewise the voltage V' across the input resistance of the amplifier is: $i = V'/R_i$. Putting these two expressions together gives

$$V' = \frac{R_i}{R_i + R_e} V = \frac{1}{1 + (R_e/R_i)} V.$$

Thus, the voltage applied to the input of the measuring device is always *smaller* than the voltage between the measuring electrodes. The deviation $V - V'$ will eventually become insignificant when $R_i \ll R_e$, which accounts for the value above of $R_i \approx 10^{10}\Omega$. A large value of R_i is also an advantage as it reduces the current that is drawn from the object during the measurement. To avoid the deleterious effects of this current flow in biological cells it is important that the load on the cell is reduced as much as possible. So in that respect the largest obtainable value of R_i is beneficial*.

A faithful recording of fast membrane potential changes with a high resistance microelectrode may be upset by too large a value of the input capacitance. The proper countermeasures to reduce the input capacitance to a minimum and to obtain a sufficiently high input impedance were due to Nastuk & Hodgkin (1950).

The microelectrode m intended for impalement into the cell is attached to a micro-manipulator, which allows precise movements down to displacements of 1 μm. With both electrodes in the extracellular medium one may observe a potential difference of less than a few millivolts, if a symmetric electrode arrangement is employed. No matter its magnitude we adjust this as our zero potential. When the cell membrane is impaled by the microelectrode we observe a new potential difference between the microelectrodes (Fig. 3.23): the potential jumps *suddenly* to a new value – the *membrane potential* – of −90 to −95 mV ($V_m = \psi^{(i)} - \psi^{(o)}$). Now, pulling the microelectrode out to leave both electrode tips again in the extracellular medium, the potential recording returns to its initial position (0 V). A new impalement of the membrane results again in the

---

* Loading a cross-striated muscle fiber with a microelectrode of resistance 10 MΩ connected to the extracellular medium draws current of about $0.1 \text{ V}/10^7 \ \Omega = 10^{-8}$ A. This current will lower the membrane potential in the region around the microelectrode by several millivolts.

3. Membrane potentials

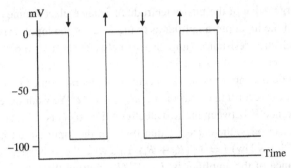

Fig. 3.23. Recording of the membrane potential in a cross-striated muscle fiber from the frog. Ordinate: membrane potential (mV). Abscissa: time (s). Symbols: ↓, impalement of the membrane by the microelectrode; ↑, withdrawal of the microelectrode to the extracellular medium.

sudden jump of the potential whose value is not significantly different from that obtained previously. Although the impalement invariably pricks a hole 1–2 μm in diameter in a membrane of thickness 80 Å, i.e. less than 1/100 of the hole diameter, the withdrawal of the microelectrode leaves no serious leakage as the membrane apparently seals the hole spontaneously, like the merging of an oil film. The duration of the impalement can be prolonged for a long time ($\frac{1}{2}$–1 h) without bringing along any appreciable change in the membrane potential. Thus, during the impalement the membrane probably also seals around the microelectrode. However, this kind of experiment requires a very effective mechanical shielding of the experimental set up to prevent transmission of mechanical vibrations from the environment to the electrode tip that might tear open the membrane beyond its self-sealing capacity.

3.8.2 The origin of the membrane potential

We shall demonstrate in this section that the resting membrane potential of a muscle fiber can – both as regards to polarity and magnitude – be accounted for as the result of a passive transport of ions through the fiber membrane. Our point of departure is the result in Section 3.7.4 that we write in the form

$$V_m = \frac{G_{Na}}{G_{tot}} V_{Na}^{(eq)} + \frac{G_K}{G_{tot}} V_K^{(eq)} + \frac{G_{Cl}}{G_{tot}} V_{Cl}^{(eq)}, \qquad (3.8.2)$$

where $G_{tot} = G_{Na} + G_K + G_{Cl}$ is the conductance of the membrane due to Na^+, K^+ and Cl^- taken together. Thus, the equation states that the membrane potential adapts itself as the result of the ions' concurrent passive diffusion through the membrane to a value that is equal to the sum of the separate contributions of each ion species to membrane potential. Each contribution equals

3.8. The membrane potential of a biological cell

Table 3.4. *Intra- and extracellular ion concentrations.*
Cross-striated muscle fiber from frog; 20 °C; membrane;
potential = −95 mV

Species	Intracellular medium (mM)	Extracellular medium (mM)	Equilibrium potential (mV)
Na^+	13	110	+55
K^+	138	2.5	−103
Cl^-	3	112.5	−93
Ca^{2+}	≤ 0.001	1	≥ +118

the equilibrium potential $V_j^{(eq)}$ multiplied by the relative share of that ion to the total conductance, that is

$$T_j = G_j/G_{tot}. \tag{3.8.3}$$

T_j is called the *transference number* or *transport number* of the ion. These quantities will obey the relation

$$\sum G_j/G_{tot} = \sum T_j = 1. \tag{3.8.4}$$

In terms of the transport numbers, Eq. (3.8.2) then reads

$$V_m = T_{Na} \times V_{Na}^{(eq)} + T_K \times V_K^{(eq)} + T_{Cl} \times V_{Cl}^{(eq)}. \tag{3.8.5}$$

Thus, although the equilibrium potential $V_j^{(eq)}$ for a particular ion may be very large, its contribution to the membrane potential V_m may be insignificant if the permeability to that particular ion is much less than those of the other ions ($T_j \ll 1$).

Table 3.4 shows typical intracellular concentrations of Na^+, K^+, Cl^- and Ca^{2+} ions from a cross-striated muscle fiber from a frog together with their equivalents in the extracellular artificial electrolyte solution (Ringer's solution), that was used during the experimental measurements of the membrane potentials. The table also shows the membrane potential $V_m = \psi^{(i)} - \psi^{(o)}$ and the *calculated* equilibrium for the ions Na^+, K^+, Cl^- and Ca^{2+}. Were the membrane permeable *solely* to K^+ ions, an equilibrium would adapt itself as described in Section 3.4.1 on selectively permeable membranes, and the membrane potential would have attained the value equal to the *equilibrium potential* for the K^+ ions, that is at −103 mV. Were the membrane permeable exclusively to the Na^+ ions we should expect a membrane potential equal to the equilibrium potential $V_{Na}^{(eq)}$, which in the present example is +55 mV. If only the Cl^- ions could penetrate the membrane, the membrane potential would correspond to the equilibrium potential $V_{Cl}^{(eq)}$ for the Cl^- ions, that is -93 mV in the present example. A membrane potential $V_m = -95$ mV and an equilibrium potential for Na^+ ions of $V_{Na}^{(eq)} = +55$ mV is absolutely incompatible with the concept of a membrane that is exclusively permeable to Na^+ ions. Similar considerations hold for the Ca^{2+} ions. As to the K^+ and Cl^- ions, we note that the equilibrium potentials of the two ions (-103 mV and -93 mV) are positioned on either side of the membrane potential (-95 mV), which is suggestive of – and compatible with – the situation where the contribution of both ions to the membrane potential is of the same order of magnitude at the same time as the permeabilities in the membrane largely exceed the permeability of the Na^+ ions, i.e. ($T_K \simeq T_{Cl} \gg T_{Na}$). Tentatively, let us put $T_K = T_{Cl} = 0.5$ and $T_{Na} = 0$ and insert these values in Eq. (3.8.3). This gives

$$V_m = 0.5 \times (-103) + 0.5 \times (-93) = -98 \text{ mV}.$$

It follows from Eq. (3.8.2) that the value of the evaluated membrane potential will always be located somewhere between the extreme values of equilibrium potentials that enter into the formula, that is

$$\left[V_j^{(eq)} \right]_{max} \geq V_m \geq \left[V_j^{(eq)} \right]_{min}. \tag{3.8.6}$$

Considerations such as those above about the nature of the membrane potential would therefore have been without any meaning if the measured membrane potential V_m had a value that oversteps the barriers set by the equilibrium potentials of the diffusing ions (in the above situation $V_{Ca}^{(eq)} = +118$ mV and $V_K^{(eq)} = -103$ mV). Had the membrane potential – with no flaws in the experimental procedure – instead turned out to be e.g. $+150$ mV or -150 mV, one would have had to conclude that mechanisms other than the passive transfer of these three ions through the membrane were participating in the build up of the membrane potential.

   The estimate of a membrane potential of -98 mV belongs to the category of "educated guesses" that may be useful as an incentive to provide the missing relevant data – in the present case the numerical values for the permeabilities in the membrane for K^+, Na^+ and Cl^- ions. In general the data are obtained by experiments using radioisotopes and employing an analysis of exchange as described in Chapter 2, Section 2.7.2. In other cases, electric methods are employed by measuring the connection between an impressed change in the membrane potential ΔV_m and membrane current ΔI_m, provided the current can be ascribed to the movement of one ionic species, e.g. K^+.

   Typical values for the membrane conductances and membrane permeabilities for Na^+, K^+, and Cl^- ions in a cross-striated muscle fiber from a frog are shown in Table 3.5. The permeability of the fiber membrane to K^+ and Cl^- ions is about two order of magnitudes larger than the permeability to Na^+ ions. One should

Table 3.5. *Membrane conductances* G_j *and membrane permeabilities* P_j

Species	G_j (S m^{-2})	P_j (m s^{-1})
Na$^+$	0.8×10^{-2}	0.02×10^{-8}
K$^+$	85×10^{-2}	2.0×10^{-8}
Cl$^-$	170×10^{-2}	4.0×10^{-8}

Cross-striated muscle fiber from frog; membrane potential = -95 mV; 20 °C
Source: Hodgkin & Horowicz, 1959.

therefore expect a membrane potential in the neighborhood of the equilibrium potentials for the K$^+$ and Cl$^-$ ions. We have from Table 3.5:

$$G_{tot} = G_{Na} + G_K + G_{Cl} = 255.8 \times 10^{-2},$$

resulting in the following transport numbers

$$T_{Na} = \frac{0.8}{255.8} = 0.003, \quad T_K = \frac{85}{255.8} = 0.332, \quad T_{Cl} = \frac{170}{255.8} = 0.665.$$

These values and the values for the equilibrium potentials in Table 3.4 are inserted in Eq. (3.8.3) giving the following value for the membrane potential

$$V_m = 0.003 \times 55 + 0.332 \times (-103) + 0.665 \times (-93) = -95.9 \, \text{mV},$$

which agrees well with the value determined experimentally.

We shall now examine whether the values of the membrane permeabilities fits with this result. The starting point is now the Goldman equation (Eq. (3.7.30)). Taking the values for the permeabilities in Table 3.5 and the respective intracellular and extracellular concentrations from Table 3.4 and insert into Eq. (3.7.30) we obtain

$$V_m = \frac{\mathcal{R}T}{\mathcal{F}} \ln \frac{0.02 \times 10^{-8} \cdot 110 + 2.0 \times 10^{-8} \cdot 2.5 + 4.0 \times 10^{-8} \cdot 3.0}{0.02 \times 10^{-8} \cdot 13 + 2.0 \times 10^{-8} \cdot 138 + 4.0 \times 10^{-8} \cdot 112.5}$$

$$= 25.7 \ln \frac{19.2}{726.3} = 25.7 \ln 0.026 \, 435 = -93.4 \, \text{mV}.$$

Hardly any importance can be attached to the difference between this and the above result.

3.8.3 *Membrane potential and ionic concentrations in the extracellular medium*

So far the evidence we have collected about the nature of the membrane potential is not at variance with the concept that we are dealing with a diffusion potential whose value is determined by the distributions extracellularly and intracellularly – or by the equilibrium potentials – of those ions that can most easily penetrate the membrane. Experiments that are designed to examine the correlation network, the equilibrium potential and the membrane potential will now be presented.

3.8.3.1 Sudden changes of both $V_K^{(eq)}$ and $V_{Cl}^{(eq)}$

The equilibrium potentials for the K^+ and Cl^- ions in the previous section were of almost equal value in contributing to the membrane potential. We see from Table 3.4 that the equilibrium potentials $V_K^{(eq)}$ and $V_{Cl}^{(eq)}$ can be reduced partly by increasing the content of K^+ ions in the extracellular medium and partly by reducing that of the Cl^- ions. An increase in the K^+ concentration from the normal 2.5 mM to, for example, 102.5 mM – with a simultaneous reduction of the Na^+ concentration by 100 mM to preserve the tonicity – will push the equilibrium potential for potassium to $V_K^{(eq)} = -8$ mV. However, the concentration of the Cl^- ions is unchanged in the immediate continuation of the change of the K^+ ions and, with an equilibrium potential of $V_{Cl}^{(eq)} = -93$ mV, this distribution of Cl^- ions will counteract the pull arising from altered K^+ concentration. A redistribution of the Cl^- ions will take place in time, and the membrane potential will settle at the value set by the new conditions. We shall not discuss this in detail since it is possible to change *both* equilibrium potentials

$$V_{Cl}^{(eq)} = \frac{\mathcal{R}T}{\mathcal{F}} \ln \frac{C_{Cl}^{(i)}}{C_{Cl}^{(o)}} \quad \text{and} \quad V_K^{(eq)} = \frac{\mathcal{R}T}{\mathcal{F}} \ln \frac{C_K^{(o)}}{C_K^{(i)}} \qquad (3.8.7)$$

by the *same* amount and thus simplifying the experimental condition. This is done by changing the composition of Ringer's solution in the particular manner that the *product of* $C_{Cl}^{(o)} \times C_K^{(o)}$ *is kept constant*. This procedure is due to Hodgkin & Horowicz (1959)[*].

If, as an example, we reduce the extracellular Cl^- concentration by a factor k from $C_{Cl}^{(o)}$ to $C_{Cl}^{(o)}/k$, the equilibrium potential for the Cl^- ions changes by the amount

$$\Delta V_{Cl}^{(eq)} = \frac{\mathcal{R}T}{\mathcal{F}} \ln \frac{C_{Cl}^{(i)}}{C_{Cl}^{(o)}/k} - \frac{\mathcal{R}T}{\mathcal{F}} \ln \frac{C_{Cl}^{(i)}}{C_{Cl}^{(o)}}$$

$$= \frac{\mathcal{R}T}{\mathcal{F}} \ln k + \frac{\mathcal{R}T}{\mathcal{F}} \ln \frac{C_{Cl}^{(i)}}{C_{Cl}^{(o)}} - \frac{\mathcal{R}T}{\mathcal{F}} \ln \frac{C_{Cl}^{(i)}}{C_{Cl}^{(o)}} = \frac{\mathcal{R}T}{\mathcal{F}} \ln k.$$

[*] Hodgkin, A.L. & Horowicz, P. (1959): *J. Physiol.*, **148**, 127.

Table 3.6. *Ringer's solution with [K⁺][Cl⁻]*
product of 300 ((mM)²)

	Na⁺ (mM)	K⁺ (mM)	Cl⁻ (mM)	Ca²⁺ (mM)	SO_4^{2-} (mM)	Sucrose (mM)
A	111.5	2.5	120	3	0	0
B	104	10	30	3	45	45
C	84	30	10	3	55	55
D	34	75	4	3	58	58

A simultaneous increase in the K⁺ concentration from $C_K^{(o)}$ to $kC_K^{(o)}$ changes the equilibrium potential for the K⁺ ions, since

$$\Delta V_K^{(eq)} = \frac{\mathcal{R}T}{\mathcal{F}} \ln \frac{kC_K^{(o)}}{C_K^{(i)}} - \frac{\mathcal{R}T}{\mathcal{F}} \ln \frac{C_K^{(o)}}{C_K^{(i)}} = \frac{\mathcal{R}T}{\mathcal{F}} \ln k,$$

whereas the product $C_{Cl}^{(o)} \times C_K^{(o)}$ has the same value before and after the concentration changes. Table 3.6 shows an example of such a Ringer's solution. The solution A is the normal frog–Ringer. In the solutions B, C and D the K⁺ concentrations are raised from the normal 2.5 mM to 19, 30 and 75 mM, respectively, while the Cl⁻ ions are reduced correspondingly to preserve a constant product of $C_K^{(o)} \times C_{Cl}^{(o)} = 300$ (mM)². In addition, the three solutions must have the same water activity (tonicity)* as solution A. An increase in the K⁺ concentration must therefore entail a reduction of the Na⁺ concentration by an equal amount. To comply with the condition of electronegativity a decrease in Cl⁻ concentration will have to be compensated for by the addition of an equivalent amount of a different ion species, which, however, must not affect the size of the membrane potential of the fiber. The SO_4^{2-} ion is for all practical purposes unable to penetrate the membrane of the muscle fiber, and will therefore not contribute to the size of the membrane potential. For this reason SO_4^{2-} is well suited as the compensating inactive anion. As SO_4^{2-} is divalent, an addition of each SO_4^{2-} ion will compensate for a decrease of two Cl⁻ ions. To keep up the same tonicity an electrolyte (e.g. sucrose) has been added in the same amounts as SO_4^{2-} in the three solutions B, C and D. A point of criticism could be raised against the composition of the three solutions because they do not have the same Na⁺ concentrations and, therefore, a complete replacement of Na⁺ with a impermeable cation such as choline ($HOCH_2CH_2N^+(CH_3)_3$) would have been preferable. This precaution has not been taken, however, knowing in advance that their permeability P_{Na} is several orders lower than those of K⁺ and Cl⁻

* See Chapter 2, Section 2.8.2.4.

(compare Table 3.5), and for this reason a change in the Na^+ concentration is not expected to influence the value of the membrane potential appreciably. A change of the equilibrium potentials for Na^+, K^+ and Cl^- by the amounts $\Delta V_{Na}^{(eq)}$, $\Delta V_K^{(eq)}$ and $\Delta V_{Cl}^{(eq)}$ should, according to Eq. (3.8.1), result in a change in the membrane potential ΔV_m that is

$$\Delta V_m = T_{Na}\Delta V_{Na}^{(eq)} + T_K\Delta V_K^{(eq)} + T_{Cl}\Delta V_{Cl}^{(eq)}$$

$$\approx T_K\Delta V_K^{(eq)} + T_{Cl}\Delta V_{Cl}^{(eq)}, \qquad (3.8.8)$$

since $T_{Na} \ll T_K \approx T_{Cl}*$. As the product $C_K^{(o)} \times C_{Cl}^{(o)}$ is constant in the solutions we have: $\Delta V_{Cl}^{(eq)} = \Delta V_K^{(eq)}$, and therefore

$$\Delta V_m = (T_K + T_{Cl})\Delta V_{Cl}^{(eq)} = (T_K + T_{Cl})\Delta V_K^{(eq)}.$$

In general we have: $T_{Na} + T_K + T_{Cl} = 1$. But in this special case where $T_{Na} \ll 1$, we put $T_K + T_{Cl} \approx 1$ and obtain

$$\Delta V_m = \Delta V_{Cl}^{(eq)} = \Delta V_K^{(eq)}.$$

As the changes in equilibrium potential result solely from changes in the extra-cellular concentrations, we have according to Eq. (3.8.7)

$$\Delta V_{Cl}^{(eq)} = -\frac{\mathcal{R}T}{\mathcal{F}}\Delta \ln C_{Cl}^{(o)} \quad \text{and} \quad \Delta V_K^{(eq)} = \frac{\mathcal{R}T}{\mathcal{F}}\Delta \ln C_K^{(o)},$$

or[†]

$$\Delta V_{Cl}^{(eq)} = -58.2\,\Delta \ln C_{Cl}^{(o)} \quad \text{and} \quad \Delta V_K^{(eq)} = 58.2 \ln C_K^{(o)}.$$

Thus, we should expect that the change in membrane potential should be proportional to the change of the logarithm of the concentration of the K^+ or Cl^- ions of the extracellular fluid, and the factor of proportionality that is expected theoretically should be 58.2 mV at 20 °C. Figure 3.24 shows an example of the correlation between the membrane potential of the muscle fiber and the content of the Ringer's fluid of K^+ and Cl^- at a fixed ion product $C_K^{(o)} \times C_{Cl}^{(o)} = 300$ (mM)^2. The points indicated by the symbols + are the values of the membrane potential measured 10–60 min after the fiber was surrounded by the new solution. The other symbols o, −o and o− show the values measured 10–60 s after a sudden replacement with a Ringer's solution of new

---

* Equation (3.8.8) requires one moment's reflection as the application of sudden changes in the equilibrium potentials $V_j^{(eq)}$ implies a sudden change in the ratio of concentrations $C_j^{(i)}/C_j^{(o)}$, whereas the T_j refers to the values of the transport numbers *before* inflicting the concentration changes. Thus, the equation refers to the instantaneous value of ΔV_m that is valid before any redistribution of ions takes place in the membrane. Compare the remarks with Eq. (3.6.31).

† At 20 °C we have $2.3026 \cdot \mathcal{R}T/\mathcal{F} = 58.2$ mV. See Section 3.4.2, Table 3.1.

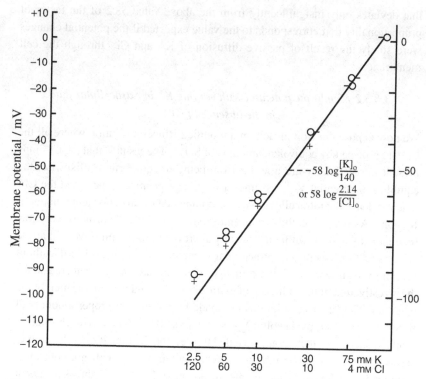

Fig. 3.24. The relation between the membrane potential and $\log C_{\mathrm{K}}^{(o)}$ or $-\log C_{\mathrm{Cl}}^{(o)}$ for solutions with $C_{\mathrm{K}}^{(o)} \times C_{\mathrm{Cl}}^{(o)} = 300\,(\mathrm{mM})^2$. Ordinate: membrane potential, (mV). Abscissa: K^+ and Cl^- concentrations (mM), log scale. Temperature is 20 °C. Symbols $+$ are potentials measured after 10–60 min equilibration in the solution referred to; $\circ$ are potentials measured 20–60 s after a sudden replacement with a Ringer's solution of new composition; $-\circ$ after an increase in $C_{\mathrm{K}}^{(o)}$; $\circ-$ after a reduction of $C_{\mathrm{K}}^{(o)}$. (After Hodgkin & Horowicz, 1959.)

composition: $-\circ$ after an increase in $C_{\mathrm{K}}^{(o)}$, and $\circ-$ after a reduction in $C_{\mathrm{K}}^{(o)}$. It is seen that the new values of the membrane potential are established quickly and in a reversible manner after the application of a Ringer's solution of altered composition. Furthermore, the relation between the membrane potential and the logarithm of the K^+ or Cl^- concentrations in the Ringer's fluid is nearly linear, especially in the ranges $C_{\mathrm{K}}^{(o)} > 10$ and $C_{\mathrm{Cl}}^{(o)} < 30$. The straight line that is fitted to the experimental data in the range $C_{\mathrm{K}}^{(o)} > 10$ and $C_{\mathrm{Cl}}^{(o)} < 30\,\mathrm{mM}$ has the slope, i.e. a change in V_{m} for a tenfold change in $C_{\mathrm{K}}^{(o)}$ or $C_{\mathrm{Cl}}^{(o)}$, of

$$\alpha = \frac{\partial V_{\mathrm{m}}}{\partial \log C_{\mathrm{K}}^{(o)}} = -\frac{\partial V_{\mathrm{m}}}{\partial \log C_{\mathrm{Cl}}^{(o)}} = 58\,\mathrm{mV},$$

that deviates only insignificantly from the above value 58.2 of the factor of proportionality that corresponds to the value expected if the potential changes were solely the result of passive diffusion of K^+ and Cl^- through the cell membrane.

3.8.3.2 Membrane potential with varying K^+ in extracellular fluid in the absence of Cl^-

We now expose the frog muscle in a modified Ringer's solution where all the Cl^- are replaced by equivalent amounts of SO_4^{2-}. The result is that the Cl^- ions present intracellularly are now far from being in equilibrium – likewise, the equilibrium potential $V_{Cl}^{(eq)}$ assumes a very large positive value – and the Cl^- ions will leave the intracellular phase accompanied by an equivalent amount of K^+ ions. As the intracellular concentration of Cl^- ions is normally very low (compare Table 3.4), within a relatively short time (30–60 min) we will arrive at the situation where the intracellular content of Cl^- is largely nil without invoking any discernible volume change. Thus, a situation has been created for the investigation of the influence of extracellular K^+ and Na^+ on the membrane potential. The Cl^--free solutions are made by mixing the proper amounts of K_2SO_4 and Na_2SO_4 but keeping $\sum\{[K^+] + [Na^+]\} = 140$ mM. To each solution is added sucrose in the amounts needed to preserve the tonicity. As the permeability of SO_4^{2-} in the membrane is very low, these ions will not contribute significantly to the size of the membrane potential. Figure 3.25 shows the result of such an experiment, where the membrane potential is plotted against the logarithm of the K^+ concentration, which varies between 0.5 and 140 mM. The membrane potential *decreases** as the external K^+ concentration increases and vanishes at about $C_K^{(o)} = 140$ mM. In the range $C_K^{(o)} > 10$ mM the membrane potential changes linearly with increasing values of $\log C_K^{(o)}$ with a slope

$$\frac{\partial V_m}{\partial \log C_K^{(o)}} = 60\,\text{mV}, \quad \text{for} \quad C_{Cl}^{(o)} = 0,$$

that is, corresponding to a tenfold increase of $C_K^{(o)}$ from 10 mM to 100 mM. At $C_K^{(o)} \approx 140$ mM the membrane potential is zero, and the equilibrium potential for the K^+ ion, $V_K^{(eq)}$, is zero, since $C_K^{(o)} = C_K^{(i)}$ (compare Table 3.1). It appears

---

* Note the particular idioms in this context. As biological cell membrane potentials invariably are negative – $V_m < 0$ – a large membrane potential means a large negative value ($V_m \ll 0$), and a *decrease* in membrane potential means a decrease of the numerical value of V_m that may approach zero. In other words: the membrane becomes *depolarized*. If by some means the membrane potential is driven more negatively relative to the existing membrane potential its effect is said to be a *hyperpolarization* of the membrane.

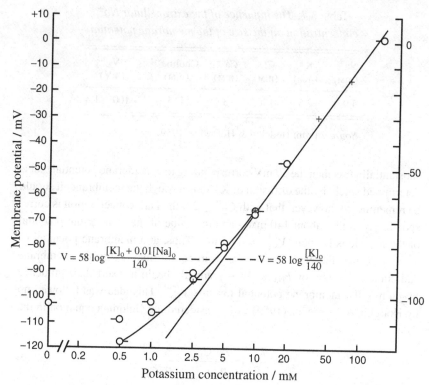

Fig. 3.25. Dependence of resting membrane potential on K^+ concentration in the extracellular fluid. Ordinate: membrane potential (mV). Abscissa: K^+ concentration (mmol dm^{-3}, log scale). Straight line: potential calculated from Nernst's equation with $C_K^{(i)} = 140$ mM. Curved line: potential calculated form Eq. (3.8.8) with $C_K^{(i)} = 140$ mM and $P_K = 100 \cdot P_{Na}$. (After Hodgkin & Horowicz, 1959.)

from Fig. 3.25 that the membrane potential is well reproduced by the Nernst equation

$$V_m = \frac{\mathcal{R}T}{\mathcal{F}} \ln \left(\frac{C_K^{(o)}}{C_K^{(i)}} \right),$$

in the concentration range $10 < C_K^{(o)} < 140$. Thus, in this situation the membrane behaves as a perfect K^+-selective permeable membrane.

For extracellular K^+ concentrations below 10 mM, the observed values of V_m deviate from the extrapolated straight line by assuming smaller values than those predicted from Nernst's equation. At $C_K^{(o)} = 2.5$ mM the slope of the curve corresponds to a change of 40 mV per tenfold change of $C_K^{(o)}$ – that is

Table 3.7. *The influence of the extracellular Na⁺*
concentration on the size of the membrane potential

Na⁺ (mM)	K⁺ (mM)	Cl⁻ (mM)	Ca²⁺ (mM)	Choline⁺ (mM)	V_m (mV)
5.0	2.5	120	3	115	-100 ± 1.4

Source: From Hodgkin & Horowicz, 1959.

substantially less than the 59 mV corresponding to a membrane potential that is determined solely by the diffusion of K⁺ ions through the membrane. It should be remembered, however, that with $C_K^{(o)} \leq 2.5$ the Na⁺ concentration is correspondingly high – about 140 mM – and the value of the equilibrium potential of the Na⁺ ions is about $V_{Na}^{(eq)} = +55$ mV. Thus, at a membrane potential of $V_m = -90$ mV the Na⁺ ions are far from equilibrium across the membrane and could – in spite of $P_{Na} \ll P_K$ – gradually begin to exert their influence by pulling the membrane potential towards $V_{Na}^{(eq)}$. This idea was followed up by Hodgkin & Horowicz (1959) who considered the Goldman equation in the form*

$$V_m = \frac{\mathcal{R}T}{\mathcal{F}} \ln \frac{P_{Na} C_{Na}^{(o)} + P_K C_K^{(o)}}{P_{Na} C_{Na}^{(i)} + P_K C_K^{(i)}}, \tag{3.8.9}$$

since $C_{Cl}^{(i)} = C_{Cl}^{(o)} = 0$ and therefore in this particular type of experiment the Cl⁻ ions do not contribute to the membrane potential. The fully drawn curve on Fig. 3.25 corresponds to values from Eq. (3.8.8) computed for a value of $C_K^{(i)} = 140$ mM and of $P_{Na} = 0.01 P_K$ shows a far better agreement with the experimentally observed values.

3.8.3.3 Membrane potential with varying Na⁺ in the extracellular fluid

To substantiate the above explanation for the deviation from the Nernst equation it would be desirable to be able to alter the value of the equilibrium potential $V_{Na}^{(eq)}$ by changing the Na⁺ concentration in the extracellular fluid. Choline $(HOCH_2CH_2N^+(CH_3)_3)$ is an organic cation that is impermeable to the muscle fiber membrane, but otherwise without deleterious effects. Table 3.7 shows results by Hodgkin & Horowicz (1959) on 11 muscle fibers where the normal extracellular Na⁺ concentration of 120 mM has been reduced to 5 mM and

---

* Actually this equation is more general than the Goldman equation, as it can be derived without assuming the presence of a constant field. See Section 3.7.2, Eq. (3.7.16).

replaced by choline ions in the equivalent amount of 115 mM, and thus maintaining the concentrations of K^+, Cl^- and Ca^{2+} ions at their normal values. In contrast to the drastic changes of the membrane potential seen in Fig. 3.25 and Fig. 3.26 resulting from large changes in the equilibrium potentials for the K^+ and Cl^- ions, similar changes in the equilibrium potential $V_{Na}^{(eq)}$ for the Na^+ ions influence the membrane potential to a much smaller extent, as a change in extracellular Na^+ concentration from 120 mM to 5 mM – that is reducing $V_{Na}^{(eq)}$ from +55 mV to about −15 to −20 mV – causes only an additional displacement in membrane potential of about 5 mV to −100 mV. The smallness of the change agrees well with the interpretation of the data in Fig. 3.26 of assigning a minor role to the contributions of the Na^+ ions to the membrane potential. However, it is worth noting that the *hyperpolarization* of the membrane that results from a removal of the normal value of $V_{Na}^{(eq)} = +55$ mV is consistent with Eq. (3.8.8) and adds support to the concept that the resting membrane potential in the muscle fiber owes its origin to the tendency to equalization of the dissimilar distributions between the cell interior and the extracellular fluid of the permeable, small ions K^+, Na^+ and Cl^- that move according to the direction prescribed by the electrochemical difference across the membrane for each ionic species.

3.8.4 Membrane potential and active Na^+/K^+ transport

The experiments above have demonstrated that the membrane potential in a striated muscle fiber belongs to a class of diffusion potential, i.e. a non-equilibrium phenomenon that will invariably come to an end owing to the passive leakage of the Na^+, K^+ and Cl^- ions unless some means of restoring the losses exists. As mentioned in Section 2.1, evolution's answer to this demand is the energy-requiring *active membrane transport*, that is the development of special protein molecules – or transport molecules – that are incorporated in all cell membranes specifically to transport a single, perhaps two or several, species simultaneously, e.g. Na^+, or Na^+ and K^+, *unidirectionally* through the membrane. The active transport system that occurs most frequently is the so-called *sodium pump* where Na^+ ions are driven *out* of the cell to the extracellular fluid – that is to a region with *higher* electrochemical potential for Na^+ – while K^+ ions are also transported actively to the cell interior*. Since this metabolic Na^+/K^+ pump carries an electric current – the pump is not *electroneutral* but *electrogenic* – we may expect the pump to exert an influence, the magnitude of

---

* The mechanisms used by the cell to keep these concentration differences at a stationary level are discussed in, e.g.: P. Laüger: *Electrogenic Ion Pumps*, Sinauer Associates Inc., 1991.

which is the membrane potential in addition to that owing to its rôle of preserving the ionic gradients at their normal levels. We shall illustrate this aspect by means of a few simple examples.

We consider a frog muscle fiber *in vivo*. Since Cl^- is distributed passively it is reasonable to presume that the Cl^- ions are in equilibrium and $J_{Cl} = 0$. The ionic current through the membrane is therefore assumed to be dominated by Na^+ and K^+, which enter partly as passive currents $I_K = \mathscr{F}J_K$ and $I_{Na} = \mathscr{F}J_{Na}$ and partly as active $I_K^{(a)} = \mathscr{F}J_K^{(a)}$ and $I_{Na}^{(a)} = \mathscr{F}J_{Na}^{(a)}$. Furthermore, we assume that the active outwardly transported Na^+ ions are tightly coupled with the number of inwardly transported K^+ ions, that is

$$I_K^{(a)} = -\Gamma I_{Na}^{(a)}, \tag{3.8.10}$$

where Γ – the coupling ratio – indicates the number of K^+ ions that are pumped inwards for each Na^+ ion pumped outwards. The minus sign in Eq. (3.8.10) indicates that the two fluxes are oppositely directed. When $\Gamma = 1$ the pump does not carry any net charge and the pump is electroneutral. When $\Gamma = 0$ the Na^+ transport through the pump system is not linked to a simultaneous inward active transport of K^+. A value of the coupling ratio of $\Gamma = \frac{2}{3}$ is frequently observed.

3.8.4.1 Absolute stationarity

This situation was first considered by Mullins & Noda (1963)*. The Cl^- ions are assumed to be in equilibrium. Furthermore, it is assumed that the intracellular concentrations of Na^+ and K^+ ions are stationary, owing to the function of the Na^+/K^+ pump. This implies that the *net fluxes* of Na^+ and K^+ vanish separately, that is

$$I_{Na}^{(a)} = -I_{Na},$$

and

$$I_K^{(a)} = -I_K.$$

However, these two equations are not mutually independent because of the constraints caused by coupling between the two active fluxes. Invoking Eq. (3.8.10), the connection between the two passive stationary fluxes is

$$\Gamma I_{Na} + I_K = 0. \tag{3.8.11}$$

* Mullins, L.J. & Noda, K. (1963): *J. Gen. Physiol.* **47**, 117–132.

The general expression for the current density of a cation is

$$I_j^+ = -\mathcal{R}T \frac{d}{dx}(u_j^+ C_j^+) - \mathcal{F}(u_j^+ C_j^+)\frac{d\psi}{dx}, \qquad (3.6.21)$$

which applied to Eq. (3.8.10) gives

$$0 = \Gamma\mathcal{R}T \frac{d}{dx}(u_{Na}C_{Na}) + \mathcal{R}\mathcal{F}(u_{Na}C_{Na})\frac{d\psi}{dx}$$
$$+ \mathcal{R}T \frac{d}{dx}(u_K C_K) + \mathcal{F}(u_K C_K)\frac{d\psi}{dx},$$

or

$$\frac{\mathcal{R}T}{\mathcal{F}} \frac{d}{dx}(\Gamma u_{Na}C_{Na} + u_K C_K) + (\Gamma u_{Na}C_{Na} + u_K C_K)\frac{d\psi}{dx} = 0,$$

which after rearrangement gives

$$-\frac{d\psi}{dx} = \frac{\mathcal{R}T}{\mathcal{F}} \frac{d}{dx}(\ln\{\Gamma u_{Na}C_{Na} + u_K C_K\}).$$

This equation is multiplied by dx and integrated from $x = 0$, where $\psi = \psi^{(i)}$ and $C_{Na} = C_{Na}^{(i)}; C_K = C_K^{(i)}$ to $x = h$, where $\psi = \psi^{(o)}$ and $C_{Na} = C_{Na}^{(o)}; C_K = C_K^{(o)}$. This gives

$$\psi^{(o)} - \psi^{(i)} = V = \frac{\mathcal{R}T}{\mathcal{F}} \ln \frac{\Gamma u_{Na}C_{Na}^{(o)} + u_K C_K^{(o)}}{\Gamma u_{Na}C_{Na}^{(i)} + u_K C_K^{(i)}},$$

As $u_{Na} \propto P_{Na}$ and $u_K \propto P_K$ the above expression can also be written in the form

$$\psi^{(o)} - \psi^{(i)} = V = \frac{\mathcal{R}T}{\mathcal{F}} \ln \frac{\Gamma P_{Na}C_{Na}^{(o)} + P_K C_K^{(o)}}{\Gamma P_{Na}C_{Na}^{(i)} + P_K C_K^{(i)}}, \qquad (3.8.12)$$

which is the formula of Mullins & Noda (1963). If the pump does not carry any net current – the pump is electroneutral – then $\mathcal{R} = 1$ and the above expression takes the form of Eq. (3.8.8) where the membrane potential is determined solely by the passive ionic fluxes, the Na^+/K^+ pump, however, still having the decisive role of sustaining the internal ionic milieu with respect to that of the extracellular fluid. At the other extreme, where only the Na^+ ion is actively transported, we have $\Gamma = 0$, and the expression for the membrane potential is reduced to that of the equilibrium potential $V_K^{(eq)}$ for the K^+ ions. If $\Gamma = 2/3$, i.e. still corresponding to a positive *outward* current of the pump, the membrane potential will, with the same values of P_{Na} and P_K, be driven more negatively towards the value of $V_K^{(eq)}$.

3.8.4.2 Total current is zero

We now consider the less-restrictive situation where both intracellular concentrations of Na^+ and K^+ change slowly with time. In *in vitro* preparations, an impaired pump function may involve a gradual, but slow, loss of cellular K^+ ions together with a gain in Cl^- ions, but in such a way that the total membrane current remains practically *zero*, i.e. the membrane potential remains constant when regarded from a quantitative point of view. The condition that now holds is

$$I_{Na} + I_{Na}^{(a)} + I_K + I_K^{(a)} = 0.$$

Introducing the coupling ratio Γ yields

$$(1 - \Gamma)I_{Na}^{(a)} + I_{Na} + I_K = 0. \tag{3.8.13}$$

Collecting the expressions for the passive currents gives

$$(1 - \Gamma)I_{Na}^{(a)} - \mathcal{R}T\frac{d}{dx}(u_{Na}C_{Na} + u_K C_K) - \mathcal{F}(u_{Na}C_{Na} + u_K C_K)\frac{d\psi}{dx} = 0.$$

We divide this equation by $u_{Na}C_{Na} + u_K C_K$, rearrange and multiply on both sides by dx. This gives

$$-d\psi = \frac{\mathcal{R}T}{\mathcal{F}}d\left(\ln\{u_{Na}C_{Na} + u_K C_K\}\right) - \frac{dx}{\mathcal{F}(u_{Na}C_{Na} + u_K C_K)}(1 - \Gamma)I_{Na}^{(a)}.$$

Integrating from $x = 0$, where $\psi = \psi^{(i)}$ and $C_{Na} = C_{Na}^{(i)}; C_K = C_K^{(i)}$ to $x = h$, where $\psi = \psi^{(o)}$ and $C_{Na} = C_{Na}^{(o)}; C_K^{(o)}$ gives

$$\psi^{(i)} - \psi^{(o)} = \frac{\mathcal{R}T}{\mathcal{F}}\ln\frac{u_{Na}C_{Na}^{(o)} + u_K C_K^{(o)}}{u_{Na}C_{Na}^{(i)} + u_K C_K^{(i)}}$$

$$- (1 - \Gamma)I_{Na}^{(a)}\int_0^h \frac{dx}{\mathcal{F}(u_{Na}C_{Na} + u_K C_K)}.$$

The integral on the right-hand side is the *integral* resistance* $R^{(int)}$. The ion mobilities are then replaced by corresponding ion permeabilities, and $\psi^{(i)} - \psi^{(o)}$ is replaced by the membrane potential V. Hence, the above expression can also be written as

$$V = \frac{\mathcal{R}T}{\mathcal{F}}\ln\frac{P_{Na}C_{Na}^{(o)} + P_K C_K^{(o)}}{P_{Na}C_{Na}^{(i)} + P_K C_K^{(i)}} - (1 - \Gamma)I_{Na}^{(a)}R^{(int)}. \tag{3.8.14}$$

---

* See Section 3.3.1, Eq. (3.3.26).

Thus, for the electroneutral pump ($\Gamma = 1$) the expression for the membrane potential is identical to that of Eq. (3.8.8) considering only passive fluxes, whereas the electrogenic pump ($0 \leq \Gamma < 1$) causes an additional hyperpolarization that is equal to the net outward active current multiplied by the integral resistance of the membrane.

3.8.4.3 The mosaic membrane

To study the effect of an electrogenic Na^+/K^+ pump on the membrane potential of a membrane having a mosaic structure, we take the passive currents* of Na^+ and K^+ (assuming Cl^- ions to be in equilibrium)

$$I_{Na} = G_{Na}\left[V - V_{Na}^{(eq)}\right] \quad \text{and} \quad I_K = G_K\left[V - V_K^{(eq)}\right].$$

Assuming again that the total current is zero we have, using Eq. (3.8.13),

$$(1 - \Gamma)I_{Na}^{(a)} + G_{Na}\left[V - V_{Na}^{(eq)}\right] + G_K\left[V - V_K^{(eq)}\right] = 0.$$

Solving with respect to V gives

$$V = \frac{G_{Na}V_{Na}^{(eq)} + G_K V_K^{(eq)}}{G_{Na} + G_K} - (1 - \Gamma)\frac{I_{Na}^{(a)}}{G_{Na} + G_k}. \tag{3.8.15}$$

The sum of the partial conductances G_{Na} and G_K now enters into this expression for the "electrogenic term", and it is the resistance $R = (G_{Na} + G_K)^{-1}$ that is polarized by the outgoing active net current $(1 - \Gamma)I_{Na}^{(a)}$.

The main significance of these results is that of imposing an upper limit to the influence of the Na^+/K^+ pump on the membrane potential. Otherwise, the calculations are somewhat formal, as the Na^+/K^+ pump is regarded as a "black box". The description of the active Na^+/K^+ transport by introducing an active current generator is naturally arbitrary. An "electromotive transport force" to represent the active transport would have served the purpose equally well.

3.9 Flux ratio analysis

The first unambiguous demonstration of active Na^+ transport in living cells was given by H.H. Ussing & K. Zerahn (1951) who developed the so-called *short-circuit technique*[†], that ever since has been a useful tool in the study of active

---

* See Section 3.7.4, Eq. (3.7.52).

[†] H.H. Ussing (1911–2000), was Professor of biological chemistry, 1958–1981, at the University of Copenhagen. He was one of the pioneers in the introduction and use of radioactive isotopes in biological research and a leading authority on epithelial transport.

transport across biological cell membranes and cell structures. An isolated frog skin[‡] was placed in a chamber and surrounded by two frog Ringer's solutions of *identical* composition. Two pairs of electrodes corresponding to those of Fig. 3.6 were placed in the chamber. One pair was placed close to the skin and connected to the input of a potentiometer of some kind. The other set was connected to a variable current source that also included the current-measuring device. A living frog skin develops a potential difference between the two solutions fairly quickly. Typically the potential $\psi^{(i)}$ of the solution in contact with the inside of the skin *exceeds* the potential $\psi^{(o)}$ of the outside solution by about 60 mV. By adjusting the current flow through the frog skin the potential difference across the skin is forced to assume the value zero ($\psi^{(o)} = \psi^{(i)}$). The frog skin is then said to be short-circuited. In this condition the skin is surrounded by two solutions that are of identical composition and at the same electric potential, pressure and temperature. That is, the electrochemical potential $\bar{\mu}_j$ of each component j in the two solutions will have the *same value* on the two sides. Thus

$$\bar{\mu}_j^{(o)}(T, P) = \bar{\mu}_j^{(i)}(T, P)$$

implies that a passive net transfer of any component j between the solutions (i) and (o) will not occur unless some irreversible process *inside* the living frog skin is capable of moving one or several species from an equilibrium state in one solution to the same equilibrium state in the other solution. Accordingly, Ussing & Zerahn (1951) concluded that the supply and removal of electric charges to the two solutions from the current generator – or, for that sake, the appearance of the skin potential difference with the system in the current-free state – reflected an action of a specific cellular transport mechanism that moved one or several ionic species *unidirectionally* through the skin, thereby creating an electric current. To identify the current carrier, radioactive sodium isotopes ^{22}Na and ^{24}Na were placed in the respective solutions (o) and (i) and the net transfer M_{Na^+} of Na^+ in a given time in the direction (o) → (i) was determined. The corresponding amount of charge $\mathscr{F}M_{Na^+}$ was found to be identical in size to the charge carried through the frog skin by the short-circuit current in the same span of time. This observation, combined with various controls to exclude other possible candidates, allowed Ussing & Zerahn to conclude that Na^+ ions were actively transported by the frog skin in the direction (o) → (i).

To decide whether a given mass transfer should be classified as a passive or an active process is not always a simple matter. In most cases the experimental data are examined to see if the process can be described satisfactorily by the theory

---

[‡] Frog skin is a complex structure consisting of several layers of cells, each having different functions. Hence, certain modifications are required in applying the transport equations in their simple forms – e.g. those of the previous section to a multilayered structure such as an epithelium.

of passive transport. If the process takes place across a single cell membrane it is often possible to determine the relevant transport parameters and to predict the system's behavior. On the other hand, if the process takes place across a layer of epithelial cells we have to consider a structure where several layers of cell membranes alternate with cytoplasm and interstitial fluid, and the simplifying assumption of homogeneity no longer holds. In certain cases this structure can, from a functional point of view, be reduced to two or perhaps three homogeneous structures arranged in series. In other cases the number of relevant details in the structure makes it impracticable to formulate a realistic model that also is mathematically manageable. In this context the so-called flux ratio criterion enters as an useful tool. In this analysis one considers the *ratio* between the fluxes of the component in question resulting from two different boundary conditions, usually those corresponding to the two *unidirectional* fluxes of the radioactive isotope

$$\overset{i \to o}{J_j} \quad \text{and} \quad \overset{i \leftarrow o}{J_j},$$

and usually used in such small amounts that their presence does not perturb the transport conditions – e.g. concentration and potential profiles, etc. – for the component j in question. As the transport route in the two directions (i) $\to$ (o) and (o) $\to$ (i) is the same for $\overset{i \to o}{J_j}$ and $\overset{i \leftarrow o}{J_j}$, which only differ as regards to boundary conditions, the transport parameters of the system will enter in the same way into the flux equations for $\overset{i \to o}{J_j}$ and $\overset{i \leftarrow o}{J_j}$ as a common factor. Therefore, taking the ratio

$$\frac{\overset{i \to o}{J}}{\overset{i \leftarrow o}{J}}$$

between the two unidirectional fluxes the common factor in the flux equation – which may contain inaccessible quantities because of our incomplete knowledge – will cancel by the division, leaving only the well-defined quantities that can be determined experimentally. Originally the flux ratio concept was developed by Behn (1897)[*], but for a long time his results remained relatively unnoticed until the introduction of radioactive isotopes gave them new importance. The flux ratio criterion was rediscovered[†] almost simultaneously by Ussing (1949) and Teorell (1949).

[*] Behn, U. (1897): Ueber wechselseitige Diffusion von Elektrolyten in verdünnten wässerigen Lösungen, insbesondere über Diffusion gegen das Concentrationsgefälle. *Ann. Physik N.F.*, **62**, 54.

[†] Ussing, H.H. (1949): The distinction by means of tracers between active transport and diffusion. *Acta Physiol. Scand.*, **19**, 43. Teorell, T. (1949): Membrane electrophoresis in relation to bio-electrical polarization effects. *Arch. Sci. Physiol.*, **3**, 205.

We shall illustrate the procedure applied to two situations, where the isotope in question is assumed to be transported passively through a cell layer either by *electrodiffusion* or by *diffusion–convection*.

3.9.1 Flux ratio and electrodiffusion

This case has been handled in several different ways. Within the framework of the present account the simplest way of derivation is to take the Nernst–Planck equation[*]

$$J_j = -D_j \frac{dC_j}{dx} - z_j e B_j C_j \frac{d\psi}{dx}, \qquad (3.9.1)$$

and then write it in the form according to Kramers' equation[†]

$$J_j \left(\frac{e^{z_j \Psi(x)}}{D_j(x)} \right) = -\frac{d}{dx} \left(C_j(x) e^{z_j \Psi(x)} \right), \qquad (3.9.2)$$

or

$$J_j \left(\frac{e^{z_j \Psi(x)}}{D_j(x)} \right) dx = -d \left(C_j(x) e^{z_j \Psi(x)} \right), \qquad (3.9.3)$$

where $\Psi(x)$ is the value of the normalized potential ($\psi/(kT/e)$) in the position x within the structure and $C_j(x)$ is the concentration of component j in the same place. The presence of inhomogeneities along the transport route is formally taken into account by considering the diffusion coefficient $D_j(x)$ as a function of the position x without the need for going into explicit details. The product $C_j(x)e^{z_j \Psi(x)}$ in the parentheses on the right-hand side – often called the *electrochemical activity* of the ion of type j – enters as a total differential of a function. Thus, this function will depend only upon the values of the variables $C_j(x)$ and $\Psi(x)$ at the initial and final states, and *not* upon the details of the path that connects the two states. We integrate Eq. (3.9.3) from a position $x = 0$ in the surrounding medium with concentration $C_j(x) = C_j^{(i)}$ and potential $\Psi^{(i)}$ through the structure to the other surrounding medium with $C_j(x) = C_j^{(o)}$ and potential $\Psi^{(i)}$ at the position $x = h$. Furthermore, we consider the process as stationary, i.e. $\partial J_j/\partial x = 0$ in the range $0 \leq x \leq h$. Hence, the fluxes J_j are put outside the

[*] See Section 3.3.2.1, Eq. (3.3.31).
[†] See Section 3.7.3, Eq. (3.7.20).

integration sign. We then have

$$J_j \int_0^h \left(\frac{e^{z_j \Psi(x)}}{D} \right) dx = \int_{\substack{\Psi^{(i)} \\ C_j(0)}}^{\substack{\Psi^{(o)} \\ C_j(h)}} d \left(C_j(x) \, e^{z_j \Psi(x)} \right)$$

$$= - \left[C_j(x) \, e^{z_j \Psi(x)} \right]_{\Psi^{(i)}, C_j^{(i)}}^{\Psi^{(o)}, C_j^{(o)}}$$

or

$$J_j = \frac{\left[C_j^{(i)} e^{z_j \Psi^{(i)}} - C_j^{(o)} e^{z_j \Psi^{(o)}} \right]}{\int_0^h \left(\frac{e^{z_j \Psi(x)}}{D_j(x)} \right) dx}$$

$$= \left[C_j^{(i)} e^{z_j \Psi^{(i)}} - C_j^{(o)} e^{z_j \Psi^{(o)}} \right] \times \mathcal{A}, \tag{3.9.4}$$

where $\mathcal{A}$ stands for

$$\mathcal{A} = \frac{1}{\displaystyle\int_0^h \left(\frac{e^{z_j \Psi(x)}}{D_j(x)} \right) dx}.$$

To evaluate the denominator, the potential profile $\Psi(x)$ and the functional dependence $D_j(x)$ must be known explicitly. To facilitate the typography we put the potential difference across the membrane $\psi^{(i)} - \psi^{(o)} = V$ and similarly $\Psi^{(i)} - \Psi^{(o)} = v$, and write Eq. (3.9.4) as

$$J_j = \left[C_j^{(i)} \, e^{z_j v} - C_j^{(o)} \right] e^{z_j \Psi^{(o)}} \times \mathcal{A},$$

or as

$$J_j = \left[C_j^{(i)} \, e^{z_j v} - C_j^{(o)} \right] \times \mathcal{A}, \tag{3.9.5}$$

if we arbitrarily put $\Psi^{(o)} = 0$.

It is assumed that the experimental set up is designed to study the transport of the *radioactive isotope* of the ion j that is added to the system in small amounts. Ideally two different tracers *j and °j of the same ion species j should be placed in the two surrounding fluids, but this is not feasible in an actual experiment*. Instead we imagine the experiment performed in two stages. Initially the tracer species is placed in phase (i) at the concentration $^*C_j^{(i)}$, while the concentration in phase (o) is kept at zero all the time. Hence, the tracer flux through the system is an *unidirectional* flux $\overset{i \to o}{J}$ in the direction (i)→(o). Equation (3.9.5) then takes the form

$$\overset{i \to o}{J} = {}^*C_j^{(i)} \, e^{z_j v} \times \mathcal{A}. \tag{3.9.6}$$

---

* Unless the two tracers have widely different half-lives.

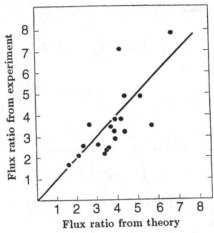

Fig. 3.26. Unidirectional flux ratios for Cl$^-$ transport through frog skin. $\overset{i \to o}{J} / \overset{o \to i}{J}$ is calculated from Eq. (3.9.8) and compared with the experimentally determined values. Nine different frog skins were used with two data sets from each skin. (After Koefoed Johnsen, Levi & Ussing, 1952*.)

We consider now the reverse situation where the radioactive tracer is placed in the phase (o) with the concentration $^*C_j^{(o)}$ and the concentration in phase (i) is maintained at zero. Thus, a *unidirectional* flux $\overset{o \to i}{J}$ flows in the direction (o) → (i) of magnitude

$$\overset{o \to i}{J} = {}^*C_j^{(o)} \times \mathcal{A}. \tag{3.9.7}$$

Taking the ratio of these two unidirectional fluxes gives

$$\frac{\overset{i \to o}{J}}{\overset{o \to i}{J}} = \frac{{}^*C_j^{(i)}}{{}^*C_j^{(o)}} \, e^{z_j v} = \frac{{}^*C_j^{(i)}}{{}^*C_j^{(o)}} \, e^{z_j \mathcal{F} V / \mathcal{R} T}. \tag{3.9.8}$$

This is the celebrated expression for the ratio between two unidirectional fluxes for an ion that moves exclusively by means of electrodiffusion through the cell structures in question[†]. Equation (3.9.8) offers an effective and often-used criterion to decide whether an ion's transport through the cell layer occurs as the result of *passive transport by electrodiffusion* or if other mechanisms of transport – in particular active transport – are in operation. Figure 3.26 shows an

---

[*] Koefoed Johnsen, V., Levi, H. & Ussing, H.H. (1952): *Acta Physiol. Scand.*, **25**, 150.
[†] Equation (3.9.8) has – apart from a missing activity coefficient – the same form as that derived by Ussing (1949) using a somewhat different approach.

example of an application of Eq. (3.9.8) in the study of Cl^- transport through frog skin. All the quantities in Eq. (3.9.8) are measured and the values of $\overset{i\to o}{J} / \overset{o\to i}{J}$ are plotted against the values of the left-hand side of Eq. (3.9.8). Complete agreement between experiment and theory would involve all the data points belonging to a straight line with the slope of 1.0. As seen from the figure, the data points are distributed partly on this line and partly more or less on both sides, but apparently in a random manner with no predominance to either side. The reasonable conclusion from such a set of experiments is that the Cl^- transport through the frog skin takes place solely by electrodiffusion. The remarkable feature of an expression such as Eq. (3.9.8) is that it does not contain adjustable parameters – in contrast to most formulas used in biophysics – but is determined solely by the conditions in the media that surround the cell structure under consideration. Equation (3.9.8) is derived assuming that stationarity holds, but the derivation has been extended to include non-stationary conditions[*], which are also in the time where the concentration profile of the tracer in question is building up to the stationary level.

3.9.2 Flux ratio and convective diffusion

If the potential difference V across the cell structure is zero, or if we are considering neutral molecules, Eq. (3.9.8) becomes

$$\frac{\overset{i\to o}{J}}{\overset{o\to i}{J}} = \frac{{}^*C_j^{(i)}}{{}^*C_j^{(i)}}, \tag{3.9.9}$$

which naturally could be derived using simpler mathematics. It may sometimes happen that influences other than an electric field cause the component j in question to move through the cell layer with a steady velocity in one direction, e.g. if the transport through the system takes place as a convection overlaid by diffusion. In Chapter 2, Section 2.7.4.2, the flux ratio for this kind of transfer across a single membrane was evaluated as

$$\frac{\overset{i\to o}{J}}{\overset{o\to i}{J}} = \frac{C(0)}{C(h)} e^{hv/D}, \tag{3.9.10}$$

where D is the diffusion coefficient and v the velocity of convection. If the domain is inhomogeneous and can be divided into n layers each of thickness

---

[*] Sten-Knudsen, O. & Ussing, H.H. (1981): The flux ratio under non stationary conditions. *J. Membrane Biol.*, **63**, 233.

$\Delta^{(i)}$ and diffusion coefficient $D_j^{(i)}$, the above expression takes the modified form

$$\frac{\overset{i \to o}{J}}{\underset{o \to i}{J}} = \frac{{}^*C_j^{(i)}}{{}^*C_j^{(o)}} \exp\left\{ v \sum \frac{\Delta^{(i)}}{D_j^{(i)}} \right\} = \frac{{}^*C_j^{(i)}}{{}^*C_j^{(o)}} e^{v/\langle P \rangle}, \qquad (3.9.11)$$

where

$$P_j^{(i)} = \frac{D_j^{(i)}}{\Delta^{(i)}}$$

is the permeability of the ith layer of thickness $\Delta^{(i)}$ and diffusion coefficient $D_j^{(i)}$, and where $\langle P \rangle$ is the *average permeability* of the whole layer which is defined as[†]

$$\frac{1}{\langle P \rangle} = \sum_{i=1}^{i=n} \frac{1}{P^{(i)}}.$$

[†] See Chapter 2, Section 2.5.4.1(ii), Eq. 2.5.24.

Chapter 4
The nerve impulse

Wonderful as are the phenomena of electricity when they are made
evident to us in inorganic or dead matter, their interest can bear
scarcely any comparison with that which attaches to the same force
when connected with the nervous system and with life ...
*(M. Faraday: Experimental Researches in Electricity,
15th Series, 1838.)*

4.1 Introduction

4.1.1 Excitability

The electric potential difference between the inside and outside of the cell
membrane – the membrane potential – plays an important part in the special
functions that are found with certain cell types. Thus, it is changes in the membrane potential that act as the *initiator* or *controller* for the secretion in glands
and for the contraction process in muscle cells. Certain cells – receptor cells –
have developed a particularly high sensitivity to the influence of certain external
physical or chemical agents. For instance, one cell type is extremely sensitive
to the action of light, another for heat and others again are sensitive to mechanical actions of various kinds, etc. However, one feature is common to all these
cells types in that the response to the stimulus takes the form of a localized
membrane potential change. In other cells special mechanisms are built into
the membrane that permit it to transmit potential changes without attenuation
along the length of the whole cell. This *propagated potential change* constitutes
an important link in the communication between the different cells. This property is developed to an extreme degree in the axons of nerve cells, but is also
present in, for example, skeletal muscle fibers, heart muscle cells and smooth
muscle cells. Collected under one designation – *excitable cells* – all these cells
share the common property of *excitability*, where an adequate stimulus induces
a change in the electrical properties of the cell membrane that successively leads

411

to a transient change in the membrane potential. This category of cells, among which are the nerve cells, muscle fibers and receptor cells, comprises a considerable proportion of the total mass of the body. The properties of excitable cells – and here in particularly those of nerves and muscles – have attracted the attention of many outstanding physiologists ever since 1786, when Galvani discovered that nerves and muscles produced electric current during activity. To this may be added several reasons that belong together: a frog nerve is a very robust preparation, whose function need not alter for many hours and which offers favorable conditions for developing precise methods of measurement of even very fast time courses. Finally, simple and clear experimental conditions can be established that give relatively unambiguous answers to the proposed questions, which can be synthesized in theories formulated precisely in terms of physics and chemistry, thus allowing for a quantitative verification.

4.1.2 The communication system

All living organisms owe their continued existence to the ability to adjust to and counteract the perpetual changes in the environment. In animals above a certain size and complexity, such an adaptability requires means of communication between the different parts of the body. Such a possibility is to some extent represented by the circulation system; it also serves as the transport system for regulating substances like hormones. This type of regulation suffers from two shortcomings: its action is slow and its capability of precise localization is limited. As regards processes like growth, digestion and equilibrium, in general this limitation does not represent a serious obstacle in the cells' conditions. But to counteract sudden changes in the surroundings a rapid and precisely acting system of communication is required. The development of excitable cells – comprising especially the nerve cells – is the response of evolution to this demand. The nerve cells are elongated cells that have developed with the special property that the proper action of the cell body causes a response to be propagated rapidly and with very little expenditure of energy along the elongated part, *the axon*, of the cell. This propagated signal is called the *nerve impulse*. The special, and only, function of the axon is to act as a transmission line conveying information between the different parts of the body. The mechanisms that underlie the excitability of the axons are also developed purposely to this aim; that is, to secure that a nerve impulse – once it is started – propagates along the whole length of the axon until it reaches its proper destination, i.e. the synaptic connection to another cell. In other instances the impulses travel from the axonal ending towards the body of the nerve cell. For instance, this happens when the nerve terminal is connected to specialized sensors, *receptor cells*, that are designed to receive and transmit information about events outside the body by initiating a propagated impulse in the nerve terminal in response to an adequate external physical stimulus. These nerve impulses may eventually terminate and activate, for example, a muscle or a gland. However, a direct connection from the receptor cells to the muscle fibers would not be of use to give a precise, well-coordinated movement in response to the external action. To this end the organism is supplied with a *central nervous system* that *sorts* and *coordinates* the incoming signals from the sense organs, whereupon the proper sequence of signals are sent out to the muscle to attain the required reaction. Whereas the nerve signal that propagates along the nerve axon represent a somewhat stereotyped phenomenon, serving only as a means of transmission, the events are far more complex at the transition from one nerve cell to the other. At the nerve terminals, structures with special functions – *synapses* – are developed to provide the transition between the cells. It is the interplay of activity in these synapses that controls the state of excitability on the membrane of the nerve cell's surface and hence is decisive for whether the impact of an incoming (afferent) impulse to the cell surface will result in the start of a new impulse from the cell body along its outgoing (efferent) axon.

4.2 Historical background

Over the centuries the properties of excitable tissue have been a subject of fascination to natural scientists. Two things have been known for a long time. (1) Nerves and muscles could be stimulated by an electric shock. Contributing to this experience was the use of the Leyden jar (the electric plate capacitor), invented by Petrus van Musschenbroek (1692–1761), professor of physics in Leyden, that allowed storage for later use of the electric charges produced by the electrification machine. (2) Certain animals (electric fishes like *Torpedo*, *Electrophorus* and *Malapterurus*) were able to cause painful shocks similar to those that was produced by the electrification machine. Furthermore, nerves and muscles were able to produce weak electric currents. The precise date for this discovery is known to be 20th September 1786. Luigi Galvani (1737–1798), professor of anatomy at the university of Bologna, had discovered a few years earlier that a frog muscle would contract when he discharged the electric tension from the electrification machine through the muscle nerve. After this discovery, Galvani took up the study of the influence of static and atmospheric electricity on nerves. On the day in question, Galvani had dissected the hind parts of frogs, pierced them by pressing thin brass needles into their spinal canal, and attached these preparations by the brass needles onto an iron railing outside his laboratory. Galvani then discovered that any time one of the frog legs accidentally

4. The nerve impulse

touched the iron railing, for example by a gust of wind, a violent contraction was triggered off in the muscles of the leg. From this observation and a series of supplementary experiments, Galvani could demonstrate that the frog's muscles contracted in the same moment a metallic connection was established between the nerve and the muscle. Galvani interpreted the phenomenon by assuming that the muscles and nerves produced electricity of the same kind that could be stored in the Leyden jar from the electrification machine and that this electricity of the tissues was discharged by connecting the metallic conductor. One group of researchers even considered this kind of electricity to be peculiar to living organisms and hence designated it as "animal electricity". Some even considered this electricity to be the "vital force" itself that would serve not only to account for the functions of nerve and muscle but also as the basic agent in the treatment of nervous diseases of all kinds. However, the bubble of these physiologists burst in 1792 when Alessandro Volta (1745–1827), professor of physics at the University of Pavia, showed that the contact of muscle to two different types of metal was decisive for Galvani's experiment and, furthermore, that insertion of animal tissue between the two types of metal was altogether unnecessary to produce an electric current: a few drops of saline water would do the job. From the end of 1793 Volta denied completely the existence of animal electricity and argued that the only merit of the animal tissue was to provide the necessary saline humidity between the metals and to serve as the indicator that an electric current was flowing in the system. Volta's investigations led to his invention of the electric battery (the Volta Pile) and formed the basis for the rapid development of electrochemistry and electrodynamics in the nineteenth century. Given this background it is hardly surprising that the controversy between Galvani and Volta immediately ended in favor of the viewpoints of the latter, in spite of Galvani's demonstration of the needlessness of the metal "contraction without metal", and that physicists in general had little sympathy for such phenomena as "animal electricity". Nevertheless, several outstanding physiologists and other scientists of the first half of the nineteenth century continued to study the "animal electricity", considering it to be crucial for the understanding of the function of nerve and muscle in spite of the physicist's denial of its distinctive character*. About 60 years after Galvani's original experiment it was demonstrated that his interpretation was also correct, as the "Galvani process" and the "Volta process" both occurred in Galvani's experiment. Carlo Matteucci (1811–1865), professor of physics at the University of Pisa, and Emil

* In this context it may be of interest also to note that the evidence and formulation of the First Law of Thermodynamics – one of the cornerstones of physics that has withstood the need for modification – was not due to professional physicists but to a brewer (*Joule*) and to two primarily medically trained people (*Mayer* and *Helmholtz*).

Du Bois–Reymond (1818–1896) in Berlin, demonstrated independently in the years 1840–43 that an isolated muscle, under certain conditions, produces an electric current and, furthermore, that the strength of this current changes when the muscle is stimulated to contract. In 1842, Du Bois–Reymond also demonstrated that a nerve can produce an electric current and that stimulation of the nerve causes a transient change of this current. From these and other observations it was concluded that a resting nerve possesses electric charges that are arranged in such a manner that the outer surface is electrically positive relative to the internal protoplasm of the nerve fiber, i.e. the surface of the fiber is *polarized* (constituting an electric double layer). An important addition to this view came 20 years later when Julius Bernstein (1839–1917) in Heidelberg showed that when a nerve impulse propagates along the length of the nerve it travels along a region that is negatively charged relative to the other regions ahead and after the impulse. That the nerve impulse travels along the nerve with a *final* velocity had already been demonstrated in 1850 by Hermann von Helmholtz (1821–1894), at that time a military surgeon in Berlin; later he became one of the leading physicists of the latter half of the nineteenth century. Helmholtz also developed the physical theory for the electric double layer and its application for describing the distribution of the current flow in the nerve.

The most fertile hypothesis to account for the propagation of the nerve impulse was that about *local currents*, which was proposed late in the nineteenth century by Ludimar Hermann (1838–1914), professor of physiology at the University of Königsberg. Hermann proposed that the mechanism of propagation contained two steps. (1) The stimulated part of the nerve's surface membrane produces an electric current. Owing to the cable-like electrical properties of the nerve, a fraction of this current will cross the neighboring region of the nerve membrane that was not part of the region stimulated initially. (2) Because of the current flow this new region becomes stimulated and also produces a nerve current, which in turn stimulates the next adjacent region, etc. The result is that a "wave of excitability" propagates along the nerve like a fuse with a certain velocity. To many electrophysiologists Hermann's idea became an incentive to test the feasibility of the proposed mechanism for nerve conduction. Two questions were chiefly in focus. How could it be imagined that a current flow through the nerve at one spot should cause the additional production of current at the same place? Does the design of nerve structure favor the spread of the current flow (the action current) to the neighboring regions and, if so, will the current have sufficient strength to stimulate the neighboring region?

In this connection, the theoretical and experimental results that engineers and physicists had previously obtained about the function of the submarine telegraph cable played an important role. A submarine cable consists of an inner high-conducting wire – the core – that is surrounded by a good insulator and is designed to transmit an electric signal over a large distance with minimum distortion and loss of amplitude. The efficiency of the transmission depends on the *cable properties*, that is the resistance of the internal core, the resistance of the insulator and the capacitance between the core and the ocean. To prevent the leak of charge inside the cable to the surrounding ocean, both the conductance of the surrounding insulator and the capacitance from the internal core to the ocean must be small, and the core should offer the least possible resistance to the longitudinal current flow. The mode of operation of the submarine cable (with disregard of the inductance of the core conductor) and its feasibility of transmitting electric signals across long distances (the Atlantic Ocean) was worked out for the first time in 1855 by Lord Kelvin (1824–1907). The calculations showed that a short voltage pulse that is impressed at the sending end of the cable becomes both smaller in amplitude and broadens out as the signal propagates towards the receiving end of the cable. The equations derived for the operation of the Atlantic cable were primarily analogous in form to those describing a one-dimensional diffusion process.

It was now convenient to ascribe to the conducting nerve properties similar to those of a cable and to measure the corresponding electric properties of the nerve. However, it turned out that from this point of view the nerve would function as a very ineffective cable with a predominant leak of current between the inner and outer side of the nerve that would make the spread of current in response to the injection of a current pulse rather short in the longitudinal direction. This observation, although making a robust cable theory for the nerve function untenable, was compatible with the slow speed of nerve conduction (10–20 m s^{-1}) – as demonstrated by Helmholtz – and was large enough to allow for the presence of the local currents as proposed by Hermann. These also remained one of the basic elements in later considerations about the nature of conduction in nerves. Experimentally, the correctness of Hermann's conceptions was first demonstrated in 1938 by A.L. Hodgkin from Cambridge.

The first attempt to account for the polarization of the nerve membrane on the basis of the difference between the ionic content in the nerve's axoplasm and the surrounding fluid was made in 1902 by Julius Bernstein – by now in Halle. The explanation was roughly as follows. In the resting state the nerve membrane is selectively permeable to K^+ ions but impermeable to Na^+, Cl^- and other ions. This assumption would not only account for the large differences between the same kinds of ions across the membrane but also for potential jumps across the membrane due to the tendency of the K^+ ions to diffuse to the extracellular fluid. Additionally, Bernstein assumed that an electric stimulation of the nerve with the current flow crossing the nerve membrane caused a change in the membrane that made it now permeable to *all* the ions in the interior axoplasm.

This removal of the barrier in the membrane that was responsible for the resting membrane potential would in consequence lead to a neutralization of the separated charges and thus cause the membrane potential to diminish – eventually to disappear completely (the membrane became *depolarized*). On account of the cable properties of the nerve, a current would flow from this locally depolarized area to the neighboring region and make a loop by crossing the nerve membrane and returning via the external fluid. This current would also cause a breakdown of the diffusion barrier across the membrane and its subsequent depolarization and production of a new loop of current flow into the next adjacent part of the nerve and so on, with the end result that the membrane depolarization propagates from the initial point of stimulation to the end of the nerve. Thus the Bernstein theory comprised essentials such as the resting membrane potential, nerve stimulation and impulse propagation, together with its electric manifestation the *action current*, whereas it still left open the questions relating to the origin and maintenance of the uneven distribution of Na^+, K^+ and Cl^- ions across the membrane. In the proposal of his theory, Bernstein pioneered work in making use of the results of electrochemistry that physical chemists had obtained a short time previously (W. Nernst, 1888; W. Ostwald, 1889). By making this leap ahead, Bernstein's theory remained accepted by the majority of physiologists for the next 40 years as the most satisfactorily explanation for the propagation of the nerve impulse. Another reason for the relative irresistibility of Bernstein's theory was undoubtedly the lack of the technical facilities that were needed to test it in detail. This required sensitive, precise and fast measuring and recording systems that could be used with flexibility. And such instrumentation first became available late in 1930 with the use of the electron tube in the construction of electric amplifiers and electronic recording instruments.

4.3 The nerve signal recorded with external electrodes

4.3.1 The nerve signal

What does the nerve signal look like when recorded with a modern electronic technique? A sketch of such a measuring arrangement is shown in Fig. 4.1. For simple demonstration experiments the n. ischiaticus from the frog is most frequently used. This nerve contain about 1000 single nerve fibers (axons). The nerve is dissected out from the frog and two pairs of electrodes are placed upon its surface. (1) The pair of *stimulation electrodes* is connected to the stimulator that enables the experimenters to apply a shortlasting electric pulse of variable strength to the nerve. (2) The pair of *recording electrodes* is connected to an amplifier of proper design, whose output is connected a suitable recording instrument, e.g. an oscilloscope, for recording the electric signal that is picked up by the electrodes when the impulse travels past the electrodes.

4. *The nerve impulse*

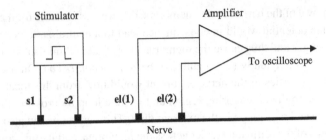

Fig. 4.1. Diagram of the arrangement to record from the surface of a nerve the electric signal that is associated with the impulse propagation. An electric stimulator is connected to the two stimulating electrodes s1 and s2. Two recording electrodes el(1) and el(2) are connected to an electronic amplifier whose output signal is transferred to an oscilloscope to record the signal.

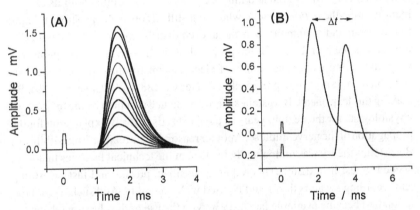

Fig. 4.2. Electric recording of the nerve signal from the surface of n. ischiaticus from frog. (A) Dependence on stimulus strength. (B) Demonstration of the signal's propagation along the nerve. The signals obtained from two pairs of electrodes are recorded on a dual beam oscilloscope. The time between the peaks (1.85 ms) represents the time of travel of the impulse from the first electrode pair to the other pair, giving in this example a conduction velocity of 54 mm/1.85 ms $= 29$ mm ms^{-1} $= 29$ m s^{-1}.

Figure 4.2(A) shows examples of recordings obtained by different strengths of stimulation. When the stimulation current is very weak and below a certain level, no signal appears at the recording electrodes in response to the stimulus. As the stimulus strength gradually increases a shortlasting *potential swing* suddenly appears on the recording screen. The stimulus strength that is just capable of eliciting a signal is called the *threshold stimulus*. An increase in the stimulus strength above this level causes the recorded diphasic signal to grow to a certain height above which the stimulus strengths have no further influence (see Fig. 4.2(A)). The underlying reason behind this dependency will be dealt with in Section 4.3.2.1.

4.3.1.1 The diphasic action potential

The transient potential swing is called the extracellularly recorded *diphasic action potential*. The recording shown in Fig. 4.2 is carried out in conformity with the "classical" convention of sign, that is a display in the record *above* the baseline of the record means that electrode (1) in Fig. 4.1, which is nearest to the place of stimulation and therefore first to meet the front of the approaching impulse, is at a *lower potential* than electrode (2), i.e. the potential difference $\psi^{(1)} - \psi^{(2)}$ is *negative*. Hence, a display above the baseline means that there is a current flow between the electrodes in the direction (2)→(1), whereas a display *below* the baseline indicates a current flow in the direction (1)→(2). Thus, the size of the display – usually a few millivolts – depends partly on the strength of the current flow and partly on the resistance between the recording electrodes. The duration of the action potential – in general a few milliseconds – will, within a certain limit, increase with increasing distances between the electrodes. The action potential appears after a delay that roughly corresponds to the time taken for the impulse to propagate from the point of stimulation to the recording electrodes. A simple way to demonstrate the actual impulse propagation is to make a series of recordings with different distances between the stimulation and recording electrodes as illustrated in Fig. 4.2(B), which shows the recordings when the impulse propagates past two pairs of recording electrodes. It is seen that the two action potentials are almost identical in shape but that the signal recorded from the more-distant electrode pair appears later than that recorded from the first pair. This delay Δt corresponds to the time required for the impulse to travel the distance Δx from the first pair of recording electrodes to the second pair. The conduction velocity of the impulse is then determined as $\Delta x / \Delta t$. Alternatively the conduction velocity is determined by making records with only one pair of recording electrodes that are placed at varying distances from the stimulating electrodes*. The time delay from the moment of stimulation to the arrival of the action potential at the recording electrode is found to vary linearly with the distance between the stimulating and recording electrode pairs. Furthermore, this line, when extrapolated to $t = 0$, intersects the distance axis at a position corresponding to the stimulation electrode that is the *cathode*. This means that the *start of the impulse* occurs at that position on the nerve where the stimulating current *leaves* the nerve.

How shall we interpret the diphasic action potential from Fig. 4.2? Is it so that this potential configuration – with a negative front and a positive tail

---

* This was essentially the method used by Helmholtz, who could not record the electric signal from the frog's nerve, but used the contraction in the attached muscle as the indicator for the arrival of the impulse at the muscle, which then was stimulated at varying distances from the muscle.

relative to those regions on the nerve that are at rest – propagates along the nerve with a constant velocity? By recording the diphasic action potential in Fig. 4.1 we measure the electric potential difference between electrode (1) and electrode (2) when the "active region" embracing the impulse travels past the electrode complex, i.e. first by affecting electrode (1) alone, next electrode (1) and electrode (2), and subsequently electrode (2) alone. As mentioned above, a display above the base line means that the potential at electrode (1) is negative with respect to electrode (2). When the front of the impulse invades electrode (1), the deflection is upwards. As long as electrode (2) is not invaded by the front of the impulse, electrode (1) will record correctly the potential variations on the surface of the membrane during the impulse propagation. This corresponds to the first part of the rising phase of the action potential (see Fig. 4.2 (A)). If the distance between the recording electrodes (1) and (2) is shorter than the extent of the propagating active region on the nerve both electrodes will in turn be involved simultaneously. This will make the conditions less easy to see as it is invariably the *potential difference* between electrode (1) and electrode (2) that the recording displays.

4.3.1.2 The monophasic action potential

To measure the spatial extent of the electric activity on the nerve's surface during the passage of the impulse, the distance between the two recording electrodes must be large enough to allow the whole active region to pass electrode (1) *before* it invades electrode (2). Unfortunately the sciatic nerve of the frog is not long enough to fulfil this demand*. But the above difficulty can be circumvented by the following artifice. The impulse conduction of the nerve is interrupted (e.g. by mechanical crushing or by a chemical blockade) at some place between the two recording electrodes (1) and (2) (see Fig. 4.3(B)), while the passive electric conductivity between the two electrodes is retained. An impulse that propagates towards the recording electrodes will pass electrode (1) but will stop at the block. Thus, the whole extent of the active region has passed electrode (1) but without involving electrode (2), which is now at a constant potential. Hence, the potential variation now recorded arises from potential variations on electrode (1) alone and, therefore, represents that potential configuration on the nerve surface that propagates with a constant velocity along the nerve as one of the nerve impulse's manifestations. The potential recording shown in Fig. 4.3(B) is called the *monophasic action potential*. This recording, which displays a time course, shows the electric potential change that occurs at a definite point on the surface of the nerve relative to positions that are kept at a fixed potential all the time

---

* N. phrenicus from a giraffe serves this end perfectly.

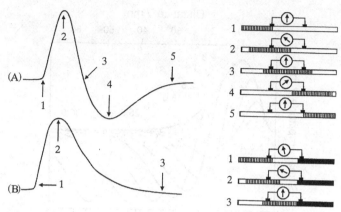

Fig. 4.3. Action potentials from frog nerve. (A) Diphasic action potential. (B) Monophasic action potential. The duration of (B) is 2.5 ms and amplitude is 1.5 mV. The schematic presentations to the right illustrate the ways of emergence of the two configurations (A) and (B). The region on the nerve that comprises the extent of the active region (the propagating nerve impulse moving from left to right) is hatched; the region with blocked impulse conduction is black; the graphs 1, 2, 3, etc., show snapshots from (A) and (B) of the course of propagation. For further explanations, see text and Fig. 4.4.

during the passage of the impulse past the point in question. The monophasic action potential in Fig 4.3.(B) also depicts the spatial potential configuration that occupies a certain length of the nerve at a given time. The length of this region – the "wave packet" – is easily found. The propagation velocity of the nerve impulse in Fig. 4.2.(B) is $v = 54$ mm/1.85 ms = 29 mm ms^{-1}. The duration of the monophasic action potential of Fig. 4.2.(B) is $\Delta t = 2.6$ ms. Hence, the length L of the region on the nerve whose electric potential differs from that of the non-active regions and through which there is a current flow has an extent – or wave length – that is $L = vt = 2.65 \times 29 = 125$ mm. Thus, replacing the time Δt by the distance L gives a "frozen" display of the spatial configuration that moves along the length of the nerve without a change in shape.

The diphasic action potential can be looked on as the difference between two monophasic action potentials that are recorded simultaneously from electrodes (1) and (2). This is illustrated schematically in Fig. 4.3.(A). The rising portion of the diphasic action potential corresponds to the invasion of the front of the monophasic action potential into electrode (1) (snapshots 1 and 2). This course of registration continues until the monophasic potential configuration after a delay Δt reaches and moves past electrode (2), whereupon both electrodes (1) and (2) are now exposed to time-variant potentials whose difference is transferred by the amplifier to the recording device. The rise of the potential at electrode (2) concurrently with the further travel of the impulse will cause the

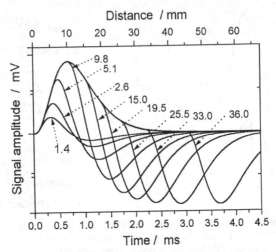

Fig. 4.4. A family of recordings of diphasic action potentials showing the influence of the distance between the recording electrodes on the form and size of the diphasic action potential. The conduction velocity is 20 m s^{-1}. The numbers attached to the curves represent the electrode distance (cm).

diphasic potential to decrease and the display will cross the baseline when the potentials at (1) and (2) are equal (see Fig. 4.3(A)) (snapshot 3) to take a transient swing below (snapshot 4) and finally to the baseline (snapshot 5) as the impulse has left electrode (1) and now leaves electrode (2) behind. It is evident that the distance between the two recording electrodes exerts an influence on the shape and amplitude of the diphasic action potential. As long as the electrode distance is less than the extent of the rising phase of the monophasic configuration, the ascending amplitude of the diphasic action potential will increase with increasing electrode distance until it contains the whole rise of the monophasic potential. The descending amplitude will, in contrast, continue to increase and the situation is attained where the *whole* extent of the monophasic configuration is contained between the two electrodes. The diphasic action potential will now be a display of two monophasic action potentials having the opposite polarity. Examples of recordings obtained with different electrode distances are shown in Fig. 4.4.

4.3.2 Some elementary properties of the nerve signal

The action potentials of Fig. 4.2 and Fig. 4.3 were recorded from a whole frog nerve (n. ischiaticus). This nerve contains about 1000 axons that separately contribute to the shape and size of the action potential that can be recorded from the surface of the nerve. Therefore recordings obtained from this nerve preparation cannot provide completely unambiguous information about the nerve signal in

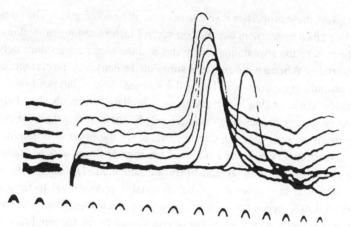

Fig. 4.5. Illustration of the "all or nothing" property of the start of the impulse propagation in a single, isolated nerve fiber from a frog. Monophasic recording. The stimulus strength in relative units are 1.00, 1.05, 1.50, 2.00, 2.50, 3.00 and 3.50. The two lower recordings, one with and one without an action potential, were taken with the same stimulus strength. Time marking: 0.2 ms. (After Tasaki, 1953*.)

the individual axon because of the interactions between the other axons and the surrounding connective tissue. Avoiding these interfering causes requires experiments on the single isolated axons. In the coastal crab *Carcinus maenas* the leg muscles are innervated by a bundle of unmyelinated axons with a diameter of about 30 μm, from which it is relatively easy to isolate a single axon about 5–6 cm long to be placed in a set up with electrodes for stimulation and recording. It is also possible – although far more difficult technically – to dissect and isolate single myelinated nerve fibers from frogs and toads.

4.3.2.1 "All or nothing" law

In the whole nerve the size of the action potential depends – within certain limits – on the strength of the stimulus (see Fig. 4.2). This pattern of behavior, however, is observed from the action potentials that are recorded from the isolated axon. Figure 4.5 shows the monophasic action potentials recorded from a single isolated axon from a crab (i.e. unmyelinated fibers). The nerve is stimulated with shortlasting (0.1 ms) electric pulses of increasing strengths, their mutual magnitude being the ratios 1.0, 1.05, 1.50, 2.00, 2.50, 3.00 and 3.50. As seen both the sizes and the time courses of the monophasic action potentials are altogether *independent of the stimulus strength*. No impulse can be elicited with a stimulus strength that is *below* a certain critical value (1.00 in the present case). This value is called the *threshold stimulus* or the *threshold value*. A nerve

---

* Tasaki, I. (1953): *Nervous Transmission*, Thomas, Springfield.

that displays these properties is said to obey an *all or nothing law*. The threshold value in a given experiment depends on several factors, among other things the arrangement of the stimulating electrodes and the electric resistance between the electrodes. When a whole nerve is stimulated electrically the current density of the stimulating current will be on the surface close to the positions of the electrodes. When gradually increasing the stimulation strength, the fibers that are nearest the electrodes (the cathode) will be the first targets to be crossed by a current stronger than the threshold value for the fibers in question. With increasing strengths more and more of the remote fibers will be stimulated above the threshold until all the fibers are activated. The amplitude of the action potential is – other things being equal – proportional to the current that flows in the nerve between the recording electrodes. Since this current grows concurrently with the number of conducting fibers, the amplitude of the summated action potential will increase with increasing stimulation strength until all the fibers in the nerve are conducting. This phenomenon is dealt with in more detail in Section 4.4.5.4 and in Appendix Q. Another cause for the gradation may be differences in thresholds among the individual nerve fibers.

4.3.2.2 Subliminal stimuli

As mentioned before, there exists a borderline – the threshold value – between the strength of the electric stimuli that elicit a nerve impulse and those that are unable to do so. Those stimuli that belong to the latter class we denote *subliminal*. Even if a subliminal stimulus does not elicit a propagated impulse in the nerve it does produce a transient, subsiding change in the state of the nerve at the electrode position in such a way that a lesser stimulation current is now required to start an impulse. We say that an *increase* in *excitability* of the nerve has occurred at the position of the electrode. To demonstrate such a change in excitability we apply a so-called double-pulse stimulator, i.e. a stimulator that produces two pulses in succession, with control of the interval between the pulses and separate controls of the strength of the two stimuli. The first stimulus, the *conditioning stimulus* $S^{(con)}$ is subliminal, e.g. 80% of the threshold value S^0 for a single stimulus. The other stimulus, the *test stimulus* is applied after a delay Δt after $S^{(con)}$, and the strength is adjusted to the threshold value $S^{(test)}_{\Delta t}$, at the time Δt after $S^{(con)}$. We will the observe that

$$S^{(test)}_{\Delta t} \leq S^0,$$

where $S^{(test)}_{\Delta t}$ approaches the original value S^0 as the time of delay Δt between the two stimuli increases. Thus, the subliminal stimulus $S^{(con)}$ has produced a transient change in the state of the nerve resulting in a reduction of the threshold value $S_{\Delta t}^{(\text{test})}$ for the stimulus that is applied after a delay of Δt ms after S^0. When the strength of the subliminal stimulus is near the threshold value, e.g. $S^{(\text{con})} = 0.95 \cdot S^0$, one could observe that 1 ms later the threshold stimulus $S_{\Delta t}^{(\text{test})} = 0.2 \cdot S^0$. A natural interpretation of this observation is that the subliminal stimulus of strength 0.95 has led to a transient increase in the excitability of the nerve such that 1 ms later it behaves as being exposed to the action of a subliminal stimulus of strength 0.8. A convenient measure for the reduction in threshold – or increased irritability – is the difference

$$\Delta S_{\Delta t} = S^0 - S_{\Delta t}^{(\text{test})},$$

which indicates the extra stimulation strength one must add to the stimulus $S_{\Delta t}^{(\text{test})}$ at the time Δt to produce a stimulus that is equal to the threshold stimulus S^0 corresponding to the single stimulation. Figure 4.6 shows an example of the time courses of the excitability increase $\Delta S_{\Delta t}$ that follows the stimulation of a frog nerve with a series of subliminal stimuli of varying strengths. It appears that even a subliminal stimulus $S^{(\text{con})}$ as weak as $0.15 \cdot S^0$ causes a transient – although slight – increase in the irritability of the nerve, which subsides nearly monoexponentially. When the subliminal stimulus approaches the threshold value a marked deviation from the exponential decline sets in, in that the initial hyperirritability becomes more and more protracted before the

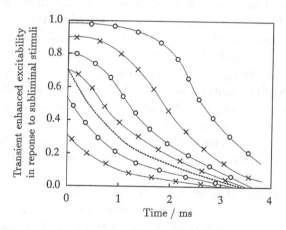

Fig. 4.6. Time course of the transient change in excitability that follows the stimulation of the frog nerve with subliminal conditional stimulus $S^{(\text{con})}$. The threshold value of this stimulus is $S^0 = 1.0$. Abscissa: time Δt (ms) after stimulation with the conditional stimulus $S^{(\text{con})} < S^0 = 1.0$ where the threshold value $S_{\Delta t}^{(\text{test})}$ is obtained. Ordinate: relative increase of the excitability expressed as: $(S^0 - S_{\Delta t}^{(\text{test})})/S^0$. The strength of the conditional stimulus (in relative units $S^{(\text{con})}/S^0$) is given by the ordinate of each curve for $t = 0$. The stippled line represents a monoexponential decay.

exponential decline begins. It will be seen later that the exponential decline can be accounted for by the *passive* cable properties of the nerve (discharge of the membrane capacitance through the membrane resistance of the charge supplied by the stimulation current), whereas the deviation from the exponential course is a reflection of the faint, uncompleted attempt on the part of the nerve to start these processes that at a somewhat greater stimulus strength results in a full-blown propagation of the impulse. This subliminal *active* reaction is called the *local response*.

4.3.2.3 The refractory period

The effectiveness of a transmission line depends among other things on the ability to conduct two impulses in succession. To examine the nerve in that respect we use double-stimulation. At the time $t = 0$ the nerve is stimulated with a stimulus $S(0)$ that is substantially greater than the threshold value S^0. Next we determine the threshold value $S^{(test)}_{\Delta t}$ for a second stimulation pulse that is applied at a time Δt after $S(0)$. By making a number of such determinations for different values of Δt one will find that marked changes in irritability occur in the wake of the action potential; there exists a short time interval $0 \leq t \leq t_{abs}$ where it is *impossible* – even when using the strongest non-injurious stimuli – to elicit another propagated action potential. The duration for which the nerve is *absolutely refractory* corresponds largely to $3/4$ of the duration of the monophasic action potential. In the interval that follows, the threshold stimulus strength $S^{(test)}_{\Delta t}$ is first very high relative to S^0, but gradually the excitability returns to the normal level. Within this interval $t_{abs} \leq t \leq t_{rel}$ the nerve is said to be in a *relatively refractory state*. An example of the *refractoriness* in a nerve that follows the application of a stimulus of superthreshold strength is shown in Fig. 4.7. When applying repetitive stimulation to a nerve, the response will in general be a train of impulses. On the face of it the duration of the absolute refractory period should indicate the shortest possible interval between successive impulses in the train. Strictly, this is not correct as the impulse propagates somewhat more slowly when the stimulus is applied in the relative refractory period of the nerve. Furthermore, after stimulation in the relative refractory period the take-off of the impulse is slowed down. Taken together this implies that the minimum interval between two impulses in a continuous train is somewhat longer that the absolute refractory period. Mammalian nerves with an absolute refractory period of 0.5 ms can produce a maximum frequency of about 1000 Hz in an impulse train. But such a train is seen only rarely in the intact organism. The acoustic nerve in mammals may sometime conduct a train of impulses with a frequency of 1000 Hz, but otherwise the sense organs and muscle nerves gave a normal range of activity amounting to 5–100 impulses per second and only rarely produce trains of a frequency larger than 200 Hz.

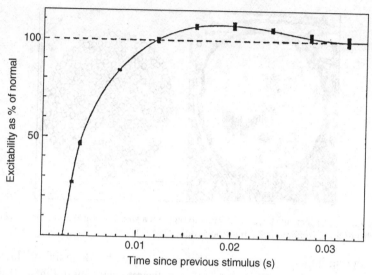

Fig. 4.7. Illustration of the absolute and relative refractory period. Abscissa: time (ms). Ordinate: reciprocal of the threshold value $S_t^{(\text{test})}$ for the second stimulus relative to $S^0 = S_{t=\infty}^{(\text{test})}$. The phase with supernormal excitability is not always observed (Adrian & Lucas, 1912).*

4.4 Results from the giant axon of the squid

In 1936, the British anatomist J.Z. Young brought the physiologists' attention to the "giant axons" that innervate the mantle muscle of the North Atlantic squid (*Loligo pealeii*). These axons have a diameter of 0.5–1 mm. Figure 4.8 shows a cross section of such a giant axon and by way of comparison a collection of myelinated nerves from a frog's n. ischiaticus.

4.4.1 Recording the resting membrane potential and action potential

Naturally, a single axon of such dimensions gives far better possibilities for us to examine directly events happening across the nerve membrane during the impulse propagation. Figure 4.9 shows that it became possible to introduce a long glass capillary of a diameter of about 0.1 mm into the one end of the axon and then push it 10–30 mm further in without touching the membrane. When successfully introduced, the presence of the glass capillary inside the axoplasm does not appear to affect the impulse conduction. The glass capillary is filled with a suitable electrolyte, e.g. sea water, and serves as the internal electrode to record the potential difference *across the membrane* both at rest

---

* Naturally, this experiment has been repeated many times ever since. This old result is reproduced as a reverence to two great physiologists of the past century. Adrian, E.D. & Lucas, K. (1912): *J. Physiol.*, **44**, 68.

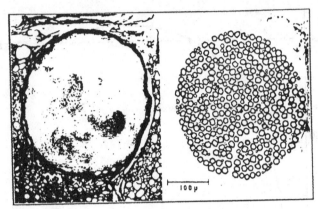

Fig. 4.8. Cross section (left) of the giant axon from a squid (*Loligo*) and a cross section (right) of an n. ischiaticus from a frog (Young, 1951*).

and during impulse propagation. The set up – used by Hodgkin[†] & Huxley[‡] (1939) in their first recordings – is shown diagrammatically in Fig. 4.10. The axon is hanging vertically surrounded by sea water. The glass capillary, filled with sea water, is connected to one of the preamplifier inputs via an Ag–AgCl electrode, the second being connected to another Ag–AgCl electrode that is placed in the bath of sea water that surrounds the nerve and – owing to its high conductivity – serves as a short circuit from the electrode to the nerve membrane. Two platinum electrodes are placed on the nerve leads via isolated wires to the electric stimulator. An example of a recording of a resting membrane potential and the signal – the action potential – that appears when the impulse travels past the tip of the glass capillary is shown in Fig. 4.11. It is seen that

---

* Young, J.Z. (1936): The giant nerve fibre and epistellar body of cephalopods. *Quart. J. Micro. Sci.*, **78**, 367.

[†] Alan Lloyd Hodgkin (1914–1998) was educated at Trinity College, Cambridge (1932–1936). After one year's study at the Rockefeller Institute he returned to Cambridge where he started collaboration with A.F. Huxley resulting in the intracellular recording of the nerve signal in the giant squid axon (1939). This collaboration was taken up after the War, leading to the physicochemical analysis of the basis mechanism of the nerve impulse in the squid axon. Hodgkin became a Royal Society Research Professor (1951–1969) and subsequently Professor of biophysics at Cambridge University. He was awarded the Nobel Prize for Physiology and Medicine in 1963, became President of The Royal Society 1970–75 and was knighted in 1972.

[‡] Andrew Fielding Huxley (b. 1917) was educated at Trinity College, Cambridge. He joined A.L. Hodgkin in 1939 in carrying out the intracellular recording of the nerve signal from the squid giant axon. From 1946 to 1952 he worked in collaboration with Hodgkin on the physicochemical analysis of the basic mechanism underlying the nerve impulse in the squid axon, holding in the course of time several college and university posts in Cambridge. After 1952 Huxley turned to muscle physiology yielding fundamental contributions to the understanding of the process of contractility. He became Jodrell Professor of physiology at the Department of Physiology, University College, London, in 1960, and from 1969 was Royal Society Research Professor. He was awarded the Nobel Prize for Physiology and Medicine in 1963, became President of The Royal Society 1980–85, and was knighted in 1974.

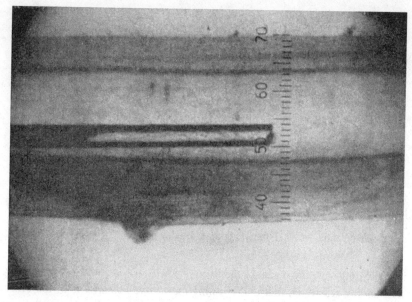

Fig. 4.9. Part of the giant axon from *Loligo* with a glass capillary (diameter 0.1 mm) inside, which functions as the inner electrode (Hodgkin & Huxley, 1945*).

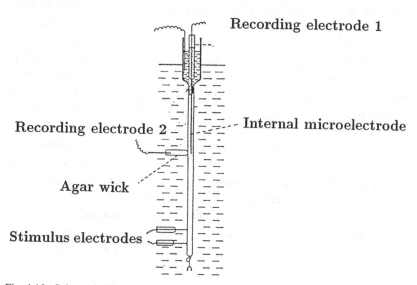

Fig. 4.10. Schematic illustration of the experimental set up to measure the resting membrane potential and the action potential from a giant axon of the squid (*Loligo*) (Hodgkin & Huxley, 1945*).

* Hodgkin, A.L. & Huxley, A.F. (1939): *Nature, Lond.*, **144**, 710; (1945): *J. Physiol.*, **104**, 176.

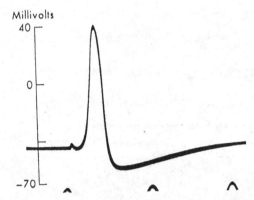

Fig. 4.11. Recording of the resting potential and action potential from the giant axon of the squid *Loligo*. Abscissa: time, distance between two marking peaks is 2 ms. Ordinate: membrane potential (mV). (After Hodgkin & Huxley, 1945.)

in the resting state the inside of the axon is 50–70 mV *negative* relative to the surrounding sea water, that is the *resting membrane potential* is of the order −50 to −70 mV.

When a propagating impulse moves past the internal electrode tip the recorded signal appears as a shortlasting change in the membrane potential by which the *interior potential of the axon transiently becomes* 40–50 mV *positive* relative to the surrounding sea water. This propagating change in the membrane potential lasting about 1.5 ms (18 °C) and having a maximum swing of 100–129 mV is called the *intracellularly recorded action potential* or just the *nerve action potential* as this recording gives the most relevant picture of the potential changes associated with the propagating impulse*. The record of Fig. 4.11 was obtained before the appearance of the Ling & Gerard (1949) glass micropipette[†]. This recording technique applies equally well, and both methods have been used in investigations on the giant axon of *Loligo*. Figure 4.12 shows recordings using both methods. In (A) the tip of a glass microelectrode is run through the membrane of the intact, undissected axon in the mantle muscle of *Loligo* and thus the nerve is stimulated *in situ*. In (B) the axon is freely dissected and placed in a set up like that of Fig. 4.10 with the axial glass capillary serving as electrode. The two recordings are essentially similar in shape but differ in duration because the two recordings were obtained at different temperatures.

* The potential configuration is highly temperature dependent both as regards the maximal potential swing and velocity of impulse conduction (Hodgkin, A.L. & Katz, B. (1949): The effect of temperature on the electrical activity in the giant axon of the squid. *J. Physiol.*, **109**, 240). For a comparison of recorded and computed responses see Huxley, A.F. (1959): Ion movement during nerve activity. *Ann. N.Y. Acad. Sci.*, **81**, 221.
[†] See Section 3.8.1.

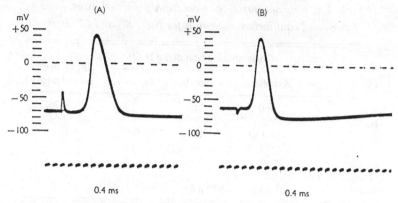

Fig. 4.12. Action potentials recorded from giant axons of a squid (*Loligo*). (A) the axon intact in the animal (temperature 8.5 °C); (B) the isolated axon (temperature 12.5 °C). Time marking is 0.4 ms between two points (Hodgkin, 1958*).

4.4.2 The resting membrane potential

In Chapter 3, Section 3.7 the membrane potential in a striated muscle fiber from the frog was accounted for as a diffusion potential whose magnitude and polarity are determined by the intracellular and extracellular distribution of the ions that most easily move passively across the membrane. Similar considerations have been applied to account for the resting membrane potential of the squid axon. The results of analysis of the axoplasm of the giant axon obtained from various sources are put together in Table 4.1[†]. The relative intracellular and extracellular distributions of K^+, Na^+ and Cl^- ions agrees well with the characteristic distributions of these ions in living systems in general, but the table does reflect that the animal is an invertebrate that lives in the sea. The last column in Table 4.1 indicates the value for the equilibrium potentials $V_j^{(eq)}$ for K^+, Na^+ and Cl^- respectively that are calculated from the table's value of intracellular and extracellular concentrations. These potentials also indicate the membrane potentials one should expect, if the membrane were in turn selective permeable to the ion in question. Thus, if the membrane were permeable to all three types one should expect a membrane potential located somewhere between the extreme values of -75 mV and $+54$ mV[‡].

The resting membrane potential that is observed experimentally is in the range of -70 to -50 mV; values that suit a diffusion regime where the membrane permeabilities P_K and P_{Cl} for K^+ and Cl^- were far greater than the permeability

---

* Hodgkin, A.L. (1958): Ionic movements and electrical activity in giant nerve fibers. *Proc. R. Soc. Lond.*, B**148**, 1.

† Hodgkin, A.L. (1964): *The Conduction of the Nervous Impulse*, Liverpool University Press, Liverpool.

‡ See Chapter 3, Section 3.8.2, Eq. (3.8.2).

Table 4.1. *Ion concentrations from freshly isolated axons from Loligo and equilibrium potentials for Na^+, K^+ and Cl^- ions*

Ion	Concentration (mmol/kg H_2O)			Equilibrium potential (mV)
	Axoplasm	Blood	Sea water	
K^+	400	20	10	-75
Na^+	50	440	460	$+54$
Cl^-	40–150	560	540	-65 to -32
Ca^{2+}	0.4	10	10	
Mg^{2+}	10	54	53	
Organic anions	350			
Water	865 g kg^{-1}	870 g kg^{-1}	966 g kg^{-1}	

(After Hodgkin, 1964.)

P_{Na} for the Na^+ ion. This view is further substantiated by the data in Fig. 4.13, which show the connection between the K^+ concentration in the external fluid – plotted on logarithmic scale – and the membrane potential, where the experimentally obtained data points indicated as (■). It appears from the graph that the membrane potential V_m – or rather its numerical value – decreases with increasing potassium concentration $C_K^{(o)}$, but that the relation between V_m and $\log C_K^{(o)}$ is not a linear one according to the behavior of a membrane that is selectively permeable to K^+. In the concentration range 2–100 mmol dm^{-3} the rate of depolarization starts with a low value that gradually increases with increasing values of $C_K^{(o)}$ until it attains the value of nearly 60 mV per decade change in $C_K^{(o)}$ and continues with that constant value (i.e. corresponding to the slope of the Nernst potential) for the remaining values of $C_K^{(o)}$ up to 500 mmol dm^{-3}. The dotted line in Fig. 4.13 shows the expected behavior of the Nernst potential in this upper range of values for $C_K^{(o)}$.

4.4.2.1 Membrane potential and the Goldman regime

The full-drawn curve in Fig. 4.13 shows the connection between the membrane potential and the K^+ concentration in the external fluid that is obtained from the Goldman–Hodgkin–Katz equation*

$$V_m = \frac{\mathcal{R}T}{\mathcal{F}} \ln \left(\frac{P_K C_K^{(o)} + P_{Na} C_{Na}^{(o)} + P_{Cl} C_{Cl}^{(i)}}{P_K C_K^{(i)} + P_{Na} C_{Na}^{(i)} + P_{Cl} C_{Cl}^{(o)}} \right), \qquad (4.4.1)$$

* See Chapter 3, Section 3.7.3, Eq. (3.7.30). Hodgkin & Katz derived the above version of the Goldman equation (Hodgkin, A.L. & Katz B. (1949): The effect of sodium ions on the electrical activity on the giant axon of the squid. *J. Physiol.*, **108**, 37) and applied it to data similar to those shown in Fig. 4.13 in their considerations about the nature of the resting membrane potential and the action potential.

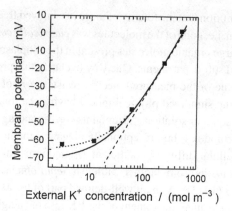

Fig. 4.13. Effect of external potassium concentration on resting membrane potential in a squid axon *Loligo*. Abscissa: $C_K^{(0)}$ in mM, logarithmic scale. Ordinate: membrane potential (mV). Experimental data ($\times$). Fully drawn curve: membrane potential calculated from the Goldman equation with concentrations in Table 4.1 and $C_{Cl}^{(axon)} = 52$ mM, assuming $P_K : P_{Cl} : P_{Na} = 1 : 0.2 : 0.025$. In the curve ($\cdots$) P_K is reduced by a factor 2.5 in the range: $2 \leq C_K^{(0)} \leq 60$ mM. The straight line (---) represents the equilibrium potential for K^+ ions, i.e. the relation to be expected if the membrane were exclusively permeable to K^+ ions. (Data ($\blacksquare$) from Hodgkin & Keynes, 1955*.)

when the ratio between the permeabilities is taken as $P_{Cl}/P_K = 0.2$ and $P_{Na}/P_K = 0.025$ and $C_{Cl}^{(axo)} = 52$ mM for the Cl^- concentration in the axoplasm (corresponding to a membrane potential of $V_m = -60$ mV). This simple assumption of different, but constant, permeabilities for the three ions fits the experimental data to a certain extent along the road ($60 \leq C_K^{(0)} \leq 500$ mM). However, the conditions appear to be somewhat more complicated as the permeability P_K for potassium decreases when the membrane potential in the resting state becomes more negative than -50 mV, which results in a lesser preponderance of the influence of the K^+ ions on the membrane potential. The broken line in Fig. 4.13, which follows the experimental data along the rest of the road, is calculated under the assumption that P_K is reduced by a factor of 2.5 in the range $2 \leq C_K^{(0)} \leq 60$ mM, whereby the balance of the ionic movements across is changed in the direction of an increase in the tendency of the Na^+ ions in the external fluid to pull the membrane potential towards the value of the equilibrium potential of the Na^+ ions ($V_{Na}^{(eq)} = 54$ mV).

4.4.2.2 Mobility of K^+ ions in axoplasm

The equilibrium potentials $V_j^{(eq)}$ for Na^+, K^+ and Cl^- ions were calculated from concentration values that were obtained by chemical analysis and used

---

* Hodgkin, A.L. & Keynes R.D. (1955): *J. Physiol.*, **132**, 592.

434   4. The nerve impulse

with the tacit assumption that the totality of these ions moved freely around as in an aqueous solution, i.e. none of the molecules was *chemically combined*, e.g. by an association to large organic molecules present in the axoplasm or confined to membrane-covered sub-departments. One way to clarify this problem has been to compare the value of the measured electric resistance r_a of the axoplasm* with the value of the simulated physical model having the ionic mobilities of u_K, u_{Na} and u_{Cl} in aqueous solution and amounts corresponding to the chemical analysis[†]. An unorthodox – but exceptionally elegant – method of measuring directly the intracellular diffusion coefficient D_K and ionic mobility u_K in the axoplasm of giant axons from another cuttlefish *Sepia officinalis* were due to Hodgkin & Keynes (1953)[‡]. A small well-defined part of the axon was charged with ^{42}K and the subsequent movement of the ^{42}K into the neighboring regions was determined by moving the axon horizontally over a Geiger counter that was provided with a suitable screening to allow the counts from a small spot only to be determined. By surrounding the axon with a film of oil, the $^{42}K^+$ ions that initially entered the axoplasm is forced to remain inside the axon. The axial concentration profile of $^{42}K^+$ was recorded partly when the $^{42}K^+$ spread from the spot of loading by *simple diffusion* and partly by diffusion with *superimposed migration*. In the latter case a pair of electrodes was arranged at each of the axon's ends serving to establish a longitudinal electric field of known strength E inside the axoplasm, which in turn gave the diffusing $^{42}K^+$ ions a unidirectional migration velocity $v = u_K E$ in the direction of the field. The results of such experiments are shown in Fig. 4.14, where the spread of the $^{42}K^+$ ions takes place by simple diffusion, and in Fig. 4.15, where an electric field of 0.548 V cm^{-1} is impressed axially in the axoplasm. Hodgkin & Keynes assumed that the spread of the substance axially could be described as originating from the location of an instantaneous matter source in the position x_0 at the time $t = 0$, and determined by a least-squares procedure the values for D_K and u_K that best fitted the experimental data (the full-drawn curves in Fig. 4.14 and Fig. 4.15, which adequately fit the data). It appears from the figures that the effect of the field in the time t is to displace the concentration in the direction of the field, whereas the change in configuration – reduced height and greater stretching out – relative to the configuration at the time of application of the field is only due to the movement of $^{42}K^+$ ions by diffusion in the time t[§]. Hodgkin & Keynes arrived at the following values for the diffusion coefficient

* See Appendix P.
[†] See Chapter 3, Section 3.3.1, Eq. (3.3.21).
[‡] Hodgkin, A.L. & Keynes, R.D. (1954): The mobility and diffusion coefficient of potassium in giant axons from *Sepia*. *J. Physiol.*, **119**, 513.
[§] Besides, compare the two figures with the analogous theoretical curves shown in Chapter 2, Section 2.6.1.1, Eq. (2.6.26).

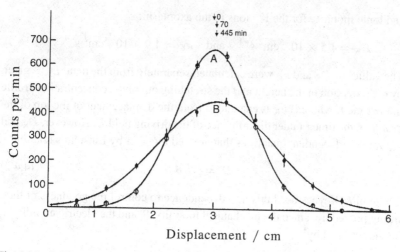

Fig. 4.14. Axial distribution of radioactive $^{32}K^+$ in a giant axon of *Sepia officinalis* after loading at a spot on the axon. Ordinate: counts per minute. Abscissa: distance (cm). Curve A initial distribution at the end of loading. Curve B distribution after 445 min (Hodgkin & Keynes, 1953).

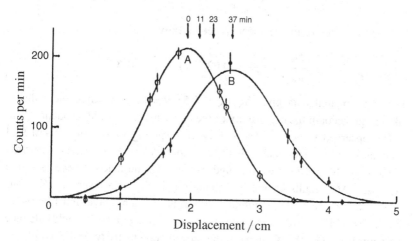

Fig. 4.15. Axial distribution of radioactive $^{42}K^+$ from a giant axon from *Sepia officinalis*. Ordinate: counts per minute. Abscissa: distance (cm). Curve A, axial distribution initially observed. Curve B, final. Between the measurements of curve A and curve B, an axial field of 0.548 V cm^{-1} was impressed for 37 min. The arrows ↓↓↓ indicate the positions of the profiles' maxima before and after application of the field. The anode was placed at the position $x = -0.05$ cm and the cathode at the peripheral end at $x = 4.2$ cm. The full-drawn curves are computed from a diffusion–migration equation of the same form as that in Chapter 2, Section 2.6.1.1, Eq. (2.6.10) after fitting the values for D_K and u_K from the experimental data (Hodgkin & Keynes, 1953).

and ionic mobility for the K^+ ions in the axoplasm:

$$D_K = 1.5 \times 10^{-5} \text{cm}^2 \text{ s}^{-1} \quad \text{and} \quad u_K = 4.9 \times 10^{-4} \text{cm}^2 \text{ s}^{-1} \text{ V}^{-1}.$$

The values of D_K and u_K were calculated separately from the data. D_K is given by the reduction in the height and the stretching out of the concentration profile in the time t, whereas u_K is calculated from the displacement of the top of the profile in the time t under the influence of the driving field E. However, D_K and u_K are interdependent quantities that are tied together by Einstein's relation*

$$D_j = k T B_j. \tag{4.4.2}$$

Hodgkin & Keynes used this interdependence to control the goodness of their fitting procedure, since the mechanical mobility B_j and the electric mobility u_j are interrelated by[†]

$$u_j = z_j e B_j. \tag{4.4.3}$$

Combining Eq. (4.4.2) and Eq. (4.4.3) gives[‡]

$$D_K = \frac{kT}{e} e B_K = \frac{\mathscr{R} T}{\mathscr{F}} u_K, \tag{4.4.4}$$

and the ratio D_K/u_K to be expected theoretically

$$\frac{D_K}{u_K} = \frac{\mathscr{R} T}{\mathscr{F}} = 0.025 \text{ (V)} \quad \text{at} \quad 20 \,^\circ\text{C}.$$

The experimental data give $D_K/u_K = 1.5/49 = 0.03$; an agreement that – taking into account the complexity of the experiment – should be adequate.

The observed values for D_K and u_K correspond to those of a 0.5 M KCl aqueous solution. This makes it rather improbable that any greater fraction of the tagged $^{42}K^+$ ions were bound, e.g. to organic macromolecules in the axoplasm. The results cannot ignore the possibility of a layer of $^{42}K^+$ ions fixed to the inside of the membrane. If that is so, such a layer must be rather thin. Thus, taking the observed value for u_K, Hodgkin & Keynes could calculate the contribution of the K^+ ions to the specific conductivity of the axoplasm, since[§]

$$\kappa_K = \mathscr{F} u_K C_K,$$

* See Chapter 2, Section 2.6.4.1, Eq. (2.6.54).
† See Chapter 3, Section 3.3.1, Eq. (3.3.5).
‡ This relation was originally derived by W. Nernst in 1888 using a different procedure.
§ See Chapter 3, Section 3.3.1, Eq. (3.3.17).

where $C_K = 400 \times 10^{-6}$ mol cm^{-3} is the concentration of K$^+$ ions in the axoplasm. This gives the value

$$\kappa_K = 96\,500 \times 4.9 \times 10^{-4} \times 400 \times 10^{-6} = 0.019\,\text{mho cm}^{-1} = (52\,\Omega\,\text{cm})^{-1},$$

a value to be compared with the total conductivity of the axoplasm, which is $\kappa_{axon} = 0.033$ mho cm$^{-1} = (30\,\Omega\,\text{cm})^{-1}$, that is, more than 60 % of the total conductivities of the axoplasm is carried by the K$^+$ ions.

4.4.3 The action potential

When a nerve is stimulated electrically by means of two surface electrodes the nerve impulse turned out to start from that electrode that served as *cathode* or in other words from that place on the nerve where the stimulation current *leaves* the nerve.

4.4.3.1 Stimulation with an intracellular microelectrode

This feature is better illustrated if one of the surface electrodes is replaced by a microelectrode that is introduced into the interior of the axon as illustrated in Fig. 4.16. The other electrode, e.g. an Ag–AgCl electrode, is placed somewhere in the external fluid. A recording microelectrode is introduced – under proper optical control – as near as possible to the stimulating electrode to record the changes in the membrane potential caused by the current flow from the stimulating electrode.

(i) The internal electrode is a cathode

We start with the situation where the external current electrode is the positive electrode, that is an *anode*, relative to the intracellularly positioned microelectrode. During the electric pulse current flows from the external fluid through the membrane to the tip of the electrode. The inward current flow through the membrane's resistance will produce an additional potential drop across the membrane directed *from the outside* to *the inside*. This potential drop adds to the prevailing membrane potential, which also drops from the outside to the inside. Thus, the result of the current flow is an *increase* in the potential drop across the membrane that drives the axoplasm *more negatively* relative to the membrane's outside, or in other words: the inwardly directed current *hyperpolarizes* the membrane. Figure 4.16 shows the effect of applying a series of shortlasting (2 ms), rectangular current pulses. The response of the membrane potential is not the exact replica of the driving current as the potential grows with a delay towards its new stationary value. This characteristic response is due

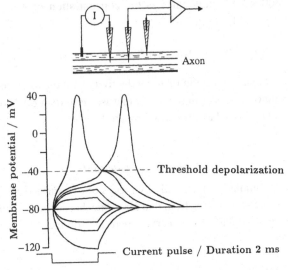

Fig. 4.16. Recordings of the changes in the membrane potential when one stimulation electrode is a microelectrode that is placed in the axoplasm. The potential-recording electrode is placed close to the tip of the current electrode. The axon is stimulated with a current pulse of 2 ms duration. Note: currents that flow in the direction that *depolarizes* the membrane are capable of *eliciting an action potential*, whereas *hyperpolarizing* currents are *ineffective* in that respect. Note also the differences in the times it takes to re-polarize the membrane after the hyperpolarizing and depolarizing currents, respectively, are switched off.

to the particular passive electric properties of the membrane as the membranes possess an ohmic *membrane resistance* and a *membrane capacitance* that are arranged in parallel. Until the new stationary potential level is reached, only part of the impressed current will flow through the resistance as another part serves to charge the membrane capacitance as the potential changes. Three conditions characterize the effect of the hyperpolarizing stimulus current. (1) The displacement of the membrane potential ΔV is proportional to the current i, and (2) the growth of the membrane potential towards the stationary value and its return after the close of the current flow take place with the same time constant, which equals the product of the membrane resistance and the membrane capacitance. (3) The hyperpolarizing currents do *not elicit* an action potential*.

(ii) The internal electrode is an anode
Reversing the direction of the stimulation current, the internal electrode becomes the anode, and the stimulation current now flows through the membrane

---

* This statement is only partly true, as very strong currents do so.

from the axoplasm to the external fluid, that is in the *outward* direction. The current that flows through the membrane resistance now causes an additional potential drop that is directed *from the inside* to *outside*, i.e. opposite to the direction-prevailing membrane potential drop. Thus, the result of the current flow is a reduction of the potential drop across the membrane. We say that the membrane is subjected to a *depolarization*, and that the current flow in the *outward direction* has a *depolarizing* action on the membrane potential. The effect of applying a series of short depolarizing current pulses of increasing strength is shown in Fig. 4.16: the weak currents produce a membrane depolarization whose transient time course – apart from the polarity – is identical to those produced by hyperpolarizing currents of the same strength. By using stronger depolarization currents that push the membrane potential to values lower than about −60 mV, the shape of the depolarizing transient changes drastically. The level of depolarization grows increasingly out of proportion with the increment in stimulus strength, and the initial exponential recovery after the current pulse becomes more and more drawn out and loses all its resemblance to an exponential decay. When exerting careful control of the stimulator one will eventually observe that there exists a stimulus strength which brings the membrane depolarization to a *critical level*, where the membrane potential – if left to itself – will linger for a certain time until one of two things happens: (1) the membrane potential returns monotonically to the initial value, and (2) the membrane depolarization continues – apparently spontaneously – with increasing speed, and the final result becomes an action potential of normal size that propagates along the axon. A further increase in current intensity – leading to a greater membrane depolarization – *always* elicits an action potential, but in such a way that the greater the depolarization, the earlier the action potential arises. Thus, the experiment demonstrates that there exists a critical level of membrane depolarization, the *threshold depolarization*, and a corresponding value of membrane potential, the *threshold potential*, where (1) all membrane depolarization *above* this level *always* leads the membrane spontaneously to run through the characteristic sequence of potential changes where the inside of the axon is temporarily charged to about +40 mV relative to the outside, whereas (2) for depolarization less than the threshold depolarization the membrane potential will repolarize to the initial resting level after the close of the current flow. The closer the displacement of the membrane potential approaches the threshold polarization, the more prolonged is the time course of the repolarization. This is in contrast to the repolarization configuration following a hyperpolarizing current, which is always in monoexponential decline. Therefore, the difference between the time course of repolarization after a depolarizing step and that of the hyperpolarizing step for equal current pulses represents activation of

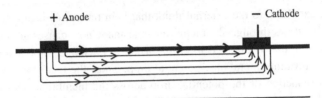

Fig. 4.17. Illustration of the current flow between two stimulation electrodes that are placed on the surface of a single axon. The bold line between the two electrodes represents the (larger) fraction of the current flowing in the external surface layer of the axon without crossing the membrane. Note: at the *anode* the current flows in the direction *outside→inside* and causes a *hyperpolarization*, whereas at the *cathode* the flow is *inside→outside* and causes a *depolarization*.

an entirely new process in the membrane that – in excess of the pure passive processes due to the presence of membrane resistance and membrane capacitance – now takes part in the control of the membrane potential. This additional configuration that doubtlessly follows a membrane depolarization is called the *local response* of the membrane.

In Section 4.3.2.1 it was emphasized that the nerve impulse started from the electrode that served as the cathode if surface stimulation was used. Figure 4.17 shows a sketch of the current flow between the two surface electrodes. Starting from the electrode that is the anode the current flows to the other electrode (the cathode) partly through the thin surface layer surrounding the nerve and partly by crossing the membrane in the direction from the outside to the axoplasm up to a spot where the cathode is localized. From this position the axial current again crosses the membrane but this time from the *inside* to the *outside* and in doing so *depolarizes* the membrane below the cathode. With sufficiently strong currents the depolarization will exceed the threshold and start a propagated action potential from the membrane region at the cathode. At the anode, on the other hand, the stimulus current has the opposite effect, namely a hyperpolarization of the membrane. The local response can also be observed by placing the two sets of electrodes on the external surface of the nerve. In fact the first direct recording of the local response was obtained by A.L. Hodgkin[*] before the era of the glass microelectrode by placing an unmedullated axon isolated from the limb nerve of the shore crab *Carcinus maenas* on three platinum hooks used as stimulation and recording electrodes with the middle one serving as both stimulus and recording electrode. The result of such an experiment is shown in Fig. 4.18.

---

[*] Hodgkin, A.L. (1938): The subthreshold potential in a crustacean nerve fibre. *Proc. R. Soc. Lond.*, B**126**, 87.

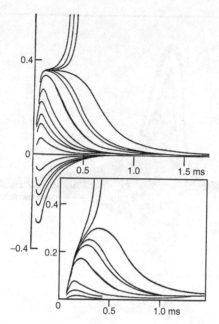

Fig. 4.18. Electric changes at stimulating electrode produced by shocks with relative strengths, successively from above, 1.0 (upper six curves), 0.96, 0.85, 0.71, 0.57, 0.43, 0.21, −0.21, −0.43, −0.57, −0.71, −1.0. The ordinate scale gives the potential as a fraction of the propagated spike, which was about 40 mV amplitude. The 0.96 curve is thicker than the others, because the local response had begun to fluctuate very slightly at this strength. (Insert) Responses produced by shocks of increasing strengths, successively from above, 1.0 (upper five curves), 0.96, 0.85, 0.71, 0.57; obtained from curves in the main part of Fig. 4.18 by subtracting anodic changes from corresponding cathodic curves. Two of the anodic curves necessary for this analysis were recorded but not shown in Fig. 4.18. Ordinate, as above. (Hodgkin, 1938.)

4.4.3.2 Conductance changes attending the action potential

A set up like Fig. 4.10, that primarily was constructed to measure resting membrane potential and action potential in the giant axon of the squid, is also applicable to measure changes in the conductance of the membrane (or rather its impedance) during the passage of the action potential past the tip of the internal electrode. In principle this is done by removing the amplifier and connecting the two input leads to an impedance-measuring device. This type of experiment was first performed by K.S. Cole & H.J. Curtis* in 1938. One of their results is shown in Fig. 4.19. The fully drawn curve in Fig. 4.19 is the action potential and the width of the dark band is proportional to the value of the

---

* Cole, K.S. & Curtis, H.J. (1938): Transverse electrical impedance of the squid giant axon. *J. gen. Physiol.*, **21**, 757.

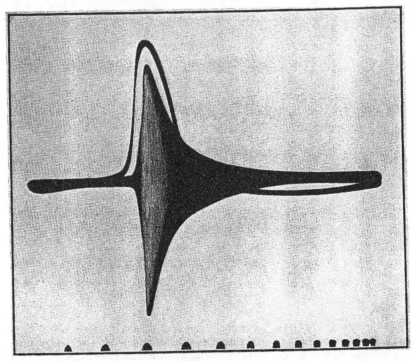

Fig. 4.19. Simultaneous recording of action potential (AP) and increase in membrane conductance (envelope) during the passage of the action potential past the extracellular electrode. (From Curtis & Cole, 1939*(slightly retouched).)

membrane conductance. A further analysis showed that the membrane resistance *decreased* from its resting value of about 1000 Ω cm^2 to about 25 Ω cm^2 at the peak of the action potential. Neither the membrane capacitance nor the axoplasm resistance changed significantly.

4.4.3.3 The effect of extracellular sodium concentration
on the action potential

An electric current through the axon's membrane is carried by *ions and not free electrons*. Hence, an increase in membrane conductance reflects an increase in the membrane permeability for one or several ionic types. At the time of its first appearance the result of Curtis & Cole (1939)* was accepted as a very satisfactory support of Bernstein's theory[†] that assumed an all-over increase in the membrane's permeability for small ions during the impulse activity. However, the overshoot of action potential of about 40–45 mV that was observed

---

* Curtis, H.J. & Cole, K.S. (1939): Electrical impedance of the squid giant axon during activity. *J. gen. Physiol.*, **22**, 649.
[†] See Section 4.2.

soon afterwards by Hodgkin & Huxley (1939) made the original concept of the Bernstein scheme untenable: whether the membrane is assumed suddenly to be absent or the permeabilities/conductances are assumed to have the same values, the result will be a displacement of the membrane potential to about −20 to −10 mV, but *never* to something about +40 mV*.

On the other hand, the observed overshoot could be obtained within the framework of an electrodiffusion model if it were assumed that the membrane change initially *only* consists of a *selective increase in the membrane's permeability to sodium* that is large enough to dominate the diffusion regime for a short time. In the extreme case one might expect an overshoot of about +54 mV – that is the equilibrium potential $V_{Na}^{(eq)}$ for the sodium ions[†] – but *never* a substantially higher value. This assumption, *the sodium hypothesis*, was first formulated by Hodgkin & Huxley (1947) and later confirmed experimentally by Hodgkin & Katz in 1949[‡] on the basis of their investigations that focused on the role of Na^+ ions in the impulse conduction. Primarily they demonstrated that replacement of all the extracellular Na^+ by choline, which does not penetrate the membrane, resulted immediately in a complete conduction block[§]. The main result of Hodgkin & Katz is Fig. 4.20, which shows that the amplitude of the action potential decreases concurrently with a reduction of the Na^+ ions in the extracellular fluid, whereas the resting membrane potential remains almost unaffected. The size of the overshoot of the action potential varies roughly linearly with the logarithm of the external Na^+ concentration within the upper range of concentrations $\frac{1}{2}C_{Na}^{(normal)} < C_{Na} < C_{Na}^{(normal)}$, behavior to be expected if the axon's membrane were chiefly permeable to Na^+. Hodgkin & Katz also showed that replacement of the normal extracellular fluid by a hypertonic solution having an excess of sodium resulted in an increase in the overshoot of a magnitude that fitted with that predicted from the Nernst equation. The results of such an experiment are shown in Fig. 4.21.

4.4.3.4 Perfused axons

In 1961, P.F. Baker & T.I. Shaw developed a technique that opened up the possibility of a study of the effects of changes in the ionic composition in the

---

* To make this estimate use either the Henderson equation (Chapter 3, section 3.6.3.2, Eq. (3.6.38)), the Goldman equation (Chapter 3, Section 3.7.3, Eq. (3.7.22)) or the Millman equation (Chapter 3, Section 3.7.4, Eq. (3.7.53)).

† See Chapter 3, Section 3.4.2, Eq. (3.4.7).

‡ Hodgkin, A.L. & Huxley, A.F. (1947): Potassium leakage from an active nerve fibre. *J. Physiol.*, **106**, 341; Hodgkin A.L. & Katz, B. (1949): The effect of sodium ions on the electrical activity of the giant axon of the squid. *J. Physiol.*, **108**, 37.

§ The English physiologist E. Overton showed in 1902 that nerves and muscles became unexcitable when placed in an Na-free medium; a remarkable result that nevertheless remained unnoticed for about 45 years, but then gave rise to something like a feud as to the reliability of Overton's result.

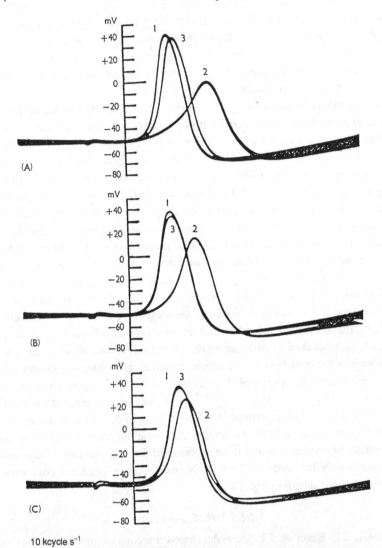

Fig. 4.20. Effect on the amplitude of the action potential by replacing the normal external fluid (sea water) by sodium-deficient solutions. All recordings labeled 1 and 3 were with the axon in normal sea water before and after the axon was exposed to the know sodium solution (recording 2): A2, 1/3 sea water and 2/3 isotonic dextrose; B2, 1/2 sea water and 1/2 isotonic dextrose; C2, 0.7 sea water and 0.3 isotonic dextrose (from Hodgkin & Katz, 1949).

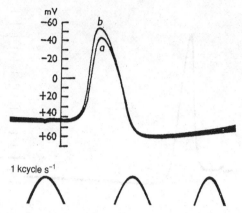

Fig. 4.21. Effect of a sodium-rich external solution on the action potential from a squid axon. Recording *a*, normal sea water; recording *b*; 50 s after exposing the axon to an external solution with sodium 1.56 times normal (from Hodgkin & Katz, 1949).

axon's interior. The method is to insert a syringe in the one end of the axon and then squeeze the axoplasm from the other end of the axon by means of a small rubber roller, and then reinflate the flattened axon by injecting a fluid, e.g. isotonic K_2SO_4. It came as a great surprise that such an artificially perfused axon was capable of conducting an action potential of normal appearance for several hours. Figure 4.22 shows an action potential from a perfused axon (A) compared with that from an intact axon (B). With this technique it became possible to study the membrane's excitability under conditions where the ionic compositions of the fluids on *both* sides of the membrane were under control. It turned out that to maintain excitability, the internal fluid had to be *isotonic* with the surrounding sea water, and it should contain K^+ ions rather than Na^+ ions, whereas the nature of the anions was not decisive as Cl^-, SO_4^{2-}, $H_3SO_4^-$ and isethionate will do equally well. Naturally, it is also possible to study the effect of changing the mutual amounts of Na^+ ions and K^+ ions in the perfusion fluid. Figure 4.23 shows an example of such an experiment. In trace A the Na^+ ions are absent in the perfusate. The action potential is now larger than normal, but not infinitely large as predicted theoretically from the Nernst equation if the membrane were exclusively permeable to Na^+ ions. Increasing the Na^+ ions at the expense of the K^+ ions as in trace B ($\frac{1}{4}K^+$ replaced) and trace C ($\frac{1}{2}K^+$ replaced) leads to a successive reduction in the amplitude of the action potential. In C, the overshoot was less than +10 mV. A further increase in the internal Na^+ concentration results in a complete *conduction block*. The results of these experiments agree well with the experiments that involve changing the Na^+ concentration in the external fluid, and both types of experiment support

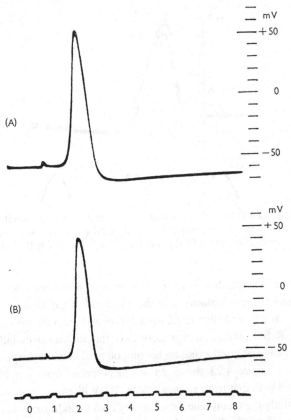

Fig. 4.22. Comparison between action potentials from (A) an axon having its axoplasm extruded and afterwards reinflated by perfusion with isotonic KCl, and (B) from an intact axon. (From Baker, Hodgkin & Shaw, 1961*.)

the *sodium hypothesis of the action potential* formulated by A.L. Hodgkin & B. Katz (1949).

(1) The action potential is caused by a temporary, selective increase in the Na^+ permeability of the axon's membrane, the result of which is that Na^+ ions flow from the extracellular fluid to the interior of the axon and in the process drive the membrane potential towards the equilibrium potential $V_{Na}^{(eq)}$ of the sodium ions.

(2) The immediate energy source for driving the Na^+ ions inward is the electrochemical potential difference for Na^+ between the value at the outside of the membrane and that on the inside.

* Baker, P.F., Hodgkin, A.L. & Shaw, T. (1961): *Nature, Lond.*, **190**, 885.

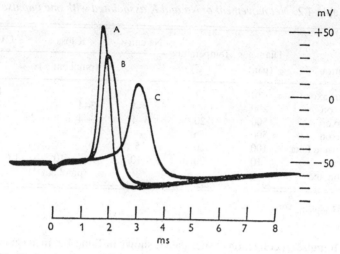

Fig. 4.23. Perfused axon. Effect on action potential of changes in the sodium concentration in the internal solution. The anion is sulfate. Traces: A, isotonic potassium sulfate; B, $\frac{1}{4}$ K$^+$ is replaced by Na$^+$; C, $\frac{1}{2}$ K$^+$ is replaced by Na$^+$. The recordings were performed in the sequence B, C and A. (From Baker, Hodgkin & Shaw, 1961.)

(3) The K$^+$ ions, which in the resting state are near equilibrium, are at the top of the action potential – with positively charged axoplasm – found in a state that is far away from equilibrium, with the result that K$^+$ ions now leave the axoplasm at a higher rate, that – eventually assisted by an increase membrane permeability P_K – after the restoration of the sodium permeability P_{Na} to normal pulls the membrane potential back to its original value.

4.4.4 The experimental substantiation of the sodium hypothesis

We shall now present the experimental evidence that consolidated the sodium hypothesis and led to a consistent quantitatively formulated theory for the mechanism behind the impulse conduction, which until now has not required any major revisions.

4.4.4.1 Net movement of radioactive sodium and potassium during electric activity

The amounts of Na$^+$ and K$^+$ ions that enter/leave the axoplasm during the activity of the membrane can be determined by the use of radioactive tracers. To improve the accuracy, a long train of impulses is generated, keeping track of the total number of impulses transmitted. The amount of Na or K, respectively, that is exchanged is usually expressed as the ionic movement associated with a

Table 4.2. *Net movement of Na and K associated with one impulse*

Preparation	Diameter (μm)	Temperature (°C)	Na entry	K loss	$CV/\mathscr{F}$*
			(pmol cm^{-2})		
Carcinus axon	30	17	—	2	1.2
Sepia axon	200	15–20	3–4	3–4	1.2
Squid axon	500	20	3–4	3–4	1.2
Squid axon	500	6	—	9	1.2
Frog muscle fiber	100	20	15	10	6
Frog nerve myelinated	10	20	5×10^{-5}	3–5×10^{-5} (pmol cm^{-1})	1.6×10^{-5}

(After Hodgkin, 1964[†].)

single impulse. A collection of such data is shown in Table 4.2. In the giant axon each impulse is associated with a *net entry of sodium* of about 4 pmol cm^{-2} per impulse[‡] and, by and large, the *same amount* of potassium ions of 4 pmol cm^{-2} per impulse *leaves* the interior of the axon. Perhaps it may be more instructive to convert these figures by saying that each impulse is associated with an *inward movement* of about 20 000 Na$^+$ ions across an area of 1 μm^2 of the axon's surface. The entrance of 4 pmol of Na$^+$ ions cm^{-2} is enough to account for the size of the action potential: the electric charge ΔQ that is required to change the voltage across a capacitor with the capacitance C by an amount ΔV is equal to $C\Delta V$. We know that the capacitance of the axon's membrane – the membrane capacitance C_m – is about 1 μF cm^{-2}. Taking the amplitude of the action potential to $\Delta V = 0.12$ V we have

$$\Delta Q = C_m \Delta V = 10^{-6} \cdot \frac{F}{cm^2} \times 0.12 \text{ V} = 0.12 \,\mu\text{C cm}^{-2}.$$

Dividing the charge by the Faraday $\mathscr{F} \approx 10^5$ C mol^{-1}, the excess of free positive charges expressed in amount of monovalent cations ΔM becomes

$$\Delta M = \frac{\Delta Q}{\mathscr{F}} = \frac{C_m \Delta V}{\mathscr{F}} = \frac{0.12 \,\mu\text{C cm}^{-2}}{10^5 \text{C mol}^{-1}} = 1.2 \times 10^{-12} \text{ mol cm}^{-2}$$

$$= 1.2 \,\text{pmol cm}^{-2},$$

* This column gives the number of monovalent cations, $CV/\mathscr{F} =$ mol cm^{-2}, that are required to charge 1 cm^2 axonal membrane capacity C to a potential $V = 0.12$ V that is equal to the amplitude of the action potential. $\mathscr{F}$ is the Faraday. C is taken as 1 μF cm^{-2} for unmyelinated axons: 5 μF cm^{-2} for muscle and 13 pF cm^{-1} for myelinated axons. The unit in the last row is mol cm^{-1}.
† Hodgkin, A.L. (1964): The *Conduction of Nervous Impulses*, Liverpool University Press.
‡ 1 pmol = 10^{-12} mol.
§ $\mathscr{F} = 96\,492$ C mol^{-1}.

as the theoretical minimal value. This is about 1/3 of the amount actually observed. The direction of this discrepancy is acceptable because inflow of Na^+ ions does not serve solely to charge the membrane capacitance to the required voltage during the progressing action potential, but also as an exchange for part of the K^+ ions that are leaving the axoplasm.

4.4.4.2 The temporal resolution of the separate ionic currents: voltage clamp technique

The movement of tagged Na^+ and K^+ ions during activity was achieved by firing a large number of action potentials (e.g. 10 000) and afterwards referring the movements to one single impulse. For the squid axon the result was an Na entry of about 4 pmol cm^{-2} and a K loss of the same magnitude. However, these experiments do not provide any information on the *temporal course* of the inward and outward flows of sodium and potassium. However, if these two oppositely directed ionic movements should generate a change in the membrane potential like the action potential, the *first event* must be a charging of the membrane's inside by an *inwards-directed* current to a positive value (e.g. to +45 mV) that is *followed* by an *outwards-directed* current that leads to the repolarization of the membrane. If these currents are carried by Na^+ and K^+ ions there must be a *time lag* between the Na entry and the K outflow. An attempt to demonstrate such a split-up of the two currents on the propagating action potential would involve almost insuperable obstacles since the membrane potential changes with time all along the length of the axon in the manner of a wave packet. Consequently the membrane current is composed partly of ionic currents crossing the membrane and partly of a component used to charge the membrane capacity C_m to a changing membrane potential, whose value is equal to the membrane capacitance multiplied by the rate of change dV/dt of the membrane potential. Therefore, to measure the ionic currents one must devise experimental equipment that eliminates the complications due to the presence of the membrane capacity. This purpose is obtained by making use of the so-called "voltage clamp" technique*. By means of this technique, the membrane potential can be kept/clamped to an arbitrary, pre-chosen value and then changed almost instantaneously to a new chosen value and kept at this value irrespective of the changes in the ionic currents that might follow from the displacement of the membrane potential as the result of (1) changes in the driving force on the ions in the membrane and (2) changes in the membrane's

---

* The voltage clamp technique was developed almost simultaneously in K.S. Cole's laboratory in the USA by Marmont, G. (1949): Studies on the axon membrane. *J. Cell. Comp. Physiol.*, **34**, 351; and in England by Hodgkin, A.L., Huxley, A.F. & Katz, B. (1949): Ionic currents underlying activity in the giant axon of the squid. *Arch. Sci. Physiol.*, **3**, 129.

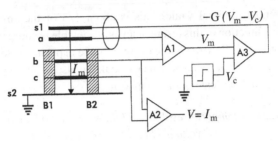

Fig. 4.24. Diagram of a voltage-clamp set up. For further description, see text.

permeability to one or several ionic species. Because of the instantaneous potential displacement from one level to another, the membrane capacity changes charge only at the instant of the membrane potential change and, therefore, the currents that may be measured during the voltage clamp (where $dV/dt = 0$) are *exclusively ionic currents* that flow through the membrane. Actually, the voltage clamp technique represents just a refined way of connecting the axon's inside and outside with a *constant voltage generator* whose output voltage is controlled by the experimenter. In practice this is done by inserting in the axoplasm an extra electrode – a so-called current electrode – that is connected to an electronic feed-back circuit, that despite changes in membrane permeabilities supplies a current of just the strength and direction to ensure that membrane potential remains at a given predetermined level. Changes in the membrane current with time at this clamped membrane potential will provide information about the changes in the membrane permeability to the surrounding ions.

(i) The control system

A flow sheet of the voltage-clamp set up is shown in Fig. 4.24.

Two electrodes are now inserted in the axoplasm. In addition to the familiar internal *potential recording electrode* (a) there is a *current electrode* (s1) that is connected to the output of an amplifier A3, which provides the electric voltage/current to clamp the membrane potential at the chosen value. Near to the outer surface of the axon is placed the external *potential recording electrode* (b), which together with electrode (a) is connected to the input of a *difference amplifier* A1 with amplification 1.0. The output signal is therefore $V^{(a)} - V^{(b)} = V_m$, where V_m is the *membrane potential*. This is only fully correct if there is no current flow between the two current electrodes (see the description in the connection with the description of Fig. 4.25). Electrode (b) and another electrode (c) are connected to another difference amplifier A2, which measures the voltage drop between electrodes (b) and (c) $= I_m R_s$, where R_s is the resistance of the liquid layer between the two electrodes. With R_s known, the output from A2 is a measure of the *membrane current* I_m that flows

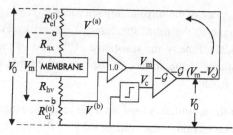

Fig. 4.25. Diagram of the feed-back circuit in a voltage clamp set up. For further details, see text.

radially from the membrane between the two barriers B1 and B2 to end at the ground electrode (s2). A3 is another difference amplifier to amplify the voltage input $V_m - V_c$, where V_c a variable voltage (the control voltage), that can be changed suddenly in well-defined steps. The gain $\mathcal{G}$ of amplifier A3 is high, e.g. 1000. The output V_0 relative to ground, with polarity *opposite* to the input signal $V_m - V_c$, is connected to the internal current electrode s1, thereby providing the current path through the membrane from the output from A3 and back to ground (s2). Let the system primarily be adjusted so that $V_m = V_c$. Hence, the output from A3 is $V_0 = 0$. If the ionic permeability of the membrane changes and ions move through the membrane, the membrane potential tends to change from the value $V_m = V_c$. If, for example, the input signal V_m changes relative to the control voltage V_c, an input signal $\Delta V = V_m - V_c > 0$ will appear at amplifier A3. The amplified output $V_0 = \mathcal{G}(V_c - V_m)$, having a polarity *opposite* to that of the input signal, will tend to move the voltage V_m back to the initial value $V_m = V_c$, that is it will force the membrane potential to be clamped again at the value V_c. The membrane current, whose action on the membrane potential V_m is now compensated for by means of the electronic feed-back, flows away in the external chamber and is recorded as the voltage drop between (b) and (c). The recording of the membrane current I_m is done by the difference amplifier A2. The basic elements that enter in the feed-back circuit of the voltage clamp set up are shown in Fig. 4.25. In the analysis of the feed-back system the path of the control current from A3 to ground requires the split-up in the following resistances in series.

(1) The resistance $R_{el}^{(i)}$ from the surface of the current electrode to the axoplasm along with the resistance of the axoplasm to the tip of the potential recording electrode (a) and in series the resistance R_{ax}, that is the resistance of the axoplasm from the tip of (a) to the inner surface of the membrane.

(2) The membrane resistance R_m of the axon in parallel with the membrane capacity C_m. Across the axon's membrane there is in the resting state a membrane potential $V_m^{(rest)}$.

(3) The axial resistance of the surrounding sea water R_{sw} from the outer surface of the membrane to the tip of the external potential recording electrode ((b) in Fig. 4.24). Finally the resistance of the rest of the sea water $R_{el}^{(0)}$ plus the resistance at the surface to the external current electrode that is connected to ground.

It is assumed that the amplifier A1 has an input resistance that is large enough to allow that the connection of the voltage $V^{(a)} - V^{(b)} = V_m$ to the input of the amplifier does not draw any current from the two measuring points (a) and (b). The difference between V_m and the control voltage V_c (clamping voltage) is amplified by the difference amplifier A3, whose output forms one of the terminals of the loop of the control current. This current flows as shown in Fig. 4.25 through the whole electrode complex and returns to the amplifier A3 via the ground wire. Let V_o represent the output voltage with respect to ground. We then have

$$V_o = -\mathcal{G}(V_m - V_c),$$

where $\mathcal{G}$ is the gain of amplifier A3 and the minus sign indicates that the polarity of the output signal is the opposite of the polarity of the input signal. By inspection of Fig. 4.25 we have

$$V_o - V_m = I\left(R_{el}^{(i)} + R_{el}^{(0)}\right) = IR,$$

where I is the control current. Elimination of V_o from these two equations gives

$$-\mathcal{G}(V_m - V_c) - V_m = -V_m(1 + \mathcal{G}) + \mathcal{G}V_c = IR.$$

Hence, we have

$$V_m = \frac{\mathcal{G}}{1 + \mathcal{G}}V_c - \frac{R}{1 + \mathcal{G}}I = \frac{1}{1 + 1/\mathcal{G}}V_c - \frac{R}{1 + \mathcal{G}}I.$$

Thus, the clamped membrane potential V_m appears as the result of an equivalent electromotive force $V_c/(1 + 1/\mathcal{G})$ having an internal resistance $(R_{el}^{(i)} + R_{el}^{(0)})/(1 + \mathcal{G})$, and forces the current I through the membrane between the points at (a) and (b) causing a voltage drop V_m. With a value of $\mathcal{G} = 500$ the effective internal resistance $(R_{el}^{(i)} + R_{el}^{(0)})/(1 + \mathcal{G})$ is less than 1 Ω and V_m will be clamped at a value that does not deviate from V_c by more than c. 0.2%, since

$$V_m = \frac{V_c}{1 + 1/500} = \frac{V_m}{1.002} = 0.998 \cdot V_c.$$

When there is a current flow through the membrane the recorded value of V_m will differ from the "current free membrane potential" by the value $I(R_{ax} + R_{hv})$. This error, which results from the presence of R_{ax} and R_{hv}, can be minimized by

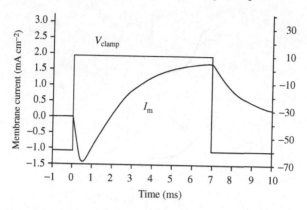

Fig. 4.26. Time course of membrane current during "voltage clamp" where the membrane potential is changed "instantaneously" from a resting potential of −60 mV to +10 mV. Inwardly-directed current is negative. The initial outwardly-directed charging current of the membrane capacity is not shown.

using a supplementary feed-back system. Nowadays, glass microelectrodes are often used to measure V_m. Since these electrodes can be placed very near to the membrane's inner and outer surfaces, R_{ax} and R_{hv} can be almost eliminated*.

Figure 4.26 shows the course of the membrane current when the axon is suddenly changed from a resting membrane potential of −60 mV to +10 mV and kept there for a period of 7 ms, i.e. exposed to a depolarization of 70 mV. The current consists of *three phases*. (1) First is a strong, short-lasting *outward-directed* pulse that lasts the few microseconds it takes to push the membrane potential to the new level and which represents the charge that must be removed from the membrane capacitor to change the membrane potential from −60 mV to +10 mV. This capacitative current peak is not visible in the figure. (2) Then comes a transient *inwardly-directed* current lasting about 1.5 ms, which (3) is followed by a delayed *outwardly-directed* current that − if the depolarization is sustained for a sufficiently long time − approaches its maximum value asymptotically and remains there as long as the depolarization is maintained. As the membrane potential is constant during phases (2) and (3) the current flow must result from an *ionic movement* across the membrane; that is, an inwardly-directed current of cations or an outwardly-directed current of anions during phase (2), and during phase (3) an outwardly-directed flow of cations or an inward flow of anions. As we shall see below, Hodgkin & Huxley succeeded in the elucidation of the nature of these ionic currents and showed that the *early inwardly-directed current was carried by sodium ions*, which moved into

---

* A modified voltage clamp technique was later introduced by B. Frankenhaeuser in Stockholm that avoided the direct insertion of the internal electrodes. This made it possible to study myelinated nerves from frogs and also unmyelinated nerves from lobsters and crabs.

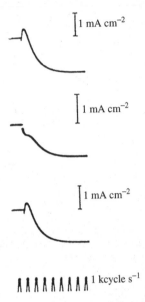

Fig. 4.27. The effect on the membrane current of changing the sodium content in the external fluid. The initial capacitative current peak is erased. For further explanation see text. (After Hodgkin & Huxley, 1952*.)

the axoplasm from the external fluid, whereas the *delayed outwardly-directed current was carried by potassium ions.*

(ii) Identification of the early inwardly-directed current
According to the sodium hypothesis, the first event in the processes leading to the action potential is the solitary entry of Na^+ ions into the interior of the axon. Therefore, to Hodgkin, Huxley & Katz it was natural to associate the early inwardly-directed current observed during the voltage clamp experiment with an entry of Na^+ ions. Figure 4.27 shows the effect of changing drastically the normal environment for the Na^+ ions in the external fluid. The membrane potential is displaced from −50 mV to +15 mV (i.e. the membrane is depolarized by 65 mV). In the upper trace the axon is surrounded by normal sea water and the course of the current flow is seen to be identical with that of Fig. 4.26 (but note that the sign for inward current flow is taken as opposite to that shown in Fig. 4.26). In the second trace all sodium in the outer fluid has been reduced to zero by replacing the Na^+ ions by choline ions, which are unable to penetrate the membrane. It is seen that the original inwardly-directed current has disappeared and is *replaced* by an *initial outwardly-directed hump* in

* Hodgkin, A.L. & Huxley, A.F. (1952): *J. Physiol.*, **116**, 449.

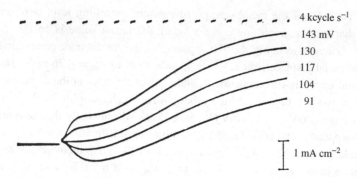

Fig. 4.28. The membrane current at different values of the clamp potential. The numbers to the right indicate the value (in mV) of the displacement from the resting membrane potential. For further explanation see text. (After Hodgkin, Huxley & Katz, 1952*.)

membrane current, which is followed by the delayed outward current similar to that of phase (3) in Fig. 4.26. In the third trace the axon is again back in its normal sea water environment. The initial inwardly-directed current has now returned to about the same size as in the upper record. The absence of the initial inwardly-directed current in the middle trace is incompatible with a notion as proposed above that this inward current flow is associated with an outward flow of anions, e.g. chloride, whereas the absence of the inward current and its replacement with an outwardly-directed hump is consistent with the idea that the initial event consists exclusively of an increase of the membrane permeability P_{Na} to sodium ions that allows them to move passively across the membrane in the direction prescribed by the driving forces that are at its disposal. In normal sea water the Na^+ ions will tend to move inwards owing to the concentration difference $C_{Na}^{(O)} - C_{Na}^{(i)} \gg 0$. This is also the case at a clamping potential $V^{(clamp)} = +15$ mV, being less than the equilibrium potential $V_{Na}^{(eq)}$, whose expected value is about $+55$ mV. In the middle trace, where $C_{Na}^{(O)} = 0$, the sodium ions will move passively outwards by a diffusion process that is assisted by the electric field due to the clamping of the interior at $+15$ mV. It is natural to ascribe the initial outwardly-directed hump of the current curve as the manifestation of this flow of Na^+ ions that results from an initial, transient increase in the sodium permeability of the membrane. Another type of experiment that also points at an association of the sodium ions to the initial early current is illustrated in Fig. 4.28, which shows a set of time courses of the membrane current in response to large displacements of the membrane potential, with the axon kept in normal sea water. The numbers to the right of each trace indicate the displacement

---

* Hodgkin, A.L., Huxley, A.F. & Katz, B. (1952:) *J. Physiol.*, **116**, 424.

$\Delta V^{(c)}$ (in mV) of the membrane potential from its resting level (-65 mV). Note that in this graph any current *below* the initial value is an *inwardly-directed current*, as in Fig. 4.26. It appears that the current trace corresponding to a potential displacement $\Delta V^{(c)} = 93$ mV – i.e. to $V_m = +26$ mV – shows an initial inwardly-directed current, and so does the trace at the displacement $\Delta V^{(c)} = 104$ mV ($V_m \equiv +39$ mV), but with a smaller amplitude. With larger displacements $\Delta V^{(c)} = 130$ mV ($V_m \equiv +65$ mV) the inwardly-directed current has disappeared and is replaced by an initial *outwardly-directed* hump, as in the middle trace of Fig. 4.27. The amplitude of the hump increases for higher values of displacement $\Delta V^{(c)} = 143$ mV ($V_m \equiv 78$ mV). By a displacement $\Delta V^{(c)} = 112$ mV ($V_m \equiv +52$ mV), however, *no current response* occurs initially until the current rises as the delayed, outwardly-directed current also seen in the other traces. An estimation of the value of the equilibrium potential $V_{Na}^{(eq)}$ for the sodium ion at the conditions of the experiment gave about the same value as the above clamping value. At a membrane potential $V_m = V_{Na}^{(eq)}$, the passive *net flux* for Na$^+$ ions is *zero* irrespective of the magnitude of the membrane permeability P_{Na}. Therefore, the most natural interpretation of this experiment was to assign the inwardly- and outwardly-directed *currents occurring initially* to the *flow of Na$^+$ ions* that are driven passively across the membrane in the direction prescribed by the concentration difference and the potential difference across the membrane, whereas the solitary S-shaped course at $V_m = +52$ mV corresponds to *delayed, outwardly-directed current potassium ions*.

Hodgkin & Huxley consolidated their interpretation by repeating the above experiment with different Na$^+$ concentrations in the external fluid. By this procedure the validity of their interpretation could be tested quantitatively. Consider two experiments: one with the axon in normal sea water with Na$^+$ concentration $C_{Na}^{(0)}$ and the other with Na$^+$ concentration reduced to $C_{Na1}^{(0)}$ by replacing part of the sodium with choline. The equilibrium potentials for sodium in the two cases are

$$V_{Na}^{(eq)} = \frac{\mathscr{R}T}{\mathscr{F}} \ln \frac{C_{Na}^{(0)}}{C_{Na}^{(i)}} \quad \text{and} \quad V_{Na1}^{(eq)} = \frac{\mathscr{R}T}{\mathscr{F}} \ln \frac{C_{Na1}^{(0)}}{C_{Na}^{(i)}}.$$

Hence, by changing the external Na$^+$ concentration from $C_{Na}^{(0)}$ to $C_{Na1}^{(0)}$ the expected change in equilibrium potential is

$$\Delta V_{Na}^{(eq)} = V_{Na}^{(eq)} - V_{Na1}^{(eq)} = \frac{\mathscr{R}T}{\mathscr{F}} \ln \frac{C_{Na}^{(0)}}{C_{Na}^{(i)}} - \frac{\mathscr{R}T}{\mathscr{F}} \ln \frac{C_{Na1}^{(0)}}{C_{Na}^{(i)}}$$

$$= \frac{\mathscr{R}T}{\mathscr{F}} \left\{ \ln \frac{C_{Na}^{(0)}}{C_{Na}^{(i)}} - \ln \frac{C_{Na1}^{(0)}}{C_{Na}^{(i)}} \right\} = \frac{\mathscr{R}T}{\mathscr{F}} \ln \frac{C_{Na}^{(0)}}{C_{Na1}^{(0)}},$$

Table 4.3. *Comparison of observed and theoretical change in sodium potential when the fluid surrounding the axon is changed from sea water to a low sodium solution*

Axon no.	Temp. (°C)	$C_{\mathrm{Na1}}^{(0)}/C_{\mathrm{Na}}^{(0)}$	$\Delta V^{(c)}$ (mV)	$\Delta V_1^{(c)}$ (mV)	$V_\mathrm{m} - V_\mathrm{m1}$ (mV)	$\Delta^{(\mathrm{obs})}$ (mV)	$\Delta^{(\mathrm{theor})}$ (mV)
20	6.3	0.3	105	78	3	30	28.9
20	6.3	0.1	96	45	4	55	55.3
21	8.5	0.1	100	48	4	56	55.6
21	8.5	0.1	95	45	4	54	55.6

(After* Hodgkin & Huxley, 1952.)[†]

which can be evaluated as $C_{\mathrm{Na}}^{(0)}$ and $C_{\mathrm{Na1}}^{(0)}$ are known. Now let $\Delta V^{(c)}$ and $\Delta V_1^{(c)}$ be the values of the displacements, which give rise neither to an early inward current nor to an early outward current. According to the assumption, the potential has been displaced to the values of the equilibrium potentials in the two cases. So we have

$$\Delta V^{(c)} = V_{\mathrm{Na}}^{(\mathrm{eq})} - V_\mathrm{m} \quad \text{and} \quad \Delta V_1^{(c)} = V_{\mathrm{Na1}}^{(\mathrm{eq})} - V_{\mathrm{m1}},$$

where V_m and V_m1 are the values of the membrane potentials at rest in the two cases that are known from the experiment. We then have

$$\Delta V^{(c)} - \Delta V_1^{(c)} = V_{\mathrm{Na}}^{(\mathrm{eq})} - V_\mathrm{m} - V_{\mathrm{Na1}}^{(\mathrm{eq})} + V_{\mathrm{m1}},$$

or by rearrangement

$$\Delta V^{(c)} - \Delta V_1^{(c)} + (V_\mathrm{m} - V_{\mathrm{m1}}) = V_{\mathrm{Na}}^{(\mathrm{eq})} + V_{\mathrm{Na1}}^{(\mathrm{eq})} = \frac{\mathcal{R}T}{\mathcal{F}} \ln \frac{C_{\mathrm{Na}}^{(0)}}{C_{\mathrm{Na1}}^{(0)}}.$$

All the terms in this expression can be determined experimentally. The left-hand side represents the shift in the equilibrium potential for sodium that is measured directly in the experiment, and the right-hand side represents the shift that is expected theoretically. Hodgkin & Huxley's summary from such experiments is shown in Table 4.3, where the last two columns give the comparison between the observed ($\Delta^{(\mathrm{obs})}$) and expected ($\Delta^{(\mathrm{theor})}$) values of the shift in the equilibrium potentials of the Na$^+$ ions as the result of diluting the Na$^+$ content in the external fluid. It is seen that the two kinds of data agree well, and thus provide strong

---

* The polarity of the membrane potential is changed to conform to the convention $V_\mathrm{m} = \psi^{(\mathrm{i})} - \psi^{(\mathrm{o})}$ used in this text.

[†] Hodgkin, A.L. & Huxley A.F. (1952): Currents carried by the sodium and potassium ions through the membrane of the giant axons of *Loligo*. *J. Physiol.*, **116**, 228.

evidence that when the membrane potential is displaced to the value of the equilibrium potential $V_{Na}^{(eq)}$ for sodium, the membrane current consist only of *delayed outward current*, that is due to K^+ ions[*], and that the early rise and fall in the recorded ionic current is carried by sodium ions. The results suggest that in the membrane there are two distinct non-interacting ionic currents arranged in parallel: (1) an early transient inward current carried by Na^+ ions and (2) a delayed current carried by K^+ ions, that lasts as long as the membrane potential is held at the displaced value (the clamp value). Taking this result as their starting point, Hodgkin & Huxley (1952) then designed experiments to split up the current $I^{(clamp)}$ in the inward Na^+ current $I_{Na}^{(in)}$ and the delayed outward K^+ current $I_K^{(out)}$ when the membrane potential was displaced to a given value $V^{(clamp)}$. Basically the procedure involved two steps.[†] With the axon placed in a solution with part of the sodium replaced by choline, a series of current recordings at different membrane potential displacements like those in Fig. 4.28 were obtained. The membrane potential corresponding to an initial flat current course was taken as the sodium equilibrium potential corresponding to the new ionic environment. The current – now directed solely outwards – was taken as representing the current $I_K^{(out)}$ carried by the K^+ ions at the membrane potential $V_m^{(clamp)}$. The low-sodium fluid was then replaced by normal sea water and the membrane potential was again displaced to $V_m^{(clamp)}$ and the current – now with an initial inward component – was recorded. The difference between this current and the above current $I_K^{(out)}$ would then represent the sodium current alone that is brought about at a membrane displacement to $V_m^{(clamp)}$[‡].

An example of such a final separation procedure is shown in Fig. 4.29. The giant axon from *Loligo* kept in normal sea water had the resting membrane potential $V_m^{(rest)} = -65$ mV. This was then displaced by the voltage clamp to $V_m = -9$ mV. Trace A shows the membrane current in normal sea water. Trace B is the membrane current when the main portion of sodium was replaced by choline to give an equilibrium for sodium of -9 mV. This current represents *the sole K^+ current* that flows in response to a membrane depolarization of 56 mV

---

[*] To substantiate this statement Hodgkin & Huxley referred in their paper to their own data from the *Sepia* axon, at that time unpublished. See Section 4.4.4.2 (i).

[†] The actual procedure was more complex. See the description on pp. 457–460 in the original paper.

[‡] The basic assumption that underlies the comparison of the currents with the axon in a low-sodium solution with those in the normal sea water, when the membrane potential was displaced to the same value during the voltage clamp, were:

(1) the time course of the K^+ current is the same in both ionic environments;
(2) the time scale and the form of the time course of the Na^+ current is similar in the two cases, although the amplitude and sometimes also the direction changes; and
(3) the increase in K^+ current did not start after the application of the clamp voltage until after a period of about one third of the time taken by I_{Na} to reach its maximum.

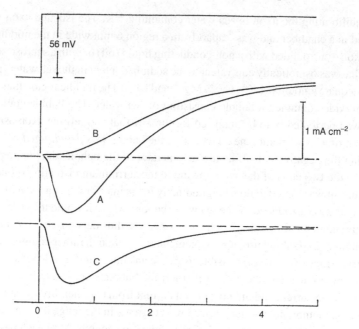

Fig. 4.29. Separation of the membrane current during a displacement of the membrane potential of +56 mV into two components carried by Na$^+$ and K$^+$ ions. Inward current downwards. Trace A, current $I_{Na} + I_K$ in sea water. Trace B, current with most external sodium replaced by choline. Trace C, difference between A and B to produce I_{Na}. Temperature 8.5 °C. For further explanation, see text. (After Hodgkin & Huxley, 1952*.)

from the resting membrane potential of −65 mV to a membrane potential of −9 mV. The difference between the current in A and that in B gives the Na$^+$ current. This is shown in trace C.

(iii) Identification of the outwardly-directed current
Considering that cations in the axoplasm are largely dominated by potassium ions, there was hardly any alternative to the assumption that the delayed outward current was carried by K$^+$. However, to get the matter settled, Hodgkin & Huxley (1953) confirmed the truth of their assumption by a series of experiments where the axon's membrane was depolarized by a strong outward current and the amount of transported electric charge was put in relation to the amount of radioactive potassium that concurrently left the axoplasm. The experiments were performed on isolated giant axons from *Sepia* with diameters of 170–260 μm, whose interior was "loaded" prior to the experiment with ^{42}K

* Hodgkin, A.L. & Huxley, A.F. (1952): *J. Physiol.*, **116**, 449.

by equilibrating the axon in sea water containing ^{42}K. The isolated axon was placed in a chamber and was – apart from a region 6 mm wide in the middle of the axon – surrounded with a non-conducting liquid (oil) or with a sucrose solution that was osmotically equivalent to the sodium content in the sea water (plus the adequate amounts of K^+, Ca^{2+}, Mg^{2+} and Cl^-). The middle region that was 6 mm wide contained a bubble consisting of sea water. The bubble could be renewed periodically to be analyzed for the ^{42}K that had left the axoplasm in a given time. The membrane contained in the bubble *was depolarized* by connecting the negative pole of a current generator to the bubble and the positive pole to the two ends of the axon. Owing to the surrounding isolating oil layer, the current was forced to flow longitudinally in the interior of the axon to the region that was surrounded by the sea water bubble, where it would then cross the membrane in the direction *from within outwards*, thus depolarizing the axon's membrane. By making use of a supplementary – although rather complicated – electric arrangement it was possible to secure an almost uniform depolarization of the membrane, and the current that left the bubble stemmed only from the depolarizing current and did not contain current from external, irrelevant leaks. The axon membrane was depolarized with currents in the range 0.1–1.0 μA in about 9 min, followed by a period of rest of the same length. After each passage of current the bubble was collected and analyzed for the activity of ^{42}K. By relating this value to the content of ^{42}K in the axoplasm and the total content of potassium one could determine the total amount M_{K^+} of K^+ that had moved during the current flow from the axoplasm into the bubble. The equivalent amount of electric charge is then $Q = \mathscr{F} M_{K^+}$, after conversion of M_{K^+} to moles. This charge was then compared with the charge that had actually moved into the bubble from the axoplasm during the passage of the current i in the time Δt (s), the magnitude of which is $Q_1 = i\,\Delta t$. Finally the dimensions were measured of that part of the axon that was contained inside the bubble. The results of such experiments performed on six axons are summed up in Fig. 4.30. The quantities mentioned above M_{K^+} and Q_1 are converted to flux J_{K^+} (in pmol cm^{-2} s^{-1}) and to current density I in (μA cm^{-2}). The figure shows that there is a *linear relation* between the current density I and the potassium flux J_{K^+}. The slope obtained by inspection is

$$\frac{\Delta I}{\Delta J_{K^+}} = \frac{100 \times \mu\text{C cm}^{-2}\text{ s}^{-1}}{1000 \times \text{pmol cm}^{-2}\text{ s}^{-1}} = \frac{10^{-6} \times \text{C}}{10^{-11} \times \text{mol}} = 10^5 \text{ C mol}^{-1}.$$

Since $\mathscr{F} = 96\,488$ C mol$^{-1} \approx 10^5$ C mol^{-1}, this slope is what one should expect if the depolarizing *outwardly-directed* current was carried *exclusively* by K^+ ions.

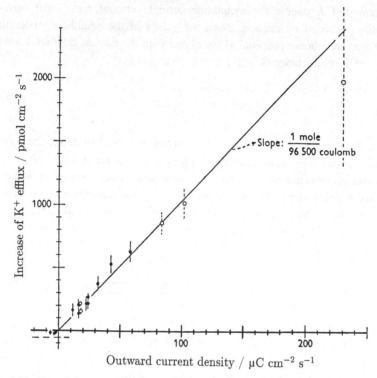

Fig. 4.30. Correlation between the depolarizing membrane current and the concurrent outflow of K^+ ions. Abscissa: mean outward membrane current density (= total current in bubble divided by the membrane area in the bubble). Ordinate: mean efflux of K^+ associated with the current flow (= efflux of K^+ divided by the membrane inside the bubble). The open points refer to results of a technique that allows use of the largest current densities. (Hodgkin & Huxley, 1953*.)

4.4.4.3 Calculation of the partial membrane conductances for sodium and potassium

Having succeeded in splitting the membrane current into the two currents carried by Na^+ and K^+ ions respectively, it is a relatively simple matter to determine the partial conductances for sodium and potassium.

(i) The time course of the changes in sodium and potassium conductance during a voltage clamp
Consider an ion of type j with an equilibrium potential $V_j^{(eq)}$. At the membrane potential $V \neq V_j^{(eq)}$ this ion will carry a component of the total membrane

---

* Hodgkin, A.L. & Huxley, A.F. (1953): Movements of radioactive potassium and membrane current in a giant axon. *J. Physiol.*, **121**, 403.

current. Let I_j denote the membrane current per unit area – the current density – carried by the ion. Since the values of the equilibrium potential $V_j^{(eq)}$ and membrane potential V associated with the current flow are known, the partial conductance G_j of the ion is calculated as

$$G_j = \frac{I_j}{V - V_j^{(eq)}}. \tag{4.4.5}$$

In Section 3.4.3.2 we introduced the above equation* to define the partial conductance – the chord conductance – for a nonlinear relation between V and I_j. In the present case, the current $I_j(V, t)$ changes as a function of the time t during the maintenance of the clamping potential V. We shall consider Eq. (4.4.1) to hold also in this case so that a value $G_j(V, t)$ of the conductance can be calculated corresponding to each value of the current $[I_j(V, t), t]$. For Na^+ and K^+ ions we have

$$G_{Na} = \frac{I_{Na}}{V - V_{Na}^{(eq)}} \quad \text{and} \quad G_K = \frac{I_K}{V - V_K^{(eq)}}. \tag{4.4.6}$$

In Fig. 4.29 the curves C and B represent the Na^+ and K^+ currents that result from a displacement of the membrane potential by +56 mV to $V = -9$ mV and with the value of $V_{Na}^{(eq)} = +52$ mV. The conversion of these current courses into the respective conductances G_{Na} and G_K is shown in Fig. 4.31. The sodium conductance starts off from a low value but rises rapidly within the next $\frac{3}{4}$–1 ms to a maximum (in the present case to about 25 mS cm^{-1}) whereupon it turns and declines more or less exponentially towards its original value *although* the clamping potential is kept unchanged. This phase of decay at a sustained clamping potential that follows the initial increase in G_{Na} is called the *inactivation of the sodium conductance*. The potassium conductance also starts off with a small value – although at a considerably higher level than that of G_{Na} – and rises more slowly than the sodium conductance along an S-shaped curve to a stationary value – provided the potential is kept displaced long enough – and remains there until the membrane potential is switched back to the previous resting level, whereupon G_K returns exponentially to the original level (the dotted line in Fig. 4.31). The same pattern of reversible decay also holds for G_{Na} when the clamping voltage is restored before the inactivation is complete, and with a rate of decay ten times higher than for G_K (compare the dotted line in the curve for G_{Na}).

* See Chapter 3, Section 3.4.3.2, Eq. (3.4.21).

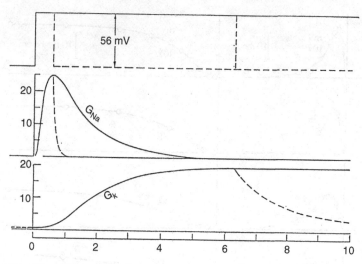

Fig. 4.31. Time course of sodium conductance G_{Na} and potassium conductance G_K in response to a sudden step of depolarization of the membrane by 56 mV. (After Hodgkin, 1958*, Hodgkin & Huxley, 1952.)

(ii) The dependence of the conductance changes on the displacement of the membrane potential

The procedure that led to the result illustrated in Fig. 4.31 was carried out for a number of different depolarizations covering the range of the action potential. The results are shown in Fig. 4.32. The time courses for the whole set of conductance changes for sodium and potassium are in principle similar in shape to those illustrated in Fig. 4.31. The characteristic features of the family are that: (1) the conductances depend strongly on the value to which the membrane potential is displaced, in particular an additional depolarization of $\Delta V = 4-6$ mV in the range around the firing threshold causes an e–fold increase ($\times 2.718$) in conductance; and (2) the time courses of the two conductances are different as the *sodium conductance rises steeply* to its maximum value whereas the *increase* in *potassium conductance* begins later and rises more slowly along an S-shaped curve to a stationary value. Moreover, the sodium conductance increase is only temporary because – in spite of a sustained holding voltage – it gradually returns to the original value. The kinetics of the time courses are all rather complete in the sense that the values of the associated time constants depend upon the value of the holding voltage.

---

* Hodgkin, A.L. (1958): Ionic movements and electrical activity in a giant nerve fibre. *Proc. R. Soc. Lond.*, B**148**, 1.

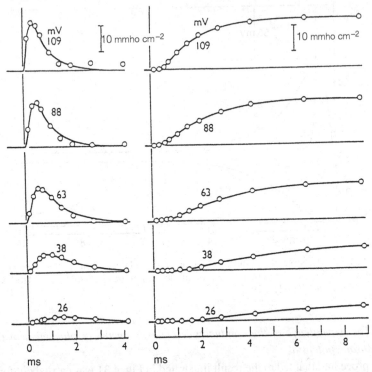

Fig. 4.32. Time course of sodium and potassium conductance for different displacements of membrane potential at 6 °C, the membrane given the depolarization indicated by the number associated with each curve. The circles are experimental estimates and the smooth curves are solutions of equations given in Section 4.7 to describe the dependence of the conductances on time and membrane displacement (Hodgkin & Huxley, 1952*).

(iii) The inactivation process

Hodgkin & Huxley soon realized that the kinetics of the time course of the sodium conductance change following a sudden step of depolarization was of such complexity that it was profitable to consider the time course of the sodium conductance as the reflection of two separate, *opposing* processes whose properties both depends on the value V of the membrane potential: (1) an initial rapid rise, the turning "on" process $G_{Na}^{(on)}$, aiming at its stationary state within less than 1 ms; and (2) an interacting, suppressing process $G_{Na}^{(supr)}$ that develops somewhat more slowly. Thus, the actual conductance change could formally be written as

$$\Delta G_{Na}(t) = G_{Na}^{(on)} \Leftarrow G_{Na}^{(supr)}, \qquad (4.4.7)$$

* Hodgkin, A.L. & Huxley, A.F. (1952): *J. Physiol.*, **117**, 500.

Membrane potential   Membrane current

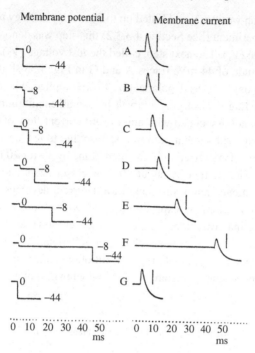

0 10 20 30 40 50   0 10 20 30 40 50
ms   ms

Fig. 4.33. Development of "inactivation" during application of a constant small depolarizing voltage. The left-hand column shows the time course of the displacement (in mV) of the membrane potential from the resting value. The test voltage $V_2 = 44$ mV was applied at different time intervals after the conditioning voltage $V_1 = 8$ mV. The right-hand column shows the time course of the membrane current density. Inward current is shown as an upward deflection – the vertical bars show the "sodium current" expected in the absence of a conditioning voltage step. (After Hodgkin & Huxley, 1952*.)

where $\Leftarrow$ symbolizes an interaction of $G_{Na}^{(supr)}$ upon $G_{Na}^{(on)}$. Hodgkin & Huxley called the suppressive process the *inactivation process* and denoted it with the letter h. The dual effect of the membrane potential change on the sodium conductance can be observed by applying an initial depolarization that is too small to produce any measurable increase in the Na^+ current. Nevertheless, if allowed to persist until a larger step of depolarization is subsequently impressed, it is found that the maximum current that is now produced is *less* than the current that was obtained with no preceding depolarization. This is illustrated in Fig. 4.33[†] where the membrane depolarization is carried out in two steps. The first step

---

* Hodgkin, A.L. & Huxley, A.F. (1952): The dual effect of membrane potential on sodium conductance in the giant axon of *Loligo. J. Physiol.*, **116**, 497.

[†] Note that our convention for the polarity of the membrane potential is the opposite to that used by Hodgkin & Huxley and the direction of the inward current is the same as that in Fig. 4.26.

466   4. The nerve impulse

of depolarization was 8 mV and lasted up to 50 ms. By analogy to a subliminal
stimulation experiment (See Section 4.3.3.2) this step was denoted the "conditioning" voltage (V_1). The next step, called the test voltage (V_2), was kept at a
constant amplitude of 44 mV. Traces A and G in Fig. 4.33 show the currents
observed when only the test depolarization V_2 was applied, whereas traces B–F
show that preceding V_2 at varying intervals by a small conditioning voltage V_1 –
that by itself is not associated with any inward current flow of significance –
alters strongly the subsequent current response due to test depolarization V_2.
Thus, a sustained depolarization of the membrane by V_1 for 20 ms reduces the
peak of the inward current by about 40%. A closer examination of the data
shows that the attenuation of the peak inward current develops smoothly with
the duration of V_1 along an exponential curve with a time constant of about
7 ms. Thus, the inactivation process evolves rather slowly compared with the
rapid rise in sodium conductance at the onset of depolarization.

   The effect of using a hyperpolarizing conditioning voltage instead is shown in
Fig. 4.34. It appears that the inward current following the potential displacement

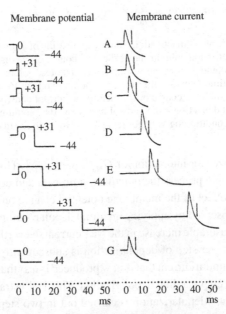

Fig. 4.34. Removal of "inactivation" by using a hyperpolarizing conditioning voltage
$V_1 = -31$ mV that makes the membrane potential more negative than the resting potential by 31 mV. The test voltage $V_2 = 44$ mV, as in Fig. 4.33. The vertical bars show the
"sodium current" expected in the absence of a conditioning voltage step. (Hodgkin &
Huxley, 1952.)

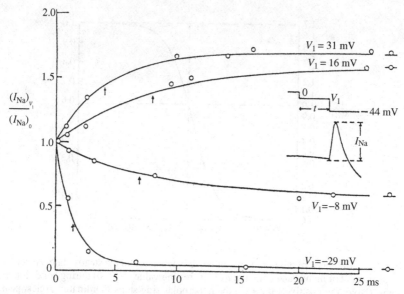

Fig. 4.35. Time course of inactivation at four different membrane potentials. Ordinate: circles, sodium current (measured as insert) relative to normal sodium current; smooth curve $y = y_\infty - (y_{\infty-1})\exp[-t/\tau_h]$, where y_∞ is the ordinate at $t = \infty$ and τ_h is the time constant (shown by arrows). Experimental details as in Fig. 4.33 and Fig. 4.34. (Hodgkin & Huxley, 1952.)

of the test voltage V_2 is now larger than the response with no conditioning voltage applied, and that this increase grows the longer the membrane is held hyperpolarized before the application of the test voltage V_2. Thus, at a holding time of 15 ms the increase is about 70%.

A more direct representation of the inactivation behavior is given in Fig. 4.35, where the peaks of the inward currents – as indicated by the inset diagram – are plotted against the duration of the conditioning voltages. Two of the traces are derived from the family of curves of Fig. 4.33 and Fig. 4.34. The other two are from families obtained from the same axon. It appears that inactivation and removal of inactivation developed in an approximately monoexponential manner with time constants that depend upon the membrane potential, having their maximum value at around the normal resting value. As the four curves of Fig. 4.35 were all obtained with the same – either positive or negative – displacements V_2, the four stationary values $y_1 \cdots y_4$ should be proportional to the inherent capability of the membrane to produce an inward current in response to a depolarization at a given value of the membrane potential. Hodgkin & Huxley followed up this idea with further experiments in which the conditioning voltage V_1 lasted long enough to allow the inactivation to attain its final level at

468	4. *The nerve impulse*

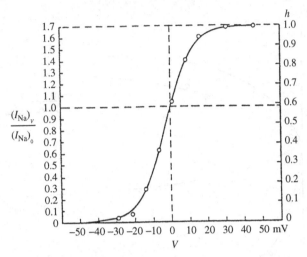

Fig. 4.36. Influence of membrane potential on inactivation in the steady state. Abscissa: displacement of membrane potential from its resting value ($V = 0$) during conditioning steps. (Note: the positive values are the hyperpolarizing steps.) Ordinate: circles, peak sodium current during test step $(\Delta I_{Na})_{cond}$ relative to peak sodium current in unconditioned test step $(\Delta I_{Na})_{uncond}$ (left-hand scale) or relative to maximum sodium current obtainable, i.e. as h (right-hand scale). The smooth curves were drawn with a value of -2.5 mV for V_h. (Hodgkin & Huxley, 1952.)

all displacements of the membrane potential (maximum hyperpolarization and depolarization being 44 mV and 29 mV, respectively). To represent the ability of the axon to respond with an increase in sodium conductance quantitatively at a steady value of the membrane potential, a quantity h, which they called the *degree of inactivation*, was introduced

$$h = \frac{\text{maximum obtainable increase in } G_{Na} \text{ at membrane potential } V}{\text{maximum obtainable increase in } G_{Na}},$$

(4.4.8)

which ranges from 0 to 1.0. The influence of the membrane potential on the final level of inactivation is shown in Fig. 4.36, where the peak of sodium current produced by the test voltage – relative to the peak value produced at the resting membrane potential V_r, i.e. $((\Delta I_{Na})_V/(\Delta I_{Na})_{V_r})$ – is plotted against the membrane potential during the conditioning period. The relation is a smooth curve that bends towards the maximum and minimum with the same curvature. It appears from Fig. 4.36 that the inactivation is complete when the membrane potential is displaced to about 0 mV*, and is nearly absent when the membrane is hyperpolarized by about 30 mV. Using this value of $(\Delta I_{Na})_V/(\Delta I_{Na})_{V_r} = 1.7$ in the denominator of Eq. (4.4.4), the left-hand ordinate of Fig. 4.36 is recalculated in terms of the inactivation factor h which is written as the right-hand ordinate. The smooth curve of Fig. 4.36 was obtained by fitting the equation

$$(h)^{(stat)} = \frac{1}{1 + e^{-(V_h - V)/b}} \qquad (4.4.9)$$

where $V_h = -2.5$ mV is the value for $h = \frac{1}{2}$ and the constant $b = 7$ mV is connected to the slope at $(V_{h,\frac{1}{2}})$ by $(dh/dV)_{V_h} = 1/4b$. It appears from Fig. 4.36 that the ability of the axon to respond with an increase in sodium conductance is not at a maximum at the resting membrane potential ($h = 0.6$ at $V = 0$). It appears strange if the membrane never utilizes as large a fraction as $\frac{4}{10}$ of its power to produce a sodium current. A likely explanation to evade this dilemma is that membrane potentials of the axons used for experiments were already lower than normal because of the procedures preceding the actual experiment[†]. Figure 4.36 shows the stationary values only, $h^{(stat)}$, of a time-dependent inactivation process $h(V, t)$, which we substitute for $G_{Na}^{(off)}$ in Eq. (4.4.3) and rewrite it as

$$\Delta G_{Na}(t) = G_{Na}^{(on)} \times h(V, t). \qquad (4.4.10)$$

Thus, when the membrane is suddenly depolarized from a membrane potential V by a step ΔV, the membrane's ability to respond with an increase in the inward sodium current – or rather an increase in sodium conductance $\Delta G_{Na}(t)$ – is the result of two opposing processes that both depend upon the membrane potential but have different time courses. (1) The fast effect, i.e. the turning on of a depolarizing step ΔV, induces a rapid increase – within 1 ms – in the sodium conductance, and an even faster return of the sodium conductance to the initial level on the removal of ΔV (see the dotted line in Fig. 4.31). The peak of the rise in conductance increases strongly with ΔV. (2) The late effect is a gradually developing suppression (inactivation) of the early conductance increase if the duration of ΔV is maintained. The initial growth of the inactivation process proceeds much more slowly than the rapid increase in the sodium conductance $G_{Na}^{(on)}$ and the peak value of $\Delta G_{Na}(t)$ is very little influenced by the inactivation. Therefore, it appears reasonable to regard the fast "on" and "off" effects above

* In Fig. 4.36 this corresponds to about $V = -50$ mV.
† In Fig. 4.22 the membrane potential in (A) recorded from the intact animal is 10 mV more negative than that obtained from the isolated axon (B). Furthermore, it is conceivable that animals in their natural surroundings might present even more optimal conditions for the axons.

as involving only the component $G_{Na}^{(on)}$ of the sodium conductance mechanism in the membrane. The final level of inactivation and the rate at which this level is attained increases with the reduction of the membrane potential. Thus, the ability of the membrane to produce an inward current in response to a large step of depolarization becomes less and less with a reduction in the steady level of the membrane potential and is almost abolished at zero membrane potential. To erase the state of inactivation and reset the power of the sodium conductance mechanism, the membrane depolarization must be removed by restoring the resting membrane potential. The fast responding part $G_{Na}^{(on)}$ of the sodium conductance mechanism returns rapidly to the initial low value of conductance, whereas the persistent inactivation is stopped in its further development and left to decay within the next 1–2 ms to return the membrane completely to its normal state.

This may appear terribly complicated. But, if speed and safety of impulse transmission were the ultimate requirement, it was apparently not within Nature's compass to devise a simpler mechanism – not even in an animal so low in the hierarchy as *Loligo*.

4.4.4.4 The membrane action potential: qualitative synthesis of the voltage clamp experiments

Before actually starting a voltage clamp experiment on the squid axon, Hodgkin & Huxley controlled the excitability of the membrane by using the internal current electrode to supply a short current pulse through the membrane to check if the membrane response was a normal sized action potential*. In this way the membrane is uniformly depolarized over a comparatively long section (the distance between the barriers B1 and B2 in Fig. 4.24). The result of applying stimulations of increasing strength is shown in Fig. 4.37. It appears that the responses are essentially similar to those of Fig. 4.16, i.e. displays of subliminal local responses and a fully blown transient swing of membrane potential of the same amplitude as a normal propagating action potential. In the present case *no propagation* occurs, but instead the membrane potential changes simultaneously along the patch of axon membrane that is confined to be clamped later in the experiment. This type of response is called a *membrane action potential* in contrast to the travelling nerve signal (*the propagated action potential*). Below we shall give an account of the appearance of the responses shown in Fig. 4.37, using the final results of the voltage clamp analysis, which

---

* To this end, the feed-back loop shown in Fig. 4.24 was interrupted by cutting off the connection between A1 and A3 and by earthing the input to A3 of the control amplifier, which was only used to supply a short stimulation current through the membrane, whereas the output from A1 recorded the response of the membrane potential.

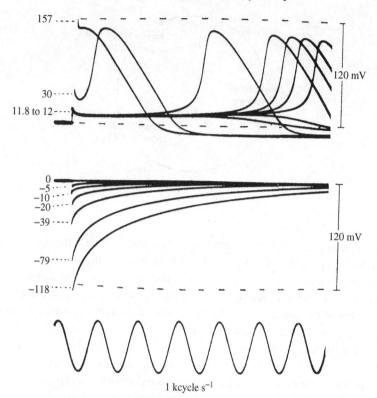

Fig. 4.37. Membrane responses resulting from applying a short current stimulus uniformly across the stretch of axon mounted in the voltage clamp assembly. Upper traces: current is directed outwards (membrane depolarized). Lower traces: inwardly-directed current flow (membrane hyperpolarized). The numbers attached to the curves give the strength of the shock (charge per unit area in μC cm^{-2}). (Hodgkin, Huxley & Katz, 1952*.)

may be summarized as follows. The membrane's permeability or conductances to Na$^+$ and K$^+$ ions are controlled by the magnitude of the membrane potential of the axon. A hyperpolarization produces a *decrease* of the conductance of both ions, whereas a depolarization results in an *increase* of the conductances of both ions that grows in a highly nonlinear manner in step with the amount of membrane depolarization. The two conductance changes differ in their time courses, as the increase in G_{Na} starts abruptly with a short delay, while the increase in G_K sets in after a delay and grows more slowly.

* Hodgkin, A.L., Huxley A.F. & Katz, B. (1952): Measurements of voltage–current relations in the membrane from the giant axon of *Loligo. J. Physiol.*, **108**, 37.

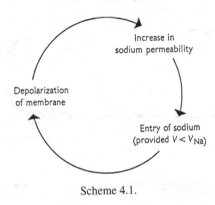

Scheme 4.1.

(i) The action potential

In the resting state, e.g. at a membrane potential of -70 mV, the tendency for the Na^+ ions to enter the axoplasm is strong because the drops of concentration and potential are both directed inwards. The actual inward movement of the Na^+ ions is held in check by the low value of the G_{Na} at rest in comparison with that of G_K and G_{Cl}. Now let the membrane become depolarized at some spot (e.g. by electric stimulation as in Fig. 4.16) above the critical level – the threshold depolarization – e.g. from -70 mV to -55 mV. The ensuing, almost instantaneous, increase in G_{Na} causes more Na^+ ions to move inward through the membrane in the direction of the drop of the electrochemical potential $\mathcal{E}_{Na} = V - V_{Na}^{(eq)}$ for the Na^+ ion. This extra supply of positive charges on the inside of the membrane leads to an additional depolarization of the membrane. And this additional depolarization of the membrane in turn causes a further increase in G_{Na}. Thus, the membrane depolarization and the conductance increase for sodium are interlocked by means of a self-amplifying mechanism. A sketch of this regenerative coupling* between the membrane depolarization and the increase in sodium conductance is shown in Scheme 4.1. By this regenerative process, where the entry of the Na^+ ions is increasingly facilitated because the membrane depolarization that progresses in step with the excess sodium entry, the size of membrane potential tends to be dominated by the permeability of the Na^+ ions. Correspondingly, the membrane potential tends to assume the value where an electrochemical equilibrium prevails for the Na^+ ions on the two sides of the membrane, i.e. the value of the equilibrium potential $V_{Na}^{(eq)}$ for sodium, which is about $+57$ mV for the squid axon. This tendency, however, is shortlasting as two counteracting influences now tend to reverse the process.

---

* Hodgkin, A.L. (1964): *The Conduction of the Nervous Impulse*, Liverpool University Press, Liverpool.

(1) The state with highly increased Na$^+$ conductance is *transient* and short-lasting. After $\frac{3}{4}$ to 1 ms the mechanism of *inactivation of* G_{Na} sets in, causing suppression of the Na$^+$ conductance that gradually returns it to the original resting value. Consequently, as the membrane returns to the state where its permeability again is dominated by the K$^+$ ion, the membrane potential becomes repolarized to assume in the end the original resting potential.

(2) With the membrane potential displaced to about $+45$ mV, the quasi-equilibrium state for the K$^+$ ions* at a resting membrane potential of about -65 mV is displaced to a state that strongly favors an increased outward movement of the K$^+$ ions, removing in the process an excess of positive charges from the axoplasm. Moreover, the outward movement of the K$^+$ ions is facilitated because depolarization also causes the membrane conductance G_K to increase, although with a delay and more slowly than that of G_{Na}.

Thus, the inactivation of the Na$^+$ conductance and the increase of the K$^+$ conductance contribute to making the membrane potential swing quickly back towards the resting value ($V_m \simeq -65$ mV), and by so doing extinguishing any conductance increase of sodium that might remain after the inactivation. Likewise, the conductance for potassium, G_K, returns to the state of rest. Sometimes this process takes place via a phase of hyperpolarization of several millivolts that reflects the return to the normal value of the resting state. The net result of this sequence of changes in the ionic permeabilities that has resulted in a transient shortlasting change in the membrane potential of about 110–120 mV is that a minute amount of sodium has entered the axoplasm (about 4 pmol cm^{-2}) and a similar amount of potassium has moved from the axoplasm to the external medium.

(ii) Threshold phenomena

The operation of the regenerative coupling between the membrane depolarization and the increase in sodium conductance as described above relates to the situation where the initial step of depolarization passes beyond a certain critical level, the *threshold depolarization*. The existence of this sharp borderline between levels of depolarization that are effective and those that are not arises because the sodium conductance increases strongly (approximately exponentially) with the size of the depolarizing displacement from the value of the resting membrane potential. For instance, a change of the membrane potential from -70 to -65 mV causes an increase in G_{Na} by a factor of about 3,

---

* The equilibrium potential for potassium $V_K^{(eq)}$ in the squid axon is about -75 mV. See Table 4.1.

4. *The nerve impulse*

whereas the factor is about 50 and 1000 at depolarization to −50 mV and −20 mV. Thus, a depolarization of, say, 5 mV from −70 mV will, owing to a relatively small increase of G_{Na}, cause only a small and rather slow increase in the inward-directed Na^+ current. The ensuing additional depolarization rises slowly enough to allow the delayed increase of the potassium conductance to make itself felt as an increased outwardly-directed potassium current that eventually exceeds the sodium current and then drives the membrane potential back to its original value – the resting membrane potential – and both sodium and potassium conductances return to their initial values. This prolonged re-polarization – "local response" (see Fig. 4.18) – which follows a subliminal depolarizing stimulus reflects the struggle between the inwardly-directed Na^+ current and the outwardly-directed K^+ current to dominate over the direction of the net current through the membrane. In the present case I_K got the upper hand and the end result was a complete repolarization of the membrane. A slight increase of the stimulus strength may prolong this competition for control of the membrane current, and it may even be possible to hit just upon the situation where the inwardly-directed Na^+ current and the outwardly-directed K^+ current are equal. Thus, even at a fairly long time after the cessation of the stimulus the situation lingers, with $I_{Na} + I_K = 0$, until a fluctuation in either I_{Na} or I_K becomes decisive as to whether the membrane potential develops further in a full-blown action potential or there is a return to the value of the resting membrane potential. This critical level of membrane depolarization where the Na^+ and K^+ conductances have just those values that result in equal inwardly-directed Na^+ current and outwardly-directed K^+ current is the *threshold depolarization*.

4.4.4.5 Investigations on single Na^+ channels

In the quantitative theory for the ionic flow in the axon membrane that Hodgkin & Huxley (1952) developed, they ascribed the origin of the Na^+ and the K^+ current to the activity of two types of ionic transferring sites – or *channels* – in the membrane. Each channel type was specifically permeable either to Na^+ or K^+ ions and possessed conductances having the characteristic voltage and time dependence shown in Fig. 4.32. The full-drawn lines of the figure are voltage and time dependence predicted by the theoretical model that Hodgkin & Huxley developed to account for the behavior of the two conductances[*]. The agreement between the theory and experiment is seen to be adequate but, moreover, the theory was also able to account satisfactorily for the shape of the propagating action potential and its velocity of conduction together with a

---

[*] Nowadays called the Hodgkin–Huxley equations, or just the H–H equations. See also Section 4.7.

number of phenomena associated with nerve excitation*. As expected the theory provoked much discussion and objections of both general and more-detailed substance. Among the latter was the assumption of separate ionic channels for Na^+ and K^+ ions. An alternative proposal put forward was that of a single channel that initially became permeable predominantly to Na^+ ions but closed after a while to become permeable instead to K^+ ions[†]. However, in its classical form, the voltage clamp technique measures the resultant flow of Na^+, K^+ and possibly Cl^- currents through a fairly large membrane area that contains an exceedingly large number of ionic channels. Therefore, it is not possible with this technique to distinguish between alternative functions and arrangements of channels in the membrane. A final clarification of this complex of problems became possible after the development of the so-called *patch clamp* technique[‡] that was developed by Erwin Neher from Göttingen. With this technique a small membrane area of about 2–10 μm^2 – which contains a few or just a single ion channel – is sealed off from the rest of the membrane by means of a micropipette that is designed for the purpose. The microelectrode is connected to a suitable electronic equipment that – by analogy with the device described in Section 4.4.4.2 – monitors a well-defined potential difference ΔV across the membrane patch and records the current I that slips through the ionic channel during the polarization. As the composition of the surrounding solutions can be chosen arbitrarily it is a relatively simple matter to determine the specific selectivity of the channel in question.

The patch clamp investigations have shown unambiguously that the excitable membrane is equipped with single ionic channels that selectively are transporters for Na^+ and K^+ ions, respectively. Furthermore, although a voltage step applied across the membrane patches elicits a rather complex current response from the single channel, the joint current responses following the simultaneous activation of large number of channels have time courses that are similar to those of the Na^+ and K^+ currents that are obtained from the larger membrane area by voltage clamp measurements. Figure 4.38(c) shows an example of nine successive recordings of the currents from a single Na^+ channel when a depolarizing voltage step of 20 mV amplitude and 20 ms duration (line *a*) was applied to the isolated membrane patch with an interval of 1 s. The response from the channel consists of one or several shortlasting current impulses of the *same* amplitude but of varying widths. The time for the occurrence of the

---

* See Section 4.4.5.3.
† Mullins, L.J., Adelman, W.J. Jr. & Sjodin, R.A. (1962): Sodium and potassium effluxes from squid axons under voltage clamp conditions. *Biophys. J.*, **2**, 257. Hoyt, R.C. (1963): The squid giant axon. Mathematical models. *Biophys. J.*, **3**, 399.
‡ Neher, E. & Sakmann, B. (1976): *Nature*, **260**, 799.

4. *The nerve impulse*

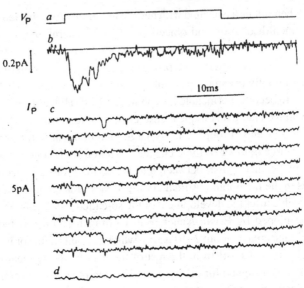

Fig. 4.38. Examples of current recordings from a single Na$^+$ channel from the membrane of myoblast fiber from rats. (*a*) Depolarization of 10 mV, which is brought about by holding the membrane patch hyperpolarized at 30 mV, and then once a second applying a depolarizing voltage step of 40 mV and 20 ms duration. (*c*) Nine successive recordings of the shortlasting inwardly-directed current pulses that are carried by Na$^+$ ions. The average current amplitude is -1.6 pA and the mean opening life time is 0.7 ms. (*b*) Sum of 300 single recordings as in (*c*). (*d*) Example of current record obtained in another experiment where two-thirds of the Na$^+$ in the pipette was replaced with tetraethylammonium ions. Temperature 18 °C. (After Sigworth & Neher, 1980*.)

current impulse appears to happen in a purely random manner. To disclose any hidden regularity in the firing pattern a large number of single recordings were summed. The effect of this procedure is that the deflections – e.g. electric noise – that occur at random in *both* directions and with the *same probability* will contribute very little to the sum, whereas the sum of the signals, having consistently a deflection in just *one* direction, appears as a distinct picture that also reveals any regularity in the arrival pattern of the signals. Figure 4.38 (line *b*) shows the result of such "noise filtration" by summing 300 current recordings of the same type as in (*c*). It appears clearly that the majority of the shortlasting current impulses occur in the time just after the membrane depolarization is switched on and peaking at about 1 ms, after which the number of occurrences decreases gradually to be absent after a course of 5–6 ms.

* Sigworth, F.J. & Neher, E. (1980): *Nature*, **287**, 447.

We can also consider Fig. 4.38 from a different point of view. Imagine the membrane patch to be large enough to contain 300 Na^+ channels – i.e. an area of perhaps 1000–3000 μm^2. Application of a *single* depolarization of 10 mV will activate the channels, each of which produces in turn their short current impulses. Figure 4.38(b) will now represent the time course of the total Na^+ current that the membrane turns on following the depolarization. The time course appears to be almost identical to that of Fig. 4.29, leaving out of account the irregularities of curve (b). However, Fig. 4.29 was the result of clamping a larger membrane area that contained at least several thousands of Na^+ channels yielding most likely a much smoother course. It should be noted that the *inactivation* – whose occurrence in the nine recordings of (c) is in no way obvious – appears when a sufficiently large number of channels are activated simultaneously.

By analysis of a large number of current recordings such as those of Fig. 4.38 (c) released at different levels of polarizations, it became possible to uncover the statistical regularity that controls the behavior of the single Na^+ channel.

(a) The probability of the opening (*activation*) of a channel increases rapidly within the first millisecond after the application of the depolarization (compare Fig. 4.38 (line b)) and with the magnitude of the membrane depolarization as well.
(b) The opening and closing of the channel – following a step of depolarization – has the character of an *all-or-nothing* event (i.e. either non-conducting or else maximum conductance), and where
(c) the open channel has a certain *lifetime*, whose duration is related to the size of depolarization.
(d) The process that leads to inactivation is likewise an *all-or-nothing* event that sets in after the opening of the channel, and is broken only by repolarizing the membrane to the original level of membrane potential. The probability of the occurrence of the inactivation is likewise dependent upon the amount of depolarization. The process of inactivation in presumably separated spatially from the region in the channel that regulates the activation.

A schematic presentation of the complex behavior of the Na^+ channel following a step of depolarization is shown in Fig. 4.39. The model consists of two openings, each provided with gates: a *gate of activation* (a) that is localized at the external surface of the membrane, and a *gate of inactivation* (in) at the inside of the membrane*. The gates are assumed to be found in one of

* This spatial localization of the two gates is the natural result described in Section 4.4.4.6.

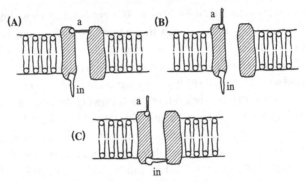

Fig. 4.39. Schematic presentation of the three states for a Na^+ channel. The gate of activation (a) is localized at the external surface of the membrane and the gate of inactivation is on the inside. (A) At the resting membrane potential (a) is closed and (in) is open. (B) A short time after depolarization, (a) is open and there is now passage for Na^+ ions but *not* for other ionic species. (C): Later in the depolarization phase the gate of inactivation (in) is closed and will only reopen at repolarization of the membrane. (From Byrne & Schultz, 1988*, modified.)

two positions: either completely closed or open. The situation at the resting membrane potential is shown in (A). In this state the gate of activation (a) is closed and the channel does not allow Na^+ ions to slip through even if the gate of inactivation (in) is open. In (B) the membrane depolarization is applied and the gate of activation (a) flies open. As the gate of inactivation (in) is in the open state the channel becomes open for a flow of Na^+ ions to pass according to the direction of their electrochemical gradient. This state with both gates fully open is only preserved for a short time, as the depolarization increases the probability for the closure of the gate of inactivation (in). When this happens the channel is again impermeable for the Na^+ ions although the activation gate (a) still remains open. Moreover the closing gate of inactivation is a definitive event[†] as a repolarization is required to reopen the gate of inactivation (in). Furthermore, the repolarization leads to a greater probability for a closure of the gate of activation (a) and thus to reconstitute the state (A).

Whether these properties of a single Na^+ channel also lead to the time and voltage dependence $G_{Na}(V, t)$ for the Na^+ conductance that is observed experimentally in a membrane area containing a very large number of Na^+ channels is a matter that cannot be settled just by plausible reasoning, but requires a

* Byrne, J.H. & Schultz, S.G. (1988): *An Introduction to Membrane Transport and Bioelectricity*, Raven Press, New York.
† One could imagine that the gate of inactivation was provided with a latch.

mathematical model to manage the interconnection between the variables. To this end the occurrence of the Na^+ current impulses was regarded as a stochastic process* with a probability density function

$$\varphi(\Delta V, t, \tau)$$

that specifies the probability

$$dP = \varphi(\Delta V, t, \tau)\, dt,$$

that the Na^+ channel, in response to a depolarization ΔV, opens at the time between t and $t + dt$ and has a lifetime τ. In this construction the time course of $\varphi(\Delta V, t, \tau)$ must be proportional to the time course of the sum of the Na^+ currents from a large number of Na^+ channels that are activated simultaneously by a step of depolarization ΔV as in Fig. 4.32. A function $\varphi(\Delta V, t, \tau)$, having the desired properties, can be constructed[†] and, therefore, the time and voltage dependence of the Na^+ conductance results from summing the apparent chaotic behavior that the stochastic model predicted for the behavior of the single ionic Na^+ channel. Furthermore, the mathematical formalism contains in principle the same elements that enter into the Hodgkin–Huxley equations[‡].

4.4.4.6 Selective effects on the Na^+ and K^+ channels

A variety of substances that influence the function of the impulse conduction have in the course of time become well known. Many became directly applicable in practical medicine and have obtained elevated status as pharmacological drugs. The arrival of the voltage clamp technique made it possible to pinpoint the mode of action on the axon membrane to some extent of these drugs. In some cases it turned out that the action of the drug was localized to just a single functional element of the process of impulse conduction. For that reason the use of the drug in question could contribute to providing new and essential knowledge about the function of the excitable membrane. In the following sections we shall describe experiments where three substances – each in itself injurious to the environment – exert a selective influence on the Na^+ and K^+ channels.

---

[*] *Stochastic* = "random" but calculable statistically with an approximate accuracy. Stochastic processes: systems that change in accordance with probabilistic laws. In contrast, deterministic processes are governed by unambiguous laws of cause and effect.

[†] A comprehensive account of the current problems is found in Hille, B. (2001): *Ion Channels of Excitable Membranes*, 3rd edition, Sinauer Associates Inc., Sunderland, MD, 722 pages.

[‡] See Section 4.7.

(i) Blocking the Na^+ channel

The ovaries of the Japanese puffer-fish *Spheroides rubripes* contain an extremely poisonous substance, which is now is produced in a pure form and marketed under the name of Tetrodotoxin (TTX). The substance, whose action has been known in Japan for the past 2000 years, annually kills several hundreds of Japanese, who like to eat the raw fish, but by some mishap also consume the ovaries. TTX is also present in the skin of a South American salamander (*Taricha torosa*). TTX belongs to the most deadly group of poisons known. Thus, mice die from a dose as small as 5 μg kg^{-1} bodyweight, whereas in comparison with a respectable dangerous poison such as KCN the lethal dose is 10 mg kg^{-1}. The cause of death is respiratory paralysis, and the underlying cause is conduction block of the nerve impulses. Voltage clamp experiments have shown that TTX acting from the *external medium* blocks the increase of the Na^+ conductance that normally results from membrane depolarization; but no effect is observed on the subsequent delayed increase in the K^+ conductance displaying a time and voltage dependence identical to that observed under normal conditions*. Figure 4.40 shows such an example. However, TTX injected *inside* the axoplasm has no effect on the impulse. This means that the site of action of TTX on the Na^+ channel must be localized somewhere at or near the outside part. The great toxicity of TTX is also reflected by its extreme binding affinity to the membrane, which also suggests that only one TTX molecule becomes bound to each Na^+ channel. On this basis the density of the Na^+ channels had been estimated, giving 5–25 channels μm^{-2} in myelinated motor nerves and unmyelinated nerves from crabs and lobster, whereas in the giant axon of the squid *Loligo* the number becomes as high as 525 channels μm^{-2}.

Saxitoxin (STX) is another substance with the same toxicity and mechanism of action as TTX. It is produced by a sea alga (*Gonyaulax catenella*, a dinoflagellate) and is concentrated in certain types of edible filtering mussels in Alaska so that consumption of just one of these mussels may be lethal for humans. The chemical formula for SXT is $C_{10}H_{17}N_7O_4 \cdot 2HCl$ and it differs from that of TTX, which is $C_{11}H_{17}N_3O_8$. There are also minor differences in action, as the nerves of the Japanese puffer-fish and the South American salamander are resistant to TTX but not to SXT.

The action of TTX on the axon membrane described above is naturally a further argument in support of the fundamental importance of the Na^+ ion for the impulse conductance, where the membrane in response to a depolarization opens up the Na^+ channels for the inward Na^+ current.

---

* Experiments carried out on axons from lobster (Narahashi, Moore & Scott, 1964), on squid giant axons (Nakamura, Nakajima & Grundfest, 1965) and on frog nerves (Hille, 1968).

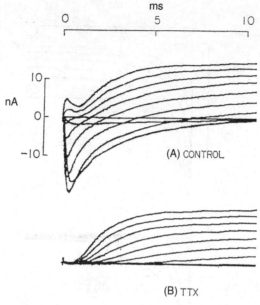

(A) CONTROL

(B) TTX

Fig. 4.40. Voltage clamp currents in a node of Ranvier in the axon from n. ischiaticus from the frog (*Rana pipiens*) in response to 19 different voltage steps from the resting potential at -67.5 mV to $+67.5$ mV with increments of 7.5 mV between each step. Abscissa: time (ms). Ordinate: membrane current (nA). Inward-directed current is negative. (A) Normal Ringer. (B) Ringer $+3 \times 10^{-7}$ M Tetrodotoxin. Note the disappearance of the Na^+ current. (Hille, 1970*.)

(ii) Blocking the K^+ channel

It had long been known that the tetraethyl ammonium (TEA) ion $N^+(C_2H_5)_4$ affects both the synaptic transmission and the impulse conduction where TEA prolongs the duration of the action potential and also provokes a repetitive impulse firing in response to a single electric stimulus. The latter observations are now understood, as the voltage clamp investigations have shown that the specific action of TEA is a *selective inhibition* – eventually a total blockage – of the membrane's ability to produce an increase in the K^+-conductance in response to a membrane depolarization.

The effect of TEA on the membrane current in the node of Ranvier is shown in Fig. 4.41. It will be seen that the K^+ current is completely absent whereas the initial Na^+ current appears to be totally unaffected by TEA. This action is in contrast to that of TTX which – as shown in Fig. 4.40 – totally eliminates the early Na^+ current but leaves the delayed outward K^+ current unaffected.

* Hille, B. (1970): *Prog. Biophys. Mol. Biol.*, **21**, 1.

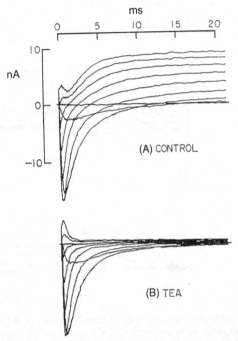

Fig. 4.41. Voltage clamp currents in a node of Ranvier in an axon from n. ischiaticus
from the frog (*Rana Pipiens*) in response to eight different steps of potential distributed
over the range of membrane potentials from −50 mV to +50 mV. Abscissa: time (ms).
Ordinate: membrane current (nA). Inwardly-directed current is negative. (A) Normal
Ringer. (B) Ringer +5 mM TEA. (Hille, 1968*.)

Whereas TEA affects myelinated nerves when applied from the outside, even
concentrations of TEA as high as 100 mM in the external fluid have no effect on
the giant axon from *Loligo*. However, perfusion of the axon with TEA results
immediately in a total block of the outward K^+ current. On the other hand, TEA
has no effect on the initial inwardly-directed Na^+ current, for which reason
one should not expect any appreciable effect on the rising phase of the action
potential. As the inactivation of the Na^+ channels is also unaffected by the
application of TEA, the membrane will repolarize in step with the development
of the inactivation. Thus, as expected, the axon is still capable of conducting
an action potential, but as seen in Fig. 4.42 the shape of the action potential
is influenced by TEA: the phase of repolarization is prolonged, and no final
phase of hyperpolarization appears as under normal conditions. This agrees
with the new conditions where TEA has blocked the ability of the depolarized

* Hille, B. (1968): *J. Gen. Physiol.*, **51**,199.

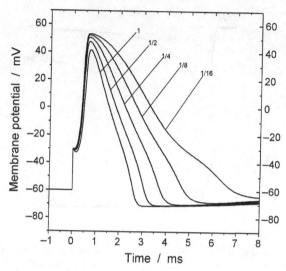

Fig. 4.42. To simulate the action of TEA a calculation from the Hodgkin–Huxley equations, Section 4.7, shows the effect on the membrane action potential resulting from a stepwise reduction of the delayed increase in K^+ conductance. The numbers attached to the curves give the reduction of the peak value of the conductance increase relative to the normal value.

membrane to respond with an increase of the K^+ conductance and with that allowing the fastest efflux of the positively charged K^+ ions.

(iii) Destruction of Na$^+$ inactivation

In connection with investigations on the action of TEA on the giant axon, Armstrong, Bezanilla & Rojas (1973) made the interesting observation that a perfusion of the axon with the proteolytic enzyme pronase* made the *inactivation of the* Na$^+$ *current disappear*. This is illustrated in Fig. 4.43. To display the pronase effect as distinctly as possible, the outward K^+ current was blocked by adding TEA to the perfusate. The recording of the inward Na$^+$ current in (A) corresponds therefore completely to the inward currents of Fig. 4.42, which all show inactivation. Figure 4.43(B) shows that the sole action of pronase is to remove a mechanism – by cutting off a component – of the Na$^+$ channel, to which the process of inactivation is normally attached. Adding pronase to the external fluid only has no effect on the inactivation. This shows that inactivation mechanism is localized to the inward-looking part of the Na$^+$ channel. This observation is also the motivation for placing the gate of inactivation on the membranes inside in Fig. 4.39.

* The substance actually consists of a mixture of different proteolytic enzymes.

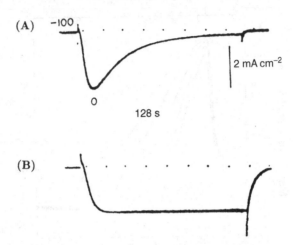

Fig. 4.43. Destruction of Na inactivation in the squid axon by perfusion with pronase (1 mg dm^{-3}). The perfusate contained also 15 mM TEA to block the K^+ current. (A) Voltage clamp current in response to a voltage step from $V = -100$ mV to $V = 0$ mV recorded just after the start of the perfusion. (B) Same potential step applied 12 min later. Note: in (A) the Na^+ current displays the usual inactivation. In (B) there is no inactivation. (Armstrong, Bezanilla & Rojas, 1973*.)

The experiments (i), (ii) and (iii) above have shown that it is possible *selectively* to dissect *each* of the three fundamental elements that respond to a depolarization of the membrane. Apart from their inherent interest, these experiments also point back to support the original assumption of Hodgkin & Huxley where the Na^+ and K^+ ions flow through their respective channels and not – as was suggested by other research groups – through one channel that sequentially shifted during the course of activation the access of the Na^+ ions to that of the K^+ ions.

4.4.5 The propagated action potential

4.4.5.1 The local current loops

The sequence of events hitherto described and summarized in Fig. 4.38 were localized to the region on the membrane that was depolarized uniformly above the threshold value to trigger off the response that we called the membrane action potential. This configuration will – once it is established – move away to both sides with a constant velocity as a *propagated action potential*. The basic mechanism underlying this impulse propagation is that the membrane depolarization at this activated region – caused by the surge of inward Na^+

---

* Armstrong, C.M., Bezanilla, F. & Rojas, E. (1973): *J. Gen. Physiol.*, **62**, 375.

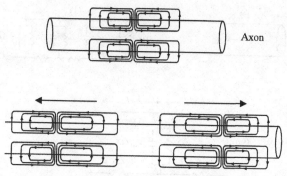

Fig. 4.44. Illustration of the propagation of the nerve impulse from the stimulation site on account of the depolarization of the adjacent membrane regions caused by the local currents. (Top) Snapshot of the local current loops that arises from the site of stimulation immediately after cessation of the stimulus. (Below) The local current loops have stimulated the neighboring regions and propagated impulses that move in both directions away from the stimulation site.

current – spreads into the two neighboring negatively charged "resting" * membrane regions. This spreading of depolarization is caused by flow of electric current from the activated region – which is at a more positive potential with respect to the rest of the axoplasm – to the neighboring axoplasm. On its flow axially along the axoplasm, the currents will successively leave the interior by crossing the membrane – in the *inside → outside* direction – and return via the external medium to the activated region, which has a positive charge deficit due to the inflow of Na$^+$ ions. This kind of electric circuit was originally suggested by Hermann[†] who coined the term *local current loops*. As can be seen in Fig. 4.44 the membranes of these two adjacent regions will be *depolarized* by the local currents that leave the axoplasm. Close to the activated region this depolarization may be sufficient to exceed the threshold depolarization of the membrane, with the result that this membrane region now undergoes the same sequence of conductance and potential changes as indicated in Fig. 4.36. That is, the active region has moved a small distance away from the original site of stimulation. In turn, this new activated region will then activate its adjacent membrane region, and so on. Thus, a wave of depolarization will propagate in both directions away from the stimulation site, whereas the activity behind the impulse will go out because of the setting in of the inactivation process[‡].

---

[*] That is, non-activated membrane regions, whose membrane potential, and Na$^+$ and K$^+$ conductances are until now in the same state as in the unstimulated axon.
[†] See Section 4.2, p. 415.
[‡] See Section 4.4.5.3.

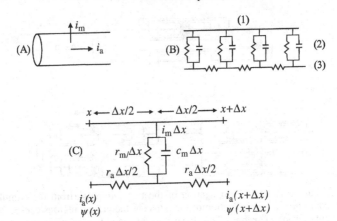

Fig. 4.45. Illustration of a one-dimensional cable (A), with the axial current i_a and the membrane current i_m. The diagram (B) shows the equivalent electric circuit. (1) External medium. (2) Membrane with membrane resistance r_m in parallel with the membrane capacitance c_m. (3) Interior with the axial resistance r_a. (C) Illustration of the current flow in the elementary axon length Δx used in deriving the cable equation.

4.4.5.2 The nerve as an electric cable

To account quantitatively for the factors that determine the spreading of the *local current loops* it is helpful to use a mathematical model called the *cable equation*. A simple version of this model is given below.

(i) Derivation of the cable equation

We model the passive electrical properties of the nerve membrane* as a cylindrical membrane of uniform cross section, which surrounds an intracellular liquid phase (axoplasm) and that is immersed in a large volume of extracellular liquid (Fig. 4.45(A)). The resistances of the axoplasm and membrane are assumed to be pure ohmic resistances. Inside the axoplasm is a (micro)electrode, that is connected to an external current source. The current from the electrode runs in the direction of the cylinder axis and is assumed to be distributed uniformly over the cross-sectional area (no radial current component inside the axoplasm). The distribution of current and potential in the model is therefore described by a *single* coordinate x that measures distance along the cylinder axis. For that reason the model is also known as a one-dimensional cable. The electrical resistance of the surrounding medium is assumed to be so small[†] that local current in the external medium will have no effect on the potential along the external

---

* That is, the electrical properties of the membrane when the voltage dependence of the Na^+ and K^+ conductances are blocked.
[†] The assumption is a simplification only as there are no principal difficulties involved in introducing a finite external resistance.

surface of the cylinder. Thus, potential in the external medium $\psi^{(o)}$ – and that along the cylinder surface – is regarded as constant; to simplify matters we set $\psi^{(o)} = 0$, and the membrane potential is $V = \psi^{(i)} - \psi^{(o)} = \psi^{(i)}$. The current electrode is placed in the position $x = 0$, sending the initial intracellular axial current I_0, whereby the potential at $x = 0$ is raised to the value V_0. If the membrane resistance were infinitely large all of the electrode current would run axially with a constant value and the voltage V in the axoplasm would decrease linearly in accordance to Ohm's law*. If, on the other hand, the membrane has a *finite resistance*, a fraction of the current would leak through the membrane for as long as there remains a potential difference to the external medium. Consequently the intracellular axial current will *no longer* be constant over any distance δx of finite length. The potential V and axial current i_a at any distance x away from the current electrode will therefore vary as a function $V(x, t)$ and $i_a(x, t)$ at any x and time t. In addition we must also consider the *membrane current* $i_m(x, t)$ that leaks through the membrane as a function of x and t.

To proceed further we need to derive the laws[†] that govern the relationship between the intracellular axial current i_a, the membrane current i_m and the potential V as we move a distance x away from the current electrode at $x = 0$. We consider two closely adjacent positions x and $x + \Delta x$, with potentials $V(x, t)$ and $V(x + \Delta x, t)$, respectively (Fig. 4.45(C)). Let r_a represent the axial resistance *per unit length* (dimension ohm per meter ($\Omega\ m^{-1}$)), such that the resistance in the cable element of length Δx is $r_a \cdot \Delta x$. Further, let the positive direction of i_a be that of the x-axis. From Ohm's law we have

$$V(x, t) - V(x + \Delta x, t) = r_a \Delta x\, i_m(x', t),$$

where $i_a(x', t)$ is the mean of the axial current over the length of axoplasm Δx, i.e. the axial current at a position x' between x and $x + \Delta x$. We rearrange this equation as

$$\frac{V(x + \Delta x, t) - V(x, t)}{\Delta x} = -r_a i_a(x', t),$$

and let $\Delta x \to 0$. On the right-hand side $i_a(x', t) \to i_a(x, t)$, and on the left-hand side the ratio tends towards the partial derivative of $V(x, t)$ with respect to x[‡]. We therefore have

$$i_a(x, t) = -\frac{1}{r_a}\frac{\partial V(x, t)}{\partial x}, \tag{4.4.11}$$

which is another form of Ohm's law.

* Provided that the end of the cable is open to external medium.
† Originally developed by the great physicist William Thomson (Lord Kelvin), 1824–1907, who developed the theory to account for the function of the transatlantic cable.
‡ See Chapter 1, Section 1.8.2, Eq. (1.8.3).

4. *The nerve impulse*

Changes in $i_a(x, t)$ along x result from a "leak" across the plasma membrane. As the axoplasm resistance r_a is defined as the resistance per unit length it is a natural thing to define the membrane current i_m as the current through the membrane per unit length of axon (units ampere per meter ($A\,m^{-1}$)). The current will be regarded as *positive* when it crosses the membrane in the *inside→outside* direction. Similarly it is natural to define the membrane conductance g_m (or membrane resistance $r_m = 1/g_m$) as the conductance of the membrane ring which is contained by one unit length of axon (the units for g_m and r_m are siemens per meter ($S\,m^{-1}$) and ohm meter ($\Omega\,m$)). The membrane capacitance of biological membranes is about $1\ \mu F\,cm^{-2}$. But in the present situation it is useful to define the capacitance c_m of the axon membrane as the capacitance per unit length of axon (unit c_m farad per meter ($F\,m^{-1}$)).

To relate i_a and i_m we consider the axon element between x and $x + \Delta x$ at the time t. The axial currents are $i_a(x, t)$ and $i_a(x + \Delta x, t)$, while the mean value of the membrane current over the length Δx is $i_m(x', t)$, where $x \le x' \le x + \Delta x$. The current that leaves the membrane ring of length Δx is $i_m(x', t)\,\Delta x$, and as the charges are conserved* it follows that

$$i_a(x, t) = i_m(x', t)\,\Delta x + i_a(x + \Delta x, t).$$

Dividing on both sides by Δx and rearranging gives

$$i_m(x', t) = -\frac{i_a(x + \Delta x, t) - i_a(x, t)}{\Delta x}.$$

When $\Delta x \to 0$, $i_m(x', t) \to i_m(x, t)$, the right-hand side tends towards the partial derivative of $i_a(x, t)$ with respect to x. The expression for the charge conservation at the position x therefore becomes

$$i_m(x, t) = -\frac{\partial i_a(x, t)}{\partial x}. \tag{4.4.12}$$

Combining Eq. (4.4.7) and Eq. (4.4.8) we obtain

$$i_m(x, t) = -\frac{\partial}{\partial x}\left(-\frac{1}{r_a}\frac{\partial V(x, t)}{\partial x}\right) = \frac{1}{r_a}\frac{\partial^2 V(x, t)}{\partial x^2}. \tag{4.4.13}$$

Thus, the *membrane current i_m is proportional to the second derivative of the membrane potential $V(x, t)$ with respect to the distance x.*

* Also called Kirchoff's second law in veneration of Gustav Kirchoff (1824–1907), Professor of physics in Breslau 1850, in Heidelberg 1854 and from 1884 in Berlin. He discovered the laws for the distribution of electric currents in branched conductors and was the founder, together with Bunsen, of spectral analysis and formulated the laws for the ratio between radiation and absorption of light at a given temperature and wavelength. He also made important contributions to the mathematical formulation of wave propagation and of diffraction of light.

Having obtained relations for the dependence of the axial current i_a and membrane current i_m on the membrane potential $V(x, t)$, we are in a position to derive the basic differential equation that governs the evolution of the membrane potential $V(x, t)$ as a function of space and time. The presence of the membrane capacitance has the effect that a step in electrode current I_0 will not be felt fully just as the current is turned on. This is because the membrane capacitance has to be charged via the ohmic resistances. At any position x it will therefore take some *time* before the capacitances are fully charged and the membrane potential V has attained its stationary value. Therefore, membrane potential, axial current and membrane current are all functions of position x and time t.

We consider a volume element of the cable between x and $x + \Delta x$. Let $V(x', t)$ denote the mean value of the potential along Δx, where $x \leq x' \leq x + \Delta x$. In the time between t and $t + \Delta t$ an amount of charge

$$\Delta q_1 = i_a(x, t) \Delta t$$

flows into the volume at x, whereby $V(x, t)$ changes from the value

$$V(x', t) \quad \text{to} \quad V(x', t + \Delta t).$$

The membrane capacitance $\Delta x \, c_m$ of the axon segment carries the charges at t and $t + \Delta t$

$$q_c(t) = \Delta x \, c_m V(x', t) \quad \text{and} \quad q_c(t + \Delta t) = \Delta x \, c_m V(x', t + \Delta t),$$

respectively. Hence, the change in charge Δq_c during the time interval Δt is $\Delta q_c(t + \Delta t) - q_c(t)$, or

$$\Delta q_c = (V(x', t + \Delta t) - V(x', t)) \, c_m \, \Delta x.$$

In the same time interval some charge Δq_{leak} will leak across the membrane resistance r_m (membrane conductance g_m). The magnitude of this charge is

$$\Delta q_{leak} = g_m \Delta x \, V(x', t') \Delta t,$$

where $V(x', t')$ is the mean value of the potential during the time interval Δt with $t \leq t' \leq t + \Delta t$.

Finally, an amount of charge Δq_2 will leave the axon element at $x + \Delta x$ in time Δt; its magnitude is

$$\Delta q_2 = i_a(x + \Delta x, t) \Delta t.$$

Note that Δq_1, Δq_c, Δq_{leak} and Δq_2 are tied together according to the principle of charge conservation

$$\Delta q_1 = \Delta q_{leak} + \Delta q_c + \Delta q_2,$$

or

$$i_a(x, t) \, \Delta t = g_m \Delta x \, V(x', t') \, \Delta t + (V(x', t + \Delta t) - V(x', t)) \, c_m \, \Delta x$$
$$+ i_a(x + \Delta x, t) \, \Delta t.$$

Dividing both sides by $\Delta x \, \Delta t$ and rearranging gives

$$-\frac{i_a(x + \Delta x, t) - i_a(x, t)}{\Delta x} = c_m \frac{V(x', t + \Delta t) - V(x', t)}{\Delta t} + g_m V(x', t').$$

When $\Delta x \to 0$ and $\Delta t \to 0$, then $x' \to x$ and $t' \to t$. The left-hand side tends towards the partial derivative of $i_a(x, t)$ with respect to x. On the right-hand side the fraction tends towards the partial derivative of $V(x, t)$ with respect to t. Accordingly, we have

$$-\frac{\partial i_a(x, t)}{\partial t} = i_m(x, t) = g_m V(x, t) + c_m \frac{\partial V(x, t)}{\partial t} \qquad (4.4.14)$$

$$= i_m(x, t) = \frac{V(x, t)}{r_m} + c_m \frac{\partial V(x, t)}{\partial t}, \qquad (4.4.15)$$

where the right-hand side denotes the *total membrane current* per unit axon length – that is, the conductive leak current through the membrane resistance r_m *plus* the charging current i_c of the membrane capacitance c_m.

Eliminating i_m in Eq. (4.4.11) by means of Eq. (4.4.9) we finally have[*]

$$\frac{r_m}{r_a} \frac{\partial^2 V}{\partial x^2} = r_m c_m \frac{\partial V}{\partial t} + V. \qquad (4.4.16)$$

This is the one-dimensional cable equation for the case where the conductance of the external medium is so high, relative to that of the axoplasm, that it does not enter the derivations of Eq. (4.4.12). The fraction $r_m/r_a = \lambda^2$ and the product $r_m c_m = \tau$ have the dimensions[†] meter squared and second, and we can define two material constants that characterize the electrical behavior of the cable. The first material constant λ is defined as

$$\lambda = \sqrt{\frac{r_m}{r_a}}. \qquad (4.4.17)$$

It is called *the length constant* or *the characteristic length* of the cable for reasons to be explained. The other material constant is defined as

$$\tau = r_m c_m. \qquad (4.4.18)$$

---

[*] For typographical reasons $V(x, t)$ is replaced by V.
[†] Dimensions ohm m/(ohm m^{-1}) = m^2, and $r_m c_m$ = ohm m (farad m^{-1}) = ohm farad = volt ampere^{-1} coulomb volt^{-1} = (coulomb s^{-1})$^{-1}$ coulomb = s.

It is called the *time constant* of the membrane. Introducing these material constants Eq. (4.4.12) reads

$$\lambda^2 \frac{\partial^2 V}{\partial x^2} = \tau \frac{\partial V}{\partial t} + V, \qquad (4.4.19)$$

which is the usual form of the *time-dependent cable equation* that governs the passive spread of the electric potential change in time and space when an electric disturbance across the membrane is introduced at some place (e.g. at $x = 0$).

(ii) The stationary state
When the electrode current has been turned on long enough the capacitative charging process of the axonal membrane will cease, i.e. $\partial V/\partial t = 0$. When the latter condition is satisfied the polarization caused by the current injection has become *stationary*, meaning that V is a function only of x. Invoking the condition $\partial V/\partial t = 0$ in Eq. (4.4.15) leads to the equation for the stationary potential distribution

$$\lambda^2 \frac{d^2 V}{dx^2} - V = 0 \quad \text{stationary}, \qquad (4.4.20)$$

which is an ordinary second-order differential equation.

To determine the stationary potential profile $V(x)$, following a constant current injection I_0 at $x = 0$ beginning at $t = 0$ $x = 0$, one must obtain a solution of Eq. (4.4.16) that satisfies the two boundary conditions

$$V(x) \to 0 \quad \text{for} \quad x \to \infty, \qquad (4.4.\,B1)$$

and

$$i_a(x) = I_0 \quad \text{for} \quad x = 0. \qquad (4.4.\,B2)$$

According to Chapter 1, Section 1.7.2, an equation of the form of Eq. (4.4.16) that satisfies Eq. (4.4. B1) has the solution

$$V(x) = Be^{-x/\lambda}, \qquad (4.4.21)$$

where the constant κ has been replaced by $1/\lambda$. Next one must determine the value of B that satisfies Eq. (4.4. B2). Inserting Eq. (4.4.17) into Eq. (4.4.7) gives

$$i_a(x) = -\frac{1}{r_a}\frac{d}{dx}(Be^{-x/\lambda}) = -\frac{1}{r_a} \cdot B(-1)\frac{1}{\lambda}e^{-x/\lambda} = B\frac{1}{r_a\lambda}e^{-x/\lambda}.$$

Setting $x = 0$ and making use of Eq. (4.4. B2) yields

$$i_a(0) = I_0 = B\frac{1}{r_a\lambda} = B\frac{1}{r_a\sqrt{r_m/r_a}} = B\frac{1}{\sqrt{r_m r_a}}$$

or

$$B = I_0\sqrt{r_m r_a}.$$

Inserting this value of B into Eq. (4.4.17) gives the following equation for the stationary potential profile as a function of x

$$V(x) = I_0\sqrt{r_m r_a}\, e^{-x/\lambda} = V_0\, e^{-x/\lambda}. \qquad (4.4.22)$$

That is, the potential decreases *exponentially* as a function of the distance x from the current electrode. The potential $V(0) = V_0$ is given by

$$V_0 = I_0\sqrt{r_m r_a} = \lambda I_0 r_a. \qquad (4.4.23)$$

Thus, as visualized from the position of the electrode, one could disregard the whole cable and replace it by a resistance R^* of magnitude

$$R^* = \sqrt{r_m r_a}, \qquad (4.4.24)$$

which is the total resistance in the longitudinal direction presented by a long cable[†]. R^* is called the *characteristic resistance of the cable*.

Figure 4.46 shows an example of the membrane potential profile according to Eq. (4.4.22). The length constant $\lambda(= \sqrt{r_m/r_a}\,)$ is a measure for how far the membrane polarization spreads away from the current injection point: at the distance $x = \lambda$ the potential is $e^{-1} = 1/e \approx 0.37$ of the initial value V_0. For the axon of *Loligo*, $\lambda \approx 3.5\text{–}5$ mm[‡].

(iii) Time-dependent solutions

We calculate the spread of the membrane potential in space and time by solving the cable equation, Eq. (4.4.15), for the two situations where at the time $t = 0$ either an instantaneous voltage or a current is imposed by the electrode at $x = 0$.

To avoid mixing up the material constants when solving we make it *dimensionless*[§] by introducing the new variables

$$T = \frac{t}{\tau} \qquad (4.4.25)$$

[†] The cable length L should be large enough to make $e^{-L/\lambda} \approx 0$.
[‡] See Section 4.4.5.5.
[§] A step which both facilitates the writing and helps to avoid trivial errors of calculation.

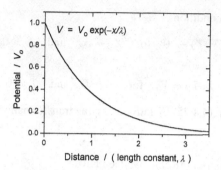

Fig. 4.46. Illustration of the length constant λ as the parameter that determines the extent of spread of membrane polarization following an injection of an axial current I_0 in the axoplasm in the position $x = 0$ Ordinate: membrane potential with the initial polarization V_0 as unit. Abscissa: distance from the current electrode in units of λ.

and

$$X = \frac{x}{\lambda}, \qquad (4.4.26)$$

i.e. time t and distance x are now expressed in units of the membrane time constant τ and the characteristic length of the cable λ. First

$$\frac{\partial V}{\partial t} = \frac{\partial V}{\partial T}\frac{dT}{dt} = \frac{1}{\tau}\frac{\partial V}{\partial T},$$

or

$$\tau\frac{\partial V}{\partial t} = \frac{\partial V}{\partial T}. \qquad (4.4.27)$$

Second

$$\frac{\partial V}{\partial x} = \frac{\partial V}{\partial X}\frac{dX}{dx} = \frac{1}{\lambda}\frac{\partial V}{\partial X},$$

or

$$\frac{\partial^2 V}{\partial x^2} = \frac{\partial}{\partial x}\left(\frac{1}{\lambda}\frac{\partial V}{\partial X}\right) = \frac{\partial}{\partial X}\left(\frac{1}{\lambda}\frac{\partial V}{\partial X}\right)\frac{\partial X}{\partial x} = \frac{1}{\lambda^2}\frac{\partial^2 V}{\partial X^2}. \qquad (4.4.28)$$

Inserting Eq. (4.4.23) and Eq. (4.4.24) in Eq. (4.4.16) transforms this equation to the following dimensionless form with respect to X and T

$$\frac{\partial^2 V}{\partial X^2} = \frac{\partial V}{\partial T} + V. \qquad (4.4.29)$$

We seek a solution for $V(X, T)$ in the domain $0 \leq X < \infty$ that satisfies the initial condition

$$V(X, T) = 0 \quad \text{for} \quad 0 \leq X < \infty \quad \text{and} \quad T = 0, \qquad (4.4.25\text{A})$$

and the boundary conditions

$$V(X, T) \to 0 \quad \text{for} \quad X \to \infty \quad \text{and} \quad T > 0, \qquad (4.4.25\text{B}1)$$

and

$$V(X, T) = V_0 \quad \text{for} \quad X = 0 \quad \text{and} \quad T > 0. \qquad (4.4.25\text{B}2)$$

We simplify Eq. (4.4.25) by introducing the transformation

$$V = e^{-T} W, \qquad (4.4.30)$$

which is due to Fourier (1822)*. We then have

$$\frac{\partial V}{\partial T} = -e^{-T} W + e^{-T} \frac{\partial W}{\partial T}, \quad \frac{\partial V}{\partial X} = e^{-T} \frac{\partial W}{\partial X}, \quad \frac{\partial^2 V}{\partial X^2} = e^{-T} \frac{\partial^2 W}{\partial X^2},$$

which, when inserted in Eq. (4.4.25), gives

$$e^{-T} \frac{\partial^2 W}{\partial X^2} = -e^{-T} W + e^{-T} \frac{\partial W}{\partial T} + V$$

$$= -e^{-T} W + e^{-T} \frac{\partial W}{\partial T} + e^{-T} W = e^{-T} \frac{\partial W}{\partial T}.$$

The transformation $V = e^{-T} W$ thus transforms the original equation Eq. (4.4.25) into a simpler one in W

$$\frac{\partial W}{\partial T} = \frac{\partial^2 W}{\partial X^2} \quad \text{for} \quad 0 \le X \le \infty \quad \text{and} \quad T > 0, \qquad (4.4.31)$$

which formally is identical to the diffusion equation Eq. (2.5.5), and we can take over results already obtained by solving Eq. (2.5.5).

The conditions that W must satisfy are found by inserting the original conditions Eq. (4.4.25A), Eq. (4.4.25 B1) and Eq. (4.4.25B2) in $V = e^{-T} W$. Thus, the initial condition becomes

$$W(X, T) = 0 \quad \text{for} \quad 0 \le X \le \infty \quad \text{and} \quad T = 0, \qquad (4.4.27\text{A})$$

and the boundary conditions become

$$W(X, T) \to 0 \quad \text{for} \quad X \to \infty \quad \text{and} \quad T > 0, \qquad (4.4.27\text{B}1)$$

and

$$W(X, T) = e^{T} V_0 \quad \text{for} \quad X = 0 \quad \text{and} \quad T > 0. \qquad (4.4.27\text{B}2)$$

---

* J.B.J. Fourier (1768–1830). His epoch-making work on heat conduction was published in 1822 in Paris. Fourier also distinguished himself as an algebraist, engineer and scholar on the history of Egypt. "Fourier's *Théorie analytique de la chaleur* is the Bible of the mathematical physicists. It contains not only an exposition of the trigonometrical series and integrals named after Fourier, but the general boundary value problem is treated in exemplary fashion for the typical case of heat conduction" (Arnold Sommerfeld, 1949).

Thus, the $V \to W$ transformation results in a simpler differential equation, but at the cost of more complicated boundary conditions. Only a comparison of the original set of equations with the transformed set will help decide which set will be the more advantageous. The solution of Eq. (4.4.31) with time-dependent boundary condition Eq. (4.4.31B2) is, according to Duhamel's theorem* (see Eq. (2.5.118)),

$$W(X, T) = \frac{X}{2\sqrt{\pi}} \int_0^T e^\tau \frac{1}{(T - \tau)^{3/2}} e^{-X^2/4(T-\tau)} \, d\tau. \quad (4.4.32A)$$

This integral is – after some algebraic manipulation[†] – brought into a form where it can be integrated in terms of known mathematical functions. The result is

$$W(X, T) = \tfrac{1}{2} V_0 \, e^T \left\{ e^{-X} \operatorname{Erfc}\left(\frac{X}{2\sqrt{T}} - \sqrt{T} \right) + e^X \operatorname{Erfc}\left(\frac{X}{2\sqrt{T}} + \sqrt{T} \right) \right\}.$$
$$(4.4.32B)$$

Making use of the transformation $V = e^{-T} W$ we then obtain the solution of the original problem in the form[‡]

$$V(X, T) = \tfrac{1}{2} V_0 \left\{ e^{-X} \operatorname{Erfc}\left(\frac{X}{2\sqrt{T}} - \sqrt{T} \right) + e^X \operatorname{Erfc}\left(\frac{X}{2\sqrt{T}} + \sqrt{T} \right) \right\},$$
$$(4.4.33)$$

which describes how the potential spreads in time and space, T, X from $X = 0$, where a voltage step of amplitude V_0 is impressed at the time $T = 0$.

Figure 4.47 shows how the potential spreads as a function of X for various values of T. With increasing T the potential invades further and further out, but more and more slowly until the potential profile becomes stationary for large values of T, where it will decrease exponentially with the distance from the electrode (cf. Eq. (4.4.22))

$$V(x) = V_0 e^{-x/\lambda} = V_0 e^{-X}.$$

This result also follows from Eq. (4.4.33) by letting $T \to \infty$

$$V(X, \infty) = \tfrac{1}{2} V_0 \left(e^{-X} \operatorname{Erfc}[-\infty] + e^X \operatorname{Erfc}[\infty] \right)$$
$$= \tfrac{1}{2} V_0 \left(e^{-X} \times 2 + e^X \times 0 \right) = V_0 e^{-X}.$$

---

* See Chapter 2, Section 2.5.5.3.
† See Appendix J.
‡ About the function $\operatorname{Erfc}(x)$, see Chapter 2, Section 2.5.5.1.

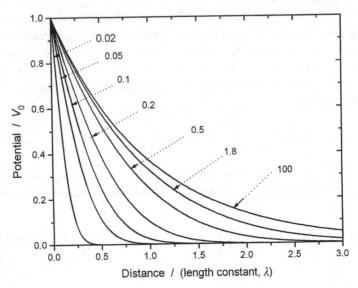

Fig. 4.47. Potential changes across the membrane in an infinitely long one-dimensional cable as the response to an impression of an instantaneous voltage step V_0 in the position $X = 0$ at time $T = 0$ (Eq. (4.4.33)). Ordinate: potential in units of V_0. Abscissa: distance along the cable in units of the length constant λ. The numbers attached to the curves give the time T in units of $\tau = r_m c_m$ after the turning on of the step.

Figure 4.48 shows the time course of $V(X, T)$ at various values of X, i.e. the time course of the potential profile and its growth towards the stationary value at a fixed position X. The time courses of the growth of $V(X, T)$ are not similar in shape for different values of X. The longer the distance is from the electrode position, the longer is the time needed for the potential to reach its stationary value. Experimentally, it is difficult to impress an instantaneous, constant voltage step V_0 across the membrane. It is easier to inject a constant current into the axoplasm. This can be done by inserting a microelectrode that is connected to a constant current generator, and the other pole is placed in the surrounding bath. If a current of magnitude $2 I_0$ is suddenly injected, a current I_0 will flow in each direction. Let $X = 0$ be the position of the microelectrode and let the current be turned on at $T = 0$. The membrane potential $V(X, T)$ caused by this current injected is found by solving the cable equation in X and T (Eq. (4.4.33))

$$\frac{\partial^2 V}{\partial X^2} = \frac{\partial V}{\partial T} + V,$$

in the domain $0 \leq X < \infty$ that satisfies the initial condition

$$V(X, T) = 0 \quad \text{for} \quad T = 0, \tag{A}$$

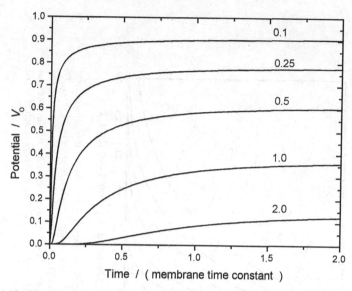

Fig. 4.48. The potential changes – calculated from Eq. (4.4.33) – across the membrane of an infinitely long cable in response to the application of the instantaneous voltage step V_0 across the membrane at $X = 0$ to the time $T = 0$. Ordinate: potential in units of V_0. Abscissa: time in units of membrane time constant τ. The number attached to the curves give distance along the cable in units of the length constant λ.

and the boundary conditions

$$V(X, T) \to 0 \quad \text{for} \quad X \to \infty \quad \text{and} \quad T > 0, \tag{B}$$

and

$$\left(\frac{\partial V}{\partial X}\right)_{X=0} = -\lambda r_a I_0 \quad \text{for} \quad X = 0 \quad \text{and} \quad t > 0, \tag{C}$$

as we have from Eq. (4.4.11) that

$$I_0 = i_a(0) = -\frac{1}{r_a}\left(\frac{\partial V}{\partial x}\right)_{x=0} = -\frac{1}{r_a}\left(\frac{\partial V}{\partial X}\frac{dX}{dx}\right)_{X=0}$$
$$= -\frac{1}{\lambda r_a}\left(\frac{\partial V}{\partial X}\right)_{X=0}.$$

The solution of this problem can be obtained from Eq. (4.4.33) already found*, and is

$$V(X, T) = \tfrac{1}{2}\lambda r_a I_0 \left\{ e^{-X} \operatorname{Erfc}\left(\frac{X}{2\sqrt{T}} - \sqrt{T}\right) - e^{X} \operatorname{Erfc}\left(\frac{X}{2\sqrt{T}} + \sqrt{T}\right) \right\},$$

$$\tag{4.4.34}$$

* See Appendix K.

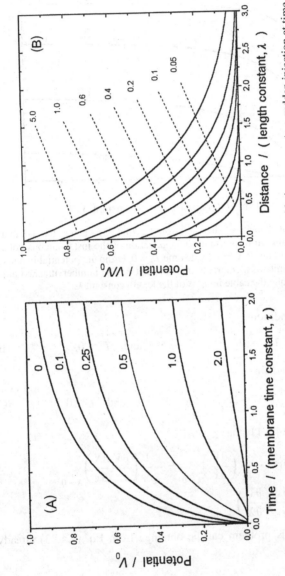

Fig. 4.49. Potential changes across the membrane of an infinitely long cable in response to a sudden injection at time $T = 0$ of a constant axial current I_0 at $X = 0$ – calculated from Eq. (4.4.34). Ordinate: potential in units of $V_0 = \lambda r_a I_0$. (A) Potential as a function of time T in units of membrane time constant τ. The numbers attached to the curves give the distance X in units of the length constant λ from the plane of current injection. Note: It takes some time for the potential to attain its stationary value even at $X = 0$. (B) Potential as a function of the distance X. The numbers attached to the curves indicate the time after the start of injection.

498

which has been derived among others by Hodgkin & Rushton (1946)*, and by Davis & Lorente de No (1947)[†]. Formally, Eq. (4.4.34) differs from Eq. (4.4.33) by having a minus between the two terms inside the curly brackets, which is a reflection of the finite charging time of the axon membrane.

Figure 4.49(A) shows the time course of the growth of the potential at different distances from the point of current injection at $X = 0$. A comparison with Fig. 4.48, where an instantaneous voltage step V_0 was impressed, shows that the potential grows somewhat *more slowly* towards the stationary value. This delay occurs because with the constant current injection, the membrane capacitance adjacent to the current electrode will be charged at a finite rate[‡]. In contrast, the application of an instantaneous voltage step at $X = 0$ as in Fig. 4.48 requires an instantaneous charging of the membrane capacitance at $X = 0$, i.e. a current surge of infinite strength. Consequently, the time to establish the potential gradient that drives a longitudinal current through the axoplasm will be shorter than when charging occurs with a finite strength. This gradual rise of the potential even at the plane of current injection caused by the time lag in charging the membrane capacitance appears even more clearly from the curves of Fig. 4.49(B).

It may be of use to recall the range of application of the solutions Eq. (4.4.22), Eq. (4.4.33) and Eq. (4.4.34) that were obtained by integration of the cable equation that satisfied the boundary conditions at $X = 0$ and $X = \infty$, therefore making no allowance for including position coordinates $X < 0$. In other words, the cable behaved as a *semi-infinite cable* stretching from $X = 0$ towards infinity along the positive direction but not permitting any oppositely directed current flow from the origin. In practice, however, the application of a potential or a current source results in currents that nearly always flow in both directions from the source. In the simplest case a current of magnitude $2I_0$ is injected into a cable structure of sufficient length to be regarded as infinite. For reasons of physical symmetry this current will divide into two equal parts of oppositely directed currents of magnitude I_0. The stationary potential profile in the positive direction will then be described by the solution Eq. (4.4.22)

$$V(x) = \sqrt{r_a r_m}\, I_0 \exp\left[-\frac{x}{\lambda}\right], \quad \text{for} \quad x \geq 0.$$

The current that flows in the negative direction follows the same pattern and, therefore, the potential profiles caused by the two oppositely directed current

---

* Hodgkin, A.L. & Rushton, W.A.H. (1946): *Proc. R. Soc. Lond.*, **B133**, 444, by using a method based upon Heaviside's's operator method for solving partial differential equations.

[†] Readers familiar with the use of integral transforms to solve partial differential equations would probably prefer to solve the two boundary problems above by applying a Laplace transform or a Fourier transform to the cable equation and the initial and boundary conditions.

[‡] See Appendix I, Eq. (I3) and Appendix J, Eq. (J6).

flows will decline in exactly the same way with the distance from the current electrode. Thus, viewed as a whole, each of the the two profiles can be regarded as the *mirror image* of the other profile with respect to the y-axis. The mirror image can be obtained from the above expression by changing the variable x to its absolute (numerical) value $|x|$, i.e.

$$V(x) = \sqrt{r_a r_m} I_0 \exp\left[-\frac{|x|}{\lambda}\right], \quad \text{for} \quad -\infty < x < \infty,$$

which works equally well in the positive domain of the cable. The same trick applies equally well to create the mirror image of the potential profile of Eq. (4.4.34). In Appendix L we illustrate the use of mirror images to handle a cable of a finite length, i.e. comparable to that of its characteristic length.

4.4.5.3 Reconstruction of the action potential

On the basis of the time and voltage dependence obtained by Hodgkin & Huxley we previously described quantitatively the events that led to a membrane action potential. However plausible such an account may have appeared, it involved a complicated interaction between several processes, and it is in no way certain that these interactions will always result in stable transient behavior such as that shown in Fig. 4.37 and, in particular, leading to a propagated action potential. This problem cannot be resolved by using more plausible reasoning, but requires a mathematical model to handle the interconnections between the variables. The starting point for setting up such a model is the cable equation Eq. (4.4.16) which is now written as

$$\frac{1}{r_a}\frac{\partial^2 V}{\partial x^2} = c_m \frac{\partial V}{\partial t} + \frac{V}{r_m} = c_m \frac{\partial V}{\partial t} + i_{Na} + i_K + i_L, \qquad (4.4.35)$$

where the last term on the right-hand side contains the actual ionic current of sodium, potassium and a leak current i_L – predominantly due to chloride – that flows through the membrane during the action potential. The currents are expressed in units of ampere per unit length of the axon. It is preferable to express these currents in ampere per unit area of the membrane. The membrane area A of one unit length of axon of radius a is $A = 2\pi a \cdot 1$. Hence $i_{Na}/2\pi a$ is the sodium current I_{Na} per unit area, etc. Likewise $c_m/2\pi a$ is the membrane capacitance C per unit area. Dividing Eq. (4.4.35) by $2\pi a$ yields

$$\frac{1}{2\pi a\, r_a}\frac{\partial^2 V}{\partial x^2} = \frac{c_m}{2\pi a}\frac{\partial V}{\partial t} + \frac{i_{Na}}{2\pi a} + \frac{i_K}{2\pi a} + \frac{i_L}{2\pi a}. \qquad (4.4.36)$$

The resistance r_a of the axoplasm can be written as $r_a = \rho_a/\pi a^2$, where ρ_a is the resistivity of the axoplasm*. Thus, $1/2\pi a\, r_a = a/2\rho_a$, and Eq. (4.4.31) takes the alternative form

$$\frac{a}{2\rho_a}\frac{\partial^2 V}{\partial x^2} = C\frac{\partial V}{\partial t} + I_{Na} + I_K + I_L, \qquad (4.4.37)$$

Equation (4.4.37) is of little use for further developments unless the dependence of V and t in I_{Na} and I_K is known explicitly. For that reason Hodgkin & Huxley, 1952[†], developed a set of empirical relations that reproduced with great precision their voltage clamp data describing the time courses of the changes in sodium and potassium at different membrane potentials V (Fig. 4.32). The equations resulted from the development of physical models to account for the behavior of the sodium and potassium channels. However, the real issue was not the reality of these physical models but whether the empirical relations being built into a membrane model like the infinite cable just described could reproduce an action potential of normal appearance.

(i) The Hodgkin–Huxley equations
The equations for the membrane currents

$$\left.\begin{aligned} I_{Na} &= G_{Na}\left[V - V_{Na}^{(eq)}\right] \\ I_K &= G_K\left[V - V_K^{(eq)}\right] \\ I_L &= G_L\left[V - V_L^{(eq)}\right] \end{aligned}\right\} \qquad (4.4.38)$$

include in addition the *leak* current I_L, which corresponds to currents other than sodium and potassium – e.g. chloride ions – whose conductance G_L is regarded as constant, whereas both G_{Na} and G_K depend on voltage and time.

The voltage and time dependence of G_K is represented by

$$G_K(V, t) = \overline{G}_K\, n^4, \qquad (4.4.39)$$

where $\overline{G}_K$ is the maximum value of G_K – presumably corresponding to the state with all K^+ channels open. The expression $n(V, t)$ is an activation function that is governed by

$$\frac{dn}{dt} = \alpha_n(1 - n) - \beta_n n. \qquad (4.4.40)$$

The expressions $\alpha_n(V)$ and $\beta_n(V)$ are rate constants depending upon V – dimension $(ms)^{-1}$ – where their particular dependence and the value of $\overline{G}_K$ are

---

* See Appendix K.
[†] Hodgkin, A.L & Huxley, A.F. (1952): A quantitative description of membrane current and its application to conduction and excitation in nerve. *J. Physiol.*, **117**, 500.

adjusted empirically to make Eq. (4.4.39) correctly reproduce the experimental data obtained from the squid axon*.

The sodium conductance is written[†]

$$G_{Na}(V, t) = G_{Na}^{on} h = \overline{G}_{Na} m^3 h,   \tag{4.4.41}$$

where $\overline{G}_{Na}$ is the maximum value of G_{Na}, m is the activation function, and h is the inactivation function. The values of m and h are given by

$$\frac{dm}{dt} = \alpha_m(1 - m) - \beta_m m,   \tag{4.4.42}$$

and

$$\frac{dh}{dt} = \alpha_h(1 - m) - \beta_h h,   \tag{4.4.43}$$

where α_m, β_m and α_h, β_h are the corresponding rate constants that also depend on the membrane potential V.

The set of equations Eq. (4.4.39)–Eq. (4.4.43) are the famous Hodgkin–Huxley equations that make up the basic elements in the reconstruction of the propagated action potential. Inserting Eq. (4.4.38) into Eq. (4.4.37) gives

$$\frac{a}{2\rho_a} \frac{\partial^2 V}{\partial x^2} = C\frac{\partial V}{\partial t} + G_{Na}\left[V - V_{Na}^{(eq)}\right] + G_K\left[V - V_K^{(eq)}\right] + G_L\left[V - V_L^{(eq)}\right].$$
$$\tag{4.4.44}$$

To simulate the behavior of the axon membrane from the squid the equations for G_{Na}, G_K and G_L above are inserted in Eq. (4.4.44) giving

$$\frac{a}{2\rho_a} \frac{\partial^2 V}{\partial x^2} = C\frac{\partial V}{\partial t} + \overline{G}_{Na}\, m(V, t)^3\, h(V, t)\left[V - V_{Na}^{(eq)}\right]$$
$$+ \overline{G}_K\, n(V, t)^4\left[V - V_K^{(eq)}\right] + G_L\left[V - V_L^{(eq)}\right],   \tag{4.4.45}$$

sometimes called the Hodgkin–Huxley modified cable equation. The solution gives the spread of $V(x, t)$ along the excitable membrane in response to an initial disturbance as a function of the variables x and t. The familiar configuration of $V(x, t)$ is the time course of the action potential that is recorded from a fixed point on the axon. To study the spread in *time alone* of a propagated action potential, the above partial differential equation can be transformed to an ordinary differential equation, using the fact that the action potential propagates along the x-axis with a constant velocity θ and with the *same* configuration. Let $V = F(x)$ represent the action potential *profile* along the distance x of the axon

---

* See Section 4.7.
[†] See Eq. (4.4.6).

at some time, e.g. $t = 0$ measured from a fixed point x_0. At a later time t this profile will be identical with that at $t = 0$ except that the profile has moved a distance θt in the positive direction of the x-axis. Taking the point at $x = \theta t$ as the new origin and denoting the distances measured from this origin as X so that $x = X + \theta t$, then the equation of the action potential profile referred to this new origin will be

$$V = F(X),$$

or, referred instead to the original fixed origin x_0,

$$V = F(x - \theta t).$$

Thus, the solution of the propagated action $V(x, t)$ depends *only* on a single variable $u = x - \theta t$, namely

$$V(x, t) = V(x - \theta t).$$

It follows that

$$\frac{\partial V}{\partial x} = \frac{\partial V}{\partial t}\frac{\partial t}{\partial x} = \frac{\partial V}{\partial t}\left(-\frac{1}{\theta}\right),$$

and

$$\frac{\partial^2 V}{\partial x^2} = \frac{\partial}{\partial x}\left(-\frac{\partial V}{\partial t}\frac{1}{\theta}\right) = \frac{\partial}{\partial t}\left(-\frac{\partial V}{\partial t}\frac{1}{\theta}\right)\frac{\partial t}{\partial x} = -\frac{\partial^2 V}{\partial t^2}\frac{1}{\theta}\left(-\frac{1}{\theta}\right) = \frac{1}{\theta^2}\frac{\partial^2 V}{\partial t^2}.$$

This implies that Eq. (4.4.45) can be written as the following ordinary differential equation

$$\frac{a}{2\rho_a\theta^2}\frac{d^2V}{dt^2} = C\frac{dV}{dt} + \overline{G}_{Na}\, m(V, t)^3\, h(V, t)\left[V - V_{Na}^{(eq)}\right]$$

$$+ \overline{G}_K\, n(V, t)^4\left[V - V_K^{(eq)}\right] + G_L\left[V - V_L^{(eq)}\right]. \quad (4.4.46)$$

This equation is highly nonlinear as the functions m, n and h are complex functions of V. To find $V(t)$ one has to solve the connected set of equations Eq. (4.4.40), Eq. (4.4.42), Eq. (4.4.43) and Eq. (4.4.44). Viewed mathematically this is a terribly complicated demand, and explicit solutions of $V(t)$ are obtainable only by use of numerical methods*.

(ii) The propagated action potential

To solve the propagation of the action potential for a particular axon the actual numerical values of a, ρ_a and θ must be used in Eq. (4.4.46). However, the value

---

* In Section 4.7.4 the five coupled first-order equations are written in the order to be handled by a computer algorithm.

504

4. The nerve impulse

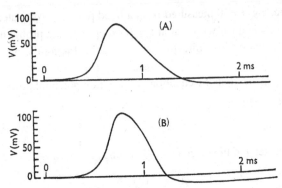

Fig. 4.50. Propagated action potentials: theoretical model (A) and squid axon at 18.5 °C (B). The calculated velocity was 18.8 m s⁻¹ and the experimental velocity was 21.2 m s⁻¹. (From Hodgkin & Huxley, 1952.)

of θ is unknown and must be guessed. In carrying out the solution procedure Hodgkin & Huxley found that the choice of the value of θ gave rise to stability problems as $V \to \pm\infty$ according to whether θ is chosen too large or too small. The unique value of θ that resulted in a stable solution by bringing the value of V back to the resting value after displaying a full-blown action potential was taken as the velocity of propagation of that particular axon.

The plot of such a calculation of $V(t)$ is shown in Fig. 4.50(A) with $a = 0.238$ mm, $\rho_a = 35.4$ ohm cm and $\theta = 18.8$ m s⁻¹, together with the record of the action potential from the appropriate squid axon, Fig. 4.50(B). The form of the reconstructed action potential agrees well with the experimentally determined action potential with a conduction velocity of 21.2 m s⁻¹, as did the calculated total Na⁺ and K⁺ currents with those determined experimentally. Thus, the result of the reconstruction must – the possible uncertainties taken into account – be considered as a highly satisfactory support of the, indeed revolutionary, theory of the impulse conduction. The time course of the Na⁺ and K⁺ conductances – which are the backbone of the impulse propagation – together with the propagating action potential are shown in Fig. 4.51. These data will also serve to give a quantitative description of the sequence of events that enter into the propagation of the action potential, which is subject to less guesswork than that we gave in the description of the membrane action potential. The impulse in Fig. 4.51 advances in the direction from right to left. The first part of the depolarization (also called the "foot" of the action potential) is caused by the local currents that flow from the active region of the axon – i.e. the membrane patch with the high sodium conductance that causes a strong inward current of Na⁺ ions to flow in the adjacent regions and by crossing the

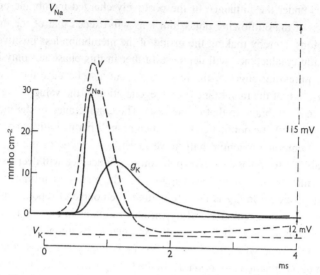

Fig. 4.51. Theoretical solution for propagated action potential and conductances at 18.6 °C. Total entry of sodium $= 4.33$ pmole cm^{-2}; total exit of potassium $= 4.26$ pmole cm^{-2}. (From Hodgkin & Huxley, 1952.)

membrane returning via the external fluid. Owing to the rapid change of V with respect to t this current is predominantly capacitative. The depolarization in the region just ahead will exceed the threshold depolarization of the membrane and cause an increase in the Na$^+$ conductance, and Na$^+$ ions move into the axoplasm in a regenerative manner, partly depolarizing the membrane further and partly supplying current to activate the next region. The sodium conductance increases rapidly and tends to approach the membrane potential to the equilibrium potential $V_{Na}^{(eq)}$ of sodium. However, before the reaching the action potential the depolarization starts the *slower* membrane processes: the *inactivation process* rises to the level where it will suppress any further increase in the sodium conductance, afterwards causing G_{Na} to decline. The potassium ions – now at a membrane potential far from their equilibrium potential $V_K^{(eq)}$ – will move out from the axoplasm to the external fluid at an increased rate, which is enhanced by the slower *rise in potassium conductance* following the depolarization. At the peak of the action potential ($dV/dt = 0$) no charge transfer takes place across the membrane capacity, and the sum of the ionic current will likewise be zero. As the conductances for sodium and potassium are now much larger than the chloride conductance, the currents carried by sodium and potassium are of equal magnitude but oppositely directed. As the sodium conductance is still being suppressed since h still increases and the potassium conductance still rises under the influence of the positively charged membrane potential, the speed of the outflowing potassium ions will exceed that of the inflowing sodium ions, thereby making the inside of the membrane less positive. Thus, the sodium conductance will decrease further in this phase not only because of the suppressing effect of the inactivation but also because the developing repolarization of the membrane (i.e. a regeneration of the value of the electric field in the membrane to that of the rest). The conductance of the membrane will again chiefly be dominated by the membrane's permeability to potassium, and the membrane potential will move towards the value of the equilibrium potential $V_K^{(eq)}$ for potassium. The potassium conductance will decrease in step with the membrane potential as it approaches the resting membrane potential, but if the decrease in G_K is lagging behind, the course of repolarization may even display a transient phase of hyperpolarization as in Fig. 4.11, where the axon's membrane potential has finally returned to its initial state after about 5 ms. At the time where the increase in G_K is finally put out a considerable amount of inactivation still remains. In the falling phase of the action potential the sodium conductance is increasingly suppressed by the inactivation, whereas the potassium conductance is still high. Therefore, in this phase there will be a state where the application of not even the strongest additional depolarization is able to reactivate the Na$^+$ channels to a conductance increase. In this phase the axon is *absolutely refractory*. As the inactivation is fading away late in the falling phase a transitional situation arises where the sodium mechanism of activation G_{Na}^{on} will reoperate, although with less intensity. As the membrane conductance is still elevated a greater membrane current is required to bring the sodium mechanism into the regenerative state. In this phase the axon is *relatively refractory*.

An electric stimulus applied to the axon away from the ends will start impulses that move away from the point of stimulation as illustrated in Fig. 4.44, because local current loops spread from the point of activation into regions having intact* sodium conductance. However, this does not apply to the current loops that belong to the "tail" of the propagating action potentials. This region is refractory since the sodium conductance is suppressed by the inactivation process.

4.4.5.4 Nerve impulse recorded with external electrodes

With the axon immersed in a large volume of external fluid no restrictions will be imposed on the spatial distribution of the returning parts of the local current loops of the propagating action potential. The calculation of the exact spatial distribution of the external current loops in an extended volume conductor is a

---

* That is, Na$^+$ channels without inactivation. Compare Section 4.4.4, Fig. 4.29, trace A.

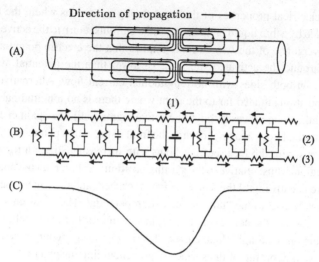

Fig. 4.52. The origin of the monophasic action potential. (A) A snapshot of the distribution along the length of the axon the local currents that are associated with the propagation of the action potential. (B) Electrical model to simulate the passive electric components of the membrane. The indicated current flow originates from a single battery with a polarity that imitates the Na^+ entry at the active region. (C) Potential variations in the thin surface layer surrounding the axon measured relative to a distant position on the axon.

difficult undertaking*, requiring use of mathematical techniques that are beyond the scope of this book. However, the current density in the external fluid will be small and as the electric resistance is small the potential variations in the external medium due to the local currents are negligible compared with the voltage swing of the action potential of about 130 mV across the axon's membrane. Thus, for all practical purposes, the axon can be regarded as surrounded by a short-circuiting layer, as we did in deriving the cable equations (Section 4.4.5.2).

It appears from Fig. 4.1 and Fig. 4.2 that it is possible in practice to record the small potential variations on the surface of the nerve during the impulse propagation. To increase the potential variations that are associated with the current loops, the resistance of the fluid in which the current flows is increased by reducing the external fluid to only a thin liquid layer, into which the external currents are now forced to flow. As the resistance of the external fluid can be made fairly high, the currents may now give rise to potential variations of the order of several millivolts[†]. A sketch of the local current loops during the action potential is shown in Fig. 4.52(A) together with a model (B) that illustrates the

---

* Clark, J. & Plonsey, R. (1968): The extracellular potential field of the single nerve fiber in a volume conductor. *Biophys. J.*, **8**, 842. Rosenfalck, P. (1969): *Intra- and Extracellular Potential Fields of Active Nerve and Muscle Fibers*, Akademisk Forlag, Copenhagen.
[†] A quantitative treatment is given in Appendix L.

passive electrical properties of the axon during conditions where the axon is surrounded by a thin liquid layer in which the currents return to the active region. The convergence of the two currents implies that the electric potential at the surface around the activated site must be lower than the potential at distant positions on both sides with no longitudinal current flow. Alternatively, one takes a position situated far to the right where there is no longitudinal current flow and then one moves through the active region until one again encounters no longitudinal flow. Moving towards the front of the action potential one enters the region where the local current flows with increasing strength towards the active region. Thus, relative to the starting position, the potential becomes more and more negative until the active region is reached, and one enters the current loops that belong to the "tail" of the action potential. These currents are also directed towards the active region and, in moving further to the left, potential rises again until a longitudinal current flow is no longer encountered. A sketch of the potential profile of the surface region encircling the external flow of the local currents is shown in Fig. 4.52(C).

The decisive role of the local currents for the impulse propagation was demonstrated experimentally in 1937 by A.L. Hodgkin*, by studying the effect of blocking the impulse propagation over a small, but well-defined length of segment of the nerve. Local cooling or local anaesthetics were used to produce the block. This treatment did not destroy the cable properties in the region of the block. Therefore, when the nerve impulse was stopped at the proximal end of the block and local currents were flowing in front of the impulse one should expect these currents to be transmitted passively into – and possible past – the blocked region. The effect of forwarding propagating impulses towards the proximal end of the block and recording at various distances from the distal end of the block is shown in Fig. 4.53. It can be seen that the result is a non-propagated signal of more or less the shape of the propagating monophasic signal that went into extinction at the block. However, the signal that is recorded at the border to the block is a good deal smaller than the propagating signal and its size attenuates in an approximately exponential manner with distance from the block. This observation agrees well with the notion of an electrical signal being transmitted passively through the region of block in a cable-like structure. The same type of recordings were also obtained when the nerve signal was replaced by an artificial electrical signal that simulated the propagated signal both in shape and size and was applied to the proximal end of the block. An example is shown in the right-hand column of Fig. 4.53. The two sets of traces agree well with respect to form and size, but to this end the size of the artificial signal had to be about 50% larger than that of the nerve impulse. However,

* Hodgkin, A.L. (1937): Evidence for electrical transmission in nerve. Parts i and ii. *J. Physiol.*, **90**, 183 and 212.

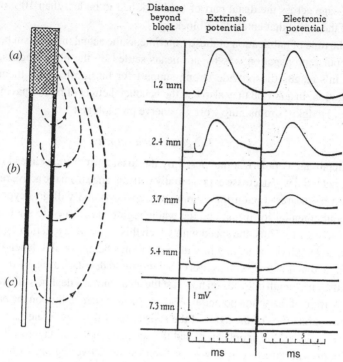

Fig. 4.53. On the left side is a sketch of the local current loops that appears beyond a complete block (c) when a nerve impulse (a) enters the proximal part of the blocked region (b) and goes into extinction. The middle column: recordings obtained beyond the block at distances (in mm) given by the numbers attached to the traces. The right-hand column: recordings obtained from the same positions in response to application of an artificial signal that simulates the nerve signal in shape and duration. Note: in the lowest recording the amplification is increased by a factor of 10×. (Hodgkin, 1937.)

taking into account the different origins of the current, the exact duplication of the two types of signal can hardly be expected. Hodgkin also showed that there was a transient increase in the electrical excitability of the region just beyond the block that followed immediately in continuation of the propagating impulse entering into the block. Furthermore, both the excitability and potential change had the same spatial and temporal configuration. Thus, provided the blocked region is not too long, the impulse is capable of transmitting currents passively through and past the blocked region that exerts weak stimulating actions on the nerve beyond the block. Finally, reducing the length of the blocked region revealed a critical length where the propagation impulse, on entering the block, simply jumped across the whole length of the block and continued to travel as a propagating impulse in the unaffected normal nerve beyond the block. It turned out that to obtain a block, the size of the passively transmitted signal at the passage across the distal part of the block had to be less than 10% of the size of the signal that entered the block.

At the time of publication of these experiments the actual mechanism behind the impulse propagation was by no means settled by the physiologists. But Hodgkin's results did provide strong support for the local circuit theory of propagation. In addition they also had the beneficial effect of finally paving the way for Hodgkin's remaining work as a nerve physiologist.

4.4.5.5 The conduction velocity

The membrane depolarization caused by the local current loops ahead of the action potential will decrease exponentially with distance from the active region. In the very nearest region the membrane depolarization will go beyond the threshold potential for firing, and the whole region is activated to elicit a new site of activation. Thus, the speed with which this "wave of excitation" spreads along the length of the axon will – all other things being equal – depend upon the length that is activated in excess of the threshold depolarization by the local currents. This length depends on how fast the local currents decline with the distance in front of the action potential. And this distance depends upon the passive electrical properties of the axon: if the resistance of the membrane r_m is large relative to that of the axoplasm r_a and the surrounding liquid layer r_o the current loops will spread a long way in front of the active region. Conversely, a small value of r_m relative to r_a and r_o has the effect of short circuiting the membrane, and the spreading of the loops will be short. A quantitative measure for the extent of the passive currents is given by the characteristic length λ

$$\lambda = \sqrt{\frac{r_m}{r_a}} \qquad (4.4.47)$$

which is the parameter for the exponential decline with distance from a membrane potential perturbation of some kind, e.g. the active region of the nerve impulse*. The extent of the region that is depolarized beyond the threshold by the local currents is proportional to λ. Therefore, the conduction velocity is also proportional to λ. The magnitude of λ depends upon the diameter d of the axon, as r_a and r_m have different dependences on d. Let ρ_a denote the resistivity of the axoplasm (unit: Ω m) and R_m be the resistance of one unit area of the axon membrane (unit: Ω m^2). We have then

$$r_a = \frac{\rho_a}{(\pi/4)d^2} \quad \text{and} \quad r_m = \frac{R_m}{\pi d},$$

---

* See Section 4.4.5.2, Eq. (4.4.13) and Eq. (4.4.17).

where the numerator in the expression for r_a is the cross-sectional area of the axon and the numerator in the expression for r_m is the membrane area per unit length of an axon with diameter d. Inserting the expressions above for r_a and r_m in Eq. (4.4.43) gives

$$\lambda = \frac{1}{2} \sqrt{\frac{\rho_m}{\rho_a}} \sqrt{d}. \tag{4.4.48}$$

It follows that λ is proportional to the square root of the axon diameter d, which again implies that one should expect the conduction velocity to be *proportional to the square root* of the axon diameter. Another factor of importance for the conduction velocity is the intensity and rate of the increase with time of the inwardly-directed sodium current at the site of activation. The greater the intensity of switching on the sodium currents, the greater will be the strength of the local currents starting off from the activated region and therefore the extent of the adjacent membrane region that is depolarized beyond the threshold.

Hodgkin (1939)* showed that an increase of the resistance per unit length r_0 of the surrounding fluid caused a reduction of the conduction velocity. Qualitatively this is in keeping with the above consideration: when the external resistance r_0 cannot be disregarded relatively to r_a it contributes to the length constant λ as r_a, i.e.

$$\lambda = \sqrt{\frac{r_m}{r_a + r_0}}.$$

It appears that an increase in r_0 makes λ diminish and through that leads to a decrease in conduction velocity. The connections between the factors that determine the conduction are treated quantitatively by Hodgkin (1954)[†].

4.5 Myelinated nerves

In unmyelinated nerves the nerve impulse propagates along the nerve as a *continuous* process, because local currents from the active region shift the site of activation to a small region in the immediate vicinity, from which new local currents in turn will flow to push forward the region of activation. This process has sometimes been compared with the motion of a burning fuse, where the temperature rise at one point causes the ignition of the powder at the adjacent portion of the fuse. Naturally, the comparison ends there, as the nerve – in contrast to a fuse – regenerates within a few milliseconds.

---

[*] Hodgkin, A.L. (1939): The relation between conduction velocity and electrical resistance outside the nerve fibre. *J. Physiol.*, **94**, 560.
[†] Hodgkin, A.L. (1954): A note on conduction velocity. *J. Physiol.*, **148**, 127.

In myelinated nerves the Schwann cells have developed a specialized structural arrangement by winding 10–20 times round the axon. As a consequence of this process very little cytoplasm remains, leaving a spirally coiled structure of layers, where each layer consists of two cell membranes packed closely together. Thereby, the axon membrane is surrounded by a structure – the *myelin sheath* – that from an electric point of view represents 20–40 membrane resistances and membrane capacitors that are arranged in series. This means that the myelin sheath offers a much higher resistance and smaller capacitance than the corresponding values of the naked axon membrane. Thus, in a frog nerve one finds values for r_m of 160 000 Ω cm and C_m of 0.0025 μF cm^{-2}. The myelin sheath is discontinued at each 1–2 mm length, making the short gaps called the *nodes of Ranvier*. Thus, the axon of myelinated nerve is surrounded by an effective isolator caused by the myelin sheet, except at the nodes of Ranvier, where the external fluid makes direct contact with the external surface membrane of the axon, where one finds values of 20 Ω cm^2 for the membrane resistance and 3 μF cm^{-2} for the membrane capacitance. Therefore, a current injected into the axoplasm of a myelinated nerve will on the whole leave the nerve at the location of the nodes of Ranvier. The consequence of this structural arrangement is that the complex of processes belonging to the impulse propagation need not proceed by involving every length segment of the axon, but can be confined only to the regions around the nodes of Ranvier. Owing to the large radial resistance of the myelin sheath the local current generated from *one* active node will flow forwards with almost no radial leak until it reaches the position of the low membrane resistance part of the *next* node of Ranvier, providing an easier return path for part of the current loop. Having sufficient strength to depolarize the axon membrane of the node above the threshold, the outwardly-directed passive current will be replaced by the inwardly-directed sodium current that charges the inside positively and thereby creates a new local current loop. Thus, the impulse conduction is discontinuous or *saltatory*, i.e. jumping from the one Ranvier node to the next (see Fig. 4.54).

The first experimental demonstration of the saltatory impulse conduction in myelinated nerves from a frog was given by Tasaki & Takeuchi (1941, 1942)* by measuring the radial current at different points along an isolated myelinated fiber from the frog. With saltatory conduction, the sequence of current

---

* Tasaki, I. & Takeuchi, T. (1941): Der am Ranvierschen Knoten entstehenden Aktionsstrom und seine Bedeutung für die Erregungsleitung. *Pflüg. Arch.*, **244**, 696. Tasaki, I. & Takeuchi, T. (1942): Weitere Studien über den Aktionsstrom mark haltigen Nervenfaser und über die elektros-altatorische Übertragung des Nervenimpulses. *Pflüg. Arch.*, **244**, 696. The publication of these papers in a German journal during World War II made them unavailable to English-speaking physiologists for some time. It goes without saying that for a time the results gave rise to sometimes overheated debates among specialists.

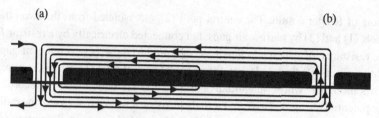

Fig. 4.54. Illustration of the local current flow in a myelinated nerve. At the node on the left (a) the membrane is activated and open for the inward sodium current charging the underlying axoplasm to a positive value. Owing to the large resistance of the myelin sheath and small capacitance, most of the axial current will reach the next node of Ranvier (b), where part of it will depolarize the axon membrane in crossing it in the direction inside → outside, and return to the activated node via the cap and the external fluid.

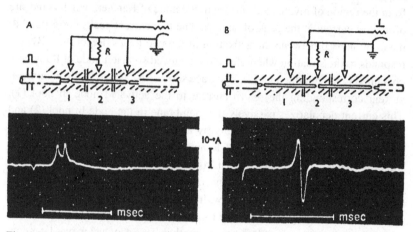

Fig. 4.55. Diagram for recording radial currents drawn from the surface of the isolated myelinated fiber placed in three pools of Ringer's fluid (1) (2) and (3) separated by two air gaps. A three-polar electrode set is used with recording of the current flow through the central low-resistance electrode to be placed in the middle pool. (A) Currents drawn from a 1 mm long myelinated region of internode. (B) Currents drawn with the node of Ranvier placed in the central pool. Outward currents upwards. (From Tasaki, 1959*.)

flows representing a membrane depolarization following repolarization that characterizes the excitable membrane should be observed only at the node of Ranvier. A diagram of the recording arrangement that was designed to measure the radial current that could be drawn from the isolated fiber during impulse conduction is shown in Fig. 4.55, together with the decisive recordings. A single myelinated fiber containing at least three nodes was placed in three miniature

* Tasaki, I. (1959): Conduction of the nerve impulse. Chapter 3 in *Handbook of Physiology, Section I Neurophysiology, Vol. I*, Williams & Williams, Baltimore.

pools of Ringer's fluid. The central pool (2) was isolated from the two other pools (1) and (3) by narrow air gaps, but connected electrically by a resistor R, the resistance of which was small as compared to those between the air gaps. With a potential of pool (2) that differs from those of pool (1) and (3), a current flows though R whose magnitude and direction are determined by recording the potential changes across R. Thus, the resistor R serves to measure the radial current flow from the isolated pool (2) to the pool (1) and (3). Recording (A) Fig. 4.55 shows the time course of the radial current when pool (2) contains only the myelinated part of the axon. The recording shows only *outwardly*-directed current, but separated by two peaks. This course is to be expected for pool (2) if activation of the node in pool (1) is later followed by an activation of the node in pool (3). In both cases there will be an external flow of the current loops from the region of myelinated surface to the node of Ranvier, which is the site of *inward* current at the peak of activity. The two current peaks correspond to the time interval between the activation of the two nodes. Recording (B) corresponds to the situation where the axon is moved so that a node of Ranvier is contained in the middle pool (2). The trace shows first a phase with an *outward* current corresponding the return current to the activated node in pool (1). This current depolarizes passively the membrane in the node in pool (2) and declines to be followed immediately by an *inwardly*-directed current arising from the activation of the node of Ranvier in pool (2). The occurrence of a diphasic radial current course drawn from the region that contains the node of Ranvier (B) and its absence when current is drawn only from the myelinated region (A) strongly supports the view that an inward current is associated with activity localized only at the nodes of Ranvier, and that transmission occurs in jumps from the one node of Raniver to the other. This conclusion was later confirmed and extended by Huxley & Stämpfli (1949)[*] using a different technique where the *longitudinal* current flow on the surface of the isolated myelinated nerve fiber was measured during the passage of an impulse at various positions along the length of the axon. The fiber was threaded through a fine glass capillary about 0.5 mm long, that was filled with Ringer's fluid. The fiber ends were then mounted on the two arms of a micromanipulator by means of which the fiber was pulled through the capillary that in turn would contain only the myelin sheath or myelin plus a node of Ranvier. Recordings of the potential difference across the capillary measured by electrodes placed on either side gave the external longitudinal current i_e that flowed in the capillary as a function of time during the impulse. Pulling the fiber stepwise through the capillary and making a recording at each step gave the longitudinal current flow at the various

[*] Huxley, A.F. & Stämpfli, R. (1949): Evidence for saltatory conduction in peripheral myelinated nerve-fibers. *J. Physiol.*, **108**, 315. Huxley, A.F. & Stämpfli, R. (1949): Saltatory transmission of the nervous impulse. *Arch.-Sci. physiol.*, **3**, 435.

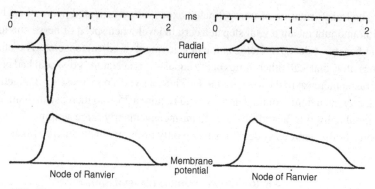

Fig. 4.56. Time courses of radial current (upper recordings) and membrane potential (lower recordings) over a node of Ranvier and a point on the fiber only covered by myelin sheath. Outward current upwards. (Huxley & Stämpfli, 1949.)

positions x along the fiber. The radial current through the myelin sheath or through a node of Ranvier was determined as

$$i_r = -\frac{\partial i_e}{\partial x},$$

i.e. by differentiation of the longitudinal current with respect to distance. Similarly, the potential variation along the length of the fiber was obtained by integration of i_e with respect to x. Figure 4.56 shows an example of the time courses of radial current and membrane potential calculated in this way[*]. The time course of the radial current across the myelin sheet region does not differ essentially from the direct measurements of the radial current according to Tasaki & Takeuchi. The current across the node of Ranvier shows initially an outwardly-directed, depolarizing current that immediately is flooded out by a much larger inwardly-directed current that most reasonably is ascribed – in analogy with the processes seen in unmyelinated nerves – to the opening of the sodium channels.

The development of the myelin sheath with its recurring interruptions results in an impulse transmission that travels with greater speed and with less expenditure of energy than in the unmyelinated axon where the cycles of excitation run through each point of the membrane. Instead the active part of the signal transmission is confined to the nodes of Ranvier, each of which functions as a relay station that boosts the signal up to full size before it is transmitted to the next relay station. Thus, compared with an unmyelinated fiber, a myelinated nerve fiber of the same diameter propagates the signal about 10 times faster and with about $\frac{1}{10}$ of the energy consumption.

---

[*] The membrane potential can be evaluated because i_e and the longitudinal axonal current i_a are numerical equal but oppositely directed. See also Appendix L.

The transformation of unmyelinated axons into myelinated nerve fibers has without doubt meant a vital step at a certain level when speed of nerve conduction became a decisive factor in the development of certain species. The other alternative that – all other things being equal – was at Nature's disposal had been a drastic increase in the axon diameter. Thus, a five-fold increase in conduction velocity in an unmyelinated nerve would require a 25-fold increase in diameter. Undoubtedly this would necessitate an inconveniently large volume to make room for the nervous system consisting only of unmyelinated nerve fibers.

4.6 Repetitive impulse transmission

In the preceding sections we have described some of the processes and mechanisms behind the generation of the impulse and its subsequent propagation following a single, shortlasting electric shock. In the intact living organism nerve impulses originate either in the periphery, where a sensory nerve fiber makes contact with a sensory receptor cell, or from the soma of the neuron. It is also characteristic that the nerve signals that are conducted along the separate axons almost always consist of a *train of impulses* of varying duration, where the interval between the individual impulses contains an essential part of the transmitted information. Thus, in a sensory axon the impulse frequency increases with the intensity of the influence on the receptor cell. In the axons that innervate muscle fibers the impulse frequency determines the strength of contraction. To produce a train of impulses an excised nerve is placed in the set up of Fig. 4.1, as most electric stimulators are designed also to produce a train of shocks. However, the repetitive firing of impulses in the intact organisms does not appear as the result of a set of discrete stimuli but of a sustained depolarization of the membrane region where the impulses start. The ability to set up a repetitive firing is not common to all excitable membranes, but appears as a property that is localized to particular membrane regions. A single nerve impulse is the response from a peripheral nerve from a vertebrate to a long-lasting current stimulus above the threshold. The same applies in general to a squid axon. But crustacean nerve fibers differ in that a constant current stimulation sets up a train of impulses as illustrated in Fig. 4.57, where action potentials are recorded with surface electrodes from an isolated nerve fiber from a crab in response to a constant current stimulation of 1–2 s duration. At the lowest strength of current (denoted 1.0 in relative units) the impulse frequency is as low as five impulses per second. This frequency grows to about 90 impulses per second as the depolarizing current at the cathode increases by a factor of 4.5.

As mentioned above a similar repetitive activity is never released in the peripheral nerve fibers of vertebrates. But just in that region where the motor

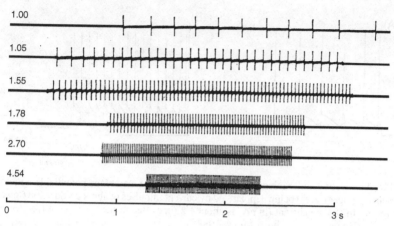

Fig. 4.57. Repetitive response in a single nerve fiber from the crab *Carcinus maenas*. Electrical changes at the cathode produced by application of constant currents with strengths shown relative to the rheobase. The beginning and end of the applied current are marked by the slight artifact. (From Hodgkin, 1948*.)

axon leaves the soma of the neuron (the initial segment) the axon membrane has developed the ability to display repetitive activity, as illustrated in Fig. 4.58. A microelectrode is inserted into the cell body of a motor neuron to record the activity in the cell near the initial segment. A second microelectrode is also placed intracellularly to provide current to polarize the cell membrane. The upper traces (A) and (B) are recordings of the membrane potential in response to the applied current whose magnitude and duration is indicated by the two lower traces. Owing to the low resistance of the cytoplasm as compared with that of the cell membrane, the current flow from the electrode tip spreads more or less uniformly over the inner surface of the cell membrane. It is seen that application of a constant current of duration of about 2 s causes a *depolarization* of the membrane upon which is superimposed a train of rhythmic activity of action potentials. This activity continues as long as the membrane depolarization is maintained. In (B) the stimulation current is about doubled. The result is again a repetitive impulse activity but – associated with the greater steady level of depolarization – firing at a much higher frequency.

The details of the processes and mechanism behind the repetitive emission of impulses may vary between the different types of excitable tissue. The description that follows touches only on the more general mechanisms that may be involved when a sustained membrane depolarization results in the transmission

---

* Hodgkin, A.L. (1948): The local electric changes associated with repetitive action in a non-medullated axon. *J. Physiol.*, **107**, 165.

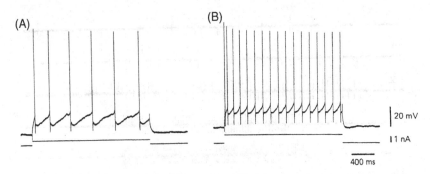

Fig. 4.58. Intracellular microelectrode recordings of the membrane potential in a motor neuron (upper traces) being depolarized by current impulses of about 2 s duration (lower traces). In (B) the current (in nA) is about 1.2 times that of the duration of (A) (lower trace). (Hounsgaard, Kiehn & Mintz, 1988[†].)

of either a single nerve impulse or a train of impulses. These two situations are considered in Fig. 4.59, showing the responses to the application of an electric current of some length that depolarizes the membrane enough to trigger off an action potential. The upper recording in Fig. 4.59 shows the response from, for example, a squid axon. The initial phase (a–b) represents partly the passive depolarization of the membrane by the current flow and the charging process of the membrane capacity, and partly a local response[‡]. At (b) the membrane potential reaches the level of threshold depolarization V^* and inphase (b–c) the sodium and potassium conductance grows and a propagated action potential is triggered off. The maintenance of the current stimulation results in a perturbation of the final level of repolarization to a membrane potential V due to the current flow, which is somewhat higher than the initial threshold depolarization V^*. Therefore, one should expect that the membrane should again respond by building up a propagating action potential anew. The non-occurrence of the action potential although $V > V^*$ means that the membrane is in a refractory state, probably because the inactivation is not turned off completely because the stimulating current maintains a depolarization of the resting membrane potential by 10–15 mV. To remove this state of refractoriness the current must be switched off to allow the membrane to repolarize to the normal value of the resting membrane potential, whereby the inactivation fades away within the next few milliseconds. The middle trace of Fig. 4.59 shows the analogous events in a membrane that responds with an impulse train to a constant membrane depolarization. Compared with the upper trace, the characteristic feature is a more

[†] Hounsgaard, J., Kiehn, O. & Mintz, I. (1988): *J. Physiol.*, **398**, 575.
[‡] See Section 4.4.3.1 (ii).

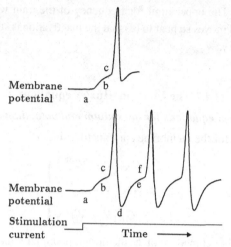

Membrane
potential

Membrane
potential

Stimulation
current

Time ⟶

Fig. 4.59. To illustrate with greater time resolution that a membrane depolarization above the threshold value elicits in one type of nerve only a single impulse (upper trace), whereas in another cell type the result is a repetitive impulse firing (middle trace). The lower trace shows the start of the depolarizing current impulse.

pronounced phase of repolarization (d) that is of longer duration and almost reaches the level of the resting potential. Such a course can be expected from a prolongation of the phase with high potassium. With the sodium conductance turned off the membrane permeability is dominated by the potassium ions and the membrane potential will approach the value of the equilibrium potential $V_K^{(eq)}$ for potassium, and at the same time suppress the polarizing action of the stimulating current. The further action of the hyperpolarization is the removal of any remaining inactivation. The potassium conductance – and the membrane resistance – gradually return to the initial value. The course from (d) to (e) in Fig. 4.59 reflects the combined action on the membrane potential of the potassium conductance returning to the resting value and the depolarization due to the stimulating current. The effect of the latter continues after attainment of the resting membrane potential. The sodium conductance that is no longer under the influence of inactivation begins to rise again at (e) and to cause a new action potential at (f). The sequence of events will go on as long as the depolarizing current is sustained. The frequency of the emission of impulses will depend on the length of the time course from (d) to (f), which again depends on the level of depolarization caused by the stimulating current. The more the membrane is depolarized passively, the faster the state (f) will be reached after each repolarization and the higher will be the frequency of the impulse train. (Compare the traces (A) and (B) in Fig. 4.58, which show the different slopes of the course from (d) to (f).) The upper limit for frequency of the train will be set by the phase (e→f) and moves so near to (d) that the inactivation is still active and the nerve is refractory.

4.7 The Hodgkin–Huxley equations

4.7.1 Empirical equations for the sodium and potassium conductances

In the equations for the membrane currents (densities)

$$
\left.\begin{aligned}
I_{Na} &= G_{Na}\left[V - V_{Na}^{(eq)}\right] \\
I_{K} &= G_{K}\left[V - V_{K}^{(eq)}\right] \\
I_{L} &= G_{L}\left[V - V_{L}^{(eq)}\right]
\end{aligned}\right\}
\tag{4.7.1}
$$

G_{Na} and G_{K} are both functions of the membrane potential V and time t. Hodgkin & Huxley* succeeded in developing a set of empirical equations, the Hodgkin–Huxley equations, which reproduced with great precision the dependence of G_{Na} and G_{K} on V and t.

The voltage and time dependence of G_{K} were represented by

$$
G_{K}(V, t) = \overline{G}_{K}\, n^{4},
\tag{4.7.2}
$$

where $\overline{G}_{K}$ is the maximum attainable value for G_{K} – the state with all K channels open at the same time – and $n(V, t)$ is an activation function, whose dependence on V and t is given by the empirical equations

$$
\frac{dn}{dt} = \alpha_{n}(1 - n) - \beta_{n}n,
\tag{4.7.3}
$$

$$
\alpha_{n} = -0.01\,(V + 50)\left/\left(\exp\left[-\frac{V + 50}{10}\right] - 1\right)\right.,
\tag{4.7.4}
$$

and

$$
\beta_{n} = 0.125\,\exp\left[-\frac{V + 60}{80}\right].
\tag{4.7.5}
$$

Note that $\alpha_{n}(V)$ and $\beta_{n}(V)$ are the rate constants in Eq. (4.7.3) depending on V (dimension ms^{-1}). The numerals – conforming to the units and polarity chosen for V – were also selected by trial and error to make the best possible fit of

* Hodgkin, A.L. & Huxley, A.F. (1952): A quantitative description of membrane current and its application to conduction and excitation in nerve. *J. Physiol.*, **117**, 500.

Eq. (4.7.2) to the experimental data that were obtained from the experiments on the squid axon.

The sodium conductance was written as

$$G_{Na}(V, t) = \overline{G}_{Na} m^3 h, \tag{4.7.6}$$

where $\overline{G}_{Na}$ is the maximum attainable sodium conductance, and m is the function of activation, whose dependence on V and t is given by the empirical equations

$$\frac{dm}{dt} = \alpha_m (1 - m) - \beta_m m, \tag{4.7.7}$$

$$\alpha_m = -0.1 (V + 35) \bigg/ \left(\exp\left[-\frac{V + 35}{10} \right] - 1 \right), \tag{4.7.8}$$

and

$$\beta_m = 4.0 \exp\left[-\frac{V + 60}{18} \right], \tag{4.7.9}$$

where α_m and β_m are the rate constants to match.

The function of inactivation h is given by the equations

$$\frac{dh}{dt} = \alpha_h (1 - h) - \beta_h h, \tag{4.7.10}$$

with rate constants

$$\alpha_h = 0.07 \exp\left[-\frac{V + 60}{20} \right], \tag{4.7.11}$$

and

$$\beta_h = 1 \bigg/ \left(\exp\left[-\frac{V + 30}{10} \right] + 1 \right). \tag{4.7.12}$$

At any time the total membrane current I – or rather current density – is

$$
\begin{aligned}
I &= C \frac{\partial V}{\partial t} + I_{Na} + I_K + I_L \\
&= C \frac{\partial V}{\partial t} + G_{Na} \left[V - V_{Na}^{(eq)} \right] + G_K \left[V - V_K^{(eq)} \right] + G_L \left[V - V_L^{(eq)} \right] \\
&= C \frac{\partial V}{\partial t} + \overline{G}_{Na} \, m(V, t)^3 \, h(V, t) \left[V - V_{Na}^{(eq)} \right] \\
&\quad + \overline{G}_K \, n(V, t)^4 \left[V - V_K^{(eq)} \right] + G_L \left[V - V_L^{(eq)} \right],
\end{aligned}
\tag{4.7.13}
$$

where C is the membrane capacitance per unit area of axon membrane. Three types of solution will now be discussed.

4.7.2 Voltage clamp

In this situation the membrane potential changes instantaneously from an initial value V_1 to a new value V_2 and is held there for a certain time. In connection with the potential jump there is a flow of current to charge the membrane capacitor until the new constant value V_2 is attained. From now on $\partial V/\partial t = 0$ and the membrane current is determined by

$$
\begin{aligned}
I &= +I_{Na} + I_K + I_L \\
&= +G_{Na}\big[V_2 - V_{Na}^{(eq)}\big] + G_K\big[V_2 - V_K^{(eq)}\big] + G_L\big[V_2 - V_L^{(eq)}\big] \\
&= \overline{G}_{Na}\, m(V_2, t)^3\, h(V_2, t)\big[V_2 - V_{Na}^{(eq)}\big] \\
&\quad + \overline{G}_K\, n(V_2, t)^4\big[V_2 - V_K^{(eq)}\big] + G_L\big[V_2 - V_L^{(eq)}\big].
\end{aligned}
\tag{4.7.14}
$$

To calculate the ionic current I requires knowledge of the time courses of the three functions $n(t)$, $m(t)$ and $h(t)$ in response to the stepwise change of $V_1 \to V_2$. These solutions can be obtained without making use of numerical methods. By way of illustration we shall solve the differential equation

$$
\frac{dn}{dt} = \alpha_n(V) - [\alpha_n(V) + \beta_n(V)]n,
\tag{4.7.15}
$$

where the rate constants $\alpha_n(V)$ and $\beta_n(V)$ are the functions of V given by Eq. (4.7.4) and Eq. (4.7.5).

To begin, we consider the state of the function n where it has become stationary, i.e. $dn/dt = 0$, at the value of $V = V_1$. Let $n(0-)$ denote the value of n at time $t = 0-$, i.e. just at the negative side of $t = 0$. This value is obtained from Eq. (4.7.15) as

$$
n(0-) = \frac{\alpha_n(V_1)}{\alpha_n(V_1) + \beta_n(V_1)}.
\tag{4.7.16}
$$

At time $t = 0+$, the potential V changes stepwise to the new constant value $V = V_2$ making the rate constants assume the new values $\alpha_n(V_2)$ and $\beta_n(V_2)$. In response to this, $n(t)$ will move in time towards a new stationary value where the time course $n(t)$ for the transition is found by solving the differential equation

$$
\frac{dn}{dt} = \alpha_n(V_2) - [\alpha_n(V_2) + \beta_n(V_2)]n,
\tag{4.7.17}
$$

which satisfies the initial condition

$$n(0) = \frac{\alpha_n(V_1)}{\alpha_n(V_1) + \beta_n(V_1)}, \tag{4.7.18}$$

since $n(t)$ must be continuous at the transition $t(0-) \to t(0+)$. Putting $dn/dt = 0$ in Eq. (4.7.17) gives

$$n(\infty) = \frac{\alpha_n(V_2)}{\alpha_n(V_2) + \beta_n(V_2)}, \tag{4.7.19}$$

which is the new stationary value for $n(t)$. The general solution of Eq. (4.7.17) is*

$$n(t)\,e^{t/\tau_n} = \int e^{t/\tau_n}\,\alpha_n(V_2)\,dt + A$$

$$= \alpha_n(V_2)\,\tau_n e^{t/\tau_n} + A, \tag{4.7.20}$$

where

$$\tau_n = \frac{1}{\alpha_n(V_2) + \beta_n(V_2)},$$

is the *time constant* for the growth of $n(t)$ towards $n(\infty)$. Inserting the initial condition Eq. (4.7.18) into Eq. (4.7.20) gives

$$\frac{\alpha_n(V_1)}{\alpha_n(V_1) + \beta_n(V_1)} = \frac{\alpha_n(V_2)}{\alpha_n(V_2) + \beta_n(V_2)} + A,$$

or[†]

$$n(0) = n(\infty) + A.$$

This value for A inserted into Eq. (4.7.20) gives

$$n(t) = n(\infty) + [n(0) - n(\infty)]\,e^{-t/\tau_n}. \tag{4.7.21}$$

The same procedure applied to Eq. (4.7.7) *et seq.* yields

$$m(t) = m(\infty) + [m(0) - m(\infty)]\,e^{-t/\tau_m}, \tag{4.7.22}$$

where

$$\left.\begin{array}{l} m(0) = \dfrac{\alpha_m(V_1)}{\alpha_m(V_1) + \beta_m(V_1)} \\[3ex] m(\infty) = \dfrac{\alpha_m(V_2)}{\alpha_m(V_2) + \beta_m(V_2)} \end{array}\right\} \tag{4.7.23}$$

---

* See Chapter 1, Section 1.9.1, Eq. (1.9.16).
[†] Compare Eq. (4.7.18) and Eq. (4.7.19).

and

$$\tau_m = \frac{1}{\alpha_m(V_2) + \beta_m(V_2)}. \qquad (4.7.24)$$

Likewise Eq. (4.7.10) *et seq.* give

$$h(t) = h(\infty) + [h(0) - h(\infty)]\,e^{-t/\tau_h}, \qquad (4.7.25)$$

where

$$\left.\begin{aligned} h(0) &= \frac{\alpha_h(V_1)}{\alpha_h(V_1) + \beta_h(V_1)} \\[2ex] h(\infty) &= \frac{\alpha_h(V_2)}{\alpha_h(V_2) + \beta_h(V_2)} \end{aligned}\right\} \qquad (4.7.26)$$

and

$$\tau_h = \frac{1}{\alpha_h(V_2) + \beta_h(V_2)}. \qquad (4.7.27)$$

Using the following data for the squid axon (FitzHugh, 1960)

$$\begin{aligned} V_{Na}^{(eq)} &= 55.0\ \text{mV} & \overline{G}_{Na} &= 120.0\ \text{mS cm}^{-2} \\ V_{K}^{(eq)} &= -72.0\ \text{mV} & \overline{G}_{K} &= 36.0\ \text{mS cm}^{-2} \\ V_{L}^{(eq)} &= -49.4\ \text{mV} & \overline{G}_{L} &= 0.3\ \text{mS cm}^{-2}, \end{aligned}$$

with $V_{rest} = V_1 = -60$ mV and $V_2 = +10$ mV, results in

$$\begin{aligned} n(0) &= 0.317\,68, & n(\infty) &= 0.920\,28, & \tau_n &= 0.215\,87 \\ m(0) &= 0.052\,93, & m(\infty) &= 0.982\,33, & \tau_m &= 1.529\,99 \\ h(0) &= 0.596\,12, & h(\infty) &= 0.002\,15, & \tau_h &= 1.016\,13 \end{aligned}$$

and the time courses for $n(t)$, $m(t)$ and $h(t)$ become

$$\left.\begin{aligned} n(t) &= 0.920\,28 - 0.602\,60\,e^{-t/0.21587} \\ m(t) &= 0.982\,33 - 0.929\,40\,e^{-t/1.530} \\ h(t) &= 0.002\,15 + 0.593\,97\,e^{-t/1.01613} \end{aligned}\right\}. \qquad (4.7.28)$$

Let the membrane now be maintained at a constant value during the time V_2 for $0 \le t \le t_0$. At $t = t_0$, where the functions of activation have assumed the values $n(t_0)$, $m(t_0)$ and $h(t_0)$, the potential V_2 is again switched back to the initial value V_1. The time courses of n, m and h for $t > t_0$ are found by solving the set of equations

$$
\left.
\begin{aligned}
\frac{dn}{dt} &= \alpha_n(V_1) - [\alpha_n(V_1) + \beta_n(V_1)]n, & n &= n(t_0) & \text{for} \quad t &= t_0 \\[2mm]
\frac{dm}{dt} &= \alpha_m(V_1) - [\alpha_m(V_1) + \beta_m(V_1)]m, & m &= m(t_0) & \text{for} \quad t &= t_0 \\[2mm]
\frac{dh}{dt} &= \alpha_h(V_1) - [\alpha_h(V_1) + \beta_h(V_1)]h, & h &= h(t_0) & \text{for} \quad t &= t_0
\end{aligned}
\right\}
$$

$$(4.7.29)$$

where the rate constants $\alpha_n, \beta_n, \alpha_m, \beta_m$ and α_h, β_h are now those corresponding to $V = V_1$.

The solutions of Eq. (4.7.29) for $t > t_0$ are:

$$
\left.
\begin{aligned}
n(t) &= n(t_0)\,e^{-(t-t_0)/\tau_n} + n(0)\big[1 - e^{-(t-t_0)/\tau_n}\big] \\[2mm]
m(t) &= m(t_0)\,e^{-(t-t_0)/\tau_m} + m(0)\big[1 - e^{-(t-t_0)/\tau_m}\big] \\[2mm]
h(t) &= h(t_0)\,e^{-(t-t_0)/\tau_h} + h(0)\big[1 - e^{-(t-t_0)/\tau_h}\big]
\end{aligned}
\right\}
$$

$$(4.7.30)$$

where the values of the new time constants are

$$
\left.
\begin{aligned}
\tau_n &= \frac{1}{\alpha_n(V_1) + \beta_n(V_1)} = 0.236\,77 \\[3mm]
\tau_m &= \frac{1}{\alpha_m(V_1) + \beta_m(V_1)} = 5.458\,56 \\[3mm]
\tau_h &= \frac{1}{\alpha_h(V_1) + \beta_h(V_1)} = 8.516\,01
\end{aligned}
\right\}.
$$

$$(4.7.31)$$

Insertion of these values of $n(t)$, $m(t)$ and $h(t)$ in Eq. (4.7.2) and Eq. (4.7.6) gives the time course of the conductances G_K and G_{Na} for the potassium and sodium ions, and the individual currents I_K, I_{Na} and I_L follow, making use of Eq. (4.8.1). Using the standard methods of solution has the additional merit of showing the exponential nature of the three functions $n(t)$, $m(t)$ and $h(t)$ and the dependence of the time constants τ_m, τ_n and τ_h on the membrane potential V.

   Naturally, nothing can be said against calculating the courses of $n(t)$, $m(t)$ and $h(t)$ by means of *numerical integration* of the equations. In principle the procedure runs as follows. A suitable range of integration $0 \le t \le t_{max}$, e.g. 10 ms, is chosen. This range is divided into an adequate – in general, large – number of intervals, e.g. 400, giving an interval length of $\Delta t = 0.025$ ms. The initial value $n(0)$ is inserted in the right-hand side of the equation for dn/dt and the increment $\Delta n = (dn/dt)\,\Delta t$ is calculated and added to $n(0)$ (first run). The new value $n_1 = n(0) + \Delta n$ is inserted in the expression for dn/dt

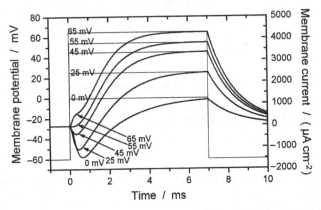

Fig. 4.60. Calculated membrane current in the Hodgkin–Huxley axon during voltage clamp conditions. Abscissa: time (ms). Left ordinate: membrane potential (mV). Right ordinate: membrane current (mA cm^{-2}). The membrane potential is displaced instantaneously from $V_m^{(rest)} = -60$ mV to the values corresponding to the five horizontal lines and maintained for 7 ms. Note the current course for $V_{clamp} = +55$ mV $= V_{Na}^{(eq)}$ and for $V_{clamp} > V_{Na}^{(eq)}$. The calculations result from a numerical integration of Eq. (4.7.14) according to the procedure of Runge–Kutta.

and the next value for Δn is calculated (second run) and added to n_1. This iteration procedure continues until the value $t = t_0$ is attained. At this point the membrane potential is switched back to $V = V_1$ and the calculation continues for $t_0 \leq t \leq t_{max}$ according to Eq. (4.7.29) and with the initial condition given by Eq. (4.7.28) with $t = t_0$. Calculations of $m(t)$ and $h(t)$ are done in the same way. This method of calculation (Euler's method) works satisfactorily only for small values of Δt and provides only a reasonably good approximation of the first order if the derivatives do not vary too fast in the interval Δt. However, procedures exist to accomplish numerical integration with great accuracy. The so-called Runge–Kutta fourth-order procedure provides us with $\Delta t = 0.025$ ms numerical values of $n(t)$, $m(t)$ and $h(t)$ that agree to six significant figures with those values obtained from the analytical solutions. The price for this precision is an increasing complexity of the numerical computations. Therefore, before the era of personal computers one had to think first before embarking upon a numerical procedure as modern as Runge–Kutta. But today these complexities cause hardly any impediment.

Figures 4.60, 4.61 and 4.62 show solutions of the voltage clamp equation Eq. (4.7.14) and the attached equations for $n(V, t)$, $m(V, t)$ and $h(V, t)$ when V changes instantaneously from $V_1 = V_m^{(rest)} = -60$ mV to a chosen value V_2 and is maintained there for 7 ms. In the calculations we have used a Runge–Kutta

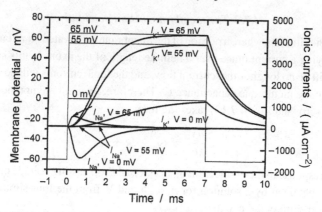

Fig. 4.61. The individual ion currents I_{Na} and I_K in the Hodgkin–Huxley axon during voltage clamp conditions. Abscissa: time (ms). Left ordinate: membrane potential (mV). Right ordinate: ionic current (mA cm^{-2}). The membrane potential is displaced instantaneously from $V_m^{(rest)} = -60$ mV to $V_{clamp} = 0$ mV, +55 mV and +65 mV and maintained there for 7 ms. Note the current courses for $V_{clamp} = +55$ mV $= V_{Na}^{(eq)}$.

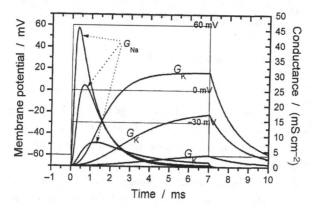

Fig. 4.62. Conductance changes of G_{Na} (fast, but transient) and G_K in the Hodgkin–Huxley axon during voltage clamp conditions. Abscissa: time (ms). Left ordinate: membrane potential (mV). Right ordinate: conductance (mS cm^{-2}). Membrane potential is displaced instantaneously from $V_m^{(rest)} = -60$ mV to $V_{clamp} = -30$ mV, 0 mV and +60 mV and maintained there for 7 ms.

fourth-order procedure and utilized the numerical data for the squid axon given above. The similarity in shape should be noted between the essential features of the theoretical time courses of total membrane current I, the individual ionic currents I_{Na} and I_K and conductances G_{Na} and g_K and the experimental courses in Fig. 4.28, Fig. 4.29 and Fig. 4.31.

4.7.3 The membrane action potential

This is the situation where a portion of the membrane is activated simultaneously by applying a current pulse over a definite length of the axon. In this situation there will be no longitudinal current flow, and the ionic currents will be spent in charging the membrane capacitance C. Therefore, the total membrane current will be zero, and Eq. (4.7.13) takes the form

$$-C\frac{\partial V}{\partial t} = I_{Na} + I_K + I_L.$$

The response to the application of an external short-lasting voltage or current pulse is obtained by a solution for n, m, h and V from the four simultaneous first-order differential equations

$$\left.\begin{aligned}
\frac{dn}{dt} &= \alpha_n(V) - [\alpha_n(V) + \beta_n(V)]n \\
\frac{dm}{dt} &= \alpha_m(V) - [\alpha_m(V) + \beta_m(V)]m \\
\frac{dh}{dt} &= \alpha_h(V) - [\alpha_h(V) + \beta_h(V)]h \\
\frac{\partial V}{\partial t} &= -\frac{1}{C}\big(\overline{G}_{Na}\, m(V,t)^3\, h(V,t)\big[V - V_{Na}^{(eq)}\big] \\
&\quad + \overline{G}_K\, n(V,t)^4\big[V - V_K^{(eq)}\big] + G_L\big[V - V_L^{(eq)}\big]\big)
\end{aligned}\right\}. \quad (4.7.32)$$

These equations are extremely nonlinear, as the one independent variable V behaves in a most nonlinear manner in each of the four equations. The system is only solvable by using a numerical procedure of integration. This system differs from that above by also containing the membrane potential V as a variable, whereby each run in the computation gives a new value for V. Before the next run this value must then be inserted into six equations for the voltage dependence of the rate constants α_n, β_n, α_m, β_m and α_h, β_h (Eq. (4.7.5), Eq. (4.7.6), Eq. (4.7.8), Eq. (4.7.9), Eq. (4.7.11) and Eq. (4.7.12)) to obtain the new values to be used in the next run. An excellent algorithm for a Runge–Kutta fourth-order procedure is found in the "yellow volumes"*. The program is started by a short displacement of the potential (e.g. 0.5 ms) by a chosen amount. This causes the conductances G_{Na} and G_K to change as in the "clamp experiment", and in response to this initial push the program begins to compute the time courses of n, m, h and V and in turn the remaining quantities G_{Na} and

* Press, W.H., Flaherty, B.P., Teukolsky, S.A. & Vettering, W.T. (1989): *Numerical Recipes. The Art of Scientific Computing*, Cambridge University Press.

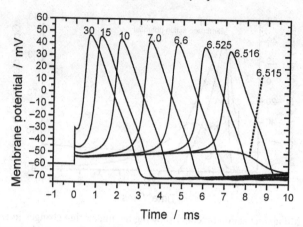

Fig. 4.63. Computed membrane action potential from the Hodgkin–Huxley axon in response to short (0.5 ms) displacements of the membrane potential to different levels. Abscissa: time (ms). Ordinate: membrane potential (mV). $V_m^{(rest)} = -60$ mV. Values of V_{stim} (mV) are given next to the curves.

G_K together with the ionic currents I_{Na}, I_K, I_L and I_{total}. All the data may then be transferred to a suitable plotter program to be drawn on the screen or some other device.

We will now present examples of computer solutions of the four simultaneous equations in Eq. (4.7.32) with their coupling to the equations for the rate constants. Figure 4.63 shows the effect of the stimulus strength. With a duration of the depolarizing voltage pulse of 0.5 ms there is a threshold depolarization at $V = -53.48$ mV, which causes the membrane potential to linger around this value for about 7 ms before it returns to the resting value. An increase of the stimulus strength shortens the time for the appearance of the action potential. At a depolarizing step to $V = -30$ mV the action potential starts off with almost no delay. Figure 4.64 shows the underlying conductance changes. The sodium conductance – starting off from a very low level – rises rapidly within the next 0.5 ms where the suppressive action of the inactivation enters to cause a return in G_{Na} to normal within the following 1.5 ms. The potassium conductance starts with a delay and increases more slowly within the next 1.5 ms and declines slowly towards its initial value causing the membrane potential to enter into a small phase of hyperpolarization before levelling at the value of the initial resting membrane potential. The essence of the flow of the ionic currents I_{Na} and I_K that are associated with the changes in G_{Na} and G_K appears clearly in Fig. 4.65: although the total charge that is carried by the two oppositely directed sodium and potassium currents is very nearly zero, it is the initial solitary

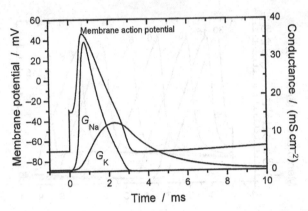

Fig. 4.64. Membrane action potential showing the underlying changes in G_{Na} and G_K caused by a stimulus pulse of 0.5 ms duration and 30 mV amplitude. Abscissa: time (ms). Right ordinate: conductance (mS cm^{-2}). Left ordinate: membrane potential (mV). $V_m^{(rest)} = -60$ mV.

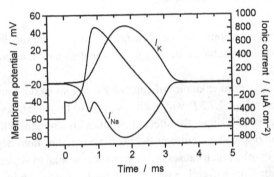

Fig. 4.65. Similar to Fig. 4.64 but showing the currents I_{Na} and I_K. Abscissa: time (ms). Left ordinate: membrane potential (mV). Right ordinate: ionic current (mA cm^{-2}).

inwardly-directed sodium current that drives the membrane potential towards the value of $V_{Na}^{(eq)}$. The behavior of the Hodgkin–Huxley axon when stimulated just below the threshold is shown in Fig. 4.66 as an illustration to the text in Section 4.4.4.4 (ii). It is seen that the essential feature of the threshold phenomenon discussed, i.e. $|I_{Na}| = |I_K|$, is reproduced by the Hodgkin–Huxley equations, which in addition predict that the prolonged threshold state is associated with a steady increase of equal size in both currents until the state is ended by one of the two conductances taking the lead. The effect of removing the inactivation function h from Eq. (4.7.6) is illustrated in Fig. 4.67. When the value of $\overline{G}_{Na}$ is maintained at the value of 120 mS cm^{-2} used up to now, the

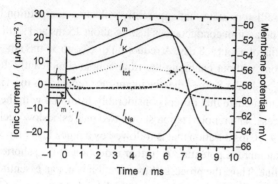

Fig. 4.66. Computed membrane action potential in the Hodgkin–Huxley axon showing membrane potential V and I_{Na}, I_K and total current I_{tot} using a stimulation just below the threshold value (6.5151 mV). Abscissa: time (ms). Right ordinate: membrane potential (mV). Left ordinate: ionic current (mA cm^{-2}).

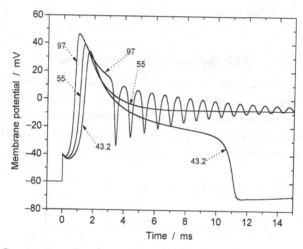

Fig. 4.67. Computed membrane action potential in the Hodgkin–Huxley axon with no inactivation function. Abscissa: time (ms). Ordinate: membrane potential (mV). The numbers attached to the curves indicate the values of $\overline{G}_{Na}$, compared with the standard value of 120 mS cm^{-2}.

stimulation causes the membrane potential to approach rapidly a value near to that of $V_{Na}^{(eq)}$ and it is kept there with little change. To make the membrane action potential respond in the usual transient manner the inward flux of sodium must be diminished by reducing the value of $\overline{G}_{Na}$. Figure 4.67 illustrates the effect of such a reduction. For $\overline{G}_{Na} = 97$ mS cm^{-2} the membrane action potential shows a steady depolarization by about 30 mV where the membrane suddenly starts a damped oscillatory state that apparently reflects a competition between the sodium and potassium conductances lasting about 15 ms and settles at a membrane potential of about -8 mV. A reduction of $\overline{G}_{Na}$ to 55 mS cm^{-2} eliminates the oscillations without changing the time course significantly. It is first at a value of $\overline{G}_{Na} = 43.2$ mS cm^{-2} that the membrane repolarizes to its resting value but with a time course that differs considerably from that with the intact inactivation process. The repolarization shows two phases: a slow decline to about -20 mV of 9–10 ms duration that is followed by a much faster repolarization to the resting membrane potential. A further reduction of $\overline{G}_{Na}$ shortens the phase of slow decline. Thus, the process of inactivation is not an essential function of the operation of the Hodgkin–Huxley axon, but a large value of $\overline{G}_{Na}$ is important for a high speed of the initial regenerative membrane depolarization. But to secure an equally high speed of repolarization, an inactivation of the sodium conductance is required to allow for the effects of the succeeding increase in the potassium conductance.

4.7.4 The propagated action potential

As previously mentioned (Section 4.4.6.3) Hodgkin & Huxley succeeded in reconstructing the propagating action potential by solving the Hodgkin–Huxley equations for the case where a longitudinal current also flows from the active region to the adjacent regions. As the starting point one can take the Hodgkin–Huxley equation for the membrane current density I (Eq. (4.7.13))

$$
\begin{aligned}
I &= C\frac{\partial V}{\partial t} + I_{Na} + I_K + I_L \\
&= C\frac{\partial V}{\partial t} + \overline{G}_{Na}\, m(V,t)^3\, h(V,t)\big[V - V_{Na}^{(eq)}\big] \\
&\quad + \overline{G}_K\, n(V,t)^4\big[V - V_K^{(eq)}\big] + G_L\big[V - V_L^{(eq)}\big],
\end{aligned}
\qquad (4.7.33)
$$

and combine it with the equation for the membrane current

$$
i_m = \frac{1}{r_a}\frac{\partial^2 V}{\partial x^2},
\qquad (4.7.34)
$$

where r_a is the resistance per unit length of axoplasm. Equation (4.7.34) expresses the condition of conservation of electric charge at any position x along the length of the cable. Thus, Eq. (4.7.34) remains valid no matter how complex the course of the ionic currents through the membrane may be in Eq. (4.7.33). The current unit (ampere per unit length) is changed to conform to that of Eq. (4.7.33): the membrane area A of one unit length of the axon is

$A = 2\pi a \cdot 1$, where a is the radius of the axon. Hence, $I = i_m/2\pi a$ or

$$I = \frac{1}{2\pi a r_a}\frac{\partial^2 V}{\partial x^2}. \tag{4.7.35}$$

Entering the resistivity of axoplasm ρ_a it follows that

$$r_a = \rho_a \frac{1}{\pi a^2}.$$

Hence

$$I = \frac{a}{2\rho_a}\frac{\partial^2 V}{\partial x^2}. \tag{4.7.36}$$

Combining Eq. (4.4.37) and Eq. (4.7.33) yields the following partial differential equation that governs the spreading of the membrane potential as a function of distance x and time t:

$$\frac{a}{2\rho_a}\frac{\partial^2 V}{\partial x^2} = C\frac{\partial V}{\partial t} + \overline{G}_{Na}\, m(V, t)^3\, h(V, t)\big[V - V_{Na}^{(eq)}\big]$$
$$+ \overline{G}_K\, n(V, t)^4\big[V - V_K^{(eq)}\big] + G_L\big[V - V_L^{(eq)}\big]. \tag{4.7.37}$$

To solve the above nonlinear partial differential equation poses rather formidable problems. To circumvent some of the difficulties Hodgkin & Huxley introduced from the outset the property that the action potential propagates along the axon with a *constant* velocity θ and with the *same configuration* of potential profile. This condition of constraint makes it possible to transform the partial differential equation into an ordinary differential equation. Let $V = F(x)$ represent the potential profile along the distance x of the axon at some time, e.g. $t = 0$. At a later time t this profile has moved a distance θt in the positive direction of the axis but with *no change* in configuration. Hence, taking the point at $x = \theta t$ as the new origin and denoting the distance measured from this origin as X so that $X = x + \theta t$, then the equation of the action potential profile referred to this origin will be

$$V = F(X),$$

or referred to the original fixed origin x_0 it will be

$$V = F(x - \theta t).$$

Thus, the solution for $V(x, t)$ extends only on a single variable $u = x - \theta t$, namely

$$V(x, t) = V(x - \theta t).$$

It follows that

$$\frac{\partial V}{\partial x} = \frac{\partial V}{\partial t}\frac{\partial t}{\partial x} = \frac{\partial V}{\partial t}\left(-\frac{1}{\theta}\right)$$

and

$$\frac{\partial^2 V}{\partial x^2} = \frac{\partial}{\partial x}\left(-\frac{\partial V}{\partial t}\frac{1}{\theta}\right) = \frac{\partial}{\partial t}\left(-\frac{\partial V}{\partial t}\frac{1}{\theta}\right)\frac{\partial t}{\partial x} = -\frac{\partial^2 V}{\partial t^2}\frac{1}{\theta}(-\frac{1}{\theta}) = \frac{1}{\theta^2}\frac{\partial^2 V}{\partial t^2}.$$

This implies that Eq. (4.7.37) can be written instead as the following ordinary differential equation in time t:

$$\frac{a}{2\rho_a\theta^2}\frac{d^2 V}{dt^2} = C\frac{dV}{dt} + \overline{G}_{Na}\,m(V,t)^3\,h(V,t)\left[V - V_{Na}^{(eq)}\right]$$
$$+ \overline{G}_K\,n(V,t)^4\left[V - V_K^{(eq)}\right] + G_L\left[V - V_L^{(eq)}\right], \quad (4.7.38)$$

which is the differential equation for the squid axon action potential that propagates with the constant velocity θ.

Inclusion of the propagation results in a second-order equation for V. However, this does not lead to any complication in the numerical integration, since a second-order differential equation

$$\frac{d^2 V}{dt^2} = a\frac{dV}{dt} + F(V,t)$$

can always be written as originating from two coupled first-order differential equations. Putting $dV/dt = U$ we have $d^2V/dt^2 = dU/dt$, and the equation above can be written as

$$\left.\begin{array}{l}\dfrac{dU}{dt} = a\,U + F(V,t)\\[2mm]\dfrac{dV}{dt} = U\end{array}\right\}.$$

The necessity for using this transformation arises if the above-mentioned Runge–Kutta algorithm is the preferred tool to be used in the numerical computations since this algorithm is worked out for handling simultaneous equations of *first order* only. The time course of the propagating action potential of the Hodgkin–Huxley axon is then obtained by solving the following five simultaneous differential equations

$$
\left.
\begin{aligned}
\frac{dn}{dt} &= \alpha_n(V) - [\alpha_n(V) + \beta_n(V)]n \\[2mm]
\frac{dm}{dt} &= \alpha_m(V) - [\alpha_m(V) + \beta_m(V)]m \\[2mm]
\frac{dh}{dt} &= \alpha_h(V) - [\alpha_h(V) + \beta_h(V)]h \\[2mm]
\frac{dU}{dt} &= \theta^2 \frac{2\rho_a}{a} \big(C\, U + \overline{G}_{\mathrm{Na}}\, m(V,t)^3\, h(V,t) \big[V - V_{\mathrm{Na}}^{(\mathrm{eq})} \big] \\
&\quad + \overline{G}_{\mathrm{K}}\, n(V,t)^4 \big[V - V_{\mathrm{K}}^{(\mathrm{eq})} \big] + G_{\mathrm{L}} \big[V - V_{\mathrm{L}}^{(\mathrm{eq})} \big] \big) \\[2mm]
\frac{dV}{dt} &= U
\end{aligned}
\right\}
\qquad (4.7.39)
$$

subject to a short-lasting voltage/current pulse above the threshold of depolarization. The solution that Hodgkin & Huxley (1952) obtained agreed well with the experimentally determined action potential and was associated with a conduction velocity of the magnitude found experimentally. Apart from the inherent complications involved in the numerical computational procedure, the process of solution was not without problems. The presence of the second-order term d^2V/dx^2 in Eq. (4.7.38) and the coupling between the two first-order equations in Eq. (4.7.39) implies that the system now has a built-in possibility of getting into a state of *instability**. The system contains four constants of interest: the membrane capacitance C, the resistivity of the axoplasm and the radius a of the axon have values that are known in advance with reasonable accuracy. With these three values fixed beforehand it is the fourth, the conduction velocity θ – whose value is unknown – that enters as the parameter that determines the behavior of the system. The crucial value of θ – which corresponds to the natural value of the velocity of propagation of the Hodgkin–Huxley axon – is that which makes the solution of V swing upwards towards a value for V of about $+45$ mV and then brings the potential back to the resting value similarly to the membrane action potential shown in Fig. 4.63. Hodgkin & Huxley handled this problem by a "trial and error" procedure, where they started the numerical computation by assigning a reasonable numerical value to θ conforming to those obtained experimentally. The system's instability then showed up by yielding solutions where $V \to \pm\infty$ according to whether the values of θ were chosen either too high or too low. The selected value for θ was that associated

---

* This instability may occur when the changing variable is dependent on two or more very different scales of the independent variable. Such a set of systems is, in computer terminology, called a *stiff* set of systems. This tendency to instability is well known in systems possessing elements of positive feedback and forms a prominent part of the modern theory of the dynamics of nonlinear systems (chaos).

with a continuous course of V until about half-way down the falling phase of V, where the tendency to instability could no longer be oppressed owing to computational difficulties. The final phase of V had to be calculated by using a different procedure based upon solving the set of Eq. (4.7.32) corresponding to $I = 0$.

For the theory it was decisive – but naturally also very satisfying – that this value of θ was in the range of conduction velocities observed experimentally on the squid axon and, likewise, that the form of the reconstructed action potential agreed so well with the one observed (see Fig. 4.50).

Chapter 5
Impulse transmission across cell boundaries

> ... the study of the end-plate potential and of its chemical origin gives us important clues to the much wider problem of biological 'chemo-electric transducers' ...
>
> *(Bernard Katz: The Sherrington Lectures X, 1969.)*

5.1 General characteristics of impulse transmission

5.1.1 The synapse

The nervous system in higher animals is usually classified into two categories: the *central* nervous system and the *peripheral* nervous system. The latter group contains the long ramifications *nerve-fibers/axons* from the centrally localized cell body called *the neuron*[*]. The peripheral nerve's solitary function is to transmit signals quickly and without error to and from the *"central computer"*.

The central nervous system receives from the *sensory receptor organs* the *afferent signals* containing information about events that occur inside and outside the organism. The signals are analyzed and processed, whereupon *efferent signals* are transmitted peripherally to the *effector organs* (muscles, glands, etc.) that require to be activated. An example of such a single structural element from this communication system is shown in Fig. 5.1. Any neuron receives afferent signals from many other neurons through the fine terminal contacts, or *synapses*[†], that are distributed all over the surface of the cell body and its dendrites. The individual neuron functions like a *mini-computer*, that is capable of integrating incoming signals to a single signal, which eventually gives rise to the start of a nerve impulse in the efferent axon that is conducted to the points

---

[*] Nowadays the convention of the neurocytologists is to use the word *neuron* for the integral cellular structure consisting of a cell *soma* with one or more ramifications, *dendrites* and *axons*.
[†] Gr. *synapsis*≡ interlocked, closely bound up.

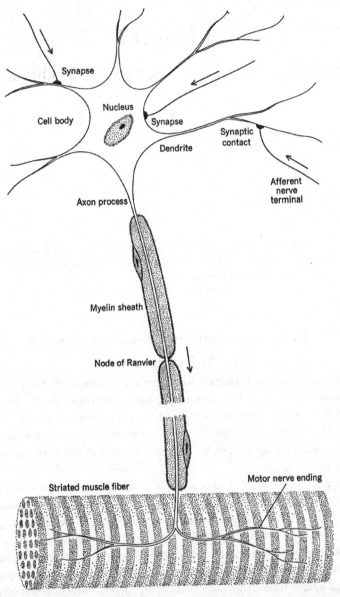

Fig. 5.1. Diagram of a nerve cell (spinal motor neuron) with some of its central and peripheral contacts (synapses). (After Katz*, 1966[†].)

---

* Bernard Katz was born in 1911 in Leipzig, and became an M.D. in 1934. He pursued advanced studies at University College, London, gaining a Ph.D. in 1938. In collaboration with A.L. Hodgkin he gave the first experimental evidence in 1949 for the Sodium Hypothesis formulated by Hodgkin & Huxley (1947). He was Professor and Head of the Biophysics Department of University College, London, from 1952 to 1978. He and his associates made numerous discoveries on the subject of neuromuscular transmission, among them the quantal release of neurotransmitter substances. Katz was knighted 1969 and awarded the Nobel Prize for Physiology and Medicine in 1970.

[†] Katz, B. (1966): *Nerve, Muscle and Synapses.* McGraw-Hill, Inc. New York.

of contact of the next station (the *target cell*), e.g. a cross-striated muscle fiber, where the synapse has a structure specially elaborated and has been given its own name: the *neuromuscular end-plate*.

5.1.2 Mechanisms of transmission

Over the course of time the synaptic region has been the subject for much discussion both as to its structure and to the mechanism that underlies the transmission of the impulse from the one cell to the other. In the main, two opposing points of view have been in the center of interest. (1) Adherents of the *theory of continuity* maintained that the growing axon invaded the cytoplasm of the target cell, and considered in reality the whole nervous system as one enormous syncytium. According to this theory the mechanisms of transmission that are at work at the synaptic junction should be the same as those that operate behind the impulse transmission before and after the synapse, i.e. the local current associated with the nerve impulse must flow uninterrupted across the synaptic junction to exert its action on the membrane of the target cell. Hence, the type of impulse transmission across the synapse is called an *electric transmission* and the synapse in which this mechanism operates is called an *electric synapse**. (2) The adherents of the contact theory (Ramon y Cajal)[†] maintained that each nerve cell – no matter how close the contact to the target cell might be – constituted a well-partitioned structural unit, i.e. the cytoplasm of the two cells that make the synapse are separated structurally from each other at the junction. This implies that the local currents that are advancing ahead of the nerve impulse – and also the impulse itself – are brought to *a stop* at the presynaptic membrane. A continuation of the impulse requires the release of a *special transmitter substance* that diffuses to the membrane of the target cell and in some way causes a new impulse to start again. This is the mechanism of *chemical transmission*. In accordance with these two views one often talks about an *electric synapse* and a *chemical synapse*.

The first convincing demonstration of the presence of a chemical transmission over a synapse was given in 1921 by the German biochemist Otto Loewi. Electric stimulation of the nerve fibers to the heart from the vagus nerve leads to a lowering in the frequency of the heart beat and a reduction in contraction force

---

[*] Some supporters of the electric transmission did not consider the continuum between the cells necessary, but regarded the presynaptic nerve ending to function as a closely placed external electrode.

[†] We owe to Santiago Ramon y Cajal (1852–1934), Professor of histology and pathological anatomy in Barcelona from 1887, fundamental work on the microscopic anatomy of the nervous system. He was the inventor of a special staining technique, and was awarded the Nobel Prize in 1906, together with the Italian histologist Camillo Golgi (1843–1919), Professor at the University of Pavia.

that eventually results in a total heart stop*, i.e. causing an inhibition of the pump function of the heart. Loewi showed now that stimulation of the fibers from the vagus nerve to the heart isolated from a frog led to a release in the surrounding fluid of a substance that had a *similar inhibitory action* on a second unstimulated isolated frog heart that was beating normally. This was demonstrated simply by transferring a small amount of the fluid from the stimulated heart to the fluid surrounding the unstimulated heart, which then showed the same pattern of inhibition as in the first electrically stimulated heart. This change could only come about if the transferred fluid contained an inhibitory substance that was released from the first heart as a result of the stimulation of the vagus fibers. Loewy called this substance "Vagusstoff" and showed that its actions were undistinguishable from those of *acetylcholine*[†]. Shortly afterwards it was shown that the liberation of acetylcholine was not localized solely to the vagus fibers to the heart since fibers belonging, for example, to the autonomic nervous system causing contraction of the pupil and accommodation of the lens, secretion of saliva, bronchial constriction, contraction of the intestinal muscle cells, etc., all transmitted the effect of the nerve signal to the respective effector organs via a release of acetylcholine. The action of the liberated acetylcholine is short-lasting because it is rapidly hydrolysed by a group of enzymes, *cholinesterases*, that are localized around the single nerve terminals. Loewi and his coworkers also showed that the strong pharmacological action of a well-known pharmacon *physostigmine*, or *eserine*, is due to an inhibition of the cholinesterase, thereby making the action of the liberated acetylcholine more potent and long lasting than normal.

The heart receives impulses partly from parasympathetic nerve fibers (n. vagus) that inhibit the pump function of the heart, and impulses partly from sympathetic nerve fibers whose action is to increase the frequency of the beats and force of contraction. Thus, the two nervous systems are antagonistic in their action on the function of the heart. Loewi also showed that impulses from the sympathetic fibers to the heart caused a release of an "accelerating" substance, whose action can be transferred to another isolated heart as in the case of acetylcholine. This substance, also called sympathin, was originally identified as the already-known hormone *adrenalin*, which is secreted by the adrenal glands. Later investigations showed that this substance is the trans-mitter substance in large parts of the sympathetic nervous system. However,

---

* This result – at that time surprising – was first observed in 1845 by the two German physiologists, the brothers Ernst Heinrich and Eduard Weber. It was the first example of a nerve stimulation causing an *inhibition* of an activity.

† The English physiologist and pharmacologist Sir Henry H. Dale had in 1914 demonstrated the vasodilation effect of acetylcholine. Dale was awarded the Nobel Prize for medicine in 1938, together with Otto Loewi, for their discoveries of the chemical transmission of nerve impulses.

the Swedish physiologist Ulf v. Euler showed in 1946 that the true transmitter substance was not adrenalin, which acts only via liberation from the adrenals, but the closely related substance *noradrenalin*. The English pharmacologists H.H. Dale, W. Feldberg & M. Vogt showed, in 1936, by using the perfusion technique of Otto Loewi, that acetylcholine is released from a skeletal muscle after stimulation of its motor nerve. A small amount of eserine was added to the perfusion fluid to delay the hydrolysis of acetylcholine. But, even then, the amount of transmitter substance that could be retrieved in the perfusion fluid was too small to be identified by ordinary chemical analysis. Instead one had to rely on the use of biological methods of identification* – in this case release of contraction in smooth muscles of leech. The results of these tests were that the substance that was released to the perfusion fluid had properties similar to those of acetylcholine. The appearance in the perfusion fluid of acetylcholine could be due to a release of the substance either from the motor nerves or from sensory nerve fibers from the muscle. The problem was settled by the following experiment. (1) To stimulate selectively the sensory fibers to the muscle, the respective dorsal root was cut proximally to the position of the stimulating electrodes. This caused no release of acetylcholine. (2) Stimulation of the motor nerves to the muscle via the ventral root resulted in a liberation of acetylcholine to the perfusing fluid. (3) Direct stimulation of a denervated[†] muscle did not give a release of acetylcholine. Thus, the acetylcholine in the perfusion fluid must have been released as a result of the stimulation of the motor nerves to the muscle. As a final step in this investigation, G.L. Brown, H.H. Dale & W. Feldberg showed that injection of very small amounts (2 μg) of acetylcholine into an artery belonging to the muscle caused a contraction of the same strength as a single contraction elicited by nerve stimulation. The contraction caused by this "close arterial injection" was prevented by a preceding injection of *curare*, used by the South American Indians as an arrow poison, which acts by blocking selectively the impulse transmission from nerve to muscle[‡]. It was also shown that injection of eserine prior to the nerve stimulation caused a larger and longer-lasting contraction. Taking together all these observations point to *the essential role that acetylcholine plays in the transmission of the impulse from nerve to muscle.*

---

* The technical name for this type of identification and measurement of potency is a **bioassay**.
† The nerves to the muscle were cut 10 days prior to the experiment to allow the peripheral nerve ends to degenerate – observation atrophy.
‡ This action was discovered in 1857 by Claude Bernard (1813–78), Professor of general physiology at The Sorbonne from 1854, and of experimental medicine at Collège de France. Bernard was the founder of modern physiology and experimental pathology. His work on diabetes, function of pancreas and liver, the innervation of the blood vessels and the action of toxic substances like CO and curare have given him the position as the greatest nineteenth-century physiologist.

5.2 Neuromuscular transmission: impulse transmission from nerve to muscle

In this section an account is given of the sequence of events that take place upon the arrival of the nerve impulse to the nerve terminals on the muscle fiber surface together with the processes that are associated with the acetylcholine release and lead to the transmission of the impulse to the surface membrane of the muscle fiber.

5.2.1 *Structure of the end-plate*

Figure 5.2 shows a schematic drawing from a light microscopic picture of the neuromuscular connections on a living sartorius muscle from a frog. The figure illustrates two points. (1) Each axon supplies several fibers – usually far more than shown on the figure – as the axon branches off just before it reaches the muscle fibers. (2) When the branches reach the surface of the muscle fiber they lose their myelin sheath and the unmyelinated terminals branch further and spread out along the surface of the fiber and form the presynaptic part of the motor end-plate. This arrangement is drawn schematically in Fig. 5.3. The terminal branching that is covered on the surface by a Schwann cell is partially buried in a groove made in the surface membrane of the muscle fiber. That part of the axon membrane that lies in the bottom of the groove is called the *presynaptic membrane*. Similarly, the region of the muscle membrane that forms the groove is called the *postsynaptic membrane*. The space between the

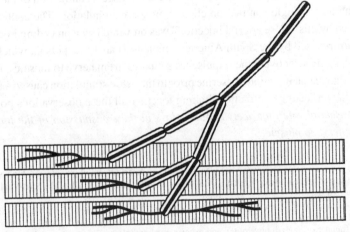

Fig. 5.2. Diagram of a motor axon and its terminals on the skeletal muscle fiber from a frog.

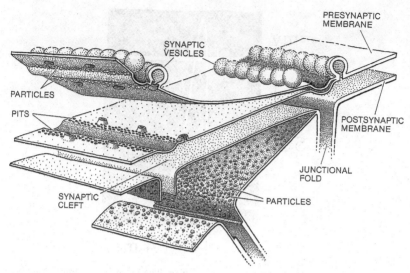

Fig. 5.3. Structure of the motor end-plate. A three-dimensional reconstruction of the pre- and postsynaptic membranes, the localization of the synaptic vesicles in the active zone, the synaptic cleft and the postsynaptic foldings. (From U.J. McMahan in Kuffler, S.W. & Nicholls, J.G. (1976): *From Neuron to Brain.* Sinauer Associates, Inc. Sunderland, Massachusetts.)

pre- and postsynaptic membranes has a thickness of about 10–20 nm, and is called the *synaptic cleft*. The postsynaptic membrane displays characteristic foldings that radiate with regular intervals perpendicular to the axis of the axon. In the presynaptic region near to the presynaptic membrane are localized *synaptic vesicles* with a diameter of 40–60 nm. The vesicles were first observed in the motor end-plate, but they are found in all nerve terminals that operate by a chemical transmission. Many investigations have now supported the view that the vesicles contain that substance that is characteristic for the synapse in question, i.e. acetylcholine in the motor end-plate. The vesicles are distributed such that the majority are localized close to the positions of the synaptic foldings. These regions are also called *active zones*, as it is assumed that it is from these regions the transmitter is released into the synaptic cleft. Besides the vesicles the presynaptic terminal contains numerous *mitochondria*, as an indication of a high rate of energy transformation, together with the enzymatic apparatus *choline acetyltransferase*, that takes part in the synthesis of acetylcholine. In particular in the region around the postsynaptic membrane there is localized the enzyme *cholinesterase**, which hydrolyses acetylcholine.

* See Section 5.2.3.3.

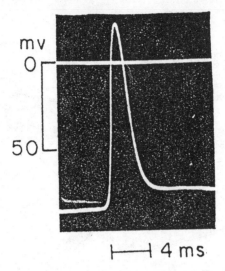

Fig. 5.4. Action potential from a frog muscle fiber. The recording is obtained at about 5 mm from the neuromuscular region. Resting membrane potential: −93 mV. Abscissa: time, ms. Ordinate: membrane potential, mV. (Gertrude Falk, 1961*.)

5.2.2 *The course of the electrical transmission*

Muscle fibers from skeletal muscles possess excitability and conduct impulses. The underlying events in the fiber membrane are in principle similar to those that operate in the axon of the squid. Thus, membrane depolarization above a critical level, the threshold depolarization, starts an action potential – almost similar in shape to that of the squid axon – that propagates along the whole length of the muscle fiber. An example of an action potential recorded with the tip of the glass microelectrode piercing the fiber membrane is shown in Fig. 5.4. In the frog muscle fiber the resting membrane potential is about −95 mV and the peak is at about +30 mV. The repolarization takes place in two phases. The first is fast – as in the squid axon – and covers most of the course of repolarization; the repolarization of the final 15 mV – the so-called tail – is much slower and may last 10–15 ms. The initial depolarization to start the impulse may be an electrical stimulus that is applied either by means of surface electrodes or microelectrodes which impale the myoplasm. This mode of stimulation is called *indirect stimulation*, in contrast to the normal stimulation of the muscle via its motor nerve, which is called *direct stimulation*. To start a propagated action potential in a muscle fiber the membrane must be depolarized from its resting membrane potential of about −90 mV to about −50 mV, i.e. by a displacement of about 40 mV. As shown below the electric currents that are associated with

* Page 269 in *Biophysics of Physiological and Pharmacological Actions*, Shanes, A.M. (Ed.), American Association for the Advancement of Science, Washington, D.C., 1961.

the propagating action potential in the nerve terminal of 1–2 μm thickness where it is "half buried" in the membrane of the muscle fiber, are far from sufficient to provide the amount of charge that is required to depolarize the muscle fiber of thickness 100 μm by more than 40 mV, even if all the current flowed in the interior of the fiber. In a myelinated frog nerve fiber the longitudinal axon current i_a during the impulse passage has approximately the form of a current pulse of strength 3×10^{-9} A and 0.5 s duration (Huxley & Stämpfli, 1949). We imagine this current injected into the interior of a muscle fiber of diameter $2a = 100$ μm. In Section 2 of Appendix I it is shown that an electric current, $2I_0$, that is injected into the interior of a cylindrical cell in the position $x = 0$, causes a membrane depolarization with the time course

$$V(0, T) = \lambda r_a I_0 \mathrm{Erf}(\sqrt{T}), \tag{5.2.1}$$

where λ is the length constant of the muscle fiber, r_a is the internal longitudinal resistance (Ω m^{-1}) and $T = t/\tau$ is the time in units of the membrane time constant $\tau = r_m c_m$. We use the following values* for the membrane constants of the muscle fiber: $\lambda = 0.2$ cm, $\rho_a = 200$ Ω cm, $\tau = 10$ ms $= 10^{-2}$ s. These values give

$$r_a = \frac{\rho_a}{\pi a^2} = \frac{200}{\pi (5 \times 10^{-3})^2} = 2.55 \times 10^6 \ \Omega \ \text{cm}^{-1}.$$

The current flows into the muscle fiber within 0.5 ms. Hence, $T = 0.5/10 = 0.05$, and $\sqrt{T} = 0.2236$. From Eq. (5.2.1) one should expect the muscle fiber to be depolarized by the amount

$$V(0, T) = 0.2 \times 2.55 \times 10^6 \times 1.5 \times 10^{-9} \times \mathrm{Erf}(0.2236)$$
$$= 7.65 \times 10^{-4} \times 0.248 = 1.9 \times 10^{-4} \ \text{V} = 0.2 \ \text{mV},$$

which is smaller by a factor of about 200 times than the 40 mV referred to above that is required to start off a propagated action potential.

The above calculation – despite its inherent uncertainties – points to the necessity for an interposed mechanism that provides the required amplification of the action of the presynaptic nerve signal to depolarize the muscle fiber membrane enough to start a propagated impulse. This mechanism of amplification results from a secretion of the chemical transmitter due to the arrival of the nerve impulse to the presynaptic region of the end-plate. That acetylcholine is the transmitter in the present case was shown by Dale, Feldberg & Vogt in 1936 in experiments on whole animals. We shall now account for the sequence of events that on the cellular level is associated with this transmitter release.

---

* See Appendix P, Table P.1.

5.2.2.1 The synaptic delay

To account for the course of electric events that are connected with the impulse transmission across the motor end-plate we refer to the sketch of Fig. 5.5(A). A frog muscle (e.g. a sartorius muscle with its nerve) is placed in a suitable chamber and the proximal end of the nerve is lifted above the bath and placed on a pair of electrodes that are connected to the stimulator S. Two recording electrodes a_1 and a_2 are placed on the nerve and connected to an amplifier. The input of a second amplifier is connected to two microelectrodes: a reference electrode that is placed somewhere in the Ringer's fluid and a recording electrode, the tip of which is carefully pressed down upon the terminal of the synaptic region. With this technique of recording (focal surface recording) one measures only potential changes from a small delimited region around the electrode tip. Figure 5.5 also shows the signals that are recorded from amplifiers (a) and (b) during the impulse passage to the muscle fibers. The first deflection in trace (B_1) is the result of a small part of the stimulation current that has spread (stimulation artifact) to the adjacent recording electrodes a_1 and a_2. The next deflection is the diphasic action potential recorded from the first amplifier with the usual convention for the polarity of surface recordings (an upward deflection where a_1 is negative with respect to a_2). The first deflection in trace (B_2) is again the stimulation artifact from the stimulation of the nerve. The following deflection is the focal surface recording from the microelectrode tip, which is placed directly upon the synaptic region. The deflections consist of two separate events. First a shortlasting, mainly negative, excursion of a magnitude of about 0.1 mV, that reflects the passage of the impulse past the electrode tip. Then after a delay of 0.5–0.8 ms there follows a diphasic deflection of several millivolts with the first peak going negative. The delay between these two signals – which is observed without exception – reflects electrically the separation between the events in the pre- and postsynaptic membrane, a separation that is confined not only to time. This appears among other things because the diphasic deflection is inhibited selectively by *curare*, or by concurrently lowering the content of Ca^{2+} in the Ringer's solution or by raising the content of Mg^{2+}. For that reason it is customary to talk partly about a *presynaptic potential*, which is localized across the presynaptic membrane, and partly about of a *postsynaptic potential*, which occurs across the postsynaptic membrane.

5.2.2.2 The postsynaptic potential

The actual nature of the diphasic deflection, whose origin evidently is postsynaptic, appears more clearly from Fig. 5.5(C), where the microelectrode that was used for the focal surface recording in the end-plate region is now pierced through the muscle fiber membrane to record the potential changes that occurs

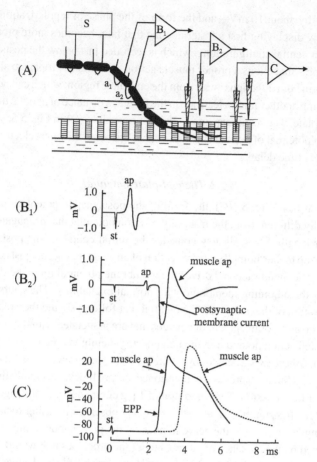

Fig. 5.5. Recordings of the electric events during the impulse transmission from nerve to muscle. (B₁) Recordings with surface electrodes on the nerve just before the end-plate region. (B₂) Recordings with a microelectrode placed directly on the nerve terminal in the end-plate region (focal recording). (C) Fully drawn line: intracellular recording from the place of focal recording. Broken line: intracellular recording 2 mm from the end-plate region. Note the synaptic delay. For further description see text.

across the postsynaptic membrane as a result of the passage of the nerve impulse along the nerve terminal, i.e. the presynaptic membrane. It appears that after the arrival of the nerve impulse to the presynaptic membrane it takes about 0.7 ms before a change in the presynaptic membrane occurs as a local depolarization that develops rapidly with a slight S-shaped course that ends with a decrease in curvature (a hump) at the threshold level for the regenerative excitation of the membrane, from where a new process of depolarization that strongly resembles the propagated action potential (see Fig. 5.4), even though the amplitude is smaller – by about 10 mV – and the form of the phase of repolarization differs slightly by displaying first a faster decline that later becomes more protracted. It is this potential configuration – which is recorded just below the postsynaptic membrane – that elicits a proper muscle action potential to propagate along the fiber axis in both directions away from the end-plate region, which can be shown by inserting another microelectrode in the fiber at a distance of about 2 mm from the end-plate. The result appears as the last recording (C) in Fig. 5.5, showing an action potential of normal appearance that arrives to the microelectrode after the correct time delay.

5.2.3 The end-plate potential

As is seen from Fig. 5.5(C), the front of the postsynaptic potential is characteristically different from the front of the action potential that propagates along the fiber length. We shall now consider the initial course of the postsynaptic potential up to the "hump" as the reflection of an event that occurs just across the postsynaptic membrane and drives the membrane potential up to the threshold level for the adjoining membrane with normal excitability. The course of the configuration of the postsynaptic potential that follows should then represent a passive invasion of the potential changes that are associated with the start of the propagated action potential in the neighboring membrane region. It is possible to separate these two processes by addition of an appropriate amount of curare to the Ringer's fluid. As previously mentioned, curare blocks reversibly the neuro-muscular transmission. Thus, addition of 1 μg curare per cm^3 of Ringer's solution blocks transmission in a frog muscle. As observed by using focal surface recording as in Fig. 5.5, the nerve impulse in the presynaptic terminal remains unaffected by the presence of curare, but all the processes that would normally occur after the synaptic delay are absent. However, this blockade does not have the character of being an "all or nothing" phenomenon, as there appears to exist a gradual transition both of the size and the form of the postsynaptic potential that is determined by the amount of curare added. This is illustrated in Fig. 5.6, which shows the effect on the postsynaptic potential of adding different doses of curare to the Ringer's fluid. Addition of progressively increasing amounts of curare in (a), (b) and (c) makes the initial course of the postsynaptic potential sluggish. The "hump" becomes more and more pronounced and appears with a greater delay. In (d) the "hump" just reaches the level for release of an action potential. A further addition of curare (e) makes the initial depolarization too small to elicit an action potential.

The results from Fig. 5.5 and Fig. 5.6 are most naturally explained as follows. The initial part of the postsynaptic potential has, under normal conditions, i.e. with an end-plate region unaffected by curare, the same protracted course as recorded in Fig. 5.6(e). But the amplitude of this depolarization always exceeds

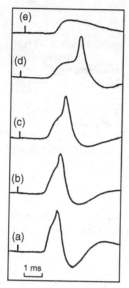

Fig. 5.6. Effect on the end-plate potential of adding curare in progressive amounts to a single nerve-muscle fiber preparation from frog. (a) Before adding curare; (b), (c) and (d) during progressive curarization showing the reduction of the initial course (the end-plate potential) with the "hump" and the progressive prolongation of the latency for the start of the muscle action potential; (e) clean end-plate potential, but reduced in size and therefore inadequate to elicit an action potential. Note: the experiment is performed on a single nerve muscle fiber, but the recordings are not obtained with intracellular microelectrodes but with fine surface electrodes, one put on the end-plate and the other on the surface of the muscle fiber. The recordings reflect reasonably correctly the course of the change in membrane potential, but with a smaller amplitude. (Modified from Kuffler, 1942.)*

the level for eliciting the propagated action potential. This level will be attained within 0.5 ms and the depolarization spreads passively into the neighboring normally excitable membrane region where a muscle action potential is initiated, that not only propagates along the length of the fiber away from the end-plate, but also spreads passively into the region of the postsynaptic membrane where it superposes and partly obscures the initial course of depolarization of the postsynaptic potential. This initial course of depolarization of the postsynaptic membrane that is only observed in its entirety – although with a reduced amplitude – when the muscle fiber is partly curarized is called the *end-plate potential*.

5.2.3.1 End-plate potential and acetylcholine

As mentioned previously, Dale, Feldberg & Vogt showed that stimulation of a motor nerve releases acetylcholine from the motor nerve endings and also that

---

* Kuffler, S.W. (1942): Electrical potential changes in an isolated nerve-muscle junction. *J. Neurophysiol.*, **5**, 18.

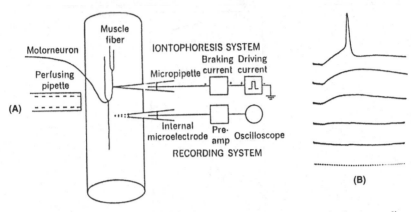

Fig. 5.7. (A) Schematic diagram showing experimental set up used when recording changes in membrane potential in the end-plate region in response to an ionophoretic microdose of acetylcholine. An inward "brake current" flows through the microelectrode to prevent the spontaneous diffusion of acetylcholine out of the pipette when ionophoresis is not applied. (B) Intracellular recordings of depolarization of the postsynaptic membrane in response to microdoses of acetylcholine in progressively increasing doses. (After Nastuk, 1955[†].)

a "close arterial injection" of acetylcholine causes a twitch-like contraction both in normal and denervated muscles. It lay near at hand to connect the appearance of the end-plate potential to the release of acetylcholine from the presynaptic region in which is present the chemical machinery for the synthesis of acetylcholine. If this assumption were correct it should also be possible to show that direct exposure of acetylcholine to the end-plate region should result in electric changes there that more or less resembled those that were observed during electric stimulation of the motor nerve fiber. In 1953, Nastuk[*] developed an ingenious technique that allowed a well-defined release lasting 1–2 ms of acetylcholine from a point source that could be placed accurately on the end-plate region within a distance of a few micrometers. At a pH = 7 acetylcholine is ionized and the acetylcholine ion is positively charged and consequently will move by electrophoresis in the direction of the field. This property is utilized by filling a glass microelectrode of the type that is used for intracellular recording with a solution containing acetylcholine. By sending a short pulse of electric current through the microelectrode it is possible to produce a short spurt of acetylcholine from the tip of the electrode. Figure 5.7(A) shows schematically

* Nastuk, W.L. (1953): Membrane potential changes at a single muscle end-plate produced by transitory application of acetylcholine with an electrically controlled microprobe. *Federation Proc.*, **12**, 102.
† Nastuk, W.L. (1955): Neuromuscular transmission: Fundamental aspects of the normal process. *Amer. J. Med.*, **19**, 663.

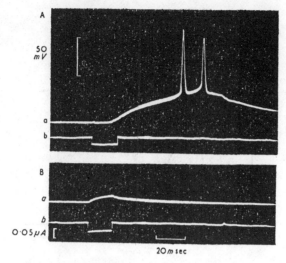

Fig. 5.8. External and internal iontophoretic microdosage of acetylcholine (ACh).
(A) Trace *b* shows the current flow through the micropipette indicated by the downward deflection. Trace *a* shows the membrane potential change observed when the micropipette is placed *just above* the end-plate. A depolarization develops after a diffusion delay that culminates in eliciting two propagating impulses. (B) Trace *b* is the same as above. In trace *a* the ACh-containing micropipette is pierced into the muscle fiber. A small displacement of the membrane potential follows the current flow from micropipette but otherwise *no effects* of the acetylcholine *release intracellularly* are observed (del Castillo & Katz, 1956*).

the experimental set up used for the application of this technique – also denoted ionophoretic microdosage of acetylcholine, and Fig. 5.7(B) shows the changes in the membrane potential across the postsynaptic membrane in response to ionophoretic jets of acetylcholine of about 2 ms duration. It appears that a progressive increase in the acetylcholine dose results in a similar graded increase in depolarization until the level that triggers off a propagated action potential. The same effects is shown in Fig. 5.8 (del Castillo & Katz, 1956), which illustrates in addition the lack of effect that is associated with an ionophoretic application of acetylcholine *intracellularly* underneath the postsynaptic membrane. The ionophoretic technique has been extensively used and refined. Thus Krnjevic & Miledi (1958)[†] succeeded in reducing the distance of diffusion from the microelectrode tip to the postsynaptic membrane to less than twice the width of the synaptic cleft with the result (Fig. 5.9) that the acetylcholine-evoked potential

* del Castillo, J. & Katz, B. (1956): Biophysical aspects of neuromuscular transmission. *Prog. Biophys. Biophys. Chem.*, **6**, 121.
[†] Krnjevic, K. & Miledi, R. (1958): Failure of neuromuscular propagation in rats. *J. Physiol.*, **140**, 440.

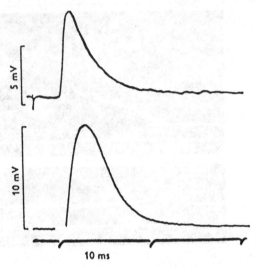

Fig. 5.9. Postsynaptic membrane potentials elicited by stimulation to the nerve of a rat diaphragm and by iontophoretic microdosage of acetylcholine. Upper recording: the synaptic potential is reduced by low Ca^{2+} and high Mg^{2+} content in the Ringer's fluid. Lower recording: acetylcholine applied ionophoretically. Note the almost identical time courses. (After Krnjevic & Miledi, 1958.)

attains a rise time that is only half the rise time of the normal end-plate potential that is evoked by nerve stimulation.

Dale, Feldberg & Vogt showed that a curare dose sufficient to cause a block of neuromuscular transmission did not to any appreciable extent affect the amount of acetylcholine that was released by stimulation of the motor nerves. It turned out that the curare added in progressively increasing doses reduces the course of end-plate potentials in essentially the same way whether evoked either by nerve stimulation or by ionophoretic microdosages. This indicates strongly that curare does not affect the amount of acetylcholine that is released during the nerve stimulation. Furthermore, there are several ways to influence the acetylcholine release from the nerve terminals on the end-plate. As mentioned above, a reduction of the Ca^{2+} concentration and an increase in the Mg^{2+} content in the surrounding fluid leads to an inhibition – eventually a total block – of the acetylcholine release. The same effect on the end-plate is seen in the substance *Botulinus toxin** (produced by anaerobic bacteria *Clostridium botulinum*, particularly in tainted canned meat), that causes a reduced or totally eliminated end-plate potential following nerve stimulation. The depolarization

---

* Like Tetrodotoxin (Chapter 4, Section 5.4.4.6 (i)), *Botulinus toxin* is fatally poisonous in extremely small doses.

of the postsynaptic membrane, however, is quite normal if acetylcholine instead is applied ionophoretically to the end-plate. The technique of ionophoresis has provided yet another important result. Injection of acetylcholine anywhere in the myoplasm causes neither a depolarization of the membrane nor a start of a propagated action potential. This indicates that the receptors to which acetylcholine is bound when causing the depolarization of the postsynaptic membrane are localized at the *outer side* of the postsynaptic membrane and not on its inner side. Additional support for this view is the observation that curare, which competes with acetylcholine in the binding to the receptor sites, is not able to inhibit the development of the receptor potential if it is applied at the inner side of the postsynaptic membrane. The above observations taken together lead to the conclusion that the end-plate potential is due to a reaction between a *specific receptor and acetylcholine* on the *outside* of the postsynaptic membrane.

5.2.3.2 The time course of the end-plate potential

It will be seen from Fig. 5.4 and Fig. 5.5(C) that the "hump" on the rising phase of the action potential is recorded only when the intracellular microelectrode is placed just in the end-plate region. In the section above, this configuration was interpreted as the combination of (i) the end-plate potential exceeding the threshold potential of the nearest excitable region to start a regenerative propagating response, which also (ii) passively invades the end-plate region. According to this interpretation the end-plate potential is – in contrast to the propagated action potential – regarded as a localized, non-regenerative, process. This view is supported from the results of Fig. 5.10. A muscle fiber – partly curarized to prevent release of a propagated action potential – is stimulated via its motor nerve. The responses are recorded with a glass microelectrode partly from the end-plate region (recording 0) and partly at various distances from the end-plate (the numbers attached to the curves in Fig. 5.10 give the distance of recording in mm). It appears that the end-plate potential spreads axially from the end-plate region into the neighboring membrane region. This distribution, however, is a passive, non-self-increasing process, as the signal attenuates and becomes more and more distorted with increasing distance and almost vanishes at a distance of 4 mm from the end-plate. The time course of the end-plate potential proper (curve 0 in Fig. 5.10), which decreases quasi-exponentially, consists of two phases. (1) A charging of the postsynaptic membrane as a result of a short-lasting depolarizing current pulse of ions through the postsynaptic membrane of 1–2 ms duration. (2) A passive spread into the neighboring regions that decays to the value of the resting membrane potential in about 20 ms. The shape of this configuration has been compared to the result obtained from passive cable theory of applying an instantaneous point charge Q at the position $x = 0$ on

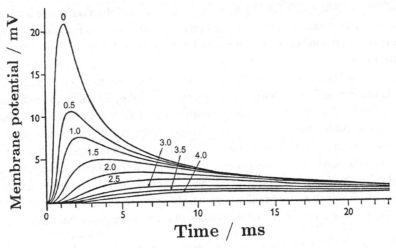

Fig. 5.10. End-plate potentials recorded with an intracellular microelectrode partly just underneath the end-plate region (recording 0) and partly at different distances from the end-plate (the distances in mm are attached to each recording). For further explanation see text. (Fatt & Katz, 1951[*].)

the cable. The result[†] is

$$V(X, T) = \frac{Q}{2c_m\lambda\sqrt{\pi T}} \exp\left\{-\frac{X^2}{4T} - T\right\}, \qquad (5.2.2)$$

where c_m is the membrane capacity, $\lambda^2 = r_m/r_a$ is the characteristic length, $X = x/\lambda$ is the distance in units of λ and $T = t/\tau$ is the time in units of the membrane time constant τ. A family of curves of $V(X, T)$ as a function of X for different value of T is shown in Appendix H. It will be seen that the curves are essentially of the same shape as those of Fig. 5.10, and when the membrane parameters of the muscle fiber of a frog are used to express Eq. (5.2.2) in terms of absolute values the equation predicts the experimentally obtained recordings reasonably well.

5.2.3.3 Cholinesterase and anti-cholinesterase

The indications of the above result are that the generation of the end-plate potential is localized solely at the postsynaptic membrane of the end-plate region. The primary process here is the binding of acetylcholine to specific receptors on the

[*] Fatt, P. & Katz, B. (1951): An analysis of the end-plate potential recorded with an intracellular electrode. *J. Physiol. (London)*, **115**, 320.
[†] Derived by A.L. Hodgkin in the above paper by Fatt & Katz (1951), appendix. A somewhat different procedure of deriving this equation is shown in Appendix M.

postsynaptic membrane; a reaction that gives rise to an inwardly-directed (depolarizing) current of 1–2 ms duration. The reason that the acetylcholine only acts for such a short time is because of the action of the enzyme *cholinesterase* that rapidly splits (hydrolyzes) the acetylcholine into the two inactive components *choline* and *acetic acid*. Acetylcholine esterase is an extremely effective enzyme, since each molecule has a maximal capacity to hydrolyze 5000 acetylcholine molecules per second, to which there corresponds a turnover time of 200 μs per acetylcholine molecule. The presence of cholinesterase in large amounts in the end-plate region has been demonstrated by specific staining methods. Thus, in each end-plate the amount of acetylcholine esterase is sufficient to hydrolyze 0.01 pmol (10^{-14} mol) acetylcholine within 5 ms, which is more than enough to inactivate the amount of acetylcholine that is released by each nerve impulse. Furthermore, acetylcholine is also present in the circulating blood, causing the splitting of the acetylcholine that diffuses from the synaptic cleft into the extracellular space and into the circulating blood. Several substances are known to *inhibit* the activity of the acetylcholine esterase. Before the Second World War only the native alkaloids – the so-called reversible inhibitors – were known, among which the most common were *Physostigmine* (Eserine), *Neostigmine* (Prostigmine) and *Edrophonium*. During the War a series of extremely toxic substances – so-called organophosphates – like tetraethylpyrophosphate and diisopropylfluorophosphate were developed, originally intended to be used as insecticides but later also as potential weapons in a war situation. The action of all these "nerve gases" is an irreversible inactivation of the acetylcholine esterase. As can be seen from Fig. 5.5 the time course of the end-plate potential is interfered with by the propagating action potential that also invades the end-plate region. To study the action on the end-plate potential of a choline esterase inhibitor like physostigmine it is necessary to block the propagation of the muscle fiber action potential. This can be done by curare, but this implies a study of changes in the end-plate potential when the end-plate region is exposed to the simultaneous action of two different substances. A more unambiguous response would be obtained by blocking the propagation of the action potential, by substituting the majority of the Na^+ ions in the external fluid with choline or sucrose. The action of prostigmine on the end-plate potential under this condition appears in Fig. 5.11(B), showing an end-plate potential that is drastically prolonged, as compared with the normal end-plate potential (A), with a plateau lasting 30–40 ms and a total time of decline of more than 200 ms. The prolonged time course undoubtedly results from the protracted action on the end-plate by the released acetylcholine whose rate of inactivation is greatly reduced because of the inhibition of the acetylcholine esterase by the prostigmine.

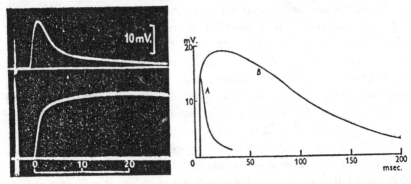

Fig. 5.11. Left-hand traces: effect of an acetylcholine esterase inhibitor (neostigmine) on the end-plate potential in a muscle fiber where the nerve impulse is blocked with a reduction of Na^+ in the external fluid (4/5 of Na^+ amount is replaced by sucrose). End-plate potential in a low sodium medium before (upper recording) and after addition of neostigmine to the Ringer's fluid (lower recording). Right-hand traces: the same, but shown with a smaller time resolution (200 ms against 20 ms). (A) Before addition of prostigmine. (B) After addition. (Fatt & Katz, 1951).

5.2.3.4 Ionic movements associated with the course of the end-plate potential

We have now presented a number of experiments that support the conception that the synaptic transmission between the motor nerve terminal and the muscle fiber is due to the release from the nerve terminal of the chemical transmitter acetylcholine, which diffuses through the synaptic cleft and exerts its action on the postsynaptic membrane, the end result being the end-plate potential. What is the nature of this action of acetylcholine?

The diffusion of the positively charged acetylcholine ions through the post-synaptic membrane has been considered as a underlying mechanism for the end-plate potential. However, a depolarization resembling an actual end-plate potential requires an entry of ions that exceeds by a factor of several thousands the amount of acetylcholine actually released from the nerve terminal. Therefore, the proposal was put forward that the acetylcholine, during its binding to specialized spots (receptors) in the postsynaptic membrane, also modified those spots by opening up new channels for small ions (e.g. Na^+, K^+ and Cl^- ions) to change the local pattern of ionic permeabilities in such a way that the post-synaptic membrane potential transiently moves towards a new stationary value having – like the original resting membrane potential – a zero membrane current[*]. This potential value could be found from the current–voltage characteristic of the postsynaptic membrane under the influence of acetylcholine, if one were available. Pursuing these lines del Castillo & Katz (1954) devised

---

[*] See Chapter 3, Section 3.6.3.1, Eq. (3.6.28) and the analysis at the end of this section.

an ingenious experiment which showed that during the acetylcholine release there was a definite value of the postsynaptic membrane potential – *the reversal potential* – that was associated with no membrane current flow: the potential across the postsynaptic membrane was displaced by the invasion of a propagating action potential that was elicited by electric stimulation of the fiber far away from the end-plate. By stimulating the muscle nerve – taking the appropriate measurements for the time it takes the impulse to reach the end-plate and the time for the propagating action potential to enter the same region – acetylcholine could be released corresponding to different levels of potential across the postsynaptic membrane ranging from e.g. -95 mV to $+20$ mV.

The effect of making the propagating action potential collide on the postsynaptic membrane with the release of acetylcholine is shown in Fig. 5.12, where each of the illustrations (a), (b), (c), (d) and (e) shows the effect of the collision together with the control. It appears that when the potential across the postsynaptic membrane $V_m^{(post)}$ is more negative than -15 mV the released acetylcholine causes an upward deflection, i.e. a *further depolarization* indicating an inward membrane current, but when the potential is less negative than -15 mV the deflection is downwards, indicating the flow of an outward membrane current. At a membrane potential of -15 mV the release of acetylcholine does not give rise to a current flow across the postsynaptic membrane. This value of membrane potential – also called the *reversal potential* for acetylcholine – must therefore be regarded as the stationary value which the postsynaptic membrane potential tends to attain during the transient action of acetylcholine. Del Castillo & Katz also observed an increase in the size of the end-plate potential as a result of a hyperpolarization of the postsynaptic membrane and a decrease following a depolarization (see also Fig. 5.13(B)). This observation lends further support to the view that the origin of the end-plate potential results from an increase in the conductances of some of the small ions that surround the postsynaptic membrane, but which ones the experiments cannot tell. That the sodium ion is likely to be involved is seen from Fig. 5.10, which – apart from the action of neostigmine – also shows that the size of the end-plate potential is reduced when the muscle fiber is surrounded by a Ringer's fluid with low sodium content.

To determine the mutual weights played by the Na^+, K^+ and Cl^- ions in the generation of the end-plate potential, Takeuchi & Takeuchi developed a modified "voltage clamp" technique (i.e. measurement of the membrane current when the membrane potential is held constant) to measure the end-plate current when acetylcholine was applied to the postsynaptic membrane that was held electronically at a predetermined value*. The flow sheet for the set up is shown

---

* See Chapter 4, Section 4.4.4.2, for an account of the voltage clamp technique of Hodgkin & Huxley.

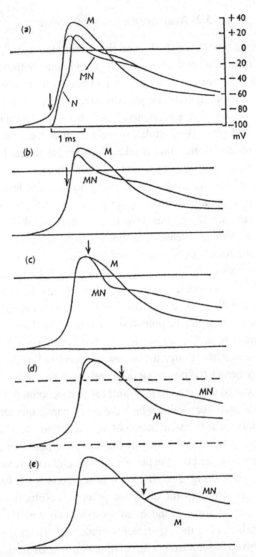

Fig. 5.12. Collision experiment of del Castillo & Katz. Recordings from an intracellular microelectrode placed in the end-plate region. The three superimposed curves M, N and MN in (a) are obtained as follows. M, the muscle fiber is stimulated far away from the end-plate. N, stimulation of the nerve to the end-plate region. The end-plate potential is followed by a propagated action potential. NM, both muscle and nerve are stimulated. The ↓ indicates the beginning of the end-plate potential. In (b), (c), (d) and (e) (only the recordings M and MN are superimposed) the nerve stimulation is delayed stepwise relative to the direct fiber stimulation. Recordings MN show how the action potential M is modified by the release of acetylcholine, indicated by the ↓. At membrane potentials lower than −15 mV (e) the potential is pulled upwards towards this level. Potential values higher than −15 mV tend to pull the membrane potential downwards. At a value (the reversal potential) of about −15 mV (d) acetylcholine has no other effect but to prolong the duration of this potential level (del Castillo & Katz, 1954).

558

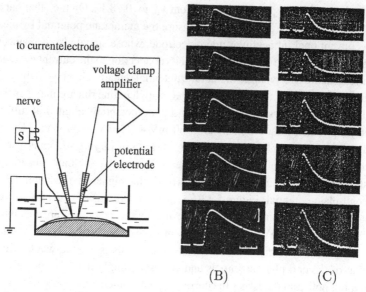

(B)   (C)

Fig. 5.13. (A) Principle for establishing a voltage clamp in the motor end-plate region. (B) End-plate potentials recorded with different membrane potentials (-60 mV, -75 mV, -90 mV, -100 mV and -112 mV) in the sequence from above downwards that are obtained by passing currents of varying strengths from the current electrode through the end-plate region. Time scale: 3 ms. Voltage scale: 10 mV. (C) End-plate currents recorded during voltage clamp conditions with the membrane potential held at -55 mV, -65 mV, -85 mV, -95 mV and -110 mV. Current scale: 0.1 mA (Takeuchi & Takeuchi, 1959[*]).

in Fig. 5.13(A). It is not possible to introduce longitudinally a current electrode of a final length in a muscle fiber as could be done in the squid axon. But owing to the small longitudinal extent of the end-plate region it is technically possible to clamp the end-plate by means of a glass micropipette that is connected to a current-controlling generator. To estimate the width of the potential profile that is due to passing a current I_0 from a point electrode placed intracellularly we make use of the result of Chapter 4, Eq. (4.4.18), which shows that the polarization decreases exponentially with distance x from the position of current injection at $x = 0$ according to

$$V(x) = V_0\,e^{-x/\lambda},$$

where V_0 is the potential at $x = 0$ and $\lambda = \sqrt{r_m/r_a}$ is the length constant of the muscle fiber, with r_m and r_a being the membrane resistance and internal longitudinal resistances both reckoned per unit length. For a frog muscle fiber we have $\lambda = 2$ mm (see Appendix K). Thus in the region ± 0.5 mm around the

---

[*] Takeuchi, A. & Takeuchi, N. (1959): Active phase of frog's end-plate potential. *J. Neurophysiol.*, **22**, 395.

current electrode the potential varies from V_0 to $0.78\ V_0$. On the other hand it does not present any difficulties to measure the membrane potential by means of a conventional intracellular microelectrode, whose signal also is connected to the electronic control system that, via its connection to the current generator, ensures that the postsynaptic membrane potential remains clamped to the preset value, irrespective of the current flow that acetylcholine release may generate. Figure 5.13(B) shows recordings of end-plate potentials elicited at different postsynaptic membrane potentials (-60 mV, -75 mV, -90 mV, -100 mV and -112 mV), brought about by passing currents of varying strengths from the current electrode through the membrane. It appears that the more negative the postsynaptic membrane potential is driven, the higher is the amplitude of the end-plate potential. Figure 5.13(C) shows recordings of end-plate currents that are elicited when the postsynaptic membrane potentials are maintained at the values -55 mV, -65 mV, -85 mV, -95 mV and -110 mV during the acetylcholine release. The shape of the currents is, on the whole, similar to that of the end-plate potentials and correspondingly the more negative the clamping potential the larger the currents' peak value.

The relevant examples that illustrate the mutual contributions Na^+, K^+ and Cl^- ions make to the end-plate current are shown in Fig. 5.14, where the peak values of the end-plate current I_{ep} are related to the values $V^{(clamp)}$ at which the postsynaptic membrane were clamped during the acetylcholine release. The range of clamping values used were -50 mV $> V^{(clamp)} > -150$ mV, as higher values of $V^{(clamp)}$ were not practicable since membrane depolarization to about -50 mV initiates undesired contraction in the contractile filament system of the fiber.

Figure 5.14(A) shows the relation between the peak end-plate current I_{ep} and the value $V^{(clamp)}$ obtained from the same end-plate with two different concentrations of tubocurarine ($3\ \mu g\ cm^{-3}$ and $4\ \mu g\ cm^{-3}$) added to the otherwise normal Ringer's fluid. As should be expected, the higher concentration causes an extra overall reduction of the end-plate current, but in both cases the relation between I_{ep} and $V^{(clamp)}$ is *linear*. Moreover, the data points in the range of clamping potentials from -50 mV to -150 mV are assembled closely enough around the linear fit to justify an extrapolation of this line to an intersection with the voltage axis to seek out that value $V_{I=0}$ of the postsynaptic membrane potential that corresponds to an end-plate current of zero. It is seen that the values of intersection for the two lines are more or less the same, namely $V_{I=0} \approx -15$ mV, which may be taken as the *reversal potential* for the postsynaptic membrane under the action of acetylcholine. That is, at all membrane potentials $V_m < -15$ mV one should expect the end-plate current to be directed inwards, whereas the current flow should reverse for $V_m > -15$ mV. The results

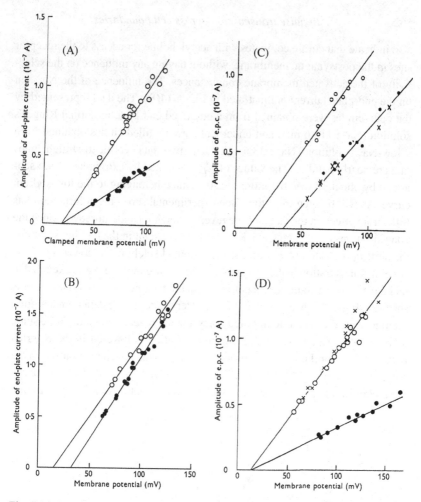

Fig. 5.14. Relationship between the peak value of the end-plate current (e.p.c) and the voltage clamped postsynaptic membrane potential from frog. Conditions are as follows. (A) Normal Ringer's solution, showing the effect of different concentrations of tubocurarine: $\bigcirc$, 3 μg cm^{-3}; $\bullet$, 4 μg cm^{-3}. (B) Effect of two different Na$^+$ concentrations in the external bathing fluid: $\bigcirc$, 113.6 mM; $\bullet$, 33.6 mM. (C) Effects of two different K$^+$ concentrations: $\bullet$, obtained in 0.5 mM K$^+$; $\bigcirc$, after soaking in 4.5 mM K$^+$; and $\times$, after return to 0.5 mM K$^+$. (D) Effect of two different Cl$^-$ concentrations: $\bullet$, obtained in normal Ringer's fluid; $\bigcirc$, after soaking in a glutamate Ringer's solution for 2–3 min; and $\times$, after soaking in glutamate Ringer for 15 min. When using glutamate Ringer the concentration of tubocurarine concentration was reduced to $\frac{2}{3}$ of that used in the normal Ringer's solution, and the Ca^{2+} concentration was raised to 5 mM. The change in tubocurarine concentration was introduced for reasons of aesthetics to prevent the merging of three sets of data. Note! Changes of Na$^+$ and K$^+$ concentrations in the bathing fluid caused a change in the reversal potential $V_{I=0}$ that – on the other hand – was uninfluenced by changes in the content of tubocurarine and Cl$^-$ ions. (From Takeuchi & Takeuchi, 1960*.) (To conform to the convention used in this book, the values on the abscissas should have a minus sign.)

* Takeuchi, A. & Takeuchi, N. (1960): On the permeability of the end-plate membrane during the action of the transmitter. *J. Physiol. (London)*, **154**, 52.

also indicate that curare competes with acetylcholine for access to the receptor sites in the postsynaptic membrane, without having any influence on the selectivity of the activated membrane conductances. The influence of the Na^+ ion on the end-plate current is illustrated in Fig. 5.14(B). The data represented by the open circles were obtained from a curarized end-plate in normal Ringer's solution (Na^+: 113.6 mM) and underneath are the filled circles obtained from a low-NaCl solution (Na^+: 33.6 mM). The two data sets fit to straight lines that are nearly parallel. The values of $V_{I=0}$ obtained by extrapolation are separated by about 17 mV, the more negative value belonging to the low-sodium curve. As will be seen from the same experimental series the chloride ion has little or no effect on the value of the reversal potential and, moreover, since the concentrations of K^+ and Ca^{2+} ions are held constant in the bathing solutions the shift in the value of $V_{I=0}$ must be attributed solely to the lowering of the sodium concentration in the Ringer's solution. The experiment was extended to include a series of sodium concentrations in the bathing solutions. The result is shown in Fig. 5.15. Irrespective of the greater scatter of the data points around the linear fit, there is a definite tendency for the reversal potential to assume more-negative values as the sodium concentration is lowered in the external bathing fluid. Similar experiments with changes in the K^+ concentrations in the bathing medium are illustrated in Fig. 5.14(C), yielding as in (B) two nearly parallel linear fits that are displaced such that the reduction of K^+ concentration

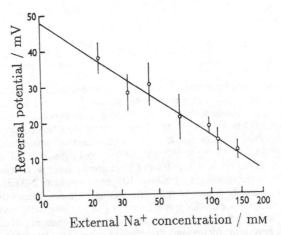

Fig. 5.15. Reversal potential $V_{I=0}$ (in mV) plotted against external sodium concentration on a semilogarithmic scale (mean ±standard deviation about the mean). The line is drawn according to $[99 + 1.29 \times 58 \log(15.5/C_{Na}^{(0)})]/2.29$. (From Takeuchi & Takeuchi, 1960.) (To conform to the convention used in this book, the values on the abscissas should have a minus sign.)

in the bathing medium results in a more-negative value of the reversal potential $V_{I=0}$. Figure 5.14(D) shows that replacement of the chloride ions by glutamate in the external medium does not affect the value of the reversal potential, even for a 10-fold reduction of the Cl^- ions.

To interpret the results of Fig. 5.14 we assume that the binding of acetylcholine to the postsynaptic membrane causes a transient change in the conductances for Na^+, K^+ and Cl^- ions to the values of $\hat{G}_{Na}$, $\hat{G}_K$ and $\hat{G}_{Cl}$ that are associated with the peak values of the end-plate current which is elicited at a membrane potential V. Furthermore, it is assumed that the ionic currents flow in their separate channels. The value $I_{ep} = I_{Na} + I_K + I_{Cl}$ of the instantaneous current is

$$I_{ep} = \hat{G}_{Na} \left(V - V_{Na}^{(eq)} \right) + \hat{G}_K \left(V - V_K^{(eq)} \right) + \hat{G}_{Cl} \left(V - V_{Cl}^{(eq)} \right),$$

or, after rearrangement

$$I_{ep} = [\hat{G}_{Na} + \hat{G}_K + \hat{G}_{Cl}]V - \left[\hat{G}_{Na} V_{Na}^{(eq)} + \hat{G}_K V_K^{(eq)} + \hat{G}_{Cl} V_{Cl}^{(eq)} \right]. \quad (5.2.3)$$

Experimentally the end-plate current I_{ep} and the clamp potential V were linearly related under all conditions. Equation (5.2.3) shows that this relation will be observed provided that the acetylcholine activated conductances $\hat{G}_j$ are constant – i.e. independent of V. The value of the slope is then expected to be

$$\hat{G}_{Na} + \hat{G}_K + \hat{G}_{Cl}.$$

The reversal potential $V_{I=0}^{(ach)}$ is located at the intersection of this straight line with the V-axis. Putting $I_{ep} = 0$ in Eq. (5.2.3) gives

$$\begin{aligned}
V_{I=0}^{(ach)} &= \frac{\hat{G}_{Na} V_{Na}^{(eq)} + \hat{G}_K V_K^{(eq)} + \hat{G}_{Cl} V_{Cl}^{(eq)}}{\hat{G}_{Na} + \hat{G}_K + \hat{G}_{Cl}} \\
&= \frac{\hat{G}_{Na}}{\hat{G}_{tot}} V_{Na}^{(eq)} + \frac{\hat{G}_K}{\hat{G}_{tot}} V_K^{(eq)} + \frac{\hat{G}_{Cl}}{\hat{G}_{tot}} V_{Cl}^{(eq)},
\end{aligned} \quad (5.2.4)$$

where $\hat{G}_{tot} = \hat{G}_{Na} + \hat{G}_K + \hat{G}_{Cl}$. It was found experimentally that the reversal potential $V_{I=0}^{(ach)}$ is unaffected by even a 10-fold change of $C_{Cl}^{(0)}$ or, in other words, to large changes in the equilibrium potential $V_{Cl}^{(eq)}$ of the chloride ions. This entails that

$$\frac{dV_{I=0}}{dV_{Cl}^{(eq)}} = \frac{\hat{G}_{Cl}}{\hat{G}_{tot}} \approx 0.$$

Therefore, we disregard the contribution from $V_{\text{Cl}}^{(\text{eq})}$ in Eq. (5.2.4) and write

$$V_{I=0}^{(\text{ach})} = \frac{\hat{G}_{\text{Na}} V_{\text{Na}}^{(\text{eq})} + \hat{G}_{\text{K}} V_{\text{K}}^{(\text{eq})}}{\hat{G}_{\text{Na}} + \hat{G}_{\text{K}}} = \frac{V_{\text{Na}}^{(\text{eq})} + (\hat{G}_{\text{K}}/\hat{G}_{\text{Na}})V_{\text{K}}^{(\text{eq})}}{1 + (\hat{G}_{\text{K}}/\hat{G}_{\text{Na}})} = \frac{V_{\text{Na}}^{(\text{eq})} + \hat{K} V_{\text{K}}^{(\text{eq})}}{1 + \hat{K}},$$
(5.2.5)

where

$$\hat{K} = \frac{\hat{G}_{\text{K}}}{\hat{G}_{\text{Na}}}$$
(5.2.6)

is the ratio of the acetylcholine-activated potassium conductance to the sodium conductance. At the resting membrane potential V_{m} the Cl^- ion is in equilibrium. Therefore

$$V_{\text{m}} = \frac{G_{\text{Na}} V_{\text{Na}}^{(\text{eq})} + G_{\text{K}} V_{\text{K}}^{(\text{eq})}}{G_{\text{Na}} + G_{\text{K}}} = \frac{V_{\text{Na}}^{(\text{eq})} + (G_{\text{K}}/G_{\text{Na}})V_{\text{K}}^{(\text{eq})}}{1 + (G_{\text{K}}/G_{\text{Na}})} = \frac{V_{\text{Na}}^{(\text{eq})} + K V_{\text{K}}^{(\text{eq})}}{1 + K},$$
(5.2.7)

where

$$K = \frac{G_{\text{K}}}{G_{\text{Na}}},$$
(5.2.8)

is the counterpart to Eq. (5.2.6) corresponding to the resting values of the conductances. Division of Eq. (5.2.8) by Eq. (5.2.6) yields

$$\left(\frac{\hat{G}_{\text{Na}}}{G_{\text{Na}}}\right) \bigg/ \left(\frac{\hat{G}_{\text{K}}}{G_{\text{K}}}\right) = \frac{K}{\hat{K}}$$
(5.2.9)

which allows a determination of the ratio of the relative increments of G_{Na} and G_{K}. The values of the ratios $\hat{K}$ and K are determined from Eq. (5.2.5) and Eq. (5.2.7) as

$$\hat{K} = \frac{V_{\text{Na}}^{(\text{eq})} - V_{I=0}^{(\text{ach})}}{V_{I=0}^{(\text{ach})} - V_{\text{K}}^{(\text{eq})}} \quad \text{and} \quad K = \frac{V_{\text{Na}}^{(\text{eq})} - V_{\text{m}}}{V_{\text{m}} - V_{\text{K}}^{(\text{eq})}}.$$

Using the values $V_{\text{m}} = -95$ mV, $V_{I=0}^{(\text{ach})} = -15$ mV, $V_{\text{Na}}^{(\text{eq})} = 50$ mM and $V_{\text{K}}^{(\text{eq})} = -100$ mV gives

$$K = \frac{50 - (-95)}{-95 - (-100)} = \frac{145}{5} = 29, \quad \hat{K} = \frac{50 - (-15)}{-15 - (-100)} = \frac{65}{85} = 0.75.$$

Inserting these values in Eq. (5.2.9) we obtain

$$\left(\frac{\hat{G}_{\text{Na}}}{G_{\text{Na}}}\right) \bigg/ \left(\frac{\hat{G}_{\text{K}}}{G_{\text{K}}}\right) = \frac{29}{0.75} = 38.$$

Thus, from the experiments of Takeuchi & Takeuchi (1960) it follows that the binding of acetylcholine to the end-plate causes an increase of the conductances for Na^+ and K^+ ions, but *not* for Cl^- ions. The ratio G_{Na}/G_K increases from a resting value of $1/K = 1/29 = 0.0345$ to $1/\hat{K} = 1/0.75 = 1.333$ at the peak of the end-plate current, that is by a factor of 38*. This increase is independent of the value at which the membrane potential V was clamped during the acetylcholine release. Thus, the postsynaptic membrane does *not contain* a regenerating conductance mechanism like that which is characteristic for the development of the nerve and muscle action potential[†]. For that reason the two ion channels are sometimes classified as *ligand-gated* and *voltage-gated*, respectively. Curare reduces the end-plate current but does not change the ratio G_{Na}/G_K, which indicates that curare competes with acetylcholine for the binding sites on the postsynaptic membrane.

5.2.4 *Quantal release of acetylcholine*

It was mentioned in Section 5.2.1 that the presynaptic region near the presynaptic membrane contained vesicles of a diameter of 40–60 μm. We shall now give an account of the observations that have led to the view that the vesicles contain acetylcholine and that the arrival of the nerve impulse to the presynaptic region causes these vesicles to incorporate into the membrane in order subsequently to empty their contents into the synaptic cleft.

5.2.4.1 *Miniature end-plate potentials*

In 1952, Fatt & Katz[‡] made the observation that the unstimulated end-plate region was not in a state of complete rest. High gain recordings, i.e. of fractions of millivolts – from a microelectrode that was impaled in the unstimulated end-plate region showed small fluctuations of the membrane potential that by and large had the same shape and duration as a normal end-plate potential elicited by nerve stimulation but having an amplitude of only about 0.5 mV. So for that reason Fatt & Katz called these small potential variations *miniature end-plate potentials* – abbreviated to MEPPs. An example of such a recording is shown in Fig. 5.16. The time interval between two succeeding MEPPs varied apparently purely at random, but a closer analysis of a great number ($N > 800$) of intervals showed that the spectrum of interval durations could be described

---

* This estimation differs from that of Takeuchi & Takeuchi (1960), who evaluated the value of the ratio $\Delta G_K / \Delta G_{Na}$.

[†] Thus, the behavior of the postsynaptic membrane has points in common with Bernstein's model for the excitable membrane.

[‡] Fatt, P. & Katz, B. (1952): Spontaneous subthreshold activity at motor nerve endings. *J. Physiol.*, **117**, 109.

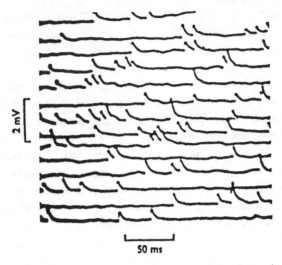

Fig. 5.16. Miniature end-plate potentials recorded from the end-plate region. The recordings were obtained consecutively from the oscilloscope where the trace after each run was lowered slightly before the start of the next passage of the track. *Note*: the voltage calibration is 2 mV. (After Fatt & Katz, 1952.)

by a decreasing monoexponential function of the interval length t as

$$n(t) = Ne^{-t/\lambda}, \tag{5.2.10}$$

where λ is a parameter that characterizes the distribution. Division on both sides by $\mathcal{N}$ gives the normalized distribution

$$\phi(t) = e^{-t/\lambda}. \tag{5.2.11}$$

This distribution can be derived from probability theory by the following argument: let t be the time passed after the occurrence of an MEPP, and let $S(t)$ represent the probability that *no* MEPP occurs within the time t. The probability for the occurrence of a new MEPP in the small time interval of duration Δt is assumed to be proportional to Δt, if $\Delta t \ll 1$. This probability is put equal to $\Delta t/\lambda$, where λ is a factor of proportionality. The probability $S(t + \Delta t)$ that no new MEPP will occur in the time $t + \Delta t$ is equal to the probability *both* for absence of an MEPP in the interval t *and* for absence in the succeeding interval Δt. The first probability is equal to $S(t)$, and the other is $1 - \Delta t/\lambda$, since the probability for either an occurrence or an absence of an MEPP in the interval Δt is equal to 1. We have then from the law of composite probabilities

$$S(t + \Delta t) = S(t)[1 - \Delta t/\lambda].$$

We write the left-hand side as $S(t) + (dS/dt)\,\Delta t^*$. Hence

$$\frac{dS}{dt} = -\frac{1}{\lambda}S,$$

or

$$\frac{1}{S}\frac{dS}{dt} = -\frac{1}{\lambda}.$$

Integrating gives

$$S(t) = A_1 e^{-t/\lambda},$$

where A_1 is an integration constant, whose value is determined by the condition $S(0) = 1$. This gives

$$S(t) = e^{-t/\lambda}, \tag{5.2.12}$$

which is the probability that *no new* MEPP occurs within the time t. The distribution of Eq. (5.2.12) is – not surprisingly – called the *exponential distribution*. The probability that a new MEPP occurs in the time between t and $t + dt$ is proportional to dt, and can be written as

$$f(t)\,dt,$$

where $f(t)$ is denoted the **probability density** for the exponential distribution. The above probability is equal to the probability for *both* absence of MEPPs in the interval t *and* occurrence of an MEPP in the interval dt, i.e.

$$f(t)\,dt = e^{-t/\lambda}\left(\frac{1}{\lambda}\right)dt.$$

Thus, the probability density for the exponential distribution for the exponential distribution is

$$f(t) = \frac{1}{\lambda}e^{-t/\lambda}. \tag{5.2.13}$$

Since an MEPP will occur sometime in the interval $0 < t < \infty$ we must have

$$\int_0^\infty f(t)\,dt = 1.$$

Inserting Eq. (5.2.13) gives

$$\frac{1}{\lambda}\int_0^\infty e^{-t/\lambda}dt = \frac{1}{\lambda}\left[-\lambda e^{-t/\lambda}\right]_{t=0}^{t=\infty} = [0 - (-1)] = 1,$$

as a control that the expression for $f(t)$ is dimensioned correctly.

---

* See Chapter 1, Section 1.2.2, Eq. (1.2.14).

The *mean value* or *expected value* of the exponential distributed time intervals is calculated as follows

$$\langle t \rangle = \int_0^\infty t f(t)\, dt = \int_0^\infty \frac{t}{\lambda} e^{-t/\lambda} dt.$$

Introducing a new variable $x = t/\lambda$, where $dt/dx = \lambda$, gives

$$\langle t \rangle = \lambda \int_0^\infty x\, e^{-x}\, dx.$$

Putting $u = x$ and $dv = e^{-x} dx$, that is $du = dx$ and $v = -e^{-x}$, partial integration* gives

$$\langle t \rangle = \lambda [-x\, e^{-x}]_0^\infty - \lambda \int_0^\infty (-1) e^{-x}\, dx$$

$$= \lambda \cdot 0 + \lambda [-e^{-x}]_0^\infty = \lambda [0 - (-1)] = \lambda. \qquad (5.2.14)$$

This shows that the parameter λ in the exponential distribution $\exp(-t/\lambda)$ is identical to the mean value $\langle t \rangle$ of the times t. Note that λ has the character of a usual time constant, as λ is the time where the (normalized) exponential distribution assumes the value $e^{-1} = 0.3679$.

The variance of the distribution is $\sigma^2 = \int_0^\infty (t - \langle t \rangle)^2 f(t) dt$. Insertion of Eq. (5.2.13) and Eq. (5.2.14) gives

$$\sigma^2 = \frac{1}{\lambda} \int_0^\infty (t - \lambda)^2 e^{-t/\lambda} dt = \frac{1}{\lambda} \int_0^\infty (t^2 - 2t\lambda + \lambda^2) e^{-t/\lambda} dt$$

$$= \frac{1}{\lambda} \int_0^\infty t^2 e^{-t/\lambda} dt - 2 \int_0^\infty t\, e^{-t/\lambda} dt + \frac{1}{\lambda} \int_0^\infty e^{-t/\lambda} dt$$

$$= \lambda \int_0^\infty \left(\frac{t}{\lambda}\right)^2 e^{-t/\lambda} dt - 2\lambda \int_0^\infty \left(\frac{t}{\lambda}\right) e^{-t/\lambda} dt + \lambda^2 [-e^{-t/\lambda}]_0^\infty$$

$$= \lambda^2 \int_0^\infty x^2 e^{-x} dx - 2\lambda^2 \int_0^\infty x e^{-x} dx + \lambda^2,$$

where the last line comes out by the substitution $x = t/\lambda$ and $dt = \lambda dx$. Putting $u = x^2$ and $dv = e^{-x} dx$ implies $du = 2x dx$ and $v = -e^{-x}$. Integration by parts then gives

$$\sigma^2 = \lambda^2 [x^2(-e^{-x})]_0^\infty - \lambda^2 \int_0^\infty 2x(-e^{-x}) dx - 2\lambda^2 \int_0^\infty x e^{-x} dx + \lambda^2 = \lambda^2,$$

which shows that the *standard deviation* of the exponential distribution

$$SD = \sqrt{\lambda^2} = \lambda, \qquad (5.2.15)$$

has the same numerical value as the mean value λ of the distribution.

* See Chapter 1, Section 1.7.2, Eq. (1.7.5).

Besides the probability density $f(t)$ for the occurrence of an MEPP *at* the time t, we can gain knowledge of the **distribution function** $F(t)$ or the accumulated probability for the occurrence of an MEPP in the course of the time t. As the probability dP for the occurrence of an MEPP in the interval between t and $t + dt$ is equal to $f(\tau)\,d\tau$, we obtain the distribution function $F(t)$ by summing all the differential probabilities dP for occurrence of an MEPP that are contained in the time interval T.

$$F(t) = \int_0^t f(\tau)\,d\tau = \frac{1}{\lambda}\int_0^t e^{-\tau/\lambda}d\tau$$

$$= \frac{1}{\lambda}\big[-\lambda e^{-\tau/\lambda}\big]_{\tau=0}^{\tau=t}$$

$$= 1 - e^{-t/\lambda}. \tag{5.2.16}$$

This result could, in the present case, be obtained more easily, since we already know that the probability for no occurrence of an MEPP in the interval t is $S(t) = \exp(-t/\lambda)$. Hence the probability for the occurrence of an MEPP in this interval is $1 - S(t)$. The function

$$\mathcal{N}(t) = N\big[1 - e^{-t/\lambda}\big], \tag{5.2.17}$$

that is Eq. (5.2.16) multiplied by the number N of observations, represents the absolute number of durations that are equal to or less than a given value T between two successive MEPPs. Figure 5.17 shows the observed number $\mathcal{N}_t^{(obs)}$ of intervals of duration less than t plotted as a function of t (Fatt & Katz, 1952). There appears to be a close agreement to the course of the distribution function $F(t) = 1 - e^{-t/\lambda}$ for the exponential distribution with a value of the mean interval $\langle t \rangle = 0.22$ s.

The miniature end-plate potentials were observed only in the end-plate region. Surgical denervating atrophy – that is, sectioning the motor nerve to the muscle which quickly leads to atrophy of the distal nerve end – causes the end-plate function and the MEPP to disappear. Likewise, application of curare reduced the amplitude of the MEPP whereas anticholine esterases, e.g. neostigmine, caused an increase.

There are several ways to depolarize or hyperpolarize the presynaptic membrane. One method is to pass an electric current through the nerve terminal from an external electrode that is placed upon the nerve as close as possible to the nerve terminal. An example is shown in Fig. 5.18. Another method consists of changing the potassium concentration in the bathing solution. In both cases it was observed that a depolarization of the *presynaptic membrane* led to an increase in the *frequency* of the occurrence of the MEPPs, while the frequency was reduced during a hyperpolarization. None of these procedures had any

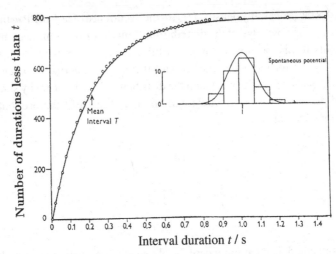

Fig. 5.17. Total number of observed intervals between two MEPPs of duration less than time t as a function of t. Open circles: observed data. Continuous line: the distribution $\mathcal{N}$ corresponding to the theoretical distribution for the exponential distribution $F(t) = 1 - e^{-t/\lambda}$. Total number of observations $N = 800$. The mean interval λ – here denoted by T – is shown by ↑. Insert diagram: histogram (published by del Castillo & Katz, 1954*) of the distribution of the MEPP amplitudes. (After Fatt & Katz, 1952.)

effect on the *amplitude* of the MEPPs, which appeared to be controlled by the state of the *postsynaptic* membrane.

Fatt & Katz interpreted the appearance of these miniature end-plate potentials as the reflection of an unceasing, irregular, intermittent release of packets of acetylcholine of similar action on the postsynaptic membrane as that associated with the release due to the arrival of a nerve impulse, but differing only by the number of molecules in the packets being several orders less than that contained in the massive release. Naturally, this raised the question as to whether the action was due to a single molecule or whether a number of molecules were involved. The amount of acetylcholine liberated per MEPP was estimated by experiments applying acetylcholine iontophoretically to the postsynaptic membrane (Krnjević & Mitchell, 1961)[†] and comparing the response with an end-plate potential that was evoked normally. The result of this estimate was that each nerve impulse was associated with a release of about 10^{-17} mol acetylcholine giving rise to an end-plate potential of about 50 mV. Then, an MEPP of 0.5 mV could – roughly estimated – be expected to contain 1/100

---

[*] del Castillo, J. & Katz, B. (1954): Quantal components of end-plate potential. *J. Physiol.*, **124**, 560.

[†] Krnjević, K. & Mitchell, J.F. (1961): The release of acetylcholine in the isolated rat diaphragm. *J. Physiol.*, **140**, 440.

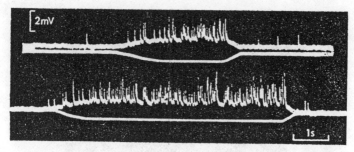

Fig. 5.18. Illustration of the effect of electrical depolarization of the presynaptic membrane on the frequency of MEPPs. In each recording the upper trace shows the MEPP, and the lower one the current that flows through the terminal part of the motor axon (del Castillo & Katz, 1954a*).

of the above amount, or at any rate a number far exceeding one molecule. Thus, in accordance with the explanation of Katz & Fatt, the occurrence of an MEPP is due to the spontaneous release from the presynaptic membrane of a well-defined "packet" of concentrated acetylcholine. These results, and others that pointed in the same direction, suggested to Katz the line of thought that the smallest observed MEPP reflected the unit of acetylcholine release. Thus, the arrival of the nerve impulse to the end-plate region did not lead to release of acetylcholine as a continuous – however short-lasting – current of separate molecules, but to an almost synchronous release of a large number of multi-molecular packets (*quanta*) of acetylcholine, each having the effect corresponding to an MEPP.

5.2.4.2 End-plate potential and miniature potentials

As previously mentioned, an increase of the Mg^{2+} concentration in the Ringer's solution causes a neuromuscular block; del Castillo & Engbaek (1954)[†] showed that the blockade was due to a reduction of the release of acetylcholine from the nerve terminals. The action of Ca^{2+} is antagonistic as an increase in the Ca^{2+} concentration diminishes the blocking effect of the Mg^{2+} ions, whereas a reduction of Ca^{2+} concentration alone causes a neuromuscular block, which completely resembles that due to excess of magnesium. On the other hand, the excess of magnesium has little effect on frequency and amplitude of the MEPPs. The progressive blockade of the neuromuscular transmission in a high-Mg^{2+} Ringer's solution can be observed by recording the membrane potential from

---

[*] del Castillo, J. & Katz, B. (1954a): Changes in the end-plate activity produced by presynaptic polarization. *J. Physiol.*, **124**, 586.

[†] del Castillo, J. & Engbaek, L. (1954): The nature of the neuromuscular block produced by magnesium. *J. Physiol.*, **124**, 553.

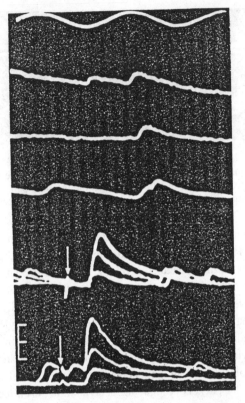

Fig. 5.19. Stepwise reduction in the amplitude of the end-plate potentials in a low-Ca^{2+} Ringer. The three upper recordings show MEPPs. The two lower ones are superimposed responses from nerve stimulation. (Stimulus marked with ↓. Time: 50 Hz. Amplitude: 1 mV. Fatt & Katz, 1952.)

a microelectrode impaled in the end-plate region and stimulating the motor nerve continuously with an interval of, for example, 1 s. As seen in Fig. 5.19, the decline in the amplitudes of the end-plate potentials does not take place smoothly and continuously; instead, the stimulus response *diminishes in small steps*, and, just before completion of the neuromuscular block, the postsynaptic response is *identical in amplitude and duration* to the spontaneously occurring *miniature end-plate potentials*. With the Mg^{2+} concentration properly adjusted to bring the end-plate region near to the entry of a total block and with a nerve stimulus of e.g. one per second one will observe the following. (1) A substantial number of stimulations do *not evoke* an end-plate potential. (2) The evoked end-plate potentials *fluctuate in amplitude* from the one stimulus to the other. An example of such fluctuations in amplitude is shown in Fig. 5.20. It is seen that

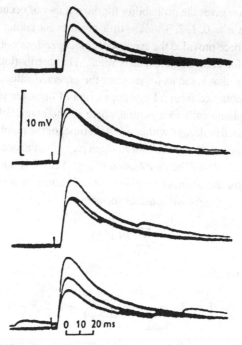

10 mV

0 10 20 ms

Fig. 5.20. Quantal fluctuations in amplitude of end-plate potential in a muscle fiber during the action of high magnesium concentration (10 mM) but with normal Ca^{2+} concentration (1.8 mM). Miniature potentials occurring spontaneously are seen in the tail of some of the traces (del Castillo & Katz, 1954[*]).

the amplitude of the end-plate potentials *changes stepwise*, as a multiple of a smallest "quantal jump". (3) The number ($n = 0, 1, 2, 3, \ldots$) of such jumps contained in a given end-plate potential varied at random with time. On this basis, del Castillo & Katz thought it natural to examine whether the number of quanta that occurred as a result of the nerve stimulation possessed a statistical behavior that could be described as Poisson distribution[†]

$$P(n;\mu) = \frac{\mu^n}{n!}\,e^{-\mu}, \quad \text{where} \quad n = 0, 1, 2, 3, \ldots, \tag{5.2.18}$$

[*] del Castillo, J. & Katz, B. (1954): *J. Physiol.*, **124**, 550.
[†] Derived in 1837 by S.D. Poisson (1781–1840) from Bernoulli's binomial distribution

$$S_n(p) = \binom{N}{n} p^n (1-p)^{N-n},$$

which is the probability that an event with probability p occurs exactly n times from a total of N observations. From this distribution Eq. (5.2.18) is obtained by letting $N \to \infty$ in such a manner that the mean value $\mu = Np$ of the distribution has a finite but small value. Equation (5.2.18) can also be derived from the distribution's own premises.

as this distribution gives the probability for the number of occurrences – given by the sequence $n = 0, 1, 2, 3, \ldots,$ – of a given event taking place within a certain time or space provided the events occur *independently* of each other and the probability for each single event is *small*. The distribution contains only one parameter μ, that is the *mean value* or the *expected* values of the number of occurrences obtained from a large population of the same process, e.g. the number of telephone calls to a central office in the course of 3 min, or the number of radioactive decays within a certain time, or the number of persons – however, not father, mother and children – that pass simultaneously through one element of a swing door. The distribution of Eq. (5.2.18) is normalized, that is the probability for the occurrence of one of the frequencies $n = 0, 1, 2, 3, \ldots,$ is equal to 1. As a control we consider the sum

$$\sum_{n=0}^{\infty} P(n; \mu).$$

and insert the expression for $P(n; \mu)$. This gives[*]

$$\sum_{n=0}^{\infty} \left(\frac{\mu^n}{n!} \right) e^{-\mu} = e^{-\mu} \sum_{n=0}^{\infty} \frac{\mu^n}{n!} = e^{-\mu} e^{\mu} = 1.$$

That the parameter μ of the distribution represents the mean value may be shown as follows

$$\langle n \rangle = \sum_{n=0}^{\infty} n P(n; \mu) = \sum_{n=0}^{\infty} n \left(\frac{\mu^n}{n!} \right) e^{-\mu} = \mu e^{-\mu} \sum_{n=1}^{\infty} \frac{\mu^{n-1}}{(n-1)!}$$

$$= \mu e^{-\mu} \sum_{n=0}^{\infty} \frac{\mu^n}{n!} = \mu e^{-\mu} e^{\mu} = \mu.$$

With the aim of examining whether the number of quanta composing an end-plate potential could be described as a Poisson process, del Castillo & Katz made a statistical analysis, partly of a large number of end-plate potentials of the type shown in Fig. 5.20, and partly of the amplitude and frequency of the MEPP. The main result was that the amplitude of the recorded end-plate potentials could be described by assuming that each end-plate potential is composed of the sum of a number ($n = 0, 1, 2, 3, \ldots$) of units ("quanta"), that have a mean amplitude and standard deviation that are identical with the mean amplitude and standard deviation of the miniature end-plate potentials occurring spontaneously. Let

$$\mu = \frac{\text{mean amplitude of EPP}}{\text{mean amplitude of MEPP}} \tag{5.2.19}$$

---

[*] See Chapter 1, Section 1.6.

denote the number of units that on average enter a single end-plate potential as a result of nerve stimulation. If the numbers of units that enter into the end-plate potential are distributed as a Poisson process, the mean value above must also be retrieved as the mean value μ of the Poisson distribution. When a set of data follows a Poisson process the mean value μ can be obtained by determining the probability that the event in question does *not occur*, since we have from Eq. (5.2.18)

$$P(0; \mu) = e^{-\mu} = \frac{\text{number of absent responses}}{\text{number of nerve impulses}}. \tag{5.2.20}$$

This gives

$$\mu = \ln\left(\frac{\text{number of nerve impulses}}{\text{number of absent responses}}\right). \tag{5.2.21}$$

If a Poisson process forms part of the distribution of units, a determination of μ by use of the two methods gives the same result; that is, the validity of the following relation must hold

$$\frac{\text{mean amplitude of EPP}}{\text{mean amplitude of MEEP}} = \ln\left(\frac{\text{number of nerve impulses}}{\text{number of absent responses}}\right). \tag{5.2.22}$$

Figure 5.21 shows the results of 10 experiments summed in this manner, where the values of μ obtained from Eq. (5.2.19) are plotted against the corresponding values of μ obtained from Eq. (5.2.21). The data points are seen to lump closely around the straight line of unit slope and thereby justify the assumption of the presence of a Poisson process[*]. On this the basis it was natural for del Castillo & Katz to examine whether all the end-plate responses were built up statistically by units whose mean and amplitudes were similar to those of the spontaneous potentials, whose occurrences are governed by the same Poisson process $P(n; \mu)$ with the values $n = 1, 2, 3, \ldots$, and μ obtained as the value corresponding to $P(0 : \mu)$ (Eq. (5.2.20) and Eq. (5.2.21)). This part of the analysis became less straightforward as the amplitude of the released EPPs did *not* grow in equidistant steps but displayed a certain variability. Hence it was not possible to place the labels $n = 1$, $n = 2$, etc., upon the single recordings, and a more thorough statistical analysis was required. An example of this method is illustrated in Fig. 5.22, showing the results of Boyd & Martin (1956)[†],

---

[*] The same result is obtained from investigations on end-plates on mammalian muscles (Boyd & Martin, 1956; Liley, 1965).

[†] Boyd, I.A. & Martin, A.R. (1956): End plate potentials in mammalian muscle. *J. Physiol. (London)*, **132**, 74.

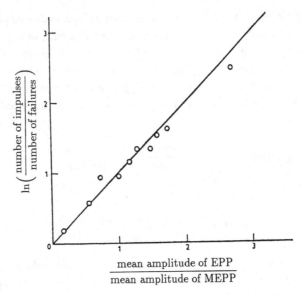

mean amplitude of EPP
―――――――――――――――――
mean amplitude of MEPP

Fig. 5.21. Test of the agreements between the two methods Eq. (5.2.19) and Eq. (5.2.21) to determine the mean value μ. Ordinate: $\ln\left(\frac{\text{number of nerve impulses}}{\text{number of absent responses}}\right)$. Abscissa: $\frac{\text{mean amplitude of EPP}}{\text{mean amplitude of MEPP}}$. The straight line corresponds to the same value of μ obtained by the two methods (del Castillo & Katz, 1954*).

obtained from mammalian muscle end-plates[†]. The inset diagram of Fig. 5.22 shows in the form of a histogram the amplitudes of 78 recorded MEPPs. From these data the probability density of the corresponding normal (Gaussian) distribution is evaluated and drawn. The mean value of the MEPP amplitudes was $m^{(\text{spont})} = 0.4$ mV with a standard deviation $\sigma_0 = 0.086$ mV[‡]. The nerve was stimulated successively 198 times with an interval of 2 s. Of these stimulations 18 gave no response. The 180 EPPs released varied in amplitude between 0.3 and 3.0 mV with a mean value $m^{(\text{stim})} = 0.933$ mV. Thus we have from Eq. (5.2.19)

$$\mu = 0.933/0.4 = 2.33,$$

and from Eq. (5.2.21)

$$\mu = \ln(198/18) = \ln 11 = 2.34.$$

---

* del Castillo, J. & Katz, B. (1954): *J. Physiol.*, **124**, 586.
† For our purpose of illustrating the quantitative procedure we prefer to reproduce the handling of the more extensive data by Boyd & Martin to that originally described by del Castillo & Katz.
‡ This value is given on p. 134 in Katz, B. (1996): *Nerve, Muscle and Synapses*, McGraw-Hill, Inc., New York.

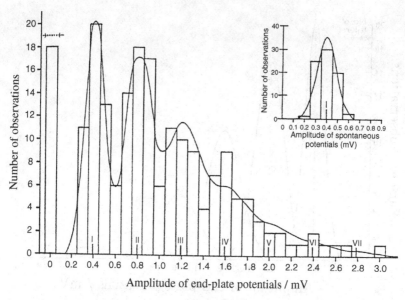

Fig. 5.22. Distribution of amplitude of EPP responses elicited from nerve stimulation and spontaneously occurring MEPP (inset diagram) from the end-plate of a mammalian muscle under the action of high Mg^{2+} concentration in the Ringer's solution. The spontaneous MEPPs are distributed around a mean amplitude of $m^{(spont)} = 0.4$ mV and the corresponding normal probability density is calculated and plotted. The histogram for the EPP displays peaks corresponding to the amplitudes 1, 2, 3 and $4 \times m^{(spont)}$. The arrows denote the number of absent responses. The fully drawn curve indicates the expected theoretical distribution of the histogram, which is calculated by making use of the values for $m^{(spont)}$ and the variance σ^2 for the spontaneously occurring MEPPs and assuming that the distribution of the amplitudes of the MEPPs is a Poisson process with the unit of amplitude equal to the mean amplitude of the MEPPs. (From Boyd & Martin, 1956.)

The agreement between the two values is adequate to justify the assumption of the quantum release as a Poisson process.

The amplitudes of the 180 EPPs due to nerve stimulation are presented in Fig. 5.22 as a histogram with an interval width of 0.1 mV. The distribution of EPP amplitudes in the histogram shows several peaks. The first is located at 0.4 mV and fits closely to the mean value $m^{(spont)} = 0.4 = m_1$ of the spontaneously occurring MEPPs shown in the inset diagram of Fig. 5.22. The second and third peaks (0.8 mV and 1.2 mV) correspond closely to $2 \times m^{(spont)} = 0.8 = m_2$ and $3 \times m^{(spont)} = 1.2 = m_3$, and the peaks that follow can – with a little ingenuity – be identified as $4 \times m^{(spont)} = 1.6 = m_4$ and $5 \times m^{(spont)} = 2.0 = m_5$. The fully drawn line indicates the theoretically expected distribution of the histogram if the basic unit in the histogram was an MEPP with mean amplitude 0.4 mV and a Gaussian distribution with a variable σ_0^2 like that of the inset diagram in

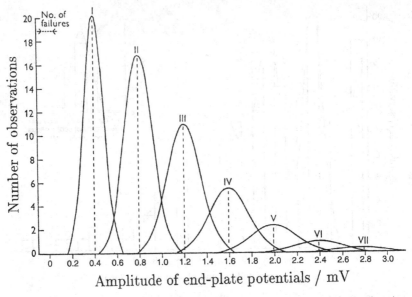

Fig. 5.23. Illustration of the method used to produce the continuous distribution function for the histogram for the EPPs in Fig. 5.21. For further explanation see text. (From Boyd & Martin, 1956.)

Fig. 5.22. The tactics used – due to del Castillo & Katz – in the construction of the theoretical curve are illustrated in Fig. 5.23. The basic assumption is that to each of the peaks in the histogram of Fig. 5.22 is assigned a class of EPP – composed respectively by 1 MEPP, 2 MEPPs, 3 MEPPs, etc. – whose distribution around the mean values

$$m_n = nm^{(\text{spont})} = 0.4 \times n, \quad \text{where} \quad n = 1, 2, 3, \ldots,$$

can be described by the normal probability density*

$$f_n(x) = \frac{\mathcal{N}_n}{\sqrt{2\pi}\sigma_n} e^{-(x-m_n)^2/2\sigma_n^2}, \tag{5.2.23}$$

where x is the amplitude of the EPP and σ_n^2 is the variance of the distribution in question. $\mathcal{N}_n$ is a constant that equals the total number of EPPs that belong to the class n concerned (e.g. corresponding to that of amplitude 2 MEPPs). It

---

* Also called a Gaussian distribution or a Gauss curve.

is assumed that this number can be calculated from the Poisson distribution as

$$\mathcal{N}_n = 198 \times P(n; \mu)$$

$$= 198 \times e^{-\mu}\, \mu^n/n! = 198 \times \frac{(2.335)^n}{n!} e^{-2.335} = 19.17 \times \frac{(2.335)^n}{n!}$$

which yields

$$\mathcal{N}_0 = 19, \quad \mathcal{N}_1 = 42, \quad \mathcal{N}_2 = 52, \quad \mathcal{N}_3 = 41,$$
$$\mathcal{N}_4 = 24, \quad \mathcal{N}_5 = 11, \quad \mathcal{N}_6 = 4.$$

Faced with the problem of selecting the numerical values of the individual Gauss curves, we start with the population of EPP corresponding to the class $n = 1$, that is with $\mathcal{N}_1 = 42$ elements and a mean amplitude equal to $m_1 = 0.4$ mV. According to the hypothesis, this population consists exclusively of EPPs, whose amplitude and duration are identical to those of the spontaneously occurring MEPPs. The variance of the distribution σ_1^2 is assumed to be identical to the variance σ_0^2 of the spontaneous MEPPs, the distribution of which is shown as the bell-shaped curve around the value of $x = 0.4$ in the insert diagram of Fig. 5.22. The next distribution corresponding to $n = 2$ has the mean value $m_2 = 0.8 = 2 \cdot m_1$. Since the quantum content is two packets the distribution can be considered as the sum of two normal distributions corresponding to $n = 1$. It follows from the probability theory that the variance of the sum of two distributions is equal to the sum of the variances of the two distributions concerned, thus

$$\sigma_2^2 = \sigma_1^2 + \sigma_1^2 = 2\sigma_0^2,$$

and with a similar assumption about the quantum content in a distribution corresponding to $n = k$ it follows that the variance is

$$\sigma_k^2 = k\sigma_0^2, \quad \text{and} \quad \sigma_k = \sqrt{k}\,\sigma_0,$$

where $\sigma_0 = 0.086$ is the variance of the distribution of the inserted diagram in Fig. 5.22. The peak values $\mathcal{N}_n$ of the individual Gaussian curves for $n = 1, 2, 3, \ldots$ can now be calculated from Eq. (5.2.23) since the number of observations that are contained inside a given interval equals the width of the interval Δx multiplied by the value obtained from the probability density function (Eq. (5.2.23)) taken at the mean value of x in the interval in question. Denoting this number as $N_i^{(\Delta x)}$, Eq. (5.2.23) gives

$$N_i^{(\Delta x)} = \frac{\mathcal{N}_n}{\sqrt{2\pi}\sigma_n} \int_{x_1 - \Delta x/2}^{x_1 + \Delta x/2} e^{-(x - m_n)^2/2\sigma_n^2}\, dx,$$

or, by making use of the mean value theorem for the definite integral*

$$N_i^{(\Delta x)} = \frac{\mathcal{N}_n \cdot \Delta x}{\sqrt{2\pi}\sigma_n} e^{-(x_i - m_n)^2/2\sigma_n^2} = f_n(x_i)\,\Delta x.$$

In the histogram of Fig. 5.22 the width of the interval is $\Delta x = 0.1$ mV. To the first Gauss curve ($n = 1$) the mean amplitude is $m_1 = 0.4$, and to the peak of $x_i = 0.4$ there corresponds the value of N_1, which is

$$N_1 = f_1(m_1)\,\Delta x = 0.4 \cdot \left(\frac{42}{0.086}\right) \cdot 0.1 = 19$$

since $1/\sqrt{2\pi} = 0.3969 = 0.4$.

For the remaining peaks we obtain

$$N_2 = 0.4 \cdot \frac{52}{1.4142 \cdot 0.86} = 17, \qquad N_3 = 0.4 \cdot \frac{41}{1.7320 \cdot 0.86} = 11,$$

$$N_4 = 0.4 \cdot \frac{24}{2.0 \cdot 0.86} = 6, \qquad N_5 = 0.4 \cdot \frac{11}{2.236 \cdot 0.86} = 2.$$

Figure 5.23 shows the individual Gauss curves for $n = 1, 2, 3, \ldots, 7$, whose algebraic summation is shown in Eq. 5.2.21 by the continuous curve, whose relative maxima appear to follow fairly well the peaks of the histogram.

This statistical analysis – which is given here in some detail out of consideration for readers who are not familiar with this kind of reasoning – strongly supports the hypothesis that under the present circumstances (low Ca^{2+} and high Mg^{2+}) the end-plate potential that is released from a nerve impulse is composed of single units (quanta) that are identical with the miniature end-plate potentials that occur spontaneously and are distributed with a mean amplitude of 0.4 mV and a standard deviation of ± 0.086 mV. Furthermore, the probability that the end-plate potential is composed of $n = 1, 2, 3, \ldots$, quanta is described as a Poisson process.

Later experiments have shown that the hypothesis of quantum liberation of acetylcholine also can be extended *beyond* the rather restrictive experimental conditions described here, provided that the basic statistical analysis (the Poisson process) is modified. This is to be expected for the reason alone that the Poisson distribution, which is an approximation from a more general probability distribution – the binomial distribution – is only valid for events having a low value p for the probability of occurrence[†]. For that reason it must be expected

---

* See Chapter 1, Section 1.3.4, Eq. (1.3.16).
[†] Note that in the Poisson distribution enters *only* the mean value μ of the distribution as the parameter but *not* the probability p for the occurrence of the event.

that under less restrictive experimental conditions – where the Ca^{2+} concentration gradually approaches normal values – must be described by Bernoulli's binomial distribution*

$$B(n; N, p) = \frac{N!}{(N - n)! \, n!} \, p^n (1 - p)^{(N-p)},$$

where $B(n; N, p)$ is the probability that an event with probability p occurs exactly n times out of a total of N trials. It can be shown that the mean of μ in this distribution is $\langle n \rangle = \mu = Np$ and a variance $\sigma^2 = Np(1 - p) = \mu(1 - p)$. If the quantal liberation after $\mathcal{N}$ nerve stimulations instead follows a binomial statistic the expected number of observations N_n containing $0, 1, 2, \ldots, N$ quanta released is

$$N_n = \mathcal{N} \frac{N!}{(N - n)! \, n!} \, p^n (1 - p)^{(N-p)}.$$

A statistical analysis based upon binomial statistics follows in principle the procedure described above, that is the mean value μ is determined, as before, by Eq. (5.2.21). A determination of the σ^2 for the quantal fluctuations permits a determination of the average probability p for the liberation of quanta since $p = 1 - \sigma^2/\mu$, from which one finds $N = \mu/p$. A series of investigations[†] with larger values of p have shown that the quantal fluctuations are reproduced satisfactorily by a model based upon the binomial statistics.

Thus, the investigations support the concept of a release of acetylcholine in the form of definite quanta that takes place as a stochastic process. The importance of this result is, among other things, due to the underlining of the necessity of describing the synaptic transmission as a combination of a *probability* for the liberation of acetylcholine quanta and the *number* of active sites on the nerve terminal from which the quanta of acetylcholine can be released.

As mentioned previously, the frequency of the MEPPs increases when the presynaptic membrane is depolarized (Fig. 5.18). A quantitative examination of the relationship shows that the frequency of the MEPPs increases by a factor of ten for a depolarization of 15 mV of the presynaptic membrane (Liley, 1956)[‡]. The result has been interpreted as a reflection of a ten-fold increase of the number N of active sites in the membrane of the terminal axon. A depolarization of the

---

* Derived by Jacob Bernoulli (1654–1708), Professor of mathematics in Basel, and published posthumously in 1713 in his great work on probability theory *Ars Conjectandi*. The Bernoulli family produced at least nine outstanding mathematicians, among others Nicolaus (1687–1759), Johannes (1887–1748) and Daniel (1700–1782), the latter called "the father of mathematical physics".

[†] Christensen, B. & Martin, A.R. (1970): *J. Physiol.*, **210**, 933. Wernig, A. (1970): *J. Physiol.*, **226**, 751. Zucker, R.S. (1973): *J. Physiol.*, **229**, 787.

[‡] Liley, A.W. (1956): *J. Physiol.*, **134**, 427–443.

axon corresponding to the arrival of a nerve impulse under normal conditions is expected to produce a very considerable increase in the number of active sites. A simultaneously occurring increase in the probability p for the quantal release will imply that even a short-lasting depolarization of the presynaptic membrane will result in an almost synchronous liberation of a "shower" of acetylcholine quanta (200–300). To produce a miniature potential, a liberation of 1000–10000 molecules of acetylcholine is required. Accordingly, a single nerve impulse should cause the release of about 2×10^6 acetylcholine molecules from the end-plate region, which agrees well with the results from application of acetylcholine ionophoretically, according to which each nerve terminal releases 10^{-17} mol acetylcholine per impulse, i.e. about $6 \times 10^{23} \times 10^{-17} = 6 \times 10^6$ molecules.

5.2.4.3 *The vesicle hypothesis*

As mentioned in Section 5.2.1, the nerve terminal, in particular close to the presynaptic membrane, contains a great number of vesicles of a diameter of about 50 nm. (Such vesicles have now been observed in all presynaptic terminals that release a chemical transmitter substance.) It was proposed by del Castillo & Katz that the vesicles contained acetylcholine and that the discharge of the contents of one vesicle into the synaptic cleft would correspond to the release of one quantum of the transmitter substance (acetylcholine). In support of this view there is the observation that from the brains of pigs it has been possible to isolate a subcellular fraction containing vesicles with a very high content of acetylcholine. Furthermore, it has been shown that application of a certain type of spider venom to the end-plate region leads to a drastic transient increase in the frequency of miniature potentials, which subsequently almost disappears. Examinations by electron microscopy of end-plates treated in this way show that the vesicle content in the nerve terminal is now very low. A similar result is also obtained by using a sustained depolarization of the nerve terminal – for example, by using a Ringer's solution where all the NaCl is substituted by potassium propionate.

Little is known about the details of the mechanism by means of which a depolarization of the presynaptic membrane causes an increase in the probability of the incorporation of the vesicles into the membrane and the subsequent draining of their contents into the presynaptic cleft. As mentioned above, the release of acetylcholine depends upon the Ca^{2+} concentration in the Ringer's solution, as sufficient lowering of the Ca^{2+} content causes a neuromuscular block. This calcium-dependent function is not peculiar to the motor end-plate, as many excitation–secretion processes in glands where neurotransmitters are

Table 5.1. *Summary of events at the motor end-plate during neuromuscular transmission*

PRESYNAPTIC MEMBRANE	SYNAPTIC CLEFT	POSTSYNAPTIC MEMBRANE
Nerve impulse		Propagated muscle
⇓		**action potential**
Depolarization		⇑
⇓		Depolarization of
Opening of Ca^{2+} channels		excitable adjacent region
⇓		⇑
Inflow of Ca^{2+}	⇐ Ca^{2+}	Local depolarization*
⇓		⇑
Discharge af **ACh**		Increase of G_K and G_{Na}
from vesicles		⇑
⇒	⇒ Diffusion of ACh ⇒	⇒ **Binding to receptor**
		⇓
		Hydrolysis of ACh
		(Cholinesterase)[†]
		⇓
		Choline +
Restorable in vesicles		CH_3COOH
⇑		⇓
Synthesis ⇐	⇐ Reuptake ⇐	⇐ Return by diffusion

* End-plate potential.

[†] Inhibits reversibly by the naturally-occurring alkaloids such as physostigmine, neostigmine, etc., or irreversibly by synthetic organophosphates such as tetratriphosphate, diisopropylfluorophosphate, etc. ("nerve gases").

involved are calcium-dependent. There are observations which indicate that it is the intracellular concentration of Ca^{2+} ions that controls the liberation of acetylcholine from the nerve terminal. Furthermore, experiments by Katz & Miledi point in the direction of a depolarization of the postsynaptic membrane that induces an opening of an ion channel to make the interior more accessible for Ca^{2+}, which by an as yet unknown mechanism contributes to an increase in the number of vesicles that discharge their contents into the synaptic cleft.

The sequence of events that take part in the neuromuscular transmission is summarized in Table 5.1.

Appendix A

About the functions Erf$\{x\}$, Erfc$\{x\}$ and calculation of the integral $\int_{-\infty}^{\infty} e^{-x^2}\, dx$

It is impossible to evaluate* the *indefinite integral* $\int e^{-x^2}\, dx$ in terms of elementary functions. A practicable way to bypass this difficulty is to develop the integrand in a power[†] series, thus

$$e^{-x^2} = 1 - [x^2] + \frac{[x^2]^2}{2!} - \frac{[x^2]^3}{3!} + \cdots + \frac{[-x^2]^n}{n!}$$

and then integrate each term separately

$$F(x) = \int \left(1 - x^2 + \tfrac{1}{2}x^4 - \tfrac{1}{6}x^6 + \cdots\right) dx$$

$$= x - \tfrac{1}{3}x^3 + \tfrac{1}{10}x^5 - \tfrac{1}{42}x^7 + \cdots + \tfrac{1}{(n+1)n!}[-x^2]^{n+1} + \text{constant}.$$

On the other hand, it is possible to calculate the value of the definite integral

$$I = \int_{-\infty}^{\infty} e^{-x^2}\, dx, \qquad (A.1)$$

making use of the following ingenious artifice, which at least can be traced back to Laplace. It runs as follows.

Since x in Eq. (A.1) is only a dummy variable which does not enter into the final result one could equally well use another variable, e.g. y, and write the integral as

$$I = \int_{-\infty}^{\infty} e^{-y^2}\, dy. \qquad (A.2)$$

We now write

$$I^2 = \left(\int_{-\infty}^{\infty} e^{-x^2}\, dx\right)^2 = \left(\int_{-\infty}^{\infty} e^{-x^2}\, dx\right)\left(\int_{-\infty}^{\infty} e^{-y^2}\, dy\right),$$

---

* To evaluate the above integral we shall make use of double integrals, and thus go beyond the mathematical requirements that were laid down in Chapter 1. But no more elementary methods of solution exist. Readers without experience in dealing with double integrals can easily skip the details of the derivation and just accept the result.

[†] See Chapter 1, Section 1.6.

584

and let x and y represent the independent variable on the x- and y-axis in a three-dimensional rectangular coordinate system (x, y, z). We have then

$$\left(\int_{-\infty}^{\infty} e^{-x^2} \, dx \right) \left(\int_{-\infty}^{\infty} e^{-y^2} \, dy \right) = \int_{-\infty}^{\infty} \int_{-\infty}^{\infty} e^{-x^2} e^{-y^2} \, dx \, dy^*$$

$$= \int_{-\infty}^{\infty} \int_{-\infty}^{\infty} e^{-(x^2+y^2)} \, dx \, dy. \tag{A.3}$$

The integration of Eq. (A.3) – which implies a summation of volume elements with surface elements $dx \, dy$ and height $e^{-(x^2+y^2)}$ – is extended over the whole x–y plane*. The form of the integrand in Eq. (A.3) suggests that the integration over the entire x–y plane is more easily carried out by using the polar coordinates r and θ, which are defined by the transformation $x = r \cos\theta$ and $y = r \sin\theta$. We then have $x^2 + y^2 = r^2$, while the surface element in polar coordinates is $r \, d\theta \, dr$. To cover the whole x–y plane the values of the new variables θ and r must cover the ranges $0 \le \theta \le 2\pi$ and $0 \le r < \infty$. It follows then

$$I^2 = \int_0^{\infty} \int_0^{2\pi} e^{-r^2} r \, dr \, d\theta = \left(\int_0^{2\pi} d\theta \right) \left(\int_0^{\infty} r \, e^{-r^2} \, dr \right) = 2\pi \int_0^{\infty} e^{-r^2} r \, dr$$

where the volume elements with surface $2\pi r \, dr$ now are summed. We have

$$\frac{d}{dr} \left(e^{-r^2} \right) = -e^{-r^2} \left(\frac{dr^2}{dr} \right) = -e^{-r^2} 2r,$$

whence

$$I^2 = -\pi \int_0^{\infty} \left(\frac{d}{dr} \left(e^{-r^2} \right) \right) dr = -\pi \left[e^{-r^2} \right]_0^{\infty} = -\pi(0 - 1) = \pi,$$

or

$$I = \sqrt{\pi},$$

which compared with Eq. (A.1) gives the celebrated result

$$\int_{-\infty}^{\infty} e^{-x^2} \, dx = \sqrt{\pi}. \tag{A.4}$$

Since the function e^{-x^2} is an even function ($f(x) = f(-x)$ like, e.g., $y = x^2$) we have

$$\int_{-\infty}^{\infty} e^{-x^2} \, dx = 2 \int_0^{\infty} e^{-x^2} \, dx,$$

and therefore also

$$\int_0^{\infty} e^{-x^2} \, dx = \tfrac{1}{2} \sqrt{\pi}. \tag{A.5}$$

---

* The theory of multiple integration provides the rule: *if the function f(x,y) can be written down as a function of x alone $\phi(x)$ and a function of y alone $\psi(y)$ then the double integral of f(x,y) over the rectangular domain $\mathcal{R}(a \le x \le b; c \le y \le d)$ in the x–y plane is equal to the product of two simple integrals:*

$$\iint_{\mathcal{R}} f(x, y) \, dx \, dy = \left(\int_a^b \phi(x) \, dx \right) \left(\int_c^d \psi(y) \, dy \right).$$

Substituting in this expression x^2 with $u^2 = \alpha x^2$ gives

$$\int_0^\infty e^{-\alpha x^2}\, dx = \int_0^\infty e^{-u^2} \left(\frac{dx}{du}\right) du = \frac{1}{\sqrt{\alpha}} \int_0^\infty e^{-u^2}\, du = \frac{1}{2}\sqrt{\frac{\pi}{\alpha}}. \tag{A.6}$$

The *error function* Erf $\{x\}$ is defined as

$$\text{Erf}\,\{x\} = \frac{2}{\sqrt{\pi}} \int_0^x e^{-t^2}\, dt. \tag{A.7}$$

The error function Erf $\{x\}$ is of importance in many branches of mathematical physics and in probability theory, and was tabulated a long time ago[*]. We consider only the limiting values. From Eq. (A.7) it follows that

$$\text{Erf}\,\{0\} = 0, \tag{A.8}$$

and

$$\text{Erf}\,\{\infty\} = 1, \tag{A.9}$$

since for $x = \infty$ the value of the integral is $\sqrt{\pi}/2$.

To evaluate Erf $\{-x\}$ we introduce a new variable of integration $-u = t$ in Eq. (A.7). Hence

$$\text{Erf}\,\{-x\} = \frac{2}{\sqrt{\pi}} \int_0^{-(-x)} e^{-u^2} \left(\frac{dt}{du}\right) du = -\int_0^x e^{-u^2}\, du = -\text{Erf}\,\{x\}. \tag{A.10}$$

The *complementary error function* Erfc $\{x\}$ is defined as

$$\text{Erfc}\,\{x\} = 1 - \text{Erf}\,\{x\}. \tag{A.11}$$

It follows from Eq. (A.7) that

$$\text{Erfc}\,\{0\} = 1, \tag{A.12}$$

and from Eq. (A.9) that

$$\text{Erfc}\,\{\infty\} = 0, \tag{A.13}$$

which combined with Eq. (A.10) gives

$$\text{Erfc}\,\{-x\} = 1 - \text{Erf}\,\{-x\} = 1 + \text{Erf}\,\{x\}, \tag{A.14}$$

and

$$\text{Erfc}\,\{-\infty\} = 1 + \text{Erf}\,\{\infty\} = 2. \tag{A.15}$$

The graphs of the functions Erf $\{x\}$ and Erfc $\{x\}$ are shown in Fig. A.1. For the numerical calculations used in this book the author has used a remarkably simple

---

[*] A four-figure table of Erf x can be found in, for example: Milne-Thomson, L.M. & Comrie, L.T. (1957): *Standard Four-Figure Mathematical Tables*, MacMillan & Co Ltd, London.

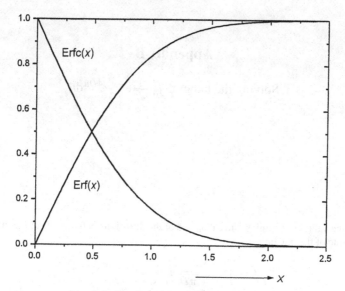

Fig. A.1. The functions Erf $\{x\}$ and Erfc $\{x\}$.

algorithm* with a precision better than 1.5×10^{-7}. In our program[†], which returns Erf $\{x\}$ and Erfc $\{x\}$ for each entered value of x, the algorithm reads as follows.

```
        DOUBLE PRECISION X,T,B1,B2,B3,B4,B5,Z,ERF
1       WRITE(*,'(//,10X,A )') 'Enter value of X: '
        READ(5,*,END=110)X
        Z=ABS(X)

        T= 1.0/( 1.0+ 0.3275911*Z)
        B1=0.254829592
        B2=-0.28449674
        B3=1.421413741
        B4=-1.453152027
        B5=1.061405429

        ERF= 1.0 - (T*(B1 +T*(B2 +T*(B3 +T*(B4 +T*(B5)))))) * DEXP(-X*X)
        IF(X.GE.0.0)THEN
            WRITE(*,'(10X,A,F8.6)')')'ERF(X)=', ERF
            WRITE(*,'(10X,A,F8.6)')')'ERFC(X)=',1-ERF
        ELSE
            WRITE(*,'(10X,A,F8.6)') 'ERF(X)=' ,-ERF
            WRITE(*,'(10X,A,F8.6)') 'ERFC(X)=' ,1+ERF
        END IF
        GO TO 1
110     END
```

* From Hastings, Jr. (1955): *Approximations for Digital Computers*, Princeton University Press, Princeton, N.J., but taken from Abramowitz, M. & Stegun, Irene A. (1965): *Handbook of Mathematical Functions*, p. 299, Dover Publications, Inc., New York.
† In FORTRAN 77.

Appendix B

Solving the integral $\int_0^t \frac{1}{\sqrt{u}} e^{-x^2/4Du} du$

With the values of charging current J_0 and of the diffusion coefficient D of the medium the value of the integral

$$C(x, t) = \frac{J_0}{\sqrt{\pi D}} \int_0^t \frac{1}{\sqrt{u}} e^{-x^2/4Du} du, \qquad (B.1)$$

depends upon the values (x, t) of position and time. To facilitate the typography we write Eq. (B.1) as

$$C(x, t) = A \int_0^t \frac{1}{\sqrt{u}} e^{-a^2/u} du, \qquad (B.2)$$

where

$$A = \frac{J_0}{\sqrt{\pi D}} \quad \text{and} \quad a^2 = \frac{x^2}{4D}. \qquad (B.3)$$

Introducing a new variable of integration

$$\mu^2 = \frac{a^2}{u} \quad \text{or} \quad u = \frac{a^2}{\mu^2},$$

the integral transforms* into

$$C(x, t) = A \int_{\mu(u=0)}^{\mu(u=t)} (\cdots) \frac{du}{d\mu} d\mu, \qquad (B.4)$$

where $(\cdots)$ symbolizes the "μ substituted" integrand. We have

$$\frac{du}{d\mu} = -2a^2 \mu^{-3} \quad \text{and} \quad \mu = \begin{cases} \infty & \text{for} \quad u = 0 \\ a/\sqrt{t} & \text{for} \quad u = t, \end{cases}$$

and hence

$$C(x, t) = A \int_\infty^{a/\sqrt{t}} \frac{\mu}{a} e^{-\mu^2} (-2a^2 \mu^{-3}) d\mu = 2aA \int_{a/\sqrt{t}}^\infty \mu^{-2} e^{-\mu^2} d\mu,$$

---

* See Chapter 1, Section 1.7.1, Eq. (1.7.2) and Section 1.7.2, Eq. (1.7.5).

which is an alternative form of Eq. (B.2). To evaluate the integral

$$I = \int_{a/\sqrt{t}}^{\infty} \mu^{-2} e^{-\mu^2} d\mu$$

by integration by parts we put

$$f(\mu) = e^{-\mu^2}, \quad \text{i.e.} \quad f'(\mu) = -2\mu e^{-\mu^2},$$

and

$$g'(\mu) = \mu^{-2}, \quad \text{i.e.} \quad g(\mu) = \int \mu^{-2} d\mu = -\mu^{-1}.$$

This gives

$$I = \int_{a/\sqrt{t}}^{\infty} \mu^{-2} e^{-\mu^2} d\mu = \left[-\frac{1}{\mu} e^{-\mu^2} \right]_{a/\sqrt{t}}^{\infty} - \int_{a/\sqrt{t}}^{\infty} -\frac{1}{\mu}(-2\mu) e^{-\mu^2} d\mu$$

$$= \frac{\sqrt{t}}{a} e^{-a^2/t} - 2 \int_{a/\sqrt{t}}^{\infty} e^{-\mu^2} d\mu = \frac{\sqrt{t}}{a} e^{-a^2/t} - 2 \frac{\sqrt{\pi}}{2} \left(\frac{2}{\sqrt{\pi}} \int_{a/\sqrt{t}}^{\infty} e^{-\mu^2} d\mu \right)$$

$$= \frac{\sqrt{t}}{a} e^{-a^2/t} - \sqrt{\pi} \, \text{Erfc} \left\{ \frac{a}{\sqrt{t}} \right\}.$$

Now $C(x, t) = (2aA) I$. Thus, inserting the values of a and A from Eq. (B.3) yields finally

$$C(x, t) = 2J_0 \sqrt{\frac{t}{\pi D}} e^{-x^2/4Dt} - \frac{J_0 x}{D} \, \text{Erfc} \left\{ \frac{x}{2\sqrt{Dt}} \right\}. \tag{B.5}$$

In connection with the actual computation and plot of the function $C(x, t)$ it may be of advantage to introduce the variables

$$X = \frac{x}{2\sqrt{Dt}} \quad \text{and} \quad T = \frac{t}{D},$$

by means of which Eq. (B.5) takes the form

$$C(X, T) = 2J_0 \sqrt{T} \left[\frac{1}{\sqrt{\pi}} e^{-X^2} - X \, \text{Erfc}\{X\} \right], \tag{B.6}$$

where the distance now is given in units of $2\sqrt{Dt}$ and the time t is replaced by the new variable $T = t/D$.

Appendix C

Evaluation of $\dfrac{N!}{[(N+m)/2]!\cdot[(N-m)/2]!}\left(\frac{1}{2}\right)^N = \sqrt{\dfrac{2}{\pi N}}\,e^{-m^2/2N}$

It was shown in Section 2.5.6.3 that a particle executing a one-dimensional "random walk", made up of N steps with equal forward and backward probability, has the probability

$$W(N,m) = \frac{N!}{\left(\dfrac{N+m}{2}\right)! \cdot \left(\dfrac{N-m}{2}\right)!} \cdot \left(\frac{1}{2}\right)^N, \qquad (2.5.156)$$

to be located finally in a position corresponding to exactly the sequence of m unidirectional steps, where $m \leq N$.

We shall now derive an approximation for this expression that for all practical purposes is correct when $N \gg 1$ and whose form is more applicable if the above probability enters into further mathematical operations. The tactic is to evoke Stirling's approximation

$$n! = \sqrt{2\pi n}\left(\frac{n}{e}\right)^n. \qquad (C.1)$$

To simplify the typography we put

$$p = \frac{N+m}{2} = \frac{N}{2}\left(1 + \frac{m}{N}\right) \quad \text{and} \quad q = \frac{N-m}{2} = \frac{N}{2}\left(1 - \frac{m}{N}\right). \qquad (C.2)$$

Substitution into Eq. (2.5.156) gives

$$W(N,m) = \frac{N!}{p! \cdot q!}\left(\frac{1}{2}\right)^N. \qquad (C.3)$$

We have, using Eq. (C.1),

$$N! = \sqrt{2\pi N}\left(\frac{N}{e}\right)^N, \quad p! = \sqrt{2\pi p}\left(\frac{p}{e}\right)^p, \quad q! = \sqrt{2\pi q}\left(\frac{q}{e}\right)^q,$$

which leads to the following expression for Eq. (C.3)

$$W(N,m) = \frac{\sqrt{2\pi N}\left(\frac{N}{e}\right)^N}{\sqrt{2\pi p}\left(\frac{p}{e}\right)^p \cdot \sqrt{2\pi q}\left(\frac{q}{e}\right)^q} \left(\frac{1}{2}\right)^N = \frac{1}{\sqrt{\pi}} \frac{N^{\frac{1}{2}}N^N}{p^{\frac{1}{2}}p^p q^{\frac{1}{2}}q^q} \left(\frac{1}{2}\right)^{\frac{1}{2}} \left(\frac{1}{2}\right)^N$$

$$= \frac{1}{\sqrt{\pi}} \frac{N^{N+\frac{1}{2}}}{p^{p+\frac{1}{2}} q^{q+\frac{1}{2}}} \left(\frac{1}{2}\right)^{N+\frac{1}{2}} = \frac{1}{\sqrt{\pi}} \frac{(N/2)^{N+\frac{1}{2}}}{p^{p+\frac{1}{2}} q^{q+\frac{1}{2}}}.$$

Insertion of Eq. (C.2) gives

$$W(N,m) = \frac{1}{\sqrt{\pi}} \frac{(N/2)^{N+\frac{1}{2}}}{\left(\frac{N}{2}\right)^{\frac{N}{2}+\frac{m}{2}+\frac{1}{2}}\left(1+\frac{m}{N}\right)^{\frac{N}{2}+\frac{m}{2}+\frac{1}{2}}\left(\frac{N}{2}\right)^{\frac{N}{2}-\frac{m}{2}+\frac{1}{2}}\left(1-\frac{m}{N}\right)^{\frac{N}{2}-\frac{m}{2}+\frac{1}{2}}}$$

$$= \frac{1}{\sqrt{\pi}} \frac{(N/2)^{N+\frac{1}{2}}}{\left(\frac{N}{2}\right)^{N+1}\left(1+\frac{m}{N}\right)^{\frac{N}{2}+\frac{m}{2}+\frac{1}{2}}\left(1-\frac{m}{N}\right)^{\frac{N}{2}-\frac{m}{2}+\frac{1}{2}}}$$

$$= \sqrt{\frac{2}{\pi N}} \left(1+\frac{m}{N}\right)^{-\frac{N}{2}-\frac{m}{2}-\frac{1}{2}}\left(1-\frac{m}{N}\right)^{-\frac{N}{2}+\frac{m}{2}-\frac{1}{2}}$$

$$= \sqrt{\frac{2}{\pi N}} \left(1+\frac{m}{N}\right)^{-\left(\frac{N}{2}+\frac{1}{2}\right)}\left(1-\frac{m}{N}\right)^{-\left(\frac{N}{2}+\frac{1}{2}\right)}\left(1+\frac{m}{N}\right)^{-\frac{m}{2}}\left(1-\frac{m}{N}\right)^{\frac{m}{2}}$$

$$= \sqrt{\frac{2}{\pi N}} \left[1-\left(\frac{m}{N}\right)^2\right]^{-\left(\frac{N}{2}+\frac{1}{2}\right)} \cdot \left[1+\frac{m}{N}\right]^{-\frac{m}{2}} \cdot \left[1-\frac{m}{N}\right]^{\frac{m}{2}}.$$

As $N/2 \gg \frac{1}{2}$, we can write

$$W(N,m) = \sqrt{\frac{2}{\pi N}} \left[1-\left(\frac{m}{N}\right)^2\right]^{-\frac{N}{2}} \left[1+\frac{m}{N}\right]^{-\frac{m}{2}} \left[1-\frac{m}{N}\right]^{\frac{m}{2}}.$$

Before going any further we seek an approximation for

$$\left(1+\frac{x}{N}\right)^b,$$

when $x/N \ll 1$. We put

$$y = \left(1+\frac{x}{N}\right)^b,$$

and take the logarithm on both sides

$$\ln y = b \ln\left(1+\frac{x}{N}\right).$$

When $|x/N| < 1$ one has*

$$\ln y = b\left\{\frac{x}{N} - \frac{1}{2}\left(\frac{x}{N}\right)^2 + \frac{1}{3}\left(\frac{x}{N}\right)^3 - \cdots\right\} = b\frac{x}{N}\left[1 - \frac{1}{2}\left(\frac{x}{N}\right) + \frac{1}{3}\left(\frac{x}{N}\right)^2 - \cdots\right].$$

* See Chapter 1, Section 1.6.

Hence

$$y = \exp\left\{b\frac{x}{N}\left[1 - \tfrac{1}{2}\left(\frac{x}{N}\right) + \tfrac{1}{3}\left(\frac{x}{N}\right)^2 - \cdots\right]\right\}.$$

For $x/N \ll 1$ the term inside the brackets approaches 1. Therefore

$$\left(1 + \frac{x}{N}\right)^b \approx \exp\left\{b\frac{x}{N}\right\}, \quad \text{for} \quad \frac{x}{N} \ll 1.$$

Making use of this approximation in the above expression gives

$$W(N,m) = \sqrt{\frac{2}{\pi N}}\exp\left\{\left(-\frac{m^2}{N^2}\right)\left(-\frac{N}{2}\right)\right\} \cdot \exp\left\{\frac{m}{N}\left(-\frac{m}{2}\right)\right\} \cdot \exp\left\{\left(-\frac{m}{N}\right)\frac{m}{2}\right\}$$

$$= \sqrt{\frac{2}{\pi N}}\exp\left\{\frac{m^2}{2N}\right\} \cdot \exp\left\{-\frac{m^2}{2N}\right\} \cdot \exp\left\{-\frac{m^2}{2N}\right\},$$

or

$$W(N,m) = \sqrt{\frac{2}{\pi N}}\, e^{-m^2/2N}, \tag{C.4}$$

which is the alternative expression – holding for $N \gg 1$ – for the above probability $W(N,m)$.

The derivation of Eq. (C.4) rested on the assumption that $N \gg 1$ and $m \ll N$. It may be instructive to examine how good is the above approximation. In the table below are shown the values for the probabilities derived from the correct expression Eq. (2.5.156) and from the approximation Eq. (C.4) for $N = 20$ together with the relative deviations Δ_W in percentages between the two expressions.

Table C.1. *Comparison between Eq. (2.5.156) and Eq. (C.4)*

m	Eq. (2.5.156)	Eq. (C.4)	Δ_W
0	0.176 20	0.178 41	−1.25
2	0.160 18	0.161 43	−0.78
4	0.120 13	0.119 59	0.45
6	0.073 93	0.072 54	1.88
8	0.036 02	0.036 02	2.54
10	0.014 79	0.014 64	1.01
12	0.004 62	0.004 87	−5.41
14	0.001 09	0.001 33	−22.0
16	0.000 18	0.000 30	−66.7
18	0.000 02	0.000 05	−150
20	0.000 000 9	0.000 008	−750

From this it appears that even for a relatively small number as $N = 20$ the deviation between the two expressions is small provided the condition $m/20 \ll 1$ holds.

Appendix D

To demonstrate that the mean value $\langle \xi^2 \rangle$ of the displacements squared is proportional to the number N of the displacements

In Chapter 2, Section 2.5.6.5, probabilistic arguments were used in predicting the behavior of a swarm of particles that executed Brownian motion. The end result was the formula

$$\frac{\partial C}{\partial t} = \frac{\langle \xi^2 \rangle}{2\tau} \frac{\partial^2 C}{\partial x^2}$$

where

$$\langle \xi^2 \rangle = \int_{-\infty}^{\infty} \xi^2 \varphi(\xi, \tau) d\xi$$

is the mean value of the square of the displacements at the time τ. This formula becomes identical with Fick's law if the fraction

$$D = \frac{\langle \xi^2 \rangle}{2\tau},$$

which was identified with the *diffusion coefficient* D of the particles, is a constant quantity. This will be the case provided it holds in general that the mean value of the displacements squared at the time τ is proportional to τ. To examine this proposition we consider a particle that – as described in Chapter 2, Section 2.5.6.1 – executes a "random walk", consisting of N steps, where the individual displacements δ_i vary in a completly random manner as regards to size and direction. This gives rise to a displacement

$$\xi = \delta_1 + \delta_2 + \delta_3 + \cdots \delta_r + \cdots \delta_N = \sum_{i=1}^{i=N} \delta_i,$$

where $\delta_1, \ldots, \delta_r, \ldots, \delta_N$ are the lengths of the individual steps – positive as well as negative. This random walk consisting of N steps is now repeated a large number of times making at our disposal the set of $j = 1, 2, \ldots, n$ displacements ξ_j, all consisting of N steps with step lengths δ_s. Squaring each displacement ξ_j gives

$$\xi_j^2 = \left(\sum_{i=1}^{i=N} \delta_{i,j} \right) \cdot \left(\sum_{i=1}^{i=N} \delta_{i,j} \right)$$

$$= \sum_{i=1}^{i=N} \delta_{i,j}^2 + \underbrace{\sum_{r=1}^{r=N} \sum_{s=1}^{s=N} \delta_{r,j} \delta_{s,j}}_{s \neq r}.$$

593

Adding the n squared displacements ξ_j^2 and dividing by n we obtain the mean square displacement

$$\langle \xi^2 \rangle = \frac{1}{n} \sum_{j=1}^{j=n} \xi_j^2$$

or

$$\langle \xi^2 \rangle = \frac{1}{n} \sum_{j=1}^{j=n} \left\{ \sum_{i=1}^{i=N} \delta_{i,j}^2 + \underbrace{\sum_{r=1}^{r=N} \sum_{s=1}^{s=N} \delta_{r,j}\, \delta_{s,j}}_{s \neq r} \right\}$$

$$= \frac{1}{n} \sum_{j=1}^{j=n} \left[\sum_{i=1}^{i=N} \delta_{i,j}^2 \right] + \frac{1}{n} \sum_{j=1}^{j=n} \left[\underbrace{\sum_{r=1}^{r=N} \sum_{s=1}^{s=N} \delta_{r,j}\delta_{s,j}}_{s \neq r} \right].$$

We then apply the rule that the mean value of a sum of elements is equal to the sum of the mean values of each individual element. This gives

$$\langle \xi^2 \rangle = \sum_{i=1}^{i=N} \langle \delta_i^2 \rangle + \underbrace{\sum_{r=1}^{r=N} \sum_{s=1}^{s=N} \langle \delta_r\, \delta_s \rangle}_{s \neq r},$$

where $\langle \delta_i \rangle$ represents the mean value of the square of the displacements δ_i that all are located in the position i, their total number being n. The last term on the right-hand side consists of the sum of cross-products $(\delta_s\, \delta_r)_{r \neq s}$. As positive and negative displacements occur with the same probability the term is made up of pairs of the form $\delta_r \delta_s$ and of $-\delta_r \delta_s$ which add to zero. Thus, the sum of cross-products vanishes and we have

$$\langle \xi^2 \rangle = \sum_{i=1}^{i=N} \langle \delta_i^2 \rangle.$$

As each step δ_i is subject to the *same* probability as regards to size and direction the mean value of the sum of each of the n squared steps $\langle \delta_i^2 \rangle$ will have the *same* value $\langle \delta^2 \rangle$ for each of the $i = 1, 2, \ldots, N$ steps. Therefore, the above expression can be written as

$$\langle \xi^2 \rangle = \sum_{i=1}^{i=N} \langle \delta^2 \rangle = N \langle \delta^2 \rangle,$$

i.e. the mean value of the squares of the displacements $\langle \xi^2 \rangle$ at time τ is proportional to the number N of displacements. Let each single displacement δ last for time $\Delta\tau$. We have then $\tau = N \, \Delta\tau$. Hence

$$\langle \xi^2 \rangle = \frac{\langle \delta^2 \rangle}{\Delta\tau} \tau.$$

Thus, the quantity $\langle \xi^2 \rangle$ is proportional to the time τ, taken to make the displacement ξ, and hence $\langle \xi^2 \rangle / \tau$ is independent of τ and equal to a constant, which may be identified as the diffusion coefficient of the particles that constitute the swarm.

Appendix E

Evaluation of the integral Eq. (2.5.191)

In the equation

$$M_\tau = \overrightarrow{M}_\tau - \overleftarrow{M}_\tau$$

$$= \int_{x_0}^{\infty} \left(C(x) \int_{x_0-x}^{\infty} \Phi(\xi, t)\, d\xi \right) dx - \int_{x_0}^{\infty} \left(C(x) \int_{-\infty}^{x_0-x} \Phi(\xi, t)\, d\xi \right) dx,$$

$$(2.5.191)$$

the concentration $C(x)$ is replaced by its Taylor expansion

$$C(x) = C(x_0) + (x - x_0)\left(\frac{\partial C}{\partial x} \right)_{x_0} + \frac{1}{2}(x - x_0)^2 \left(\frac{\partial^2 C}{\partial x^2} \right)_{x_0}$$

$$+ \cdots \frac{1}{n!}(x - x_0)^n \left(\frac{\partial^n C}{\partial x^n} \right)_{x_0} + \cdots$$

for $C(x)$. This gives

$$M_\tau = C(x_0)\left\{ \int_{x_0}^{\infty} \left(C(x) \int_{x_0-x}^{\infty} \Phi(\xi, t)\, d\xi \right) dx - \int_{x_0}^{\infty} \left(C(x) \int_{-\infty}^{x_0-x} \Phi(\xi, t)\, d\xi \right) dx \right\}$$

$$+ \left(\frac{\partial C}{\partial x} \right)\left\{ \int_{x_0}^{\infty} \left((x - x_0) \int_{x_0-x}^{\infty} \Phi(\xi, t)\, d\xi \right) dx \right.$$

$$\left. - \int_{x_0}^{\infty} \left((x - x_0) \int_{-\infty}^{x_0-x} \Phi(\xi, t)\, d\xi \right) dx \right\}$$

$$+ \frac{1}{2}\left(\frac{\partial^2 C}{\partial x^2} \right)\left\{ \int_{x_0}^{\infty} \left((x - x_0)^2 \int_{x_0-x}^{\infty} \Phi(\xi, t)\, d\xi \right) dx \right.$$

$$\left. - \int_{x_0}^{\infty} \left((x - x_0)^2 \int_{-\infty}^{x_0-x} \Phi(\xi, t)\, d\xi \right) dx \right\} + \cdots$$

$$+ \frac{1}{n!}\left(\frac{\partial^n C}{\partial x^n} \right)\left\{ \int_{x_0}^{\infty} \left((x - x_0)^n \int_{x_0-x}^{\infty} \Phi(\xi, t)\, d\xi \right) dx \right.$$

$$\left. - \int_{x_0}^{\infty} \left((x - x_0)^n \int_{-\infty}^{x_0-x} \Phi(\xi, t)\, d\xi \right) dx \right\},$$

$$(E.1)$$

where the subscripts x_0 of the derivatives $\partial^n C / \partial x^n$ are omitted.

The first integral in the lowest line

$$I_1(n) = \int_{x_0}^{\infty} \left((x - x_0)^n \int_{x_0-x}^{\infty} \Phi(\xi, t) \, d\xi \right) dx$$

is evaluated by partial integration by putting

$$u = \int_{x_0-x}^{\infty} \Phi(\xi, t) \, d\xi = -\int_{\infty}^{x_0-x} \Phi(\xi, t) \, d\xi \quad \text{and} \quad dv = (x - x_0)^n dx,$$

which gives

$$v = \frac{1}{n+1} (x - x_0)^{n+1}$$

and

$$\frac{du}{dx} = \frac{d}{d(x_0 - x)} \left\{ -\int_{\infty}^{x_0-x} \Phi(\xi, t) \, d\xi \right\} \cdot \frac{d(x_0 - x)}{dx}$$

$$= -\phi_\tau(x_0 - x)(-1) = \phi_\tau(x_0 - x) = \phi_\tau(x - x_0),$$

as, according to assumption (a) in Section 2.5.6.4 (ii), $\Phi(\xi, t)$ is symmetric around $\xi = 0$. We have then

$$I_1(n) = uv - \int v \, du$$

$$= \left[\frac{(x - x_0)^{n+1}}{n+1} \right]_{-\infty}^{x_0} \cdot \left[\int_{x_0-x}^{\infty} \Phi(\xi, t) \, d\xi \right]_{-\infty}^{x_0}$$

$$- \frac{1}{n+1} \int_{-\infty}^{x_0} (x - x_0)^{n+1} \phi_\tau(x - x_0) \, dx.$$

Consider the first term on the right-hand side. At the upper limit $x = x_0$ we have

$$0 \cdot \int_0^{\infty} \Phi(\xi, t) \, d\xi = 0 \cdot \frac{1}{2} = 0,$$

as the integral has the value $\frac{1}{2}$ according to assumption (c) in Section 2.5.6.4 (ii), Eq. (2.5.169). At the lower limit $x \to -\infty$ and the term is of the form

$$\lim_{x \to \infty} \left(x^n \int_x^{\infty} \Phi(\xi, t) \, d\xi \right),$$

which according to assumption (b) in Section 2.5.6.4 (ii) takes the value zero. Thus, the first term vanishes and we have

$$I_1(n) = -\frac{1}{n+1} \int_{-\infty}^{x_0} (x - x_0)^{n+1} \phi_\tau(x - x_0) \, dx.$$

The second integral in the lowest row

$$I_2(n) = \int_{x_0}^{\infty} (x - x_0)^n \left(\int_{-\infty}^{x_0-x} \Phi(\xi, t) \, d\xi \right)$$

is likewise evaluated by integration by parts by putting

$$u = \int_{-\infty}^{x_0 - x} \Phi(\xi, t)\,d\xi \quad \text{and} \quad dv = (x - x_0)^n dx,$$

which yields

$$\frac{du}{dx} = \frac{d}{d(x - x_0)}\left\{ -\int_{\infty}^{x_0 - x} \Phi(\xi, t)\,d\xi \right\} \cdot \frac{d(x_0 - x)}{dx}$$
$$= \Phi_\tau(x_0 - x)(-1).$$

Hence

$$I_2(n) = uv - \int v\,du$$
$$= \frac{(x - x_0)^n}{n + 1}\left[\int_{-\infty}^{x_0 - x} \Phi_\tau(\xi)\,d\xi \right]_{x_0}^{\infty} + \frac{1}{n + 1}\int_0^{\infty} (x - x_0^{n+1})\Phi(\xi, t)\,d\xi,$$

or

$$I_2(n) = \frac{1}{n + 1}\int_0^{\infty} (x - x_0^{n+1})\Phi(\xi, t)\,d\xi,$$

as the first term vanishes for the same reasons as mentioned in connection with the evaluation of $I_1(n)$. Thus integrals in the lowest row of Eq. (E.1) can be written

$$I(n) = I_1(n) - I_2(n)$$
$$= -\frac{1}{n + 1}\int_{-\infty}^{x_0} (x - x_0)^{n+1}\Phi_\tau(x - x_0)\,dx + \frac{1}{n + 1}\int_0^{\infty} (x - x_0^{n+1})\Phi_\tau(x - x_0)\,dx,$$
$$= \frac{1}{n + 1}\int_{-\infty}^{\infty} (x - x_0^{n+1})\Phi_\tau(x - x_0)\,dx.$$

Putting $x - x_0 = \xi$ we can also write

$$I(n) = -\frac{1}{n + 1}\int_{-\infty}^{\infty} \xi^{n+1}\,d\xi. \tag{E.2}$$

Inserting Eq. (E.2) into Eq. (E.1) gives

$$M_\tau = -C(x_0)\int_{-\infty}^{\infty} \xi\,d\xi - \frac{1}{2}\left(\frac{\partial C}{\partial x}\right)_{x_0}\int_{-\infty}^{\infty} \xi^2\Phi(\xi, t)\,d\xi$$
$$- \frac{1}{6}\left(\frac{\partial^3 C}{\partial x^3}\right)_{x_0}\int_{-\infty}^{\infty} \xi^3\Phi(\xi, t)\,d\xi - \frac{1}{24}\left(\frac{\partial^4 C}{\partial x^4}\right)_{x_0}\int_{-\infty}^{\infty} \xi^4\Phi(\xi, t)\,d\xi \ldots$$
$$- \frac{1}{(n + 1)n!}\left(\frac{\partial^n C}{\partial x^n}\right)_{x_0}\int_{-\infty}^{\infty} \xi^{n+1}\Phi(\xi, t)\,d\xi - \ldots.$$

According to assumption (a) in Section 2.5.6.4 (ii), $\Phi(\xi, t)$ is symmetric around $\xi = 0$. It follows then

$$\int_{-\infty}^{\infty} \xi^k \Phi(\xi, t)\,d\xi = 0$$

for k odd, i.e. $k = 2n + 1$. Thus, all integrals having odd powers of ξ^n will vanish, hence

$$M_\tau(x_0) = -\frac{1}{2}\left(\frac{\partial C}{\partial x}\right)_{x_0} \int_{-\infty}^{\infty} \xi^2 \Phi(\xi, t)\,\mathrm{d}\xi$$

$$-\frac{1}{24}\left(\frac{\partial^3 C}{\partial x^3}\right)_{x_0} \int_{-\infty}^{\infty} \xi^4 \Phi(\xi, t)\,\mathrm{d}\xi - \frac{1}{720}\left(\frac{\partial^5 C}{\partial x^5}\right)_{x_0} \int_{-\infty}^{\infty} \xi^6 \Phi(\xi, t)\,\mathrm{d}\xi$$

$$-\cdots\frac{1}{(n+1)n!}\left(\frac{\partial^n C}{\partial x^n}\right)_{x_0} \int_{-\infty}^{\infty} \xi^{n+1} \Phi(\xi, t)\,\mathrm{d}\xi, \tag{2.5.193}$$

which is the formula used in Section 2.5.6.4 (ii).

Appendix F

Evaluation of the integral $\int_\infty^x e^{mX-(X+x_0)^2/B}dX$

In Section 2.6.1.2, Eq. (2.6.24), the function $w(x, t)$ entered into the equation

$$w(x, t)\, e^{mx} = -2mA \int_\infty^x \exp\{mX\} \exp\{-(X + x_0)^2/B\}dX \qquad \text{(F.1)}$$

with

$$m = \frac{v}{2D}, \quad \text{and} \quad B = 4Dt, \quad \text{and} \quad A = \frac{\mathcal{N}}{2\sqrt{\pi Dt}}. \qquad \text{(F.2)}$$

To evaluate definite integrals of the type

$$\int_\infty^x \exp\{mX - (X + x_0)^2/B\}\, dX$$

the standard procedure is to "complete the square". We rewrite the exponent

$$
\begin{aligned}
mX - \frac{(X + x_0)^2}{B} &= \frac{mBX - X^2 - 2Xx_0 - x_0^2}{B} = -\frac{X^2 + 2Xx_0 + x_0^2 - mBX}{B} \\
&= -\frac{X^2 + 2X(x_0 - mB/2) + x_0^2}{B} \\
&= -\frac{X^2 + 2X(x_0 - mB/2) + (x_0 - mB/2)^2 - (x_0 - mB/2)^2 + x_0^2}{B} \\
&= -\frac{(X^2 + (x_0 - mB/2))^2 - (x_0 - mB/2)^2 + x_0^2}{B} \\
&= -\frac{(X + x_0 - mB/2)^2 + mBx_0 - (mB)^2/4}{B} \\
&= -\frac{(X + x_0 - mB/2)^2}{B} - mx_0 + m^2B/4.
\end{aligned}
$$

Thus, the integral above can also be written as

$$I = \exp\{-mx_0 + m^2B/4\} \int_\infty^x \exp\left\{-\frac{(X + x_0 - mB/2)^2}{B}\right\} dX.$$

599

Equation (F.1) then reads

$$w(x, t) = -2m A e^{-m(x+mx_0)+m^2 B/4} \int_{\infty}^{x} \exp\left\{-\frac{(X + x_0 - mB/2)^2}{B}\right\} dX.$$

We have from Eq. (F.2)

$$mB/2 = (v/2D)4Dt/2 = vt; \quad m^2 B/4 = (v^2/4D^2)4Dt/4 = v^2/4D.$$

Inserting these values and $m = v/2D$ in the above equation gives

$$w(x, t) = -2m A \exp\left\{\frac{v^2 t}{4D} - \frac{v}{2D}(x + x_0)\right\} \int_{\infty}^{x} \exp\left\{-\frac{(X + x_0 - vt)^2}{4Dt}\right\} dX.$$

We introduce a new variable of integration

$$\xi = \frac{X + x_0 - vt}{2\sqrt{Dt}},$$

so that

$$dX/d\xi = 2\sqrt{Dt},$$

and

$$\xi \to \infty \quad \text{for} \quad X \to \infty, \quad \text{and} \quad \xi = (x + x_0 - vt)/2\sqrt{Dt} \quad \text{for} \quad X = x.$$

With these substitutions we obtain

$$w(x, t) = -2m A 2\sqrt{Dt} \exp\left\{\frac{v^2 t}{4D} - \frac{v}{2D}(x_0 + x)\right\} \int_{\infty}^{(x+x_0-vt)/2\sqrt{Dt}} e^{-\xi^2} d\xi.$$

The factors on the right-hand side are redressed as

$$-2m A 2\sqrt{Dt} = -2\frac{v}{2D}\frac{\mathcal{N}}{2\sqrt{\pi Dt}}2\sqrt{Dt} = -\frac{v}{D\sqrt{\pi}}\mathcal{N}.$$

Hence

$$w(x, t) = -\mathcal{N}\frac{v}{D\sqrt{\pi}} \exp\left\{\frac{v^2 t}{4D} - \frac{v}{2D}(x_0 + x)\right\} \int_{\infty}^{(x+x_0-vt)/2\sqrt{Dt}} e^{-\xi^2} d\xi$$

$$= \mathcal{N}\frac{v}{2D} \exp\left\{\frac{v^2 t}{4D} - \frac{v}{2D}(x_0 + x)\right\} \frac{2}{\sqrt{\pi}} \int_{(x+x_0-vt)/2\sqrt{Dt}}^{\infty} e^{-\xi^2} d\xi *$$

$$= \mathcal{N}\frac{v}{2D} \exp\left\{\frac{v^2 t}{4D} - \frac{v}{2D}(x_0 + x)\right\} \mathrm{Erfc}\left\{\frac{x + x_0 - vt}{2\sqrt{Dt}}\right\},$$

which is the expression referred to previously.

---

* As $\mathrm{Erfc}(z) = \frac{2}{\sqrt{\pi}}\int_z^\infty e^{-\xi^2} d\xi$. See Appendix A.

Appendix G

Example of the application of the theory of Brownian motion

The theory of Brownian motion was examined experimentally mainly by Perrin and Svedberg. One of the most remarkable results was the determination of Avogadro's number. Below we shall give an example of these investigations.

A spherical particle of radius $r = 4 \times 10^{-7}$ m executes Brownian motion in water at temperature 300 K, which has the viscosity $\eta = 10^{-3}$ N s m^{-2}. Microphotographs of the position of the particle were taken at times t and $(t + 2)$ s, and the net displacement resulting from the numerous elementary Brownian steps executed by the particle in moving from the first observed position $x(t)$ to the final one at $x(t + 2)$ in the intervening time interval was determined as $\Delta = x(t + 2) - x(t)$. The result of a large number of such measurements ($n = 401$) is shown in the table below.

Table G.1. *Displacements executed by a Brownian particle in the course of 2 s*

$\Delta(\mu m)$			Frequency
Between	0.5 and	−0.5	111
Between	0.5 and	1.5	87
Between	−0.5 and	−1.5	95
Between	1.5 and	2.5	47
Between	−1.5 and	−2.5	32
Between	2.5 and	3.5	8
Between	−2.5 and	−3.5	15
Between	3.5 and	4.5	3
Between	−3.5 and	−4.5	2
Between	4.5 and	5.5	0
Between	−4.5 and	−5.5	1
Between	5.5 and	6.5	0
Between	−5.5 and	−6.5	0

Temperature 300 K. After P.M. Morse (1962)*.

* Morse, P.M. (1962): *Thermal Physics*, W.A. Benjamin, Inc., New York, p. 430.

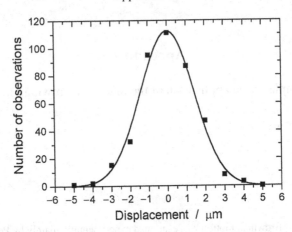

Fig. G.1. Displacements of a Brownian particle in the course of 2 s. Abscissa: displacement from the initial position (in μm). Ordinate: number of displacements to a given position $x = -5, -4, \ldots, + 4, +5$. Data from Table G.1 are given by ■. Full line: theoretical course for the displacements calculated from Chapter 2, Section 2.5.6.3, Eq. (2.5.157) with $D = 5.063 \times 10^{-13}$ m^2 s^{-1}, $t = 2$ s and $n = 401$.

From these observations one finds

$$\langle \xi \rangle = \frac{(87 - 95) + (47 - 32) \cdot 2 + (8 - 15) \cdot 3 + (3 - 2) \cdot 4 - 1 \cdot 5}{401} \approx 0 \; \mu\text{m},$$

as expected according to the theory, and

$$\langle \xi^2 \rangle = \frac{(87 + 95) + (47 + 32) \cdot 4 + (8 + 15) \cdot 9 + (3 + 2) \cdot 16 + 1 \cdot 25}{400} \; \mu\text{m}^2$$

$$= 2.025 \times 10^{-12} \; \text{m}^2.$$

From this value the diffusion coefficient of the particle can be determined from the Einstein–Smoluchowski equation*

$$D = \frac{\Delta^2}{2\tau} = \frac{2.025 \times 10^{-12}}{4} = 5.063 \times 10^{-13} \; \text{m}^2 \cdot \text{s}^{-1}.$$

As the value of the gas constant $\mathcal{R} = 8.314$ J mol^{-1} K^{-1} is known, it is possible to determine the numerical value of Avogadro's constant using the Einstein–Stokes relation[†]

$$D = \frac{\mathcal{R} T}{6\pi \eta r \, \mathcal{N}_A}.$$

This gives

$$\mathcal{N}_A = \frac{8.314 \times 300}{6 \times 3.1416 \times 10^{-3} \times 4 \times 10^{-7} \times 5.06 \times 10^{-13}} = 6.53 \times 10^{23},$$

which only deviates by about 8% from the now accepted value for $\mathcal{N}_A = 6.023 \times 10^{23}$.

---

* See Chapter 2, Section 2.5.6.1, Eq. (2.5.162).
[†] See Chapter 2, Section 2.7.4.2, Eq. (2.6.58).

Figure G.1 shows graphically the result from Table G.1. To compare theory with experiment, the full line represents the distribution of displacements that one should expect from simple "random walk" considerations, by inserting the values of the parameters in the expression for the probability density*

$$\varphi(x, t) = \frac{1}{2\sqrt{\pi Dt}} \, e^{-x^2/4Dt},$$

that are obtained experimentally: observation time $t = 2$ s, $D = 0.5063 \times 10^{-12}$ m^2 s^{-1} and $n = 401$. Hence the expected distribution of the 401 data points becomes

$$\Phi(x, 2) = \frac{401}{2\sqrt{\pi\, 0.5063 \times 2}} e^{-x^2/4\times0.5063\times2} = 112.4\,e^{-0.2469\,x^2}.$$

The agreement is adequate enough to leave out further tests of the goodness of fit.

* See Chapter 2, Section 2.5.6.3, Eq. (2. 5.157).

Appendix H

A note on the physical meaning of the pressure gradient

In Section 2.8.2.6 it was stated that a body of volume V that is immersed in a fluid with a hydrostatic pressure changing in a given direction x is subject to a force X of magnitude

$$X = -V \frac{dP}{dx}$$

in that direction.

We consider an incompressible fluid in which its pressure varies solely in the direction indicated by the x-axis (i.e. $\partial P / \partial y = \partial / \partial z = 0$). A prism of length h and end-surfaces of area A is placed as shown on the figure at the planes at x and $x + h$ where the pressures have the values P and $P(x + h)$. The two oppositely directed forces acting on the end-surfaces at x and $x + h$ are $\overrightarrow{X}(x) = P(x)A$ and $\overleftarrow{X}(x + h) = -P(x + h)A$. Hence the net force X acting on the prism in the direction of the x–axis is

$$X = \overrightarrow{X}(x) + \overleftarrow{X}(x + h)$$
$$= [P(x) - P(x + h)]\,A = \left[P(x) - \left(P(x) + h\frac{dP}{dx} \right) \right] A,$$

where it is assumed that h is sufficiently small to make the linear first-order approximation of $P(x + h)$ valid. It follows then

$$X = -hA\frac{dP}{dx} = -V\frac{dP}{dx}$$

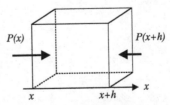

Fig. H.1. Calculation of the hydrostatic force acting on a prism of volume $V = h\,A$.

604

since $V = hA$ is the volume of the prism. Dividing both sides by V we obtain

$$\frac{X}{V} = X_{\mathrm{V}} = -\frac{\mathrm{d}P}{\mathrm{d}x}.$$

Thus, the hydrostatic force X_{V} that acts on the immersed body per unit volume is equal to the negative pressure gradient.

The density of the immersed body is $\rho = M/V$. Hence, the hydrostatic force X_{M} that acts on the immersed body per unit mass is

$$X_{\mathrm{M}} = -\frac{1}{\rho}\frac{\mathrm{d}P}{\mathrm{d}x}.$$

Appendix I

About hyperbolic functions

In Chapter 3, Section 3.5.2, hyperbolic functions appeared in several of the formulas. For those readers who might be less familiar with these functions, an account is given below for the derivation of the relations that occur in the text.

1 Analytical definition

From the middle of the eighteenth century mathematicians realized that certain combinations of the exponential functions e^x and e^{-x} had much in common with the trigonometric $\sin x$, $\cos x$, $\tan x$ and $\cot x$ – also called circular functions – and that these combinations came up so frequently in the integral and differential calculus in both pure and applied mathematics that it became profitable to introduce the following functions, which are called *hyperbolic functions*

$$\sinh x = \frac{e^x - e^{-x}}{2} \quad \text{and} \quad \cosh x = \frac{e^x + e^{-x}}{2}, \tag{I.1}$$

where $\sinh x$ and $\cosh x$ are called the hyperbolic sine and hyperbolic cosine of x, respectively. It is seen that $\cosh x = \cosh(-x)$, i.e. $\cosh x$ is an *even function* with $\cosh x = 1$ for $x = 0$, whereas $\sinh x = -\sinh(-x)$ is an *odd function* with $\sinh x = 0$ for $x = 0$. Similarly

$$\tanh x = \frac{\sinh x}{\cosh x} = \frac{e^x - e^{-x}}{e^x + e^{-x}} = \frac{e^{2x} - 1}{e^{2x} + 1} \tag{I.2}$$

and

$$\coth x = \frac{\cosh x}{\sinh x} = \frac{e^x + e^{-x}}{e^x - e^{-x}} = \frac{e^{2x} + 1}{e^{2x} - 1}, \tag{I.3}$$

are called the **hyperbolic tangent** and **hyperbolic cotangent** of x. When x takes the values from $-\infty$ to $+\infty$ then $\tanh x$ varies from -1 to $+1$ and assumes the value zero $x = 0$, whereas $\coth x$ is not defined for this value as $\coth(0_-) \rightarrow -\infty$ and $\coth(0_+) \rightarrow +\infty$.

A large number of relations can be derived among the hyperbolic functions with an appearance analogous to the circular functions. However, only the relations needed in this book will be derived.

606

From Eq. (I.1) we obtain by addition

$$e^x = \cosh x + \sinh x, \tag{I.4}$$

and by subtraction

$$e^{-x} = \cosh x - \sinh x. \tag{I.4A}$$

From these two equations we obtain

$$e^x e^{-x} = [\cosh x + \sinh x][\cosh x - \sinh x]$$

from which the remarkable relation follows

$$\cosh^2 x - \sinh^2 x = 1, \tag{I.5}$$

which also has a bearing upon the named hyperbolic function: putting $\xi = \cosh x$ and $\eta = \sinh x$ it is seen that $\cosh x$ and $\sinh x$ are the connected points on the rectangular hyperbola

$$\xi^2 - \eta^2 = 1.$$

For the similar circular functions $\xi = \cos x$ and $\eta = \sin x$ we have

$$\xi^2 + \eta^2 = 1,$$

and the points ξ and η are coordinates lying on the circle with radius 1.

2 Addition theorems

We consider the exponential expression $e^{(u+v)}$, where u and v are two arbitrary positive numbers. We have then from Eq. (I.4)

$$e^{(u+v)} = \cosh(u+v) + \sinh(u+v).$$

The left-hand side can also be written as

$$e^u \cdot e^v = (\cosh u + \sinh u) \cdot (\cosh v + \sinh v)$$
$$= \cosh u \cosh v + \sinh u \sinh v + \cosh u \sinh v + \sinh u \cosh v.$$

Hence

$$\cosh(u+v) + \sinh(u+v) = \cosh u \cosh v + \sinh u \sinh v$$
$$+ \cosh u \sinh v + \sinh u \cosh v.$$

Replacing in this expression u with $-u$ and v with $-v$ gives

$$\cosh(u+v) - \sinh(u+v) = \cosh u \cosh v + \sinh u \sinh v$$
$$- \cosh u \sinh v - \sinh u \cosh v,$$

as $\cosh(-x) = \cosh(x)$ and $\sinh(-x) = -\sinh(x)$. Addition and subtraction respectively of these two equations result in the *equations of addition* below

$$\cosh(u+v) = \cosh u \cosh v + \sinh u \sinh v. \tag{I.6}$$

$$\sinh(u+v) = \cosh u \sinh v + \sinh u \cosh v. \tag{I.7}$$

Replacing here $v = -v$ yields the *equations of subtraction**

$$\cosh(u - v) = \cosh u \cosh v - \sinh u \sinh v. \qquad (I.8)$$
$$\sinh(u - v) = \sinh u \cosh v - \cosh u \sinh v. \qquad (I.9)$$

Next we put $u = v = x$ in Eq. (I.5). In combination with Eq. (I.7) this gives

$$\cosh 2x = \cosh^2 x + \sinh^2 x = 2\cosh^2 x - 1, \qquad (I.10)$$

and correspondingly with Eq. (I.7)

$$\sinh 2x = 2 \sinh x \cosh x. \qquad (I.11)$$

3 Differentiation and integration

The analogy to the trigonometric functions also shows in the formulas for differentiation. For the exponential function we have $d(e^x)/dx = e^x$. From this it follows that

$$\frac{d}{dx}(\sinh x) = \frac{e^x - (-1)e^{-x}}{2} = \cosh x, \qquad (I.12)$$

and

$$\frac{d}{dx}(\cosh x) = \frac{e^x + (-1)e^{-x}}{2} = \sinh x, \qquad (I.13)$$

and by making use of the rule for differentiation of a fraction (Chapter 1, Section 1.2.1.1(2))

$$\frac{d}{dx}(\tanh x) = \frac{\cosh x \cosh x - \sinh x \sinh x}{\cosh^2 x} = \frac{\cosh^2 x - \sinh^2 x}{\cosh^2 x} = \frac{1}{\cosh^2 x}. \qquad (I.14)$$

Proceeding in the same way we find the counterpart

$$\frac{d}{dx}(\coth x) = -\frac{1}{\sinh^2 x}. \qquad (I.15)$$

To calculate the integral

$$\int \frac{1}{\sinh x}\, dx$$

we make use of Eq. (I.11) and write

$$\int \frac{1}{\sinh x}\, dx = \int \frac{1}{2 \sinh\left(\frac{x}{2}\right)\cosh\left(\frac{x}{2}\right)}\, dx = \int \frac{1}{\sinh\left(\frac{x}{2}\right)\cosh\left(\frac{x}{2}\right)}\, d\left(\frac{x}{2}\right).$$

* Note the analogy between these formula and the corresponding trigonometric formulas.

Putting $u = x/2$ and multiplying numerator and denominator by $1 = \cosh u / \cosh u$ we can write

$$\int \frac{1}{\sinh x}\, dx = \int \frac{1}{\tanh u} \cdot \frac{1}{\cosh^2 u}\, du = \int \left(\frac{1}{Z}\right)\left(\frac{dZ}{du}\right) du,$$

where $Z = \tanh(x/2) = \tanh u$. The integral is now in the form of a function $F(Z(u))$ of a function $Z(u)$ multiplied by the derivative dZ/du of the inner function. Hence*

$$\int \frac{1}{\sinh x}\, dx = \int \frac{1}{Z}\, dZ = \ln Z = \ln \tanh\left(\frac{x}{2}\right) + C_1 \qquad (I.16)$$

which is the result that was used in Chapter 3, Section 3.5.3.2.

4 Inverse hyperbolic functions

To the hyperbolic functions $\sinh x$, $\cosh x$, $\tanh x$ and $\coth x$ there correspond *inverse* hyperbolic functions. Very often they are written as

$$t = \operatorname{ar\,sinh} x, \quad t = \operatorname{ar\,cosh} x, \quad t = \operatorname{ar\,tanh} x, \quad \text{and} \quad t = \operatorname{ar\,coth} x,$$

as the argument to a hyperbolic function is an *area sector* – contrary to the circular functions whose argument is an angle – that is partly delimited by the rectangular hyperbola $\xi^2 - \eta^2 = 1$.

The inverse functions can be expressed in terms of logarithms. The equation $t = \operatorname{ar\,sinh} x$ is equivalent to

$$x = \sinh t = \frac{e^t - e^{-t}}{2} = \frac{u - 1/u}{2},$$

where we have put $e^t = u$. This expression can be written as

$$u^2 - 2xu - 1 = 0.$$

Solving with respect to u gives

$$u = \frac{2x \pm \sqrt{4x^2 + 4}}{2} = x \pm \sqrt{x^2 + 1} = e^t.$$

As $e^t = u$ it follows that

$$t = \operatorname{ar\,sinh} x = \ln(x + \sqrt{x^2 + 1}), \qquad (I.17)$$

where the value $-\sqrt{x^2} + 1$ is disregarded as a negative argument of $\ln x$ makes no sense in the theory of functions of a real variable. The above expression is valid for all $-\infty < x < \infty$. It may be convenient to split Eq. (I.17) in this way

$$\operatorname{ar\,sinh} x = \begin{cases} \ln(x + \sqrt{x^2 + 1}), & \text{for } x \geq 0 \\ -\ln(\sqrt{x^2 + 1} - x), & \text{for } x \leq 0 \end{cases} \qquad (I.18)$$

as $\operatorname{ar\,sinh}(x) = -\operatorname{ar\,sinh}(-x)$, since $\sinh x$ is an odd function.

* See Chapter 1, Section 1.7.1.

Finally we consider the function $t = \operatorname{ar} \tanh x$ and its inverse function

$$x = \tanh t = \frac{e^t - e^{-t}}{e^t + e^{-t}} = \frac{e^{2t} - 1}{e^{2t} + 1},$$

and put again $e^t = u$, i.e. $e^{2t} = u^2$. We then have

$$x = \frac{u^2 - 1}{u^2 + 1}.$$

Solving with respect to u gives

$$u^2 = \frac{1 + x}{1 - x} = e^{2t}.$$

Hence it follows that

$$t = \operatorname{ar} \tanh x = \frac{1}{2} \ln \frac{1 + x}{1 - x} \text{ in interval } -1 < x < 1. \tag{I.19}$$

The derivative of $\operatorname{ar} \tanh x$ is*

$$\frac{d}{dx}(\operatorname{ar} \tanh x) = \frac{dt}{dx} = \frac{1}{2}\left(\frac{1 + x}{1 - x}\right)^{-1} \cdot \frac{d}{dx}\left(\frac{1 + x}{1 - x}\right)$$

$$= \frac{1}{2}\left(\frac{1 - x}{1 + x}\right)\frac{(1 - x)(1) - (1 + x)(-1)}{(1 - x)^2}$$

$$= \frac{1}{(1 + x)(1 - x)} = \frac{1}{1 - x^2}. \tag{I.20}$$

We have then

$$\int \frac{1}{1 - x^2}\, dx = \operatorname{ar} \tanh x + C_1, \tag{I.21}$$

corresponding to Chapter 3, Section 5.5.2, Eq. (3.5.52).

5 Differential equations

The second-order differential equation

$$\frac{d^2 y}{dx^2} - \alpha^2 y = 0 \tag{I.22}$$

has a general solution of the form

$$y = Ae^{\alpha x} + Be^{-\alpha x} \tag{I.23}$$

where A and B are adjustable constants (Chapter 1, Section 1.9.2.2). In Eq. (I.23) we put

$$A = B = \frac{1}{2}, \quad \text{and} \quad A = \frac{1}{2}, B = -\frac{1}{2}$$

---

* See Chapter 1, Section 1.2.5.1, Eq. (1.2.15), Section 1.4.2, Eq. (1.4.3.) and Section 1.2.3.1(e).

respectively. This yields

$$y = \cosh(\alpha x) \quad \text{and} \quad y = \sinh(\alpha x).$$

Hence the general solution of Eq. (I.22) can also be written as

$$Y = A \cosh(\alpha x) + B \sinh(\alpha x). \tag{I.24}$$

It follows from the addition theorems Eq. (I.6)–Eq. (I.9) that

$$Y = A \cosh(\alpha[L \pm x]) + B \sinh(\alpha[L \pm x]), \tag{I.25}$$

where L is a constant, is also a solution of Eq. (I.22). Certain types of boundary conditions are more easily handled by the solutions involving the hyperbolic functions, e.g. if the values of y or dy/dx are specified at $x = 0$ and $x = L$.

Appendix J

Evaluation of an integral of Duhamel's type

The integral in Chapter 4, Section 4.4.5.2 (iii), Eq. (4.4.32A)

$$W(X, T) = \frac{X}{2\sqrt{\pi}} \int_0^T e^\tau \frac{1}{(T - \tau)^{3/2}} e^{-X^2/4(T-\tau)} \, d\tau, \qquad (4.4.32A)$$

can be expressed in terms of the complementary error function Erfc$\{X, T\}$, which does not offer any numerical problems, because comprehensive tables as well as good algorithms for computer use are available.

It appears natural to introduce a new variable of integration μ

$$\mu = \frac{X}{2\sqrt{(T - \tau)}} = \frac{X}{2} (T - \tau)^{-1/2}, \qquad (J.1)$$

by means of which Eq. (4.4.32A) takes the form

$$W(X, T) = \frac{X}{2\sqrt{\pi}} \int_{\mu(\tau=0)}^{\mu(\tau=T)} (\cdots) \frac{d\tau}{d\mu} \, d\mu,$$

where $(\cdots)$ symbolizes the integrand of Eq. (4.4.32A). We have from Eq. (J.1)

$$\tau = T - \frac{X^2}{4\mu^2}.$$

Thus

$$\frac{d\tau}{d\mu} = -\frac{X^2}{4}(-2)\mu^{-3} = \frac{X^2}{2} \frac{8}{X^3}(T - \tau)^{3/2} = \frac{4}{X}(T - \tau)^{3/2}.$$

Furthermore, $\mu = X/2\sqrt{T}$ for $\tau = 0$ and $\mu = \infty$ for $\tau = T$. Hence

$$\begin{aligned}
W(X, T) &= \frac{X}{2\sqrt{\pi}} \int_{X/2\sqrt{T}}^\infty e^{T - X^2/4\mu^2} \frac{1}{(T - \tau)^{3/2}} e^{-\mu^2} \frac{4}{X}(T - \tau)^{3/2} d\tau \\
&= \frac{2}{\sqrt{\pi}} \int_{X/2\sqrt{T}}^\infty e^{T - (X^2/4\mu^2 + \mu^2)} \, d\mu \\
&= \frac{2}{\sqrt{\pi}} e^T \int_{X/2\sqrt{T}}^\infty e^{-(X^2/4\mu^2 + \mu^2)} \, d\mu, \qquad (J.2)
\end{aligned}$$

as the provisional alternative to Eq. (4.4.32A). To avoid messing up the algebra more than necessary we put

$$A = \frac{X}{2\sqrt{T}} \quad \text{and} \quad a = \frac{X}{2} \tag{J.3}$$

and write

$$I(X, T) = \int_A^\infty e^{-\left(\frac{a}{\mu}\right)^2 - \mu^2} \, d\mu, \tag{J.4}$$

which makes the relevant elements in the integral of Eq. (J.2) easier to see. The integration procedure which follows consists of three parts.

(A) By using the identity

$$\left(\frac{a}{\mu}\right)^2 + \mu^2 = \left(\frac{a}{\mu} + \mu\right)^2 - 2a,$$

Eq. (J.4) can be written as

$$I(X, T) = \int_A^\infty e^{-\left(\frac{a}{\mu} + \mu\right)^2 + 2a} d\mu = e^{2a} \int_A^\infty e^{-\left(\frac{a}{\mu} + \mu\right)^2} d\mu,$$

which invites us to replace the variable of integration μ by

$$\xi = \frac{a}{\mu} + \mu, \quad \text{i.e.} \quad d\xi = \left(1 - \frac{a}{\mu^2}\right) d\mu.$$

We now consider the integral

$$e^{2a} \int_A^\infty e^{-\left(\frac{a}{\mu} + \mu\right)^2} \left(1 - \frac{a}{\mu^2}\right) d\mu = e^{2a} \int_{\frac{a}{A} + A}^\infty e^{-\xi^2} d\xi,$$

as $\xi = a/A + A$ for $\mu = A$, and $\xi = \infty$ for $\mu = \infty$. The left-hand side is written out fully, which gives

$$e^{2a} \int_A^\infty e^{-\left(\frac{a}{\mu} + \mu\right)^2} d\mu - e^{2a} \int_A^\infty \frac{a}{\mu^2} e^{-\left(\frac{a}{\mu} + \mu\right)^2} d\mu = e^{2a} \int_{\frac{a}{A} + A}^\infty e^{-\xi^2} d\xi,$$

or

$$\int_A^\infty e^{-\left(\frac{a}{\mu} + \mu\right)^2 + 2a} d\mu - e^{2a} \int_A^\infty \frac{a}{\mu^2} e^{-\left(\frac{a}{\mu} + \mu\right)^2} d\mu = e^{2a} \int_{\frac{a}{A} + A}^\infty e^{-\xi^2} d\xi,$$

or

$$\int_A^\infty e^{-\left(\frac{a}{\mu}\right)^2 - \mu^2} d\mu - e^{2a} \int_A^\infty \frac{a}{\mu^2} e^{-\left(\frac{a}{\mu} + \mu\right)^2} d\mu = e^{2a} \int_{\frac{a}{A} + A}^\infty e^{-\xi^2} d\xi, \tag{J.5}$$

by invoking the identity: $(a/\mu + \mu)^2 - 2a = (a/\mu)^2 + \mu^2$.

(B) We use the new identity

$$\left(\frac{a}{\mu} - \mu\right)^2 + 2a = \left(\frac{a}{\mu}\right)^2 + \mu^2,$$

Appendix J

and proceed as in (A) by writing

$$I(X, T) = \int_A^\infty e^{-\left(\frac{a}{\mu}-\mu\right)^2 - 2a}\, d\mu = e^{-2a} \int_A^\infty e^{-\left(\frac{a}{\mu}-\mu\right)^2}\, d\mu,$$

and replace the variable of integration μ with the new variable

$$\xi = \mu - \frac{a}{\mu}, \quad \text{or} \quad d\xi = \left(1 + \frac{a}{\mu^2}\right) d\mu,$$

and consider then the integral

$$e^{-2a} \int_A^\infty e^{-\left(\frac{a}{\mu}-\mu\right)^2}\left(1 + \frac{a}{\mu^2}\right) d\mu = e^{-2a} \int_{A-\frac{a}{A}}^\infty e^{-\xi^2}\, d\xi,$$

as $\xi = A - a/A$ for $\mu = A$, and $\xi = \infty$ for $\mu = \infty$. The left-hand side is again split into

$$e^{-2a} \int_A^\infty e^{-\left(\frac{a}{\mu}-\mu\right)^2}\, d\mu + e^{-2a} \int_A^\infty \frac{a}{\mu^2} e^{-\left(\frac{a}{\mu}-\mu\right)^2}\, d\mu = e^{-2a} \int_{A-\frac{a}{A}}^\infty e^{-\xi^2}\, d\xi.$$

But

$$\left(\frac{a}{\mu}\right)^2 + \mu^2 = \left(\frac{a}{\mu} - \mu\right)^2 + 2a = \left(\frac{a}{\mu} + \mu\right)^2 - 2a,$$

and therefore

$$\int_A^\infty e^{-\left(\frac{a}{\mu}-\mu\right)^2 - 2a}\, d\mu + e^{2a} \int_A^\infty \frac{a}{\mu^2} e^{-\left(\frac{a}{\mu}+\mu\right)^2}\, d\mu = e^{-2a} \int_{A-\frac{a}{A}}^\infty e^{-\xi^2}\, d\xi,$$

which also can be written alternatively

$$\int_A^\infty e^{-\left(\frac{a}{\mu}\right)^2 - \mu^2}\, d\mu + e^{2a} \int_A^\infty \frac{a}{\mu^2} e^{-\left(\frac{a}{\mu}+\mu\right)^2}\, d\mu = e^{-2a} \int_{A-\frac{a}{A}}^\infty e^{-\xi^2}\, d\xi. \qquad (\text{J.6})$$

(C) Addition of Eq. (J.5) and Eq. (J.6) results in

$$2 \int_A^\infty e^{-\left(\frac{a}{\mu}\right)^2 - \mu^2}\, d\mu = e^{2a} \int_{A+\frac{a}{A}}^\infty e^{-\xi^2}\, d\xi + e^{-2a} \int_{A-\frac{a}{A}}^\infty e^{-\xi^2}\, d\xi.$$

We have

$$\frac{2}{\sqrt{\pi}} \int_y^\infty e^{-y^2}\, dy = \text{Erfc}\{y\}.$$

Thus, the above integral can also be written as

$$2 \int_A^\infty e^{-\left(\frac{a}{\mu}\right)^2 - \mu^2}\, d\mu = \frac{\sqrt{\pi}}{2} \left[e^{2a} \, \text{Erfc}\left\{ A + \frac{a}{A} \right\} + e^{-2a} \, \text{Erfc}\left\{ A - \frac{a}{A} \right\} \right].$$

Invoking $A = X/2\sqrt{T}$ and $a = X/2$ gives

$$I(X, T) = \int_{X/2\sqrt{T}}^{\infty} e^{-X^2/4\mu^2 - \mu^2} \, d\mu$$

$$= \frac{\sqrt{\pi}}{4} \left[e^X \operatorname{Erfc} \left\{ \frac{X}{2\sqrt{T}} + \sqrt{T} \right\} + e^{-X} \operatorname{Erfc} \left\{ \frac{X}{2\sqrt{T}} - \sqrt{T} \right\} \right],$$

which inserted into Eq. (J.2) gives finally

$$W(X, T) = \tfrac{1}{2} e^T \left[e^X \operatorname{Erfc} \left\{ \frac{X}{2\sqrt{T}} + \sqrt{T} \right\} + e^{-X} \operatorname{Erfc} \left\{ \frac{X}{2\sqrt{T}} - \sqrt{T} \right\} \right].$$

The function $W(X, T)$ came out as the result of Fourier's substitution

$$V(X, T) = e^{-T} W(X, T).$$

Hence the solution of the boundary problem becomes

$$V(X, T) = \tfrac{1}{2} \left[e^X \operatorname{Erfc} \left\{ \frac{X}{2\sqrt{T}} + \sqrt{T} \right\} + e^{-X} \operatorname{Erfc} \left\{ \frac{X}{2\sqrt{T}} - \sqrt{T} \right\} \right],$$

which is the formula of Eq. (4.4.33).

Appendix K

Calculation of the potential profiles resulting from injection of a constant current I_0 in the axon

The problem raised in Chapter 4, Section 4.4.5.2 (iii) required a solution of the cable equation

$$\frac{\partial^2 V}{\partial X^2} = \frac{\partial V}{\partial T} + V,$$

in the range $0 \leq X < \infty$ with the initial condition

$$V(X, T) = 0 \quad \text{for} \quad T = 0, \tag{K.1}$$

and the boundary conditions

$$V(X, T) \rightarrow 0 \quad \text{for} \quad X \rightarrow \infty \quad \text{and} \quad T > 0, \tag{K.2}$$

and

$$\left(\frac{\partial V}{\partial X} \right)_{X=0} = -\lambda r_a I_0 \quad \text{for} \quad X = 0 \quad \text{and} \quad t > 0. \tag{K.3}$$

We know already that the solution of a related problem, i.e. finding potential profile as the response to the application at $T = 0$ of a constant voltage V_0 at the position $X = 0$, is

$$V(X, T) = \tfrac{1}{2} V_0 \left\{ e^{-X} \operatorname{Erfc} \left(\frac{X}{2\sqrt{T}} - \sqrt{T} \right) + e^X \operatorname{Erfc} \left(\frac{X}{2\sqrt{T}} + \sqrt{T} \right) \right\}$$

as derived in Appendix J. This solution also satisfied the initial condition (K.1) and boundary condition (K.2) above. Therefore, it is possible to see whether this solution contains the basic elements to satisfy the boundary condition (K.3) of the above problem by proposing a solution of the form

$$V(X, T) = A e^{-X} \operatorname{Erfc} \left(\frac{X}{2\sqrt{T}} - \sqrt{T} \right) + B e^X \operatorname{Erfc} \left(\frac{X}{2\sqrt{T}} + \sqrt{T} \right), \tag{K.4}$$

that is, to examine whether the constants A and B can be adjusted such that

$$\left(\frac{\partial V}{\partial X} \right)_{X=0} = \text{a constant}, \tag{K.5}$$

that is equal to $-\lambda r_a I_0$ according to the boundary condition (K.3).

From the definition of Erfc$\{u\}$

$$Y(u) = \mathrm{Erfc}\{u\} = 1 - \mathrm{Erf}\{u\} = 1 - \frac{2}{\sqrt{\pi}} \int_0^u e^{-\xi^2}\, d\xi,$$

where $u = u(x)$ is a function of x, we have

$$\frac{dY}{dx} = \frac{d}{du}\left\{ -\frac{2}{\sqrt{\pi}} \int_0^u e^{-\xi^2}\, d\xi \right\} \frac{du}{dx} = -\frac{2}{\sqrt{\pi}} e^{-u^2} \frac{du}{dx}.$$

Thus

$$\frac{d}{dX}\left(\mathrm{Erfc}\left(\frac{X}{2\sqrt{T}} \pm \sqrt{T}\right)\right) = -\frac{2}{\sqrt{\pi}}\left[\exp\left\{-\left(\frac{X}{2\sqrt{T}} \pm \sqrt{T}\right)^2\right\}\right] \frac{d}{dX}\left(\frac{X}{2\sqrt{T}}\right)$$

$$= -\frac{1}{\sqrt{\pi T}} \exp\left\{-\left(\frac{X}{2\sqrt{T}} \pm \sqrt{T}\right)^2\right\}.$$

Hence, we have from Eq. (K.4)

$$\frac{\partial V}{\partial X} = A\left[-e^{-X}\left(\mathrm{Erfc}\left(\frac{X}{2\sqrt{T}} - \sqrt{T}\right)\right) - \frac{e^{-X}}{\sqrt{\pi T}}\exp\left\{-\left(\frac{X}{2\sqrt{T}} - \sqrt{T}\right)^2\right\}\right]$$

$$+ B\left[e^{X}\left(\mathrm{Erfc}\left(\frac{X}{2\sqrt{T}} + \sqrt{T}\right)\right) - \frac{e^{X}}{\sqrt{\pi T}}\exp\left\{-\left(\frac{X}{2\sqrt{T}} + \sqrt{T}\right)^2\right\}\right].$$

Next, putting $X = 0$ gives

$$\left(\frac{\partial V}{\partial X}\right)_{X=0} = -\lambda r_a I_0$$

$$= A\left[-\mathrm{Erfc}(-\sqrt{T}) - \frac{e^{-T}}{\sqrt{\pi T}}\right] + B\left[\mathrm{Erfc}(\sqrt{T}) - \frac{e^{-T}}{\sqrt{\pi T}}\right]$$

$$= A\left[-2 + \mathrm{Erfc}(\sqrt{T}) - \frac{e^{-T}}{\sqrt{\pi T}}\right] + B\left[\mathrm{Erfc}(\sqrt{T}) - \frac{e^{-T}}{\sqrt{\pi T}}\right]*$$

$$= (A + B)\left[\mathrm{Erfc}(\sqrt{T}) - \frac{e^{-T}}{\sqrt{\pi T}}\right] - 2A.$$

If the right-hand side has to be a constant and independent of T the terms outside the parentheses must add to zero, i.e.

$$A + B = 0,$$

from which it follows that

$$A = \tfrac{1}{2}\lambda r_a I_0 \quad \text{and} \quad B = -\tfrac{1}{2}\lambda r_a I_0.$$

Inserting these values into Eq. (K.4) gives the solution

$$V(X, T) = \tfrac{1}{2}\lambda r_a I_0 \left\{ e^{-X}\,\mathrm{Erfc}\left(\frac{X}{2\sqrt{T}} - \sqrt{T}\right) - e^{X}\,\mathrm{Erfc}\left(\frac{X}{2\sqrt{T}} + \sqrt{T}\right)\right\},$$

which is Eq. (4.4.34) of Chapter 4, Section 4.4.5.2.

---

* Since $\mathrm{Erfc}\{-u\} = 1 - \mathrm{Erf}\{-u\} = 1 + \mathrm{Erf}\{u\} = 1 + 1 - \mathrm{Erfc}\{u\} = 2 - \mathrm{Erfc}\{u\}$.

Appendix L

A note on the method of images

In Section 4.4.5.2 we derived the cable equation

$$\lambda^2 \frac{\partial^2 V}{\partial x^2} = \tau \frac{\partial V}{\partial t} + V,$$

and obtained – perhaps rather laboriously – solutions that were valid for the semi-infinite cable in the range $0 \leq x \leq \infty$ and subject to the boundary condition $V = 0$ for $x = +\infty$. When a constant current I_0 is injected at time $t = 0$ at $x = 0$ the solution became

$$V(X, T) = \tfrac{1}{2}\lambda r_a I_0 \left\{ e^{-X}\mathrm{Erfc}\left(\frac{X}{2\sqrt{T}} + \sqrt{T} \right) - e^{X}\mathrm{Erfc}\left(\frac{X}{2\sqrt{T}} - \sqrt{T} \right) \right\}, \quad (4.4.34)$$

where X and T are expressed in units of length constant λ and membrane time constant τ respectively. It may happen in many experimental conditions that the distance L from the current injecting electrode to the end of the cable is of the same order of length as the length constant $\lambda = \sqrt{r_m/r_a}$. This will change the boundary condition problem to that of having to assign a value of V or $\partial V/\partial x$ at the end of the cable at $X = L$, which in many cases may be

$$\frac{\partial V}{\partial X} = 0, \quad \text{for} \quad X = L, \tag{L.1}$$

a condition that may render a complicated problem even more complicated. If, as in the above case, the solution that is valid for the semi-infinite region is at one's disposal, there is a fairly simple way out of the difficulties known as the *methods of images*, which is based on superposing a series of segments of the solution for the semi-infinite space that are selected in such a way that their sum results in the above condition.

With a view to the illustration of the above theory we shall begin by solving a problem of the above category whose integration is straightforward, namely that of finding the stationary potential profile in a cable of finite length L with longitudinal current injection I_0 at the one end at $X = 0$ and no current flow across the other end.

This potential profile is found by solving the stationary cable equation (Eq. 4.4.20)

$$\lambda^2 \frac{d^2 V}{dx^2} - V = 0 \quad \text{or} \quad \frac{dV^2}{dX^2} - V = 0, \tag{L.2}$$

in the range $0 \leq X \leq L$ which satisfies the boundary conditions

$$\frac{dV}{dX} = 0, \quad \text{for} \quad X = L, \tag{A}$$

and

$$I_0 = -\frac{1}{r_a}\frac{dV}{dx} = -\frac{1}{r_a}\frac{1}{\lambda}\frac{dV}{d(x/\lambda)} = -\frac{1}{\lambda r_a}\frac{dV}{dX}, \quad \text{for} \quad X = 0. \tag{B}$$

To include these boundary conditions we have to take the whole of the general solution*

$$V = Ae^{-X} + Be^{X}, \tag{L.3}$$

from which it follows that

$$\frac{dY}{dX} = -Ae^{-X} + Be^{X}. \tag{L.4}$$

Inserting the two boundary conditions (A) and (B) in Eq. (L.4) gives the two connected equations

$$-Ae^{-L} + Be^{L} = 0, \tag{L.5}$$

and

$$-A + B = -\lambda r_a I_0, \tag{L.6}$$

for the determination of the constants A and B. Solving for A in Eq. (L.5) gives

$$A = Be^{2L},$$

which inserted into Eq. (L.6) yields

$$B = \frac{a}{1 - e^{2L}},$$

where $a = -\lambda r_a I_0$. Returning this value for B in the above expression for A gives

$$A = \frac{a}{1 - e^{2L}}e^{2L} = a\frac{e^{L}}{e^{-L} - e^{L}} = -a\frac{e^{L}}{2\sinh(L)}.$$

Inserting then the values of A and B in the general solution gives the potential profile as

$$V = -a\frac{e^{L}}{2\sinh(L)}e^{-X} + \frac{a}{1 - e^{2L}}e^{X} = -\frac{a}{2}\frac{e^{L}e^{-X}}{\sinh(L)} - \frac{a}{2}\frac{e^{X}e^{-L}}{\sinh(L)}$$

$$= -a\frac{e^{(L-X)} + e^{-(L-X)}}{2}\frac{1}{\sinh(L)} = -a\frac{\cosh(L-X)}{\sinh(L)},$$

and by replacing a with $-\lambda r_a I_0$

$$V = \lambda r_a I_0 \frac{\cosh(L - X)}{\sinh(L)}, \tag{L.7}$$

which is the equation for the potential profile in the regiom $0 \leq X \leq L$.

* See Chapter 1, Section 9, Eq. (1.9.21).

This result could have been obtained more directly if the general solution had been expressed directly from the start in terms of the hyperbolic functions sinh and cosh and at the same time had shifted the argument from X to $(L - X)$. The function

$$V = A \cosh(L - X), \quad \text{i.e.} \quad \frac{dV}{dX} = -A \sinh(L - X),$$

automatically satisfies the boundary condition (A) for $X = L$. Thus, it is enough to apply the boundary condition (B) to $V = A \cosh(L - X)$, which immediately results in

$$A = \frac{\lambda r_a I_0}{\sinh(L)}.$$

We shall now handle the same problem by means of the method of images. The stationary profile in the semi-infinite region $0 \leq X \leq +\infty$ is

$$V = \lambda r_a I_0 e^{-X} k \tag{L.8}$$

or

$$U(X) = e^{-X}, \tag{L.9}$$

where U is a normalized potential expressed in units of $V_0 = \lambda r_a I_0$. To simplify the numerical example further we assume that the cable length is equal to λ, i.e $L = 1$. The finite cable is contained within the region $0 \leq X \leq 1$ shown in Fig. L.1(A). Curve (a) is the course of $U(X)$ inside the cable domain. It is seen to have a definite negative slope when crossing the plane at $X = 1$, whereas the required potential profile at this point should have a gradient of value zero. To this end we add to the domain a potential profile whose value coincides with $U(X)$ at $X = 1$ but has a slope that is numerically equal but of *opposite sign* to that of $U1$. Consider now the function

$$\Im_{2+} = e^{-|X-2|}.$$

It has two legs that decline symmetrically with respect to the y-axis at $X = 2$. In the range $X \geq 2$ the course is identical to that of $U(X)$ apart from being displaced to $X = 2$. The course of the leg for $X \leq 2$ can be looked upon as the *mirror image* of $U(X)$ that is reflected at the plane containing the y-axis at $X = 2$. Curve (b) shows the course of this image through the cable region $0 \leq X \leq 1$. The position $X = 1$ lies *midway* between the origins of the function $U(X)$ at $X = 0$ and its mirror image $\Im_{2+}$ at $X = 2$. Therefore, the values of the two functions are both equal to $U(1)$ and, furthermore, the slopes of the two functions are numerically equal but of *opposite* sign. Hence, the superposition of $U(X)$ with its mirror image $\Im_{2+}$ results in a curve that enters the position at $X = 1$ with the value $2U(1)$ and with a *horizontal slope* as the boundary condition demands. This procedure, however, does not solve the problem completely because the portion (b) of the mirror image *leaves* the region at $X = 0$ with a *finite* slope. To eliminate this unwanted effect we introduce the mirror image of $\Im_{2+}$ placed at $= -2$,

$$\Im_{-2} = e^{-|X+2|},$$

by means of which the origin $X = 0$ is left in the position midway between the two sources. Thus, the superposition of $\Im_{-2}$, curve (c), causes as before a horizontal slope of the resultant profile at $X = 0$ and a finite slope at the other end at $X = 1$. This is

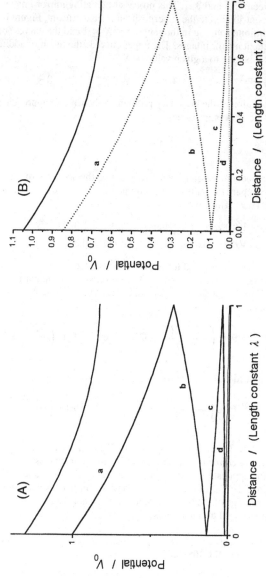

Fig. L.1. (A) Illustration of the operation of the method of images. (B) Using the transient solution of the infinite cable with $T = 1$ to solve for the short cable with boundary condition $\partial V / \partial X = 0$ for $X = 0.8$.

621

removed by introducing the next mirror image

$$\Im_{+4} = e^{-|X-4|}$$

of $\Im_{-2}$ whose origin is at $X = 4$ to cause the position of the cable end at $X = 1$ to be found midway betweem $\Im_{-2}$ and $\Im_{+4}$. This process of creating mirror images is continued until the contribution to the superposition is insignificant. Figure L.1 curve (a) shows the semi-infinite solution in the range $0 \leq X \leq 1$ and the curves (b), (c) and (d) show the subsequent mirror images. The upper curve is the result of adding each of the five ordinates that belong to a given value of X,

$$W(X) = U(X) + e^{-|X-2|} + e^{-|X+2|} + e^{-|X-4|} + e^{-|X+4|} \quad \text{for} \quad 0 \leq X \leq 1, \quad \text{(L.10)}$$

and represents the solution of the boundary problem. When $W(X)$ approaches the cable end at $X = 1$ the slope approaches

$$\left(\frac{dW}{dX}\right)_{X=1} = 0,$$

in accordance to the boundary requirement. To examine the accuracy of the approximation we compare the values at the end points with the exact values calculated from Eq. (L.7). This gives

$$V(0) = 1.313 \quad \text{against} \quad W(0) = 1.307$$
$$V(1) = 0.850 \quad \text{against} \quad W(1) = 0.843.$$

To improve the agreement additional terms would be needed.

To make the time-dependent solution Eq. (4.4.34) satisfy the boundary condition Eq. (L.1) for $X = L$ we establish a sequence of mirror images, where the first one is situated at $X = 2L$, namely

$$\Im_{+2L} = \tfrac{1}{2}\lambda r_a I_0 \left\{ e^{-|X-2L|} \text{Erfc}\left(\frac{|X-2L|}{2\sqrt{T}} + \sqrt{T}\right) - e^{|X-2L|}\text{Erfc}\left(\frac{|X-2L|}{2\sqrt{T}} - \sqrt{T}\right)\right\}.$$

The second image placed at $X = -2L$ is

$$\Im_{-2L} = \tfrac{1}{2}\lambda r_a I_0 \left\{ e^{-|X+2L|} \text{Erfc}\left(\frac{|X+2L|}{2\sqrt{T}} + \sqrt{T}\right) - e^{|X+2L|}\text{Erfc}\left(\frac{|X+2L|}{2\sqrt{T}} - \sqrt{T}\right)\right\},$$

and the next two at $X = 4L$ and $-4L$ respectively. Figure L.1(B) shows an example of the cable profile calculated for $L = 1.5$ by summing the four images.

An example of the application of the method of images can be found in the study of Hodgkin & Nakajima (1972)* of the electrical constants in frog skeletal muscle fibers, which required knowledge of the transient solution in a short cable. Naturally, the method applies equally well to other fields of physics.

* Hodgkin, A.L. & Nakajima, S. (1972). The effect of diameter on the electrical constants of frog skeletal muscle. *J. Physiol.*, **221**, 105–120.

Appendix M

Cable analysis of the end-plate potential (Fatt & Katz, 1951)

In Chapter 5, Section 5.2.3.2, we mentioned that the end-plate potentials that were recorded at various distances from the very end-plate region could be regarded as the passive spread in a cable-like structure resulting from a short-lasting depolarization occurring in the end-plate due to the release of acetylcholine from the presynaptic membrane. Fatt & Katz (1952) refer in their appendix to the mathematical formula that describes the perturbation along the fiber length of the membrane potential. The formula was derived by A.L. Hodgkin as the solution of the cable equation subject to the condition of placing an electric charge Q at the time $t = 0$ in the position $x = 0$. No details were given of the solution of the problem considered, for which several methods are available. We shall here use the resources that already are at our disposal.

We start with the cable equation Eq. (4.4.19)

$$\lambda^2 \frac{\partial^2 V}{\partial x^2} = \tau \frac{\partial V}{\partial t} + V, \tag{4.4.19}$$

where λ is the length constant and τ is the membrane time constant. Introducing the new variables

$$T = \frac{t}{\tau},$$

and

$$X = \frac{x}{\lambda},$$

brings Eq. (4.4.19) into the dimensionless form

$$\frac{\partial^2 V}{\partial X^2} = \frac{\partial V}{\partial T} + V, \tag{4.4.29}$$

which by means of Fourier's substitution

$$V = e^{-T} W,$$

takes the form

$$\frac{\partial W}{\partial T} = \frac{\partial^2 W}{\partial X^2}, \tag{4.4.31}$$

623

which is analogous to the form of the simple diffusion equation. In the present case, however, where W has the dimension of an electric potential, Eq. (4.4.31) represents the propagation of the voltage across an electric cable without leakage ($r_m = \infty$).

Green's function* for this equation is

$$G(X, T) = \frac{1}{2\sqrt{\pi T}} e^{-X^2/4T} \tag{M.1}$$

if the instantaneous source is placed at the position $X = 0$ at the time $T = 0$. In case the initial value of W is known, that is

$$W(X, 0) = f(X) \quad \text{for} \quad -\infty < X < \infty \quad \text{and} \quad T = 0, \tag{M.2}$$

we have for all times $T > 0$ that[†]

$$W(X, T) = \frac{1}{2\sqrt{\pi T}} \int_{-\infty}^{+\infty} f(\xi) e^{-(X-\xi)^2/4T} \, d\xi. \tag{M.3}$$

We now imagine that an electric charge Q is placed at the time $T = 0$ in the region $0 \leq X \leq \Delta X$, which is raised to the potential

$$\frac{Q}{\Delta C} \quad \text{for} \quad 0 \leq X \leq \Delta X \quad \text{and} \quad T = 0,$$

where ΔC is the membrane capacitance corresponding to the length element $\Delta X = \delta x / \lambda$ in dimensionless units, with Δx being the length element in absolute units (e.g. in meters). Let c_m represent the membrane capacitance per unit length of the muscle fiber. We have then

$$\Delta C = c_m \, \Delta x = c_m \lambda \, \Delta X.$$

The initial condition

$$W(X, 0) = f(X) = \begin{cases} 0, & \text{for} \quad \xi < 0; \\ \dfrac{Q}{\lambda c_m \, \Delta X}, & \text{for} \quad 0 \leq \xi \leq \Delta X; \\ 0, & \text{for} \quad \xi > \Delta X; \end{cases}$$

applied to Eq. (M.3) gives

$$\begin{aligned} W(X, T) &= \frac{1}{2\sqrt{\pi T}} \int_0^{0+\Delta X} \frac{Q}{\lambda c_m \, \Delta X} e^{-(X-\xi)^2/4T} \, d\xi \\ &= \frac{Q}{2 c_m \lambda \sqrt{\pi T}} \left\{ \frac{1}{\Delta X} \int_0^{0+\Delta X} e^{-(X-\xi)^2/4T} \, d\xi \right\}. \end{aligned}$$

Using the mean value theorem for the definite integral[‡] we obtain

$$W(X, T) = \frac{Q}{2 c_m \lambda \sqrt{\pi T}} \left\{ \frac{1}{\Delta X} \cdot \left(\Delta X \, e^{-(X-\xi'(\Delta X))^2/4T} \right) \right\},$$

---

* See Chapter 2, Section 2.5.4.5, Eq. (2.3.126).
† See Chapter 2, Section 2.5.4.5(i), Eq. (2.5.128).
‡ See Chapter 1, Section 1.3.4, Eq. (1.3.16).

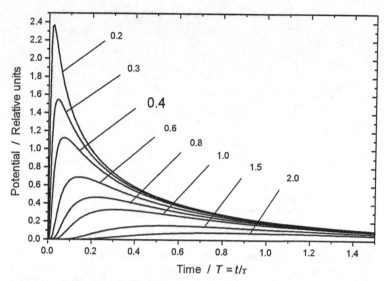

Fig. M.1. Time course of the potential distribution across the membrane in response to placing an instantaneous charge Q in the position $X = 0$. Abscissa: dimensionless time in units of the membrane time constant τ. Ordinate: dimensionless membrane potential in units of $Q/2c_m\lambda$. The numbers attached to the curves give the distance from the position $X = 0$ in units of the length constant λ of the membrane.

where $0 \leq \xi'(\Delta X) \leq \Delta X$. When $\Delta X \to 0$ then $\xi'(\Delta X) \to 0$. This makes the above expression assume the value

$$W(X, T) = \frac{Q}{2c_m\lambda\sqrt{\pi T}}e^{-X^2/4T},$$

corresponding to the charge Q is now concentrated in the position $X = 0$ at the time $T = 0$. Insertion of the substitution $V = e^{-T}W$ gives

$$V(X, T) = \frac{Q}{2c_m\lambda\sqrt{\pi T}}\exp\left\{-\frac{X^2}{4T} - T\right\}, \tag{M.4}$$

which is identical to Hodgkin's equation except that Eq. (M.4) contains the relative, dimensionless variables $X = x/\lambda$ and $T = t/\tau$ and not the absolute variables x and t, as written by Hodgkin.

Equation (M.4) reproduces the potential profile $V(X, T)$ in position X at time T after the charge Q was placed instantaneously at $T = 0$ in $X = 0$. In Fig. M.1 are shown examples of calculated time courses of $V(X, T)$ in units of $Q/2c_m\lambda$ as a function of the distance X. It appears that the time courses from this mathematical model and those recorded experimentally by Fatt & Katz as shown in Chapter 5, Section 5.2.3.2, Fig. 5.9, are similar in shape. Fatt & Katz did not make this comparison between the two plots but examined whether the time of the rising phase of the EPP to reach its maximum was related to the distance x at this point as predicted by the theory. This relation is established by finding the value of T that makes

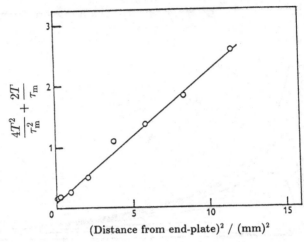

Fig. M.2. Analysis of end-plate potentials recorded outside the end-plate region in a curarized frog muscle fiber, assuming that the signals represent a passive spread in a one-dimensional cable structure. Ordinate: $4T^2 + 2T$, where $T = t/\tau$ is the time – in units of the membrane time constant τ – to reach the maximum of the EPP that is recorded at the distance x from the end-plate. Abscissa: the square of the position x where each EPP is recorded. From cable theory the relation $4T^2 + 2T = (1/\lambda^2)x^2$ is predicted. (From Fatt & Katz, 1951.)

$\partial V/\partial T = 0$ in Eq. (M.2). We take the logarithm on both sides of the equation, thus

$$\ln V = \ln \text{const} - \ln(\lambda\sqrt{T}) - \frac{X^2}{4T} - T$$

$$= \ln \text{const} - \ln \lambda - \frac{1}{2}\ln T - \frac{X^2}{4T} - T,$$

differentiate partially with respect to T and put the result equal to zero. This gives

$$\frac{\partial \ln V}{\partial T} = \frac{1}{V}\frac{\partial V}{\partial T} = -\frac{1}{2}\frac{1}{T} + \frac{X^2}{4T^2} - 1 = 0,$$

where T now represents that particular value that passes until $\partial V/\partial T = 0$. Multiplication by $4T^2$ brings about the result

$$X^2 = 4T^2 + 2T,$$

or

$$\left(\frac{x}{\lambda}\right)^2 = 4\left(\frac{t}{\tau}\right)^2 + 2\left(\frac{t}{\tau}\right), \tag{M.5}$$

which implies that a plot of $Y = 4(t/\tau)^2 + 2(t/\tau)$ against the square of the distance x corresponding to the maximum value of the EPP should result in a straight line with the slope $(1/\lambda)^2$.

Experiments on the muscle fibres in the same sartorius muscle provided Fatt & Katz with numerical values of τ and λ, which enabled them to test the validity of the prediction of λ from Eq. (M.5). An example is shown in Fig. M.2, which displays a satisfactory straight-line relationship apart from quite small values of t and x where discrepancies between the experimental situation (a final spatial extension of the end-plate region and a non-instantaneous charging process there) and the idealized model in particular come to light. In the example, a value of $\lambda = 2.5$ mm was predicted by the model analysis. Another experiment gave $\lambda = 2.4$ mm. Determinations of λ by standard cable analysis in seven fibers of similar diameter as in the above experiments gave values for λ between 2.2 and 2.6 mm (mean value: 2.4 mm). The agreements – qualitatively as well as quantitatively – of the model with the experimental data permitted Fatt & Katz, as mentioned in Chapter 5, Section 5.2.3.2, to conclude that the prolonged spread and decline of the EPP outside the end-plate region are determined by the passive electric properties of the resting muscle fiber.

Appendix N

Measuring the electric parameters in a spherical cell

After the development of the technique to measure membrane potentials by means of glass micropipettes it became a relatively simple matter to develop this technique further to include measurement the electric *resistance* and the *capacitance* of the membrane. Dealing with *spherical* cells makes the situation particularly simple. A rough outline of a measuring assembly is shown in Fig. N.1. Two microelectrodes $\mathcal{E}_V$ and $\mathcal{E}_i$ are inserted into the cell. $\mathcal{E}_V$ is connected to an electronic amplifier that measures the potential difference between the interior of the cell and the surrounding medium, i.e. the potential difference across the cell membrane. The other microelectrode, $\mathcal{E}_i$, the current electrode, is connected to one pole of a constant-current generator IGen, whose other pole is connected to the electrode el_{bath} that is placed in the bathing solution surrounding the cell. Ideally the tip of the current electrode $\mathcal{E}_i$ should have been placed in the center of the cell. However, as the electric resistance of the cell's cytoplasm is negligible as compared to the resistance presented by the membrane the current flow from $\mathcal{E}_i$ will not – irrespective of the position of the tip of $\mathcal{E}_i$ – cause any noticeable potential drop inside the cell before reaching the cell membrane, and from there the current will distribute itself evenly, i.e. with the same current density everywhere across the cell membrane. During the passage through the membrane the current divides into two components. One is flowing through the membrane resistance and causing a change in the potential difference across the membrane that is recorded by the voltage recording electrode $\mathcal{E}_V$. The other charges the membrane capacitance concurrently with changes in the membrane potential. The current that flows through the cell membrane is measured as the voltage drop across a small resistance that is placed in the wire from el_{bath} to the current generator, where a suitable feed-back arrangement ensures that the potential of the bath does not change when turning on the circuit from IGen. As the current flow across the membrane can be regarded as uniform, the physical situation can be reduced to a simple model where the evenly distributed resistances and capacitances of the membrane are lumped into *one single* resistance R and capacitance C as illustrated in Fig. N.1(B).

Figure N.2 shows an example of the time course of the potential change across the membrane in response to a constant current flow I through the tip of $\mathcal{E}_i$ that starts at time $t = 0$. It appears that the potential does not change instantaneously to the new level corresponding to the value of I, but grows gradually to approach asymptotically a new stationary value, where it remains as long as the current flow is sustained. When the current is turned off, the potential returns to the initial value with a time course that makes a counterpart to the rising phase. The delay of the potential relative to the current I is due to the presence of the membrane capacity C, since a portion – though

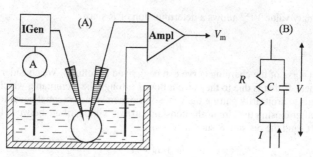

Fig. N.1. (A) Set up for the simultaneous measurement of membrane potential, membrane resistance and membrane capacitance in a spherical cell. Ampl: electronic amplifier connected to the voltage recording microelectrode $\mathcal{E}_V$. The other input lead is connected to an electrode el $_{\text{bath}}$ in the fluid that surrounds the cell. IGen is a constant current generator that injects current into the cell interior via the microelectrode $\mathcal{E}_i$. After crossing the cell membrane the current returns from the bath via the electrode el $_{\text{bath}}$ to IGen. (B) The electric equivalent circuit to imitate the electric behavior of the cell membrane is made of the *total* resistance R and capacitance C of the cell surfaces arranged in parallel.

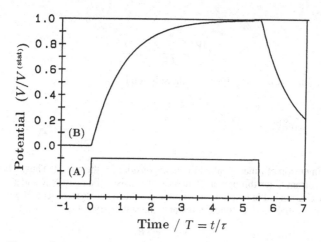

Fig. N.2. Change of membrane potential in the spherical cell in response to the application of a constant current through the membrane. (A) Recording of current that is turned on at time $t = 0$ and off at time t_1. (B) Recording of the displacement of the membrane potential V_m from the resting value as a function of time t.

alternating – of the current will enter into charging the membrane capacity C while the rest of the constant current flows through the membrane resistance R. Initially, all the current will flow into the capacitor, but as the charge grows on C – and connected with it the voltage V – more and more current will flow through R and consequently less current to C. Finally, the situation will arrive where the voltage across C has attained the value where all the current flows through R. The situation is now *stationary*, and this stationary value $V^{(\infty)}$ allows a determination of R since

$$R = \frac{V^{(\infty)}}{I}. \tag{N.1}$$

The time course of the charging curve can be derived as follows. We consider only the displacement V that is due to the current flow I through the membrane, which corresponds to arbitrarily putting the value of the membrane potential equal to zero. At any time after closing the circuit the constant current will distribute between the capacitor C and the resistor R such that

$$I_C + I_R = I, \tag{N.2}$$

where I_C is the charging current to the capacitor C and I_R is the current flowing through the resistor R. We have

$$I_C = C\frac{dV}{dt} \quad \text{and} \quad I_R = \frac{V}{R},$$

where V is the value of the membrane potential V at the time t. Inserting the expressions above for the two currents in Eq. (N.1) we obtain the differential

$$C\frac{dV}{dt} + \frac{V}{R} = I,$$

or

$$\frac{dV}{dt} + \frac{1}{RC}V = \frac{I}{C}.$$

With the initial value $V = 0$ for $t = 0$ the solution is*

$$V = IR\left[1 - e^{-t/\tau}\right], \tag{N.3}$$

where the quantity

$$\tau = RC, \tag{N.4}$$

having dimension of time, is called the *time constant* for the circuit. Thus, when a constant current flows through an element that consists of a resistor R and a capacitor C arranged in parallel, the voltage produced across this element will grow mono-exponentially asymptotically towards the stationary value

$$V^{(\infty)} = IR, \tag{N.5}$$

and since I is known and the value of $V^{(\infty)}$ is determined from the recording, the total resistance R of the membrane can be calculated. Putting $t = \tau$ in Eq. (N.3) we obtain

$$V(\tau) = V^{(\infty)}\left[1 - e^{-1}\right] = V^{(\infty)}\left[1 - 0.3679\right] = 0.632\, V^{(\infty)}. \tag{N.6}$$

Thus, the value of time constant τ is found as that time where the membrane potential has attained the value $V^{(*)} = 0.632\, V^{(\infty)}$. Since the membrane resistance is already known the membrane capacitance C can be calculated as $C = \tau/R$. Let the sphere have the diameter d. Then the surface area of the sphere is $A = \pi d^2$, and the normalized values are then, according to Appendix K,

$$C_m = \frac{C}{A} \;(\text{F m}^{-2}) \quad \text{and} \quad R_m = R \cdot A \;(\Omega\, \text{m}^2).$$

* See Chapter 1, Section 1.9.1.2, Eq. (1.9.6), but replacing K with I/C and α with $1/RC$.

Appendix O

Measuring the electric parameters in a cylindrical cell

To describe the passive spread of the potential changes in space and time in a cylindrical cell that results when an external current is injected into the cell, the three following electric parameters are required.

(1) The axial resistance per unit length r_a, S.I. unit: (Ω/m).
(2) The membrane resistance for one length unit r_m, S.I. unit: (Ωm).
(3) The membrane capacitance per unit length c_m, S.I. unit: (F m^{-1}).

In addition to this there are the three following useful quantities.

- The length constant $\lambda = \sqrt{r_m/r_a}$, S.I. unit: (m).
- The characteristic resistance $R_0 = \sqrt{r_m r_a} = \lambda r_a$, S.I. unit: (Ω).
- The membrane time constant $\tau = r_m c_m$, S.I. unit: (s).

Below we shall describe some of the methods to determine r_a, r_m and c_m that are mostly in use. In principle the methods are the same as that described in Appendix N, namely that of injecting a known current ΔI into the interior of the cell and recording the potential change ΔV that the current causes across the membrane. However, ΔV is now a function both of the time t and the position x, and as a rule it is not possible to depolarize the membrane uniformly over such a distance that it is meaningful to consider the ratio $\Delta V/\Delta I$ as was done in Appendix N. For that reason it is nearly always necessary to adapt the experimental technique with special reference to relevant interrelations that can be derived from the one-dimensional cable equation

$$\lambda^2 \frac{\partial^2 V}{\partial x^2} = \tau \frac{\partial V}{\partial t} + V. \qquad (4.4.19)$$

1 Determination of the length constant

A glass microelectrode that functions as a current electrode is inserted into the interior of the cable at the position $x = 0$. At the time $t = 0$ the current of strength $2I_0$ is turned on. Owing to symmetry conditions the current flows axially to both sides with the same strength I_0. The potential change across the membrane due to the current is recorded along the cable at distances x_1, x_2, x_3, ..., from the current electrode. The time of each recording is taken to last long enough to allow the membrane capacitances to be fully charged, i.e. the potential changes represent the *stationary*

631

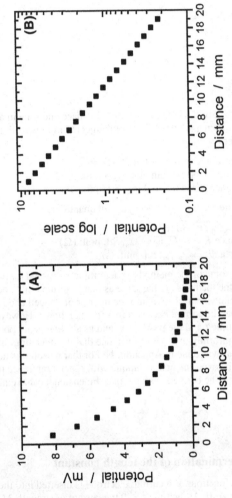

Fig. O.1. Stationary potential values V measured at various distances x from a current electrode placed internally in the position $x = 0$, from which a constant axial current $I_0 = 1\ \mu A$ flows in the positive direction of the x-axis. Abscissa: distance from the electrode tip (mm). Ordinate: potential (mV). (A) Ordinate with linear scale. (B) Ordinate with logarithmic scale. From the graphs it is found that $\lambda = 0.5$ cm, and $V_0 = 10$ mV.

values. Hence, the shape of potential profile $V(x)$ is found by solving the equation

$$\lambda^2 \frac{d^2 V}{dx^2} - V = 0, \qquad (4.4.20)$$

that satisfies the boundary conditions

$$\left.\begin{array}{ll} V(x) = 0 & \text{for } x \to \infty \\ i_a(x) = I_o & \text{for } x = 0 \end{array}\right\} \qquad (O.1)$$

is

$$V(x) = I_o\sqrt{r_a r_m}\, e^{-x/\lambda} = V_o\, e^{-x/\lambda}, \qquad (4.4.22)$$

i.e. the potential $V(x)$ declines mono-exponentially with the distance from the position of the current electrode at $x = 0$, where the initial value V_o is

$$V_o = \sqrt{r_a r_m}\, I_o. \qquad (4.4.23)$$

For $x = \lambda$ Eq. (4.4.22) gives

$$V(\lambda) = V_o\, e^{-1} = 0.3679\, V_o \approx 0.37 V_o. \qquad (O.2)$$

Thus, from the plot of the values $V(x)$ measured at the positions $x_1, x_2, x_3, \ldots$, the value of length constant λ is found from the graph as that value x where V has declined to 37% of the initial value V_o. An alternative method to determine λ consists of plotting $\log V$ against the distance x. This results in a straight line having a slope

$$\frac{d\log V}{dx} = -\frac{1}{\lambda}. \qquad (O.3)$$

Extrapolation of the straight line $\log V$ versus x to $x = 0$ the value of V_o results from intersection with the ordinate axis. When λ is found we also know the value of the ratio r_m/r_a, as

$$\frac{r_m}{r_a} = \lambda^2. \qquad (O.4)$$

As the value of I_o is known and V_o is found by the extrapolation the value of $r_m r_a$ can be obtained from Eq. (4.4.23) as

$$r_m r_a = \left(\frac{V_o}{I_o}\right)^2. \qquad (O.5)$$

From these two equations the values of r_m and r_a can be found.

EXAMPLE From the data of Fig. (O.1) where $I_o = 1\,\mu A$, a graphical determination gives $\lambda = 0.5$ cm and $V_o = 10$ mV. Hence,

$$r_m r_a = \left(\frac{10^{-2}}{10^{-6}}\right)^2 = 10^8 \quad \text{and} \quad \frac{r_m}{r_a} = 0.25,$$

yielding

$$r_m = \sqrt{0.25 \times 10^8} = 5 \times 10^3\,\Omega\,\text{cm} \quad \text{and} \quad r_a = 5 \times 10^3/0.25 = 20 \times 10^3\,\Omega\,\text{cm}^{-1}.$$

2 Determination of the membrane capacitance

Similar to the response in the spherical cell, it is the magnitude of the *membrane time constant*

$$\tau = r_m c_m = R_m C_m,$$

that determines the rate at which the membrane potential grows to its stationary value in response to a current step. Thus, if an analysis of the time-dependent potential response $V(x, t)$ provides a determination of the membrane time constant τ, then the value of the membrane capacitance c_m is computable as the membrane resistance r_m is determined as shown above from the stationary potential values $V(x)$.

As in section 1 we consider the situation with the current electrode placed in the position $x = 0$. At time $t = 0$ a constant current $2I_o$ is turned on and flows axially to both sides with equal values I_o. In the range $x > 0$ the current injection I_o causes in the position x a potential response whose expected time course is given by

$$V(X, T) = \frac{1}{2}\lambda r_a I_o \left\{ e^{-X} \operatorname{Erfc}\left(\frac{X}{2\sqrt{T}} - \sqrt{T}\right) - e^X \operatorname{Erfc}\left(\frac{X}{2\sqrt{T}} + \sqrt{T}\right) \right\}, \quad (4.4.34)$$

where $\lambda r_a I_o = V_o$ is the stationary value of potential at the position $x = 0$, i.e. corresponding to the position of the electrode tip. In Eq. (4.4.34) the distance $X = x/\lambda$ is expressed in units of the length constant λ of the cell and the time $T = t/\tau$ is in units of the membrane time constant τ. We consider first a special situation.

(i) Time course of the transient at position $X = 0$

Putting $X = 0$ in Eq. (4.4.34) gives the following simpler expression

$$V(0, T) = \tfrac{1}{2}\lambda r_a I_o \{\operatorname{Erfc}(-\sqrt{T}) - \operatorname{Erfc}(\sqrt{T})\}$$

$$= \tfrac{1}{2}\lambda r_a I_o \{2 - \operatorname{Erfc}(\sqrt{T}) - \operatorname{Erfc}(\sqrt{T})\}$$

$$= \tfrac{1}{2}\lambda r_a I_o\, 2 \{1 - \operatorname{Erfc}(\sqrt{T})\}$$

$$= \lambda r_a I_o\, \operatorname{Erf}(\sqrt{T}), \qquad\qquad\qquad (O.6)$$

as $\operatorname{Erfc}(-u) = 2 - \operatorname{Erfc}(u)$. Equation (O.6) represents the potential response obtained if the voltage recording electrode was inserted through the membrane exactly at the position of the tip of the current electrode, i.e. at $X = 0$. The time course of Eq. (O.6) is shown in Fig. (O.2). It appears to display some similarity to the charging response from the spherical cell, which for reasons of comparison is also shown as curve b. In the course of a charging time of $t = \tau$ (i.e. for $T = 1$) Eq. (O.6) yields

$$V(0, 1) = \lambda r_a I_o\, \operatorname{Erf}(1) = 0.8427\, V_o.$$

Therefore, if one can arrange these very special conditions of recording (recording electrode placed at the position $x = 0$, i.e. at the location of the tip of the current electrode) measuring the time that is required to charge the membrane to 84% of the stationary value V_o provides the value of the membrane time constant $\tau = r_m r_a$, which allows the calculation of c_m if r_m is determined as described in section 1.

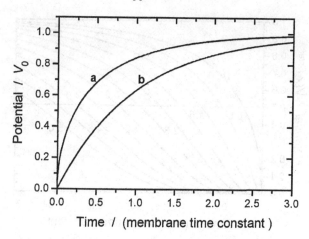

Time / (membrane time constant)

Fig. O.2. Time course of the displacement of the membrane potential (curve a) at the position $X = 0$ after a constant axial current $i_a = I_o$ is applied at the position $X = 0$ on the cable. $V(X, T)$ is calculated from Eq. (O.6). Ordinate: membrane potential in units of the stationary value $V_o = \lambda r_a I_o$. Abscissa: time $T = t/\tau$ in units of the membrane time constant τ. At the time of $\tau = 1$ the voltage has grown to 84% of the stationary value. By way of comparison the somewhat slower time course of the charging of the spherical cell is also shown (curve b), where the membrane charges to 63% of the stationary value at the time of τ.

(ii) Analysis of the transients at different distances X

It appears from Fig. (4.44)*, that the response of the potential $V(x, t)$ at the distance X from the current electrode does not only decrease in size with increasing values of X, but arrives with *a delay* that also grows with X. This pattern is illustrated in more detail in Fig. O.3 by calculating a series of time courses of the potential $V(X, T)$ at various positions X, but now being modified such that every time course is recalculated in units of the stationary value of the potential $V(X, T)$ in question, i.e. as

$$W(X, T) = \frac{V(X, T)}{V(X)_{\text{stat}}} = \frac{V(X, T)}{\lambda r_a I_o\, e^{-X}},$$

or by insertion of Eq. (4.4.34)

$$W(X, T) = \frac{1}{2}\left\{\text{Erfc}\left(\frac{X}{2\sqrt{T}} - \sqrt{T}\right) - e^{2X}\,\text{Erfc}\left(\frac{X}{2\sqrt{T}} + \sqrt{T}\right)\right\}. \qquad (O.7)$$

It has been known for a long time that the time $t^{(\frac{1}{2})}$, that it takes for the potential $V(x, t)$ in the position x to rise to half the value of the stationary value in question grows linearly with the distance x from the current electrode. Also that the coefficients of this straight line $t^{(\frac{1}{2})} = \alpha x + q$ contain both λ and τ. No theoretical justification for this relation has been given. However, this empirical number magic is illustrated in Fig. (O.3), where the point of intersection of the horizontal line $W = 0.5$ with each of the curves gives the time $T^{(\frac{1}{2})}$ for the rise of the potential $V(X, T)$ to half of its

* Chapter 4, Section 4.4.5.2.

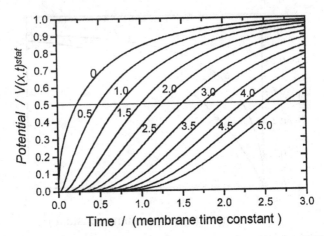

Fig. O.3. Illustration of the delayed arrival of the potential responses $V(X, T)$ with increasing distance X from the current electrode. All responses are normalized in units of their own stationary value. Ordinate: normalized potentials $W(X, T)$ calculated from Eq. (O.7). Abscissa: time $T = t/\tau$ in units of the membrane time constant. The numbers associated with each curve denote the distances $X = x/\lambda$ from the current electrode in units of the length constant. The distances from one position to the next are almost the same $\Delta X = 0.5$. The intersection of the curves with the straight line $W = 0.5$ is the time $T^{(\frac{1}{2})}$ for the response in question to rise half-way to its stationary value.

stationary value. It holds for all the curves that the distance from one position X_n to the next X_{n+1} is the same, namely $\Delta X = 0.5$. It appears that for $X \geq 0.5$ the delay ΔT from one curve to the next is the *same*. With a little ingenuity it may be seen on the figure that for a distance equal to one length constant ($\Delta X = 1$) the delay is equal to half the time constant ($\Delta T = \frac{1}{2}$). To elucidate this matter somewhat we have determined the values $T^{(\frac{1}{2})}$ by solving the equation

$$\frac{1}{2}\left\{\text{Erfc}\left(\frac{X}{2\sqrt{T}} - \sqrt{T}\right) - 2X\,\text{Erfc}\left(\frac{X}{2\sqrt{T}} + \sqrt{T}\right)\right\} = \frac{1}{2}, \qquad (O.8)$$

for different values of X. The result is shown in Fig. (O.4). It is seen that provided neither X nor T are near zero the values of $(X, T^{(\frac{1}{2})})$ belonging together are located on a straight line with the slope $\frac{1}{2}$ and intersects the T-axis for $X = 0$ at $T = \frac{1}{4}$. Thus, the empirical relation between the position X and the value of the time $T^{(\frac{1}{2})}$ for the potential to attain half the stationary value is

$$T^{(\frac{1}{2})} = \frac{1}{2}X + 0.25, \qquad (O.9)$$

or by replacing X and T by their absolute values $t = \tau T$ and $x = \lambda X$

$$t^{(\frac{1}{2})} = \left(\frac{\tau}{2\lambda}\right)x + 0.25\,\tau. \qquad (O.10)$$

Thus, plotting the absolute values of the time $t^{(\frac{1}{2})}$, to attain half the stationary value of the potential response as a function of the distance x from the tip of the current

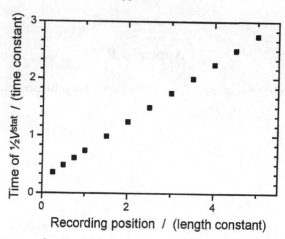

Fig. O.4. The time $T^{(\frac{1}{2})}$ to attain half of the stationary value of $W(X, T)$ as a function of the position X as calculated from Eq. (O.7). Ordinate: $T^{(\frac{1}{2})}$, where $T = t/\tau$. Abscissa: position $X = x/\lambda$.

electrode, a straight line with the slope $\tau/2\lambda$ is obtained which intersects the t-axis at $t = \tau/4$. The method allows the determination of λ and τ, but to determine the capacitance of the membrane c_m the membrane resistance r_m must also be known, e.g. by making use of the stationary values of the signals as shown in section 1.

It has been attempted to account for Eq. (O.9) more rigorously by putting $W(x, T) = \frac{1}{2}$ in Eq. (O.7) and examine whether this can be satisfied by a relation of the form

$$X = 2T + C,$$

where C is a constant. The result of these not particularly transparent arithmetic exercises is that the equation will be satisfied for larger values of T if $C = -0.5$, which agrees with Eq. (O.9). But this procedure is still an empirical demonstration of the validity of Eq. (O.9).

Appendix P

Electric parameters for a cylindrical unmyelinated axon

We consider an axon with a cylindrical cross section and a constant diameter. The cable properties of the axon are described by the following basic and derived constants.

Fundamental constants

a = radius of axon, SI unit: (m)
ρ_a = specific resistance of axoplasm, SI unit: (Ωm)
ρ_e = specific resistance of external medium, SI unit: (Ωm)
ρ_m = specific resistance of membrane, SI unit: (Ωm)
δ = thickness of membrane, SI unit: (m)
C_m = membrane capacitance per unit area, SI unit: (F m^{-2})

Derived constants

r_a $= \rho_a/\pi a^2 =$ axial resistance of axoplasm per unit length, SI unit: (Ω m^{-1})
R_m = membrane resistance for one area unit of membrane, SI unit: (Ω m^2)
r_m $= \rho_m \delta/2\pi a = R_m/2\pi a =$ membrane resistance for one length unit of axon, SI unit: (Ωm)
c_m $= 2\pi a C_m =$ membrane capacitance per unit length of axon, SI unit: (F m^{-1})
λ $= \sqrt{r_m/r_a} = \sqrt{R_m a/2\rho_a} =$ length constant of axon, SI unit: (m)
R^* $= \sqrt{r_m r_a} = \lambda r_a = \sqrt{\rho_a R_m/2a^3}/\pi =$ characteristic resistance of axon, SI unit: (Ω).

The connection between the two types of constant can be derived as follows. Consider an element of axon of length h, radius a and membrane thickness δ. The resistance of the axoplasm in the direction x of the axis is

$$r' = \rho_a \frac{h}{\pi a^2} = \left(\frac{\rho_i}{\pi a^2}\right) \times h, \quad \text{SI unit: } (\Omega).$$

The first factor represents the resistance of axoplasm in the direction x reckoned per unit length of the axon, thus

$$r_a = \frac{\rho_a}{\pi a^2}, \quad \text{SI unit: } (\Omega \text{ m}^{-1}).$$

638

We imagine then the cylinder surface cut along a generatrix, and unfold the membrane to a plane layer with thickness δ and area $2\pi ah$. A current that flows perpendicularly to the layer encounters the resistance

$$r'_m = \rho_m \frac{\delta}{2\pi ah} = \left(\frac{\rho_m \delta}{2\pi a}\right) \times \frac{1}{h}.$$

The first factor represents the membrane resistance that the radially directed current is met with for each length unit of the axon, thus

$$r_m = \frac{\rho_m \delta}{2\pi a} = \frac{R_m}{2\pi a}, \quad \text{SI unit: } (\Omega m),$$

where

$$R_m = \rho_m \delta, \quad \text{SI unit: } (\Omega m^2),$$

is the membrane resistance for one unit area of the membrane.

Let c'_m represent the membrane capacitance of the length element. We have then

$$c'_m = 2\pi ah C_m = (2\pi a C_m) \times h, \quad \text{SI unit: } (F).$$

The first factor represents the membrane capacitance per unit length of the axon, thus

$$c_m = 2\pi a C_m, \quad \text{SI unit: } (F\ m^{-1}).$$

Typical values for a squid axon are

$$\rho_a = 30\ \Omega cm = 0.3\ \Omega m$$
$$R_m = 1\ k\Omega\ cm^2 = 0.1\ \Omega m^2$$
$$C_m = 1\ \mu F\ cm^{-2} = 0.01\ Fm^{-2}.$$

Thus, in an axon of diameter $2a = 500\ \mu m = 0.5\ mm$, i.e. $a = 0.025\ cm$, the expected values are

$$r_a = \frac{30}{\pi \times (0.025)^2} = 15.2\ k\Omega\ cm^{-1} \quad \text{and} \quad r_m = \frac{1000}{2\pi \times 0.025} = 6.37\ k\Omega\ cm^{-1},$$

$$c_m = 2\pi \times 0.025 \times 1\ \mu F\ cm^{-1} = 0.16\ \mu F\ cm^{-1}.$$

We have then

$$\lambda = \sqrt{\frac{6.37}{15.2}} = 0.65\ cm = 6.5\ mm,$$

and

$$\tau = 6.37 \times 10^3 \times 0.16 \times 10^{-6} \approx 10^{-3}\ s = 1\ ms,$$

and also

$$R_0 = \sqrt{6.37 \times 15.2} = 9.6\ k\Omega.$$

Cable constants from unmyelinated nerve fibers from three species and from a frog muscle fiber are collected in Table P.1.

Appendix P

Table P.1. *Cable constants from nerves and muscle fibers, 18 °C*

Fibre	d (μm)	λ (mm)	τ (ms)	R_m Ωcm^2	C_m μF cm^{-2}	ρ_i Ω cm	ρ_o Ω cm
Loligo	500	5	0.7	700	1	30	22
Lobster	75	2.5	2	2000	1	60	22
Crab	30	2.5	5	5000	1	60	22
Frog muscle	75	2	24 (10)	4000	6 (2.5)	200	87

(From B. Katz, 1966.)

The frog muscle fiber has two membrane capacitances, which charge and discharge with different time constants. The smaller (2.5 μF cm^{-2}) probably represents the capacitance of the muscle membrane itself (including the T-tubules); the larger is attributed to the sarcoplasmatic reticulum.

Appendix Q

Surface recorded action potential and membrane action potential

We consider a single axon where the surrounding medium is now reduced to a thin layer, whereby the external current loops that are associated with the impulse passage are forced to flow largely longitudinally and parallel to the fiber surface. Figure Q.1 shows the current loops in the active region for an impulse that propagates in the direction right → left, i.e. the positive direction of the x-axis. Let i_o and i_a represent the longitudinal currents that flow in the external medium and the axoplasm respectively, where the potentials are V_o and V_i. As the currents are forced to flow in a closed loop the relation

$$i_o + i_a = 0 \qquad (Q.1)$$

will hold for every position x on the axon. Let r_o and r_a represent the resistances of the outside medium and of the axoplasm reckoned per unit length. We have then

$$\frac{\mathrm{d}V_o}{\mathrm{d}x} = -r_o i_o \quad \text{and} \quad \frac{\mathrm{d}V_i}{\mathrm{d}x} = -r_a i_a. \qquad (Q.2)$$

We consider two positions: one at x_0, where both i_o and i_a are zero, i.e. the potentials V_o and V_i are constant, and the other at $\hat{x}$ where the inward-directed membrane current is maximal. Integration of Eq. (Q.2) from x_0 to $\hat{x}$ gives

$$\int_{V_o(x_0)}^{V_o(\hat{x})} \mathrm{d}V_o = -r_o \int_{x_0}^{\hat{x}} i_o(x)\,\mathrm{d}x = V_o(\hat{x}) - V_o(x_0), \qquad (Q.3)$$

and

$$\int_{V_i(x_0)}^{V_i(\hat{x})} \mathrm{d}V_i = -r_a \int_{x_0}^{\hat{x}} i_a(x)\,\mathrm{d}x = V_i(\hat{x}) - V_i(x_0). \qquad (Q.4)$$

Subtraction of Eq. (Q.4) from Eq. (Q.3) gives

$$V_i(x_0) - V_o(x_0) - (V_i(\hat{x}) - V_o(\hat{x})) = r_a \int_{x_0}^{\hat{x}} i_a(x)\,\mathrm{d}x - r_o \int_{x_0}^{\hat{x}} i_o(x)\,\mathrm{d}x.$$

Invoking $i_a = -i_o$ from Eq. (Q.1) in the above equation gives

$$V_i(x_0) - V_o(x_0) - (V_i(\hat{x}) - V_o(\hat{x})) = -(r_a + r_o) \int_{x_0}^{\hat{x}} i_o(x)\,\mathrm{d}x.$$

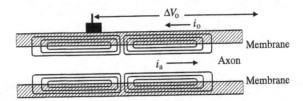

Fig. Q.1. Schematic illustration of the local current loops that flow in the axoplasm and the external medium during the impulse passage in the active region. The external medium consists of a thin liquid layer, in which electrodes are placed to record the surface action potential. The direction of propagation is right → left.

Now

$$V_m = V_i(x_0) - V_o(x_0),$$

is equal to the *resting membrane potential* V_m and

$$\hat{V}_m = V_i(\hat{x}) - V_o(\hat{x}),$$

is the membrane potential corresponding to the peak of the membrane action potential. Therefore

$$V_m - \hat{V}_m = -(r_a + r_o) \int_{x_0}^{\hat{x}} i_o(x)\,dx.$$

Furthermore

$$\Delta V_i = \hat{V}_m - V_m = (r_a + r_o) \int_{x_0}^{\hat{x}} i_o(x)\,dx, \qquad (Q.5)$$

represents the potential drop from the position $\hat{x}$ corresponding to the peak of the action potential $\hat{V}_i$ to the region where the potential is constant and equal to the resting membrane potential. Similarly the potential drop on the outside of the axon is

$$\Delta V_o = V_o(\hat{x}) - V_o(x_0) = -r_o \int_{x_0}^{\hat{x}} i_o(x)\,dx, \qquad (Q.6)$$

which is the maximal potential difference that can be measured between the external electrodes under the given conditions. The sum of these two contributions is equal to the electromotive force V_{emf} of the "battery" in the membrane that produces the local current loops, and whose magnitude corresponds to the amplitude ΔV_{ap} of the *action potential* as measured with an intracellular microelectrode, when the fiber is placed as usual in an extended medium. Thus, we have

$$\Delta V_{ap} = \Delta V_i + \Delta V_o = (r_a + r_o) \int_{x_0}^{\hat{x}} i_o(x)\,dx - r_o \int_{x_0}^{\hat{x}} i_o(x)\,dx$$

$$= r_a \int_{x_0}^{\hat{x}} i_o(x)\,dx.$$

This expression could also be obtained from Eq. (Q.5) when $r_a \gg r_o$. The connection between the surface action potential ΔV_o that is recorded with a given surface

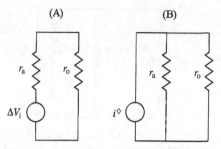

Fig. Q.2. Electric equivalent diagram to illustrate the potential division of the action potential between the axoplasm with resistance r_a and the surrounding liquid layer with resistance r_o. (A) Potential equivalent consisting of an EMF ΔV_i in series with its internal resistance r_a to which is connected the external resistance r_o. (B) Current equivalent consisting of a current generator of strength $i^\circ = \Delta V_i / r_a$ with r_a and r_o in shunt connection.

resistance and the membrane action potential ΔV_{ap} itself is then[*]

$$\frac{\Delta V_o}{\Delta V_{ap}} = -\frac{r_o}{r_a}. \qquad (Q.7)$$

It appears that the surface recorded action potential is related to the membrane action potential as the ratio between the surface resistance r_o per unit length and the axial resistance of the axon r_a per unit length.

Dividing Eq. (Q.5) into Eq. (Q.6) yields[†]

$$\frac{\Delta V_o}{\Delta V_i} = -\frac{r_o}{r_a + r_o}. \qquad (Q.8)$$

Thus, the recorded surface voltage V_o appears as the potential division between the surface resistance r_o and the axoplasm resistance r_a of the longitudinal axoplasmatic voltage drop ΔV_i.

Figure Q.2(A) shows an electric circuit that simulates Eq. (Q.8). It consists of a potential source with an electromotive force ΔV_i in series with its internal resistance r_a. A resistance r_o is connected to the battery. The resulting current is

$$i = \frac{\Delta V_i}{r_a + r_o}. \qquad (Q.9)$$

The similar equivalent circuit that is based upon a *current source* is shown in Fig. Q.2(B). It consists of a current source, whose strength i° is equal to the current that is drawn from circuit (A), when the output is short circuited ($r_o = 0$), thus

$$i^\circ = \frac{\Delta V_i}{r_a} = g_a \Delta V_i. \qquad (Q.10)$$

---

[*] A more rigorous – but also mathematically more complex – derivation is due to P. Rosenfalck (1969): *Intra- and Extracellular Potential Fields of Active Nerve and Muscle Fibres*, Akademisk Forlag, Copenhagen.

[†] The derivation is also found in B. Katz (1966): *Nerve, Muscle and Synapses*, McGraw-Hill, Inc., New York.

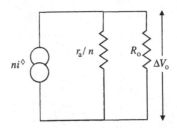

Fig. Q.3. Electric equivalent diagram to illustrate the dependence of the size of the surface recorded action potential on the number $n < N$ of activated fibers in a whole nerve consisting of a total of N fibers. Each activated nerve fiber is represented with a current generator of strength i°. The n current generators acting in parallel produce a total current $n\,i^\circ = n\Delta V_i/r_a$, which is shunted by the resistance r_a/n and the total resistance R_o of the external liquid surrounding the nerve fibers.

Shunting this current source with r_a results in the current equivalent, whose behavior is identical to the potential equivalent of Fig. Q.2(A), as loading with the external resistance r_o produces the same current i as given by Eq. (Q.9). The generator current i° divides between r_a and r_o, giving $i^\circ = i_a + i_o$ or

$$i^\circ = g_a\Delta V_i = \frac{\Delta V_o}{r_a} + \frac{\Delta V_o}{r_o} = \Delta V_o(g_a + g_o),$$

from which we obtain

$$\Delta V_o = \frac{g_a}{g_a + g_o}\,\Delta V_i = \frac{1/r_a}{(r_a + r_o)/r_a r_o}\,\Delta V_i = \frac{r_o}{r_a + r_o}\,\Delta V_i. \qquad (Q.11)$$

We consider now a whole nerve consisting of N identical fibers. We assume that the number of fibers that are activated increases with increasing stimulus strength until all the N fibers are activated. On the basis of the above equivalent models we shall now give the connection between the size of the summated action potentials that are recorded from the nerve's surface and the number n of activated fibers. We regard each activated fiber as functioning as a current generator with strength i° and with the internal shunt resistance r_o, that sends a current i out to the surrounding medium, which consists of the total liquid layers that surround the fibers in the nerve. Let this resistance per unit length be R_o. The stimulus strength is assumed to be submaximal and thus causing a certain number $n < N$ nerve fibers to become activated. Instead of dealing with only one current generator with the strength $i^\circ = \Delta V_i/r_a$ we have now n current generators each of strength i° that are arranged in parallel. The current that the activated fiber sends into the resistance R_o rises concurrently with an increase in the number n of activated fibers and with that also the longitudinal voltage change in the liquid surrounding the fibers. The electric model of this situation is shown in the current equivalent of Fig. Q.3. The n current generators that are functioning in parallel are lumped together to a single current generator of strength $i^\circ = n\,\Delta V_i/r_a = ng_a\,\Delta V_i$ which is shunted by n resistances r_a connected in parallel, i.e. by a total shunt resistance amounting to r_a/n or a shunt conductance of $n g_a^*$. The voltage drop ΔV_o

---

* The surface membranes of the unactivated fibers are considered to function as isolators, so that the membrane resistances r_m are assumed not to contribute to R_o.

along the external resistor R_o (with conductance $G_o = 1/R_o$) due to this current generator is, according to Eq. (Q.11), equal to

$$\Delta V_o = \frac{n\,i^\circ}{G_o + n\,g_a} = \frac{n\,g_a}{G_o + n\,g_a}\,\Delta V_i. \tag{Q.12}$$

Since

$$G_o + n\,g_a = \frac{1}{R_o} + \frac{n}{r_a} = \frac{r_a}{R_o r_a} + \frac{nR_o}{R_o r_a} = \frac{r_a + nR_o}{R_o r_a},$$

Eq. (Q.12) is also written as

$$\Delta V_o = \frac{nR_o}{nR_o + r_a}\,\Delta V_i = \frac{R_o}{R_o + r_a/n}\,\Delta V_i. \tag{Q.13}$$

  In the present model analysis the effect of increasing the number of activated fibers is that a greater amount of current is pressed into the extra-axonal space, thereby causing a greater longitudinal voltage drop. This viewpoint is probably more in conformity with what actually happens physically. But inspection of the above formula might invite us to take a different point of view that would also arise if we replaced the current source by the equivalent potential source. This would lead to a collection of n sources each consisting of an EMF ΔV_i having the internal resistance r_a arranged in parallel. This is equivalent to a single battery with the same EMF but now having an internal resistance r_a/n. This lowering of the internal resistance by the factor n results – as it should – in a greater external current flow where the battery is loaded with the resistance R_o and resulting again in a voltage drop ΔV_o across R_o equal to that above. However, although identical equivalent circuits must return identical end-results, the latter approach is perhaps less plain from a physical point of view.

Appendix R

About concentration scales

The *mole fraction* X, the *molality* m and the *molarity* C are the concentration scales most widely used when dealing with solutions. The definitions of these quantities and their interconnections are given below.

We consider a binary solution that contains n_s mole of a substance in n_w mole water of molecular weight M_w and density ρ_w (g cm^{-3}). The total volume of the solution is V and its density is ρ. In theoretical calculations in physical chemistry the *mole fraction* is the preferred measure of concentration. For the above solution we have

$$X_w = \frac{n_w}{n_w + n_s} \quad \text{and} \quad X_s = \frac{n_s}{n_w + n_s}. \tag{R.1}$$

The *molality* of the dissolved substance is the number of gram formula weights per $1 \text{ kg} = 1000$ g solvent (water). The solution above contains $M_w n_w$ g H_2O. Per g H_2O the solution contains $n_s / M_w n_w$ mole of the dissolved substance. Hence, the molality m is

$$m_s = \frac{1000 \, n_s}{M_w n_w}. \tag{R.2}$$

Relating Eq. (R.1) to Eq. (R.2) gives

$$\frac{m_s}{X_s} = \frac{1000 \, (n_w + n_s)}{M_w n_w}. \tag{R.3}$$

When $n_s \to 0$; $m_s / X_s \to 1000 / M_w$. Thus, if our solution is very diluted we have

$$m_s \approx \frac{1000 \, X_s}{M_w}, \tag{R.4}$$

i.e. the mole fraction is proportional to the molality. There are two reasons to prefer the use of molality as a measure of concentration. It is easier to weigh out a substance with precision than to measure out volumes. Furthermore, the molality is independent of the temperature of the solution. In biology this property is rarely a major problem.

The *molarity* C or the volume concentration is defined as the number of moles that are dissolved per volume unit of the solution – in SI units as mol dm^{-3}, i.e.

$$C_s = \frac{n_s}{V}. \tag{R.5}$$

The weight in grams of the volume V of solution is $1000\,\rho V$ g. Hence

$$1000\,\rho V = M_w n_w + M_s n_s,$$

where M_s is molecular weight in grams of the dissolved substance. Thus

$$V = \frac{M_w n_w + M_s n_s}{1000\,\rho},$$

which together with Eq. (R.5) gives

$$C_s = \frac{1000\,\rho n_s}{M_w n_w + M_s n_s}. \tag{R.6}$$

Combining this with Eq. (R.1) gives

$$\frac{C_s}{X_s} = \frac{1000\,\rho(n_w + n_s)}{M_w n_w + M_s n_s}. \tag{R.7}$$

When $n_s \to 0$, then $C_s/X_s \to 1000\,\rho/M_w$. Thus, for very dilute solutions we have

$$C_s \approx \frac{1000\,\rho X_s}{M_w}, \tag{R.8}$$

showing that the molarity is proportional to the mole fraction. It appears from Eq. (R.7) that the molarity of the solution will change with the temperature since this holds for the density ρ of the solution. Combining Eq. (R.3) and Eq. (R.7) gives

$$C_s = \frac{\rho}{1 + (M_s n_s)/(M_w n_w)} m_s. \tag{R.9}$$

For dilute solutions we have $(M_s n_s)/(M_w n_w) \ll 1$ and $\rho \approx \rho_w$. Hence,

$$C_s = \rho_w m_s \approx m_s, \tag{R.10}$$

i.e. the numerical values of the two concentration scales are identical.

Appendix S

Units and physical constants

The SI system is based upon the seven fundamental units in Table S.1.

By these *basic units* we can express the other physical units, which accordingly are called *derived units*. In Table S.2 are listed a number of physical quantities that occur in this book, together with the symbols used, their SI names and SI dimensions.

Two units from the previous *practical system of units* have retained a considerable power of survival, namely the *calorie* as a unit for heat energy and the *atmosphere* as a unit of pressure. The conversion factors for these units to the SI system are

$$1 \text{ cal}_{15} = 4.186 \text{ joule,}$$

and

$$1 \text{ atm} = 760 \text{ mm Hg} = 760 \text{ Torr} = 1.013\,25 \times 10^5 \text{ Pa} = 101.325 \text{ kPa,}$$

and

$$1 \text{ Torr} = 1 \text{ mm Hg} = 133.322 \text{ Pa} = 0.1333 \text{ kPa.}$$

When the greatest precision is not needed, the rule is

$$1 \text{ atm} \approx 100 \text{ kPa.}$$

In Table S.3 are listed the numerical values of the physical constants that are of relevance in the present text.

Table S.1. *Basic units in the SI system*

Quantity	Name	SI symbol
Length	meter	m
Mass	kilogram	kg
Time	second	s
Temperature	kelvin	K
Electric current	ampere	A
Amount of substance	mole	mol
Luminous intensity	candela	cd

Table S.2. *Some derived units in the SI system*

Quantity	Symbol	SI name	SI symbol	SI unit
Area	A			m^2
Volume	V			m^3
Velocity	v			$m\ s^{-1}$
Acceleration	a			$m\ s^{-2}$
Force	X	newton	N	$kg\ m\ s^{-2}$
Work	W	joule	J	newton m
Energy	U, W	joule	J	N m
Mechanical mobility	B			$m\ s^{-1}\ N^{-1}$
Electric charge	Q	coulomb	C	ampere s
Charge density	ϱ			$coulomb\ m^{-3}$
Electric potential	ψ, V	volt	V	$joule\ coulomb^{-1}$
Electric field	E			$volt\ m^{-1}$
Electric resistance	R	ohm	Ω	$volt\ ampere^{-1}$
Electric conductance	g, G	siemens	S	$ampere\ volt^{-1}$
Electric mobility	u			$m^2\ s^{-1}\ volt^{-1}$
Electric capacitance	c, C	farad	F	$coulomb\ volt^{-1}$
Current density	I			$ampere\ m^{-2}$
Pressure	p	pascal	Pa	$newton\ m^{-2}$
Diffusion coefficient	D			$m^2\ s^{-1}$
Viscosity coefficient	η			$m^{-1}\ kg\ s^{-1}$

Table S.3. *Some fundamental constants in SI units*

Constant	Symbol	Value in SI
Avogadro number	$\mathcal{N}_A$	$6.0225 \times 10^{23}\ mol^{-1}$
Elementary charge	e	$1.602 \times 10^{-19}\ C$
Faraday constant	$\mathcal{F} = \mathcal{N}_A e$	$96\,488\ C\ mol^{-1}$
Gas constant	$\mathcal{R}$	$8.3143\ J\ K^{-1}\ mol^{-1}$
		$1.987\ cal\ K^{-1}\ mol^{-1}$
		$0.082\,05\ dm^3\ atm\ K^{-1}\ mol^{-1}$
Boltzmann constant	$k = \mathcal{R}/\mathcal{N}_A$	$1.3805 \times 10^{-23}\ J\ K^{-1}$
Ideal gas volume (STP)	V_0	$2.2414 \times 10^{-2}\ m^3\ mol^{-1}$
Vacuum permittivity	ϵ_0	$8.84 \times 10^{-12}\ F\ m^{-1}$
Coulomb constant	K_c	$8.9876 \times 10^9\ N\ m^2\ C^{-2}$
Velocity of light	c	$2.9979 \times 10^8\ m\ s^{-1}$
Acceleration of gravity at equator at sea level	g	$9.7805\ m\ s^{-2}$

List of symbols

Below are listed the majority of symbols that are used in the text. The use of **bold face** type indicates the first occurrence of the symbol in the text either on the page number or as section number or equation number.

A	Arrhenius's activation energy, **Section 2.5.6.6**, **Eq. (2.5.222)**.
A	Area, **Section 2.4.2**.
A_1, A_2	Integration constants, **page 343**.
a	Axon radius, **page 500**.
B	Mechanical mobility, **Eq. (2.4.5)**.
B_1, B_2	Integration constants, **page 343**.
b	Molar mechanical mobility, **page 273**.
C	Concentration; quantity (e.g. mol) per unit volume, **Section 2.4.2**.
$C_j(x)$	Concentration of component j in position x, **page 271**.
$C_j^{(i)}$	Concentration of component j in phase (i), **Eq. (3.4.7)**.
$C_j^{(o)}$	Concentration of component j in phase (o), **Eq. (3.4.7)**.
C_m	Membrane capacitance, **page 283**.
c_m	Membrane capacitance per unit length of axon, **page 489**.
D	Diffusion coefficient, **page 74**.
$\mathrm{d}y$	Differential of the function y, **Eq. (1.2.12)**.
$\dfrac{\mathrm{d}}{\mathrm{d}x}$	Differential operator acting on succeeding term, **Eq. (1.2.5)**.
$\dfrac{\partial f}{\partial x}$	Partial derivative of function $f(x, y, z)$ with respect to x, **Eq. (1.8.4)**.
$\mathrm{Erf}(x)$	Error function, **Eq. (2.5.103)**.
$\mathrm{Erfc}(x)$	Complementary error function: $1 - \mathrm{Erf}(x)$.
E	Electric field, **Eq. (3.2.5)**.
e	$= 2.718\ 281\ 8\ldots$ base of the exponential function e^x, **Eq. (1.5.10)**.
e	Positive electric elementary charge (proton charge), **page 262**.
$\mathcal{E}$	Electron.
$F(t)$	Exponential distribution, **Eq. (5.2.12)**.
$f_n(x)$	Normal distribution density, **Eq. (5.2.23)**.
$\mathcal{F}$	Faraday's constant, **Eq. (3.3.9)**.
$G(x, t \mid \xi, 0)$	Green's function, **Eq. (2.5.126)**.
G	Electric conductance, **Eq. (3.3.14)**.
G_j	Partial conductance in layer of thickness h, **Eq. (3.3.27)**.

650

G	$G^{(i)}$, $G^{(s)}$ Gibbs free energy, **page 239.**
$\mathcal{G}$	Amplification factor, **page 452.**
g_j	Slope conductance, **Eq. (3.4.61).**
h	Synonymous with Δx.
h	Length/membrane thickness, **Section 2.5.4.1.**
h	$h(V, t)$ inactivation function for G_{Na}, **Eq. (4.4.43).**
I	Current density, **page 267.**
I_j	Current density of ion of species j, **Eq. (3.3.8).**
i	Electric current.
i_a	Longitudinal intra-axonal current, **Eq. (4.4.11).**
i_m	Membrane current per unit axon length, **Eq. (4.4.12).**
$J_\rightarrow$	Flux, **Eq. (2.2.1).**
$J_{i\rightarrow o \ i\leftarrow o}$	Flux in the arrow's direction among a particle swarm, **Eq. (2.5.214).**
J, J	Unidirectional fluxes used at flux ratio analysis, **page 405.**
K	Relative dielectric constant ($\varepsilon/\varepsilon_0$), **Eq. (3.3.46).**
K	Constant of proportionality used in Henry's law, **Eq. (2.5.40).**
K	Krogh's diffusion constant, **Eq. (2.5.44).**
K_c	Coulomb's constant, **Eq. (3.2.2).**
k	Boltzmann's constant, **Section 2.6.4.2, Eq. (2.6.53).**
k	Rate constant, **Eq. (2.7.9).**
$k^{(1)}$, $k^{(2)}$	Rate constants in connection with compartment analysis, **page 218.**
$k^{(in)}$, $k^{(out)}$	Rate constants for unidirectional fluxes, **Eq. (2.7.48).**
L_P	Hydraulic conductivity, **Eq. (2.280).**
$\ln x$	Natural logarithm of x, **Eq. (1.4.2).**
$\log x$	Logarithm (with base of 10) of x.
$M^{(x)}$	Amount of diffusing substance through the area A in position x, **Section 2.5.3.1.**
m	Mass, **Eq. (2.4.2).**
m	Number of one-way directed steps in random walk, **Section 2.5.6.3.**
m	m_0, $m^{(i)}$ and $m^{(o)}$ quantity of material, **page 214.**
m	$m(V, t)$ activation function for G_{Na}, **Eq. (4.42).**
N	Concentration: number of particles per unit volume, **Section 2.4.2.**
N_+	Positive steps in random walk, **Section 2.5.6.3.**
N_-	Negative steps in random walk, **Section 2.5.6.3.**
$N^{(\Delta t)}$	Mass transport through area A in time t, **Eq. (2.5.1).**
ΔN_τ	Increment of number of particles in a volume in time τ, **Eq. (2.5.200).**
$\mathcal{N}$	Surface density, **Eq. (2.5.125).**
$\mathcal{N}_A$	Avogadro's constant, **page 69.**
n	Number of particles in a given volume.
n	$n(V, t)$ activation function for G_K in H–H equations, **Eq. (4.4.42).**
P	Permeability coefficient, **Eq. (2.5.15), Section 2.7.1.**
$\langle P \rangle$	Equivalent permeability, **Eq. (2.5.21), Eq. (2.5.24).**
P_j	Permeability of ion of species j, **Eq. (3.4.68).**
p	Pressure, **page 281.**
p_j	Partial pressure of gas component j, **Eq. (2.5.36).**
Q	Q_1, Q_2 electric charge, **Eq. (3.2.1).**
$\mathcal{R}$	Gas constant, **page 204.**
R	Electric resistance, **Eq. (3.3.11).**
R_e	Electrode resistance, **page 387.**
R_i	Input resistance of amplifier, **page 387.**
R^*	Characteristic resistance of axon, **Eq. (4.4.24).**

r	Radius of cylinder/circle, **Section 2.5.4.2**.
r_D	Gibbs–Donnan distribution ratio, **Eq. (3.5.2)**.
r_a	Resistance of axoplasm per unit length of axon, **page 487**.
STP	Standard temperature and pressure, **page 96**.
T	Dimensionless time in units of membrane time constant τ, **Eq. (4.4.25)**.
$\mathcal{T}_j$	Gas tension, **Eq. (2.5.41)**.
ΔT_f	Freezing point depression, **Eq. (2.8.30)**, **Eq. (2.8.31)**.
T_j	Transport number for ion of species j, **Eq. (3.8.3)**.
t	Time, **Section 2.4.1**.
U	Potential energy barrier, **Section 2.5.6.6**, **Eq. (2.6.46)**.
$\mathcal{U}$	Generalized cation concentration introduced by Planck, **Eq. (3.6.23)**.
U_0	Height of potential barrier, **page 177**.
u_j	Electric mobility, **Eq. (3.3.5)**.
u_j^0	Electric mobility at infinite dilution of ion of type j, **page 269**.
V	Potential difference, **page 266**.
V	Membrane potential, **page 259**.
V_m	Resting membrane potential or reversal potential, **page 310**.
V_j^{eq}	Equilibrium potential for ion species j, **Eq. (3.3.45)**, **Eq. (3.4.7)**.
V_D	Donnan potential, **Eq. (3.5.11)**.
V_0	Potential impressed on cable in position $x = 0$, **Eq. (4.4.23)**.
$V^{(i)}, V^{(o)}$	Volume of phase (i) and phase (o), **page 213**.
$V^{(dif)}$	Diffusion potential, **Eq. (3.6.6)**.
$\overline{V}$	Molar volume, **page 239**.
v	Velocity, **Section 2.4.1**.
υ	Normalized membrane potential (V in units of $\mathcal{R}T/\mathcal{F}$), **page 367**.
v	Normalized Donnan potential, **page 323**.
$\overline{v}_j$	Partial molar volume of component j, **page 316**.
W	Transformed potential in cable equation, **Eq. (4.4.31)**.
$W(N, m)$	Probability for a position $m \le N$ in random walk, **Eq. (2.5.165)**.
$\mathcal{W}$	Generalized anion concentration introduced by Planck, **Eq. (3.6.27)**.
X	Force, **Section 2.4.1**.
X	Dimensionless length in units of length constant λ, **Eq. (4.4.26)**.
X_f	Frictional force, **Section 2.4.1**.
X_j	Mole fraction, **Eq. (2.8.5)**.
x	Position coordinate along x-axis.
$\overline{x}$	Mean displacement in one direction, **Eq. (2.5.161)**.
$\langle x \rangle$	Mean displacement in random walk, **Eq. (2.5.160)**.
$\langle x^2 \rangle$	Mean root square of displacements, **Section 2.5.6.3.(iv)**.
$\overline{x}_+$	Mean displacement in the direction of force, **page 191**.
$\overline{x}_-$	Mean displacement against the direction of force, **page 191**.
Y	Degree of rectification, **page 229**.
z_+, z_-	Charge number for cation K^{z+} and anion A^{z-}, **Eq. (3.2.7)**, **Eq. (3.2.8)**.
z_j	Charge number for ions of species j, **page 266**
α	Distribution coefficient between membrane and adjacent solution, **Eq. (2.7.3)**.
α	Bunsen's absorption coefficient, **Eq. (2.5.36)**.
α_n, β_n	$\alpha_n(V)$, $\beta_n(V)$ rate constants for n, **Eq. (4.7.4)**, **Eq. (4.7.5)**.
α_m, β_m	$\alpha_m(V)$, $\beta_m(V)$ rate constants for m, **Eq. (4.7.8)**, **Eq. (4.7.9)**.
α_h, β_h	$\alpha_h(V)$, $\beta_h(V)$ rate constants for h, **Eq. (4.7.11)**, **Eq. (4.7.12)**.
β	Frictional coefficient, **Eq. (2.4.1)**.
Γ	Ratio of coupling for active transport of Na^+ and K^+, **Eq. (3.8.10)**.

Δx	Increment of variable x, **Section 1.2.4**, **Eq. (1.2.8)**.
ΔT_f	Freezing point depression, **Eq. (2.8.30)**, **Eq. (2.8.31)**.
δ	Thickness of axon membrane, **page 638**.
δ_1, δ_2	Thickness of unstirred layer, **Eq. (2.5.25)**.
η	Coefficient of viscosity, **Eq. (2.6.56)**.
ε_0	Permittivity of vacuums, **Eq. (3.2.4)**.
κ	Specific conductivity, **Eq. (3.3.14)**.
κ_j	Partial specific conductivity, **Eq. (3.3.17)**.
λ	Width of a potential barrier, **page 177**.
λ	Debye length, **Eq. (3.5.58)**.
λ	Mean value and standard deviation of exponential distribution, **Eq. (5.2.14)**.
λ	Characteristic length (length constant) of a cable, **Eq. (4.4.17)**.
λ	Step length in random walk, **page 152**.
λ	Width of a potential barrier, **page 133**.
λ_j	Molar conductivity of ion of species j, **Eq. (3.3.18)**.
λ_j^0	Molar conductivity at infinite dilution of ion of type j, **page 269**.
μ	Chemical potential, **Eq. (2.8.7)**.
$\mu(x)$	Chemical potential at position x, **page 373**.
$\overline{\mu}$	Electrochemical potential, **page 316**.
μ	Mean value in Poisson distribution $P(n; \mu,)$, **Eq. (5.2.18)**.
ν_+, ν_-	Stoichiometric coefficients for anions and cations, **Eq. (2.8.15)**.
ξ	Boltzmann's variable of transformation ($\xi = x\, t^{-\frac{1}{2}}$), **Eq. (2.5.93)**.
ξ	Potential parameter introduced by Planck, **Eq. (3.6.33)**.
ξ	Running position variable in relation to a fixed value of x, **Eq. (2.5.126)**.
Π	Osmotic pressure, **Eq. (2.8.4)**.
τ	Running time variable, e.g. with fixed t.
τ	Time between two steps in a random walk, **page 152**.
τ	Membrane time constant, **Eq. (4.4.18)**.
τ_n, τ_m, τ_h	Time constants for n, m and h during voltage clamp, **page 523**.
ρ	Resistivity, **Eq. (3.3.23)**.
$\rho(x)$	Charge density per unit volume in position x, **Eq. (3.3.48)**.
$\rho(x)$	Mass production per unit volume in position x, **Section 2.5.3.2**.
ρ_a	Specific resistivity of axoplasm, **page 638**.
ρ_m	Resistance of an unit area of the axon membrane, **page 638**.
σ	Standard deviation of the Gauss distribution, **Eq. (2.5.145)**.
σ_j	Reflections coefficient, **Eq. (2.8.35)**.
$\varphi(x, t)$	Probability density for a random walk, **Eq. (2.5.158)**.
ϕ	Practical osmotic coefficient, **page 243**.
$\Phi(\xi, \tau \mid 0, 0)$	Probability density function, **Eq. (2.5.182)**.
ψ	Electric potential, **Eq. (3.2.9)**.
Ψ	Normalized electric potential (in terms of $\mathcal{R}T/\mathcal{F}$), **Eq. (3.3.39)**.
$\overset{\text{def}}{=\!=}$	Meaning "is defined as ...".
$\overset{\Delta}{=\!=}$	Meaning "has dimension ...".

Subscripts

j	Chemical component of type j.
j+	Cation of type j.

j–   Anion of type j.
o   Initial state.

Superscripts

(i)   Inner phase (i) – often the membrane's inside.
(o)   Outer phase (o) – often the membrane's outside.
(eq)   Equilibrium state.
+   Refers to cation.
–   Refers to anion.

References

Abramowitz, M. & Stegun, Irene A. (1965): *Handbook of Mathematical Functions*, Dover Publications Inc., New York.

Adrian, E.D. & Lucas, K. (1912): On the summation of propagated disturbances in merve and muscle. *J. Physiol.*, **44**, 68–124.

Aidley, D.J. (1989): *The Physiology of Excitable Cells*, 3rd edn, Cambridge University Press, Cambridge (468).

Alonso, M. & Finn, E.J. (1963): *Physics*, Addison-Wesley Publishing Company, Inc., Reading, Massachusetts.

Armstrong, C.M., Benzanilla, F. & Rojas, E. (1973): Destruction of sodium conductance inactivation in squid axons perfused with probase. *J. Gen. Physiol.*, **62**, 175.

Ayres, F. (1964): *Theory and Problems of Differential and Integral Calculus*, Schaum Publishing Co., New York (345).

Baker, P.F., Hodgkin, A.L. & Shaw, T.I. (1862): Replacement of the axoplasm of giant nerve fibres with artificial solutions. *J. Physiol.*, **164**, 330.

Bayliss, W. M. (1927): *Principles of General Physiology*, 4th edn, Longmans, Green, Co. (page 389, Fig.108).

Behn, U. (1897): Ueber wechselseitige Diffusuion von Elektrolyten in verdünnten wärigen Lösungen, insbesondere über Diffusuion gegen das Concentrationsge-fälle. *Ann. Physik, N.F.*, **62**, 54.

Boas, M. (1966): *Mathematical Methods of the Physical Sciences*, John Wiley and Sons, Inc., New York (778).

Bockris, J.O'M. & Reddy, A.K.N. (1970): *Modern Electrochemistry*, 2 volumes, Macdonald & Co. (Publishers) Ltd., London (1432).

Boyd, I.A. & Martin, A.R. (1956): The end-plate potential in mammalian muscle. *J. Physiol.*, **132**, 74.

Bryan, G.H. (1891): Note on the linear conduction of heat. *Proc. Camb. Phil. Soc.*, **7**, 246.

Bryan, G.H. (1891): An application of the method of images to the conduction of heat. *Proc. Lond. Math. Soc. (I)*, **22**, 424.

Butkov, E. (1968): *Mathematical Physics*, Addison-Wesley Publishing Company, Inc., Reading, Massachusetts (735).

Butler, J.A.V. (1924): Studies in heterogeneous equilibria: Part II – The kinetic interpretation of the Nernst theory of the electromotive force. *Trans. Farad. Soc.*, **19**, 729.

656   References

Byerly, W.E. (1898): *An Elementary Treatise of Fourier Series and Spherical, Cylindrical and Ellipsoidal Harmonics*, Ginn and Company, Boston (278).

Byrne, J.H. & Schultz, S.G. (1988): *An Introduction to Membrane Transport and Bioelectricity*. Raven Press, New York.

Carslaw, H.S. & Jaeger, J.C. (1956): *Conduction of Heat in Solids*, Oxford University Press, London (510).

Chandrasekhar, S. (1943): Stochastic problems in physics and astronomy, *Rev. mod. Physics*, **15**, Nr. 1.

Chandrasekhar, S. (1995): *Newton's Principia for the Common Reader.* Clarendon Press. Oxford (595).

Churchill, R.V. (1958): *Operational Mathematics*, McGraw-Hill Book Company, Inc., New York (337).

Cohen, H. & Cooley, J.W. (1965): The numerical solution of the time dependent Nernst–Planck equations. *Biophys. J.*, **5**, 145.

Cole, G.H.A. (1967): *The Statistical Theory of Classical Simple Dense Liquids*, Pergamon Press Ltd., Oxford (284).

Cole, K.S. (1965): Electrodiffusion models for the membrane of the squid axon. *Physiol. Rev.*, **45**, 340.

Cole, K.S. (1968): *Membranes, Ions and Impulses*, University of California Press, Berkeley and Los Angeles (569).

Constant, F.W. (1963): *Fundamental Laws of Physics*, Addison-Wesley Publishing Company, Inc., Reading, Massachusetts (403).

Courant, R. (1945): *Differential and Integral Calculus*, 2 volumes, Nordeman Publishing Company, Inc., New York (682 + 616).

Dainty, J. & House, C.R. (1966): An examination of the evidence for membrane pores in frog skin. *J. Physiol.*, **185**, 172.

Del Castillo, J. & Katz, B. (1954): Quantal components of the end-plate potential. *J. Physiol.*, **124**, 560.

Del Castillo, J. & Katz, B. (1954): Changes in the end-plate activity produced by pre-synaptic polarization. *J. Physiol.*, **124**, 584.

Del Castillo, J. & Katz, B. (1954): Quantal components of end-plate potentials. *J. Physiol.*, **124**, 560.

Del Castillo, J. & Katz, B. (1954): The membrane change produced by the neuromuscular transmitter. *J. Physiol.*, **125**, 546.

Del Castillo, J. & Katz, B. (1956): Biophysical aspects of neuromuscular transmission. *Progr. Biophys.*, **6**, 121.

Denbigh, K. (1957): *The Principles of Chemical Equilibrium.* Cambridge University Press (491).

Dwight, H.B. (1947): *Tables of Integrals and other Mathematical Data*, The Macmillan Company, New York (250).

Einstein, A. (1926): *Investigations on the Theory of the Brownian Movements*, edited by R. Fürth, Methuen & Co. Ltd, London; translated by A.D. Cowper from R. Fürth (Ed.) (1922): *Untersuchungen über die Theorie der Brownschen Bewgung*, in: *Ostwalds Klassiker der exakten Wissenschaften*, Akademische Verlagsgesellschaft, Leipzig.

Einstein, A. (1944): *Relativity, the special and the general theory. A popular exposition*, 12th edn, Methuen & Co. Ltd, London.

Falk, G. (1961): In *Biophysics of Physiological and Pharmacological Actions*, A.M. Shanes (Ed.), p. 269, American Association for the Advancement of Science, Washington, D.C.

Fatt, P. & Katz, B. (1951): An analysis of the end-plate potential recorded with an intra-cellular electrode. *J. Physiol.*, **115**, 320.

Fatt, P. & Katz, B. (1952): Spontaneous subthreshold activity at motor nerve endings. *J. Physiol.*, **117**, 109.

Fick, A. (1855): Ueber Diffusion. *Ann. Phys. Chem.*, **194**, 59–86.

Finkelstein, A. (1976): Water and nonelectrolyte permeability of lipid bilayer membranes. *J. Gen. Physiol.*, **68**, 127–135.

Finkelstein, A. (1987): *Water Movements through Lipid Bilayers, Pores, and Plasma Membranes. Theory and Reality*, John Wiley and Sons, New York.

Finkelstein, A. & Mauro, A. (1963): Equivalent circuits as related to ionic systems. *Biophys. J.*, **3**, 215.

FitzHugh, R. (1960): Threshold and plateaus in the Hodgkin–Huxley nerve equations. *J. Gen. Physiol.*, **43**, 867–896.

Fourier, J. (1822): *Thèorie analytique de la chakeur*. (Evres de Foutier. English translation with previous corrigenda incorporated by Freeman, A. (1878): *The Analytic Theory of Heat*, Reprinted by Dover Publications, Inc., New York, 1955.

Frenkel, J. (1946): *Kinetic Theory of Liquids*. Oxford University Press, Oxford (488).

Fürth, R. (1917): Einige Untersuchungen über Brownsche Bewegung an einem Einzelteilschen. *Ann. d. Physik*, IV. Folge, **53**, 177.

Hartley, G.S. & Crank, J. (1949): Some fundamental definitions and conceptes in diffusion processes. *Trans. Farad. Soc.*, **45**, 801.

Henderson, P. (1907): Zur Thermodynamik der Flüssigkeitsketten. *Z. physik. Chemie*, **59**, 118.

Hertz, G. (1922): Ein neues Verfahren zur Trennung von Gasgemischen durch Diffusion. *Physik. Z.*, **23**, 433.

Hille, B. (1968): Pharmacological modifications of the sodium channels of the frog nerve. *J. Gen. Physiol.*, **51**, 199.

Hille, B. (1970): Ion channels in nerve membrane. *Prog. Biophys. Mol. Biol.* **21**, 1.

Hille, B. (1992): *Ionic Channels of Excitable Membranes*, Sinauer Associates Inc. Publishers, Sunderland, Massachusetts (607).

Hobson, E.W. (1887): Synthetical solutions in the conduction of heat. *Proc. Lond. Math. Soc. (I)*, **XIX**, 279–301.

Hodgkin, A.L. (1937): Evidence for electrical transmission in nerve, Part 1. *J. Physiol.*, **90**, 189.

Hodgkin, A.L. (1939): The subthreshold potentials in a crustacean nerve fibre. *Proc. R. Soc. Lond.*, B**126**, 87.

Hodgkin, A.L. (1948): The local electric changes associated with repetitive action in non-medulled axon. *J. Physiol.*, **107**, 165.

Hodgkin, A.L. (1958): Ionic movements and the electrical activity in giant nerve fibres. *Proc. R. Soc. Lond.*, B**148**, 1.

Hodgkin, A.L. (1964): *The conduction of the nervous impulse*. The Sherrington Lecture VII, Liverpool University Press.

Hodgkin, A.L. & Horowicz, P. (1959): The influence of potassium and chloride ions on the membrane potential of a single muscle fibre. *J. Physiol.*, **148**, 127.

Hodgkin, A.L. & Huxley, A.F. (1945): Resting and action potentials in single nerve fibres. *J. Physiol.*, **104**, 176.

Hodgkin, A.L. & Huxley, A.F. (1947): Potassium leakage from an active nerve fibre. *J. Physiol.*, **106**, 341.

Hodgkin, A.L. & Huxley, A.F. (1952): Current carried by the sodium and potassium ions through the membrane of the giant axon. *J. Physiol.*, **116**, 449.

658   *References*

Hodgkin, A.L. & Huxley, A.F. (1952): The components of membrane conductance in the giant axon of *Loligo*. *J. Physiol.*, **116**, 473.

Hodgkin, A.L. & Huxley, A.F. (1952): The dual effect of membrane potential on sodium conductance in the giant axon of *Loligo*. *J. Physiol.*, **116**, 497.

Hodgkin, A.L. & Huxley, A.F. (1952): A quantitative description of membrane current and its application to conductance and excitation in nerve. *J. Physiol.*, **117**, 500.

Hodgkin, A.L. & Huxley, A.F. (1953): Movement of radioactive potassium and membrane current in a giant axon. *J. Physiol.*, **121**, 403.

Hodgkin, A.L., Huxley, A.F., & Katz, B. (1952): Measurements of current–voltage relations in the membrane of the giant axon of *Loligo*. *J. Physiol.*, **116**, 424.

Hodgkin, A.L. & Katz, B. (1949): The effect of sodium ions on the electric activity of the giant axon of the squid. *J. Physiol.*, **108**, 37.

Hodgkin, A.L. & Keynes, R.D. (1953): The mobility and diffusion coefficient of potassium in giant axons from *Sepia*. *J. Physiol.*, **110**, 513.

Hodgkin, A.L. & Keynes, R.D. (1955): The potassium permeability of a giant nerve fibre. *J. Physiol.*, **128**, 61.

Hodgkin, A.L. & Nakajima, S. (1972): The effect of diameter on the electrical constants of frog skeletal muscle fibres. *J. Physiol.*, **221**, 105–120.

Hounsgaard, J., Kiehn, O. & Mintz, I. (1988): Response properties of motorneurones in a slice preparation of the turtle spinal cord. *J. Physiol.*, **398**, 575.

House, C.R. (1974): *Water Transport in Cells and Tissues*. Edward Arnold (Publishers) Ltd. London 1974 (562).

Hoyt, R.C. (1963): The squid giant axon. Mathematical models. *Biophys. J.*, **3**, 399.

Huang, C. & Thompson, T.E. (1966): Properties of lipid bilayer membranes separating two aquaeous phases: Wayer permeability. *J. Mol. Biol.*, **15**, 539–554.

Huxley, A.F. & Stämpfli, R. (1949): Evidence for saltatory conduction in peripheral myelinated nerve fibres. *J. Physiol.*, **108**, 318.

Huxley, A.F. & Stämpfli, R. (1950): Saltatory transmission of nervous impulse. *Arch.-Sci. physiol.*, **3**, 435.

Jost, W. (1970): *Diffusion in Solids, Liquids, Gases*. Academy Press, Inc., New York (551).

Katz, B. (1939): *Electric Excitation of Nerve*. Oxford University Press, London (151).

Katz, B. (1962): The Croonian Lecture. The transmission of impulses from nerve to muscle and the subcellular unit of synaptic action. *Proc. R. Soc. Lond.* B**155**, 455.

Katz, B. (1966): *Nerve, Muscle and Synapses*. McGraw-Hill, New York (193).

Katz, B. (1969): *The release of neural transmitter substances*, The Sherrington Lecture X. Liverpool University Press.

Keynes, R.D. (Organizer) (1975): *A Discussion on Excitable Membranes. Phil. Trans. R. Soc. Lond.*, B**270**, 295.

Keynes, R.D. & Aidley, D.J. (1981): *Nerve and Muscle*, Cambridge University Press, Cambridge (163).

Kline, M. (1967): *Calculus: An Intuitive and Physical Approach*, 2 volumes, John Wiley and Sons, Inc., New York (574 + 415).

Kohn, P.G. (1865): Table of some physical and chemical properties of water. In *The State and Movements of Water in Living Organisms. Symposium of the Society for Experimental Biology, no. XIX, Cambridge*, pp. 1–16.

Kramers, H.A. (1940): Brownian movements in a field of force and the diffusion model of chemical reactions. *Physica*, **7**, 284.

Krnjević, K. & Miledi, R. (1958): Failure of neuromuscular transmission in rats. *J. Physiol.*, **149**, 440.

Kuffler, S. (1942): Electric potential changes at an isolated nerve–muscle junction. *J. Neurophysiol.*, **5**, 18.

Kuffler, S.W. & Nicolls, J.G. (1992): *From Neuron to Brain*. Sinauer Associates Inc. Publishers, Sunderland, Massachusetts, first printing (468).

Laidler, K.J. (1978): *Physical Chemistry with Biological Applications*. The Benjamin Cummings Publishing Co. Inc., Menlo Park, California (587)

Larsen, E.H., Sørensen, J.N. & Sørensen, J.B. (2002): Analysis of the sodium recirculation theory of solute coupled water transport in small intestine. 10-1013/jphysiol.2001.013248

Lassen, U.V. & Rasmussen, B.E. (1978): Use of microelectrodes for measurements of membrane potentials. In: *Membrane Transport in Biology, Volume I, Concepts and Models*, Chapter 5, Springer Verlag.

Laüger, P. (1991): *Electrogenic Ion Pumps*, Sinauer Associates Inc.

Le Claire, A.D. (1958): Random walks and drift in chemical diffusion. *Phil. Mag.*, **3**, 921–939.

Liley, A.W. (1956): The effects of presynaptic polarization on the spontaneous activity at the mammalian neuromuscular junction. *J. Physiol.*, **134**, 427–443.

Lorente de Nó, R. (1947): *A study in nerve physiology*. Rockefeller Institute Studies, **131,132**, New York.

McQuarrie, D.A. (1973): *Statistical Mechanics*. Harper & Row Publishers, New York (641).

Millman, J. (1940): A useful network theorem. *Proc. I.R.E.*, **28**, 413.

Milne-Thompson, L.M. & Comrie, L.J. (1957): *Standard Four-Figure Mathematical Tables*, Macmillan & Co. Ltd, London.

Moore, W.J. (1972): *Physical Chemistry*, 5th edn, Longman Group Ltd., London.

Morse, P.M. (1962): *Thermal Physics*, W.A. Benjamin Inc., New York.

Mullins, L.I., Adelman, W.J. & Sjodin, R.A. (1962): Sodium and potassium efflux from squid axons under voltage clamp conditions. *Biophys. J.*, **2**, 257.

Mullins, L.I. & Noda, K. (1963): The influence of sodium–free solutions on the membrane potential of frog muscle fibers. *J. Gen. Physiol.*, **47**, 117–132.

Nastuk W.L. (1955): Neuromuscular transmission. *Am. J. Physiol.*, **19**, 66.

Nastuk, W.L. & Hodgkin, A.L. (1950): The electrical activity of single muscle fibres. *J. cell. comp. physiol.*, **35**, 39.

Nernst, W. (1888): Zur Kinetik der ir Lösunger befindlischer Körper. *Zeit. physik. Chem.*, **2**, 613–637.

Nernst, W. (1889): Die elektromotorische Wirksamkeit der Ionen. *Zeit. physik. Chem.*, **4**, 129.

Northrop, J.H. & Anson, M.V. (1929): A method for the determination of dittusion constants and the calculation of the rasius and weight of the hemoglobin molecule. *J. Gen. Physiol.*, **12**, 543–554.

Ogden, D. (Ed.) (1987): *Microelectrode Techniques*, The Plymouth Workshop Hand-book, The Company of Biologists Limited, Cambridge (448).

Ostwald, W. (1890): Elektrische Eigerschaffter halbdurchlässiger Scheidewänder. *Zeit. physik. Chem.*, **6**, 129.

Overbeek, T.T. (1956): The Donnan Equilibrium. *Progr. Biophys. biophys. Chem.*, **6**, 57.

Perrin, J. (1910): *Brownian Movement and Molecular Reality*, F. Soddy (ed.), Taylor and Francis, London.

Planck, M. (1890a): Über die Erregung von Elektricitet und Wärme in Elektrolyten. *Ann. Physik. u. Chemie. Neue folge*, **39**, 161.

660   *References*

Planck, M. (1890b): Über die Potentialdifferenz zwischen zwei verdünnten
    Lösungen binärer Elektrolyte. *Ann. Physik. u. Chemie. Neue folge*, **40**, 561.
Planck, M. (1917): Über einen Satz der statistischen Dynamik und seine Erweiterung
    in der Quantentheorie. *Sitz. der preuss. Akad.* (324).
Plonsey, R. (1969): *Bioelectric Phenomena.* McGraw-Hill Book Company, New York
    (380).
Preston, C.M. & Agre, P. (1991): Isolation of the cDNA for erythorcyte integral
    membrane protein of 28 kilodalton member of an ancient channel family. *Proc.
    Nat. Acad. Sci. USA*, **88**, II 110– II 114.
Robbins, M. & Mauro, A. (1960): Experimental study of the independence of diffusion
    and hydrodynamic permeability coefficients in collodion membranes. *J. Gen.
    Physiol.*, **185**, 172.
Rosenfalck, P. (1969): *Intra- and Extracellular Potential Fields of Active Nerve and
    Muscle Fibres,* Akademisk Forlag, Copenhagen.
Schlögl, R. (1964): Stofftransport durch Membranen. Fortschritte der Physikalischen
    Chemie, Band 9. Dr. Dietrich Steinkopff Verlag, Darmstadt.
Setlov, R.R. & Pollard, E.C. (1962): *Molecular Biophysics.* Addison-Wesley
    Publishing Company Inc., Reading, Mass.
Sigworth, F.J. & Neher, E. (1980): Single Na^+ channel currents observed in cultured
    rat muscle cells. *Nature*, **287**, 447.
Smoluchowski, M.v. (1915): Über Brownsche Molekularbewegung unter Einwirkung
    äusserer Kräfte und deren Zusammenhang mit der verallgemeinerten
    Diffusionsgleichung. *Ann. d. Physik*, **48**, 1103.
Smoluchowski, M.v. (1916): Drei Vorträge über Diffusion, Brownsche
    Molekularbewegung und Koagulation von Kolloidteilchen. *Physik. Zeitschr.*,
    17, 557.
Smoluchowski, M.v. (1923): Abhandlungen über die Brownschen Bewegung und
    verwandte Erscheiningen. In: *Ostwalds Klassiker der exakten Wissenschaften.*
    Akademische Verlagsgesellschaft, Leipzig.
Snell, J.L. (1989): *Introduction to Probability.* McGraw-Hill Book Company Inc.,
    New York.
Sommerfeld, A. (1949): *Partial Differential Equations in Physics.* Academic Press
    Inc., New York (335).
Sten-Knudsen, O. (1978): *Passive Transport Processes.* I: Membrane Transport in
    Biology. Volume I, Concepts and Models, Chapter 2. Springer-Verlag, Berlin,
    Heidelberg, New York (113).
Sten-Knudsen, O. & Ussing, H.H. (1981): Flux ratio under non-stationary conditions.
    J. Membrane Biol., **63**, 233.
Takeuchi, T. & Takeuchi, N. (1858): Active phase of frog's end-plate potential.
    J. Neurophysiol., **22**, 395.
Takeuchi, T. & Takeuchi, N. (1860): On the permeability of the end-plate membrane
    during the action of the transmitter. *J. Physiol.*, **154**, 52.
Tasaki, I. (1953): *Nervous Transmission.* Thomas, Springfield.
Tasaki, I. (1959): Conduction of the nerve impulse. In: *Handbook of Physiology*,
    Section I: Neurophysiology, Chapter III, 73.
Teorell, T. (1949): Membrane electrophoresis in relation bioelectrical polarization
    effects. *Arch. Sci. Physiol.*, **3**, 205.
Teorell, T. (1953): Transport processes and electrical phenomena in ionic membranes.
    Progr. Biophys. biophys. Chem., **3**, 305.
Tranter, C.J. (1951): *Integral Transform in Mathematical Physics.* Methuen & Co. Ltd,
    London, and John Wiley and Sons Inc., New York (118).

References 661

Ussing, H.H. (1949): On the distinction by means of tracers between active transport and diffusion. *Acta Physiol. Scand.*, **19**, 43–56.

Ussing, H.H. (1952): Some aspects of the application of tracers in permeability studies. *Advanc. Enzymol.*, **13**, 21–65.

Vernig, A. (1972): Changes in the statistical paraneters during facilitation at the crayfish neuromusclar junction. *J. Physiol.*, **226**, 751–759.

Vernig, A. (1972): The effects of calcium and magnesium on statistical release parameters at the crayfish neuromuscular junction. *J. Physiol.*, **226**, 761–768.

Warburg, E.J. (1922): Studies on carbonic acid components and hydrogen ion activities in blood and salt solutions. *Biochem. J.*, **16**, 11–340.

Wax, N., (Ed.) (1954): *Selected Papers on Noise and Stochastic Processes*. Dover Publications Inc., New York.

Young, J.Z. (1936): The giant nerve fibres and epistellar body of cephalopods. *Quart. J. Micr. Sci.*, **78**, 367.

Young, J.Z. (1951): *Doubt and Certainty in Science*. Clarendon Press, Oxford.

Zeuthen, T. (2000): Molecular water pumps. *Rev. Physiol. Biochem. Pharmacol.*, **141**, 97–151.

Zucker, R.S. (1973): Changes in the statistics of transmitter release during facilitation. *J. Physiol.*, **229**, 787–810.

Index